In Quest of the Universe

Universe

FOURTH EDITION

In Quest of the Universe

Universe

FOURTH EDITION

Karl F. Kuhn
Eastern Kentucky University

Theo Koupelis
University of Wisconsin—
Marathon County

JONES AND BARTLETT PUBLISHERS

Sudbury, Massachusetts

BOSTON   TORONTO   LONDON   SINGAPORE

World Headquarters
Jones and Bartlett Publishers
40 Tall Pine Drive
Sudbury, MA 01776
978-443-5000
info@jbpub.com
www.jbpub.com

Jones and Bartlett Publishers Canada
2406 Nikanna Road
Mississauga, ON L5C 2W6
CANADA

Jones and Bartlett Publishers International
Barb House, Barb Mews
London W6 7PA
UK

Production Credits
Chief Executive Officer: Clayton Jones
Chief Operating Officer: Don W. Jones, Jr.
Executive V.P. & Publisher: Robert W. Holland, Jr.
V.P., Design and Production: Anne Spencer
V.P., Sales and Marketing: William Kane
V.P., Manufacturing and Inventory Control: Therese Bräuer
Executive Editor, Science: Stephen L. Weaver
Managing Editor, Science: Dean W. DeChambeau
Associate Editor, Science: Rebecca Seastrong
Senior Production Editor: Louis C. Bruno, Jr.
Marketing Manager: Matthew Bennett
Marketing Associate: Matthew Payne
Text and Cover Design: Anne Spencer
Composition: Graphic World, Inc.
Illustrations: Precision Graphics, Carlisle Communications, Elizabeth Morales, and Graphic World, Inc.
Printing and Binding: Courier Kendallville
Cover Printing: Lehigh Press

Library of Congress Cataloging-in-Publication Data
Kuhn, Karl F.
    In Quest of the Universe.—4th ed./Karl F. Kunn, Theo Koupelis
        p. cm.
    Includes bibliographical references and index.
    ISBN 0-7637-0810-0 (alk. paper)
    1. Astronomy. I. Koupelis, Theo. II. Title.
    QB45.2.K84 2004
    520—dc22   2003054663

Printed in the United States of America
07 06 05 04   10 9 8 7 6 5 4 3 2

About the cover: About 98 million light-years away, the majestic spiral galaxy NGC3370 looms large in this NASA *Hubble Space Telescope* image. The galaxy is similar in size to our own Milky Way. In the background is an impressive array of other far more distant galaxies. This image, taken by *Hubble's* Advanced Camera for Surveys, is roughly 95,000 light-years (3.4 arcminutes or 29,000 parsecs) wide.
[Photo Credit: NASA, The Hubble Heritage Team, and A. Riess (STScI)]

To Mandy, Jim, Ty, Parker, Patrick, Adam, Meg, Trey, Katie, MaryCarol, Jackson, and to the newest supernova in my life, Hayden.

Karl F. Kuhn

για τους γονεις μου, and for Billy, Alex, and Nancy; I am so lucky to have you in my life.

Theo Koupelis

BRIEF CONTENTS

CONTENTS

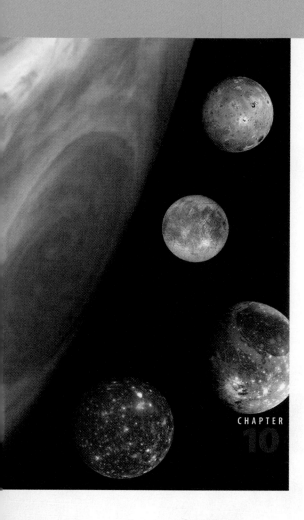

CHAPTER
10

Pluto and Solar System Debris 324

CHAPTER
11

The Sun 358

CHAPTER
17

A Diversity of Galaxies 540

CHAPTER
18

Cosmology: The Nature of the Universe 580

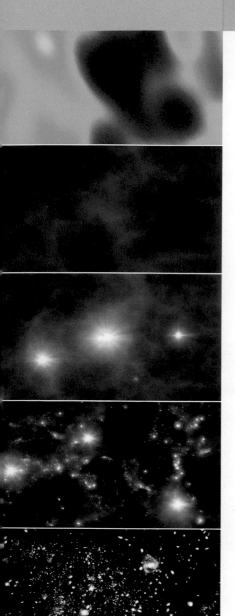

Frontmatter photo credits: pages ii–iii, NASA, The Hubble Heritage Team, and A. Reiss (STScI). **Pages iv–v,** Akira Fujii/NASA. **page viii,** NASA. **page ix,** Bill Schoening, Vanessa Harvey/REU program/NOAO/AURA/NSF. **pages x and xi,** NASA/Goddard Space Flight Center Scientfic Visualization Studio. **page xii,** NASA. **page xiii,** SOHO (ESA & NASA). **page xiv,** NOAO. **page xv,** European Southern Observatory. **page xvi,** NASA/WMAP.

BOXED FEATURES

ADVANCING THE MODEL

BOXED FEATURES

TOOLS OF ASTRONOMY

ACTIVITIES

We have written this book as an introduction to astronomy, as well as an introduction to the general methods of science. As with earlier editions, we recognize that many students using this text are not planning to become astronomers or scientists of any sort. However, we strongly believe in the importance of every student (and every person, for that matter) knowing about how science operates and how we know what we know about our world. A habit of skeptical thought and a reliance upon observation and testing to verify an idea are valuable not just for scientists but for everyone.

We present astronomy in the traditional order of solar system, stars, and galaxies. This gives us the opportunity to emphasize the historical development of science and technology as well as to proceed from the more familiar to the less familiar. It also allows students to expand their sense of astronomical dimensions gradually from the sizes of planets to the distances between the stars to the vast spaces between the galaxies. We frequently stop our tour of the universe to discuss how new scientific theories gradually replaced older ones, building upon established information by adding new observations and new insights into how the universe works.

New to the Fourth Edition

Even though we have retained the organization and basic goals of the book as presented in the third edition, this edition represents a thorough rewrite of the book that includes a re-thinking of every argument and an update of all the information presented in all its forms (text, numbers, tables, and figures). Important changes can be found throughout the book, but the following are some of the major updates.

- We have expanded the section on searches for other planetary systems.
- We have completely revised the section on Mars based on the latest data from the *Mars Global Surveyor* and *Mars Odyssey*.
- We have included information from the *NEAR* mission in the Advancing the Model box "The Mission to Eros."
- The section on the "solar neutrino problem" and neutrino oscillations includes all the latest data from the Super-Kamiokande experiment, the Sudbury Neutrino Observatory, and the KamLAND project.
- We have expanded the discussion on the stormy areas on the Sun's surface, including the interplay between magnetic fields and convection, solar flares, and sunspots.
- We have added two new Advancing the Model boxes on "Nucleosynthesis" and "Gamma-Ray Bursts."
- A new section on "Evidence for the Big Bang" has been added.
- We present a clear and in-depth description of all the latest findings on cosmology, including dark energy and dark matter, based on data from the *Wilkinson Microwave Anisotropy Probe* and other experiments. These new results have signaled the dawn of the era of precision cosmology.
- A great number of new images are included from a variety of sources that cover the entire electromagnetic spectrum and are the result of many international collaborative efforts. We have tried to select pictures that are useful as well as exciting, illustrating many of the awesome events that take place throughout the cosmos.

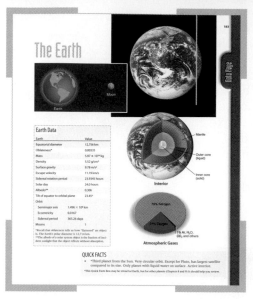

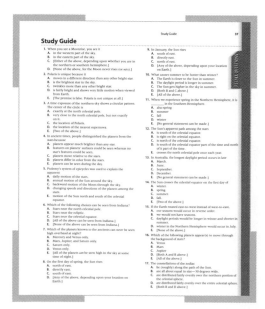

- **Data Pages** on individual planets, the Moon, the Sun, and planetary nebulae should help students in reviewing the material and focusing on important concepts. **Planetary Data Comparisons** appear at the ends of Chapters 7, 8, 9, and 10. These two-page spreads present data on the objects of the solar system in a format for easy comparison, emphasizing what we can learn by considering the variety of observed temperatures, atmospheres, interior compositions, and magnetic fields.

- **Marginal definitions and Marginal notes** are located throughout the text. Definitions appear next to key terms highlighted in the text; these definitions are collected together in the Glossary at the end of the book. Marginal notes often refer to relevant material in other chapters or provide additional facts about the topic under discussion.

- **Quotations** appear in the margin to give students an idea of how astronomers and physicists themselves thought about science or about their work. Other quotations show what people in other disciplines or other cultures think about the issues involved in astronomy.

- **Student misconceptions** are addressed by placing in the margin typical questions often asked by students. Highlighted by a question-mark icon, these questions appear in the chapter section that provide the answers to such questions. Students can use these questions as guidelines for review, making sure they understand the correct explanation for each question.

- **Advancing the Model** boxes discuss historical developments and biographies of astronomers.

- **Tools of Astronomy** boxes present more advanced concepts.

- **Try One Yourself** problems involve a more mathematical understanding of astronomy.

- Worked **Examples** illustrate how astronomers determine some of the data we now accept as accurate descriptions of celestial objects.

- A built in **Study Guide** at the end of each chapter provides three sets of exercises.
  - **Recall Questions** are mostly multiple choice but also include other review questions.
  - **Questions to Ponder** are generally more complex than the Recall Questions and sometimes ask students to relate material in the chapter to ideas from earlier chapters. Some questions ask students for personal evaluations of issues or to research topics in other resources.
  - **Calculations** pose numerical problems for students to solve, based on examples in the chapter.

Activities are also included in the Study Guides, featuring some observational astronomy and some drawing of scale models, among other things. New exercises have been added and some old ones have been revised. Answers to even-numbered exercises are included in Appendix H.

Ancillaries

Instructor's Tool Kit – CD-ROM

Compatible with Windows and Macintosh platforms, this CD-ROM provides instructors with the following traditional ancillaries:

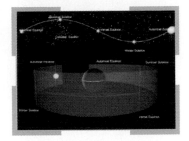

- **Over 60 animations** of key illustrations in the text have been created, expanding the reach of the figures beyond the static page. Illustrations in the text accompanied by the icon ▇ indicate that a corresponding animation is on the Instructor's ToolKit CD and on the StarLinks Student CD. Because so many concepts in astronomy involve motion, these 3D animated illustrations are an invaluable tool for helping students understand the concepts.
- The *Instructor's Manual,* provided as a text file, includes chapter summaries, complete chapter lecture outlines, learning objectives, key terms, teaching suggestions and classroom activities, and questions to test student understanding.
- The *Test Bank* is available as straight text files and as part of the Diploma™ Test Generator software included on the CD. The Test Generator software helps choose an appropriate variety of questions, create multiple versions of tests, even administer and grade tests online.
- The *PowerPoint™ Lecture Outline Slides* presentation package provides lecture notes and images for each chapter of *In Quest of the Universe.* Instructors with the Microsoft PowerPoint™ software can customize the outlines, art, and order of presentation.
- The *Kaleidoscope Media Viewer* provides a library of all the art, tables, and photographs in the text to which Jones and Bartlett Publishers holds the copyright or has digital reprint rights. The Kaleidoscope Media Viewer uses your browser—Internet Explorer or Netscape Navigator—to enable you to project images from the text in the classroom, insert images into PowerPoint presentations, or print your own acetates.
- Additional movies and animations from various NASA sites including galaxy formation, differential rotation near a black hole, a solar eclipse, Kepler's First Law, and many others.
- Additional activities, courtesy of the *Space Science Reference Guide, 2nd Edition* CD-ROM, provide extra resources to effectively learn space science.

Starlinks Student CD-ROM

Provided free with this text, this interactive learning tool contains
- An electronic study guide with chapter learning objectives, outlines, and summaries, key terms, questions to test understanding, and activities.
- Additional author-written practice questions to help students review chapter material and prepare for exams. For each chapter there are multiple choice, fill-in, short answer, and mathematical review questions.
- Flash Cards for learning the new vocabulary encountered in astronomy.
- **Over 60 animations** of key illustrations in the text.
- *Home Planet 3.1* planetarium/satellite-tracking software by John Walker.

Starlinks

Starlinks is the Jones & Bartlett web site that accompanies *In Quest of the Universe,* 4/e. This site, www.jbpub.com/starlinks, provides additional resources for problem-solving, research, and activities.
- **Explorations** are links to independent web sites that provide further information about topics in the text.
- **NetQuestions** provide additional end-of-chapter questions with links to independent web sites where students can research the questions.

- To help review, **On-line Quizzes** for each chapter provide the students with web-based, multiple-choice questions and answers.
- **Additional animations** and movies of visually reinforce phenomena discussed in the text and provide historical context.
- **Astronomy Picture of the Day** is a link to the NASA site that features daily astronomy photos and background information.

Acknowledgments

Good teachers learn from their students. Attention to students' questions helps instructors to discover which concepts are most difficult and require special explanation. The same is true for textbook writers, and we thank our students for their questions.

Although it is the authors' names that appear on the cover of a textbook, all textbook writers know that "their" book is far from theirs alone. The sweat of many brows moistens the pages of every book. We thank the Jones and Bartlett team, led by Steve Weaver, and including Anne Spencer, Louis Bruno, Dean DeChambeau, Rebecca Seastrong, and Matt Bennett.

K. Kuhn: I thank my wife Sharon, my kids, and my grandkids for teaching me things that are more important than those learned in school.

T. Koupelis: My support and guidance during the entire process of revising/updating this text came from two main sources. First, my love for my discipline, my way of thinking about science and its ways, and my biases as to how a good teacher should react to his/her students' needs and about the need for a science-literate student body; for all this, I owe a debt of gratitude to my teachers through the years, great teachers and friends—examples that have been very hard to follow. I thank them for everything they taught me and for providing me with a never-ending challenge. Second, my thanks to my parents and my family for their unspoken but ever-present support, for their love and patience, for making it all worthwhile. I hope they forgive me for not being there for them as much as they would have liked, and I hope they know that they are the stars in my sky and that I love them very much.

Reviewers play a major role in the success of a book. We would like to thank all those reviewers who have spent time commenting on all editions of *In Quest of the Universe*.

Kurt Anderson
New Mexico State University

Louis Beyer
Murray State University

William J. Boardman
Birmingham Southern College

Donald J. Bord
University of Michigan–Dearborn

Paul J. Camp
Coastal Carolina University

Emerson Cannon
Salt Lake City Community College

Peter W. Deutsch
Pennsylvania State University

Bernd Enders
Marin College

Terry L. Goforth
Southwestern Oklahoma State University

Susan Hartley
University of Minnesota–Duluth

Howard Hayden
University of Connecticut

Lon C. Hill
Broward Community College

Adam Johnston
Weber State University

Thomas M. Jordan
Ball State University

Robert C. Kennicutt, Jr.
University of Arizona

Robert J. Leacock
University of Florida

Paul-Emile Legault
Laurentian University

J. Patrick Lestrade
Mississippi State University

George E. McCluskey, Jr.
  Lehigh University

Brian Milbrath
  Eastern Kentucky University

Scott Niven
  Olympic College

Thomas C. Olien
  Humber College

John Oliver
  University of Florida

Michael O'Shea
  Kansas State University

Paul J. Thomas
  University of Wisconsin–Eau Claire

Cynthia Peterson
  University of Connecticut

Roger Ptak
  Bowling Green State University

Robert M. Rickett
  Northeast Louisiana State University

Robert D. Schmidt
  University of Nebraska–Omaha

John Silva
  University of Massachusetts

Beverley Taylor
  Miami University–Hamilton

Scott Temple
  Cuesta College

Louis Winkler
  Pennsylvania Sate University

Robert Zimmerman
  University of Oregon

   Special thanks to our colleagues at Eastern Kentucky University and at the University of Wisconsin–Eau Claire who provided valuable feedback as they used the book. Also, special thanks to Dr. Roger Angel for his input on the Tools of Astronomy box on "Spinning a Giant Mirror."

   We invite readers of *In Quest of the Universe* to write to us with comments and suggestions. In particular, if you know of better illustrations or better examples of everyday phenomena that help students understand astronomical concepts, we would appreciate hearing from you. It is you who know best what should be changed in future editions. Our addresses are

Theo Koupelis
Univ. of Wisconsin—Marathon
518 S. 7th Ave
Wausau, WI 54401
tkoupcli@uwc.edu

Karl Kuhn
Retired from
Department of Physics and Astronomy
Eastern Kentucky University
Richmond, KY 40475

The Quest Ahead

Comet Shoemaker-Levy 9 was broken by Jupiter's gravitational forces into more than 20 fragments (bottom photo), all following the same path toward a collision with Jupiter. This photo of Jupiter (top photo) shows the impact spots left by some of the fragments.

ON THE NIGHT OF MARCH 23, 1993, AMATEUR ASTRONOMER DAVID LEVY photographed part of the sky near the planet Jupiter. His friends, fellow astronomers Carolyn Shoemaker and her husband Eugene Shoemaker, spotted something unusual in the picture: a comet that had broken up into about 20 pieces. The comet became known officially as Comet Shoemaker-Levy 9 (the ninth comet discovered by these three sky-watchers) and unofficially as the "string of pearls" comet.

When astronomers announced the news of the comet, on June 1, 1993, they had traced its path closely enough to tell that it had come under the influence of Jupiter's powerful gravitational field. They had deduced that the comet was pulled apart by Jupiter's gravity in July of 1992. They predicted that the comet would crash into Jupiter on or about July 25, 1994.

By October, 1993, astronomers had refined their predictions. The comet was then stretched out over a distance of 3 million kilometers (1.8 million miles) in orbit around Jupiter. The first piece was predicted to hit Jupiter at 4:01 PM (Eastern daylight time in the U.S.) on July 16, and the last piece was predicted to hit at 4:00 AM on July 22, 1994. But nobody knew whether the impacts would be big or small, or even observable by telescopes on Earth.

By the time of the predicted impacts, the entire world was watching, linked together by the Internet and watching on television. The *Hubble Space Telescope* was trained on Jupiter, as was the *Galileo* space probe then approaching the Jupiter system, as well as most major observatories around the world and untold numbers of amateur telescopes. It has been said that more telescopes were aimed at the same spot, Jupiter, than ever before or since. And the viewers were not disappointed.

Fragment A of the comet hit Jupiter at 4:11 PM Eastern daylight time, only 10 minutes off the prediction made 10 months earlier. Traveling at 60 kilometers/second (130,000 miles/hour), each fragment created a fireball in Jupiter's upper atmosphere reaching up as high as 3000 kilometers (1900 miles), and leaving darkened areas in the atmosphere as wide across as the diameter of the Earth. Scientists estimated that some of the larger impacts, caused by fragments about 3 kilometers across, released energy equivalent to 6 million megatons of TNT. (The largest nuclear bomb ever detonated produced a blast equivalent to about 5 megatons.) The event made front-page headlines around the world, featuring photographs of fireballs and dark impact spots across the giant planet.

Six million megatons is about 600 times the estimated arsenal of the world.

How were astronomers able to predict the time of impact with such accuracy? What did they learn from observing the aftereffects of this stupendous collision? Many scientists believe that a similar collision with a comet occurred on Earth 65 million years ago, heralding the end of the age of dinosaurs. Since seeing the visible proof of such collisions on Jupiter, many people worry about the chances of further catastrophes on our own planet. Astronomers say that the chances of such an event happening here are roughly once in 300,000 years; but how can they make such a prediction?

You will learn the answers to some of these questions in this book. You will find that while collisions of comets and planets are very rare in our immediate neighborhood of space, the vast regions of the universe show tremendous activity. Stars are born in dense clouds of gas while other stars explode, spewing forth gas back into space. Galaxies collide over periods of millions of years, hurtling stars around at fantastic speeds. The universe is a wild and fabulous showcase, and we will see some of its highlights, looking out from our quiet corner called Earth.

Astronomy is the oldest of the sciences, but it is still changing rapidly, even today. It is a study rich in interesting personalities and exciting discoveries. Its long history and recent advances make it an excellent example of the progressive nature of science.

Nothing is rich but the inexhaustible wealth of nature. She shows us only surfaces, but she is a million fathoms deep.
Ralph Waldo Emerson

Science is fundamentally a quest for understanding, and in the case of astronomy, the subject of the quest is the entire universe. The study of astronomy covers the full range of matter from smallest to largest. On the small end, it includes atoms and their internal workings, for astronomers need to know about the nuclear reactions that make **stars** shine; this knowledge may one day lead to peaceful applications of thermonuclear reactions. Astronomers also study Earth's neighbors, the planets, and in doing so help us solve problems right here on Earth. For example, a study of weather patterns on Venus and Jupiter has advanced our understanding of weather on Earth. In addition, studying Mars and Venus helps us understand the past and future of our own planet. Information we collected using unmanned spacecrafts and Earth-based telescopes has revolutionized our understanding of how our solar system formed billions of years ago and has evolved since then. In particular, our neighborhood was, and still is to a lesser extent, a shooting gallery; we can see the scars of collisions on the surfaces of Mercury, the Moon, and other planetary satellites. Understanding the composition of the objects in our vicinity may one day prove extremely valuable in our efforts to find additional sources of materials for our own survival. The apparent weightlessness and very low density of space may one

star A self-luminous celestial object.

Planets shine by reflected sunlight, but stars have an internal source of energy for their light. In Chapter 11 we discuss this stellar energy source.

day lead to the manufacture of exceptional substances. In some cases astronomy's results are more trivial, but nevertheless practical: materials developed for the Apollo missions to the Moon are now used in tennis rackets.

Astronomers also study the largest groupings of billions of stars, and they ask *big* questions, the answers to which tell us something about ourselves, both as individuals and as a species. Our studies of stars and *nebulae* help us better understand the natural cycle of the formation of a star, its evolution, its death and the recycling of some of its material into a new star. Our studies of *galaxies* help us better understand the creation and evolution of our universe. Studying astronomy causes us to change our basic outlook as we better appreciate where humans fit into the magnificent universe. Our species and our planet seem to be dwarfed by the immensity of the cosmos. In a sense, astronomy teaches humility.

At the same time, one marvels that humans have been able to learn so much about the universe in which we live, and that such tiny beings are able to comprehend the huge, complex universe. From this we may decide that size is not so important after all; at least, not when compared to the intelligence and spirit of our species. We should take pride in our accomplishments but also continue pushing the frontiers of knowledge.

nebula (plural **nebulae**). An interstellar region of dust and/or gas.

galaxy A group of gravitationally linked stars, ranging from a billion to perhaps a thousand billion stars.

P-1 Science and Its Ways of Knowing

Science is a very human endeavor, and the science produced by a particular culture is in part a reflection of the characteristics of that culture. In astronomy the human aspects of science are very visible.

Science is about discovering the orderliness of nature and the rules that govern this order. It is also the body of knowledge about nature that represents the collective efforts of people throughout history. Science progresses through the interaction of observations and experiments with theory. With our theories we make predictions, while our observations and experiments support, disprove, or place constraints on these theories. This information provides a means of determining the best description of nature. The characteristics of good science include the continuous search for new knowledge, the ability to make predictions and test them against observations, and the ability to repeat and verify any experimental result. Learning from observations and experiments and, as a result, refining our theories is not a weakness in science—it is a strength.

It is said that the **scientific method** is the best approach for doing good science. This method is used to gain, organize and apply new knowledge and was first introduced in the sixteenth century. Essentially it involves the following series of steps: (a) recognize a problem; (b) make an educated guess, called a **hypothesis**, about the problem's solution; (c) predict the consequences of the hypothesis; (d) perform experiments to test the predictions; (e) find the simplest general rule that organizes the hypothesis, prediction, and experimental outcome into a **theory**. The scientific method is a never-ending cycle of hypothesis, prediction, data gathering, and verification. In reality, this method is not always used. For example, some of the most important discoveries in science were accidental, such as Becquerel's discovery of radioactivity by noting that a chunk of uranium ore caused a photographic plate to turn cloudy. The important ingredient is the continuous comparison between observations and theory.

It is important to emphasize again that a **hypothesis** is nothing but an educated guess made in explaining the results of an experiment or observation. A hypothesis must be testable and, at least in principle, it must be susceptible to being shown wrong. A useful hypothesis is one that makes predictions about nature that can be confirmed or refuted by observations. It is only considered to be a **fact** after it has been demonstrated by experiments.

Science is a way of thinking much more than it is a body of knowledge.
Carl Sagan

Science is not about truth, but the most reliable information at the time.
Freeman Dyson

We are trying to prove ourselves wrong as quickly as possible, because only in that way can we find progress.
Richard Feynman

In everyday life the word *fact* implies something that is absolute. In science, however, it simply implies a generally close agreement among most competent scientists about a series of observations of the same phenomenon. Thus, something that in science was considered a fact a few decades ago may turn out to be wrong. For example, we used to consider it a fact that the universe is static, while our current observations clearly show that we live in an expanding universe.

When a hypothesis has been tested over and over again and has not been contradicted, it may become known as a **law** or **principle.** In science we describe reality using **models,** which are hypotheses that are supported by observations and experiments. A model gives us a description of a phenomenon and allows prediction of future events, but it is not necessarily the truth or reality.

Finally, a **theory** is a synthesis of a large body of information that includes related hypotheses (which have been repeatedly tested and verified), giving us a self-consistent description of a certain aspect of the natural world. A theory can never be proven to be true, but data can prove a theory to be false. Among two or more competing theories (which explain all known observations equally well), we prefer the one using the fewest arbitrary assumptions; this philosophical choice is known as **Occam's razor**; we say that such a theory is more aesthetically pleasing. In everyday usage the word *theory* implies something that is nothing but a wild guess that has little to do with reality. In science, however, a *theory* represents a coherent framework describing an aspect of nature very well.

Before embarking on our quest to understand the universe, let us take a quick look at what lies ahead.

"But that's only a scientific **theory.** Why should I believe it?"

Science is built up with facts, as a house is with stones. But a collection of facts is no more a science than a heap of stones is a house.
Jules Henri Poincaré, Scientist and Mathematician, 1854-1912

P-2 The View from Earth

In our modern world, few of us can escape from city lights to sit under a clear sky and enjoy its splendors. Fortunately, our ancestors had the time and inclination to do so. The sky on a clear, dark night is a wondrous sight, and before electric lights and television, watching the sky must have been a very popular activity.

As your eyes adjust to the darkness, perhaps the most impressive spectacle you see is the myriad of twinkling stars in the sky, ranging from dim to bright and from lone stars to clusters of many stars (**Figure P-1**). The wondrous *Milky Way,* a

Milky Way Historically, the diffuse band of light that stretches across the sky. Today the term refers to the **Milky Way Galaxy,** the group of a few hundred billion stars of which our Sun is one. It is this group of stars that causes the diffuse band of light.

Figure P-1
The night sky illuminates a fir tree in this photo. Brighter stars appear larger in a photograph because they overexpose the film. Their apparent size is not related to their actual size. Return to this figure after you gain some experience in recognizing stars and constellations on the night sky and try to find the Southern Cross, Carina, and alpha and beta Centaurus; these are some of the most significant southern stars and constellations.

Figure P-2
This wide-angle photo shows the Milky Way [in the constellations Sagittarius (left) and Scorpius (right)].

planet Any of the nine (so far known) large objects that revolve around the Sun, or similar objects that revolve around other stars.

hazy white area, may stretch across the sky (**Figure P-2**). If the Moon is visible, it presents a magnificent view, varying from night to night but always interesting.

Besides the hazy Milky Way, you can see a few smaller patches of dim light in the sky. We call them nebulae because of their indistinct, nebulous appearance. Photographs (**Figure P-3**) show that they have much more color than the naked eye can detect.

The patient observer is likely to see a meteor, a quick flash of light across the sky. Once called "falling stars," meteors are caused by (mostly) rocky visitors from beyond the Earth that glow as they burn in our atmosphere. A few times in your life, you will be able to see a comet, which will probably appear as the largest object in the sky.

If you watch for a few hours, you will see that the stars are not stationary at all, but move across the sky, most of them rising in the east and setting in the west (**Figure P-4**). The entire sky seems to be on a bowl rotating around us. Why are we privileged to be at the center of this majestic spectacle?

As you watch the sky from night to night and begin to remember patterns of the positions of stars, you will see that a few "stars" do not stay in the same place relative to the others. These are the ***planets***—wanderers on an apparently irregular path across the sky. You will make another observation as you watch the sky through the seasons: different stars are visible at different times of the year because the entire sky shifts to the west very slightly from night to night.

If you have the opportunity to use binoculars or a small telescope to view the heavens, a myriad of sights will be available. As we proceed with our quest for understanding in this text, we will point out many of the objects that are accessible to the backyard astronomer.

Questions, Answers, and Methods

Ancient observers wondered about the sights we have just described. We wonder at the same sights today. What are these objects, and where do we humans and our Earth fit into the realm of the universe?

Through the methods of astronomy, we are answering questions asked by those ancient observers as well as questions far beyond their imagining. As we find answers to our questions, we not only are filled with awe at the splendid universe, but we find more questions. The questions seem to be endless, providing excitement to those who are fortunate enough to be interested in the search for answers.

In this text we will try to describe not only the answers but also the methods used in the search. Studying the methods used by astronomers actually means studying scientific methods in general, for although each of the sciences uses different instruments, the basic methods of inquiry are similar for all the natural sciences. Thus we

Figure P-3
This is a photo (not a painting!) of a small area of the sky. The Eagle Nebula is a very luminous cluster of stars surrounded by dust and gas. The three pillars at the center of the image were made famous in an image by the Hubble Space Telescope.

will explore the very nature of the scientific endeavor, showing how astronomers gather data, how those data are developed into theories, how various theories compete against one another, and how and why some theories are retained and others abandoned. Data and theories are the working materials of all the sciences, and examples of the interplay between data and theory will be an essential part of our study.

Figure P-4
A time exposure shows the motion of stars as they set in the west. The lines are actually parts of circles, whose center is not in the picture. You can make photographs like this with any camera capable of taking time exposures. Choose a dark, starry night and point the camera toward the part of the sky you want to photograph. Fix the camera so that it won't move, and open the shutter for an hour or two. (See margin note by Figure 1-3 on page 15.)

P-3 From Earth to Galaxies

What we now know about celestial objects would have astounded our forefathers. We know, of course, that the Moon is a solid planet-like object about one-fourth the diameter of the Earth. The planets range in size from tiny Pluto, with a diameter less than 20% of Earth's, to Jupiter, a giant more than 11 times wider than Earth.

The celestial object most important to us, the Sun, dwarfs even Jupiter. Its diameter is about 10 times Jupiter's, 109 times Earth's. The nature of the Sun and the tremendous energy it produces have been mysteries through the centuries, but now we know that the Sun is a sphere made up almost entirely of hydrogen and helium and that its energy comes from nuclear fusion reactions.

What is truly astounding is that each of the stars that we see in the night sky is simply another sun, shining by the same processes that take place in our Sun. As you sit under the stars some night, try to think of each of those stars as a sun, and imagine how far away they must be to look as dim as they are. How many stars are there? Only about 5000 can be seen with the naked eye from Earth, but telescopes reveal hundreds of billions of stars clustered in a giant disk that we call the Galaxy (**Figure P-5a**). The Milky Way that stretches across our sky is caused by our view along the disk of the Galaxy (Figure P-5b).

Nebulae (Figure P-3) are giant clouds of gas, illuminated by light from stars within them. Though we compare them to clouds, we know today that the gas within them is extremely sparse, more sparse than a laboratory vacuum here on Earth. A nebula is visible to us only because it is very large; our line of sight passes through so much material that although any given part of it provides only slight illumination, the nebula as a whole is visible.

The Moon is discussed in Chapter 6 and the planets in Chapters 7 through 10.

The Sun is described in Chapter 11.

We will discuss the stars in Chapters 12 through 15.

Galaxies are the subject of Chapters 16 and 17.

Nebulae are involved in both the birth and the death of stars.

Figure P-5.
(a) Our Sun is one of a few hundred billion stars that form a spiral disk somewhat like this simplified, face-on drawing of our Galaxy. (b) This simplified edge-on drawing of the Milky Way shows that when we see it in the sky, we are looking along its disk, and distant stars are not distinct but instead appear as a bright haze. When we look in other directions, we see right out of the Galaxy.

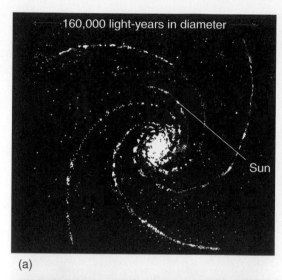

160,000 light-years in diameter

Sun

(a)

1 mile is about 1.6 kilometers.
1 yard is about 0.9 meters.

Looking out of the disk

Looking along the disk

(b)

astronomical unit A distance equal to the average distance between the Earth and the Sun. This is about 150 million kilometers or 93 million miles.

light-year The distance light travels in a year.

?

Is a light-year a unit of time or distance?

Figure P-6
This distant spiral galaxy (named NGC 2997) is in the constellation Antlia. Although this galaxy is not visible to the naked eye, a few other galaxies are. The bright stars in the photo are in our Galaxy and are much closer to us than NGC 2997, like flies on a car windshield.

Other objects that were formerly called "nebulae" are separate galaxies from ours, groups of millions to hundreds of billions of stars (**Figure P-6**). They appear to the eye as tiny, hazy splotches only because they are so very far from us.

P-4 Units of Distance in Astronomy

In everyday life we use different units of distance for different types of measurements. We might use centimeters and meters (or inches and feet in the United States) to measure distances around the house. Although it would be possible to continue to use these units when we describe distances across the country or across the Earth, it is much more convenient to use kilometers (or miles) in these cases.

To measure distances in the solar system, kilometers and miles are too small to be convenient. In this case we use the *astronomical unit* (abbreviated **AU**), which is defined as the average distance between the Earth and the Sun. Thus we are able to say that Mars is 1.5 AU from the Sun and that Venus gets as close as 0.3 AU to the Earth.

When we go from considering distances within the solar system to distances between stars, the astronomical unit is too small to be very useful. A unit that is handy in this case is the *light-year*, defined as the distance light travels in one year. The speed of light is tremendous—a bit less than 300,000 kilometers/second or about 186,000 miles/second—so a light-year is indeed a great distance. Since a year has approximately

$$365 \text{ days} \times \frac{24 \text{ hours}}{\text{day}} \times \frac{60 \text{ minutes}}{\text{hour}} \times \frac{60 \text{ seconds}}{\text{minute}} = 3.2 \times 10^7 \text{ seconds}$$

a light-year is about

$$300,000 \frac{\text{km}}{\text{second}} \times 3.2 \times 10^7 \text{ seconds} = 9.6 \times 10^{12} \text{ km}$$

or about 9.6 trillion kilometers or about 63,000 AU. (See the accompanying Tools of Astronomy box for an explanation of this notation. You see here the reason it is needed.) This is about 6 million million miles, or 6 trillion miles.

P-5 The Scale of the Universe

It is difficult for us to appreciate the sizes of objects in the universe and the distances between them (**Figure P-7**). To try to get a sense of the objects that are the subject of astronomy, let us construct an imaginary scale model of part of the universe.

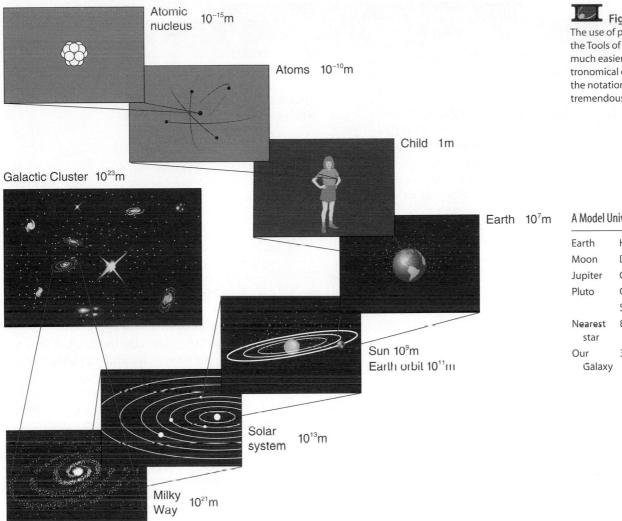

Atomic nucleus 10^{-15}m

Atoms 10^{-10}m

Child 1m

Galactic Cluster 10^{23}m

Earth 10^{7}m

Sun 10^{9}m
Earth orbit 10^{11}m

Solar system 10^{13}m

Milky Way 10^{21}m

Figure P-7
The use of powers-of-ten notation (see the Tools of Astronomy box) makes it much easier to describe the sizes of astronomical objects, but be careful that the notation does not obscure the tremendous differences in size.

A Model Universe with Sun as Basketball

Earth	Head of pin, 33 m from Sun
Moon	Dot, 8 cm from Earth
Jupiter	Grape, 170 m from Sun
Pluto	Grain of sand, 1.3 km from Sun
Nearest star	8,740 km from Sun
Our Galaxy	330,000,000 km diameter

Suppose in our model we represent the Sun by a basketball, as in **Figure P-8**. On this scale, the Earth is the head of a pin, about 107 feet away and invisible in the photo. The Moon is a dot the size of a period on this page and is about 3 inches from Earth. Jupiter, the largest planet, is the size of a grape about 186 yards away—almost 2 football fields away. Pluto, normally the farthest planet from the Sun, is a grain of sand about eight-tenths of a mile away. Thus the entire solar system occupies an area about 1.6 miles in diameter with a basketball-size Sun at the center.

Where is the nearest star (other than the Sun)? About 5400 miles away! If our model solar system is located in Halifax, Nova Scotia, the nearest star might be in Honolulu, Hawaii (**Figure P-9**). This is about the average distance between stars in our Galaxy, so we must imagine basketballs spread around randomly with approximately 5400 miles between adjacent ones. The diameter of the Galaxy on this scale would be about 206,000,000 miles. To imagine a scale model of the Galaxy, we would have to make the Sun much smaller; a 206,000,000-mile galaxy of basketballs is beyond our imagining.

By developing such models, we begin to appreciate the distances involved in the real universe. Such imaginary scale models will be constructed throughout the text. If you try to imagine the distances involved each time, the repetition will not be boring, but instead will be mind-expanding.

Here we approximate our basketball to be one foot in diameter. A "standard" basketball is more like 9.4 inches in diameter.

Figure P-8
If the Sun is a basketball, the Earth is the head of a pin 107 feet away.

TOOLS OF ASTRONOMY

Powers of Ten

Numbers in science—and particularly in astronomy—are sometimes extremely small or extremely large. For example, the diameter of a typical atom is about 0.0000000002 meter, and the diameter of the Galaxy is about 1,000,000,000,000,000,000,000 meters. Both are rather awkward numbers to use. We can avoid such clumsy numbers by using a variety of units, such as the astronomical unit within the solar system and the light-year for distances between stars. But sometimes it is inconvenient to switch units simply to avoid large and small numbers. Instead, scientists use *powers-of-ten* notation, also called *scientific* notation, or *exponential* notation. (The exponent is the power to which a number is raised.)

Scientific notation is simple because when the number 10 is raised to a positive power, the exponent is the number of zeros. For example:

$$10^1 = 10$$
$$10^2 = 100$$
$$10^3 = 1000$$

Written in meters, the diameter of the Galaxy contains 21 zeros, so it is written as 10^{21} meters.

If the number to be expressed in this notation is not a simple power of 10, it is written as follows:

$$2100 = 2.1 \times 1000 = 2.1 \times 10^3$$
$$305,000 = 3.05 \times 100,000 = 3.05 \times 10^5$$

Notice that to change a number from scientific notation to regular notation, you simply move the decimal point to the right by a number of places equal to that indicated by the exponent, filling in zeros if necessary.

Rather than explain negative exponents, we will just give some examples and let you see the pattern:

$$10^0 = 1$$
$$10^{-1} = \frac{1}{10^1} = 0.1$$
$$10^{-2} = \frac{1}{10^2} = 0.01$$
$$10^{-3} = \frac{1}{10^3} = 0.001$$
$$2 \times 10^{-8} = \frac{2}{10^8} = 0.00000002$$
$$4.67 \times 10^{-5} = \frac{4.67}{10^5} = 0.0000467$$

In this case, the same rule is followed, moving the decimal point by a number of places equal to the number indicated by the exponent, but this time moving it to the left and supplying any needed zeros.

When two numbers written in scientific notation are multiplied, to get the final answer you multiply the coefficients of the two powers of ten and add their exponents. For example:

$$(3.2 \times 10^5) \times (2 \times 10^2) = (3.2 \times 2) \times 10^{(5+2)} = 6.4 \times 10^7$$

and

$$(4.2 \times 10^5) \times (3 \times 10^{-2}) = (4.2 \times 3) \times 10^{(5+(-2))}$$
$$= 12.6 \times 10^3 = 1.26 \times 10^4$$

In the case of a division, you divide the coefficients of the two powers of ten and subtract their exponents. For example:

$$\frac{3.2 \times 10^5}{2 \times 10^2} = \frac{3.2}{2} \times 10^{5-2} = 1.6 \times 10^3$$

and

$$\frac{4.2 \times 10^5}{3 \times 10^{-2}} = \frac{4.2}{3} \times 10^{5-(-2)} = 1.4 \times 10^7$$

Figure P-9
If our solar system with its basketball Sun is located in Halifax, the nearest star is at the distance of Honolulu.

Simplicity and the Unity of Nature

It is our human nature to try to simplify things. In the sky we see a tremendous variety of objects and phenomena. Astronomy provides a method of seeing order in the apparent confusion. The more we learn about the objects that make up the universe, the more patterns we see, and the more order we find. As our knowledge of the universe expands, we become more and more aware of the unity of the cosmos.

Each of the natural sciences studies a different aspect of the universe, but because the unity exists, the various sciences overlap in many areas. Astronomy is particularly close to physics, with the two fields becoming indistinguishable at times. Likewise, astronomy and geology combine as we attempt to understand the formation, evolution, and surface features of the planets. Chemistry and biology become areas that astronomers are concerned about as they study the compositions of astronomical objects and the possibility of extraterrestrial life. And meteorology aids the astronomer in studying weather patterns on other planets.

astrophysics Physics applied to extraterrestrial objects.

The application of physics in astronomy is now so common that "astrophysics" is often used synonymously with "astronomy."

P-6 Astronomy Today

We live in exciting times. A few decades ago, humans walked and drove a vehicle on the Moon. In 1997, a robot vehicle sent us pictures from the surface of Mars. In 2004, two rovers at two different sites on the surface of Mars will start searching for and collecting information on a range of rocks and soils that holds clues to the planet's past water activity. After a 7-year journey, the *Cassini* orbiter will enter Saturn's orbit in July 2004 and, 6 months later, release its attached *Huygens* probe for descent through the thick atmosphere of Saturn's largest moon, Titan. Farther afield, we've detected planets around distant stars. The *Hubble Space Telescope* is returning amazing images to us, images of things never seen before (**Figure P-10**). Numerous new telescopes are under construction or being put into operation. Through the use of new instrumentation and new methods, we are discovering celestial objects that were undreamed of a few decades ago. The most distant galaxies are being observed and studied, and answers are being sought to questions as basic as the origin and the fate of the universe. Discoveries can be made only once, and the generations of people now alive are witnessing some of the most exciting discoveries ever.

Of what value is the science of astronomy? Will it help us advance toward worldwide prosperity? Probably not. Technological advances (sometimes beneficial and sometimes not) have followed almost every scientific advance. This, however, is not astronomy's purpose. Astronomy is a pure science rather than an applied science, and astronomers seek knowledge because knowledge is a reward in itself. Asking and answering questions is one of the things that make humans different from the other animals with whom we share our Earth. We are curious because we are human; we study the heavens because we are curious.

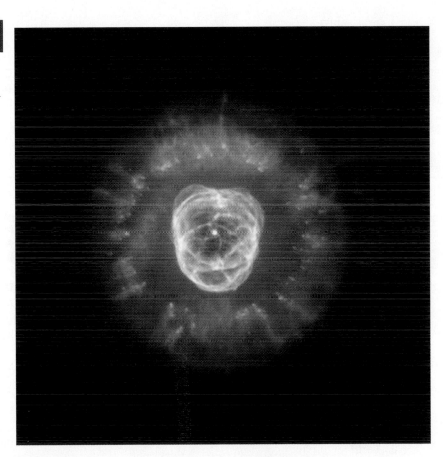

Figure P-10
The "Eskimo" nebula, first spotted by William Herschel in 1787, looks like a face in a fur parka when seen through ground-based telescopes. But the *Hubble Space Telescope* shows it to be an inner bubble of gaseous material being blown into space by a dying star, surrounded by a ring of comet-shaped objects. The nebula is about 5000 light-years from Earth. Glowing gases in the nebula produce the colors in this image: nitrogen (red), hydrogen (green), oxygen (blue), and helium (violet).

Conclusion

We know today that the universe is more wondrous than the most imaginative dreamer of old could have envisioned. To fully appreciate the wonder, however, one must understand the questions asked, the methods used, and the results produced by modern astronomy.

In our quest to understand the universe, we will journey through the solar system, to the stars, and then to distant galaxies, and we will come to appreciate that although our Earth is but a tiny rock circling an ordinary star, we humans have significance in the universe. The significance lies not in our size, but—at least in part—in the fact that our tiny brains have the ability to comprehend the spacious universe.

Study Guide

RECALL QUESTIONS

1. How does a star differ from a planet?
   A. Most stars are smaller than planets.
   B. Stars shine by their own light; planets don't.
   C. Stars move across the sky faster than planets.
   D. Stars appear dimmer than planets.
   E. Stars appear brighter than planets.

2. The Milky Way that we observe with the naked eye is
   A. the path planets take across the sky.
   B. debris left by the motions of planets.
   C. debris left by comets.
   D. the asteroid belt.
   E. stars in the galaxy of which the Sun is a part.
   F. sunlight scattered from water vapor in the atmosphere.

3. As we watch the sky during the night, most of the stars move across the sky
   A. from east to west.
   B. from west to east.
   C. from north to south.
   D. from south to north.

4. The smallest planet is _____ and the largest is _____.
   A. Earth . . . Saturn
   B. Mercury . . . Earth
   C. Pluto . . . Earth
   D. Pluto . . . Jupiter
   E. Earth . . . Jupiter

5. The diameter of the Sun is about _____ the diameter of the largest planet.
   A. one-tenth
   B. one-half
   C. the same as
   D. 10 times
   E. 100 times

6. In astronomical units, how far is the Earth from the Sun?
   A. 0.5
   B. 1.0

   C. 1.5
   D. 3.0
   E. 93,000,000

7. A light-year is defined as the
   A. time required for light to travel from Earth to Sun.
   B. distance to the nearest night-time star.
   C. distance around the Earth's orbit.
   D. distance from the Earth to the Sun.
   E. distance light travels in one year.

8. Which choice lists the objects in correct size from smallest to largest?
   A. Earth, Sun, solar system, Galaxy
   B. Sun, Earth, solar system, Galaxy
   C. Earth, Sun, Galaxy, solar system
   D. Earth, Galaxy, Sun, solar system
   E. [None of the above.]

9. The fundamental purpose of astronomy is to
   A. improve the living standards of humans by applying astronomical knowledge.
   B. ensure the safety of the human species by understanding the heavens.
   C. seek astronomical knowledge for its own sake.
   D. [All of the above about equally.]

10. Name at least seven types of celestial objects that are visible to the naked eye.

11. What is a star?

12. What causes the Milky Way we see in the sky?

13. Describe a scale model that includes the Earth, Moon, Sun, and nearest star.

14. What is a galaxy?

15. What is it about the appearance of a nebula that caused us to give it that name?

16. What causes a meteor?

QUESTIONS TO PONDER

1. The text states that astronomy is "still changing rapidly" to-day. Report on a new discovery or new hypothesis in astronomy that you find in the news during the next two weeks. Suggested sources: national television news, newspapers, and news magazines.

2. Science never arrives at a final answer. As it answers one question, it just finds more. Why then should we continue doing science?

3. Some people argue that astronomy does not produce anything useful for our lives, and therefore it does not deserve public funding. How do astronomers answer this? What do you think?

4. The ancient Greeks believed that "the universe is comprehensible" and that it can be described using mathematics. We consider this to be one of their most important contributions. What do you think we mean by saying that "nature is rational"?

CALCULATIONS

1. Using powers of ten, evaluate the following expressions:

$$(2 \times 10^{-2}) \times (3 \times 10^{3})$$

and

$$\left(\frac{4 \times 10^{-4}}{2 \times 10^{-3}}\right)^{2}$$

2. Verify that one light-year is about 63,000 AU.

3. The speed of light is about 300,000 kilometers/second and the average distance between Earth and Sun (1 AU) is about 150 million kilometers. How long does it take light to travel from the Sun to Earth? (Hint: An object moving at a speed v for a time t covers a distance d given by $d = v \times t$.)

4. Using scientific notation and data found in the appendices of the book, answer the following questions: (a) How many times larger in size is the Sun than the Earth? (b) The stars Rigel and Betelgeuse are both in the constellation Orion. Are they at the same distance from us? What does your answer suggest about the constellations? (c) How long does it take light to travel from Rigel to Earth? What does that tell you about the *current* state of this star?

5. The Crab Nebula is the remnant of a supernova explosion that was first seen by the Chinese in 1054 AD. The distance of this nebula from Earth is about 6,000 light years. In what Earth year did the explosion actually occur?

ACTIVITIES

Let us assume that we can build a *replica* of the solar system in which the Earth has a diameter of 1 mm. (Use relevant data from the appendices in your text.) In *this* replica, find the following: (a) the Sun's diameter (in cm); (b) the Earth-Moon distance (in cm); (c) the distances between the Sun and the planets (in m);

(d) the distance to the nearest star other than the Sun (in km). Consult a map of your area (for example, from a phone book), and plot on it the positions of the planets (and the Moon), assuming the Sun is located at your school.

EXPANDING THE QUEST

1. The book *The Measure of the Universe,* by I. Asimov (Harper and Row, 1983) will take you for a trip through the universe using powers of ten.

2. The book *Powers of Ten* by Philip and Phylis Morrison (W. H. Freeman, 1982) will also take you for a tour through the universe using powers of ten. There is also a video based on the book (Pyramid Film and Video, 1989).

Quest Ahead to Starlinks:
www.jbpub.com/starlinks

Starlinks is this book's online learning center. It features **eLearning,** which contains chapter quizzes and other tools designed to help you study for your class. You can also find **on-line exercises,** view numerous relevant **animations,** follow a guide to **useful astronomy sites** on the web, or even check the latest **astronomy news** updates.

STARLINKS

CHAPTER 1

An Earth-Centered Universe

This hand-colored engraving of Ursa Major appeared in *Uranometria* by Johann Bayer in 1603. This was the first attempt at a celestial atlas, and began the practice of naming stars after the constellation in which they appear.

URSA MAJOR—LATIN FOR *GREAT BEAR*—IS A PROMINENT CONSTELLATION in the northern sky, although most people see and refer to only the Big Dipper portion of it. The group of stars that form Ursa Major was identified as a bear by ancient civilizations in North America, Europe, Asia, and Egypt. Historian Owen Gingerich has suggested that the labeling of these stars as a bear may have originated in Asia or Europe as far back as the ice age and then gradually spread to other cultures. The Greeks associated it with a bear before the time of Homer, and the constellation is mentioned in the *Odyssey*. According to the Greek story of the origin of the constellation, the bear had once been a nymph who attracted the attention of Zeus, the father of the other gods. This caused Hera (the wife of Zeus) to be jealous and to change the nymph into a bear. Finally, to protect the bear from hunters, Zeus grabbed it by the tail and flung it into the sky. This is why Ursa Major has a longer tail than earthly bears do.

One legend in native American culture held that hunters had chased the big bear onto a mountain from which it leaped into the sky. The bowl of our Big Dipper is now the bear, and the handle is made up of the hunters who followed.

We currently have good explanations for what we see in the sky, and we understand the Earth's relationship to other astronomical objects. The search for this understanding began long ago, long before telescopes were invented. We will begin by examining what can be learned about the heavens without the use of modern instruments. Then we will consider a theory of the universe that is nearly 2000 years old.

Why does an astronomy text begin by looking back into history instead of plunging into today's astronomy? Furthermore, why start with an outmoded theory? There is good reason. In order to understand how astronomy—and science in general—works, we must look at how it progresses with time. We must see what led up to today's ideas. Science is a dynamic enterprise and the continual refinement of theories is one of its strengths. Every advance in science is necessarily incomplete and may well be partly inaccurate; this is a natural result of the fact that science is a human activity and thus includes people's biases. Looking back and understanding how a certain idea evolved allows us to better appreciate all the specific influences on that idea from people of different times and nationalities. We can also better understand the basis of the idea, and see how the scientific "knowledge filter" works, at least ideally, by recognizing and removing weaknesses such as personal biases during the different evolutionary stages of the idea.

Our starting place in the study of astronomy will be the same as that of people of long ago. We will consider the heavens as seen by the naked eye and will examine the observed motions of stars and planets. In the first three chapters we will consider two major theories that explain these motions, examining them for two different purposes: first, to answer the question of where the Earth fits into the scheme of things; second, to see how well these theories match the criteria for a good scientific theory. Finally, in these chapters we will show how and why one of these theories won out over the other, and how the understanding of nature provided by the successful theory eventually allowed humans to travel beyond the Earth and leave footprints on the Moon.

1-1 The Celestial Sphere

As we watch the sky during the night from any mid-latitude location on Earth, we see some stars rising in the east and setting in the west (Figure P-4). Stars above the poles of the Earth move in concentric circles, centered on a spot in the sky above each pole (**Figure 1-1**).

Another observation that can be made by even a casual observer is that the stars stay in the same patterns night after night and year after year. The Big Dipper *seems* to retain its shape through the ages as it moves around and around the North Star; in reality, its shape is very slowly changing, as we describe in the next section. It is easy to see why the ancients concluded that the stars act as if they were glued on a huge sphere that surrounds the Earth and rotates around it. **Figure 1-2** illustrates this *celestial sphere.* Its axis of rotation passes through the sphere at the center of the circles of Figure 1-1, and at the center of similar circles for the stars in the Southern Hemisphere. The photograph in Figure 1-1 was taken from the Northern Hemisphere, and the point at the center of the circles, the ***north celestial pole,*** is exactly above the North Pole of the Earth. Above the Earth's South Pole, we find a corresponding point on the celestial sphere called the ***south celestial pole.*** The motions of the stars make it appear that the Earth is sitting still at the center of the celestial sphere as it rotates around us. To picture the motion of the celestial sphere, you might think of it as spinning on rods that extend straight out from the Earth's North and South Poles. These rods would connect to the celestial sphere at the north and south celestial poles.

celestial sphere The imaginary sphere of heavenly objects that seems to center on the observer.

celestial pole The point on the celestial sphere directly above a pole of the Earth.

Figure 1-1
Notice that the stars in this time exposure of the northern sky seem to move around one point. Polaris, the North Star, formed the short, bright streak near the center of the circles. Polaris also moves around this center because our North Star is not located exactly above the Earth's rotation axis. Color differences between the streaks were caused by variations in brightness, not by the stars' actual colors.

Constellations

The word *constellation* comes from Latin, meaning "stars together."

constellation An area of the sky containing a pattern of stars named for a particular object, animal, or person.

It is natural, when we look at the sky, to look for order—for a pattern. All people feel this desire for order, and it is basic to science. Indeed, we can see (or imagine) patterns in the stars. The ancients saw similar patterns and identified them by associating them with beings in their particular mythology. **Figure 1-3a** is a photograph of the *constellation* Orion, and part (b) shows a drawing of the hunter Orion in Greek mythology. He is fighting off Taurus, the bull at the upper right. If you look at the evening sky in December, January, February, or March, you will see the stars of Orion. You may find it easy to imagine the three closely spaced stars as the belt of the hunter. And if you have a good dark sky, you can see the stars that form his sword's sheath hanging from the belt, as well as the stars forming his upraised arm.

The mythological creatures of most other constellations are much more difficult to imagine. Ursa Major, the Big Bear, is a constellation of the northern sky. (See the chapter-opening drawing.) Chances are, this group of stars will not remind you of a bear, but you will probably recognize the pattern of stars in the bear's rear and tail—the Big Dipper.

We know that as early as 2000 BC the Sumerians had defined several constellations, including a bull and a lion. As described in the story that opens this chapter, the constellation of the bear was common to many cultures. The 88 constellations that are used today were established by international agreement of astronomers. About half of these are constellations identified by the ancient Greeks, and the names we know them by are Latin translations of the original names. Today we realize that the constellations have no real identity. They are simply accidental patterns of stars, much the same as patterns you might have seen in the clouds when you were a young dreamer watching them pass overhead. The stars don't change their patterns as quickly as the clouds, however; so perhaps it was natural for ancient peoples to attribute real meaning to them. Besides, the stars were in the "heavens," and their association with the gods seemed natural.

Why do we say that the patterns are accidental? To begin with, the various stars are located at different distances from Earth. This means that if the Earth were in a greatly different position, we would see different patterns. For example, the stars of Orion are not all close to one another. Consider the two stars Rigel and Betelgeuse (Figure 1-3b). Rigel is nearly two times as far from Earth as is Betelgeuse. In addition, stars gradually move relative to one another along different directions and at different speeds. Given enough time, the shape of any constellation will change, but the distances from Earth of the stars in a constellation are so large that thousands of years must pass for a change in shape to be easily recognized. (Figure 12-9 shows how the Big Dipper is changing.)

In spite of their artificial nature, constellations are used by astronomers today to identify parts of the sky. We might say, for example, that Halley's comet was in the constellation Aquarius on Christmas Day, 1985. The ancient Greeks had no constellations in areas of the sky that did not have bright stars, nor in areas they could not see from their location in the Northern Hemisphere, so others have had to be added to their list. Today's 88 constellations fill the entire celestial sphere. In addition, the constellations of ancient cultures had poorly defined boundaries, so

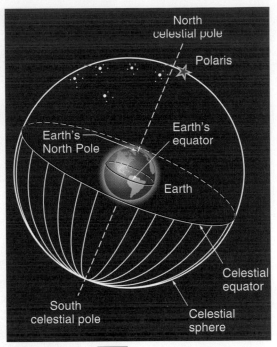

Figure 1-2
Because of their daily motion, objects in the sky appear to be on a sphere surrounding the Earth. (The celestial sphere is an imaginary sphere whose size is really much larger compared to Earth than it appears in this diagram, of course.)

(a)

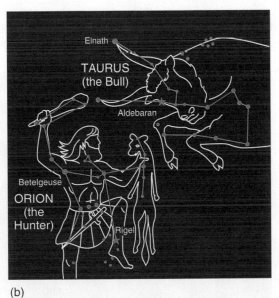

(b)

Figure 1-3
(a) The stars of the constellation Orion are visible high in the winter sky. (b) Ancient Greeks pictured Orion as a hunter warding off Taurus the Bull.

Do the shapes of constellations change through time?

Trails of stars that are on the celestial equator are straight lines. The rise and set points of such stars mark true east and west, and the angle that the trail of such a star makes with the horizon reveals the geographic latitude from which the picture was taken. A star that is almost exactly on the celestial equator is Mintaka (delta Orionis), the rightmost of the three stars in Orion's belt, as shown in Figure 1.3. (See Figure P-4, on page 5.)

Figure 1-4
The constellation Cygnus was seen as a swan, but we often call it the Northern Cross, as outlined here. The official boundaries of Cygnus are shown as white lines. Draco, Lyra, Vulpecula, and Lacerta are neighboring constellations. (You can find Cygnus in the evening sky in late summer and early fall.)

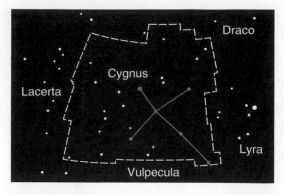

angular separation The angle between lines originating from the eye of the observer toward two objects.

astronomers have had to establish definite ones. **Figure 1-4** shows the constellation Cygnus and its boundaries.

Measuring the Positions of Celestial Objects

When people speak of objects in the sky, they sometimes talk about how far apart they are. What they are probably referring to is their *angular separation,* or the angle between the objects as seen from here on Earth. **Figure 1-5** illustrates this angle. We say, for example, that the angular separation of the two stars at the ends of the arm of the Northern Cross (Figure 1-4) is about 16 degrees. As **Figure 1-6** in-

Is an arcminute or an arcsecond a unit for measuring time?

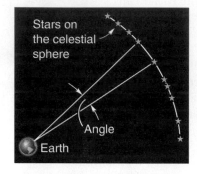

Figure 1-5
The two stars, when viewed from Earth, have an angular separation as shown.

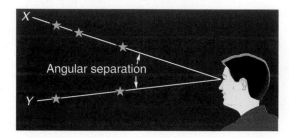

Figure 1-6
The angular separation of stars says nothing about their distances apart. All of the stars that lie along line *X* have the same angular separation from the stars that lie along line *Y*.

EXAMPLE

Mizar and Alcor are two stars in the Big Dipper that can be distinguished by the naked eye (**Figure 1-7**). They are separated by 12 arcminutes. Mizar, the brighter of the two, reveals itself in a telescope to be two stars, separated by 14 arcseconds. Express each of these angles in degrees.

Figure 1-7
The Big Dipper is part of the constellation Ursa Major. (Refer to the chapter-opening figure to see the full drawing of Ursa Major, the Big Bear.)

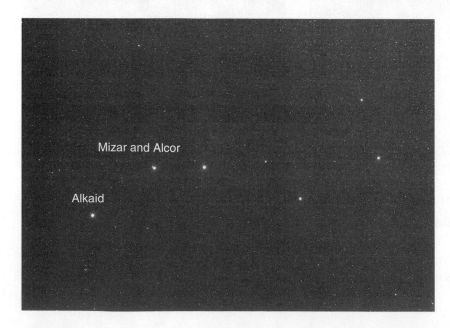

SOLUTION   Since 60 arcminutes equals one degree, we can create the conversion factors

$$\frac{1°}{60'} = 1 \qquad \text{and} \qquad \frac{60'}{1°} = 1$$

We choose the form of the conversion factor that causes the given unit to cancel, giving us the unit we want. In this case, multiply the 12 arcminutes by the first ratio, thus converting its value to degrees:

$$12' \times \frac{1°}{60'} = 0.20°$$

The second part of the problem is done in like manner, but another conversion factor is needed to convert arcseconds to arcminutes:

$$14'' \times \frac{1'}{60''} \times \frac{1°}{60'} = 0.0039°$$

TRY ONE YOURSELF

The top star in the head of Orion (Figure 1-3b) is a double star with a separation of 4.5 arcseconds. Express this angle in degrees.

The answer to each Try One Yourself is in Appendix H.

dicates, this says nothing about the actual distance between the stars. It tells us their angular separation, but not their distance apart.

It is often necessary in astronomy to discuss angles much smaller than one degree. For this purpose, each degree is divided into 60 **minutes of arc** (or 60 **arcminutes,** or 60'), and each minute is divided into 60 **seconds of arc** (or 60 **arcseconds,** or 60″). Notice that although these units are very similar in definition to units of time, they are not units of time, but units of angle. As an example of the use of these smaller units, a good human eye can detect that two stars that appear close together are indeed two stars if they are separated by about 1 arcminute or more. In Chapter 5 we will explain how the use of a telescope allows us to detect such double stars if they are separated by as little as one arcsecond—3600 times smaller than one degree.

There is an easy way to estimate angles in the sky. Make a fist and hold it at arm's length. The angle you see between the opposite sides of your fist is about 10 degrees (**Figure 1-8a**). For estimating smaller angles, the angle made by the end of your little finger held at arm's length is about one degree (Figure 1-8b). You can use these rules to estimate the angular separations of stars.

minute of arc One-sixtieth of a degree of arc.

second of arc One-sixtieth of a minute of arc.

celestial equator A line on the celestial sphere directly above the Earth's equator.

Celestial Coordinates

In order to describe the location of an object on Earth we use the coordinate system of longitude and latitude whose center is the center of the Earth. The use of these two numbers uniquely defines the position of the object on Earth for anyone. Similarly, astronomers specify locations of objects in the sky by using a coordinate system inscribed on the celestial sphere, the imaginary sphere centered on the Earth. The *equatorial coordinate system* describes the location of objects by the use of two coordinates, *declination* and *right ascension*.

The declination of an object on the celestial sphere is its angle north or south of the **celestial equator** (see

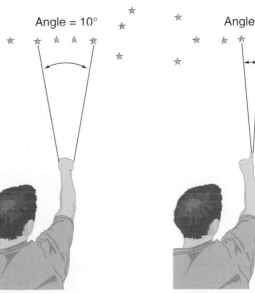

Figure 1-8
(a) Your fist held at arm's length yields an angle of about 10 degrees. (b) Your little finger held at arm's length cuts off an angle of about one degree. Both the Sun and the Moon have a diameter of about one-half degree.

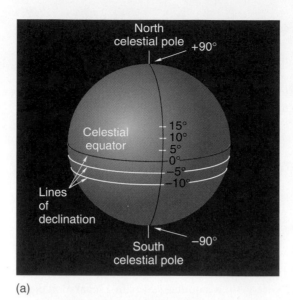

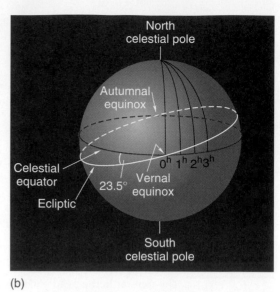

(a)   (b)

Figure 1-9
(a) Declination measures the angle of a star north or south of the celestial equator. Angles north are positive and angles south are negative. (b) Right ascension measures the angle around the celestial equator eastward from the vernal equinox, where the Sun crosses the equator moving northward. Angles are expressed in hours and parts of an hour, with 24 hours encompassing the entire circle. The ecliptic is the apparent path of the Sun on the celestial sphere; the angle between the planes of the ecliptic and the celestial equator is about 23.5 degrees.

Figure 1-9a and Figure 1-2). Angles north of the equator are designated positive and those south are negative. Thus the scale ranges from +90 degrees at the North Pole to −90 degrees at the South Pole. For example, Sirius (the second brightest star in our sky after the Sun) has a declination of −16°43′.

The right ascension of an object states its angle around the celestial sphere, measuring eastward from the vernal equinox (the location on the celestial equator where the Sun crosses it moving north). Instead of expressing the angle in degrees, however, it is stated in hours, minutes, and seconds (see Figure 1-9b). These units are similar to units of time, with 24 hours around the entire circle. Sirius has a right ascension of 6h 45m 9s, or—as it is usually written—$6^h45^m.15$, since 9 seconds is 0.15 minute.

1-2 The Sun's Motion Across the Sky

Like the stars, the Sun and Moon also seem to move around the Earth as the hours pass, rising in the east and setting in the west. Watching the Sun's motion over a few days might lead us to conclude that it is moving at the same rate as the stars, but if we carefully observe the stars in the sky immediately after sunset, we will see that they change as the weeks and months go by. It is as though the Sun were out among the stars, but not staying in the same place on the sphere. Instead, it seems to move constantly eastward among the stars. To explain: If we have a map of the stars that surround the Earth, we can locate the position of the Sun on this map by observing the stars that appear just after the Sun sets and again just before the Sun rises. The Sun is between those two groups of stars. If we do this again two weeks later, we would find that the Sun has moved and is now farther toward the east on our map. So it appears that the Sun does not participate fully in the motion of the celestial sphere. It appears to move around the Earth, but not as fast as the sphere of stars.

How long does the Sun take to get all the way around the sphere of stars? You could determine the approximate time simply by drawing the pattern of stars you see in the western sky after sunset tonight, and then waiting until you see that same pattern again. If you do that, you would probably find that the time is somewhere between 350 and 380 days. If you did this over a number of cycles, perhaps you could determine that the time is about 365 days. And, being an alert observer, you would undoubtedly notice that this cycle coincides with the cycle of the seasons here on Earth. Finally, you—or perhaps your descendants to whom you hand down your data—would decide that the time for the Sun's cycle through the stars exactly fits the cycle of the seasons, and that 365 days is the length of the year. (The time the Sun takes to return to the same place among the stars is actually not a whole number of days. It is about 365-1/4 days. Every four years we add one day to our year to make up for this difference. This is our leap year.)

See the Advancing the Model box on page 23 for more details on leap years.

Early in history, people noticed that certain star patterns appear in the sky at the same time every year. Even before the development of the calendar, the arrival of these patterns was used as an indication of a coming change of seasons. For example, the arrival of the constellation Leo in the evening sky meant that spring was coming, and this alerted people that even though the weather might not yet indicate it, warm days were on the way.

The Ecliptic

For an observer on Earth, as the Sun moves among the stars, it traces out the same path year after year. **Figure 1-10a** is a map of the stars in a band above the Earth's equator extending 30 degrees on either side of the equator. Part (b) of the figure shows the relationship of these stars to the Earth. In fact, the dashed line drawn straight across the map is directly above the Earth's equator and is called the celestial equator. The other line on the map shows the Sun's apparent path among the stars and shows that it is sometimes north of the equator and sometimes south. The apparent path that the Sun takes among the stars is called the *ecliptic,* and it is not the same as the celestial equator. (The name comes from the fact that an eclipse can occur only when the Moon is on or very near this line. This will be discussed further in Chapter 6.)

You may want to see Figure 2-5, which shows how today's model explains the apparent motion of the Sun among the stars.

ecliptic The apparent path of the Sun on the celestial sphere.

The constellations through which the Sun passes as it moves along the ecliptic are called constellations of the *zodiac.* **Figure 1-11** shows the 12 major constellations of the zodiac. (Actually, the Sun spends almost three weeks in a thirteenth constellation, Ophiuchus, which is a much longer time than it spends in Scorpius.)

zodiac The band that lies 9° on either side of the ecliptic on the celestial sphere.

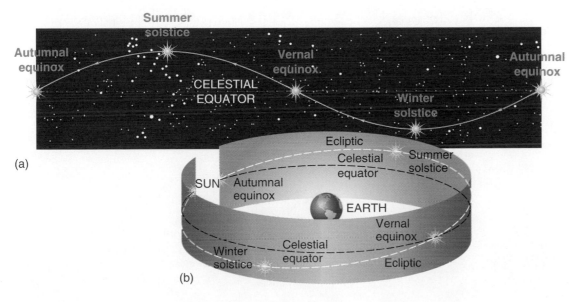

 Figure 1-10
Part (a) is a map of the stars within 30 degrees of the equator. Picture this map wrapped around the Earth as shown in part (b).

Figure 1-11
The constellations of the zodiac lie along the ecliptic. (Don't confuse the constellations of the zodiac with the astrological signs of the zodiac.) Note that in addition to the 12 zodiacal constellations that most people are familiar with, there is a thirteenth—Ophiuchus—that the Sun passes through. You may want to look at Figures 2-5 and 2-6 for an explanation of the Sun's apparent motion among the stars.

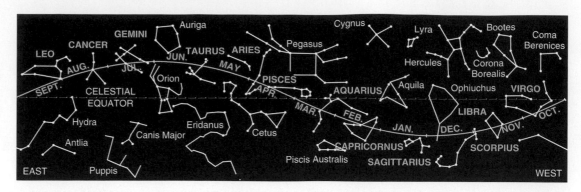

Notice the months indicated on the ecliptic. These show the Sun's locations at various times. On March 21 the Sun is in the constellation Pisces and is crossing the equator on its way north. The changing position of the Sun on the celestial sphere is related to the changing of the seasons, as we now explain.

The Sun and the Seasons

Why do we have seasons?

There are three easily observed differences between the behavior of the Sun in winter and summer:

- For an observer in the Northern Hemisphere, the Sun rises and sets farther north in the summer than in the winter. **Figure 1-12** shows the Sun's rising and setting points at various times of the year, as well as the Sun's path across the sky in each case. Although we may say that the Sun rises in the east, it actually rises exactly east only when it is on the celestial equator—at a particular time in March and again in September.

- The Sun is in the sky longer each day in summer than in winter. In December you may leave for an 8:00 AM class in near-darkness, and it may be dark again for your evening meal. Contrast this to June days, when it is more difficult to arise before the Sun, and darkness does not arrive until late in the evening. Figure 1-12 shows why this happens, for the Sun's path above the horizon is much

Figure 1-12
The Sun's apparent path across the sky of the Northern Hemisphere in (a) December, (b) March or September, and (c) June. Notice that the Sun rises and sets at different places on each date and that its noontime altitude differs in each case.

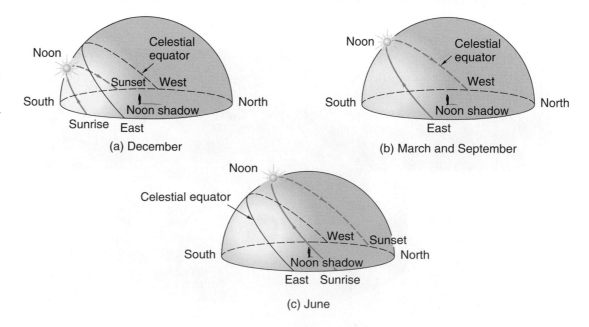

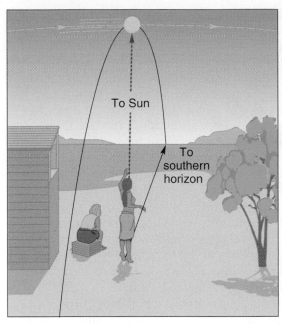 

(a) June   (b) December

Figure 1-13

The Sun's apparent path across the sky in the Northern Hemisphere in (a) summer and (b) winter. In each case we have drawn a line from south to north straight over the person's head (the meridian). The Sun moves across the sky along the yellow line and reaches a much higher position in the sky in summer than it does in winter.

Figure 1-14

In (a) the light from the flashlight shines perpendicularly onto the surface, while in (b) it strikes the surface at an oblique angle. The fact that the same amount of light illuminates more surface in (b) means that each little part of the surface is less illuminated in that case. Relate this to the noontime Sun's position in the sky in summer and winter.

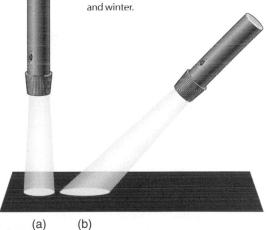

(a)   (b)

longer in June than in December. This effect is one reason for the seasons. Because the summer Sun is in the sky longer, we collect more solar energy during the daytime, and less time is available at night for our surroundings to lose the energy they have gained.

- The third difference in the Sun's observed behavior results in two more reasons for our seasonal differences. As Figure 1-12 illustrates, the Sun reaches a point higher in the sky in summer than in winter, and **Figure 1-13** emphasizes this by showing the Sun's path and its location at midday in late June and in late December. When the Sun is higher in the sky, its light hits the surface at an angle that gets closer to 90 degrees. Consider shining a flashlight directly down onto a surface in one case and shining it at an angle to the surface in another (**Figure 1-14**). In the second case, the same amount of light is spread over more surface area, and thus each portion of the surface receives less light. The same is true for sunlight, so a given area, in a given amount of time, receives more energy from the Sun in June than in December. In addition, since the Sun is never high in the sky in winter, its light must pass through more atmosphere in winter than in summer. As a result, each portion of the lit surface receives less direct light, since more of it is scattered and absorbed in the atmosphere. (See Figure 8-2 for this effect.)

If you live in the Southern Hemisphere, you know that this explanation is backward for your part of the Earth. In the Southern Hemisphere, the Sun gets higher in the sky in December than in June, and the seasons are reversed from those in the Northern Hemisphere. While people are enjoying summer in Canada, Australians are feeling the chill of winter.

It is logical for someone to believe that the difference in seasons is directly related to the change in distance of the Earth from the Sun. That is, you may believe that when we are closer to the Sun we have summer and when we are farther from the Sun we have winter. After all, you do feel warmer when getting closer to a fireplace. However, if this were the *only* reason for the seasons, then

You may want to see Figure 2-6, which shows that according to today's model, the tilting of the Earth's axis with respect to its orbital plane explains the seasons.

how could we have two different seasons, for the two hemispheres, at the same time? It turns out that the distance of the Earth from the Sun does not vary too much during a year, at least as a percentage of its average distance of one astronomical unit. (Also, the Earth is closer to the Sun in January and farther from the Sun in July.) By analogy, suppose you were standing at a distance of 1 meter from a fireplace and then walked toward or further from it by less than 2 centimeters. It obviously would not make too much of a difference in the amount of energy you received from the fireplace. Thus, even though in principle the amount of energy we receive from the Sun does depend on how far we are from it, in the case of our planet, distance is not an important factor and it does not really influence the seasons. The important factor in this case is the orientation of the Earth with respect to the Sun, as shown in the observations mentioned above.

To see how the path of the Sun along the ecliptic relates to the seasons, refer to the star map of Figure 1-11. Notice that the Sun is at its northernmost position on about June 21. If you picture the celestial sphere circling the Earth in late June, you can see that it will carry the Sun to a much higher position in the sky of North America. The Sun reaches its southernmost position on about December 21, and on that date it reaches the least ***altitude*** in the sky of the Northern Hemisphere (**Figure 1-15**).

The two dates discussed in the last paragraph are unique. During the spring the Sun gets higher and higher in the midday sky. Then about June 21—at the ***summer solstice***—it stops climbing, and after that it starts getting lower again. The reverse happens about December 21—at the ***winter solstice.*** (The word *solstice* is a conjunction of the Latin words "sol" meaning "Sun" and "sistere" meaning "to stand still.") These dates were given this name because at the solstice the Sun *stops* and reverses its direction. For example, during the spring the Sun moves farther and farther northward from day to day. At summer solstice it stops moving north and begins moving south.

The star map reveals two more unique events each year: Around March 21 the Sun crosses the celestial equator moving north, and about September 22 it crosses the equator moving south. On these dates every location on Earth experiences equal periods of day and night. The events are called the ***vernal equinox*** (or

?

Is the change of our distance from the Sun an important factor that causes seasons?

altitude The height of a celestial object measured as an angle above the horizon.

summer and winter solstices The points on the celestial sphere where the Sun reaches its northernmost and southernmost positions, respectively.

vernal and autumnal equinoxes The points on the celestial sphere where the Sun crosses the celestial equator while moving north and south, respectively.

Figure 1-15
The height the Sun reaches in the sky is explained by its position on the celestial sphere. In (a), the Sun is shown in its June position, and in (b) it is in its December position on the celestial sphere.

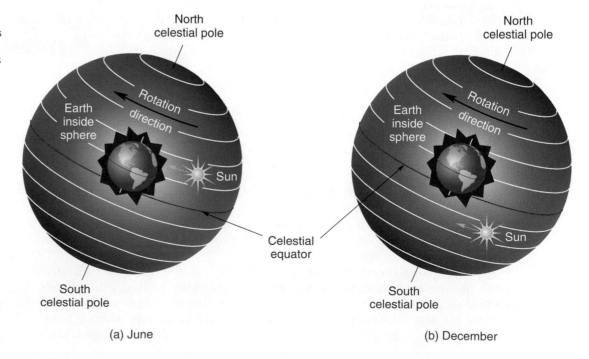

ADVANCING THE MODEL

Leap Year and the Calendar

People in early times (c. 30,000 BC) began keeping records of celestial events because they realized that certain cycles observed in the sky corresponded to familiar events in their surroundings, such as the arrival of migratory birds and the ripening of fruits. The agricultural revolution that began around 10,000 BC led to the need to better understand the celestial cycles and to predict the seasonal changes. Observational evidence collected through the years showed that there is a correlation between the seasons and the apparent motion of the Sun among the stars. As a result, the first calendars were created by the end of this period. The Indians of the American Southwest followed a lunar calendar (of 29.5 days average per lunar cycle), resulting in a 29.5 × 12 = 354-day year, which obviously needed an annual adjustment. A solar calendar (of 360 days) was used in Mesopotamia and Egypt. The 360-day year follows a Babylonian practice of using the sexagesimal (60-based) system, instead of our current decimal (10-based) system. (Our division of the circle into 360 degrees is probably a result of this Babylonian approximation for the year and the fact that 360 is divisible by 2, 3, 4, 5, and 6.)

Observations of solstices and equinoxes offered a way for early civilizations to reset the calendar every year. Archaeological evidence suggests that numerous astronomical traditions of keeping a "calendar" sprang up independently around the world; for example, at Stonehenge, England around 2800–2200 BC, and by the Plains Indians of Wyoming around 1500–1760 AD. Such evidence includes special alignments or designs of temples that mark solstices or other events, legends, and historical records of observation ceremonies carried out at the sites. Buildings designed with astronomical orientations were also built by the Egyptians (the great pyramids around 3000 BC), the Aztecs, Mayans, and Incas.

Our present calendar comes primarily from the Roman calendar. This calendar started its year in March; the Latin words for the numbers 7 through 10 are "septem," "octo," "novem," and "decem." By the time of Christ, January and February had been added, giving us the 12 months we have now. The months of the calendar were based on the Moon's period of revolution around the Earth and alternated in length between 29 and 30 days in order to match the average of 29 1/2 days between full Moons. Twelve of these lunar months totaled only 354 days, which was defined as the length of the standard year. To make up for the fact that this calendar quickly got out of synchronization with the seasons, an entire month was inserted about every three years—a sort of "leap month."

No single authority controlled the calendar, and by the time of Julius Caesar (100–44 BC), the date assigned to a specific day differed widely between different communities. In some countries, nearby communities did not even agree on what year it was. Caesar reformed the calendar in 46 BC, making the months alternate between 31 and 30 days, except for February, which had 29. Thus the *Julian calendar* had 365 days, and (almost) like ours, it added one day at the end of February every four years to make it correspond more closely to the time the Sun takes to return to the vernal equinox.* This latter time determines the seasons and is called the *tropical year*. Thus, the Julian year had an average of 365.25 days.

The tropical year is 365.242190 days long. This means that the seasons on Earth repeat after that period of time, and the difference between the tropical year and the average 365.25 days of the Julian calendar caused the calendar to gradually get out of synchronization with the seasons. The difference between the Julian and tropical years is about 0.0078 days per year, which corresponds to about 3 days for every 400 years. By the year 1582, the vernal equinox occurred on March 11 rather than March 25, as it had when Julius Caesar instituted the calendar. It was time for more reform. Pope Gregory XIII declared that 10 days would be dropped from the month of October, so that October 15 followed October 4 that year. That restored the vernal equinox to March 21, which corresponded to what the church wanted for establishing the date of Easter each year. To keep the calendar from having to be adjusted this way again, Pope Gregory instituted a new leap year rule: every year whose number was divisible by four would be a leap year—as in the Julian calendar—unless that year was a century year not divisible by 400 (such as 1800 or 1900). Thus 1700, 1800, and 1900 were not leap years, but the year 2000 was a leap year. This rule takes care of the 3 days per 400 years discrepancy mentioned above. Thus, the Gregorian year has an average of 365.2425 days.

Roman Catholic countries accepted the *Gregorian calendar*, but most other countries chose to stick with the old calendar. England changed in 1752, at which time it was necessary to omit 11 days. Russia did not adopt the Gregorian calendar until 1918.

However, the Gregorian calendar reform is also an approximation, albeit a better one than the Julian calendar reform. Since the difference between the Gregorian and tropical years is about 0.0003 days per year, there is a difference of about one day for every 3,300 years. As a result, a proposal has been made to include an exception to the Gregorian calendar: the years 4000, 8000, 12,000, and so on will not be leap years, as they would have been according to the original Gregorian calendar. Such a calendar will have an average of 365.24225 days and will be accurate enough that it will not have to be revised for about 20,000 years.

*The vernal equinox, defined in this chapter, determines the moment when spring begins.

spring equinox) and the ***autumnal equinox,*** respectively. (*Equinox* means "equal night.")

Why did we estimate each of the four dates above rather than state them exactly? Because they may vary a day or so from year to year. The fact that there are not exactly 365 days per year causes our calendars to get out of synchronization with astronomical events until a leap year allows us to catch up. In addition, your "today" may be the previous day or the next day somewhere else on the Earth.

1-3 Scientific Models

scientific model A theory that accounts for a set of observations in nature.

Thus far in this chapter we have explained the celestial motions we see by describing ***scientific models.*** The idea of the stars residing on a giant celestial sphere is a model that explains the observation of the daily motion of the stars. Likewise, the changing position of the Sun in the sky explains the changing of the seasons.

A scientific model is not necessarily a physical model, in the sense of a model car or a model airplane. A scientific model is basically a mental picture that attempts to use analogy to explain a set of observations in nature. Thus we are able to say that the stars appear as if they are on a sphere rotating around the Earth. Later in the book we will encounter some scientific models that cannot be represented well by a physical construction.

We can combine our explanation for the seasons with our model of the celestial sphere. Imagine a path along the ecliptic. As the celestial sphere rotates around the Earth, the Sun moves along this path (**Figure 1-16**), following the sphere's general motion. However, gradually the Sun creeps back eastward along the track. As the months pass, the Sun changes its position among the stars. It moves along the path with such speed that in about 365 days it is back to where it started.

The Sun in our model moves from north to south of the equator and back again. This corresponds to the observed motion of the real Sun. We are not saying that there is actually a real track on which the Sun moves, but our model provides this picture that helps us feel comfortable with the motion of the Sun. That is one of the functions of a scientific model: to allow us to make sense of a set of observations and thus feel comfortable with them. Our model does not actually explain the Sun's motion in the sense of telling us *why* it occurs, but it provides a mechanism that allows us to say, "OK, that makes sense now."

geocentric model A model of the universe with the Earth at its center.

The model we have constructed is a ***geocentric model;*** an Earth-centered model. This is not today's model of the universe, as we will explain in Chapter 2. However, in order to see how scientific models are developed, changed, and replaced, we will describe the geocentric model developed by the Greeks some 2000 years ago, and then we will examine reasons why the model has been replaced (but not entirely abandoned).

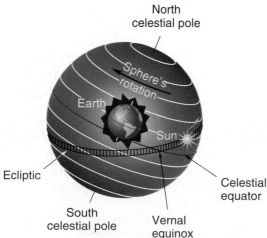

Figure 1-16
The Sun acts as if it follows a path around the celestial sphere. The sphere rotates toward the west as the Sun gradually moves along it toward the east.

The Greek Geocentric Model

To trace the evolution of our present model of the system of the heavens, we begin with the advanced Greek culture that lasted from about 600 BC until about 200 AD. Before looking directly at the Greek model, however, it is necessary to begin with a discussion of the world as seen by some of the ancient Greek philosophers, especially Aristotle. Aristotle lived from about 384 to 322 BC, but still influences today's thinking.

There is a fundamental difference between the contributions to astronomy made by the ancient Greeks and those made by the other ancient civilizations. The Babylonians studied the heavens mostly because they were interested in interpreting daily events, in the same way as some people do today with astrology. The ancient Egyptians studied the heavens because they were interested in making predictions for agricultural purposes. The ancient Chinese believed in a kind of astrology in which events in the heavens influenced events on Earth; thus they were very good at recording celestial events. However, the ancient Greeks were interested in astronomy because of a pure philosophical desire to understand how the universe works. They believed in, and looked for, a sense of symmetry, order, and unity in the cosmos. They took the first steps in creating a unified model of the universe (**Figure 1-17**).

Thales of Miletus (about 600 BC) believed that rational thought can lead to understanding of the universe. He also speculated that the Sun and stars were not gods, as was then usually thought, but balls of fire. Pythagoras (about 530 BC), one of the first experimental scientists, believed that nature can be described by numbers (that is, mathematics) and was the first to propose that Earth is spherical. His students, called the Pythagoreans, proposed around 450 BC that the universe is spherical, with a central "fire" containing a force that controls all motion. Around this fire, in order outward from the center, move Earth, the Moon, the Sun, the five planets, and the stars. This system predates by more than 2000 years Nicolaus Copernicus' revolutionary model of the planets moving around the Sun. The Pythagoreans were also the first to use the observed roundness of the Earth's shadow on the Moon as a supporting argument for the Earth being a spherical body. Plato (about 380 BC) taught the ideas of Thales and believed that the planets are spheres moving in circular orbits. He reasoned that astronomy contributed to the civilization of humanity.

Aristotle rejected the idea of the Pythagoreans that the Earth moves around a central "fire" and placed it back at the center of the solar system. Even though to us this choice seems to be obviously wrong, we must keep in mind that Aristotle's reasoning was supported by the observational evidence of his time. He argued that if Earth were moving, we ought to be able to see changes in the relative positions of the various stars in the sky, just as, if you drive down a highway, you see changes in the relative positions of nearby and distant trees. Such a shift in position due to motion is called parallax.

Had we never seen the stars, the sun, and the heavens, none of the words we have spoken about the universe would have been uttered. But now the sight of day and night, and the revolutions of the years, have created number and given us a concept of time as well as the power of inquiring about the nature of the universe: and from this source we have derived philosophy. No greater good ever was or will be given by the gods to mortal man.

Plato, in Timaeus

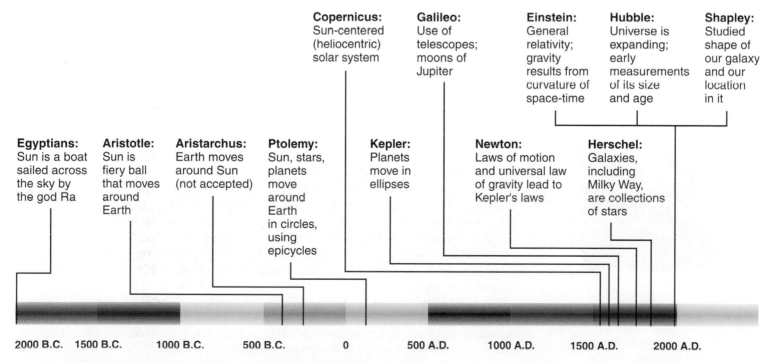

Figure 1-17
Models of the universe have changed over the past 4000 years, with most changes coming in the last 500 years. Chapter 1 takes us only to year zero.

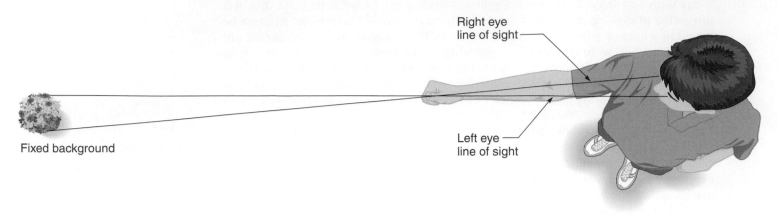

 (labels: Right eye line of sight; Left eye line of sight; Fixed background)

Figure 1-18

To observe parallax, hold one thumb in front of you and look at it first with one eye and then with the other.

parallax The apparent shifting of nearby objects with respect to distant ones as the position of the observer changes.

stellar parallax The apparent annual shifting of nearby stars with respect to background stars, measured as the angle of shift.

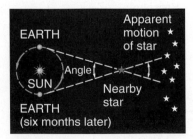

Figure 1-19

As the Earth goes around the Sun, we see parallax of a nearby star as it shifts its position against background stars. The amount of parallax is very much exaggerated here. For the nearest star, the angle shown is only about 1.5 arcseconds (0.0004 degree).

Hold your thumb in front of your face while you close one eye and look at the wall across from you (**Figure 1-18**). Now, without moving your thumb, view the wall with the other eye. Wink with one eye and then the other. You will see that your thumb seems to move from one spot on the wall to another. The explanation, of course, is that you are looking at your thumb from a different location each time you change eyes. In general, if our observing location changes, nearby objects will appear to move with respect to distant objects. This observation is given the general name *parallax.*

Figure 1-19 shows the Earth at two positions on opposite sides of its orbit. Just as your thumb shifted its apparent position when you alternately winked your eyes, if stars differ in their distances from Earth, a nearby star would be expected to shift its position relative to very distant stars. Such **stellar parallax** was first observed in 1838. All of the stars are so far away, however, that the greatest annual shift observed for any star is only 1.5 arcseconds, and this is why stellar parallax was not observed earlier.

A visual survey of the stars and constellations over time showed no evidence of such shift, so Aristotle concluded that Earth must not move. This is a great example of how a perfectly correct logical argument can lead to wrong conclusions if based on incomplete data. One can hardly blame Aristotle for not being able to observe parallax, since the stars are so much farther away from Earth than anyone of his time could imagine.

Aristotle thought the Moon is spherical, probably by noticing that the line separating the lit side of the Moon from the unlit side changes its curvature as the phases progress. He believed that the Sun is farther away than the Moon because the Moon's crescent phase shows that it passes between the Earth and Sun, and that the Sun appears to move more slowly in the sky than the Moon. Aristotle used three observations to argue that the Earth is spherical, saying that only on a sphere do all falling bodies seek the center; that as a traveler goes north, more of the northern sky is exposed while the southern stars sink below the horizon; and that during lunar eclipses, the edge of the Earth's shadow is always circular.

Aristotle's system of the world made a great distinction between earthly things and the things of the heavens. He observed that matter on Earth and matter in the sky seem to behave fundamentally differently. For example, objects on Earth have an undeniable tendency to fall to the ground; they fall *down.* In fact, we might define the word "down" as the direction things fall. Objects in the sky don't do that; instead they seem to move in circles around the Earth. Aristotle said that it is "natural" for earthly objects to move downward, that they seem *naturally* to seek the downward direction. On the other hand, it seems reasonable that heavenly objects move in circles. Something in the very nature of the objects seems to make them behave differently.

Aristotle saw another basic difference between the two types of objects: While earthly objects always seem to come to a stop, heavenly objects just keep going. Thus the "natural" motion of the two classes of objects appears to be different. This idea from Aristotle was a basic part of the world view of the Greeks, and it can be seen in their celestial models. They believed that there were two different sets of rules: one for earthly objects and one for celestial objects.

A celestial model developed by Greek thinkers before the time of Aristotle placed the stars on a sphere, similar to the model we developed earlier in this chapter. Instead of having the Sun move on a path on that sphere, however, the Sun was placed on another sphere that rotates around the Earth inside the sphere of the stars, as shown in **Figure 1-20**. To account for the fact that the Sun is sometimes seen north of the celestial equator and sometimes south, the Sun's sphere was thought to turn on a different axis from the one on which the stellar (celestial) sphere rotates. The difference between the tilts of the two spheres is shown in Figure 1-20.

There was good reason for the Greeks to use a sphere to carry the Sun, and the reason is linked to their admiration of geometry, which is traceable to the time of Pythagoras, the discoverer of the Pythagorean theorem. They sought geometric explanations for all natural phenomena. The Greeks then carried the idea of spheres much further and developed a model of the Earth, Sun, Moon, and planets using many spheres rotating around one another.

The Greek model we will discuss most thoroughly is that of Claudius Ptolemy (**Figure 1-21**), who lived around 150 AD. In his book (which we call the *Almagest*), he presented a comprehensive model that lasted for more than 1300 years after his death. Today we know this geocentric model as the ***Ptolemaic model.*** Ptolemy abandoned the spheres of earlier Greek models, for he saw no need to imagine actual physical objects carrying the Sun, Moon, and planets. He still spoke of the stars, however, as being on the celestial sphere.

In accordance with the thinking of Aristotle, people of Ptolemy's time thought that things in the sky (the "heavens") must be perfect. It seemed reasonable that the heavens would feature the circle, a symmetrical shape with no beginning and no end. Indeed, the stars seem to move in circles around the Earth. It seemed natural that they would lie in a spherical arrangement, and that the sphere would move around us at a constant speed. As you probably know from observation, the Moon changes its position among the stars from day to day just as the Sun does (but at a different speed). The motion and phases of the Moon are discussed in Chapter 6.

The Moon fits the Ptolemaic scheme perfectly. All that is necessary is that it moves around the Earth as the closest heavenly body. In fact, the Moon fit the overall Greek philosophical approach. Being the closest to the imperfect Earth, the Moon might be expected to have imperfections on it. Indeed, it has dark and light areas. It is somewhere between the imperfect Earth and the perfect heavens.

Observations of Planetary Motion

Thus far we have barely mentioned the other major class of objects in the sky: the planets. Without a telescope we can see five planets: Mercury, Venus, Mars, Saturn, and Jupiter. The Greeks, of course, knew of these planets. In fact, the word "planet" comes from a Greek word meaning "wanderer." And wander they do. Like the Sun and Moon, the planets move around among the stars on the celestial sphere. The Sun and Moon always move eastward among the stars, but the planets sometimes

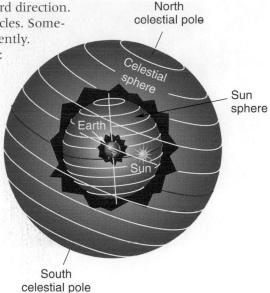

Figure 1-20
In order to account for the Sun's path around the Earth (and among the stars), the Greek model located the Sun on a sphere that moves around the stationary Earth inside the celestial sphere of stars. Notice that the axis of the Sun's sphere is tilted with respect to the axis of the celestial sphere.

Ptolemaic model The theory of the heavens devised by Claudius Ptolemy.

The Greeks listed the Sun and Moon as planets. That made seven planets, and this is the origin of our week having seven days. In fact, some of the days are named after planets.

Figure 1-21
Ptolemy.

(a)   (b)

Figure 1-22

(a) On January 8, 1999, Mars appeared as a bright star in the constellation Virgo, at upper right in the photo. East is toward the left. (b) This photo was taken on February 11, 1999. You can see that Mars moved toward the east (and a little southward) during the month. To understand why maps and photographs of the sky show east and west reversed from Earth maps (on which north is at the top, east on the right, and west on the left), think of yourself lying face down on the ground with your head toward the north. Your right arm points toward the east. Suppose you now turn over to look up to the sky; your right arm is now toward the west! Thus the difference occurs because when we look at a map of the Earth we are looking down, but when we view a map of the sky we are looking up.

? Why is East on the left on sky photographs?

stop their eastward motion and move westward for a while. They lack the simple, uniform motion of the Sun and Moon.

Figure 1-22a shows parts of Virgo and Libra on the night of January 8, 1999. Mars is the bright object at upper right. (The bright object at right center is the star Spica.) Figure 1-22b shows the same area of the sky a month later, on February 11. Notice that Mars has moved to the left and downward. These photographs show the sky with east toward the left and south at the bottom. So Mars moved eastward (and southward) with respect to the stars. A photograph taken a few weeks later would show that it was still moving eastward and southward. Things would be fine if Mars continued on in this uniform way. Uniform motions fit the idea of simple, circular motions in the heavens. (The model can account for the southward motion of Mars if the plane of Mars's motion is placed at an angle with the celestial equator. Mars moves northward again later in the year. This is similar to how the model explains the Sun's motion.)

The drawing in **Figure 1-23** shows the progress of Mars through the spring and summer of 1999, with its position marked for various dates. Notice that Mars continued its eastward motion until about March 19 and then started moving *backward,* heading west! It moved backward until about May 29 and then resumed its eastward motion again. Mars continues moving eastward until April, 2001, at which time it begins another of its backward loops. We call this backward motion ***retrograde motion.*** Retrograde motion is characteristic of planets, including those discovered in modern times.

retrograde motion The east-to-west motion of a planet against the background of stars.

Although the planets move among the stars in a seemingly irregular manner, there are limits to where they move, for they never get more than a few degrees from the ecliptic. Mercury and Venus have an additional limitation on their motion. These two never appear very far from the position of the Sun in the sky. We only see them either in the western sky shortly after the Sun has set or in the eastern sky shortly before sunrise. Mercury appears so close to the Sun that it is hard to find, even if we know where to look. This is because even when Mercury is at its maximum *elongation,* we have to look for it in the semi-bright sky during dawn or twilight. You never see Mercury or Venus high in the sky at night.

elongation The angle in the sky from an object to the Sun.

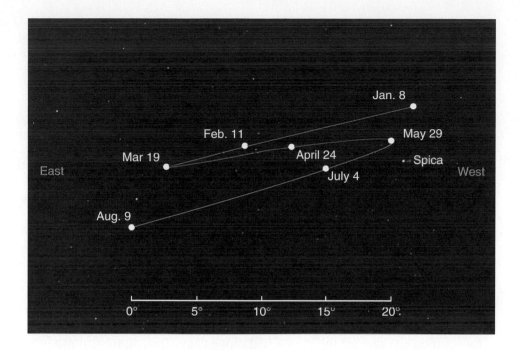

■ Figure 1-23
This shows the motion of Mars through the sky during its retrograde motion of 1999.

Any model of the planets must explain not only their observed retrograde motion, but why they stay near the ecliptic, and also the peculiar behavior of Mercury and Venus. How did the Greeks account for the planets' unique motions? Were the heavens perhaps imperfect, with the planets not moving uniformly in a smooth circle?

A Model of Planetary Motion: Epicycles

Ptolemy was indeed able to use circles to make a model that fit the planets' motions. It simply took more than one circle for each planet. (Though this idea was not originally his, he did elaborate on it, and he is generally given credit for it.) **Figure 1-24**

Having set ourselves the task to prove that the apparent irregularities of the five planets, the sun and the moon can all be represented by means of uniform circular motions, because only such motions are appropriate to their divine nature. . . . We are entitled to regard the accomplishment of this task as the ultimate aim of mathematical science based on philosophy.
Ptolemy

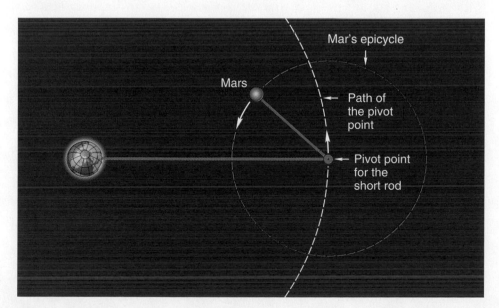

Figure 1-24
Mars seems to move in a circle on a second rod that rotates on the longer one.

shows an imaginary rod extending out from Earth with a second imaginary rod attached to its end. The second rod can swing around its connecting point on the longer rod. At the other end of this second rod, we place Mars. Now we rotate the big rod uniformly around the Earth. At the same time, we rotate the small rod around its pivot point on the large one. By a correct choice of rotation speeds, an observed motion for Mars such as that shown in **Figure 1-25** can be produced.

A person on Earth viewing the motion of Mars among the stars sees the planet moving eastward most of the time. This corresponds to the long rod turning around counterclockwise in the drawing. Occasionally, however, the planet retrogrades. This occurs when the planet on the end of the little rod passes closest to Earth, so that it appears to back up for a short time.

The rods referred to here did not appear in the Ptolemaic model because the model did not use actual physical objects to guide the planets. Instead a point moving around the Earth served as the center of motion for the smaller circle of the planet's motion. This smaller circle is called an *epicycle.* In this way, Ptolemy was able to preserve the idea of perfect heavenly circles and perfect, uniform, heavenly speeds.

How, then, does the model explain why the planets never move far from the ecliptic? The answer is that the plane of their circular motion lies very close to the plane of the Sun's motion. The greater the angle between these two planes, the farther the planet will move from the ecliptic as it moves among the stars.

The different behavior of Mercury and Venus was explained by having the centers of their epicycles remain along a line between the Earth and the Sun (**Figure 1-26**). While the center of other planets' epicycles could be anywhere on their circle around the Earth without regard to where the Sun is, Mercury and Venus are special cases.

epicycle The circular orbit of a planet in the Ptolemaic model, the center of which revolves around the Earth in another circle.

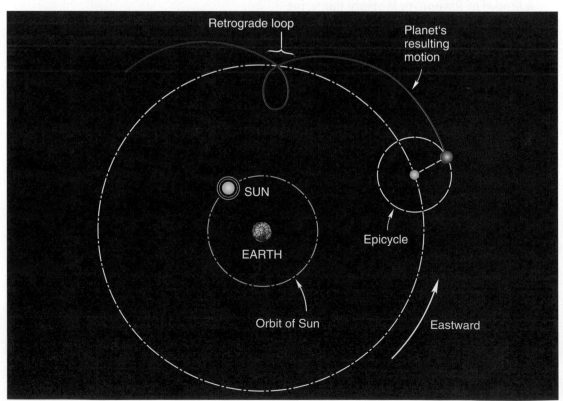

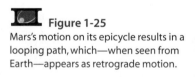 **Figure 1-25**
Mars's motion on its epicycle results in a looping path, which—when seen from Earth—appears as retrograde motion.

Retrograde loop

Planet's resulting motion

SUN

EARTH

Epicycle

Orbit of Sun

Eastward

You might want to look ahead to the middle of Chapter 2 for today's explanation of the limits on where Mercury and Venus appear.

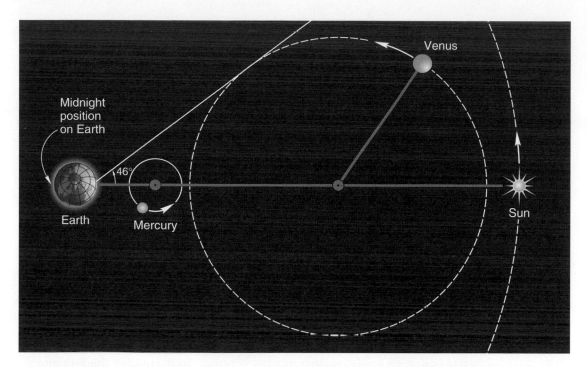

Figure 1-26
In the Ptolemaic model, the centers of Mercury's and Venus's epicycles stay between the Earth and the Sun. This accounts for the fact that the planets are never seen far from the Sun. Venus reaches a maximum of 46 degrees elongation.

1-4 Criteria for Scientific Models

The Ptolemaic model had many features that we look for today in a scientific model. We will discuss this model in light of three criteria that are applied to scientific models today. Even though these are the standards of today's science and not of the thinking of Ptolemy's times, it is instructive to apply them to Ptolemy's model.

- The first criterion is that the model must fit the data. It must fit what is observed. The Ptolemaic model fits the data pretty well in that it gives an explanation for the motions of the heavenly objects. The stars' motions were explained by the rotation of the celestial sphere on which they reside. The Sun and the Moon were theorized to move in circles around the Earth. Explanation of the planets' motions required the use of epicycles and special treatment for Mercury and Venus, but when these features were included, the planets' motions were accounted for. So the Ptolemaic model passes the first test. It fits the available data.

- The second criterion is that the model must make predictions that allow it to be tested. In other words, it must be possible to disprove the model—that is, to show that it needs to be modified to fit new data or perhaps needs to be discarded entirely. Many testable predictions are built into the Ptolemaic model. First, the model made predictions for the locations of the planets at times in the future. One could use the model to predict that Jupiter would be at a particular place in the sky at a particular time the next year. Another example: the Ptolemaic model holds that the Earth is stationary. It would predict that as knowledge advances and new methods are found to measure the motion of the Earth—either rotational motion or motion through space—no motion would be found. If further experimentation did not confirm this, the model would have to be adjusted drastically or abandoned.

"Prediction" here does not necessarily refer to the future. It means that the theory itself indicates an observation that will either support it or disprove at least

Actually, to make the model fit the data more accurately, Ptolemy either had to make the planets vary their speeds along their paths, or he had to move the Earth off-center among the planets' paths. He chose the latter method. After this and a few similar adjustments, the model fit pretty well.

part of it. Sometimes the observation has already been made and needs only to be checked against the theory. For example, one might check the Ptolemaic model's prediction of where Jupiter was 10 years ago.

- The third and final feature of a good model is that it should be aesthetically pleasing. This concept is difficult to define. It generally means that the model should be simple, neat, and beautiful. For example, Pythagoras is credited with the idea of placing the heavenly objects on a number of spheres rotating within one another. Later followers of Pythagoras wrote that the motions of these spheres resulted in "music of the spheres" that could be heard by those attuned to it. Such a claim would be quite unconventional—to say the least—in today's science. Nevertheless, the fact that people of that time spoke of such things does illustrate the necessity of a good scientific model having a pleasing quality or a beauty. Today's idea of beauty in a scientific model is more along the lines of symmetry and simplicity. A model should be as simple as possible; that is, it should contain the fewest arbitrary assumptions.

The principle that the best theory is the one that requires the fewest assumptions is often called ***Occam's razor.*** It is named for a fourteenth-century Franciscan monk-philosopher who stressed that in constructing an argument one should not go beyond what is logically required. Using Occam's razor, one "cuts out" extraneous suppositions.

Let's test Ptolemy's model against this last criterion. In his model, the Sun and Moon travel around the Earth in different paths than the stars do, but they obey basically the same rules, moving from east to west around the Earth, and they all travel in circles. This idea that all objects obey the same rules would be described by scientists as a form of symmetry.

To explain the motions of planets, epicycles were required. In using epicycles, the model retained the use of circles, a shape that was dear to Greek thinking; but somehow one wonders if a better method than epicycles couldn't be found to represent what is observed. In addition, the fact that different rules are required for Mercury and Venus makes the Ptolemaic model less pleasing. The need for different rules for certain planets means that the model lacks an aspect of simplicity. How many new special rules will be needed? Each one makes the model less appealing in the scientific sense. So we see that the Ptolemaic model begins to look less attractive as a good scientific model.

Remember, though, that the primary rule for a good model is that it fit the observations. An aesthetically pleasing model that did not come close to the real world would be almost useless. The Ptolemaic model *did* fit the data of the time, so we must judge it to be an acceptable model. It just lacks that certain neatness we would like.

We will return several times to a discussion of the features of a good scientific model.

Model, Theory, and Hypothesis

The three terms in the title of this section are words used not only by scientists, but by the general public. However, they mean something different in science from what they normally do in everyday language. We have been calling the work of Ptolemy a model and have discussed how this does not mean a physical model, but rather a developed set of ideas used to describe some aspect of nature. Instead of calling Ptolemy's creation a model, we could have called it a theory. The two words mean about the same thing in scientific usage, and we could have spoken of the "Ptolemaic theory." In science a scheme is not usually called a theory until its ideas are shown to fit observed data successfully. In nonscientific use, the word *theory* is often used to refer to ideas that are much more fanciful and less secure.

A hypothesis is a guess—perhaps a very intelligent guess—at a theory or part of a theory. Generally, a theory starts as a hypothesis. The idea of epicycles to explain

The simplicity spoken of here is not accepted as a necessary attribute of beauty in art, so we are speaking of a different type of beauty.

Occam's razor The principle that the best explanation is the one that requires the fewest unverifiable assumptions.

As pointed out in a previous margin note, Ptolemy's actual model was even more complicated and thus even less aesthetically pleasing.

ADVANCING THE MODEL

Astrology and Science

Astrology is the belief that the relative positions of the Sun, planets, and Moon affect the destiny of humans. Often astrology is used to determine the nature of a person's personality or of certain circumstances in the future, and it is traditionally used for timing or resolving problems in marriage, business, medicine, or politics. To discover the astrological effect on an individual, an astrologer designs, or "casts," a horoscope.

Casting a horoscope involves determining the positions of significant heavenly objects at the moment of the person's birth. To do so, the zodiac is divided into 12 equal-sized *signs*. Although these signs have the same names as the 12 principal constellations of the zodiac, they lack correspondence with the constellations in two ways. First, the constellations differ from one another in size, whereas all signs are the same size. The constellation Cancer, for example, is much smaller than Virgo or Pisces, but their signs are the same size. Second, the signs are at different locations in the sky from the constellations. At one time the positions corresponded fairly well, but because of *precession* (to be discussed in Chapter 3), the signs and constellations are now "out of whack" by about one position. Thus the sign of Aries is located mostly in the constellation Pisces. This means that a person born on March 22 is said by astrologers to be an Aries, while the Sun actually appeared to be in the constellation Pisces at the person's birth. (Some astrologers, called sidereal astrologers, use the constellations rather than the signs.)

Astrologers divide the sky above us into 12 *houses*, starting with the eastern horizon and proceeding around the sky. The houses do not rotate with the stars, but remain in the same position relative to Earth. Thus the first house stays above the eastern horizon. The houses are arbitrarily associated with various areas of our lives, such as children, health, and enemies. Likewise, the planets are associated with various good or evil influences. Mars, for example, is associated with war.

When the above information is determined for the moment of a person's birth, we have what is called the *natal chart* for that person. This chart could be made by anyone who knows the sky and has tables of solar, lunar, and planetary positions. It might be determined, for example, that the planet Mars was just above the eastern horizon at the time of your birth. So what? This is where interpretation by the astrologer comes in. After such interpretation, the astrologer is supposed to be able to tell you about your personality and life pattern.

Scientific Criteria Applied to Astrology

Having briefly described the nature of astrology, we will now apply to it the criteria for a good scientific theory. The first criterion: Does astrology fit the observations? It claims to be able to account (at least in part) for our personalities and our characteristics. Can it? This question has been tested many times. Invariably, the tests show that astrologers are unable to identify the personality traits of people or to predict future events. (For example, read the report in the December 5, 1985 issue of *Nature* concerning the research done by Shawn Carlson at the University of California.) After numerous tests, we must conclude that astrology fails our first criterion.

The second criterion for a good scientific theory is that it makes verifiable predictions that could possibly prove the theory wrong. There is confusion as to whether astrology does this or not. In Carlson's study, astrologers claimed that if enough cases were considered, they could make verifiable predictions. Astrologers would say that it is not fair to test the accuracy of a horoscope of an individual person because a horoscope only predicts tendencies, and people do not necessarily follow their tendencies. In other words, you might be very different from what is predicted by your horoscope and still not contradict astrological predictions. If this is the case, there is no way that astrology can be proved incorrect. It therefore fails the second criterion.

The third criterion is one of aesthetics. Astrology does not pass this test because it is not a single, unified theory at all, but instead is a group of arbitrary rules as to the effect of heavenly objects on people. One could hardly call it aesthetically pleasing.

Since astrology is a faith system, it cannot be disproved to those who believe. People who believe in it do not care whether scientists can explain the belief or not.

Scientists do not accept astrology as a valid theory. Some scientists strongly object to the publication of astrological columns in newspapers, but others claim that astrology does no harm if people use it only as a pastime and do not let it actually determine what they do with their lives.

planetary motion had been used by Hipparchus, who lived about 300 years before Ptolemy. It was then a hypothesis. The hypothesis was developed and included with others to form Ptolemy's final model. According to today's scientific use of the words *theory* and *model*, Ptolemy's plan would not even have been given those names until it was shown to be able to explain the heavenly motions.

The gods did not reveal, from the beginning, all things to us; but in the course of time, through seeking, men find that which is better. But as for certain truth, no man has known it, nor will he know it; neither of the gods, nor yet of all things of which I speak. And even if by chance he were to utter the final Truth, he would himself not know it; for all is but a woven web of guesses.

Xenophanes, Greek Historian, 6th century BC

heliocentric Centered on the Sun.

You might hear someone refer to Einstein's theory of relativity or to the theory of biological evolution and say, "Well, it is *only* a theory," meaning that you mustn't put too much faith in it. It is correct that these two concepts are only theories, but keep in mind what a scientist means in calling something a theory. Later chapters will show that Einstein's theory is well founded and has been experimentally verified many times. Although the theory of evolution will not be discussed here, virtually all biologists form the same conclusion about its validity.

As we will explain in Chapter 3, although it is called a law, Newton's law of gravity is no more a "law" and no less a "theory" than Einstein's theory.

1-5 Aristarchus's Heliocentric Model

About 400 years before Ptolemy, around 280 BC, the Greek philosopher Aristarchus had proposed a moving-Earth solution to explain the motions of the heavens. According to this model, the reason the sky seems to move westward is that the spherical Earth is spinning eastward. Although Ptolemy was aware of Aristarchus's model, he argued that if the Earth turned on an axis, it would be moving through the air around it, and therefore a tremendous wind should be observed in the opposite direction. Ptolemy refered to the moving-Earth model as being a "simpler conjecture." He saw that it was more aesthetically pleasing, but he believed it was not a good model because it presented obvious contradictions with the observation that the air stays where it is, along with everything loose on Earth. Ptolemy was unable to see that the Earth might be carrying the air and everything on it along as it rotates.

Aristarchus's model was a *heliocentric model*, constructed about 1800 years before Copernicus proposed the one that we use today (in a modified form). Aristarchus visualized the Moon in orbit around a spherical Earth and the Earth in orbit around the Sun. One could argue that Aristotle's argument, based on the lack of observable parallax for the stars, clearly shows that Aristarchus's heliocentric model is wrong. Why, then, did Aristarchus propose such a model? As we will see, the answer lies in the results he obtained about the relative distances and sizes of the Earth, Moon, and Sun.

Relative Distance to the Sun When the Moon is exactly half lit, Aristarchus correctly stated that the angle between Earth-Moon-Sun is 90 degrees (**Figure 1-27**).

Figure 1-27
Relative positions of the Earth, Moon and Sun when the Moon is exactly half lit. Aristarchus used this geometry to find the relative distances of the Moon and Sun from Earth.

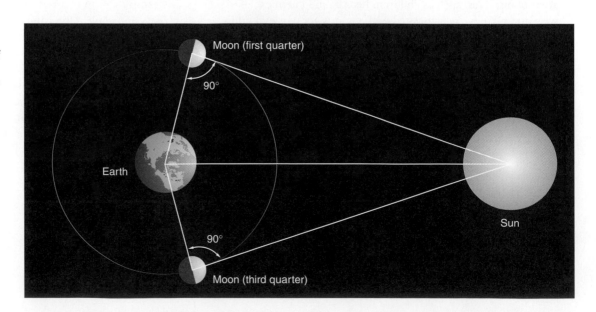

For the angle Moon-Earth-Sun Aristarchus used 87 degrees, instead of the correct value 89.85 degrees; we do not know what argument he used to defend this choice. Knowing the angles in the right triangle Earth-Moon-Sun he was then able to obtain a relationship between the distances Earth-Moon and Earth-Sun. He found that the Sun is about 20 times farther from the Earth than the Moon is, instead of the correct value of about 390. Note, however, that this large difference of a factor of about 20 is due to the difference of only about 3 degrees in the choice of the angle Moon-Earth-Sun. What is important here is not the quantitative error, but the power and simplicity of the argument he used to find, 2000 years before anybody else, the relative distance to the Sun.

A calculation at the end of the chapter offers such an argument.

Relative Sizes of the Moon and Earth From the time it takes the Moon to move through the Earth's shadow during a total lunar eclipse, Aristarchus concluded that the Earth is about three times larger than the Moon (the correct ratio is about 3.7). The geometry by which he estimated the relative sizes of the Moon and Earth is shown in **Figure 1-28.** The fact that total solar eclipses just barely occur leads to the conclusion that the angular diameters of the Moon and Sun are about the same as seen from Earth. Aristarchus measured this angular diameter to be 0.5 degree and thus he represented Earth's shadow by lines leading away from Earth and converging at an angle of 0.5 degree. These criteria specified the position and size of the Moon in the Earth's shadow.

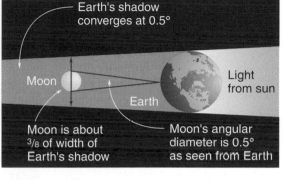

Figure 1-28
Aristarchus used the time it takes for the Moon to go through the Earth's shadow to find the relative size of the Earth and Moon.

Relative Sizes of the Moon and Sun Since the angular diameter of the Moon and the Sun are the same (0.5 degree) as seen from Earth, Aristarchus correctly concluded that

$$\frac{\text{radius of Moon}}{\text{radius of Sun}} = \frac{\text{distance Earth–Moon}}{\text{distance Earth–Sun}}$$

Using the calculations mentioned above, he found that the Sun is about 20 times larger than the Moon, instead of the correct ratio of about 390; this error is the result of the difference of only 3 degrees in the choice of the angle Moon-Earth-Sun shown in Figure 1-27.

These calculations clearly showed the Sun being much bigger than Earth. As a result, Aristarchus concluded that the Sun, not Earth, must be the central body in the solar system, thus proposing the first heliocentric system. For this, an outraged critic declared he should be charged with impiety.

Aristarchus had made a map of the solar system; he simply did not have the scale for it. For example, if he knew the radius of the Earth and the distance to the Moon, then he could accurately find how big the Moon and Sun were and their exact relative distances.

Measuring the Size of the Earth

The first person to clearly understand the shape and approximate size of the Earth, 1700 years before Columbus, was Eratosthenes (276–195 BC). His calculation of the radius of the Earth is very simple (**Figure 1-29**).

Eratosthenes heard that at noon, during the summer solstice, the Sun shone directly down a well near Syene (the present city of Aswan, in Egypt), casting no shadow. This implied that the Sun was directly overhead, or at the **zenith** at that time. In his city of Alexandria, on the same day, he noted that the Sun's direction was off the vertical by about 7 degrees. Since he was aware of Aristarchus' results, Eratosthenes could safely assume that the Sun is far away from the Earth and thus its light travels in almost parallel rays toward the Earth. Thus, he realized that the difference of 7 degrees had to be due to the curvature of the Earth. Since 7 degrees is about 1/50 of a full circle (7/360), Eratosthenes concluded that the Earth's circumference was

zenith The point in the sky located directly overhead.

Figure 1-29
Eratosthenes calculated the size of the Earth by comparing the Sun's direction off the vertical at two locations at the same time. When the Sun was overhead at Syene, it was 7 degrees from the zenith at Alexandria.

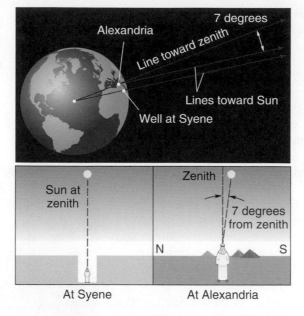

solar system The system that includes our Sun, planets and their satellites, asteroids, comets, and other objects that orbit our Sun.

about 360/7 or 50 times the distance from Alexandria to the site of the well. This distance was about 5000 stadia, where a stadium was a unit of distance equivalent to about 0.15 to 0.2 kilometers. If we assume that one stadium was 1/6 km, then the distance between the two cities is about 830 kilometers. Thus, Eratosthenes calculated the Earth's circumference to be $50 \times 830 = 41{,}500$ kilometers and thus the Earth's radius to be 6,600 km. This is amazingly close to the correct value of 6,378 kilometers!

Combining the calculations of Aristarchus and Eratosthenes, the ancient Greeks had, for the first time, measurements of the radii of Earth, Moon, and Sun, and the relative distances between them. Even though later astronomers obtained more accurate measurements, we would have to wait until 1769 AD to obtain the actual value of the astronomical unit and thus the true dimensions of the *solar system.*

What is important here is not the accuracy of the measurements. It is the fact that simple logical arguments allowed the ancient Greeks to have a very good sense of the solar system more than 2000 years ago. Sometimes an idea is so revolutionary that it takes a long time for others to become convinced. In this case, the lack of observable parallax for the stars allowed the Ptolemaic model (instead of Aristarchus') to be the accepted model for the heavens for almost 1400 years, until a little after the time of Copernicus. Parallax of stars was successfully measured around 1838 AD. As a result, Aristotle's argument convinced one of the greatest observers of all times, Tycho Brahe (1546–1601), to believe in a geocentric universe decades after the Copernican revolution.

Conclusion

There is a tendency to think that Greek science was bad science because it is not today's science. This is simply not true. It is true that the Ptolemaic model is more primitive than today's. The Greeks did not have the accurate observations and extensive data we have today. Their model, however, did fit the data they had. In fact, it fits the casual observations of most people today. Can you think of any direct evidence obtainable without a large telescope that contradicts Ptolemy's model? What observations can you personally make, without relying on reference materials, which will show the 2000-year-old Ptolemaic model to be a poor model?

Study Guide

1. When you see a Moonrise, you see it
   A. in the western part of the sky.
   B. in the eastern part of the sky.
   C. [Either of the above, depending upon whether you are in the northern or southern hemisphere.]
   D. [None of the above, for the Moon never rises (or sets).]

2. Polaris is unique because it
   A. moves in a different direction than any other bright star.
   B. is the brightest star in the sky.
   C. twinkles more than any other bright star.
   D. is fairly bright and shows very little motion when viewed from Earth.
   E. [The premise is false. Polaris is not unique at all.]

3. A time exposure of the northern sky shows a circular pattern. The center of the circle is
   A. exactly at the north celestial pole.
   B. very close to the north celestial pole, but not exactly on it.
   C. the location of Polaris.
   D. the location of the nearest supernova.
   E. [Two of the above.]

4. In ancient times, people distinguished the planets from the stars because
   A. planets appear much brighter than any star.
   B. features on planets' surfaces could be seen whereas no star's features could be seen.
   C. planets move relative to the stars.
   D. planets differ in color from the stars.
   E. planets can be seen during the day.

5. Ptolemy's system of epicycles was used to explain the apparent
   A. daily motion of the stars.
   B. annual motion of the Sun around the sky.
   C. backward motion of the Moon through the sky.
   D. changing speeds and directions of the planets among the stars.
   E. motion of the Sun north and south of the celestial equator.

6. Which of the following choices can be seen from Indiana?
   A. Stars near the north celestial pole.
   B. Stars near the ecliptic.
   C. Stars near the celestial equator.
   D. [All of the above can be seen from Indiana.]
   E. [None of the above can be seen from Indiana.]

7. Which of the planets known to the ancients can *never* be seen high overhead at night?
   A. Mercury and Venus only.
   B. Mars, Jupiter, and Saturn only.
   C. Saturn only.
   D. Venus only.
   E. [All of the planets *can* be seen high in the sky at some time of night.]

8. On the first day of spring, the Sun rises
   A. north of east.
   B. directly east.
   C. south of east.
   D. [Any of the above, depending upon your location on Earth.]

9. In January, the Sun rises
   A. south of east.
   B. directly east.
   C. north of east.
   D. [Any of the above, depending upon your location on Earth.]

10. What causes summer to be hotter than winter?
    A. The Earth is closer to the Sun in summer.
    B. The daylight period is longer in summer.
    C. The Sun gets higher in the sky in summer.
    D. [Both B and C above.]
    E. [All of the above.]

11. When we experience spring in the Northern Hemisphere, it is _____ in the Southern Hemisphere.
    A. also spring
    B. summer
    C. fall
    D. winter
    E. [No general statement can be made.]

12. The Sun's apparent path among the stars
    A. is south of the celestial equator.
    B. is right on the celestial equator.
    C. is north of the celestial equator.
    D. is south of the celestial equator part of the time and north of it part of the time.
    E. crosses the north celestial pole once each year.

13. In Australia, the longest daylight period occurs in late
    A. March.
    B. June.
    C. September.
    D. December.
    E. [No general statement can be made.]

14. The Sun crosses the celestial equator on the first day of
    A. winter.
    B. spring.
    C. summer.
    D. fall.
    E. [Two of the above.]

15. If the Earth rotated east-to-west instead of west-to-east,
    A. our seasons would occur in reverse order.
    B. we would not have seasons.
    C. daylight periods would be longer in winter and shorter in summer.
    D. winter in the Northern Hemisphere would occur in July.
    E. [None of the above.]

16. Which of the following planets appear(s) to move through the background of stars?
    A. Venus
    B. Mars
    C. Jupiter
    D. [Both A and B above.]
    E. [All of the above.]

17. The constellations of the zodiac
    A. lie (roughly) along the path of the Sun.
    B. are all about equal in size—30 degrees wide.
    C. are distributed fairly evenly over the northern portion of the celestial sphere.
    D. are distributed fairly evenly over the entire celestial sphere.
    E. [Both B and D above.]

18. The ecliptic and celestial equator intersect at two points called the
    A. equinoxes.
    B. solstices.
    C. tropics.
    D. sidereal points.
    E. poles.

19. Why have astronomers added modern constellations to the sky?
    A. To replace some that changed in shape through the years.
    B. To replace some whose meaning has changed.
    C. To combine previous small ones into large ones.
    D. To name some where none had been designated.
    E. [Three of the above answers are correct.]

20. In science, what is the difference between a *theory* and a *hypothesis?*
    A. A hypothesis is more fully developed than a theory.
    B. A theory is more fully developed than a hypothesis.
    C. A hypothesis is based on a model, while a theory isn't.
    D. There is essentially no difference. The words are interchangeable.

21. Eratosthenes calculated the size of the Earth from
    A. angles to the Sun from locations a measured distance apart.
    B. angles to the Moon from locations a measured distance apart.
    C. its angular size and distance from the Sun.
    D. its orbital speed and distance from the Sun.
    E. its calculated rotational speed.

22. A minute of arc is
    A. a measure of how far the Sun moves during one minute of time.
    B. one-sixtieth of a degree.
    C. how far the Earth turns on its axis in one minute.
    D. 60 degrees.
    E. the angular diameter of the Sun.

23. Thirty arcminutes is about _____ degrees.
    A. 0.008
    B. 0.5
    C. 180
    D. 1800
    E. [None of the above.]

24. In what century did Ptolemy live?

25. As you watch a planet during the night, in what direction does it appear to move across the sky?

26. How do stars near Polaris appear to move as we watch them through the night?

27. What is a constellation and how do astronomers use constellations?

28. What observation convinced Pythagoras that the Earth was spherical?

29. What is the ecliptic?

30. What are the approximate dates of the summer and winter solstices and what is their significance?

31. What is the origin of the word *planet?*

32. In what direction across the background of stars do the planets normally move?

33. Define *retrograde motion* of a planet.

34. Name the planets that are never seen far from the Sun in the sky.

35. What is an epicycle and what did the use of epicycles enable us to explain?

36. In what part of the sky must you look to find the planets Mercury and Venus?

37. Define *geocentric.*

38. List three criteria for a good scientific model.

39. Which of the criteria for a scientific model is the most important?

40. What is meant by the *elongation* of a planet?

41. Explain how Eratosthenes measured the diameter of the Earth.

QUESTIONS TO PONDER

1. Ancient sky-watchers were quite imaginative in attaching figures to groups of stars in the sky. Use the stars of Figure 1-3 to outline some figure other than a person and a bull.

2. Figure out a method not mentioned in this chapter that would allow you to measure the length of the year by astronomical observations.

3. Specify the two coordinates by which we describe locations in the sky. List the units of each and the maximum and minimum value of each.

4. How does our calendar adjust for the fact that the year is about 365-1/4 days long rather than an even 365 days?

5. What is meant by saying that one criterion for a good scientific model is its aesthetic quality?

6. Ptolemy's model placed the Sun on a sphere whose axis was not the same as the axis of the celestial sphere. Why was it necessary for the Sun's sphere to have a different axis?

7. Why did Ptolemy include epicycles in his model?

8. Explain why the Greeks used circles and spheres to account for heavenly motions.

9. What observation forced the Ptolemaic model to locate the center of Venus's and Mercury's epicycles on a line between the Earth and the Sun?

10. Name the three criteria for a good scientific model and discuss their relative importance.

11. Describe one observation that would disprove the Ptolemaic model and explain why it would conflict with the model.

12. The star in Orion's left leg (Rigel) is much farther away than the other bright stars in the constellation (see Figure 1-3). If you look at Orion in the sky, and then move far into space in the direction to your left, how will that star shift relative to the others?

13. We sometimes speak of the sky as being "the Heavens," and it is traditional in Christian circles to speak of a spiritual heaven as being up above us. Cite a pre-Christian historical reason for this link between the sky and heaven.

14. Identify the position of the vernal equinox, autumnal equinox, and the two solstices on Figure 1-11.

15. Explain the concept of Occam's razor by giving an example of its use in distinguishing which is the better of two hypotheses. (You may make up two hypotheses about some everyday event if you wish.)

16. Why do people who live near the equator not experience major seasonal changes? (The answer, "It is always hot there." is not sufficient.)

17. Find the Arctic Circle on an Earth globe. What is the astronomical significance of this line? Where is the Tropic of Capricorn and what is its significance?

18. Ptolemy observed parallax of the Moon against the distant stars. Explain how this can be observed. (Hint: It is not due to the motion of the Moon or of the Earth around the Sun.)

CALCULATIONS

1. Express an angle of 15 arcseconds in degrees.

2. Four degrees of angle is how many arcminutes?

3. Suppose one star is 3 arcminutes from another. What is this angle in degrees?

4. Given that the angle between the celestial equator and the ecliptic is 23.5 degrees, what is the declination and right ascension of the vernal equinox? Of the autumnal equinox? Of the winter solstice? Of the summer solstice?

5. Suppose you live on planet X. One day the Sun is directly over a place 500 km to your south, while at your position the Sun's direction is 10 degrees off the vertical. What is the radius of planet X?

6. The time between successive similar phases for the Moon is on the average about 29.53 days. (In Chapter 6, we define this time period as a *lunar month* or the *synodic period* of the Moon.) If it takes the Moon an average of 1.2 hours more to go from first to third quarter than to go from third to first quarter, show that the angle Moon-Earth-Sun in Figure 1-27 is about 89.85 degrees.

ACTIVITIES

1. Obtain one of the star charts published monthly in astronomy magazines such as *Astronomy* and *Sky & Telescope*. On a clear, cloud-free night, use the star chart to locate as many constellations of the zodiac as you can. Can you locate the celestial equator? the ecliptic? Spend some time outdoors and see if you could clearly distinguish the motion of the stars toward the west. Has Polaris moved? What is the overall pattern that the stars follow?

2. Let us assume that you would like to repeat Eratosthenes' calculation of the radius of the Earth. However, neither you nor your friend who volunteered to help you live at a place where the Sun is directly overhead at any day of the year. You decide to give it a try anyway. Both of you set a stick vertically on the ground and around noontime on the same day you measure the smallest shadow that the stick makes on the ground. You know the size of your stick and therefore, by drawing a scale of a right triangle (vertical stick and horizontal ground) you are able to determine the angle that the Sun's direction makes with the vertical. You also know the direct, north-south distance between your locations. Do you have enough information to find the size of the Earth? If yes, how would you calculate it?

EXPANDING THE QUEST

Astronomy and *Sky & Telescope* are the most popular monthly magazines covering the latest developments in astronomy for a general audience.

1. A simple description of archaeoastronomy is given in "Archaeoastronomy: Past, Present and Future" by A. Aveni (*Sky & Telescope*, Nov. 1986, p.456).

2. A recounting of Eratosthenes' experiment is given by Carl Sagan in *Cosmos* (Random House, 1980), in the chapter entitled "The Shores of the Cosmic Ocean."

3. O. Gingerich, "From Aristarchus to Copernicus" (*Sky & Telescope*, Nov. 1983, p.410).

4. An award-winning history of the development of astronomical thought is given by Timothy Ferris in *Coming of Age in the Milky Way* (Morrow, 1988).

5. The book *Introductory Readings in the Philosophy of Science, 3rd ed.* (edited by E. D. Klemke, R. Hollinger, D. Wyss Rudge, and A. D. Kline; Prometheus Books, 1998) has a number of interesting articles on science, including Paul Thagard's "Why Astrology is a Pseudoscience."

STARLINKS

Quest Ahead to Starlinks: www.jbpub.com/starlinks

Starlinks is this book's online learning center. It features **eLearning,** which contains chapter quizzes and other tools designed to help you study for your class. You can also find **on-line exercises,** view numerous relevant **animations,** follow a guide to **useful astronomy sites** on the web, or even check the latest **astronomy news** updates.

CHAPTER 2

A Sun-Centered System

"WHAT STRUCK ME MOST WAS THE SILENCE. It was a great silence, unlike any I have encountered on Earth, so vast and deep that I began to hear my own body: my heart beating, my blood vessels pulsing, even the rustle of my muscles moving over each other seemed audible. There were more stars in the sky than I had expected. The sky was deep black, yet at the same time bright with sunlight.

The Earth was small, light blue, and so touchingly alone, our home that must be defended like a holy relic. The Earth was absolutely round. I believe I never knew what the word 'round' meant until I saw Earth from space."*

These are the words of Russian cosmonaut Aleksei Leonov, describing what it was like for him to walk free in space. For a time, Leonov was a human satellite of Earth.

The motion of satellites was first described almost 400 years ago by Johannes Kepler. His laws of planetary motion apply to planets orbiting the Sun, to the Moon orbiting the Earth, and even to satellites, space stations, and solitary astronauts.

■ *Interview with Aleksei Leonov, March 18, 1987, Zvyozdny Gorodok, USSR. By permission of Michael Woods, Senior Editor, the *Toledo Blade.* Quoted in *The Home Planet,* Kevin W. Kelley, Ed. (Reading, MA: Addison-Wesley and Moscow: Mir Publishers, 1988).

Kepler's laws supported the Sun-centered model of the solar system described by Nicolaus Copernicus, providing tools for better understanding of planetary motions that the Earth-centered systems could not match.

Today's space age provides striking evidence of how the Earth fits in the Sun's planetary system, but Copernicus and Kepler published their ideas before there was any proof. They relied on elegant reasoning and data from observations that were becoming more accurate and more abundant with time. As a result, astronomy grew slowly but steadily as a science.

In Chapter 1 we described the development of Ptolemy's geocentric model of the heavens. Recall that although this model is called the Ptolemaic model, it was not entirely due to Ptolemy. He worked in Alexandria, Egypt, and had access to the great Greek library there. Ptolemy constructed his model by molding the thought and tradition of the past—much of it the work of Aristotle and Pythagoras—and adding his own insights to arrive at the final result. This is typical of scientific development. When we look with hindsight at the great advances in science, we see that they occur when the time is ripe for them, when someone is ready to bring together previous thoughts in the area with new insight to take a great step forward. It seems that all that is needed is the right person to take this step—the person who sees a little more clearly than others, is ready to take the leap of imagination into the future, and is able and willing to do the work and report the results.

Ptolemy's model, developed about 125 AD, served as the accepted model for more than 1400 years. In this chapter we present a new model, due to Copernicus, and examine the arguments for and against each of the two competing models, based on modern scientific criteria. As we will see, the new model finally won out because of adjustments made long after it was first introduced that helped it pass the primary test of a scientific model: its ability to fit the observations accurately. Finally, we present the work of Tycho Brahe, one of antiquity's best observers, and discuss how his data helped his assistant Johannes Kepler to formulate the first *mathematical* laws describing the motion of the planets around the Sun. The time period discussed in this chapter marks the beginning of a revolution in our understanding of the heavens.

2-1 The Marriage of Aristotle and Christianity

During the thirteenth century, Saint Thomas Aquinas, one of the greatest theologians and philosophers of the Christian church, incorporated the works of Aristotle and Ptolemy into Christian thinking. Aquinas insisted that there must be no conflict between faith and reason, and he blended the natural philosophy of Aristotle with Christian revelation. For the Aristotelian and Ptolemaic ideas we have discussed, the blend was an easy one. The idea of an Earth-centered universe fit comfortably with literal biblical interpretation, for it placed humans at the center of God's creation—the ultimate expression of the divine will (**Figure 2-1**).

As we pointed out in Chapter 1, the idea of a central, unmoving Earth was natural for early humans. Through the work of Aquinas, this easily accepted idea was shown to fit perfectly with Christian beliefs. So Aristotle's science—and with it the Ptolemaic

Figure 2-1
A medieval illustration from a world atlas of 1661, showing a model of the universe with the Earth at its center. This was the model approved by the Catholic Church, which had the power to punish anyone who argued against this model. Each planet is shown being carried in a chariot drawn by winged horses.

model—became even more entrenched in Western culture. It was no longer just a natural, normal way of thinking about the world, but was part of Christian thinking and religious dogma.

Why have we presented material that seems to be more closely related to church history than to science? The reason is that science does not exist apart from human culture. Science and scientists are part of the society in which they live, and it is necessary to know something of the flavor of an age to appreciate the work of the thinkers of that age—including those thinkers who today might be labeled as scientists. Thus it is important to emphasize that after the thirteenth century, the teachings of Aristotle and the Ptolemaic model were an ingrained part of Western thinking.

A very important characteristic of the Middle Ages was a great reliance on authority, particularly on authorities of the past. Today most of us have a much greater tendency to rely on our own thoughts, observations, experiences, and feelings than did people of the times we are discussing. Thus Aquinas relied on the authority of the Bible, the authority of earlier churchmen, the authority of his superiors in the church, and the authority of Aristotle. Arguments were often settled by reference to authorities rather than by personal experience or independent experimentation.

Into this world came a man who was to cause a revolution in the way people think.

2-2 Nicolaus Copernicus and the Heliocentric Model

Figure 2-2
The Sun-centered model of the Universe that appeared in Copernicus's *De Revolutionibus*. It is said that a copy of the printed book was rushed to Copernicus so that he could see it as he lay on his death-bed.

At the time when Columbus was journeying to America, a brilliant Polish student was studying astronomy, mathematics, medicine, economics, law, and theology. As was common for scholars of that time, Nicolaus Copernicus did not limit himself to studies in a particular discipline as we do today (perhaps necessarily). Later in life his primary responsibility was that of a churchman, canon of the cathedral of Frauenberg. He is known today, however, for initiating the revolution that finally resulted in replacing the geocentric system of Ptolemy with a *heliocentric*—Sun-centered—system.

Copernicus worked on his heliocentric system for nearly 40 years. Even after all this work, he was slow to publish it. It was eventually published with the title *On the Revolutions of the Heavenly Spheres,* but it is often called *De Revolutionibus,* a shortened form of the original Latin title (**Figure 2-2**).

Copernicus apparently decided to develop and publish his model for two primary reasons. First, as the centuries had passed, it was found that Ptolemy's predicted positions for celestial objects were not in agreement with the carefully observed positions. For example, Ptolemy's model could be used to predict the position of Saturn on some night in the future. The prediction of its position for a night a few years later was accurate enough, but when the prediction was made for a few centuries in the future, the predicted position might differ from the carefully observed position by as much as 2 degrees. This difference corresponds to about four times the diameter of the Moon. Ptolemy's model had been updated several times through the years so that it would be fairly accurate for a while. These corrections were not refinements in the way it worked, but were "resettings" of each planet's position, that were made to fit the latest observations. A good model would not require such updating. Copernicus recognized this and sought a more accurate model.

A second reason Copernicus sought another model was that he did not believe that the Ptolemaic model was aesthetically pleasing enough. We will return to a discussion of the criteria for a good model after we have described the model developed by Copernicus.

ADVANCING THE MODEL

Copernicus and His Times

Mikolaj Koppernigk (or Nicolaus Copernicus by his Latinized name; **Figure B2-1**) was born in Toruń, Poland, and was educated in Poland and Italy. During most of his life he worked for the church, serving as canon of the cathedral of Frauenberg, Poland. Copernicus lived during an exciting—and disturbing—time, as a list of some of his contemporaries suggests: Henry VIII, Martin Luther, Michelangelo, Leonardo da Vinci, Christopher Columbus, and Raphael.

Copernicus entered the University of Bologna (Italy) in 1496 and was liberally educated in several subjects. His great interest in astronomy can be attributed to the close link between astronomy and religion at the time. There were two reasons for this link: First, it was important to the church that its feast days be celebrated at the right time, and this required a correct calendar—a matter for astronomers. (The dates of Easter and the feasts that follow it depend even today upon accurate determination of the time of the vernal equinox.) Second, at the time of Copernicus, astrology was considered a more important study than astronomy, but astrology cannot function without accurate astronomical data. Thus astronomy was a necessary stepsister of what were considered more important subjects.

Copernicus lived about 100 years before the invention of the telescope. Measurements of the heavens had to be made with the naked eye, and yet accurate measurements were necessary. One important measurement was the exact time that a celestial object crossed the meridian. (The *meridian* is an imaginary line drawn on the sky from south to north passing directly over the observer's head.) Copernicus made this measurement by having his house constructed with a narrow slot in one of its

walls. By placing himself at the correct location and looking through the slot, he could determine accurately when a given object was crossing (or *transiting*) the meridian.

It has long been known that planets change their brightnesses as they shift positions among the stars. The Ptolemaic system accounted for this with its epicycles, for they would cause a planet to change its distance from the Sun and from Earth. Copernicus noticed, however, that Mars seemed to change in brightness even more than would be predicted by that model. The

Figure B2-1
Nicolaus Copernicus (1473–1543).

epicycles of Ptolemy were simply not large enough to explain the great changes in Mars's brightness. Thus Ptolemy's model explained the basic phenomenon, but it did not explain it very precisely when accurate data were used. It was this observation that first prompted Copernicus to reconsider the heliocentric system of Aristarchus, which had lain hidden in obscurity for 2000 years.

Copernicus studied the ancient writings and became convinced that a Sun-centered system would not only provide better data for use by the church and by astrologers, but that it would also be a more aesthetically pleasing system. He linked astronomy very closely with his religious faith, arguing that placing the Sun at the center of everything is completely logical because the Sun is the source of light and life. The Creator, he said, would naturally place it at the center.

The Copernican System

Copernicus's system revived many of the ideas of Aristarchus. (Recall the discussion near the end of Chapter 1.) Ptolemy had explained the daily motions of the heavenly objects as being due to the rotation of the celestial sphere from east to west around a stationary Earth. Like Aristarchus, Copernicus pointed out that an Earth rotating from west to east under a stationary sky produces the same observations as the Ptolemaic model (**Figure 2-3**). Ptolemy had said that such a rotating Earth would produce a great wind. To solve this problem, Copernicus stated as an assumption that the air around the Earth simply follows the Earth around. How high above the Earth does this air extend? Copernicus did not answer this question, but notice that his theory requires that the air not extend all the way to the stars, or even to the Moon.

Copernicus's system is heliocentric; that is, the Sun is at its center. The Earth assumes the role of just another one of the planets, all of them revolving around the Sun. The Earth becomes the third planet from the Sun. **Figure 2-4** shows the order of the planets that were known in the sixteenth century.

Try the first Activity at the end of this chapter to appreciate the change in outlook as one switches from a geocentric to a heliocentric system.

If the air extended to the Moon, it would tend to drag the Moon along with the Earth.

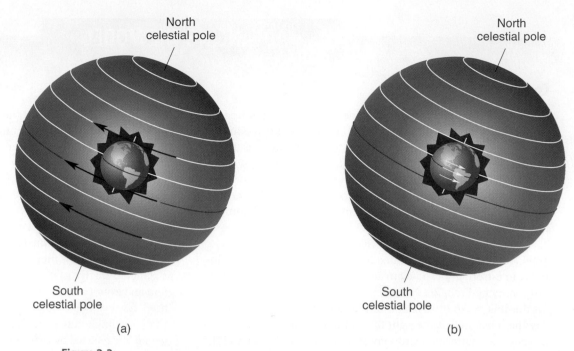

North
celestial pole

North
celestial pole

South
celestial pole

South
celestial pole

(a)

(b)

Figure 2-3
(a) The celestial sphere rotating toward the west around a stationary Earth produces the same observed motion of stars as (b) the Earth rotating toward the east under a stationary celestial sphere.

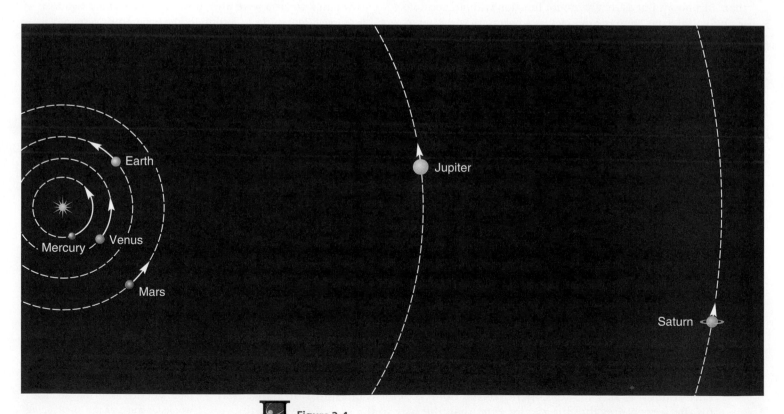

Earth

Jupiter

Mercury   Venus

Mars

Saturn

Figure 2-4
The planets that can be seen by naked-eye observations, shown in order from the Sun. Their relative orbital speeds are indicated by the lengths of the arrows.

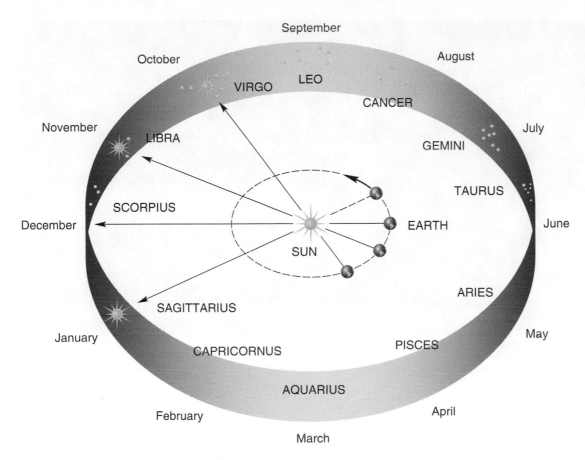

Figure 2-5
As the Earth moves around the Sun during the year, the Sun appears to move among the background stars.

If we consider the view of Figure 2-4 to be that of someone far out in space above the Northern Hemisphere of the Earth, the planets all move around in a counterclockwise direction, as shown by the arrows in this figure. Another feature of Copernicus's model is that the planets closer to the Sun move faster than the planets farther away. In the figure, the lengths of the arrows represent the speeds of the planets, with the longer arrows indicating greater speeds.

Figure 2-5 shows the Earth in a circular orbit around the Sun. Its path does not appear as a circle in the drawing because we are seeing it at an angle. If our point of view in the drawing were straight above the system, the Earth's orbit would appear as a circle and the band of stars would likewise appear as a circle. (Remember, we're describing Copernicus's model, not today's model.) Notice in the drawing that when viewed from Earth, the Sun can be thought of as located among certain stars. As the Earth moves around the Sun, the position of the Sun among those stars appears to change. The Sun seems to be moving across the background of the stars. This, of course, is what we observe. Ptolemy explained it by having the Sun revolve around the Earth independent of the celestial sphere. Copernicus explained it by having the Earth move around a stationary Sun.

Recall from Figure 1-11 that the Sun appears to move from among the stars south of the equator to those north of the equator and back again. To fit this observation, Copernicus's model had the plane of the Earth's equator tilted with respect to the plane of its orbit around the Sun (**Figure 2-6**). (Recall from Chapter 1 how and why this change in the Sun's apparent position causes the seasons.)

Look carefully at Figure 2-6 to see that the tilt of the Earth's axis would cause the ecliptic to be sometimes above and sometimes below the celestial equator.

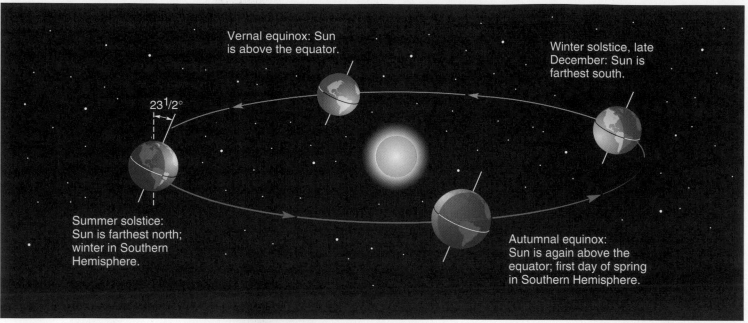

Vernal equinox: Sun
is above the equator.

Winter solstice, late
December: Sun is
farthest south.

$23\frac{1}{2}°$

Summer solstice:
Sun is farthest north;
winter in Southern
Hemisphere.

Autumnal equinox:
Sun is again above the
equator; first day of spring
in Southern Hemisphere.

Figure 2-6

The Copernican model explains the Sun's apparent motion north and south of the equator by having the Earth's equator tilted with respect to the planet's orbit around the Sun.

Motions of the Planets

We stated in Chapter 1 that someone watching the planets from Earth would see that these wanderers spend most of their time moving from west to east against the background of the stars. How does Copernicus's model handle this motion? **Figure 2-7** shows several positions of the Earth in its orbit, along with the corresponding positions of Mars. The lines from Earth to Mars illustrate the direction in which we on Earth see Mars. Look at positions 1, 2, and 3 for each planet. As time goes by, Mars appears to move among the background stars. Now imagine that you, the observer, are standing on Earth as it is moving through its different positions on its orbit. You will see that this motion of Mars (for the first three positions) is from west to east among the background stars. The model fits.

The retrograde motion of the planets presents a greater problem. The planets stop their west-to-east motion and move for a time in the opposite direction, according to a regular schedule that is different for each planet. This retrograde motion forced Ptolemy to use epicycles, but here we see a major difference between the two models. Copernicus explained retrograde motion in an entirely different way.

Locations 4, 5, and 6 show Earth and Mars as they continue their motion. Look at what happens when Earth passes Mars. Mars no longer seems to be moving among the stars in the same direction it was before. It seems to have reversed its direction. This means that while the Earth is passing Mars, Mars appears to move from east to west. Retrograde motion!

To picture how this works, think of riding along a freeway in a fast lane and quickly passing a slow-moving car. Now suppose you watch that car against a distant background—perhaps some trees across a field or even some distant mountains (**Figure 2-8**). As you are approaching the car, it will seem to move forward along the background. Then as you pass the car, you will see it appear to sweep backward against the background. In order to see this, you must be sure to concentrate on the background against which you see the car. That is what we are doing as we observe retrograde motion of a planet: we watch as it appears to move among faraway stars.

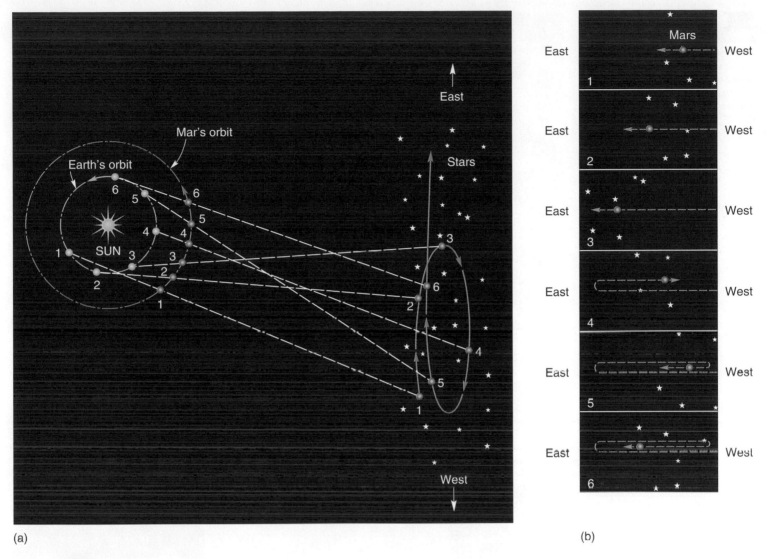

Figure 2-7

(a) While Earth moves from position 1 to position 3, Mars appears to be moving one direction among the stars. Then as Earth passes Mars, Mars appears to move backward. This is the heliocentric explanation for retrograde motion. (b) This is the view of Mars as seen from Earth for each position.

The Moon appears to complete about one revolution around the Earth each month. Copernicus accounted for this in his model by having the Moon on a sphere that revolves around the Earth in that time. For Copernicus, everything else circles the Sun but the Moon revolves around the Earth.

2-3 Comparing the Two Models

In Chapter 1 we discussed the scientific criteria for deciding the value of a proposed model. Let's use these criteria to compare the Copernican and Ptolemaic models.

1. Accuracy in Fitting the Data The heliocentric model, as we have described it, is able to account qualitatively for the observed motions of the stars, the Moon, and the planets. If we are going to compare it to the Ptolemaic model, however, we must ask how accurately it fits the data. To be a good model, it must account for the

Figure 2-8

As the car in the left lane passes the other car, its driver sees the slower car appear to move backward against the distant mountains. Part (a) illustrates the lines of sight, and (b) shows what is seen from the car in the left lane.

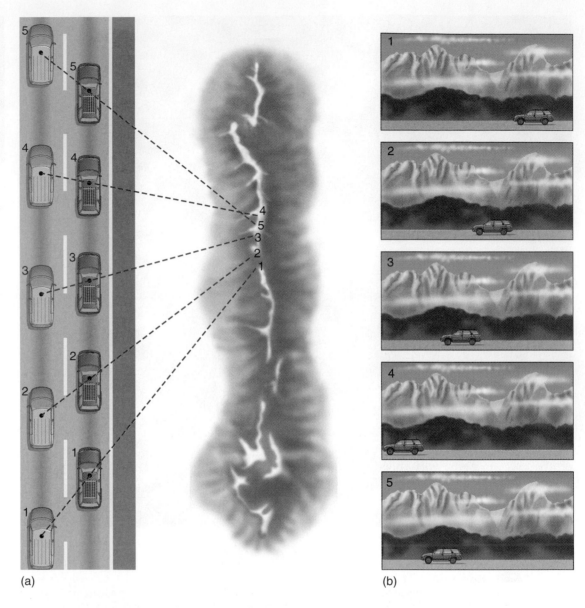

(a)

(b)

In the center of everything rules the sun; for who in this most beautiful temple could place this luminary at another or better place whence it can light up the whole at once? . . . In this arrangement we thus find an admirable harmony of the world, and a constant harmonious connection between the motion and the size of the orbits as could not be found otherwise.

Copernicus

most accurate positions recorded for the planets. Fitting a model to the observations is particularly difficult in the case of planetary retrograde motions. Did the positions of the planets according to the Copernican model correspond to accurate observations of the planets' positions? Well, not quite. There were differences between predicted positions and observed positions.

Copernicus sought to increase the accuracy of his model by inserting small epicycles for the celestial objects to move on. This was necessary because Copernicus assumed—like Ptolemy—that the planets move at a constant speed. In fact, the observed motions did not correspond exactly to planets moving at constant speed. The epicycles of Copernicus were smaller than those of Ptolemy, and the motion of the planets around them was much slower, so that each planet moved in a somewhat elongated circle.

Why did Copernicus not abandon the idea of circles altogether? The reason lies partly in the fact that the circle had such a long tradition. Beyond this, however, Copernicus felt that there was good reason for using the circle. He argued that the motions in the heavens are repetitive, and "a circle alone can thus restore the place of a body as it was."

So Copernicus was unable to abandon the epicycles of Ptolemy. And even with his epicycles, Copernicus's model had errors in predicting positions of planets that typically came to about 2 degrees over a few centuries of planetary motion. This is about the same error as in Ptolemy's model. How, then, do we choose between the two models? One way is to search for more data that might distinguish between them. If we could find an observation that could be explained by one model and not the other, we would have evidence upon which to build a case.

Both the Copernican model and the Ptolemaic model held that the stars are on a sphere a great distance from the Earth. Neither model included the idea that some stars may be much farther from Earth than others. *Parallax* provided such evidence, as we will see.

Figure 1-19 showed the Earth at two positions on opposite sides of its orbit. Just as your thumb shifts its apparent position when you alternately wink your eyes, if stars differ in their distances from Earth, a nearby star would be expected to shift its position relative to very distant stars. All of the stars are so far away, however, that the greatest annual shift observed for any star is only 1.5 arcseconds. This is why stellar parallax was not observed until 1838.

Once parallax was detected, it provided obvious evidence that the heliocentric model is the better one (although the question had been settled by that time). In addition, parallax provided evidence that stars are not all at the same distance from Earth. The Ptolemaic model held that the Earth does not move, but the observation of stellar parallax showed that it does. Although it would have surprised Copernicus to learn that the stars are at such different distances, this new idea could have easily been incorporated into his model.

No definitive observation could be made in the 1500s that would decide which of the two competing models fit the data more accurately. For that reason, using the evidence available in the 1500s, we still must call our comparison of the models a draw, based on the criterion of data fitting.

2. Predictive Power The second prerequisite of a good scientific theory is that it must make testable predictions that might allow the theory to be disproved. The Ptolemaic model predicted that stellar parallax should not exist, while the Copernican model predicted that it would, so both pass this test. (A prediction that is eventually disproved by new data is as acceptable as a prediction that is eventually supported by new data.) But wouldn't all models make this type of prediction? Well, suppose someone presents a model that says the planets appear to move the way they do because of a divine plan. Using this model, whatever we later find out about a planet can be explained by saying that it was a part of this divine plan. There would be no way to prove this theory wrong. Whatever observation is made, one could attribute it to this unverifiable divine plan. This model cannot be accepted as a good scientific model because it does not contain within itself predictions that allow it to be disproved.

This seems to be an odd criterion. It says that a theory must contain the potential seeds of its own destruction. If a theory's prediction turns out to be correct, this serves as further evidence that the theory is good, but it does not "prove" the theory. On the other hand, a theory—or at least some aspects of a theory—can always be proven wrong by further evidence. We can disprove a theory, but we can never absolutely prove one.

Before moving to the final criterion, let's look at another prediction made by the heliocentric model: the relative distances of planets from the Sun. Based on his theory, Copernicus was able to calculate such distances. We will not show here how this was done; two Activities at the end of this chapter show the calculations for Mars and Venus. By such methods, Copernicus predicted that Mars is 1.5 times farther from the Sun than the Earth is. As noted in the Prologue, to simplify discussion of distances in the solar system, the *astronomical unit* (abbreviated AU) is defined as the average distance from the Earth to the Sun. Since Mars is 1.5 times farther from the Sun than the Earth is, Mars is 1.5 AU from the Sun. Similar calculations for the other planets yielded for Copernicus the values shown in the center column of **Table 2-1**. The table also shows today's measurements.

parallax The apparent shifting of nearby objects with respect to distant ones as the position of the observer changes.

As the example of parallax shows, predictions are inherent in models. They need not be stated by the model's designer.

astronomical unit (AU) A unit of distance equal to the average distance between the Earth and the Sun.

Table 2-1

Planetary Distances: Copernicus's Values versus Today's Values

Planet	Sun-to-Planet Distances	
	Copernicus's Values	Today's Values
Mercury	0.38 AU	0.387 AU
Venus	0.72	0.723
Earth	1.00	1.000
Mars	1.52	1.524
Jupiter	5.2	5.204
Saturn	9.2	9.582

We could have defined "the Martian Unit (MU)." This would have made Mars 1.0 MU from the Sun and Earth 0.66 MU from the Sun. There is nothing special about the definition of an AU, it just makes the discussion simpler.

Copernicus had no way to determine the distances in everyday units such as kilometers; he did not know the value of the astronomical unit. How do we know the actual distances today? One way to determine distance is to bounce a radar beam from a planet and measure the time it takes for the beam to return. Knowing the speed of the radar signal, we can calculate the distance to the planet and from this the value of the astronomical unit. Since the average distance from the Earth to the Sun is about 150,000,000 kilometers (93,000,000 miles), that is the value of one astronomical unit.

The calculations of planetary distances based on the heliocentric model serve as an example of testable predictions made by the model. If later measurements had shown the distances to be wrong, the theory would have suffered a setback. The criterion that a theory must make testable (and therefore falsifiable) predictions is very important, and if this feature is lacking in a theory, the theory cannot be considered scientific.

There is a final important criterion that is used to compare scientific theories: aesthetics.

3. Aesthetics: Mercury and Venus Copernicus worked for years to develop his heliocentric model. He, and other thinkers of his time and the years following his death, preferred it to the Ptolemaic model even though it had no particular advantage as far as the criteria we have discussed so far. One reason he preferred it was that he thought it was somewhat more accurate (which really was not the case). But his—and others'—real reason for preferring this model is due to our third criterion for a good model: neatness, simplicity, beauty, and aesthetic quality.

By the very nature of aesthetics, there can be disagreement over which of two things is more aesthetically pleasing. Let us, however, look at the two models in question and try to make a judgment. To make the comparison, it is instructive to see how the two models treated the motions of Mercury and Venus.

Recall that Mercury and Venus never get very far from the Sun in the sky. In order to explain this observation, the Ptolemaic theory treated these planets as special cases by requiring that the centers of their epicycles remain along a line between the Earth and the Sun (Figure 1-26). The Copernican system explains the observation in a different manner. As **Figure 2-9** shows, neither planet can get very far from the Sun as seen from Earth simply because they orbit the Sun at a distance less than the Earth's distance. Notice that no matter where Venus is in its orbit, it can never appear farther than 46 degrees away from the Sun as viewed from Earth.

Now, referring to Figure 2-9, locate the position on Earth where the time would be midnight. Notice that there would be no way for a person to see Mercury or Venus at

Venus is often called the "morning star" or "evening star" because it is so bright and easily seen in the morning or evening sky. It outshines all surrounding stars.

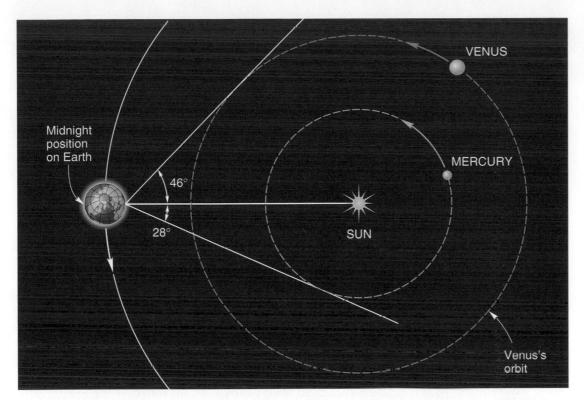

Figure 2-9
This figure illustrates the heliocentric model's explanation for the fact that Mercury and Venus never get far from the Sun and can never be seen in the sky late at night. Compare it to Figure 1-26.

midnight. Thus both models are able to explain why the observed motions of Mercury and Venus differ from those of the other planets, and both models are able to explain why Mercury and Venus are never seen high in the night sky. The Ptolemaic model, however, required a special rule to bring this about. In the Copernican model, this was a natural consequence of the fact that Mercury's and Venus's orbits are inside the Earth's orbit. They are not treated in any special manner. So the Copernican theory must be judged to be more aesthetically pleasing in this case.

Ptolemy used epicycles to explain retrograde motion. Each planet must be assigned an epicycle of a certain size, a speed for its motion around the epicycle, and another speed for the center of the epicycle moving around the Earth. The speeds that were assigned did not form a pattern, and, as we have seen, Mercury and Venus were special cases. Copernicus, on the other hand, saw a pattern in planetary speeds. According to his model, the farther from the Sun a planet is, the slower it moves in its orbit. This rule of motion was all that was needed to account for retrograde motion. Because of this simplicity, Copernicus and others felt that his was a better model for explaining the heavens.

This discussion of the two systems has almost neglected the fact that Copernicus also used epicycles in his model. They were much smaller epicycles than Ptolemy's, but he did find them necessary to make his model as accurate as he could. They were not needed to explain retrograde motion.

This leaves us with a dilemma: the basic idea of the heliocentric system would have to be judged as neater. But when one considers it in its full detail, it loses some of its beauty. So what is the final verdict? Probably in favor of the Copernican model, though the decision seems less than overwhelming.

Why were people of Copernicus's time concerned with how simple and orderly a theory is? This concern arose from a religious belief that a Creator would not have created a disorderly, overly complicated universe. Scientists today would express their desire for simple models in different terms, but the desire is still present. One reason scientists search for a simple model is that our experience since the time of Copernicus tells us that when there are two competing models, the simpler one is

A person on Mars would never see Mercury, Venus, or Earth around midnight.

Recall from a margin note in Chapter 1 (p. 31) that Ptolemy's model actually contained further complications, such as the Earth being slightly off-center of the planets' paths. This contributed to Copernicus's conviction that his own system was simpler.

Figure 2-10
Tycho Brahe (1546–1601).

◆ People of Tycho's time are often known by their first names. For example, do you know the last name of Michelangelo?

◆ One-tenth of a degree is the angle intercepted by a quarter when viewed face-on from a distance of 50 feet (15 meters).

more likely to fit new data—data from observations not yet made. This is often expressed as Occam's razor, as we discussed in Chapter 1. The belief that this pattern will continue with new models is a faith different from that of Copernicus and his supporters. It is a faith in the unity and beauty of nature, based on past experience.

Copernicus died in 1543, just as *De Revolutionibus* was being released. This single book started such an upheaval in people's thinking that the word "revolution" took on a second meaning—an upheaval, as in a social or political revolution. Copernicus began the revolution, but it was up to others to carry it forward to completion.

2-4   Tycho Brahe: The Importance of Accurate Observations

Three years after Copernicus died, Tycho Brahe (pronounced "bra" or "bra-uh") was born in Denmark (**Figure 2-10**). When he was a teenager studying law, he developed an interest in astronomy and learned that both the Ptolemaic and Copernican models were based on recorded planetary data that were inaccurate. That is, he found discrepancies between different tables of observed planetary positions. He was convinced that before a decision could be made about which model was better, or before a completely new and better model could be devised, there was a need for more accurate observations of the positions of the planets as they move across the background of stars.

At this time the telescope had not yet been invented, so all of Tycho's observations were made with the naked eye. In an observatory built for him by King Frederick II of Denmark, Tycho built the largest observing instruments yet constructed. The mural in **Figure 2-11** shows some of his equipment. Because of the large size of this equipment, Tycho was able to measure angles to an accuracy of better than 0.1 degree, far more accurate than any measurements made before his time and close to the limit the human eye can observe.

In addition to making such accurate measurements, Tycho was careful to determine the accuracy of each individual measurement. For example, he would not only state the position of a particular planet, but would also give a value indicating the amount of uncertainty in his measurement. Because of this statement of uncertainty, when such data are later compared to the predictions of a particular model in order to test the model for accuracy, we can tell how close the predictions of the model should be expected to come to the reported measurements. For example, if Tycho stated that a particular measurement was accurate to 0.1 degree of angle, then we should expect a model's predictions to come within 0.1 degree of that measurement. On the other hand, if the model did not make the prediction within 0.1 degree of the measurement, it would have to be considered not accurate enough. The inclusion of a statement concerning the accuracy of measurements is now a common practice in science, but it wasn't in Tycho's time.

Figure 2-11
This mural depicts Tycho's quadrant, which was designed to measure the time and the elevation angle of an object crossing the meridian. The quadrant itself consists of the slot at upper left and the large quarter-circle. Tycho is shown as the observer at the edge of the mural at right center. In addition, he had himself painted on the quadrant (the large figure on the wall near the center) to remind his assistants that he was watching. Tycho's quadrant had a radius of over 2 meters. The large quarter-circle in the foreground is drawn to the scale of the people in front. The person shown partially on the drawing at right center is looking past a pointer on the degree markings and through the slot in the wall at upper left.

Tycho's Model

You might expect that the quality of Tycho's observations would convince him of the correctness of the Copernican model. Exactly the opposite was true, for a reason having to do with parallax. Tycho became famous for his observations of two very important phenomena early on in his career. In 1572 he observed a bright "star," what we now call a supernova explosion, that suddenly appeared in the sky and was observable for about 18 months, while in 1577 he observed a bright comet. What is important about these observations is that Tycho tried to measure the parallax of each of these two objects, but failed. As a result, he concluded that these objects must be very far from the Earth, showing that change occurs in the heavens, a view contrary to thousands of years of traditional belief. While at his observatory, Tycho continued his observations by trying to find parallaxes for the stars. He argued that if the Earth were in motion around the Sun, as the Copernican model suggested, then nearby stars should show parallax as the Earth revolved around the Sun. Because he failed to observe any such parallax, he concluded that the Earth must be at the center of the universe and that the Copernican model was wrong.

Tycho's model of the heavens was a mix between the Ptolemaic and Copernican models. He positioned the Earth at the center with the Sun revolving around it, but he argued that the other planets were revolving around the Sun. Tycho's conclusion was wrong but his argument for placing the Earth at the center was a very good one. It was the same argument used by the ancient Greeks, based on the lack of observable parallax for the stars. However, as we now know, even the nearby stars are so far away from us that their parallax is very difficult to measure. We cannot blame Tycho, one of the best observers of all times, for reaching a wrong conclusion because he had incomplete data. However, this clearly shows the importance of having many independent observations of the same phenomenon and the difficulty in understanding the universe based on data that will always be incomplete.

Night after night for 20 years, Tycho recorded data concerning the positions of the planets, with particular emphasis on Mars. In addition to his accurate measurements, Tycho is known for something he did the year before he died: in 1600 he hired Johannes Kepler as an assistant.

In Tycho's time, it was considered bad manners to leave a formal dinner banquet before the noble host. Tycho drank too much at a dinner in 1601, compounding an existing urinary problem. He stayed at the table, however, and died two weeks later from the resulting burst bladder.

2-5 Johannes Kepler and the Laws of Planetary Motion

Johannes Kepler, who was born in what is now Germany, concentrated on the study of theology. During this study, he learned of the Copernican system and became an advocate for it.

In 1600 Kepler accepted a position as assistant to Tycho Brahe, who assigned him the job of working on models of planetary motion. After Tycho's death, Kepler took over most of Tycho's records and continued the work he began under Tycho. After four years and trying 70 different combinations of circles and epicycles, he finally devised a combination for Mars that would predict its position—when compared to Tycho's observations—to within 0.13 degrees.

Kepler's accuracy of 0.13 degrees (8 arcminutes) is about one-fourth the diameter of the Moon. Recall that prior to this, a typical error between predicted positions and observed positions was about 2 degrees of angle. Kepler, however, was not satisfied. The error of 0.13 degrees still exceeded the likely error in Tycho's measurements (about 0.1 degree). Kepler knew enough about Tycho's methods to know that an error of 8 arcminutes in the data was too much, and he sought a model that would fit the data to the limits of its precision. This is a tribute both to Kepler's persistence and to his belief in the accuracy of Tycho's careful measurements.

Finally, Kepler decided to abandon the circle as the basic motion of the planets and to try other shapes. He tried various ovals, with the speed of the planet changing in different ways as it went around the oval. After nine years of work, he found a shape that fit satisfactorily with the observed path of Mars. What's more, he found that the same basic shape worked not just for Mars, but for every planet for which he had data. The shape? An ellipse.

Figure 2-12
To draw an ellipse (a) start with two tacks, a pencil, and a string. Stick the two tacks into a board some distance apart and put a loop of string around them. (b) Then use a pencil to stretch the string as shown. Finally, keeping the string taut, move the pencil around until you have completed the ellipse. Every point (P) on the ellipse is at the same total distance from the two fixed points F_1, F_2 (the foci); that is, $PF_1 + PF_2$ = constant.

ellipse A geometrical shape of which every point (P) is the same total distance from two fixed points (the foci). (That is, in Figure 2-12, $PF_1 + PF_2$ = constant = major axis.)

focus of an ellipse One of the two fixed points that define an ellipse. (See the definition of *ellipse*.)

eccentricity of an ellipse The result obtained by dividing the distance between the foci by the longest distance across an ellipse (the *major axis*).

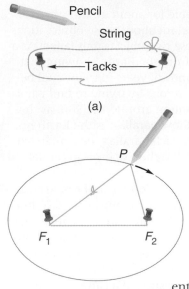

Pencil

String

Tacks

(a)

P

F_1 F_2

(b)

The Ellipse

At first glance, an *ellipse* is nothing more than an oval. But it is more specific than that; not every oval is an ellipse. **Figure 2-12** shows how to draw an ellipse. Each point where we placed a tack is called a *focus* of the ellipse. The plural of this word is "foci," so an ellipse has two foci.

Ellipses can be of various *eccentricities (e)*, from $e = 0$ to $e = 1$. The eccentricity of an ellipse is defined as the ratio of the distance between the foci to the longest distance across the ellipse (the major axis; see **Figure 2-13**). Essentially, the eccentricity tells us how "flat" the ellipse is. For example, if the tacks coincide, the ellipse would be a circle and its eccentricity would be zero (see **Figure 2-14a**). If the tacks are a small distance apart, the ellipse would have a small eccentricity and might look like Figure 2-14b. And if the tacks are farther apart or the string is shorter, the ellipse would have a greater eccentricity and might look like Figure 2-14c. Although different ellipses have different eccentricities, it is important to see that a definite rule governs the shape of an ellipse; that is, every point (P) on the ellipse is at the same total distance from the foci: $PF_1 + PF_2$ = constant = major axis. Compare this to the definition of a circle, where every point on the circle is at the same distance from a fixed point, the circle's center. In this case, F_1 and F_2 coincide and thus $e = 0$.

An example of an elliptical shape you see every day is that of a circle seen at an angle. **Figure 2-15** shows a round object seen in perspective. The shape the artist drew to represent the rim of the basket correctly is an ellipse. The eccentricity of such an apparent ellipse is changed by changing your angle of viewing the rim. Viewing it from below, you see a circle, but by viewing it at an angle, you see its apparent shape become more eccentric.

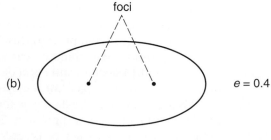

(a) $e = 0$

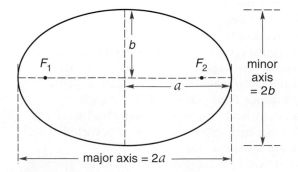

F_1 b F_2

a

major axis = 2a

minor axis = 2b

Figure 2-13
The points F_1 and F_2 are the foci of the ellipse. The eccentricity is calculated by dividing the distance between the foci (F_1F_2) by the length of the major axis 2a. Half of the major axis is called the semimajor axis (a) and is what we have previously referred to as the average distance from a planet to the Sun. The semiminor axis is usually denoted by b and the eccentricity can also be calculated by $e = \sqrt{1 - (b^2/a^2)}$.

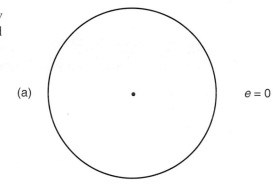

foci

(b) $e = 0.4$

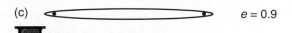

(c) $e = 0.9$

Figure 2-14
(a) An ellipse with zero eccentricity ($e = 0$), that is a circle.
(b) An ellipse with a small eccentricity ($e = 0.4$). (c) A more eccentric ellipse ($e = 0.9$).

ADVANCING THE MODEL

Johannes Kepler

Figure B2-2
Johannes Kepler (1571–1630).

Johannes Kepler (**Figure B2-2**) was born in a small town in southern Germany in 1571. His father was poor and unreliable, his mother had a violent temper, and Johannes was a very sickly child. When he was four, he contracted smallpox and nearly died. The disease left him with poor eyesight, a condition that prevented him from enjoying the astronomical sights reported to him by Galileo. So rather than becoming an observer, he made his contributions in theoretical astronomy by applying his great mathematical abilities and his insights to the observational data collected by others. The most accurate information was the nontelescopic data from Tycho Brahe, for whom Kepler worked during the last year of Tycho's life. (Tycho hired Kepler in the hope that the mathematician would be able to verify Tycho's Earth-centered model, but Kepler apparently took the position with the idea of gaining access to the accurate data he needed to test his own ideas.)

Kepler was a mathematician of some note; his work on determining volumes of objects by using approximations (inspired by trying to find the most convenient proportions for wine barrels) was an important early contribution to the development of calculus. Today we know Kepler for his three laws of planetary motion, but these laws seem to have been developed almost accidentally as he searched for other, deeper patterns in the heavens. For example, his first great endeavor was to try to find out why the planets are spaced apart as they are. When he was about 24 years old and working as a professor of mathematics, an idea occurred to him while he was lecturing. He was discussing the size of the largest circle that could be drawn inside a triangle and the smallest that could be drawn outside the triangle when he wondered if this could possibly be the basis for the spacings of the planets. After working with various shapes, he decided that it was not two-dimensional figures that were significant, but three-dimensional solids. It was known that only five symmetric solids were possible (the 4-sided tetrahedron, 6-sided cube, 8-sided octahedron, 12-sided dodecahedron, and 20-sided icosahedron). That number is exactly what would be needed to fill the spaces between six planets. **Figure B2-3** shows a construction that illustrates the plan Kepler finally worked out. Kepler might have

saved a lot of time if all of today's planets had been known then, for he would have had too many planets for the five solids.

Ptolemy and Copernicus had tried to work out a model for the heavenly motions, but Kepler went further. He also asked *why* the planets should have such motions. It was Kepler—not Isaac Newton, whom we will discuss in the next chapter—who first hypothesized that there was a force that kept the planets near the Sun: a force we now identify as gravity. In addition to this force, however, Kepler felt that there must be a force sweeping the planets around the Sun. Just like the gravitational force, this force should become weaker with distance. That is why, he argued, the more distant planets move around the Sun more slowly. It was his consideration of this sweeping force that led him to his third law. Although no such sweeping force exists, it helped Kepler find a rule that later led Newton to discover the forces that *do* exist.

Kepler remained unappreciated throughout his life. His difficulties were many: his first marriage was unhappy, his wife and one of his children died of smallpox, his mother was convicted of witchcraft, and he was forced to cast horoscopes—which he personally ridiculed—to support himself and his family. His consolation was that he believed his mathematical and scientific findings were important and would be recognized after he died.

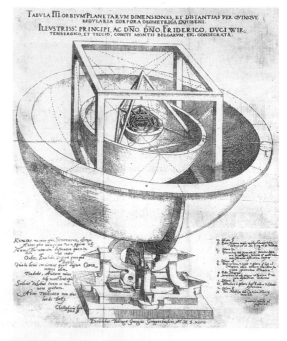

Figure B2-3
Kepler worked out a plan using the five regular three-dimensional figures to fix the spheres of the six known planets at their correct distances.

Figure 2-15
A circular shape, such as the rim of a basketball hoop, appears elliptical when viewed at an angle.

To solve the problem of how to determine Mars' unknown orbit while riding on planet Earth, whose orbit was also unknown, Kepler selected observations of Mars on the same day in different years. This effectively froze the Earth in position. Then he did the same thing in different *Martian* years to map the orbit of Earth. Clever!

Figure 2-16
Kepler's first law states that a planet moves in an elliptical orbit with the Sun at one focus of the ellipse. The orbits of actual planets are much more circular than the ellipse shown here.

www.jbpub.com/starlinks

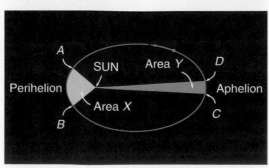 **Figure 2-17**
Kepler's second law: A line from the Sun to a planet sweeps out equal areas in equal times. In the drawing, area *X* = area *Y*.

perihelion/aphelion The closest/farthest point from the Sun on a planet's orbit.

Kepler's First Two Laws of Planetary Motion

Kepler published his first model of planetary motion in 1609 in his book *The New Astronomy*. Today we summarize his findings for planetary motion into three laws, the first of which is:

> **KEPLER'S FIRST LAW: Each planet's path around the Sun is an ellipse, with the Sun at one focus of the ellipse.**

Figure 2-16 shows an elliptical path of a planet around the Sun. We have exaggerated the eccentricity of the ellipse to make it more obvious. The elliptical paths of most planets are not very eccentric; they are nearly circles.

The second law tells us about the speed of a planet as it moves around its elliptical path. Kepler found that a planet moves faster when it is closer to the Sun and slower when it is farther away. He was able to make a more definite statement than this, however, one that would allow a calculation of the speed:

> **KEPLER'S SECOND LAW: A planet moves along its elliptical path with a speed that changes in such a way that a line from the planet to the Sun sweeps out equal areas in equal intervals of time.**

The second law requires some explanation. Suppose we consider the Earth moving in an elliptical path, such as that shown in **Figure 2-17**. Further suppose that point *A* in the figure represents the position of the Earth on December 19, and point *B* is its position on January 18. The yellow-shaded area (area *X*) represents the area "swept out" during those 30 days. Then suppose that on June 19 the planet is at point *C*. Kepler's second law tells us that in the next 30 days, the line from the Earth to the Sun must sweep out an area equal to *X*. Thus the Earth must move more slowly when it is farther from the Sun, so that it gets only to point *D* on July 19. In fact, the law tells us that during any 30 days, the area swept out must be the same.

The time that is applied in Kepler's second law need not be 30 days. The point of the law is that no matter what time interval is selected, if we compare the area swept out by the imaginary line during that amount of time at different places on the elliptical journey of the planet, we will get the same area everywhere. In Figure 2-17, we have again greatly exaggerated the eccentricity of the Earth's orbit. In actuality, the elliptical path of the Earth has a very small eccentricity, so much so that it is almost a circle. Copernicus's circle very nearly fits the path of the Earth.

In essence, Kepler's second law describes how the orbital speed of a planet changes as the planet revolves around the Sun. A planet moves fastest when it is at the nearest point to the Sun (called the *perihelion*) while it moves most slowly at its farthest point from the Sun (called the *aphelion*).

Kepler's Third Law

The first two laws proposed by Kepler tell us about the paths of each individual planet. Although they propose the same basic shape for the path of each planet (the ellipse), they do not tell us anything about how the speed of one planet along its path compares to that of another. The third law does.

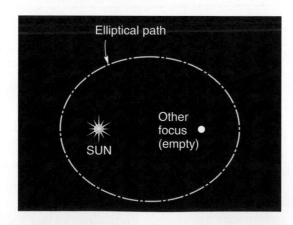

Elliptical path

SUN

Other focus (empty)

A

SUN

Area *Y*

D

Perihelion

Aphelion

Area *X*

B

C

KEPLER'S THIRD LAW: **The ratio of the cube of the semimajor axis of a planet's orbit to the square of its orbital period around the Sun is the same for each planet.**

The third law is most easily understood when it is expressed in symbols. Let

a = semimajor axis of planet's elliptical orbit =
    (approximately) average distance of a planet from the Sun

P = the planet's orbital period = the time it takes the planet to complete a
    full orbit around the Sun, relative to the stars (called the *sidereal period*)

and

C = a constant whose value depends on the physical units used for a and P

Then

$$\frac{a^3}{P^2} = C$$

Kepler's third law states that a^3/P^2 has the same value for every planet. In using Kepler's third law, any units of measure can be used for the period and the distance. We will choose units that make the calculation simple, expressing time in years and average distance in astronomical units. Thus, for the Earth, $a = 1$ AU and $P = 1$ year. Applying Kepler's third law to the Earth we get

$$\frac{a^3}{P^2} = \frac{(1\ \text{AU})^3}{(1\ \text{year})^2} = 1\ (\text{AU})^3/(\text{year})^2$$

Notice the advantage of choosing the units we did: the ratio is equal to 1 for the Earth. That is, the numerical value of the constant C in Kepler's third law is 1 for all planets in our solar system, as long as we measure the planet's orbital period in years and the semimajor axis of its orbit in astronomical units.

Kepler was very careful to use the sidereal period of a planet in his third law. This is the true orbital period of a planet as it revolves around the Sun, as seen by someone above the solar system, and it differs from its *synodic period,* which is the time it takes for the planet and Sun to be in the same configuration when seen from *Earth.*

The distance in Kepler's third law—that is, the *semimajor axis,* or half the length of the major axis of the orbit—is very nearly the same as the planet's average distance from the Sun.

sidereal period The time it takes a planet to complete a full orbit around the Sun, relative to the stars.

We will discuss the sidereal period and synodic period in more detail in later chapters.

synodic period The time between two successive identical configurations between a planet and the Sun, as seen from Earth.

EXAMPLE

Use the fact that Mars's orbital period is 1.88 years to calculate the average distance of Mars from the Sun.

SOLUTION The third law says that a^3/P^2 is also equal to 1 for Mars when the above units are used. After rewriting the equation, we substitute the value of Mars's period and solve for its distance from the Sun.

$$\frac{a^3}{P^2} = 1\ (\text{AU})^3/(\text{year})^2$$

$$\frac{a^3}{(1.88\ \text{year})^2} = 1\ (\text{AU})^3/(\text{year})^2$$

$$a^3 = 3.53\ (\text{AU})^3$$

$$a = 1.52\ \text{AU}$$

This agrees with the measured value of the average distance of Mars from the Sun.

<u>TRY ONE YOURSELF</u>

Calculate the radius of Venus's orbit using the fact that its period of revolution around the Sun is 0.615 year. (Venus's orbit is very close to circular, so its average distance from the Sun is essentially the radius of its orbit.)

Table 2-2
Testing Kepler's Third Law

Planet	Distance from Sun	Sidereal Period	Distance Cubed	Sidereal Period Squared
Mercury	0.387 AU	0.241 yr	0.0580 AU³	0.0581 yr²
Venus	0.723	0.615	0.378	0.378
Earth	1.000	1.000	1.000	1.000
Mars	1.524	1.881	3.540	3.538
Jupiter	5.204	11.862	140.9	140.7
Saturn	9.582	29.457	879.8	867.7
Uranus	19.201	84.011	7079.0	7057.8
Neptune	30.047	164.790	27,127	27,156
Pluto	39.236	247.68	60,402	61,345

Table 2-2 shows modern data for each of the planets known to Kepler. For completeness, it also includes data for Uranus, Neptune and Pluto. In each case, Kepler's third law tells us that *to the limits of the accuracy of the data*, the distance cubed is equal to the sidereal period squared. (It is not completely accurate because planets are influenced not only by the Sun but also by the gravitational tug of the other planets.)

Kepler had fairly accurate data for the orbital periods of the planets. The actual distances to the planets from the Sun were unknown to him, but remember that the Copernican system allows us to calculate the *relative* distances to the planets. For example, it was known that Mars is 1.52 times farther from the Sun than the Earth is. This, however, is all that is needed to do calculations with Kepler's third law, because 1.52 is the radius of Mars's orbit in astronomical units. So Kepler didn't need to know the actual distances. To the limits of the accuracy of Brahe's data, Kepler's third law fit.

2-6 Kepler's Contribution

In the first place, lest perchance a blind man might deny it to you, of all the bodies in the universe the most excellent is the sun, whose whole essence is nothing else than the purest light, than which there is no greater star; it is . . . called king of the planets for his motion, heart of the world for his power, its eye for his beauty, and which alone we should judge worthy of the Most High God.
Kepler

I measured the skies, now the shadows I measure. Skybound was the mind, earthbound the body rests.
Kepler's epitaph, as he composed it.

Kepler's modification to the Copernican model brought it into conformity with the data; finally, the heliocentric theory worked better than the old geocentric theory. To achieve this fit to the data, however, it had been necessary to abandon the long-held idea of perfect circles in the heavens. Kepler's only reason for proposing ellipses for planetary motion was that they worked.

Our understanding of the solar system would be very unsatisfactory if it had remained where Kepler left it. At the time of Kepler, our scientific understanding of the world was just beginning to take major steps forward, and a man who was a contemporary of Kepler—Galileo Galilei—provided what was needed for the next leap in understanding. The discussion of Galileo's contributions will be left to the next chapter, in which we will show how they led, in turn, to the crowning achievements of Isaac Newton.

Conclusion

Two major models of the heavens have been presented: the geocentric model of Ptolemy and the heliocentric model of Copernicus. Both models had simple explanations for the daily motion of the stars and for the annual motion of the Sun, but their explanations of planetary motions were more complicated. Copernicus's model had an aesthetic advantage in that it explained all planetary motions in the same way, whereas Ptolemy's model had special rules for Mercury and Venus. However, when the models were compared in the most important of scientific criteria—fitting the data—the heliocentric system of Copernicus was not significantly better than the old geocentric system. To accept Copernicus's model would mean taking an entirely different view of the world, for his model moved the Earth from its central position and reduced it to being just one of the planets. This fact, along with the model's lack of advantage in fitting the data, meant that it had little chance of acceptance.

Kepler's revision of Copernicus's model solved its data-fitting problem by using ellipses rather than circles and by presenting simple rules for the speeds of planets. Using the criteria for scientific models, we judge Kepler's model a better one than Ptolemy's. Kepler's revision suffered one flaw, however: the laws he proposed did not "make sense" in that they did not correspond to anything else in nature. They made the model fit the data, and they were reasonably "neat and clean," but they seemed to have been pulled out of thin air. Before the heliocentric system could be fully accepted, it would have to be tied successfully to other phenomena in our experience.

Study Guide

1. It is _____ to observe Venus in the sky throughout the entire night from central United States.
   A. sometimes possible
   B. always possible
   C. never possible

2. It is _____ to observe Jupiter in the sky throughout the entire night from a point on the Earth's surface.
   A. sometimes possible
   B. always possible
   C. never possible

3. If the plane of the Earth's equator were not tilted with respect to the ecliptic plane,
   A. the daylight period on Earth would be the same year-round.
   B. there would be no seasonal variations on Earth.
   C. Earth's poles would not experience six-month-long nights.
   D. [All of the above.]

4. Which of the following choices is *not* a criterion for a good scientific theory?
   A. A theory should be aesthetically pleasing.
   B. A theory should be agreed upon by all knowledgeable scientists.
   C. A theory should fit present data.
   D. [None of the above; all are criteria for a good theory.]

5. The Copernican model explained retrograde motion as due to
   A. planets moving along epicycles.
   B. planets stopping their eastward motion, moving westward a while, and then resuming their eastward motion.
   C. different speeds of the Earth and another planet in their orbits.
   D. [Both A and B above.]
   E. [None of the above; the Copernican model was unable to explain retrograde motion.]

6. Why was Copernicus forced to use epicycles in his model?
   A. To account for retrograde motion.
   B. To account for phases of the Moon.
   C. To accurately predict the position of a planet.
   D. [Both A and B above.]
   E. [All of the above.]

7. The observation that Mercury and Venus never get high in the night sky was explained by
   A. the geocentric model only.
   B. the heliocentric model only.
   C. both the geocentric and heliocentric models.

8. Which of the following choices could be calculated from the Copernican model but not from the Ptolemaic model?
   A. The length of the day on each planet.
   B. The actual distance from the Sun to each planet.
   C. The relative distance from the Sun to each planet.
   D. The approximate position of a given planet in the sky a few years in the future.

9. If you lived on Jupiter, which of the following planets might you see high overhead during the night (assuming that your sky was clear enough that you could see the stars)?
   A. Mercury and Venus only.
   B. Mercury, Venus, Earth, and Mars only.
   C. All of the planets except Mercury and Venus.
   D. All of the planets except Mercury, Venus, Earth, and Mars.

10. The Sun appears to move among the stars. The Copernican model accounts for this as being due to
    A. the Earth's rotation on its axis.
    B. the Earth's revolution around the Sun.
    C. the actual motion of the Sun against distant stars.
    D. the Earth changing speed in its orbit.
    E. different planets moving at different speeds.

11. If the Earth's diameter were double what it is, stellar parallax would be
    A. easier to observe.
    B. more difficult to observe.
    C. the same, for stellar parallax does not depend upon the Earth's diameter.

12. Why did the existence of stellar parallax not convince people of Copernicus's time that his model was better than the geocentric model?
    A. Few people understood stellar parallax.
    B. Stellar parallax was understood, but ignored.
    C. Although most people understood the phenomenon, it was not clear that it provided evidence of the heliocentric theory.
    D. The phenomenon actually provides evidence for a geocentric theory.
    E. Stellar parallax had not yet been observed.

13. According to the heliocentric model, the reason the planets always appear to be near the ecliptic is that
    A. the ecliptic is only 23.5 degrees from the celestial equator.
    B. the planets revolve around the Sun in nearly the same plane.
    C. compared to the stars, the planets are near the Sun.
    D. the planets come much nearer to us than does the Sun.

14. The apparent motion of the planet Jupiter in the sky during a single night as seen by an Earthbound observer is
   A. from west to east, then backward a while, and then west to east again.
   B. a west-to-east motion throughout the night.
   C. a east-to-west motion throughout the night.
   D. [None of the above, for no motion is observed.]

15. According to Kepler's laws, a planet moves
   A. at a constant speed through its orbit.
   B. fastest when nearest the Sun.
   C. fastest when farthest from the Sun.
   D. [None of the above, for there is no simple rule.]

16. The primary reason that it is hotter in Australia in December than it is in July is that
   A. the Southern Hemisphere is tilted toward the Sun in December.
   B. the Earth is closer to the Sun in December than in July.
   C. the Earth is moving faster in its orbit in December than in July.
   D. the Earth is on the hotter side of the Sun in December.
   E. [Both B and C above.]

17. If the Earth were in an orbit closer to the Sun, the
   A. day would be longer.
   B. day would be shorter.
   C. year would be longer.
   D. year would be shorter.
   E. [Two of the above are correct.]

18. Kepler's law of equal areas predicts that a planet moves fastest in its orbit when
   A. it is closest to the Sun.
   B. it is closest to the Earth.
   C. the Earth, Moon, and Sun are in a line.
   D. it is farthest from the Sun.
   E. [None of the above; the law makes no predictions as to speed.]

19. If there were another planet between Mercury and Venus,
   A. its "day" would be shorter than Earth's.
   B. its "year" would be shorter than Earth's.
   C. its speed in orbit would be less than Earth's.
   D. [Both A and C above.]
   E. [Both B and C above.]

20. From the law of equal areas, one can predict that the Earth
   A. moves the same distance in January as in July.
   B. moves faster when it is closer to the Sun.
   C. spins faster when it is closer to the Sun.
   D. [Both A and C above.]
   E. [Both B and C above.]

21. If a planet were found at a distance of 3 AU from the Sun, its sidereal period would be
   A. about 2.1 years.
   B. 3 years.
   C. about 5.2 years.
   D. 9 years.
   E. about 0.3 years.

22. If a new planet were found with a period of revolution of 6 years, what would be its average distance from the Sun?
   A. About 2 AU
   B. About 3.3 AU
   C. 6 AU
   D. About 9 AU
   E. 36 AU

23. Identify or define: heliocentric, geocentric, astronomical unit, *De Revolutionibus*.

24. How did the Copernican model explain the daily motion of the heavens?

25. How did the Copernican model explain retrograde motion?

26. Why was Copernicus forced to use epicycles in his model?

27. Name a discovery made in the nineteenth century that the Copernican model fits (or can be made to fit), but the Ptolemaic model does not.

28. What rule did the Copernican model have concerning the speed of one planet compared to another?

29. How did each model explain why Mercury and Venus are never seen high in the night sky?

30. Explain how to draw an ellipse, including how to draw one with more or less eccentricity.

31. State and explain Kepler's second law, the law of equal areas.

32. List the planets known in Kepler's time in order of their speeds in orbit. List them in order of their distances from the Sun.

33. Copernicus had stated that planets farther from the Sun move slower than nearer ones. What did Kepler's third law add to this statement?

QUESTIONS TO PONDER

1. Was Ptolemy's model based on his own observations? Discuss the origin of the ideas in his model.

2. According to the Copernican model, what causes the Sun to change its position periodically from south of the equator to north of the equator and back again?

3. Compare the Copernican and Ptolemaic models as to their accuracy in fitting the data available in the sixteenth century.

4. What is Occam's razor and why is it important in considering the value of a scientific theory?

5. Explain why Copernicus was dissatisfied with Ptolemy's model.

6. Compare the Ptolemaic and Copernican models as to how each explains the Sun's apparent motion among the stars.

7. What observation led Tycho Brahe to develop his own model of the solar system? Describe his model.

8. It might be possible to prove a theory false, but it is never possible to prove it true. Explain.

9. When you read how each of the models explained retrograde motion, you may have found it easier to understand the explanation of the Ptolemaic model than the explanation of the Copernican model. If so, how can Copernicus's explanation for retrograde motion be called the simpler one?

10. Suppose that you have a friend across the Atlantic with whom you converse on the telephone. Describe a measurement that the two of you might make that would allow you to calculate the distance to the Moon. (Assume that you know how far apart the two of you are.)

11. When observed from Earth, Mercury and Venus have different motions than the other planets. Describe this difference.

(Hint: We can NOT observe directly that these planets are closer to the Sun than other planets.) Compare Copernicus's explanation for this difference to Ptolemy's explanation.

12. What is meant by the "eccentricity" of an ellipse? What is the eccentricity of a circle?

13. The Earth is closest to the Sun in January. Then why do we not experience our hottest weather in January?

---

1. Calculate the radius of Jupiter's orbit (in astronomical units) from data available to Copernicus. Data: Jupiter's sidereal period is 11.86 years and quadrature occurs 87.5 days after opposition. (Hint: See the Activity "The Radius of Mars's Orbit.")

2. Jupiter's semimajor axis is 5.2 AU. What are the smallest and largest distances possible between Earth and Jupiter?

3. A comet moves around the Sun on a highly eccentric orbit. Assume that its distance from the Sun at perihelion is 0.1 AU. The comet's period is 729 years. Use Kepler's third law to find the average distance of the comet from the Sun and the largest distance between the comet and the Sun.

4. Mercury's period of revolution is 88 days, or 0.24 years. Use Kepler's third law to calculate its average distance from the Sun in astronomical units.

5. The planet Uranus was discovered in 1781. The semimajor axis of its orbit is 19.2 AU. Calculate its period of revolution around the Sun.

6. A quarter at 50 feet spans an angle of about 0.1 degrees. About how many quarters could be placed side-by-side around a circle with a radius of 50 feet?

7. The eccentricity of Pluto's orbit is 0.24. Use a drawing of an ellipse to explain what this means quantitatively.

CALCULATIONS

ACTIVITIES

---

1. The Rotating Earth

On a clear dark night, find a good observing location and draw a quick sketch of some stars high over your head. Look for a pattern so that you can remember these stars later. If you can, mark your map so that it shows which stars are toward the north, east, south, and west.

Take about two hours off; then go back to your observing location and find the stars you drew. How has their position changed? You probably know what this answer should be from having read the text. Your real job, however, is to stand under the stars and imagine that their motion is due to the rotation of the Earth. Try to "feel" the Earth turning under the stars. Which is easier to imagine, the stars on a sphere rotating around the Earth or the Earth spinning under the stars?

Finally, spend a half hour watching either a sunrise or a sunset. Or better yet, watch the Moon rise or set, particularly when it is full or nearly full. Now picture this phenomenon as explained by the heliocentric system. Try to think of the Earth turning instead of the Sun or Moon moving. Imagine yourself on a little ball that is turning so that the place where the Sun's light hits the ball changes.

2. The Radius of Venus's Orbit

In this activity we will measure the orbital radius of Venus. This is the method used by Copernicus to measure the orbital radius of the two *inferior planets,* Venus and Mercury (called this because their orbits are smaller than Earth's).

Mark a location for the Sun on the right side of a piece of paper and draw an arc of a circle about halfway across the paper, representing the orbit of the Earth. The radius of this circle represents 1 AU. Make a scale drawing and assume that the Earth and Venus move in circular orbits. Then draw a line from the Sun horizontally across the paper and mark its intersection point with the Earth's orbit as the position of the Earth. Copernicus measured the angle between the Sun and Venus, when Venus was at its *greatest elongation.* This is the farthest from the Sun that Venus is seen to be in the sky from Earth. Therefore, when

an inferior planet is at greatest elongation, it means that the line connecting it with the Earth is tangent to the planet's orbit. The angle measured by Copernicus for Venus was 46 degrees.

Center a protractor on the Earth so that the Sun is at 0 degrees. Mark a point at 46 degrees and draw a line from the Earth to that point, extending it a bit further. Draw the perpendicular to this line from the Sun. This perpendicular distance is the radius of Venus's orbit. Using a compass you can now draw the orbit of Venus. Measure the radius of Venus's orbit, divide this by the Earth's distance to the Sun and compare your result to that in Table 2-1. Are you close?

3. The Radius of Mars's Orbit

Let us determine the radius of Mars's orbit from data available to Copernicus. To do so, we start by making a scale drawing, assuming that the Earth and Mars move in circular orbits. Mark a location for the Sun on one side of a piece of paper. As in **Figure 2-18a**, draw an arc of a circle about halfway across the paper to represent the orbit of the Earth. Then draw a line from the Sun horizontally across the paper. Suppose that when the Earth crosses your horizontal line, Mars is also crossing it, so the Sun and Mars are on opposite sides of the Earth. (Mars is then said to be in *opposition.*) Mars and the Earth are both moving in the direction shown in Figure 2-18a. Now suppose that Mars is observed night after night until it is at *quadrature,* which is defined as the position when Mars is 90 degrees from the Sun in the sky (seen from Earth), as shown in Figure 2-18c. This is observed to occur 106 days after opposition. This observation, along with the sidereal period of Mars (1.88 years), is all that we need to determine the radius of its orbit.

To draw your figure to scale, you must first calculate the angle through which the Earth moves in 106 days. This is obtained by dividing 106 days by 365 days and multiplying by 360 degrees. Draw a line from the Sun at the angle you have calculated (angle *X* in Figure 2-18b), and determine the position of Earth (point *E*) when Mars is at quadrature.

Since Mars is at quadrature when the Earth reaches point *E,* you should start at *E* and draw a line toward the right 90 degrees from the line to the Sun. This line points to the planet Mars. You don't know yet where Mars is located on this line. You can determine this by calculating the angle Mars moves through in 106 days. Use its period of 1.88 years to calculate this angle. (Since 1.88 years is 686 days, Mars moves 106/686 of a complete circle in 106 days.)

Starting from the Sun, draw a line at the angle you have calculated (angle *Y* in Figure 2-18d). Since Mars lies along this line, it must be located where the two lines cross.

Measure the distance from the Sun to Mars, divide this by the Earth's distance to the Sun and compare your result to that in Table 2-1. Are you close? This was the method used by Copernicus to measure the radius not only of Mars's orbit but also the orbital radius of every **superior planet,** that is of every planet with orbit larger than Earth's.

Figure 2-18
Steps involved in determining the radius of an outer planet's orbit. This figure is not drawn to scale, as yours must be.

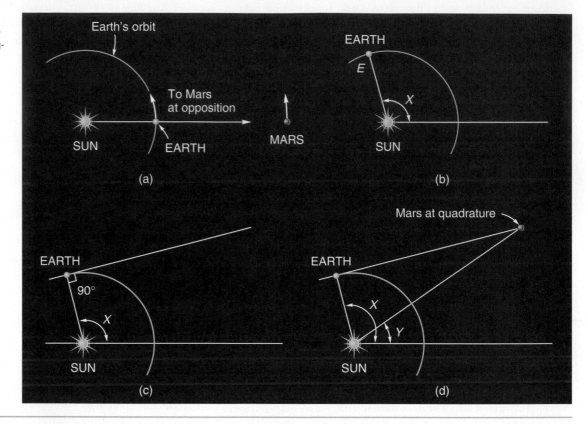

The following books and articles will give you a sense of the impact that Copernicus, Tycho and Kepler had on our understanding of the planets.

1. M. Caspar: *Kepler* (Dover, 1993).

2. O. Gingerich: *The Great Copernicus Chase, and Other Adventures in Astronomical History,* a collection of essays including the period of Copernicus and Kepler (Sky Publishing/Cambridge University Press, 1992).

3. "Observing the Occasion" by E. C. Krupp, a short biography of Tycho, in *Sky and Telescope* (December 1996).

4. "Copernicus and Tycho" by O. Gingerich, in *Scientific American* (December 1973).

EXPANDING THE QUEST

STARLINKS

Quest Ahead to Starlinks:
www.jbpub.com/starlinks

Starlinks is this book's online learning center. It features **eLearning,** which contains chapter quizzes and other tools designed to help you study for your class. You can also find **on-line exercises,** view numerous relevant **animations,** follow a guide to **useful astronomy sites** on the web, or even check the latest **astronomy news** updates.

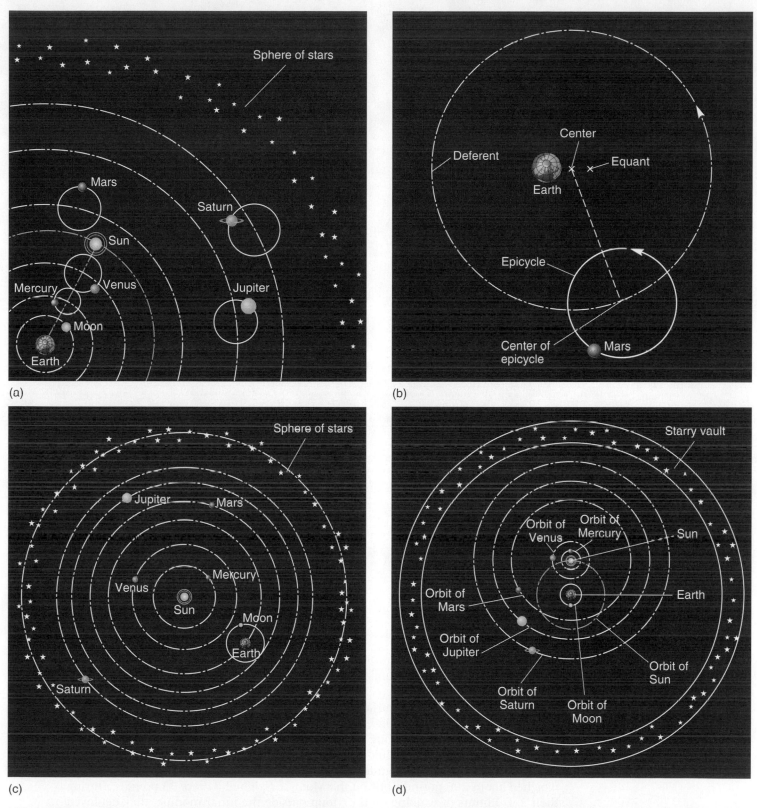

(a)

(b)

(c)

(d)

(a) Ptolemy's Earth-centered epicycle theory (around A.D. 100) of the layout of the universe, according to which the five visible planets move on epicycles around Earth. The epicycles of the two innermost planets, Mercury and Venus, are centered on the line joining Earth to the Sun. (b) In this modification of his model, Ptolemy has the Earth displaced slightly from the center of the deferent (the circle on which the center of a planet's epicycle moves). The center of the epicycle moves at constant speed as seen from the equant but not as seen from the earth or the center. (c) Copernicus's Sun-centered theory of the layout of the universe, A.D. 1543. The diagram is simplified; the planets all move on epicycles, similar to those in Ptolemy's theory, but here there are far fewer epicycles. Furthermore, Copernicus used no equants. (d) Tycho Brahe's Earth-centered model of the universe.

CHAPTER 3

Gravity and the Rise of Modern Astronomy

Earthrise over the lunar surface. This photo was taken from the command module *Columbia* during the Apollo 11 mission, less than 20 hours before the *Eagle* landed.

ON JULY 20, 1969, HUMANS LANDED FOR THE FIRST TIME on another astronomical object. From the Moon's Sea of Tranquility they reported to mission headquarters in Houston, "Houston, Tranquility Base here. The Eagle has landed." Shortly thereafter, Neil A. Armstrong stepped onto the surface of the Moon saying, "That's one small step for a man, one giant leap for mankind." (Historians might note that although the words quoted here are what Armstrong meant to say, he actually left out the word "a.")

Armstrong and Edwin (Buzz) Aldrin, Jr., stayed on the Moon for 21.6 hours. When they left the lunar module, each wore a 90-kilogram spacesuit. On Earth this weighed about 200 pounds, but on the Moon it weighed a mere 33 pounds. During their 2-1/2 hours of walking on the Moon outside the lunar module, they deployed several instruments to measure various features of the Moon and unveiled a plaque that reads "Here Men From Planet Earth First Set Foot Upon the Moon. July 1969 A.D. We Came in Peace For All Mankind." As they left the Moon to rejoin Michael Collins in the command module and return to Earth, the plaque remained behind to express for ages to come a lofty ambition of the human race.

The flight to the Moon and back was made possible by many people and many discoveries, some of which we discuss in this chapter. Perhaps the most basic discovery of all was that the laws of physics that describe motion on Earth are the same laws that describe motion on the Moon, and indeed, everywhere else in the universe. Neil Armstrong's step onto the lunar surface was the last step in a sequence of events dating back over 300 years, with contributions from some of the greatest scientific minds of all time. Their work led to the exploration of space as well as to the rise of modern astronomy.

Although Kepler's laws succeeded in adding accuracy to the heliocentric system, it was the work of Galileo Galilei and Isaac Newton that led to the final triumph of that system. We usually credit the beginning of modern science to Galileo, while the work of Newton helped describe gravity and the causes of planetary motions. Newton formulated some of the fundamental laws of physics, and his work not only helped us understand the force of gravity, but also sparked the technological developments of the industrial revolution. In the twentieth century, Albert Einstein changed our understanding of such basic concepts as space, time, and mass, refining Newton's laws to make them applicable to the vast regions and exotic objects that we find in a universe much larger and more wondrous than Newton envisioned.

In this chapter we look first at how Galileo's use of the telescope established the validity of the heliocentric system. Then we examine how Newton's new description of mechanical motion and his law of gravitation successfully explained the motions of celestial objects so that only a few unsolved problems remained. Those remaining problems were not solved until Einstein proposed an entirely new theory, the general theory of relativity.

3-1 Galileo Galilei and the Telescope

The contributions of Galileo Galilei to our understanding of the solar system consisted both of observations and theory. His observations were of particular value because he was the first person to use a telescope to study the sky. Galileo built his first telescope in 1609, shortly after hearing about telescopes being constructed in the Netherlands.

Several of Galileo's observations with his telescope affected the comparison between the geocentric and heliocentric theories:

- Mountains and valleys on the Moon
- Sunspots—dark areas on the Sun that move across its surface
- More stars than can be observed with the naked eye
- Four moons of Jupiter
- The complete cycle of phases of the planet Venus (similar to the Moon's phases)

Each of these five observations will be discussed in turn to see how they affect the choice of models.

Observing the Moon, the Sun, and the Stars

The first three observations in the list provide no data that completely rule out either theory. All three, however, cast doubt on basic assumptions of the geocentric theory. The Ptolemaic idea of the perfection of the heavens was at the heart of the geocentric system, yet Galileo's telescope (like the small telescope used for **Figure 3-1**)

Notice that "moon" is not capitalized here. In this book, when talking in general about satellites of planets, we will use lower-case letters. Earth's satellite is named "Moon," so we capitalize its name.

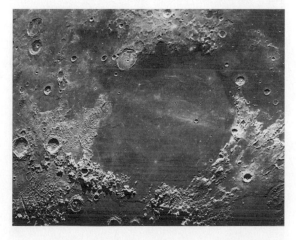

Figure 3-1
Even a small telescope reveals features such as mountains and craters on the Moon. This photo was taken in suburban Chicago with a 10-inch telescope.

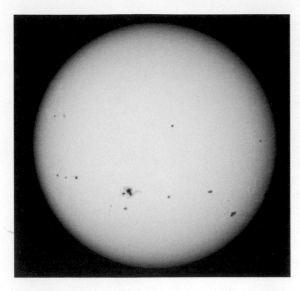

Figure 3-2
Sunspots appear as dark spots on the Sun's surface. See Chapter 11 for more details on sunspots.

Galilean moons The four natural satellites of Jupiter that were discovered by Galileo. Their names, starting closest to Jupiter, are Io, Europa, Callisto, and Ganymede.

Figure 3-3
(a) The orbits of Jupiter's Galilean satellites as seen face on. (b) The same satellites and orbits seen almost edge on. From Earth, this is how this system appears. (c) This 1979 photo from the *Voyager 2* spacecraft shows the moons Io (left) and Europa (right) in front of Jupiter.

revealed Earth-like features on the Moon. The mountains there do not appear basically different from mountains on Earth. In the case of the Sun, the existence of dark spots did not fit at all with the idea of perfection in the celestial realm (**Figure 3-2**).

Galileo's telescope also revealed that the sky contains many more stars than had previously been imagined. Recall that Thomas Aquinas had incorporated the Ptolemaic model into Christian theology. Part of this idea was the centrality of humans, not only in position, but in importance. Passages in the first book of the Bible can be interpreted to indicate that the stars were put in the sky for the exclusive purpose "to shed light on the Earth" (Gen. 1:17). Why, then, do stars exist that are so dim that they cannot even be seen by the unaided eye? What is their purpose? The existence of these stars seemed to undermine the Ptolemaic model, and with it a literal interpretation of the Bible. For this reason, many people in Galileo's time simply refused to look through a telescope at the stars.

Jupiter's Moons

In January, 1610, Galileo turned his attention to Jupiter. Near the planet, he saw three very faint stars. They were all in line, two east of the planet and one west. As he continued to observe them night after night, he noted that there were four stars, rather than three, and it became clear to him that they were not stars at all, but natural satellites—moons—of Jupiter. He concluded that they were objects that revolve around the planet, just as the planets themselves revolve around the Sun in the heliocentric model. (Today we call these four satellites the ***Galilean moons*** of Jupiter.) The fact that they never appear north or south of the planet indicated to Galileo that their orbital plane is aligned with the Earth. **Figure 3-3** shows that someone viewing Jupiter's satellites from Earth would see them moving back and forth from one side of the planet to the other.

The Ptolemaic model held that the Earth is the center of everything and that everything revolves around the Earth. Galileo's observation of these satellites indicated otherwise. In the heliocentric system, everything revolves around the Sun except for our Moon, which circles the Earth. It further holds that the Earth is just one of the planets. We see, therefore, that the heliocentric model is much more comfortable with explaining Galileo's observation of a "solar system" in miniature. In addition, the fact that Jupiter is able to move through space without leaving its

Europa

Ganymede

Io

Callisto

(a)

(b)

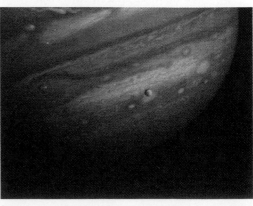

(c)

Galileo Galilei

Galileo Galilei (**Figure B3-1**) was born in Pisa, Italy, on February 15, 1564, the same year that Shakespeare was born (and Michelangelo died). He was the eldest of seven children born to Vincenzio Galilei, who was an accomplished musician and the author of works on musical theory. The family had been very influential and wealthy during the previous century, but Vincenzio was not a man of wealth. At age 12 Galileo went away to school, studying the usual course of Greek, Latin, and logic. He entered medical school when he was 17, but lost interest in it and took up mathematics. However, he was forced to leave school after four years, before graduating, because of a lack of money.

Galileo's creativity began to show when he was in medical school. While there he invented a pendulum device for measuring pulse rates. After leaving school, he continued to study mathematics, applying it to the physics of motion and of liquids. In 1589, at the age of 25, he was appointed professor of mathematics at Pisa and spent the next two decades as a college professor.

In his early years, Galileo had little interest in astronomy, but medical students whom he taught were required to learn some astronomy for use in medical astrology. He therefore became quite familiar with the Ptolemaic model. In 1597, however, he obtained a book published by Kepler concerning the Copernican theory. Galileo apparently preferred the Copernican theory from the time he learned of it, but kept these ideas secret (except from Kepler and a few others) to avoid controversy. Not until his publication of *Letters on Sunspots* in 1613, well after many of his telescopic discoveries had been made, did he openly espouse Copernicus's ideas.

Galileo's announcement of support for the heliocentric system started an uproar that would cause him, and society in general, much grief. One factor contributing to the opposition engendered by the *Letters* was that Galileo wrote them in Italian, his native language, while most scholarly writings of the time were written in the "international" language of Latin. Latin was the accepted language for scholarly papers because it was understood by well-educated people in all countries. The ordinary literate person, however, was less likely to be familiar enough with Latin to read books written in that language. Galileo chose to write in the language of the people of his country because he was convinced that people other than scholars could understand his arguments and his evidence that the Copernican system was correct.

Primarily because of this controversy, in 1616 the Roman Catholic church declared that the Copernican doctrine was "false and absurd" and issued a proclamation prohibiting Galileo from holding or defending it. Why did the church make such a strong statement? We must recall the times and the fact that the Ptolemaic theory and Aristotelian physics were a definite part of both religious doctrine and the general culture.

The Catholic church was not alone in voicing opposition to the Copernican theory. Martin Luther, who was a contemporary of Copernicus, had called Copernicus "the fool who would overturn the whole science of astronomy." John Calvin

asked, "Who will venture to place the authority of Copernicus above that of the Holy Spirit?" To appreciate why religious leaders were so concerned about this issue, we must realize that they considered the spiritual salvation of the individual of paramount importance— more important than answering the question of what is the best theory of the universe. They feared that the idea of a non-geocentric model might seem to undermine the supremacy of humans in God's plan, thereby confusing people and threatening their chance of salvation.

Figure B3-1
Galileo Galilei (1564–1642)

At the age of 70, Galileo was sentenced by an Inquisition court to house arrest for life. The court also banned his book *A Dialogue on the Great World Systems,* published in 1632, in which two fictional characters debate whether or not the Earth is at the center of the universe. Galileo was forced to abandon publicly "the false opinion" that the Sun is at the center of the universe, thus avoiding the fate of Giordano Bruno (a mystic and astronomer who also believed in the heliocentric model), who was burned at the stake 33 years earlier. In the end, the Catholic Church was not able to stop the spreading of the new ideas and in 1992, Pope John Paul II formally accepted that it was a mistake to condemn Galileo. He also proclaimed that knowledge obtained by science is soundly based on evidence and that this knowledge "reason can discover by its own power."

The important point in Galileo's story is not what the Catholic Church did, but that similar things happen in cultures and societies around the world. There are many examples in our history, in many fields, where the status quo defends a specific ideology against new ideas that turn out to be true. It takes courage to defend one's ideas and to be open to the possibility of being proven wrong.

Figure B3-2
A reproduction of one of Galileo's telescopes, looking at the sky over Florence.

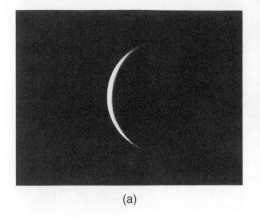

(a)

(b)

(c)

(d)

Figure 3-4
When viewed through a telescope, Venus exhibits a full set of phases, including (d) full, (c) gibbous, and (a and b) crescent. Notice how the planet's apparent size changes as it goes through its phases.

satellites behind conflicts with the Aristotelian/Ptolemaic view, which held that if the Earth moved through space it would leave the Moon behind.

The Phases of Venus

The observation that Galileo found most convincing in choosing between the contending models was his observation of a complete set of phases for the planet Venus. **Figure 3-4** shows four photographs of Venus at different times. Notice how it exhibits phases, similar to the Moon. To the naked eye, Venus appears as a dot of light, but Galileo's telescope revealed its phases. As we will see, this observation was very important evidence for the heliocentric theory.

First, some definitions are needed for terms related to phases, either of a planet or of the Moon. When the entire disk of an object is seen lit, the object is said to be in *full* phase. Venus appears full in Figure 3-4d. (A full Moon is similar; we see the entire Moon lit.) When more than half but less than the entire disk of an object is seen lit, it is said to be in a *gibbous* phase. Figure 3-4c shows a gibbous Venus. An object is said to be in a *crescent* phase when less than half of its lit disk is visible. Figures 3-4a and b both show Venus in crescent phases.

Recall that repeated observations of Venus lead us to conclude that Venus never gets far from the Sun in the sky. It always appears either in the east shortly before sunrise or in the west shortly after sunset, and it is never seen high overhead at night. The Ptolemaic model explained this by saying that the center of Venus's epicycle always remains on a line that joins the Earth and the Sun (**Figure 3-5a**).

What phases of Venus would we see according to the Ptolemaic model? Notice in Figure 3-5a that the sunlit side of Venus never faces the Earth. At times we should be able to see part of its illuminated surface, but never much of it. Venus should never get beyond a crescent phase. Galileo, however, saw Venus in a gibbous phase, an observation unexplainable by the Ptolemaic model. Let's look at the heliocentric model and see if it can account for the observation.

Figure 3-6 is an illustration of the orbit of Venus according to the heliocentric model. At point *X*, Venus would be seen in a crescent phase by an Earth-bound viewer. As it moves farther around and reaches point *Y*, it should appear in *quarter* phase. Then when it reaches point *Z*, it should appear in a gibbous shape, similar to the gibbous Moon.

Thus the heliocentric system of Copernicus (and Kepler) is able to explain Galileo's observations of gibbous phases of Venus. Here, finally, we have an observation that gives us data for choosing one model over another. A model must be able to

full (phase). The phase of a celestial object when the entire sunlit hemisphere is visible.

gibbous (phase). The phase of a celestial object when between half and all of its sunlit hemisphere is visible.

crescent (phase). The phase of a celestial object when less than half of its sunlit hemisphere is visible.

quarter (phase). The phase of a celestial object when half of its sunlit hemisphere is visible.

◆ Additional concepts such as **inferior** and **superior planets, quadrature, greatest elongation** and **opposition** are covered in Activities 2 and 3 at the end of Chapter 2.

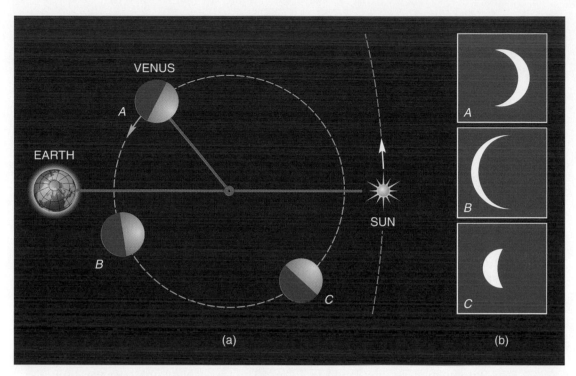

Figure 3-5
(a) Venus's motion according to Ptolemy. (b) This is how Venus would appear from Earth when it is at each of the three points shown in part (a).

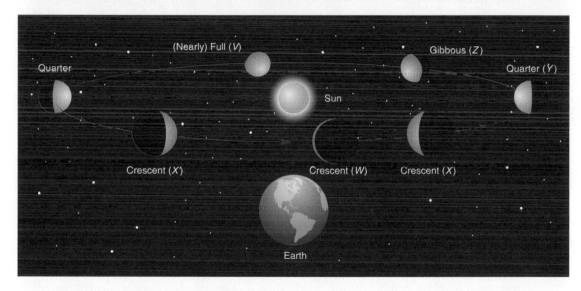

Figure 3-6
Venus at various places in its orbit, according to the Copernican, heliocentric model. This drawing shows how the model predicts that all phases are possible and correspond to what is actually observed. Compare the phases of Venus marked by V, Z, X, and W to those in Figure 3-4. Venus is brightest at positions X and X' as we explain in the Advancing the Model box on Our Changing View of Venus in Chapter 8.

explain the observations. Ptolemy's model cannot explain the gibbous phases that are observed for Venus, and thus it must be rejected. The heliocentric model explains all the phases and therefore becomes the model of choice. In addition, it also explains the correlation between the phases and the corresponding observed sizes of Venus. When Venus is in gibbous phase, it looks smaller than when it is in crescent phase. According to the heliocentric model, this happens because at gibbous phase Venus is farther from the Earth than when it is in crescent phase.

Galileo was a leader in breaking the bonds of Aristotelian and Ptolemaic thought. He and Kepler gave a new direction to the methods of stating natural laws, for their laws were stated in a manner that allowed measurement and testing by the methods of mathematics. In addition, Galileo referred constantly to experiments that would test his hypotheses. This was a new procedure in the study of nature. Even though experimentation was not a new idea (experiments were conducted by some ancient

The Holy Spirit intended to teach us in the Bible how to go to heaven, not how the heavens go.

Galileo

I do not feel obliged to believe that the same God who has endowed us with sense, reason, and intellect has intended us to forgo their use.

Galileo

Greek scholars), the synthesis of Aristotle's ideas into religious belief by Aquinas resulted in an almost complete dependence on "authoritative" opinions. Today, a reliance on observation and experimentation, rather than on authority, is a cornerstone of science. Its beginning is usually credited to Galileo.

One problem remained after Galileo's observations: there seemed to be no logic supporting the laws of Kepler, except that they work. We continue our quest for understanding the rules that govern the solar system by turning to the work of Isaac Newton, the man who "put it all together."

3-2 Isaac Newton's Grand Synthesis

The same year that Galileo died, Isaac Newton was born. Newton was the genius who took Galileo's findings and tied them together into one expansive theory of motion. Galileo had predicted this, writing that he had "opened up to this vast and most excellent science, of which my work is merely the beginning, ways and means by which other minds more acute than mine will explore its remote corners."

Newton's First Two Laws of Motion

Can a moving object slow down and stop by itself?

Today we summarize Newton's conclusions concerning force and motion in three laws, known as "Newton's laws of motion." The first of Newton's laws was built directly on conclusions (though incomplete) that Galileo had reached. Newton stated that an object at rest tends to stay at rest, while a moving object has a tendency to continue moving in a straight line at the same speed. A stationary object starts to move only when something causes it to move. A moving object changes direction or stops only because something causes it to change direction or stop. The tendency of an object to maintain whatever speed and direction of motion it has is called *inertia*, and Newton's first law is often called the *law of inertia*.

inertia The tendency of an object to resist a change in its motion.

> **Newton's First Law (The Law of Inertia): Unless an object is acted upon by a net, outside force, the object will maintain a constant speed in a straight line (if it were initially moving), or remain at rest (if initially at rest).**

The "net" force acting on an object is the sum of all forces, taking into account both their magnitudes and their directions. Equal forces in opposite directions, acting on the same object, cancel one another.

Newton's first law is a basic observation about motion and cannot be directly proven or derived. Still, the law of inertia is a powerful tool that emphasizes cause and effect. It indicates that a force is needed to change the speed and/or the direction of an object's motion; that is, to *accelerate* it. Newton's second law quantifies and extends the first law. It tells us how much force is necessary to produce a certain *acceleration* of an object.

accelerate To change the speed and/or direction of motion of an object.

Consider the brick shown in **Figure 3-7** and imagine that the wheels allow the brick to move without friction. The brick is at rest in part (a), and because it has inertia, it will stay at rest. In (b), you give the brick a push. While you are pushing, the brick accelerates. The amount of force you exert determines how great the acceleration is. A tiny force will produce little acceleration. If you exert a greater force, the brick's acceleration is greater (c). This idea gives us the first part of Newton's second law: The acceleration of an object is proportional to the force exerted on it. "Proportional to" actually goes beyond what we deduced from our thought experiment. It tells us that twice the force will cause twice the acceleration.

acceleration A measure of how rapidly the speed and/or direction of motion of an object is changing.

The question of the origin of inertia has not yet been resolved. Newton believed that inertia is an intrinsic property of matter. Some today believe that it arises from the interaction of all matter in the universe, while others are exploring alternative ideas.

Refer to **Figure 3-8**. Here we show one hand pushing on a frictionless brick as before and another hand pushing on a stack of two identical bricks. If the hands push with equal force, which hand will cause the most acceleration? It should be intuitive that the

ADVANCING THE MODEL

Isaac Newton

To give an idea of the importance of Isaac Newton (**Figure B3-3**) to the history of thought, we quote the English poet Alexander Pope, who lived at the same time as Newton:

Nature and nature's laws lay hid in night.

God said, Let Newton be! and all was light.

Newton was born prematurely on Christmas Day, 1642, in a small village in England. He was so small at birth that his mother said that he would have fit into a quart pot. His father, a fairly prosperous farmer, had died before Isaac was born, and his mother decided that he was too frail to become a farmer. He was not a particularly good student, however, "wasting" much of his time tinkering with mechanical things, including model windmills, sundials, and kites that carried lanterns and scared the local people at night. When her second husband died, Isaac's mother called him home from school to help on the farm, where he spent most of his teenage years.

At the age of 19, Newton was admitted to Cambridge University. There he studied mathematics and natural philosophy (known today as science). In 1665, after Newton had received his degree and was serving as a junior faculty member, England was swept by the bubonic plague. This incurable disease killed more than 10% of the population of London in only three months, and those who could afford to do so escaped it by moving away from population centers. Newton returned to his mother's home at Woolsthorpe. There he worked feverishly on science for the next two years. These must have been two of the most productive years in the history of science, for during this time Newton made discoveries in light and optics, in force and motion, and in gravitation and planetary motion. He also devised a theory of color. To solve a problem in gravitation, he invented calculus. During this time he outlined what would become his major book, *Philosophiae Naturalis Principia Mathematica*, usually called *The Principia*. In later years, Newton wrote, "All this was in the two plague years of 1665 and 1666, for in those days I was in the prime of my age for invention, and minded Mathematics and Philosophy more than at any time since."

Newton's manner of attacking a problem was simply to concentrate his mind on it with such intensity that he finally solved it. His ability to concentrate must have been tremendous. As a result, he often appeared to be what we might call absent-minded. It was not uncommon for him to work all day, forgetting to eat. The story is told that once when riding his horse, he got off the horse to unlatch a gate, led the horse through the gate, relatched the gate, and then led his horse home, forgetting to get back on. He was concentrating on some problem, and walking probably gave him more time to think than riding did, anyway.

Figure B3-3
Sir Isaac Newton (1642–1727) worked in many fields. In this painting he is shown experimenting with a prism to investigate the nature of light.

As a young man, Newton was not interested in publishing his work. He seemed to want to discover the mysteries of nature simply for his own curiosity, and his friends often had to persuade him to publish his findings. Besides, he did not like having to defend his views against criticism, and he wanted to avoid getting involved in arguments over who was the first to make certain discoveries. Nevertheless he *did* get involved quite fiercely in such disputes, including one with Gottfried Leibniz over which of them first developed calculus. This dispute lasted long after both had died. (Today they are both credited for discovering calculus independently.)

It is interesting to note that Newton had a very practical side as well as being a theoretician. He served as Warden of the Mint and while in this office, he began the practice of making coins with small notches around their edges. (Check the quarters in your pocket.) This was done to discourage people from illegally shaving off and retaining the valuable metal of the coins.

In many cases, a person is not recognized as a genius until after his or her death. Such was not the case for Newton. He received honors and was given positions of authority. He was elected president of the Royal Society, an organization of scientists, in 1703 and every year thereafter until his death in 1727. In 1705 he was knighted, the first scientist to receive this honor. When he died, Sir Isaac Newton was buried in Westminster Abbey after a state funeral.

---

greater acceleration will be produced by the hand pushing on the single brick. It is important to see that friction has nothing to do with this. It is tougher to accelerate two bricks than one simply because the two have more inertia. In fact, two identical bricks have twice as much inertia as one brick. This is an idea we didn't discuss above: some objects have more inertia than others. So before continuing with the second law, we must introduce the property of an object that determines its inertia—its *mass*.

Figure 3-7
(a) The wheeled brick will accelerate (b) if a force is exerted on it. (c) If twice as much force is exerted on it, it will accelerate at twice the rate.

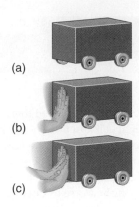

(a)

(b)

(c)

Figure 3-8
The same amount of force will give twice as much mass only half the acceleration.

mass The quantifiable property of an object that is a measure of its inertia.

 Rotational inertia—the inertia of a spinning object—involves more than just mass and will be discussed in a later chapter.

Is mass the same thing as weight?

Is a kilogram a unit of weight?

A kilogram on the Moon weighs only about one-third of a pound.

The concept of force is a fundamental one. It is reasonable to think of force as pushes and pulls, but this is subjective and not always accurate. We see the changes produced by a force and maybe the best way of describing force is as *the agent of change*.

An Important Digression—Mass and Weight

Mass, like inertia, is one of those terms that is used in everyday language but has a very specific definition in science. An object's **mass** is simply the measure of its inertia. Instead of saying that one object has twice the inertia of another, we normally say that it has twice the mass. Mass is an intrinsic property of an object and remains the same independent of where the object is located in the universe.

It is important to say what mass is not. Mass is *not* volume. Suppose that the brick in Figure 3-7 is not a true brick, but is instead a piece of styrofoam made to look like a brick. If you do not know this, however, you will have quite a surprise when you push on the brick. You will find that it accelerates much more than you expected. Why? Because a styrofoam brick has much less mass than a real brick, even though they both have the same volume.

Mass also is *not* weight. In our examples of pushing the frictionless bricks, the weight of the brick was not a factor. The brick's weight is simply the downward force experienced by the brick as a result of its gravitational interaction with the Earth. (See Section 3-4.) Weight did not oppose the pushing hand. This is a subtle distinction, but a very important one.

The worldwide unit of mass measurement is the kilogram. At the International Bureau of Weights and Measures near Paris is a platinum cylinder that has, by definition, a mass of *exactly* one kilogram. To give you an idea of what one kilogram of mass is, a kilogram weighs about 2.2 pounds at the surface of the Earth. It is not correct to say that a kilogram is about *equal to* 2.2 pounds, because a pound is a unit that expresses weight and a kilogram is a unit that expresses mass. A kilogram weighs about 2.2 pounds on the surface of the Earth, but at some other location it might weigh a different amount.

Back to Newton's Second Law

Figure 3-8 shows one hand pushing on one brick and another hand pushing on two bricks. We stated that if the two applied forces were equal, the acceleration of the two bricks would be smaller. In fact, measuring the accelerations would show that if the forces are equal, the acceleration of the two bricks is exactly half that of the single brick, and the same force applied to three bricks results in one-third the acceleration. Thus, acceleration is inversely proportional to the mass being accelerated. If the force is the same, more mass means less acceleration, in exact proportion.

We have thus far discussed the relationships between force and acceleration and between mass and acceleration. We can sum up these relationships in a single mathematical statement:

$$\text{acceleration} = \frac{\text{net force}}{\text{mass}}$$

NEWTON'S SECOND LAW: A net external force applied to an object causes it to accelerate at a rate that is inversely proportional to its mass.

The second law tells us that the greater the net external force applied to an object, the greater the object's acceleration. Also, the greater the object's mass, the less is its acceleration. The second law also makes it apparent that if the net external force is zero, there is no acceleration, which agrees exactly with Newton's first law. The expression above is usually written as

$$\text{Force} = \text{mass} \times \text{acceleration}$$

or in symbols:

$$F = ma$$

In Newton's second law, force and acceleration are always in the same direction.

In using Newton's second law we assume that the speeds involved are much less than the speed of light; otherwise, Einstein's theory of relativity comes into play.

Newton's Third Law

Newton's third law is simple to state:

NEWTON'S THIRD LAW: When object X exerts a force on object Y, object Y exerts an equal and opposite force back on X.

This law seems very simple, but its implications are great. It is sometimes stated as: "For every action there is an equal and opposite reaction." The implication is that it is impossible to have a single, isolated force. Forces always occur in pairs.

When you sit on a chair, you exert a force downward on the chair (**Figure 3-9**). The third law states that the chair exerts an equal force upward on you. We may call the force you exert on the chair the *action* force and the force of the chair on you the *reaction* force, but these labels can be assigned in an arbitrary way. The chair's force could just as well be called the action force and your force on it the reaction force. The point is that one of these forces cannot exist without the other and neither "comes first."

Notice that the statement of Newton's third law indicates that two objects (called X and Y here) are always involved in the application of forces. *Always*. A force cannot be exerted without an object to exert it and an object on which it is exerted. The word "object" here might refer to an individual atom, or it might refer to a collection of atoms, and the collection might be a gas or a liquid as well as a solid object. For example, if you hold your hand out of the window of a moving car to feel the force of air resistance, it is the air that exerts a force on your hand. The third law tells us that your hand therefore exerts a force on the air. This is not easy to see, but we know that your hand must deflect the air as it comes by and a force is necessary to deflect the air. Calling the air an object may seem odd, but air is simply a collection of objects—atoms.

Force of chair on body

Force of body on chair

Figure 3-9
When you sit, you exert a downward force on the chair and the chair exerts an upward force on you. By Newton's third law, these forces are equal in magnitude and opposite in direction, but they do not cancel each other because they are exerted on different objects.

3-3 Motion in a Circle

Recall that an object is said to accelerate if either its speed or its direction of motion is changed. Thus far, our examples of acceleration have involved only changes in speed, not direction changes. Now let us examine changes in directions.

Newton's second law says that a net external force acting on an object always produces an acceleration. Consider what happens if you whirl a rock on a string. You are exerting a force on the rock toward the center of its circular motion as you pull inward on the string (**Figure 3-10a**). If you are careful, you might be able to make the rock go around you with a constant speed. Where is the acceleration that, according to Newton's law, must be produced by the force you are exerting? If the rock moves

Does accelerating an object always cause it to speed up or slow down?

 **Figure 3-10**
(a) The string breaks as the rock is whirled in a circle. (b–e) Which way does the rock go after the string breaks?

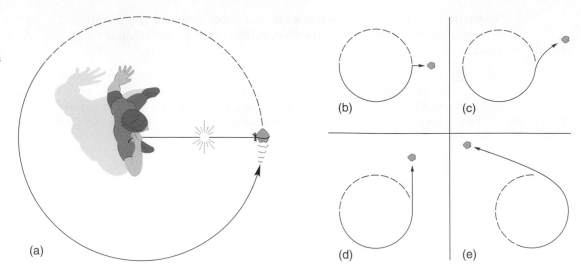

(a)   (b)   (c)   (d)   (e)

at a constant speed, the force is not causing a change in speed. Instead, it changes the *direction* of the rock's motion.

Motion of an object in a circle at constant speed is an example of acceleration causing a change in direction. It may seem odd to call this an acceleration, but notice that if acceleration is defined this way, Newton's law holds in all cases. As a result, the force you exert toward the center of the circle when you whirl a rock on a string does indeed cause an acceleration, which is in the direction of the force—toward the center of the circle.

You might try the Activity "Circular Motion" (at the end of the chapter) now.

Parts (b) through (e) of Figure 3-10 show several potential paths for the rock when the string breaks. Based on Newton's laws, which of these paths is the way you expect the object to travel?

Once the string breaks, there is no horizontal force on the rock. If this is the case, there can be no horizontal acceleration of the rock. Thus, according to Newton's first law, it will continue in a straight line (Figure 3-10d). When the rock was moving in the circle, the only horizontal force on it was exerted by the string. (The rock's weight pulls down, but we are considering only horizontal forces and horizontal motions.) The horizontal acceleration produced was therefore due only to the string.

centripetal force The force directed toward the center of the curve along which the object is moving.

A force is necessary to make something move in a curve. We give this force a name, calling it the **centripetal** (meaning "center-seeking") *force.* This is *not* a new kind of force, but rather it is the name we apply to a net force if that net force is causing something to move in a curve. For example, the centripetal force involved when you were whirling the rock was the force exerted by the string on the rock. The centripetal force on a car rounding a curve on a level roadway is the frictional force of the road acting on the tires in the direction toward the center of curvature. The roadway exerts a force on the car, keeping it in the curve. Do not think of centripetal force as another type of force, similar to a push, a pull by a string, or a friction force. It might be any of those three or other single forces or combinations of them; it is simply what we call whatever net force is responsible for an object's motion along a curved path.

Now let us consider the force that causes the Moon to move in its (almost) circular path.

3-4 The Law of Universal Gravitation

The importance to astronomy of Newton's three laws of motion becomes evident when they are combined with another of his accomplishments: the law of universal gravitation, or "the law of gravity."

Newton's first law states that an object continues at the same speed in the same direction unless some unbalanced external force acts on it. This law is applicable to objects on Earth, but what about objects in the sky? Aristotle believed that earthly laws do not hold in the heavens, but Newton sought to apply his laws of motion there, too. It was during his great, productive years of 1665 and 1666 that, in Newton's words, he "began to think of gravity extending to the orb of the Moon." The Moon follows a nearly circular path around the Earth, so if the laws of motion apply to the Moon, a force must be exerted on it toward the center of its circle—a centripetal force.

Newton hypothesized that there is a force of attraction between the Earth and the Moon and that this serves as the centripetal force. What causes this force? Newton didn't know, but he hypothesized that it is the same force that causes an object here on Earth to fall to the ground. The insight that all matter (hence the term "universal") interacts gravitationally developed gradually. Even Copernicus had made a similar, but incomplete, proposal. Newton's formulation of this idea is as follows.

THE LAW OF UNIVERSAL GRAVITATION: Between every two objects there is an attractive force, the magnitude of which is directly proportional to the mass of each object and inversely proportional to the square of the distance between the centers of the objects.

In equation form this is

$$F = G \frac{m_1 m_2}{d^2}$$

where m_1 and m_2 are the masses of the two objects, d is the distance between their centers, and G is a constant number that depends upon the units used.

Let's stop a moment and emphasize what this means. Newton was saying that *every* object in the universe attracts every other one. This includes the book you are reading and the pen in your hand. According to Newton, they attract one another. You can't feel the force pulling these two objects toward one another because they have very little mass and the strength of the force depends upon the amount of mass. When one of the attracting objects is the Earth, however, one of the masses is enormous, and the force we experience in this case is what we call *weight* (**Figure 3-11**).

Since Newton's law of gravitation applies to every two objects, you may wonder why you are not gravitationally attracted to the person sitting next to you in the classroom. If we use Newton's law and substitute appropriate numbers for your mass, the mass of your classmate, and the approximate distance between you, we get a force that is about equal to the weight of a small flea. This tiny force, combined with the presence of frictional forces, explains why people are not attracted to each other (gravitationally, anyway).

According to Newton, the attractive force of gravity not only makes objects fall to Earth, but keeps the Moon in orbit around the Earth and keeps the planets in orbit around the Sun. Newton proposed that this law, along with his laws of motion, could explain the motions of the planets as well as the falling of objects on Earth. If he could explain the planets' motions, this would clear up the mystery of why Kepler's laws worked and bring the motions of the heavenly planets within the realm of our scientific understanding.

(a)

(b)

Figure 3-11
An object's mass doesn't change from one location to another, but its weight can change. An object that weighs 12 pounds on Earth (a) would weigh 2 pounds on the Moon (b), but its mass is the same.

More than 100 years after Newton published his law, Henry Cavendish was able to measure the gravitational force between two ordinary objects in a laboratory. This experiment enabled him to determine the numerical value of G.

weight The gravitational force between an object and the planetary/stellar body where the object is located.

▮▮▮▮▮▮   EXAMPLE   ▮▮▮

Let us now apply Newton's law of gravity to compare the force required to keep the Earth orbiting the Sun, F_{ES}, to the one required to keep the Moon orbiting the Earth, F_{ME}. The masses of the three objects and their relative distances can be found in the appendices of this book.

▮ SOLUTION ▮   You might think that the force from the Sun on the Earth will be the larger one because the Sun has much more mass than the Moon. However, keep in mind that the gravitational force decreases as the *square* of the distance between two objects, and the Sun is much farther from the Earth than the Moon is. Applying Newton's law of gravity for both pairs we have

$$F_{ES} = G\,\frac{m_E m_S}{(d_{ES})^2} \quad \text{and} \quad F_{ME} = G\,\frac{m_M m_E}{(d_{ME})^2}$$

Taking a ratio of the two expressions, we get

$$\frac{F_{ES}}{F_{ME}} = \left(\frac{m_S}{m_M}\right)\left(\frac{d_{ME}}{d_{ES}}\right)^2 = 180.$$

Indeed, the force from the Sun on the Earth is much larger (by about a factor of 180) than the force from the Earth on the Moon.

<u>TRY ONE YOURSELF</u>

At the surface of the Earth, a distance of about 6400 km from the Earth's center, an astronaut's weight is about 160 pounds. This is the attractive force on the astronaut from Earth. What is the value of this attractive force on the astronaut if he were at a height of 300 kilometers above the surface of the Earth? Taking account of your answer, how do you explain the fact that astronauts in the Space Shuttle, orbiting at 300 kilometers above the Earth's surface, feel no weight?

Arriving at the Law of Universal Gravitation

How did Newton arrive at his formulation of the law of gravity? Kepler (in 1609) was the first to suggest that two objects placed in space would attract gravitationally and approach each other in proportion to each of the two masses involved. This is easy to test—in one case, anyway. A 10-kilogram object has twice the mass of a 5-kilogram object. (Remember, we are speaking here of the objects' *masses,* not their weights.) The law of gravity predicts that a 10-kilogram object has twice the weight of a 5-kilogram object, and indeed it does. Weight is proportional to mass, so the law seems to work when one of the objects is the Earth. That is, the gravitational attraction between the Earth and an object is proportional to the object's mass (say m_1). But, according to Newton's third law, the object is also attracting the Earth with an equal and opposite force. Therefore, the gravitational force between the Earth and an object is also proportional to the Earth's mass (say m_2). Thus, the force is proportional to the product of the masses of the two objects, that is, $F \propto m_1 m_2$.

> Kepler believed that planets were pushed along their orbits by a force that spread out from the Sun and therefore decreased with distance.

To test gravity's dependence on distance, we must be able to compare forces on objects at different distances from each other. The law states that the force is inversely proportional to the square of the distance between the objects' centers. For example, if the two objects are the Earth and your book and if you take your book to a location twice as far from the center of the Earth as it is now, the law says that the book will weigh only one-fourth (which is $(1/2)^2$) as much. At three times as far from the Earth, it will weigh one-ninth as much (**Figure 3-12**).

> Because of today's sensitive instruments, we can now measure differences in weight at different locations on Earth. As expected, the measurements confirm Newton's law of gravitation.

Newton, of course, was unable to change an object's distance from Earth's center significantly. He could have taken an object up on a mountain and measured its weight, but his theory predicted that the weight change in this case would be so small that it would not be measurable with the methods he had available. Instead, he used an object already in the sky: the Moon. And rather than comparing forces directly, he compared accelerations produced by the forces.

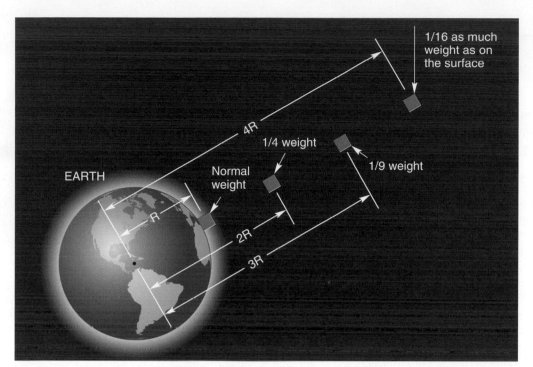

EARTH

R

2R

3R

4R

Normal weight

1/4 weight

1/9 weight

1/16 as much weight as on the surface

Figure 3-12
When an object gets farther from Earth, its gravitational force toward Earth decreases as the square of the distance to Earth's center.

The surface area of a sphere of radius R is $4\pi R^2$. Therefore, anything that diffuses outward from a central point, filling all space uniformly in all directions, becomes less concentrated by a factor of $1/R^2$.

An object at the surface of the Earth falls toward the ground with a certain acceleration. Newton knew that as an object gets higher above the Earth, this acceleration should change in the same way that the force does; that is, the acceleration should be less at greater distances from the Earth, and it should decrease as the square of the distance. The Moon's path is actually an ellipse, but it is close enough to circular that we can consider it a circle to do a rough test of the law of gravity. Newton knew that the distance from the center of the Earth to the Moon is about 60 times the distance from the center of the Earth to its surface. According to his law, then, the centripetal acceleration of the Moon should be $(1/60^2)$, or $1/3600$ of the acceleration of gravity on Earth. But how could he measure the acceleration of the Moon? He did this by using the fact that the Moon moves in (nearly) a circle and therefore has a centripetal acceleration—it is accelerating toward the Earth.

By analyzing the Moon's path, Newton calculated the acceleration of the Moon toward the Earth and found that it is indeed $1/3600$ of the acceleration of an object near the Earth's surface. Thus he checked his hypothesis concerning gravitational force. This was one of the calculations that convinced him that his law was valid and that gravitation is the force that keeps the Moon in its orbit.

As we will explain in Chapter 6, Ptolemy used parallax to determine that the distance from the Earth to the Moon is 27.3 Earth diameters, very close to today's value of 30.17 for the average distance to the Moon.

3-5 Newton's Laws and Kepler's Laws

In the previous section we showed how Newton was able to use circular motion to apply his laws to the motion of the Moon. In fact, Newton was able to show that, based on his laws of motion and his law of gravity, objects in orbit around the Sun would be compelled to move in elliptical orbits. This, of course, is what Kepler had found to be their paths. The mathematics involved in Newton's calculations will not be shown here, but it is important to emphasize that if Kepler's first law, the law of ellipses, had not been known beforehand, it could have been derived from Newton's laws.

Kepler's second law states that as a planet orbits the Sun, a line from the planet to the Sun sweeps out equal areas in equal times. Again, we will not show the

One had to be a Newton to see that the Moon is falling, when everyone sees that it doesn't.

Paul Valery (French poet and philosopher, 1871–1945)

ADVANCING THE MODEL

Travel to the Moon

I believe that this nation should commit itself to achieving the goal before this decade is out, of landing a man on the Moon and returning him safely to Earth.

—PRESIDENT JOHN F. KENNEDY, 1961

The motivation for President Kennedy's commitment to place a man on the Moon was not primarily scientific; it involved national and international politics. Nevertheless, science—particularly astronomy—certainly benefited from that program, advancing our understanding of space travel and of the Moon itself.

We might think of Earth-orbiting satellites as being a development of the second half of the twentieth century, and indeed the technology to launch satellites was not developed until that time. But 300 years earlier, Isaac Newton developed the required gravitational theory and predicted the possibility of artificial satellites. Newton presented the following argument for placing an object in orbit around the Earth, based on his laws of motion and his law of universal gravitation:

If a mountain could be found that extends above the Earth's atmosphere, and if a cannonball were fired from the summit, the ball's path would be somewhat as shown in **Figure B3-4**, line *A*. If a greater charge were used to shoot the cannonball, the ball might follow path *B*. Shoot harder yet and get path *C*. With careful adjustment (and a very powerful cannon!) the ball can be made to go around the Earth and return to where it started (line *D*). Then it will

continue around again in the same path; the cannonball will be in orbit.

The gravitational aspects of a flight to the Moon are not vastly more complicated than those of Earth orbit. The mission begins with the spacecraft orbiting the Earth. As indicated in **Figure B3-5**, a rocket then launches the spacecraft out of Earth orbit to begin its trip to the Moon. As the spacecraft gets farther from the Earth and nearer the Moon, the gravitational pull of the Moon becomes significant. Recall that the

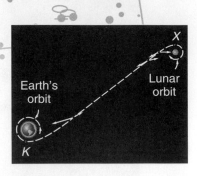

Figure B3-5
A craft sent to orbit the Moon starts out in Earth orbit. A rocket is fired at point *K* to send it toward the Moon. At point *X*, a rocket is fired to put it in Moon orbit. The arrows along the path illustrate the changing gravitational pull on the craft from the Earth and Moon. Distances are vastly out of scale here.

law of gravity tells us that every object in the universe attracts every other. This means that the Moon exerts a gravitational force on a satellite even when it is in Earth orbit (and, in fact, the Moon exerts a force on us here on the surface of the Earth). This force is very small compared to the gravitational attraction toward the Earth until the craft gets closer to the Moon. We have drawn arrows along the path of the Moon-bound craft in Figure B3-5, which are meant only as a guide to the changing strength of the gravitational forces exerted on the craft by the Earth and Moon.

At one point the gravitational pulls toward Earth and Moon exactly cancel one another. Up until this point, the force pulling the craft toward the Earth has been greater than that pulling it toward the Moon, and the spacecraft has been slowing down. After passing that point, the pull toward the Moon is greater, and the spacecraft speeds up. Finally, when the spacecraft has reached the appropriate point in its journey (at about *X* in Figure B3-5), a rocket is fired to slow it down so that it remains in orbit around the Moon. Without this firing, the craft would have so much speed that it would swing right past the Moon. It is interesting to note that for most of the trip the astronauts feel weightless, since they are in free fall, coasting along with the craft.

Now the astronauts are in orbit around the Moon. At this point the lunar module—(often pronounced "lem") disconnects from the command/service module that remains in orbit, fires its rockets to slow down, and descends to the Moon's surface. The

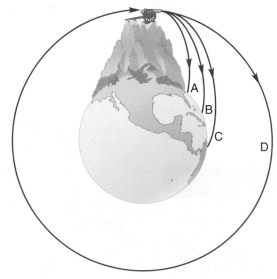

Figure B3-4
Newton drew a figure similar to this to illustrate how an artificial satellite would circle the Earth.

(Continued)

ADVANCING THE MODEL

Travel to the Moon *(Cont'd)*

lunar module does not have to be very large and does not need particularly powerful rockets. This is because the force of gravity on the surface of the Moon is only one-sixth of that on the Earth, so the fall toward the Moon is not as fast and the lift-off requires much less energy. In addition, the Moon does not have an atmosphere, so the frictional effects of an atmosphere need not be considered, as they do for takeoff and landing on Earth.

To leave the Moon, part of the lunar module is left behind when a rocket launches the small remaining craft into lunar orbit to reconnect ("dock") with the command module. Then, at the appropriate point, rockets fire again to send the craft out of lunar orbit and back toward the Earth.

Notice that very little of the trip to and from the Moon is spent with the rockets firing. They are needed only to begin and end each portion of the trip and to make minor midcourse corrections. On one of the missions to the Moon, an astronaut spoke with his son by radio. His son asked who was driving. The reply was that Isaac Newton was doing most of the driving at the time.

The first manned *Apollo* flight took place in October, 1968. In December of that year, three astronauts orbited the Moon, but did not land. Their mission provided the first whole-Earth photos ever made (similiar to the chapter opening figure). In July, 1969, astronauts Neil A. Armstrong and Edwin (Buzz) Aldrin, Jr., guided their lunar module "Eagle" onto the Moon at the Sea of Tranquillity and uttered the famous words, "Tranquillity Base here. The Eagle has landed." In all, 12 men visited the Moon between July, 1969, and December, 1972, when *Apollo 17* blasted off from the Moon.

---

mathematics that prove that this law necessarily follows from the laws of Newton, but we will show, without math, that Newton's laws at least make Kepler's second law seem reasonable. **Figure 3-13** shows the exaggerated elliptical path of a planet. Consider the planet when it is at position *A*, getting closer to the Sun as it moves along. The arrow pointing from the planet toward the Sun represents the gravitational force exerted on the planet by the Sun. Notice that when the planet is in this position, the direction of the force is not perpendicular to the motion of the planet; it is not simply a centripetal force. If it were a centripetal force, it would be exerted perpendicular to the motion of the planet, and it would have only one effect: to cause the planet to curve in its path. Instead, it has two effects: it causes the planet to curve, and it also causes the planet to speed up. This is because it pulls *forward* as well as sideways on the planet.

At location *B* in the figure, the planet is moving away from the Sun. Here the arrow representing gravitational force shows us that the force pulls both backward and sideways on the planet, thus slowing it down as well as curving its path. So we see that as the planet moves toward the Sun, it speeds up, and as it moves away, it slows down. This motion is incorporated in Kepler's second law.

We have not shown that Newton's laws would actually result in Kepler's law of equal areas, but only that both laws result in the planet going fastest when it is closest to the Sun. In fact, it can be shown mathematically that application of Newton's laws of motion and his law of gravitation *necessarily* results in the law of equal areas. Here again, if Kepler's second law had not already been known, Newton could have deduced it from his laws. **Table 3-1** summarizes the types of motion as they were analyzed by Newton.

Finally, we come to Kepler's third law, the one that relates a planet's period of revolution to its distance from the Sun. Newton showed mathematically that if nature obeys an inverse square law for gravitation, planets must necessarily have the period–distance relationship of Kepler's third law. In addition, Newton expanded Kepler's third law, showing that the masses of the objects are important in the relationship. We can now use Kepler's third law for any two objects

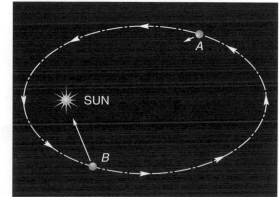

Figure 3-13
At point *A*, the planet is moving closer to the Sun, and the gravitational force on it causes it to speed up as well as curve from a straight line. At point *B*, the force of gravity slows the planet as well as curving its path.

Table 3-1

Summary of Types of Motion

Motion	Type of Acceleration	Direction of Force and Acceleration
Linear	Change in speed	Forward (to speed up) Backward (to slow down)
Circular	Change in direction	Toward center of circle
Orbital	Change in speed and direction	Toward the object being orbited

binary system A system of two objects orbiting each other due to their mutual gravitational attraction.

orbiting each other as a result of their mutual gravitational attraction (a ***binary system***). Using the same convention as in the previous chapter, Newton's formulation of Kepler's third law is

$$\frac{a^3}{P^2} = K \cdot (m_1 + m_2)$$

where a = semimajor axis of the orbit

P = period of the orbit

m_1, m_2 = the masses of the two objects

K = a constant whose value depends on the units of the other quantities

Using standard units (meters for distance, seconds for period, and kilograms for mass), the value of K is given by $G/4\pi^2$, where G is the gravitational constant. Thus, $K = 1.69 \times 10^{-12} \text{ m}^3/(\text{s}^2 \cdot \text{kg})$. However, in some cases it is easier to measure the semimajor axis a in AU and the period P in years; then K is equal to the inverse of the mass of the Sun ($K = 1/m_{\text{Sun}}$).

For the case of a planet (of mass m_1) orbiting the Sun (of mass $m_2 = m_{\text{Sun}}$), Newton's version of Kepler's third law actually simplifies to the original version by Kepler. This is the case because the mass of any planet is very small compared to the mass of the Sun; that is, $m_1 \ll m_{\text{Sun}}$, so the expression on the right side of the equation is essentially equal to one. Thus, for all objects orbiting the Sun, we get Kepler's original version: $a^3/P^2 = 1$.

EXAMPLE

Let us now examine the case of two stars orbiting each other. Suppose observations suggest that the orbital period of this system is 4 years and that the average distance between the stars is 8 AU. Substituting these numbers into Kepler's third law, we find

$$\frac{8^3}{4^2} = 32 = \frac{m_1 + m_2}{m_{\text{Sun}}}$$

Thus, the total mass of both stars is 32 times the mass of the Sun. Without any additional information, we cannot find the mass of each star. We will come back to this problem at the end of the next section.

An artificial satellite is orbiting the Moon with period $P = 11.5$ hours (or 1.3×10^{-3} years) and at an average distance from the Moon's center of 5960 kilograms (or 4×10^{-5} AU). What is the mass of the Moon? (The Sun's mass is about 2×10^{30} kilograms. Make sure you use the appropriate units for the period and average distance.)

Science seeks to show that the various phenomena we observe are not independent of one another, but are instead manifestations of a relatively few basic principles. We see here the success of Newton's work in relating Kepler's laws to the more basic ideas of gravitation and the motion and mass of objects. Once Newton's laws are known, all three of Kepler's laws necessarily follow.

Before leaving the discussion of Newton's laws, we must look at a change these laws brought in our understanding of the paths objects take as they orbit one another. Then we will discuss a particular success of the laws: the explanation of ocean tides.

3-6 Orbits and the Center of Mass

Newton's third law tells us that when one object exerts a force on another, the second object exerts an equal force back on the first. Thus we should expect that gravitational forces affect both objects. One object cannot just remain still while another object orbits around it. Instead, the two objects orbit about a point between them. This point is called the ***center of mass*** of the objects. (Even though an object's center of mass is not always the same as its center of gravity, for almost all the cases we will discuss the two points will be considered to be identical.) As a child, you probably played on a seesaw (or teeter-totter) with someone much heavier than yourself. Recall that the larger person had to sit closer to the pivot point in order to balance. In fact, both persons must position themselves in such a way that their center of mass is at the pivot point of the seesaw if they are to balance on it. Suppose one person weighs 50 pounds and the other person weighs 150 pounds. Since one person is three times heavier than the other, the lighter person must sit three times farther from the pivot of the seesaw.

Figure 3-14 shows a large ball and a small ball connected by a rod and held by a string tied to the rod. The two balls balance because the string is connected at the center of mass of the objects. If this contraption were thrown into the air with a spinning motion—like a baton—it would spin around that point, the center of mass. A planet and the Sun are somewhat like the two balls on the rod, but instead of being held together by a rod, they are held by the force of gravity.

Until now we have spoken of a planet orbiting the Sun as if the Sun stayed still and the planet moved around it. In fact, the two objects revolve around their common center of mass, or ***barycenter.*** The Sun is 330,000 times more massive than the Earth, however, so the barycenter of the Sun–Earth system is essentially in the center of the Sun. To examine a case where the center of mass has a more significant effect, let's look at the Earth–Moon system.

To determine the location of the barycenter of the Earth–Moon system, we use the fact that the Earth's mass is 81 times that of the Moon. Therefore, the Moon is 81 times farther from the center of mass of the Earth–Moon system than the Earth is. This means that the center of mass is about 4800 kilometers from Earth's center and 380,000 kilometers from the Moon. This point is inside the Earth. **Figure 3-15** illustrates that it is about 1700 kilometers below the surface.

It was the relationship between the distances of Earth and Moon from the system's center of mass that allowed us to calculate the mass of the Moon. Today we

center of mass The average location of the various masses in a system, weighted according to how far each is from that point.

barycenter The center of mass of two astronomical objects revolving around one another.

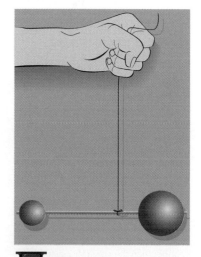
Figure 3-14
The two balls at the ends of the rod balance at the center of mass of the device, where the string is connected.

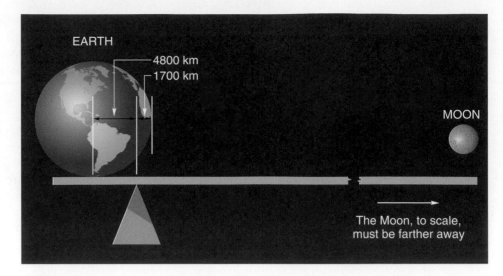

Figure 3-15
The center of mass of the Earth–Moon system is 4800 kilometers from the center of the Earth. If a model of the system were constructed to scale to sit on a weightless board, the construction would balance as shown.

Historically, the barycenter of the Earth–Moon system was determined by observing parallax of nearby planets due to Earth's motion around the barycenter.

know the mass of the Moon more accurately by observing its gravitational effect on space probes that have flown by it, but until the space age, the method described above was the most accurate. In Chapter 12 (and the example below) we show that the center-of-mass method is what allows us to calculate the masses of stars.

EXAMPLE

In the previous example we found that the total mass of the binary system is 32 solar masses. Suppose observations suggest that one star (say star X) is three times farther from the barycenter than the other (star Y). Therefore, star Y has three times the mass of star X. As a result, we find that the mass of star X is 8 solar masses and the mass of star Y is 24 solar masses.

When an entire system moves under the influence of a net external force, it moves according to Newton's second law, behaving as if all of its mass were concentrated at the center of mass.

Now consider the orbit of the Earth around the Sun. Kepler's laws would indicate that the Earth moves in an elliptical path. In fact, it is not the Earth, but the barycenter of the Earth–Moon system that follows the elliptical path. **Figure 3-16a** illustrates the path of a hammer that has been thrown with a spinning motion. The center of mass of the hammer follows a regular path; similarly, the center of mass of a planet and its

Figure 3-16
(a) The center of mass of a thrown hammer follows a smooth path as the hammer rotates around that point. (b) Likewise, it is the center of mass of the Earth–Moon system that orbits the Sun in an elliptical path. (By holding a straight edge up to the paths drawn, you can see the Earth and Moon are always curving toward the Sun during the orbit. This is because the Sun's gravitational force on either object is stronger than the gravitational force from the other object.)

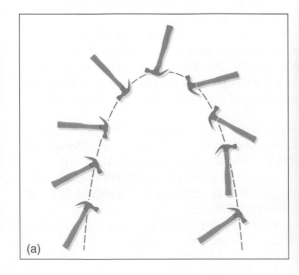

(a)

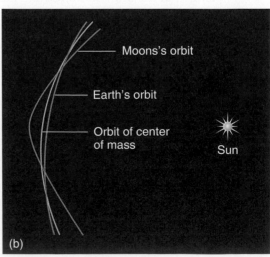

(b)

satellite is the point that follows Kepler's elliptical path (see Figure 3-16b). Kepler's law of ellipses, therefore, must be corrected as to the point that follows the ellipse.

Thus we see that the laws of Kepler, like most laws and theories in science, are approximations. The laws fit Kepler's data perfectly, but as more accurate and different observations became available, Kepler's laws had to be modified and improved. Kepler's laws as originally stated are accurate enough to justify their use in many instances, but they help remind us that scientific theories are always tentative and are subject to revision.

3-7 The Tides

If you live near the seashore or visit it often, we don't need to describe the tides to you. But "landlubbers" don't often get a chance to experience the tides. **Figure 3-17** shows a seashore at low and high tide; the difference in depth of water is obvious. Most locations on the Earth experience a high tide about every 12 hours and 25 minutes and a low tide midway between high tides. Thus on most days there are two high tides and two low tides.

Newton realized that the force of gravity between the Earth and an object is not really a single force exerted by the entire Earth, but the result of all the forces of gravitational attraction between the object and each part of the Earth—each little mass within the Earth. This idea applies to the force of gravity between the Earth and the Moon. The Moon exerts gravitational force not on the Earth as a whole, but on each individual part of the Earth. Consider only three little parts of the Earth. **Figure 3-18** uses arrows to indicate the force of gravity between the Moon and a small mass at three different locations within the Earth. The mass on the side of the Earth nearest the Moon feels the greatest lunar gravitational force, since it is closest to the Moon.

Figure 3-17
Those of us who don't often get to the coast may not be familiar with the phenomenon of tides. Observe how much deeper the water is at high tide. Part (a) shows low tide at a bay of Campobello Island, Canada, and (b) is of the same bay at high tide.

(a)

(b)

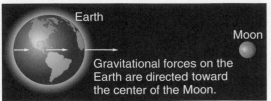

Figure 3-18
The gravitational force exerted by the Moon on a given amount of Earth's mass is greatest on the side nearest the Moon, less on the mass at the center of the Earth, and least on the mass on the side farthest from the Moon. Arrows here represent the forces, and their differences have been exaggerated.

◆ A "differential" pull means that it is different at different points on the Earth.

The mass at the center of the Earth feels less force toward the Moon, and the mass at the far side of the Earth feels the least force. Specifically, the mass on the side of the Earth nearest the Moon feels a gravitational force from the Moon that is about 3% larger than the force on the mass at the Earth's center, which in turn is about 3% larger than the force on the mass at the far side of the Earth.

Now, each of these three parts of the Earth responds to the gravitational force toward the Moon, but the part closest to the Moon feels a greater force than the other two. There is more force toward the Moon on one kilogram of mass here than there is on one kilogram of mass at Earth's center, so the mass at the surface feels pulled away from the center. Water covers most of the Earth, and since it is liquid and free to flow, some water flows to the area under the Moon. As the water becomes deeper at that point, it causes a high tide.

The high tide on the other side of the Earth occurs because the center of the Earth feels a greater force toward the Moon than water on that far side. Thus the main body of the Earth is pulled away from that water, creating another high tide there.

This *differential* gravitational pull on the various parts of the Earth results in two areas of the Earth experiencing high tides. The water that went to making those high tides has been pulled away from other parts of the Earth, so a low tide lies midway around the Earth between the areas of high tides.

You might think that since the Earth completes one rotation a day, there would be exactly two high tides each day. Instead, about 12 hours and 25 minutes pass between successive high tides, so that if a high tide occurs on my beach at 10:00 this morning, the high tide tomorrow morning will be at about 10:50 A.M. The reason for this is that the Moon is not stationary as the Earth rotates. Instead, the Moon moves through about 1/30 of its cycle each day. Therefore the Earth must turn for the additional 50 minutes, which is about 1/30th of 24 hours, before a spot on the Earth returns to the same position with respect to the Moon. This is what causes the Moon to rise (or set) about 50 minutes later each day. **Figure 3-19** illustrates the Moon's motion as the Earth turns for 24 hours and 50 minutes.

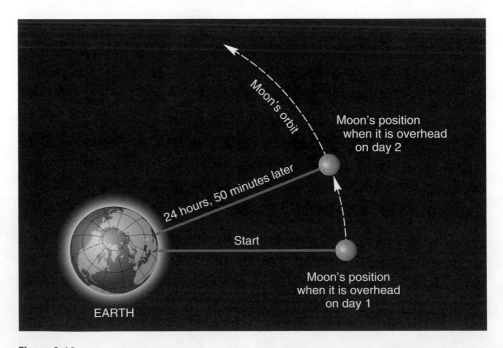

Figure 3-19
The Earth is seen here from above the North Pole. Since the Moon moves in its orbit, the Earth must make more than one complete rotation for a point on its surface to return to the same position with respect to the Moon. About 24 hours and 50 minutes pass between the two positions shown.

As you might suspect, the actual tidal phenomenon is more complicated than we have just described. The Earth is rotating and its land masses disturb the flow of water. The shape of the shoreline, the depth of the water, and the location of the Moon all play a part in determining exactly when high and low tides occur at a particular location on Earth and just how high and how low those tides are.

In addition, the Sun causes tides on the Earth. The Sun's gravitational pull on one kilogram of Earth's mass is about 180 times stronger than the corresponding pull from the Moon. However, the Sun is much farther from the Earth than the Moon is and therefore its pull is almost the same on every kilogram of Earth's mass. Specifically, the mass on the side of the Earth nearest the Sun feels a gravitational force from the Sun that is about 0.01% larger than the force on the mass at the Earth's center, which in turn is about 0.01% larger than the force on the mass at the far side of the Earth. As a result, the difference in the Moon's pull on opposite sides of the Earth is a bit more than 2 times larger than the difference in the Sun's pull. Therefore, the tides due to the Sun are small and are not noticed independently of the Moon's tides. However, when the tides due to the Sun correspond to the Moon's tides (near the times of new and full Moon), we see extreme tides on Earth; these are known as *spring tides.* On the other hand, when the Sun's tides are 90° from the Moon's (near quarter Moon), they tend to cancel partially the Moon's and this causes the change from low to high tide to be less than normal; such tides are called *neap tides.*

> In the example on page 76 we showed where the factor of 180 comes from.

> **spring tide** The greatest difference between high and low tide, occurring about twice a month, when the lunar and solar tides correspond.

> **neap tide** The least difference between high and low tide, occurring when the solar tide partly cancels the lunar tide.

Rotation and Revolution of the Moon

The period of the Moon's rotation exactly matches its period of revolution. As a result, the Moon keeps the same face toward Earth at all times. This effect is caused by tidal forces. To understand it, recall that as the Earth rotates, it interacts with tides in the water. If there were no friction between the solid Earth and its oceans, one area of high tide would stay exactly under the Moon and another would be on the opposite side of the Earth. The land masses, however, exert forces on the water as the Earth rotates, causing the point of highest tide to be pushed from directly under the Moon (and directly opposite the Moon). **Figure 3-20** illustrates this behavior.

If the land masses exert a force on the tidal bulges as the Earth turns, it follows from Newton's third law that the bulges exert an equal and opposite force on the land masses. This results in the Earth's rotation being slowed down because of the *tidal friction* exerted on it. The Earth is indeed turning more slowly today than it was years ago. Our days are very slightly longer than Shakespeare's were, increasing by about 25 billionths of a second every day.

> The next time you are at the seashore, notice that high tide occurs after the Moon is highest in the sky.

> **tidal friction** Friction forces that result from tides on a rotating object.

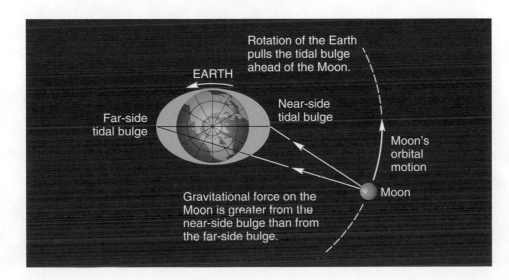

Figure 3-20
The Earth's rotation tends to drag the tides along with it, so that a high tide is not directly under the Moon but instead is farther to the east.

Tides also occur on the dry land. In this case, the rocks and dirt actually stretch to allow the surface to rise and fall. The maximum dry-land tide is about 9 inches; that is, the surface is about 9 inches farther from Earth's center at high tide than at low tide.

Just as the Moon causes tides on Earth, the Earth causes Moon tides, similar to the tides on the solid Earth. And just as tides on Earth are causing it to change its speed of rotation, the tides on the Moon have resulted in a change in its speed of rotation. At one time in the past, the Moon must have had a rotation period different from its revolution period. Through millions of years, the tides have slowed the Moon's rotation until it now keeps its same face toward the Earth. As a result of the tidal interactions between the two objects, the Moon is also pushed farther away from the Earth at a rate of about one centimeter per year.

Precession of the Earth

Think of what happens when you spin a child's top on a smooth table. The axis of the top does not stay in the same orientation (unless you were able to begin the rotation around a perfectly vertical axis). Instead, the top wobbles around. What causes the wobble? The mathematics to describe this effect is far from simple, but it boils down to the fact that the top has a tendency to fall over, and its rotation prevents this from happening. Instead, the top wobbles around, keeping the same angle with the table's surface until friction slows it down. Any time a spinning object feels a force trying to change the orientation of its axis, it will wobble. The wobble is called *precession*.

precession The conical shifting of the axis of a rotating object.

The Earth also spins on its axis. It might seem that the Earth would not precess because there is nothing below it trying to pull it over. There is, however, a force on the Earth tending to change the orientation of its axis of rotation. Refer to **Figure 3-21**. The Earth drawn there is not spherical, but is a bit "flattened," as is the real Earth, although the real Earth is much more spherical than the one in the figure. The Earth's diameter is about 26 miles greater across the equator than from pole to pole. This is caused simply by the fact that it is spinning. (Imagine what would happen if you made a ball of wet clay and set it spinning. The Earth is not exactly a ball of wet clay,

Figure 3-21
The Earth is not a perfect sphere, and as the Moon exerts a greater gravitational force on its nearer side, this force tends to twist the Earth into a different orientation. This effect causes the Earth to precess. (The lengths of the arrows are not to scale.)

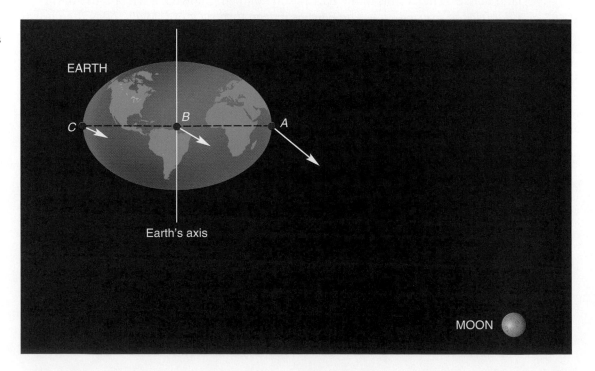

but it is still flexible enough to show the effect.) The measure of how much a planet is "flattened" *(oblate)* is called its **oblateness.**

$$\text{oblateness} = \frac{d_{\text{large}} - d_{\text{small}}}{d_{\text{large}}}$$

Finally, recall that the Moon exerts a gravitational force on each particle of the Earth. The arrows in Figure 3-21 illustrate this force at three places. Since point *A* is closer to the Moon than point *C*, the Moon exerts a greater force on a particle at *A*. This results in an overall force seeking to change the tilt of the spinning Earth's axis. Therefore we have what is needed to cause precession. The Earth's axis does indeed precess, although very slowly. While the child's top may complete a precession in about a second, the Earth requires about 26,000 years (**Figure 3-22**)!

What effect does precession have on what we see in the sky? Right now, Polaris is the closest bright star to the Earth's north celestial pole. But it will not remain so forever. The pole will gradually change, and about 12,000 years from now the star Vega will be our "North Star." A corresponding effect is that the position of the vernal equinox changes over the centuries.

3-8 The Importance of Newton's Laws

We have seen some of the value of Newton's laws. From them we can derive Kepler's laws. They allow us to explain tides and precession. In Chapter 9 an Advancing the Model box tells the story of how they were used to predict the existence of the planet Neptune. From these examples you might conclude (correctly) that Newton's laws differ in their very nature from Kepler's laws. Kepler's laws explain a particular situation—the orbiting of the planets around the Sun. Newton's laws, on the other hand, apply everywhere and to all objects. They are much more fundamental than Kepler's laws, not only explaining motion, but doing so in a measurable, mathematical manner. This is important in science. We wish to measure things and to predict events quantitatively. Newton's laws do this.

Can we say that Newton's laws are *right?* Scientists usually avoid calling a theory right or wrong. But we can say that Newton's laws work and are therefore good laws. They fit the data, they make predictions that can be checked, and they fit with other laws to make an overall theory that is a simple, unified package. (We call this package "Newtonian mechanics" or "classical mechanics," to distinguish it from the mechanics of Einsteinian relativity.)

Up until the time of Newton, Aristotle's idea of the separateness of the Earth and the heavens was the accepted world view (**Figure 3-23**). Isaac Newton was not the only one who thought differently, for from the time of Galileo the idea of the unity of nature had begun to take shape. In fact, other thinkers of Newton's time were working on theories of gravitation. It was Newton, however, who put it all together. It is difficult to overemphasize the effect that Newton's work has had on our thinking.

For one thing, Newton's laws were the first ever that could be shown to hold for both the heavens and the Earth. The idea that there were two natures, one up there and one down here, had been a part of Western culture since before Greek times. Newton showed that the ancient idea was wrong; now it was possible to look upon the universe as ONE. In fact, the word universe begins with the prefix "uni," meaning "one, single" (*unit, unify, unique, united,* and so on). Such a concept was foreign before the time of Newton. He showed us that we live in one cosmos, that nature is unified, that things up there might be expected to be like things down here. In a sense, Newton changed our world from a "duoverse" to a *universe.* Our modern science of astronomy could not exist without this basic understanding.

oblateness A measure of the "flatness" of a planet, calculated by dividing the difference between the largest (d_{large}) and smallest (d_{small}) diameter by the largest diameter.

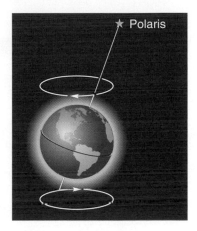

★ Polaris

Figure 3-22
As the Earth spins, its axis precesses with a period of about 26,000 years, pointing to different "pole stars" over the centuries.

I do not know what I may appear to the world; but to myself I seem to have been only like a boy playing on the seashore, and diverting myself in now and then finding a smoother pebble or a prettier shell than ordinary, whilst the great ocean of truth lay all undiscovered before me.
Newton

All things by immortal power,
Near or far,
Hiddenly
To each other linked are,
That thou canst not stir a flower
Without troubling of a star.
Francis Thompson (1859–1907)

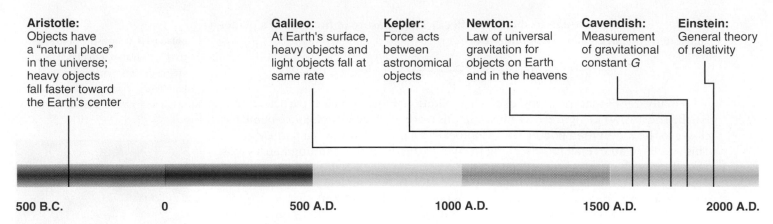

Figure 3-23
Aristotle's explanation of gravity lasted for almost 2000 years, before it was replaced by scientific models based on observation and experiment.

In addition, the work of Sir Isaac Newton confirmed the belief of the ancient Greeks that nature is explainable; that if we work hard enough at it, we have the ability to understand the many seemingly mysterious things that occur in nature. No longer do we have to fear the darkness, for the objects in the skies are part of our universe.

3-9 Beyond Newton: How Science Progresses

Mass was defined earlier in this chapter as the measure of the inertia of an object. In stating the law of gravity, however, Newton proposed that mass is also the quantity that determines the strength of gravitational attraction. Why should the same quantity be the measure of two seemingly different physical properties? It is certainly not obvious that *inertia* (the resistance to a change in motion) should have anything to do with *gravitational attraction*. Yet the measures of these two properties are not just similar; experiments show that they are identical to at least one part in 1 trillion. Can this be simply a coincidence?

Scientists do not like such coincidences. They feel that if two things are so similar, there must be a reason. The attempt to explain this apparent coincidence is what led Albert Einstein to develop his general theory of relativity. General relativity, developed more than two hundred years after Newton stated the gravitational relationship, successfully relates the two concepts of inertia and gravity so that the puzzle of the coincidence no longer exists. In addition, it makes correct predictions in cases where Newton's law of gravity has been found to be in error. Thus, to continue the story, we move suddenly to the twentieth century.

3-10 General Theory of Relativity

The general theory of relativity fundamentally changes the way we look at the phenomenon of gravity. The theory begins with a statement of the equivalence of gravity and acceleration.

The woman in **Figure 3-24a** has dropped a book on the surface of the Earth. In part (b) she drops the same book while in a spaceship that is far from Earth (and any

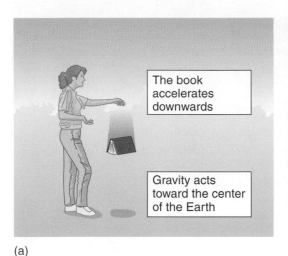

(a)

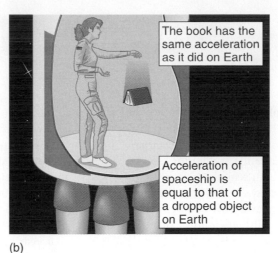

(b)

Figure 3-24
Whether on Earth or in a spaceship far from Earth and any other gravitational influences, accelerating at the acceleration due to gravity, the book will "fall" in the same way.

other gravitational influences) and accelerating in the direction shown. If her spaceship is accelerating at the same rate that falling objects accelerate here on Earth, she will observe the book falling toward her feet in exactly the same way as it did on Earth's surface. She won't be able to tell the difference between "falling" caused by the ship's acceleration and falling caused by gravity. And if she steps on a scale in such a spaceship, the scale will register the same weight as a scale on the Earth's surface. The ***principle of equivalence*** of Einstein's general theory of relativity tells us that there is no experiment whatsoever that she can do to distinguish between the two conditions.

The general theory of relativity makes other predictions than the principle of equivalence. For example, when we study black holes in Chapter 15, we will see that general relativity also predicts that light curves as it passes by an object. That is, light is affected by gravity. Einstein predicted this in 1915, and observations made during a solar eclipse in 1919 confirmed that light passing near the Sun is indeed bent by the presence of the Sun. Since then, this prediction has been confirmed in every experiment in which it has been tested.

Space Warp

Einstein's general theory of relativity explains the similarity of gravity and acceleration, as well as the bending of light when it passes by an object, by considering the curvature of space. Space warp is not easy to imagine because the world we experience is a three-dimensional one and we cannot imagine what dimension our world can curve into. We say that our world has three dimensions because we use three directions to specify the exact location of something. We can state these directions as north-south, east-west, and up-down. For example, you might state that to get from the library to your bed at home, you must go 4325 feet north, 5843 feet west, and 12 feet up to the second floor. Although other choices might be made for the three coordinates that specify the location of an object relative to your position, you must always give three pieces of information in our three-dimensional world.

Imagine a land of two dimensions, "Flatland," on which two-dimensional creatures live. The surface of a desk might represent Flatland. Imagine that these creatures know only the two dimensions of their universe; they can perceive north-south and east-west but have no conception of up-down. The location of everything in the universe of Flatland can be specified by saying how far it is along an east-west line and how far along a north-south line.

The spaceship is far enough from Earth (and other objects) that there is no observable gravitational force.

principle of equivalence The statement that effects of acceleration are indistinguishable from gravitational effects.

This idea is borrowed from "Flatland," written by Edwin A. Abbott in 1884. This book is a classic illustration of multiple dimensions in space, but even more that that, it's great fun to read.

ADVANCING THE MODEL

Albert Einstein

One of the greatest theoretical physicists of all time was considered a slow learner as a child. The young Albert Einstein (**Figure B3-6**) found the formal, disciplinary schools of Germany at the end of the last century intimidating and boring, and he dropped out before completing high school. He studied at home, however, and learned to play the violin well enough that—although he played it only for his own enjoyment and relaxation—he became an accomplished violinist. His studies of geometry and science led him to conclude, at age 12, that the Bible is not literally true. That shock implanted in him a deep distrust of authority of any kind—a distrust that he carried with him throughout life.

On his second try, Einstein was granted admission to the Swiss Federal Institute of Technology in Zurich. At the university he often failed to attend class, preferring instead to study on his own, reading the classical works of theoretical physics. He was granted a Ph.D. in 1900, but it was two years before he found regular work—as a patent examiner in the Swiss Patent Office.

Einstein was similar to Isaac Newton in his lack of success in formal school and in the timing of his scientific work as well. Like Newton, Einstein's most revolutionary work was done during a very few years when he was in his early twenties. In 1905, Einstein published four important papers in the prestigious German physics journal *Annalen der Physik.* Although the paper for which he is best known is the one introducing special relativity, he was awarded the Nobel Prize for another, concerning the photoelectric effect of light. Many physicists quickly recognized the importance of his work, but it was not until 1909 that Einstein was given a full-time academic position at the University of Zurich.

In 1903 Einstein married his college sweetheart, Mileva Maritsch, and they had two sons. When World War I began, Einstein had a position in Germany, but his wife and children were vacationing in Switzerland. They were unable to return to Berlin, and the forced separation resulted in a divorce in 1919. Later that year, Einstein married his cousin Elsa.

In 1916, Einstein published his general theory of relativity, but his relativity theories were slow to gain acceptance due to the lack of experimental verification. In 1919, however, the Royal Society of London announced that its scientific expedition to observe the solar eclipse of that year had verified Einstein's prediction of the gravitational deflection of starlight as it passes near the Sun. The international acclaim that followed changed Einstein's life, for he was suddenly considered a genius. He began to travel more and more, giving lectures throughout the world.

Figure B3-6
Albert Einstein (1879–1955) had several hobbies; one of his favorites was sailing.

Although Einstein continued his scientific work until he died in 1955, his fame also allowed him to exert influence in world affairs. Though a Jew and a critic of the political situation in Germany, he escaped the Nazis because he was visiting California when Hitler assumed power in 1933. Einstein renounced his German citizenship and never returned to his home country. The next year he became an American citizen.

It is ironic that Einstein's name is so closely linked to the atomic bomb. Although his theories predicted that mass could be converted to energy, which was the idea that led to the development of nuclear weapons, Einstein was an avowed pacifist who worked untiringly to prevent war, which he saw as the ultimate scourge of humanity. He argued that the establishment of a world government was the only permanent solution.

Einstein disliked fame and the trappings that accompany it. In his travels to the Far East he refused to ride in rickshaws, feeling that to do so would be degrading to the person pulling him along. He preferred to be treated as a common person and disliked formal attire. He regularly gave important lectures in an open-collar shirt and was often seen near his Princeton home in rumpled clothes, carrying his violin.

Now imagine some tiny fleas on an expanding balloon. Instead of regular fleas, make the creatures Flatfleas and the balloon Flatland. Flatfleas, of course, have no height and are only two-dimensional creatures. You might object that the surface of the balloon isn't flat, but if the balloon is large enough compared to the size of the Flatfleas, they would not easily perceive its curvature (**Figure 3-25**). If one Flatflea realizes that his universe is curved, how can he explain this idea to his contemporaries?

Saying that the universe is curved "downward" would have no meaning, for "up" and "down" are undefined in Flatland. We humans see that the balloon's surface is curved into the third dimension, but it would take a great stretch of the Flatfleas' imagination to think of a third dimension, for it is not part of their everyday world.

By a similar analogy, we picture the curvature of *our* space. Suppose that the presence of a massive object causes space to be warped. We can picture space near the Sun as being warped analogous to the way the surface of a waterbed is warped by a bowling ball placed in its center (**Figure 3-26**). The waterbed's surface would be distorted so that a straight line following the surface would have to follow the distortion. Similarly, a beam of light passing near the bowling ball would follow the curvature of space caused by the massive object. Let us also imagine the motion of a small ball on this waterbed. Far from the distortion, the surface of the waterbed is fairly flat and thus the ball will move in a straight line. However, the closer the ball gets to the distortion, the more it will curve toward it. If the ball's speed is chosen appropriately with respect to the distortion, the ball may end up moving in an orbit around the sides of the distortion, in the same way a planet moves around the Sun. We can say that matter (in our case, the heavy bowling ball) tells space (the surface of the waterbed) how to curve and that the space curvature (or distortion) tells matter (the small ball) how to move.

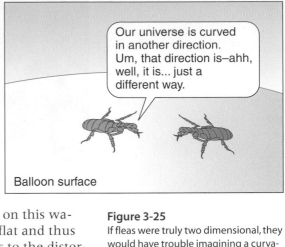

Figure 3-25
If fleas were truly two dimensional, they would have trouble imagining a curvature of their world.

We will see later in the text that space curvature also explains black holes.

 Figure 3-26
According to the General Theory of Relativity, matter curves the space around it. This curvature, in turn, influences the motion of objects in the vicinity. In order to visualize the curvature of three-dimensional space, in this figure we show how (a) a two-dimensional space (the surface of a waterbed) curves around a massive object (a heavy bowling ball (b)). The resulting distortion, in turn, affects the paths taken by light and other objects passing by (c). (Changes in the paths have been exaggerated for clarity.)

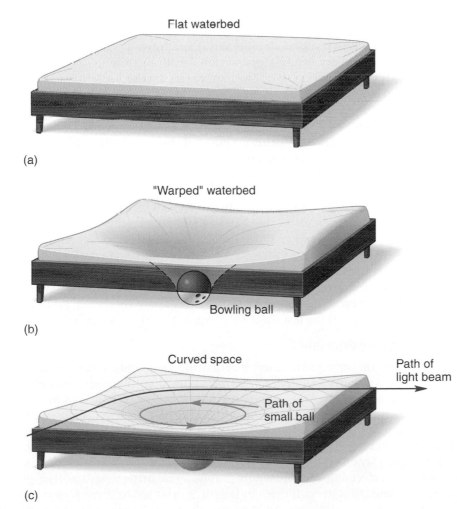

Flat waterbed

(a)

"Warped" waterbed

Bowling ball

(b)

Curved space

Path of light beam

Path of small ball

(c)

Figure 3-27
This illustration exaggerates the actual precession of the perihelion (the closest point to the Sun) of Mercury's orbit.

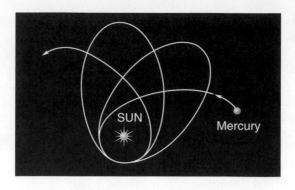

3-11 Gravitation and Einstein

Einstein proposed that instead of thinking of gravity as an attractive force between objects, we think of space being curved by the presence of mass, so that as objects move, they follow that curvature. His general theory of relativity holds that a planet in orbit around the Sun is responding to the curvature of space that results from the Sun's mass. Space curvature thereby explains not only why objects fall and why planets orbit the Sun, but also why light bends near a massive object.

Here again we have two theories that explain a set of observations. How do we choose between them? First, we ask whether one theory better fits the data. We have seen that Einstein's theory explains the bending of light near a massive object and that Newton's does not. We now consider a final observation, one concerning Mercury's orbit.

The Orbit of Mercury

precession (of an elliptical orbit)
The change in orientation of the major axis of the eliptical path of an object.

In 1859, 14 years after predicting the existence of Neptune, Urbain Le Verrier reported that Mercury's elliptical orbit *precesses;* that is, it does not keep the same orientation in space. **Figure 3-27** illustrates a greatly exaggerated precession, showing the near point in the orbit (the perihelion) gradually sliding around the Sun. Mercury's orbit precesses very slowly—at the rate of 574 arcseconds per century—which is less than a degree per century. Calculations show that Newton's theory of gravity can account for the basic effect, because gravitational pulls by other planets, particularly Venus and Jupiter, would cause it. The total precession accounted for by these gravitational tugs amounts to 531 arcseconds per century, however, and this is 43 arcseconds short of the observed 574 arcseconds.

The unaccounted-for 43 arcseconds of precession per century presented a mystery for astronomers. One hypothesis held that there is another planet in the inner solar system that is responsible for the extra precession. This planet was given the name Vulcan, and extensive searches for it were carried out, but it was never found.

The quotation is from Clifford M. Will, *Was Einstein Right?* (New York: Basic Books, 1993).

One of the first problems to which Einstein applied his new general relativity theory was the precession of Mercury's orbit. He found that the theory accounted precisely for the 574 arcseconds of precession. Einstein later wrote that "for a few days, I was beside myself with joyous excitement." Like Newton's theory, general relativity predicts the precession effect caused by the other planets, but unlike Newton's theory, it predicts additional precession due to properties of curved space. Thus the Sun itself caused the extra 43 arcseconds of precession. This was the first direct test of conflicting predictions made by the two theories, and the test showed Einstein's theory to be the better one.

The Correspondence Principle

correspondence principle The idea that predictions of a new theory must agree with the theory it replaces in cases where the previous theory has been found to be correct.

There is a general principle in science concerning the replacement of old theories by new ones. The *correspondence principle* states that the predictions of the new theory must agree with those of the previous theory where the old one yields correct results. This is entirely reasonable; in the case we are discussing, it simply says that general relativity must not disagree with Newton's gravitational theory where the older theory provided correct results. For example, Newton's theory predicts that the planets orbit the Sun in elliptical paths, but that there are minor deviations from perfect ellipses (caused by gravitational attractions to other planets). We will see later that irregularities in Uranus's orbit led to the discovery of Neptune. In fact, the discovery of Neptune after it was predicted by Urbain Le Verrier and John Adams (Advancing the Model, Chapter 9) provided a dramatic

ADVANCING THE MODEL

The Special Theory of Relativity

About 10 years before he developed the general theory of relativity, Einstein proposed the special theory of relativity. It is called *special* because it does not apply to accelerated motion, but only to the special case of uniform motion.

The special theory of relativity is based on two postulates, the first of which states:

- All laws of physics are the same for all non-accelerating observers, no matter what the speed of those observers.

Recall that people once thought that the laws of nature that govern objects here on Earth are different from those that rule the heavens. Then Newton showed that the same laws work for both—at least, the mechanical laws of force and motion. Einstein's first postulate completes the progression; Einstein begins with the assumption that *all* laws, including those of electricity and magnetism, are the same everywhere. The first postulate abolishes the idea of absolute rest and therefore absolute motion. Motion is relative, since it is not possible to distinguish experimentally between two different uniformly moving observers.

Einstein's second postulate concerns the speed of light:

- The speed of light is the same for all non-accelerating observers, no matter what their motion relative to the source of the light.

This predicted behavior of light is different from the behavior of ordinary objects in our experience. For example, when we catch a baseball, we see it coming at us faster if we are moving toward the thrower than if we are standing still. If Einstein's postulate is true, this doesn't happen for light. **Figure B3-7** shows an imaginative case in which people on fast-moving spaceships are measuring the speed of light that comes from an Earthling's flashlight. Technological advances since Einstein's time have allowed us to repeatedly confirm the second postulate; we now consider the constancy of the speed of light a law of nature.

On the basis of his two postulates, Einstein reached several startling conclusions, including the statement that Newton's laws of motion become increasingly inaccurate with increasing speed. The special theory of relativity predicts that (1) the observed length along the line of motion of a moving object becomes less than its length when measured at rest; (2) the observed passage of time becomes slower for the moving object; and (3) the observed inertia of an object becomes greater than its inertia when at rest.

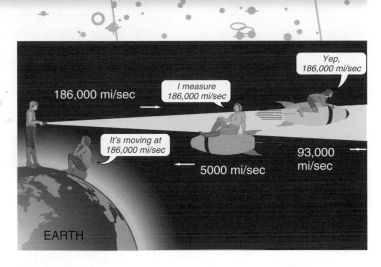

Figure B3-7
The special theory of relativity predicts that if we measure the speed of light, we will get the same result regardless of the motion of the measurer. That speed is 186,000 miles per second, or 300,000 kilometers/second.

OK, so Einstein made two assumptions (postulates) and used them to make predictions. Fine, but are the predictions correct? Does the theory fit the data? The three predictions mentioned above become significant only at speeds greater than those attained in everyday life, and they are therefore difficult to check experimentally. Nevertheless, experimental checks are possible in some cases. Every time a test has been conducted to check the special theory of relativity, the theory has passed the test. The theory *works!*

One more conclusion based on the postulates of special relativity should be mentioned: mass can be transformed to energy and vice versa. The conversion between the two is governed by the equation $E = mc^2$. (In this equation, E is energy, m is mass, and c is the speed of light.) The equation was dramatically verified in 1945 by the first explosion of a nuclear bomb, for the energy of these bombs comes from the conversion of mass to energy. A more peaceful example of mass-energy conversion is provided by nuclear power plants, which produce electrical energy based on Einstein's theory. Finally, we will see in Chapters 11 and 12 that nuclear energy is the source of the energy of the stars. The theory of special relativity is not a simple one to understand, but it is firmly established as a tool that astronomers must use if we are to advance in our quest to understand the universe.

confirmation of Newton's theory of gravitation. Einstein's theory must also predict elliptical orbits, and it must not contradict the confirmed irregularities predicted by Newton's theory.

Indeed, the general theory of relativity is in accord with the correspondence principle, for it does predict elliptical orbits with the variations that have been observed.

The correspondence principle is usually credited to Niels Bohr, whom we discuss further in Chapter 4.

The difference is that it views the orbits as being due to planets following their natural paths in a space that is warped by the presence of the Sun.

All tests of Einstein's relativity theory have confirmed it. The theory is a solid one and forms the basis of modern astronomy. Why, then, do we continue to talk about gravitation and Newton's ideas? The answer is the same as we discussed in the case of the geocentric/heliocentric question: we use the more easily understood theory when it fits. The theory of relativity is not an easy one to understand, and since most of us do not need it to explain things in our everyday lives, few of us think in terms of curved space. Thus you'll see little mention of curved space in this text until the discussion of black holes, where Einstein's theory is essential.

✦ Recall that we still speak of "sunrise" and "sunset" from the geocentric model.

Conclusion

In the previous two chapters, we introduced two primary models: the geocentric model of Ptolemy and the heliocentric model as presented by Copernicus and revised by Kepler. Galileo made observations with his telescopes that led to the conclusion that Kepler's model fit the data better and, in general, was the better theory. The theory suffered one flaw, however: Kepler's ellipses and his rules concerning the speeds of planets did not "make sense," in that they did not correspond to anything else in nature.

A more complete understanding of force and motion was necessary before Kepler's laws could be related to anything other than planetary motions. Newton's three laws of motion, along with the law of universal gravitation, gave humanity a different outlook on the world and provided scientists with a means of understanding the motions of objects, not only here on Earth but in the heavens as well.

Now we find that Newton's theories are not the final answer; they have been supplanted by Einstein's. It is interesting that during the times of Copernicus and then of Kepler and Galileo, people were confused and scandalized by the theories these men proposed. People's common sense told them that the Earth was the center of the universe, and they were convinced that anything else was nonsense. Nearly a century ago, Einstein proposed a new way of thinking about the universe. His theory of relativity might seem to some to upset the order of the universe; in fact, it adds order. To most people it seems completely nonsensical in that it does not fit their experience at all. However, general relativity is firmly established in science, and perhaps someday it will become part of our everyday thinking. It is also not only possible but likely that at some point in the future Einstein's theories will be supplanted by new theories that more accurately describe the universe.

Study Guide

1. Using his newly invented telescope, Galileo discovered all of the following except
   A. moons of Jupiter.
   B. phases of Venus.
   C. sunspots.
   D. stellar parallax.
   E. mountains on the Moon.
2. Which of the following planets can be seen (from Earth) in a crescent phase?
   A. Mercury
   B. Venus

C. Mars
D. [Two of the above.]
E. [All of the above.]

3. Which of the following observations by Galileo was most convincing in deciding between the two opposing theories of our planetary system?
   A. moons of Jupiter
   B. mountains on the Moon
   C. phases of Venus
   D. sunspots

4. Suppose you are riding as a passenger in a car when the car stops suddenly. What pushes you forward?
   A. The force of inertia.
   B. Your weight.
   C. The car's forward motion.
   D. Nothing pushes you forward.

5. If the same net force is applied to two different objects, one with a mass of 1 kilogram and the other with a mass of 2 kilograms,
   A. the 1-kilogram mass will have twice the acceleration of the other.
   B. the 1-kilogram mass will have more acceleration than the other, but not necessarily twice as much.
   C. both objects will have the same acceleration, for the force determines the acceleration.

6. If the mass of the Earth magically decreased with no change in its size, your weight (as you stand on the surface) would
   A. increase.
   B. not change.
   C. decrease.

7. According to Newton, the natural motion of an object is
   A. a circle.
   B. an ellipse.
   C. a straight line.
   D. retrograde motion.

8. If the distance between two objects is tripled, the gravitational force exerted by one on the other will be
   A. the same.
   B. one-third as much.
   C. one-ninth as much.
   D. three times as much.
   E. nine times as much.

9. The gravitational attraction between an object and the Earth
   A. stops just beyond the Earth's atmosphere.
   B. extends to about halfway to the Moon.
   C. extends about five-sixths of the way to the Moon.
   D. extends to infinity.

10. The mass of the Moon is about 1/81 that of the Earth. If you were on the Moon, though, you would weigh 1/6 of your weight on Earth. Why do you weigh more than 1/81 of your Earth-weight?
    A. The Moon is made of different materials than Earth.
    B. The Moon has a different density than Earth.
    C. The Moon has no atmosphere, while Earth does.
    D. The Moon is much smaller than Earth.
    E. [Both A and B above.]

11. Which statement best describes the relationship between Newton's laws and Kepler's laws?
    A. Newton proved that Kepler was wrong.
    B. Newton's laws and Kepler's laws are now considered equally valuable.
    C. Newton's laws are more fundamental and more powerful than Kepler's laws.
    D. Neither Newton's laws nor Kepler's laws have any modern applications.

12. When you whirl a rock around on the end of a string, centripetal force
    A. pulls outward on the rock.
    B. pulls inward on the rock.

C. pulls outward on your hand.
   D. [Both A and B above, and the two forces balance.]
   E. [All of the above, for all of the forces are equal.]

13. Kepler's third law states that the ratio of the cube of a planet's semimajor axis to the square of the planet's period of revolution is equal to a constant. Newton found that the value of that constant depends upon
    A. the sizes of the two objects.
    B. the masses of the two objects.
    C. the velocities of the two objects.

14. Newton checked his hypothesis concerning an inverse square law of gravitation by calculating
    A. the Moon's acceleration toward the Earth.
    B. the time required for the Moon to complete one orbit.
    C. the mass of the Earth.
    D. the mass of the Moon.
    E. [Both C and D above.]

15. The force of gravity is responsible for
    A. the weight of an object on Earth.
    B. the mass of an object on Earth.
    C. the tides.
    D. holding the Earth in its orbit.
    E. [All of the above.]

16. If the Moon were covered with water, the number of tidal bulges on it would be
    A. one.
    B. two.
    C. [None.]

17. Tides on the Earth are primarily due to the mutual gravitational attraction between the Earth and
    A. the Moon.
    B. the Sun.
    C. Jupiter.
    D. [None of the above.]

18. How many high tides are observed most days at most seaports on Earth?
    A. One, caused by the Moon.
    B. Two, because the Sun and the Moon each cause one.
    C. One or two, depending upon the relative positions of the Earth, Moon, and Sun.
    D. Two, roughly 12 hours apart.

19. The gravitational force due to the Moon is exerted
    A. only upon the side of the Earth nearest the Moon.
    B. only upon the point on the Earth nearest the Moon.
    C. upon the center of the Earth only.
    D. upon the entire Earth.
    E. upon the water surfaces of the Earth but not upon the land surfaces.

20. The principle of equivalence of the general theory of relativity tells us that
    A. the force of gravity is equivalent to acceleration in the opposite direction.
    B. speeds measured in any location are equivalent.
    C. the speed of light is the same for all observers.
    D. Newton's laws are equivalent to Kepler's laws.
    E. Newton's laws are equivalent to Einstein's theories.

21. Which of the following choices confirmed a prediction made by the general theory of relativity?
    A. Observations of Jupiter's satellites.
    B. Observations of phases of Venus.
    C. Calculations of the orbit of Mercury.
    D. Calculations predicting the existence of Mars's moons.
    E. [None of the above; general relativity has not been successful in astronomical applications.]

22. The correspondence principle states that
    A. predictions made by a new theory must agree with those of the theory it replaces where the old theory fit the data.
    B. all predictions made by a new theory must agree with those of the theory it replaces.
    C. no predictions made by a new theory are expected to agree with those of the theory it replaces.
    D. the force of gravity corresponds to acceleration in the opposite direction.

23. List the following men in the order in which they lived: Copernicus, Aristotle, Ptolemy, Kepler, Galileo, Brahe. (Hint: You may have a "tie" between two of them.)

24. Most of Galileo's observations argued *against* the Ptolemaic system rather than *for* a heliocentric system. What was the exception?

25. Which of the phases of Venus could not be explained by the Ptolemaic model?

26. Define inertia and give an example of its action.

27. Which is the more fundamental quantity, mass or weight? Describe an observation that confirms your answer.

28. State Newton's three laws and cite an example of each.

29. Explain why there are two areas of high tide on Earth rather than one.

30. How was the mass of the Moon (compared to Earth's mass) first determined?

31. What determines the magnitude (strength) of the gravitational force between two objects?

32. How did Newton use an astronomical object to check his hypothesized law of gravitation?

33. Explain how the law of gravitation accounts for the fact that planets move fastest when they are closest to the Sun.

34. Did Newton's laws conflict with Kepler's laws? Explain.

35. What provides the centripetal force to keep the Earth in orbit around the Sun?

QUESTIONS TO PONDER

1. What events took place during the century before Galileo that contributed to the revolutionary flavor of his times?

2. Neither the geocentric nor the heliocentric system made a direct prediction about whether Jupiter has moons. Why, then, did Galileo's discovery of moons have an impact on the choice of a model?

3. Figure 3-4 shows that Venus appears to be larger when it is in certain phases. Why does this occur?

4. Mars exhibits phases. What phase(s) would you expect to see in viewing Mars? In what phase(s) would Mars never appear when viewed from Earth?

5. Explain the distinction between mass and weight, showing that the difference is more than just a matter of what units are used.

6. Newton's laws tell us that no force is needed to keep something moving. Why, then, when we are driving on level ground, don't we turn off the engine of our car?

7. It seems presumptuous to call the law of gravitation "universal." What evidence did Newton have that the law applies beyond the Earth?

8. Explain why the work of Newton had implications beyond science.

9. Is it just coincidence that the Moon's periods of revolution and rotation are the same? If not, explain the cause.

10. Kepler held that the center of the Earth orbits the Sun in an elliptical path. Newton's laws tell us that this isn't exactly true. Explain.

11. Venus is the greatest contributor to Mercury's orbital precession because the two are neighbors. The planet causing the next most effect is not Earth, however, but Jupiter. Hypothesize as to why this occurs.

12. Why do we teach Newton's law of gravitation even though general relativity is a more up-to-date explanation of the phenomena involved?

13. What causes weight?

14. Give an example of the correspondence principle in the case of the heliocentric theory replacing the geocentric theory.

15. Some people even today oppose the theory of evolution on religious grounds. Compare this position to the conflicts of Copernicus and Galileo with the church establishment.

CALCULATIONS

1. Suppose that you move three times as far from the center of the Earth as you are now. By what factor will your weight change?

2. On Earth's surface, Big Al is about 6500 kilometers from its center. If the force of gravity on Big Al here is 250 pounds, how much will it be on him at a distance of 13,000 kilometers from the Earth's center?

3. Two planets (*A* and *B*) orbit a star S. Planet *B* is three times farther from the star than *A* is and has three times the mass of *A*. The force from the star on planet *A* is *x*. What is the force, in units of *x*, from the star on planet *B*?

4. In the preceding problem, suppose that the orbital period of planet *A* is two years and that it orbits the star at an average distance of 2 AU. What is the mass of the star compared to that of our Sun? (Assume the planets have small masses.) What is the orbital period of planet *B*?

5. Using data for Jupiter and its Galilean satellites (Io, Europa, Ganymede, and Callisto) given in the appendices of your book, show that these data agree with Newton's form of Kepler's third law.

1. Circular Motion

This activity should be done outside, away from anything breakable. Tie some object such as a shoe to the end of a fairly long (6 or 8 feet) string or rope. Now whirl the object in a horizontal circle around you. Feel the pull you must exert to keep the object in the circle. Whirl it faster. Do you have to increase the force you exert on the string?

Now let go of the string and carefully observe the path taken by the object. Forget the downward motion (caused by gravity) and concentrate on how the object travels horizontally. Figure 3-10 shows several potential paths for the object. Which of these did your object take?

2. Observing Venus

If you have a small telescope and Venus happens to be visible in the evening sky, observe the planet once a week for a month. In essence, you will be repeating Galileo's observations. Do your observations show that the observed shape of the planet is changing? If yes, can you tell if it is coming toward us or moving away from us?

The following books and articles will give you a sense of the impact that Galileo, Newton and Einstein had on our understanding of the universe.

1. "Newton's Discovery of Gravity," by I. B. Cohen, in *Scientific American* (March, 1981).

2. "Newton's *Principia:* A Retrospective," by G. Christianson, in *Sky and Telescope* (July, 1987).

3. "How Galileo Changed the Rules of Science," by O. Gingerich, in *Sky and Telescope* (March, 1993).

4. J. Fauvel, et al., *Let Newton Be!* (Oxford University Press, 1988).

5. O. Gingerich, *The Great Copernicus Chase, and Other Adventures in Astronomical History,* a collection of essays including the period of Galileo and Newton (Sky Publishing/Cambridge University Press, 1992).

6. Clifford M. Will, *Was Einstein Right?* (Basic Books, 1986).

7. Lewis Carroll Epstein, *Relativity Visualized* (Insight Press, 1985).

8. George Gamow and Roger Penrose, *Mr Thompkins in Paperback* (Cambridge University Press, reissue edition, 1993).

9. George Gamow, *The Great Physicists from Galileo to Einstein* (Dover Publications, 1988).

10. Galileo Galilei, *Discoveries and Opinions of Galileo,* Stillman Drake, translator (Anchor Books/Doubleday, 1957).

11. James Reston, *Galileo,* audio cassette, Jeff Riggenbach, narrator (Blackstone Audio Books, 1996).

12. Gale E. Christianson, *Isaac Newton and the Scientific Revolution (Oxford Portraits in Science)* (Oxford University Press, 1998).

13. John L. Heilbron, *The Sun in the Church: Cathedrals as Solar Observatories* (Harvard University Press, 1999).

Quest Ahead to Starlinks:
www.jbpub.com/starlinks

Starlinks is this book's online learning center. It features **eLearning,** which contains chapter quizzes and other tools designed to help you study for your class. You can also find **on-line exercises,** view numerous relevant **animations,** follow a guide to **useful astronomy sites** on the web, or even check the latest **astronomy news** updates.

ACTIVITIES

EXPANDING THE QUEST

STARLINKS

atomic hydrog

radio continuum (2.5 GH

molecular hydrog

infrar

mid-infrar

near infrar

optic

x-r

gamma r

CHAPTER 4

Light and the Electromagnetic Spectrum

The Milky Way seen at 10 wavelengths of the electromagnetic spectrum.

THE GREATEST PHYSICIST FROM THE TIME OF ISAAC NEWTON to Albert Einstein was probably James Clerk Maxwell. Maxwell was born into a wealthy family in Edinburgh, Scotland, in 1831. His genius was apparent early in his life, for at the age of 14, he published a paper in the proceedings of the Royal Society of Edinburgh. One of his first major achievements was the explanation for the rings of Saturn, in which he showed that they consist of small particles in orbit around the planet. In the 1860s Maxwell began a study of electricity and magnetism, and discovered that it should be possible to produce a wave that combines electrical and magnetic effects, a so-called electromagnetic wave. His analysis of this hypothetical wave showed that its speed would be 300,000 kilometers/second. Since this is the speed of light, Maxwell concluded that he had discovered the nature of light: light is an electromagnetic wave.

The story is told that on the evening after he made his discovery, Maxwell went out walking with his wife-to-be, and she pointed out the beauty of the stars.

He told her that she was with the only person in the world who understood what starlight is.

In 1888, nine years after Maxwell's death, radio waves were discovered by Heinrich Hertz and were shown to have properties similar to those of light. This verified Maxwell's prediction. The importance of Maxwell's work is indicated in the following quotation from the Nobel Prize winner Richard Feynman:

> *From a long view of human history—seen from, say ten thousand years from now—there can be little doubt that the most significant event of the 19th century will be judged as Maxwell's discovery of the laws of electrodynamics. The American Civil War will pale into provincial insignificance in comparison with this important scientific event of the same decade.*

The Feynman Lectures on Physics, vol. 2 (Reading, Mass.: Addison-Wesley Publishing Co., 1964).

An important part of your study of astronomy is to learn about the fantastic objects that make up our universe. But perhaps more important is to see how astronomy functions, by learning *how* we know what we know. How do we know that our Sun is a star like the thousands of others we see in the sky at night? How do we know what the stars are made of? How do we know their sizes? How can we measure a star's mass? In fact, how can we learn anything about objects that we cannot hold, feel, weigh, and experiment with?

The only thing we obtain from such objects is the radiation they emit. This radiation, including not only visible light but many other types of radiation, carries to us a tremendous amount of information about the objects it comes from. To understand how astronomers analyze radiation to answer questions about celestial objects, it is necessary to learn something about radiation itself.

In this chapter we will examine the nature of light and show how we measure three major properties of stars: their temperatures, their compositions (that is, the chemical elements of which they are made), and their speeds relative to the Earth. This is quite a lot to learn about the stars, but it is just the beginning. Keep in mind that we already have a tool, in the form of Kepler's third law, that allows us to find the total mass of a binary system. In addition, there are other tools, which we will describe in later chapters, that allow us to learn more about the stars (for example, how far away and how big they are).

One of the major differences between Astronomy and other disciplines is that the subject of our studies, the universe, is beyond our control. Stars are born and die continually, galaxies collide and, most important of all, when we look at an object we see it as it was when it emitted the light we now see. One of the means to learn about the cosmos is by unlocking the information in the radiation we collect. In this chapter we will follow the evolution of our ideas about light and describe some of the tools astronomers developed in trying to understand the information it carries.

4-1 Temperature Scales

Light transmits energy. We know this because we can clearly feel the warmth of sunshine. When we think of the Sun as an energy source (even though very different from other energy sources, such as a fireplace or a very hot iron bar), an obvious question to ask is how "hot" it is. What is its temperature? Even more important, what do we mean by the term "temperature" and how do we measure it? It should not be surprising that we started measuring temperatures long before we understood what temperature is. Before going any further, we must describe the temperature scales currently in use.

The scale in common use in the United States is the Fahrenheit scale, but most of us are at least somewhat familiar with the Celsius temperature scale. The freezing point of water is defined as 0°C or 32°F, while the boiling point of water is defined as 100°C or 212°F. Since 180 divisions on the Fahrenheit scale correspond to 100 divi-

Around 1592, Galileo invented the first device indicating the "degree of hotness" of an object. The invention of the thermometer, in 1631, is credited to J. Rey, a French physician.

sions on the Celsius scale, a Fahrenheit degree is smaller than a Celsius degree by a factor of $180/100 = 9/5$. Thus, the two scales are related by

$$T_F = 32 + \frac{9}{5} T_C \quad \text{or} \quad T_C = \frac{5}{9}(T_F - 32)$$

The first two thermometers of **Figure 4-1** compare the Fahrenheit and Celsius scales, from extremely low temperatures up to the boiling point of water. But let's think for a minute about what these temperature scales can and cannot tell us. What does it mean to say that object A is two (or a thousand) times hotter than object B, which is at 0°F? What is then the temperature of object A? The obvious answer is 0°F, but we know there is something wrong with this. (We could ask a similar question using 0°C.) Neither scale—nor, for that matter, any other scale that defines its zero mark arbitrarily—has anything to do with what temperature "is." Clearly, we would like to have an "absolute" temperature scale, one on which the zero mark is associated with the lowest temperature possible (absolute zero). Such a scale exists. It is called the ***Kelvin*** scale, and is the one commonly used in science.

The Kelvin scale is based on the fact that there is a coldest temperature that can exist. This temperature is about −273°C, and the Kelvin scale defines this as its zero point. The intervals on the Kelvin scale are the same as on the Celsius scale, so in Kelvin temperature, the freezing point of water is 273 K and the boiling point is 373 K. (Notice

Kelvin temperature scale A temperature scale with its zero point at the lowest possible temperature ("absolute zero") and a degree that is the same size (same temperature difference) as the Celsius degree.
$T_K = T_C + 273$

Figure 4-1
A comparison of temperature scales.

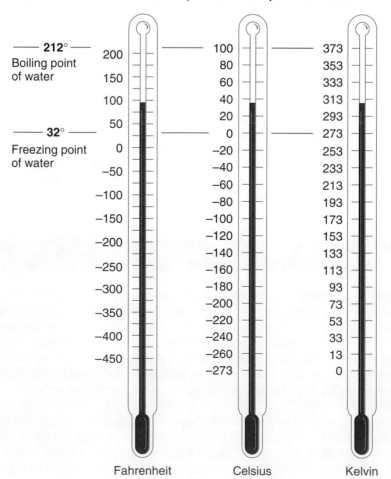

A comparison of temperature scales

that no degree symbol is included; the latter temperature is stated as "373 kelvin.") Figure 4-1 includes the Kelvin temperature scale on the right.

We now recognize that temperature is a fundamental quantity, as mass and time are. As such, it cannot be expressed in terms of other quantities. However, it is not a bad approximation for us to say that given the temperature of an object, such as a pot of water on a stove, we can calculate the average speed of each of its constituent particles; that is, temperature is a measure of their average *kinetic energy.* As the temperature of an object increases, each of its constituent particles moves faster, while as the temperature decreases, the particle speed also decreases (though not in a linear fashion). At absolute zero, we have a state of minimum atomic motion. In science we mostly use the Kelvin scale in measuring temperature.

The Kelvin scale was first proposed by William Thomson (1824–1907), a British physicist and engineer who was made a baron in 1892, taking the title Lord Kelvin. He contributed important ideas across the entire range of physics.

kinetic energy An object's energy due to motion. For an object of mass m and speed v, its kinetic energy is equal to $(1/2)mv^2$.

EXAMPLE

A star's temperature is 6000 K. What is its temperature in degrees Celsius and Fahrenheit?

SOLUTION In degrees Celsius the star's temperature is $6000 - 273 = 5727°C$. In degrees Fahrenheit it is

$$\frac{9}{5}(5727) + 32 = 10{,}341°F$$

For a large enough Kelvin temperature, we can approximate its Celsius equivalent by the same number, while its Fahrenheit equivalent is approximately twice as large in value.

TRY ONE YOURSELF

Two objects A and B are identical except that B is at 0°C and A is twice "as hot" (meaning its temperature is twice as large). What is the temperature of object A? (Hint: You must work with the Kelvin scale for both objects.)

4-2 The Wave Nature of Light

Two lines of thought about light came to us from the ancient Greeks. The first suggested that light is a stream of extremely small, fast-moving particles and that our vision is the result of the interaction between our eyes and this stream. The second idea, suggested by Aristotle, pictured light as an "aethereal motion." Aristotle added *aether* as the fifth element to his four elements of nature (fire, air, earth, and water) and imagined that it fills all space. According to this idea, our vision is the result of movement of the aether produced by the object we perceive.

Our understanding of the nature of light has changed several times over the years (**Figure 4-2**). The first step away from Aristotle's idea was taken by Isaac Newton, who theorized that light consists of tiny, fast-moving particles.

Figure 4-3 shows a beam of white light passing through a glass prism. The emerging light is separated into colors—into a *spectrum.* Isaac Newton showed that the prism does not add color to the light, as was previously thought, but rather that color is already contained in white light and that the prism merely separates the light into its colors. He showed this by using a second prism to recombine the colors produced by the first. In a separate experiment, Newton allowed one color of the spectrum created by a prism to pass through a second prism. Since each color of the spectrum remained unchanged by the second prism, he was able to show that color is a fundamental property of light. We will see in later chapters that analysis of the spectrum of light from stars is extremely important in astronomy.

spectrum The order of colors or wavelengths produced when light is dispersed.

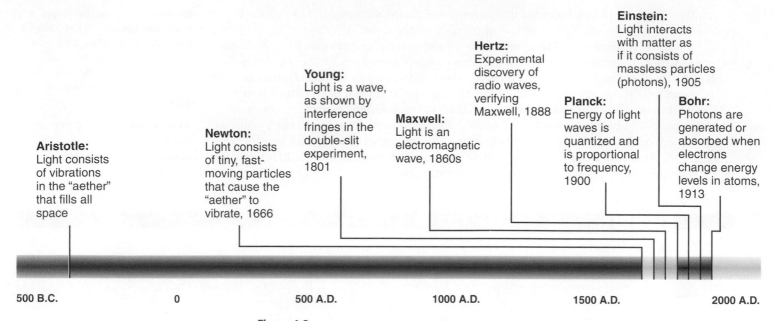

Aristotle:
Light consists of vibrations in the "aether" that fills all space

Newton:
Light consists of tiny, fast-moving particles that cause the "aether" to vibrate, 1666

Young:
Light is a wave, as shown by interference fringes in the double-slit experiment, 1801

Maxwell:
Light is an electromagnetic wave, 1860s

Hertz:
Experimental discovery of radio waves, verifying Maxwell, 1888

Planck:
Energy of light waves is quantized and is proportional to frequency, 1900

Einstein:
Light interacts with matter as if it consists of massless particles (photons), 1905

Bohr:
Photons are generated or absorbed when electrons change energy levels in atoms, 1913

500 B.C.   0   500 A.D.   1000 A.D.   1500 A.D.   2000 A.D.

Figure 4-2
Over the years, explanations for the nature of light went back and forth between wave models and particle models. As we will see later in this chapter, today we talk about the wave-particle duality of light; it propagates through space as a wave, but interacts with matter as a stream of particles.

Figure 4-3
A prism separates white light into its component colors.

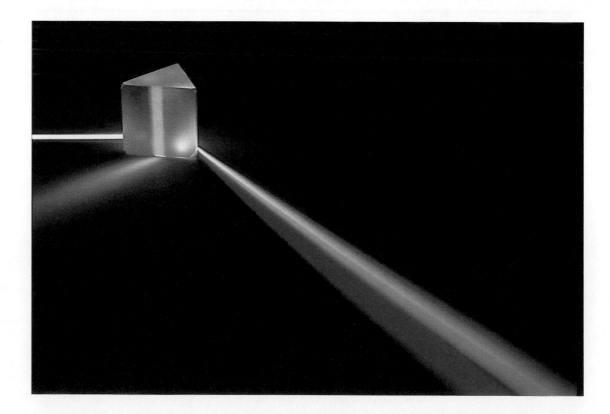

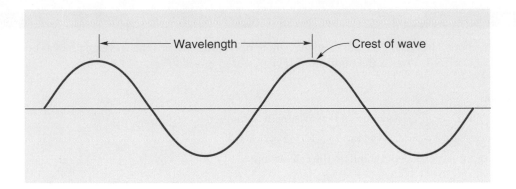

Figure 4-4
Wavelength is the distance between successive crests of a wave.

Characteristics of Wave Motion

Light acts like a wave. **Figure 4-4** is a simplified drawing of a wave. As the figure indicates, the distance between successive peaks (crests) of the wave is called the ***wavelength.*** The wave in the drawing has a wavelength of 2 1/4 inches; waves you make by dipping your hand in a swimming pool might have a wavelength of about this value.

A wave does not sit still, however. Imagine yourself fishing while sitting on a pier, watching waves pass underneath. As the waves move by, they cause the cork on the fishing line to move up and down. This indicates that the water itself moves up and down, rather than along the direction of the wave's motion. As the wave travels along the surface, the water's motion is primarily in the vertical direction and not along the direction of the wave.

Now suppose that you count the number of times the cork moves up and down in one minute. You might find that the cork moves through a complete cycle 30 times each minute. We say that the ***frequency*** of the cork's motion is 30 cycles per minute and therefore that the frequency of the wave is 30 cycles/minute. (Frequency is often reported in units of ***hertz,*** abbreviated Hz, where 1 hertz = 1 cycle/second.)

Now suppose you measure the wavelength of the waves and find it is 20 feet. You can use the fact that the wavelength is 20 feet and the frequency is 30 cycles/minute to calculate the speed with which the wave is moving. Each wave, from crest to crest, is 20 feet long, and 30 of these waves pass by you each minute. This must mean that the waves move at a speed of 600 feet/minute. We multiply wavelength by frequency to obtain the speed of the wave. In equation form,

$$\text{wave speed} = \text{wavelength} \times \text{frequency}$$

or, using symbols,

$$v = \lambda \times f$$

where $\quad v$ = wave speed

λ = wavelength

f = frequency

This equation applies not only to water waves, but to all types of waves, including light and sound waves. An application of its use for the case of sound waves is shown in the next example. If you do the suggested exercise in this example, you will see that a sound of higher frequency has a shorter wavelength, since the speeds are the same. In the example we use sound rather than light because sound waves have frequencies, velocities, and wavelengths within our everyday experience, whereas light waves do not. Now let's return to light and the spectrum produced when white light shines through a prism.

wavelength The distance from a point on a wave, such as the crest, to the next corresponding point, such as the next crest.

frequency The number of repetitions per unit time.

hertz (abbreviated **Hz**) The unit of frequency equal to one cycle per second.

λ is the lowercase Greek letter "lambda."

EXAMPLE

Sound travels at a speed of about 344 meters/second in air at room temperature. What is the wavelength of a sound that has a frequency of 262 cycles/second? (This is the frequency of the note C in music.)

SOLUTION

We start with the equation

$$\text{wave speed} = \text{wavelength} \times \text{frequency}$$

$$344 \text{ m/s} = \lambda \times 262 \text{ cycles/s}$$

We now solve the equation for the wavelength and do the calculation:

$$\lambda = \frac{344 \text{ m/s}}{262 \text{ cycles/s}} = 1.31 \text{ m}$$

So the wavelength of the sound wave is about 1 1/3 meters.

TRY ONE YOURSELF

What is the wavelength of a sound that has a frequency of 4000 Hz ? (The speed of sound is the same for all frequencies, about 344 meters/second in air at room temperature.)

Light as a Wave

It only takes light 8.3 minutes to reach us from the Sun, but for far away objects we see them the way they were millions or even billions of years ago.

?

Can anything travel faster than the speed of light in vacuum?

nanometer (abbreviated **nm**) A unit of length equal to 10^{-9} meter.

White light is made up of light of many different wavelengths, all traveling at the same speed in a vacuum (and interstellar space is essentially a vacuum). The speed of light (*c*) is tremendous, 3.00×10^8 meters/second, which is 300,000 kilometers/second or 186,000 miles/second. If a light beam could be made to travel around the Earth's surface, it would circle the globe seven times in one second. According to Einstein's special theory of relativity (see Tools of Astronomy in Chapter 3), the speed of light in vacuum is always measured as the same speed and is the fastest speed possible in the universe. (How can we measure this incredibly high speed? See the box Advancing the Model: Measuring the Speed of Light for an answer.)

The same equation that relates the speed, wavelength, and frequency of other types of waves also applies to light: $v = \lambda \times f$, or for light, $c = \lambda \times f$. We perceive light of different wavelengths as different colors. The wavelength of the reddest of red light is about 7×10^{-7} meters, or 0.0000007 meter. The wavelength decreases across the spectrum from red to violet, and the wavelength at the violet end of the spectrum is about 4×10^{-7} meters.

We have already seen that meters and kilometers are not convenient units with which to describe distances in the solar system, so we use astronomical units (AU) there. In describing the wavelengths of visible light, a meter is much too long to be convenient, so scientists use another unit—the **nanometer**. One nanometer (abbreviated nm) is 10^{-9} meters. So the shortest violet wavelength and the longest red wavelength are about 400 and 700, respectively.

When the equation $c = \lambda \times f$ is used to calculate frequencies of light waves, extremely high frequencies are obtained: 400 nanometers corresponds to 7.5×10^{14} cycles/second, and 700 nanometers corresponds to 4.3×10^{14} cycles/second. The frequency of visible light is tremendously large, larger than we can imagine.

The particular frequency or wavelength of light in vacuum determines its color. However, since color is so subjective and people are unable to distinguish between two very similar colors, scientists describe light by referring to wavelength in vacuum rather than color (**Table 4-1**). They might mention the color

Table 4-1

Approximate Vacuum Wavelength* Ranges for the Various Colors

Color	λ (nm)
Violet	380–455
Blue	455–492
Green	492–577
Yellow	577–597
Orange	597–622
Red	622–720

*From now on, whenever we refer to the wavelength of light, we mean its wavelength in vacuum. As light travels from one medium to another, its speed and wavelength change but its frequency remains the same.

The color we see is not a property of the light itself but a manifestation of the system that senses it, that is our eyes, nerves and brain.

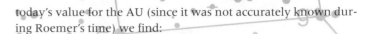

ADVANCING THE MODEL

Measuring the Speed of Light

Today we know that the speed of light c is 2.9979×10^8 meters/second. How do we know that? You can't exactly time a light beam with a stop watch. Our understanding of light and the speed of light parallels the development of the technology that was used to measure this speed.

In the early 1600's, Galileo attempted to measure the speed of light by using his pulse-beat and comparing the time between opening his lantern while on a hilltop and seeing the light from his assistant's lantern from a distant hilltop. His attempts were not successful, but he correctly concluded that light is simply too fast to be measured by the slow human reaction.

Around 1675, the Danish astronomer Ole Roemer made the first accurate measurement of the value of c. **Figure B4-1** shows his method of measurement. He observed that when the Earth–Jupiter distance is large (point A), eclipses of Jupiter's moons occur slightly later than anticipated. When the Earth is near Jupiter (point B), the eclipses occur slightly earlier than anticipated. Roemer correctly attributed this effect to the time required for light to travel from Jupiter to Earth. Light takes about 16.5 minutes to travel across the diameter of the Earth's orbit (2 AU); Roemer's actual measurement was 22 minutes. Using today's value for the AU (since it was not accurately known during Roemer's time) we find:

$$c = \frac{\text{distance}}{\text{time}} = \frac{2 \times 1.5 \times 10^{11}\,\text{m}}{16.5 \times 60} = 3 \times 10^8\,\text{m/s}$$

(Roemer's actual calculation gave $c = 2.3 \times 10^8$ meters/second.)

In 1849, French physicist Armand Fizeau measured the speed of light using a rotating toothed wheel (**Figure B4-2**). His idea was to bounce a beam of light off a mirror and place the wheel so the rotating teeth could block the light. However, when the rotation rate was just right, the light would pass between the teeth in both directions. From the rotation rate, Fizeau calculated the time for the wheel to move from one gap between teeth to the next; this was the time during which light traveled from the wheel to the mirror and back. Dividing the distance by the time, Fizeau calculated $c = 3.1 \times 10^8$ meters/second.

We can now routinely make accurate measurements of c in the laboratory. The speed of light *in vacuum* is now defined to be $c = 299,792.458$ kilometers/second. Light has a smaller speed when going through other transparent media, but unless otherwise specified we will use $c = 300,000$ kilometers/second.

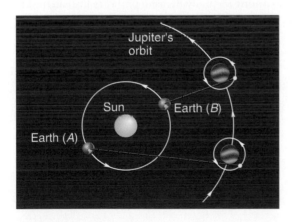

Figure B4-1
Roemer's method of measuring the speed of light. The scale of the orbits of Earth and Jupiter have been changed to exaggerate the effect.

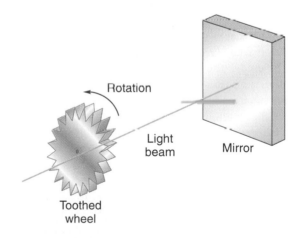

Figure B4-2
Fizeau's method of measuring the speed of light.

in some cases, but this is to help us better picture the situation. Remember that light can be described more accurately than by simply calling it "red" or "green."

An important point: As we describe the light coming from astronomical objects, we will sometimes refer to its wavelength (in a vacuum) and sometimes to its frequency. Remember that if we know one, we can calculate the other.

ADVANCING THE MODEL

Evidence for the Wave Model of Light

In the 1660s, English scientist Robert Hooke challenged Aristotle's particle model of light for the first time. He proposed a simple wave theory in which light was the manifestation of fast oscillations in the aether. He reasoned that this model better explained the color patterns observed in thin transparent films, such as soap bubbles. On the other hand, Newton favored the particle model for light and proposed a very interesting solution. He suggested that light has a dual nature; that it is a stream of particles that can induce oscillations in the aether. (Keep this idea in mind as our understanding of the nature of light continues to evolve!) Newton's preference of a particle model for light was based on misunderstandings about how waves behave. He expected that if light were a wave, it should clearly spread out after passing through an opening, instead of producing the observed narrow beams. In his opinion, the particle model for light better explained how light propagates along straight lines and why shadows are sharp. The problem is that light *does* spread out as Newton expected, but only if the opening is of the same approximate size as the wavelength of the light, which is extremely small (at least for visible light, which was what he used at the time).

Both the particle and wave models for light explained two of the most common phenomena we observe. Light reflects—it bounces like a tennis ball off a flat surface. Also, light refracts—it changes direction when going from one medium (such as air) to another (such as water). However, the particle model of light gained acceptance for more than a century, mainly due to Newton's reputation and authority. There was a need for a simple but convincing experimental test in order to distinguish between the two competing models.

Thomas Young first proposed such a test around 1801. Young accepted the wave model of light proposed by Christian Huygens (1629–1695; a contemporary of Newton's), in which light is described by the usual quantities we use to describe a wave, such as frequency (f), wavelength (λ), and speed ($c = \lambda \times f$). Young's experiment was to shine light of a specific wavelength (from a single source) through two narrow parallel slits, close to each other. The light then fell on a screen a certain distance away (**Figure B4-3**). Young observed a pattern of light and dark bands on the screen, called interference fringes, which he explained by assuming that light is a wave. As light waves go through the slits, they **diffract** (they spread out) in a series of crests and troughs. When the waves meet, they add up algebraically as shown in **Figure B4-4**.

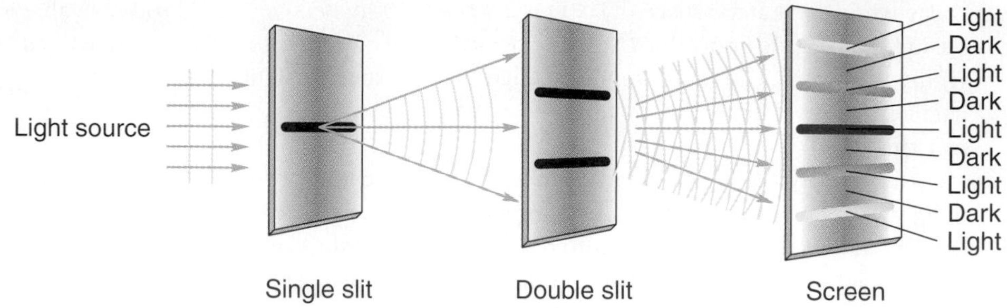

Figure B4-3
Young's double-slit experiment. Light rays interfere after passing through the two slits, forming light and dark fringes on the screen. The light areas are due to constructive interference, the dark areas to destructive interference. The colors used for the light areas are for illustration purposes only and correspond to the decreasing intensity of the light.

When the interference is such that a wave crest from one slit meets a wave crest from the other slit, we have constructive interference and a bright fringe appears on the screen. When the interference is such that a wave crest from one slit meets a wave trough from the other slit, we have destructive interference and a dark fringe appears on the screen. Young used this **double-slit experiment** (actually he used pinholes instead of slits) to measure the wavelength for violet (400 nm) and red light (700 nm).

Under everyday conditions, it is very difficult to see diffraction of visible light, which is why Newton thought he was observing sharp shadows. The experiments of the French physicist Augustin Fresnel (1788–1827) helped him put the wave model of light on a firm mathematical basis. The interference of light clearly showed its wave nature, so the wave model seemed to be the correct description for light.

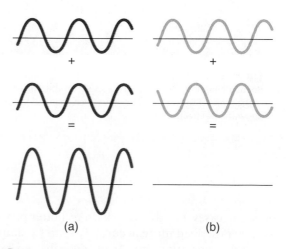

Figure B4-4
(a) Waves interfering constructively. (b) Waves interfering destructively.

ADVANCING THE MODEL

Evidence for the Wave Model of Light *(Cont'd)*

What was the nature of the waves? In the early 1860s, the Scottish mathematical physicist James Clark Maxwell (1831–1879) succeeded in unifying what was then known about electricity and magnetism into four equations, which today are named after him. He found that his equations allowed the existence of waves that combine electromagnetic effects and move at a speed that is almost the same as the then known speed of light. Also, these waves are *transverse* waves (such as the up-and-down waves traveling on a stretched rope), with the electric and magnetic fields perpendicular to each other and to the wave's direction of propagation (**Figure B4-5**). Maxwell wrote that "we can scarcely avoid the inference that light consists in the transverse modulations of the same medium which is the cause of electric and magnetic phenomena." Nine years after Maxwell died, the German physicist Heinrich Hertz (1857–1894) produced radio waves in his lab, which confirmed all the properties of light known at the time. However, the story does not end here, but continues in Section 4-6.

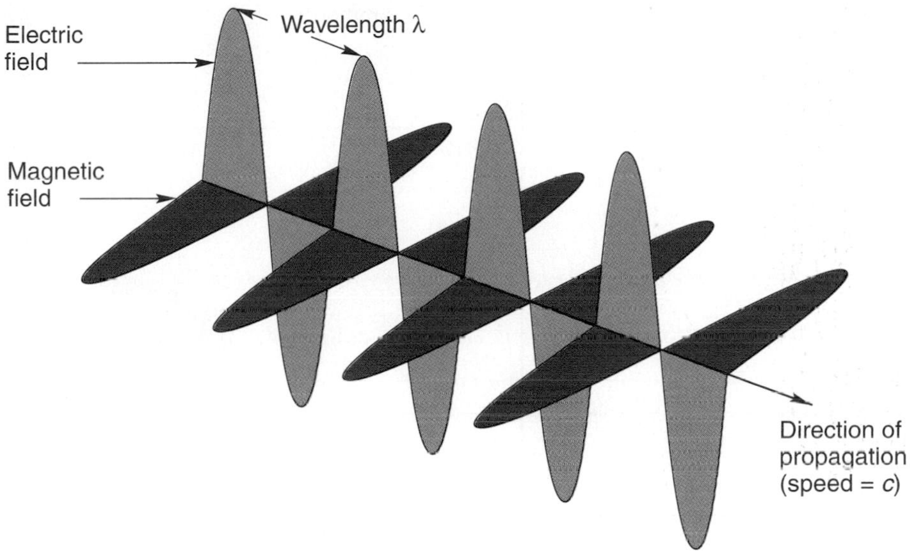

Figure B4-5
Electromagnetic radiation (from radio waves to X-rays) consists of oscillating electric and magnetic fields, perpendicular to each other, moving through space at the speed of light.

4-3 The Electromagnetic Spectrum

Can a light wave have a wavelength longer than 700 nanometers? Yes, although such a wave is not visible to us. The waves we see—visible light—are just a small part of a great range of waves that make up the ***electromagnetic spectrum.*** Waves somewhat longer than 700 nanometers (the approximate limit of red) are called *infrared* waves. **Figure 4-5** shows the entire electromagnetic spectrum. Notice that the infrared region of the spectrum goes from 700 nanometers at the border of visible light up to about 10^{-4} meters, which is a tenth of a millimeter or 100,000 nanometers. Electromagnetic waves longer than that are called *radio* waves.

Going the other way, from visible light toward shorter wavelengths, we first encounter *ultraviolet* waves, then *X-rays* and *gamma rays*.

It is important to emphasize that all of these types of waves (or rays, as certain portions of the spectrum are known) are essentially the same phenomenon. They differ in wavelength, and this causes some of their other properties to differ. For example, visible light is just that—visible. Ultraviolet is invisible to us, but it kills living cells and causes our skin to tan or burn. On the other hand, we perceive infrared as "heat" radiation. And yes, the radio waves in the spectrum are the same radio waves we use to transmit messages on Earth. They are handy for carrying messages containing

electromagnetic spectrum The entire array of electromagnetic waves.

It is probably not worth memorizing these wavelengths of light waves, but it is handy to remember that the wavelength of visible light ranges from 400 (violet) to 700 (red) nanometers.

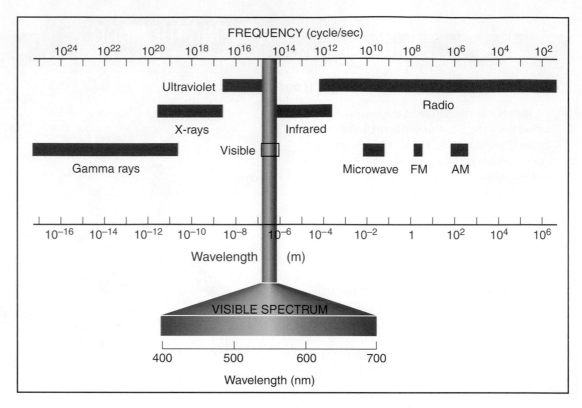

Figure 4-5
The electromagnetic spectrum is divided into several regions, depending upon the properties of the radiation. Region boundaries are not well defined. Notice the small portion of the spectrum occupied by visible light.

◆ Infrared, radio, ultraviolet, X-rays, and gamma rays are defined by their frequency and/or wavelength, as shown in Figure 4-5.

◆ The waves are called "electromagnetic" because they consist of combined, perpendicular, oscillating electric and magnetic fields that result when a charged particle accelerates (see Figure B4-5 and the Advancing the Model box on Evidence for the Wave Model of Light).

Is most of the electromagnetic spectrum made up of visible light?

sound and pictures (in the case of television) for several reasons, including the fact that they pass through clouds and bend around obstacles.

Again, we emphasize that all of these various waves are electromagnetic waves, just as visible light is. We give the various regions different names because of their properties and the uses we have for them. Nature does not see the spectrum as seven or eight or ten different regions; all of these regions are simply electromagnetic waves.

The electromagnetic spectrum is important to astronomers because celestial objects emit waves in all the different regions of the spectrum. Notice that visible light is a very small fraction of the entire spectrum. We humans tend to regard it as the important part, but this only reveals our limited outlook. Astronomers learn a great deal from the *invisible* radiation emitted by objects in the heavens.

One of our problems with this invisible radiation is that most of it does not pass well through air, so it does not reach the surface of the Earth. Our air is transparent to visible light and to part of the radio spectrum, but most of the rest of the electromagnetic spectrum is blocked to some degree. The chart in **Figure 4-6** shows the relative absorbency of the atmosphere to various regions of the spectrum. Where the graph line is highest, the least amount of radiation gets through. Notice that not much ultraviolet radiation penetrates to the surface. This is fortunate because, as you recall, this radiation damages living cells.

Astronomers, however, wish to detect and examine these nonpenetrating radiations from space. They accomplish this by using balloons to carry detectors high into the atmosphere or by using artificial satellites to take detectors completely above the atmosphere. This will be covered in a later chapter.

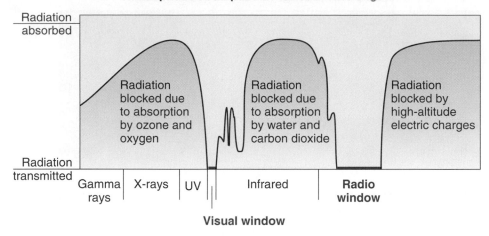

Atmospheric absorption at various wavelengths

Astronomers refer to *windows* in the atmosphere, saying that there is a visual window and a radio window. This means that our atmosphere allows radiation in these two regions of the spectrum to penetrate to the surface.

Figure 4-6
The height of the curve indicates the relative amount of radiation of a given wavelength blocked by the atmosphere from reaching Earth's surface. The atmosphere is transparent to two regions of the spectrum: visible light and part of the radio region.

4-4   The Colors of Planets and Stars

Let us return to the visible spectrum and examine how the spectrum from a celestial object is analyzed in order to determine some of the object's properties. This analysis can be divided into two parts. First, we look at the overall spectrum of light from the object. From this, we typically determine the color of the object. However, remember that when we look at the spectrum, we are really examining the actual wavelengths of light rather than just the color. The second analysis, discussed later in this chapter, involves examining individual regions of the spectrum.

Color from Reflection—The Colors of Planets

When you see a visible spectrum spread out before you—on a screen, perhaps—you see a particular color at a given location on the spectrum. This is because a wave whose wavelength is associated with that color is coming to your eye from that spot. However, the color we see in most objects does not correspond to a single wavelength. **Figure 4-7a** might be the spectrum of light from some lemons. The spectrum contains many different wavelengths of light, but notice that there is no violet or blue light and that the center of the spectrum is indeed in the yellow. Part (b) of the figure shows a graph that indicates the intensity of light of each wavelength. Where the graph is higher, the light of the corresponding color is brighter. Lemons have this spectrum because when white light strikes them, they absorb some of the wavelengths of the white light, especially blue and violet. They reflect only those wavelengths we see in their spectrum (the wavelengths centered on the color yellow). This is what determines their color.

Planets have their colors because of a process like that described for the lemons. The rusty red color of Mars, for example, occurs because the material on its surface absorbs some of the wavelengths of sunlight and reflects a combination of wavelengths that looks rusty red to us.

Figure 4-8 shows the color spectrum and graph of intensity versus wavelength for the light from the red taillight of a car. The spectrum includes not only many wavelengths in the red part of the spectrum, but also some orange. In this case, the bulb inside the taillight emits white light, but part of that light is absorbed by the plastic cover. The light that gets through the cover has the spectrum in the figure.

Figure 4-7
An example of reflected color. (a) If light reflected from the lemons were sent through a prism to reveal its spectrum, we would see that the lemons reflect mostly yellow light. (b) This graph indicates the relative intensity of the various wavelengths of light from the lemons.

Spectrum of light from lemons

(a)

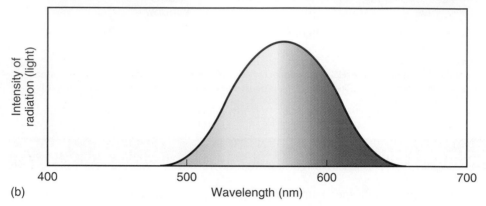

(b)

Figure 4-8
An example of transmitted light. (a) The spectrum of light from the taillight of this particular car. The light bulb emits white light, but the plastic cover over the bulb absorbs much of the light, letting pass some wavelengths in the red, orange, and yellow regions of the spectrum. (b) This graph indicates the relative intensity of the various wavelengths of light in this case.

Spectrum of light from taillights

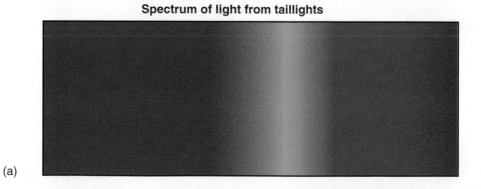

(a)

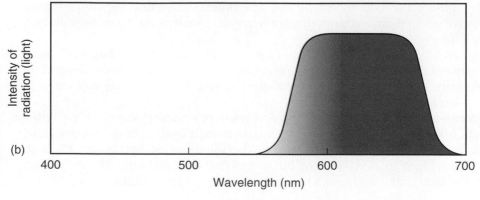

(b)

The colors of the Sun and other stars are determined in part by a process somewhat like the taillight of the car, for light from a star is produced by emission within the star and some light is absorbed as it passes through the star's outer layers. We will look at this in more detail later.

Color as a Measure of Temperature

The light emitted by the Sun and other stars can be compared to light coming from a light bulb or from the element of an electric stove in an otherwise dark room. Consider what happens when you turn the burner of an electric stove to a low setting. It glows a dull red. **Figure 4-9** is a graph of intensity versus wavelength for this case. Notice that on this graph we have included not only the visible portion of the spectrum, but quite a lot of the infrared. Recall that we experience infrared radiation as heat. The stove burner is emitting quite a bit of infrared radiation. (If you don't believe this, put your hand near it and feel its heat, the energy it emits.) In fact, the graph indicates that more infrared radiation is being emitted than visible radiation, and the graph reaches its peak in the infrared portion of the spectrum. Some red light is emitted, but very little light from the center and violet end of the visible spectrum.

Now turn up the heat on the stove. The burner begins to take on an orange glow. The second curve from the bottom in **Figure 4-10** is a graph of intensity versus wavelength for this burner. Compare its curve to the bottom curve (the red-hot burner). First, the orange burner is emitting more radiation of all wavelengths. This should correspond to your experience, for you can feel that more infrared is being emitted and you can see that more light is coming from the burner. Second, the peak of the graph has moved over toward the visible portion of the spectrum, toward the shorter wavelengths. (Remember, infrared radiation has longer wavelengths than visible light.)

This is about as far as you can go with an electric stove burner. If you have an object with temperature you can control, you can increase its temperature so that the object emits most of its light in the yellow part of the spectrum. The third curve from the bottom is a graph of such an object. Finally, suppose you increase the temperature of the object so that it is emitting light about equally in all visible regions of the spectrum. The object will appear white. The fourth curve from the bottom of Figure 4-10 is an example of such an object.

As this sequence indicates, the graph of intensity versus wavelength of an object emitting electromagnetic radiation can be used to detect its temperature. The graph in Figure 4-10 shows curves that represent widely different temperatures, including one at 10,000 K. The temperature corresponding to each curve is shown.

We want you to imagine a dark room because in a well-lit room, the lamp and stove element reflect light and you can see them even if they are turned off.

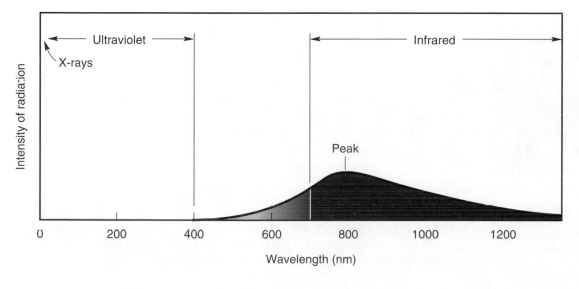

Figure 4-9
The spectrum of a red-hot stove burner shows that it emits more infrared radiation than visible light.

Figure 4-10
The bottom curve represents an object (at 3600 K with $\lambda_{max} = 805$ nm) that emits mostly infrared radiation and is the same as the curve in Figure 4-9. The next curve corresponds to an object (at 4750 K with $\lambda_{max} = 610$ nm) that emits most of its light in the orange part of the spectrum. The third curve from the bottom corresponds to a still hotter object (at 5000 K with $\lambda_{max} = 580$ nm), and the fourth one corresponds to an even hotter object. The graph is extended to temperatures found in stars. The 10,000 K object would most likely seem "blue" because it emits most of its light in the ultraviolet part of the spectrum. Recall that color is subjective.

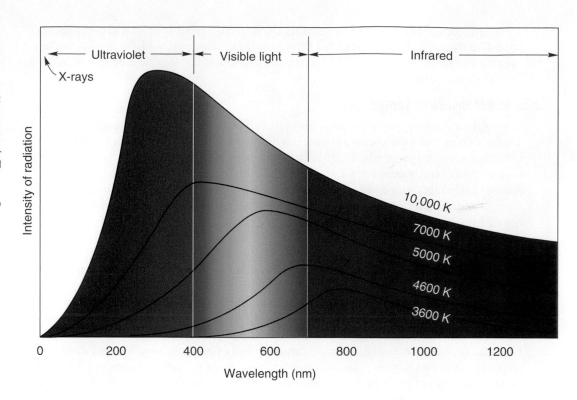

Astronomers usually call the graph of intensity versus wavelength for a star its *thermal spectrum.*

Beta Cygni (named Alberio) is the second brightest star in the constellation Cygnus. It is the brighter of a closely spaced pair of stars with obvious color differences.

Figure 4-11
The *Hubble Space Telescope* imaged one of the lowest-temperature stars ever seen (upper right). It is a companion to the dwarf star at lower left. The surface temperature of the cool star may be as low as 2300 degrees Celsius. (The white bar is an effect produced by the camera.)

The peak of the intensity-versus wavelength curve for a light-emitting object always falls at a wavelength that depends upon the object's temperature. This is important in astronomy because by plotting this graph for a star, we can know the temperature of its surface. The intensity versus wavelength graph provides astronomers with an extremely valuable tool in their attempts to understand stars.

The temperatures indicated in Figure 4-10 are typical of temperatures of stars, and the curve that reaches its peak at the shortest wavelength on the graph represents the hottest star. Notice that a star of this temperature emits more intense violet and blue light than light of the longer wavelengths. We say that it is a "blue" star. **Figure 4-11** is an image of one of the coolest stars ever found. We often refer to stars by color; this is a quick way to indicate their temperature. A "white" star is hotter than a "red" star. In practice, of course, a "red" star does not appear red like a Christmas tree bulb, but it definitely has a red tint. To the unpracticed naked eye, color differences between stars are not at all obvious. However, if you have the opportunity to use a telescope to observe pairs of closely spaced stars, you can see this color difference easily. The important point is that by examining the thermal spectrum of a star, we can determine the temperature of the surface of the star without ever visiting it!

Wien's Law In 1893, a German physicist, Wilhelm Wien, discovered the mathematical relationship between the temperature of an object and the wavelength at which it emits most of its radiation. The position of the maximum emission is easy to locate on any intesity/wavelength curve (see, for example, Figure 4-10). Wien's law can be stated as

$$\lambda_{max} = \frac{2,900,000}{T}$$

where the wavelength that corresponds to the peak of the thermal spectrum, λ_{max}, is in nanometers and the temperature T is expressed in kelvin. You can check that this equation applies to the lines of Figure 4-10.

Blackbody Radiation Wien's law was derived from theoretical calculations made in the late nineteenth century, when scientists studied the radiation that would be emitted from an object that absorbs (or emits) all wavelengths completely. Such an ideal object is called a ***blackbody,*** and the radiation it emits is called ***blackbody radiation.*** It was found that although stars are not perfect blackbodies, the theory applies very closely to them. Figure 4-10 contains examples of blackbody curves. A blackbody emits a ***continuous spectrum.*** This spectrum has a peak at a certain wavelength λ_{max} that becomes smaller as the temperature increases, but some energy is emitted at *all* wavelengths.

Figure 4-12 shows the thermal spectrum of the Sun and the corresponding blackbody curve of about 5800 K. The Sun's spectrum peaks in the visible part of the electromagnetic spectrum, where our eyes have become most sensitive.

The Stefan-Boltzmann Law Figure 4-10 indicates that the hotter an object is, the more radiation it emits. In 1879 an Austrian physicist, Josef Stefan, discovered the mathematical relationship between temperature and energy emitted, based on data published 14 years earlier. In 1884, another Austrian physicist, Ludwig Boltzmann, used a theoretical argument to show why the relationship occurs. This rule, now called the *Stefan-Boltzmann law,* is

$$F = \sigma T^4$$

where *F*, called the energy flux, is the energy emitted each second per square meter of the surface, *T* is the temperature on the Kelvin scale, and σ (the Greek letter sigma) is a constant, called the *Stefan-Boltzmann constant,* that relates the two quantities.

Consider the following example. Figure 4-10 shows a wavelength versus intensity curve marked 5000 K and another marked 10,000 K. Because the temperature of the hotter star is twice that of the cooler star, the Stefan-Boltzmann law tells us that it emits 2^4, or 16, times more energy per second per square meter than does the cooler star.

As we will see in Chapter 12, the Stefan-Boltzmann law allows us to find the radius of a star. (From observations, we can find a star's temperature and the total energy it emits each second. Recall the meaning of energy flux and the fact that the surface area of a sphere is proportional to the square of its radius.)

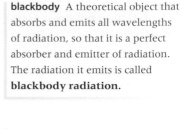

blackbody A theoretical object that absorbs and emits all wavelengths of radiation, so that it is a perfect absorber and emitter of radiation. The radiation it emits is called **blackbody radiation.**

continuous spectrum A spectrum containing an entire range of wavelengths, rather than separate, discrete wavelengths.

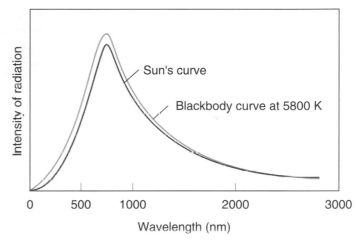

Figure 4-12
The Sun is almost an ideal blackbody at 5800 K. The measurements of the Sun's intensity were made above the Earth's atmosphere.

To get this result, we don't put the numeral "2" in for temperature, but we observe that energy is related to temperature to the fourth power. Thus twice the temperature means 16 times the energy.

4-5 Types of Spectra

The spectrum of visible light described above and shown in Figure 4-3 is called a ***continuous spectrum.*** Such a spectrum is produced when a solid object (in this case the filament of a lamp) is heated to a temperature great enough that the object emits visible light. Not all spectra are of this type, as was discovered nearly two hundred years ago.

Kirchhoff's Laws

In 1814, Joseph von Fraunhofer, a German optician, used a prism to produce a solar spectrum. He noticed that the spectrum was not a continuous spectrum at all, but had a number of dark lines across it (**Figure 4-13**). Fraunhofer had no explanation for these dark lines, but it was later discovered that the dark lines produced in the solar spectrum result from the sunlight passing through cooler gases (in the Sun's and the

Figure 4-13
The Sun's spectrum may appear at first to be a continuous spectrum, but when it is magnified, we see dark lines across it where specific wavelengths do not reach us. Thus the solar spectrum is an absorption spectrum.

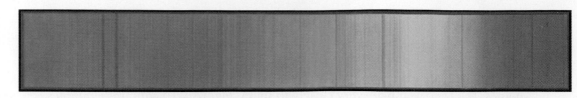

Figure 4-14
The visible bright line spectrum of four different chemical elements. Each element emits specific wavelengths of light when it is heated. This applies not only to visible light (shown here) but also to infrared and ultraviolet. Such a spectrum is called an emission spectrum.

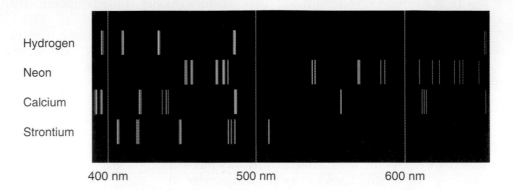

Gustav Kirchhoff (1824–1887) was a German physicist and astronomer whose primary work was in the field of spectroscopy—the study of the spectrum.

Earth's atmospheres). Then, in the mid-1800s, several German chemists discovered that if gases are heated until they emit light, neither a continuous spectrum nor a spectrum with dark lines is produced; instead a spectrum made up of *bright* lines appears. Further, they discovered that each chemical element has its own distinctive pattern of lines (**Figure 4-14**). This proved to be a very valuable way of identifying the makeup of an unknown substance and was soon developed into a standard technique that allows us to identify the chemical composition of matter.

In the 1860s, Gustav Kirchhoff formulated a set of rules, now called Kirchhoff's laws, which summarize how the three types of spectra are produced:

1. A hot, dense glowing object (a solid or a dense gas) emits a continuous spectrum.

2. A hot, low-density gas emits light of only certain wavelengths—a bright line spectrum.

3. When light having a continuous spectrum passes through a cool gas, dark lines appear in the continuous spectrum—a dark line spectrum.

The dark lines that result when light passes through a cool gas (process 3) have the same wavelengths as the bright lines that are emitted if this same gas is heated (process 2).

4-6 The Bohr Model of the Atom

Kirchhoff's laws tell us how to produce the various types of spectra, but the science behind the laws—the connection between the laws and the nature of matter—remained a mystery in the nineteenth century. The connection was finally made in 1913, when a young Danish physicist, Niels Bohr, proposed a new model of the atom.

The atomic model accepted at that time was due primarily to the New Zealand physicist Ernest Rutherford. His model described the atom as having a **nucleus** with a positive electrical charge, circled by **electrons** with a negative electrical charge. Positive and negative electrical charges attract one another, and this electrical force holds the electrons in orbit around the nucleus. But what keeps the electrons from simply spiraling into the nucleus, leading to collapse of the atom? Bohr's model attempted to answer this question.

nucleus (of atom) The central, massive part of an atom.

electron A negatively charged particle that orbits the nucleus of an atom.

The **Bohr atom,** as Niels Bohr's model is called, is based on three new ideas, or postulates:

1. Electrons in orbit around a nucleus can have only certain specific energies To imagine different energies for the electrons, imagine electrons orbiting at different distances from the nucleus. The negatively charged electrons are being attracted to the positively charged nucleus; thus, to pull an electron farther away from the nucleus requires energy. Since only certain energies are possible, we speak of "allowed" orbits for the electrons. The element hydrogen has only one electron, but many possible energy levels and therefore many allowed orbits. The drawing in **Figure 4-15a** depicts the hydrogen atom with its electron in the lowest orbit and also shows, as examples, three other orbits this electron might have. The point of Bohr's first postulate is that the electron can have only these particular energies and therefore these particular orbits. This is far different from the solar system, where there are no limitations on possible positions of orbits. We will see later that asteroids and comets seem to have random, chaotic orbits. Bohr proposed the revolutionary idea that there is a restriction on where electrons can orbit.

2. An electron can move from one energy level to another, changing the energy of the atom In terms of electron orbits, when energy is added to an atom, the electron moves farther from the nucleus. On the other hand, the atom loses energy when an electron moves from an outer orbit to an inner orbit. This lost energy leaves the atom in the form of electromagnetic radiation.

Our previous discussion of waves and light assumed that light acts as a simple long wave, similar to the waves you can make in a swimming pool. Theoretical work by Albert Einstein, Niels Bohr, and others showed that light is more complicated than this. (See the Advancing the Model box on The Evidence for the Particle Model of Light.) The Bohr model holds that light is emitted not as continuous waves, but in tiny bursts of energy; each burst is emitted when an electron moves to an orbit closer to the nucleus. According to the Bohr model, light is emitted when an electron falls from an outer to an inner orbit. (We say "falls" because an electron is attracted toward the nucleus in a manner analogous to the way we are attracted to Earth; it seems natural to think of the electron's move inward as a fall.) These tiny bursts of electromagnetic energy are called **photons.** The energy of the photon depends upon the spacing between electron orbits.

3. The energy of a photon determines the frequency of light that is associated with the photon The greater the energy of the photon, the greater the frequency of light, and vice versa. Thus a photon of violet light has more energy than a photon of red light. (Recall that violet light has a greater frequency than red light.)

The relevant equation is

$$E = hf$$

where E is the energy of the photon, h is a constant (called *Planck's constant*), and f is the frequency of the light.

Bohr atom The model of the atom proposed by Niels Bohr; it describes electrons in orbit around a central nucleus and explains the absorption and emission of light.

photon The smallest possible amount of electromagnetic energy of a particular wavelength.

Every physicist thinks he knows what a photon is. I spent my life to find out what a photon is and I still don't know it.
Albert Einstein

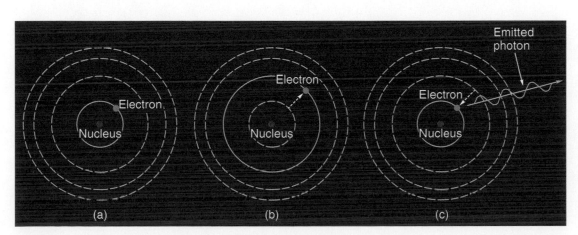

Figure 4-15
(a) The electron is in the lowest possible energy state—the lowest orbit. (b) When the atom gains energy (perhaps by collision with another atom), the electron jumps up to a higher orbit. (c) The electron then quickly falls down to its original orbit and, as it does, emits energy in the form of a photon of electromagnetic radiation.

ADVANCING THE MODEL

Niels Bohr

Niels Bohr (**Figure B4-6**) was born in 1885 into a very cultured Danish home. His father was a physiologist and university professor, and both of his parents read at least four languages and were lovers of art and music. His father's interest in science led Niels to that subject, and he became known as a student who gave his utmost to every project—a reputation that continued throughout his life.

Niels was married in 1910. He and his wife Margarethe had six sons and a number of grandchildren. He was very family oriented and greatly enjoyed his children. Margarethe and Niels remained devoted to one another until Niels's death in 1962. One son, Aage, followed him in the study of physics (Aage won the Nobel Prize in physics in 1975), and the two worked together on several projects (and rode motorcycles together—one of Niels's hobbies).

In 1922 Niels Bohr was awarded the Nobel Prize in physics "for his services in the investigation of the structure of atoms and of the radiation emanating from them." In his acceptance speech, he emphasized the limitations of his theory, and, indeed, he seems to have been more aware of its limitations than other scientists who worked with the theory.

One major project on which Niels and Aage Bohr worked together was the development of the atomic bomb. Niels's mother was Jewish, and after Hitler's army overran Denmark, Niels and his family were forced to flee their native land to avoid arrest. Niels and Aage came to the United States and helped with the Manhattan Project (the code name for the bomb development effort).

Bohr's contribution to science goes far deeper than the development of the Bohr model of the atom, as important as that is. His philosophical ideas on the nature of physical theory are perhaps his greatest contribution, but to discuss them here would re-

Figure B4-6
Niels Bohr and his sons.

quire much more space than we have. We only mention that many important ideas of modern physics were clarified through the friendly arguments between Bohr and Einstein. Einstein would try to imagine situations in which the new ideas of quantum mechanics (the modern description of matter and energy) did not work, and Bohr would always figure out how quantum mechanics did explain these situations. It is a classic example of how science can progress through the informal discussions between scientists.

Niels Bohr was very interested in people and was greatly admired by all with whom he worked. He was a very soft-spoken, gentle man, and his work will influence scientific thinking for ages to come.

Emission Spectra

The Bohr model of the atom can be used to explain why only certain wavelengths are seen in the spectrum of light emitted by a hot gas. In its normal, lowest energy state, the electron of a hydrogen atom is in its lowest possible orbit, as indicated in Figure 4-15a. If this atom is given energy (perhaps by collisions with other atoms), the electron might jump to a more distant orbit. Figure 4-15b illustrates this jump. An atom will not stay in its energized state long. Quickly, the electron falls down to a lower orbit, emitting a photon as it does, as shown in Figure 4-15c. The energy of this photon is exactly equal to the energy difference between the two orbits. Finally, since the energy of the photon determines the frequency of the radiation, the radiation coming from this atom must be of the corresponding frequency (and color).

The lowest energy state of an atom is usually called the *ground state,* and the energized states are called *excited states.*

ADVANCING THE MODEL

Evidence for the Particle Model of Light

At the close of the 19th century, the wave model for light seemed to be very successful. However, some unanswered questions still remained. The most important ones dealt with the continuous spectrum of blackbody radiation and with the emission and absorption spectra. Gustav Kirchhoff had already established his three laws on the production of spectral lines, as we discussed in the text. Astronomers were quick in utilizing these laws to analyze light from stars. Wien's law allowed astronomers to find the surface temperature of a star; the Stefan-Boltzmann equation, coupled with Wien's law, permitted a determination of the size of a star; spectroscopy and Kirchhoff's rules enabled astronomers to find the chemical composition of the atmosphere of a star. However, there was no solid theoretical foundation to analyze the observed spectra.

In 1900, while trying to explain the continuous spectrum of blackbody radiation, the German physicist Max Planck (1858–1947) discovered an equation that fit the blackbody curve. In an effort to understand its meaning, Planck was forced to accept that electromagnetic waves can only have discrete (**quantized**) energy values that are integer multiples of a minimum energy value, called a **quantum** of energy. This quantum of energy is given by hf, where f is the frequency of the wave and h is Planck's constant. Today we consider h to be a fundamental constant of nature, like the gravitational constant G and the speed of light c. It is also the basis of **Quantum Mechanics**, which is the modern description of matter and energy. However, the quantum idea did not fit at all with the prevailing wave model of light.

The quantum nature of light was taken seriously by Einstein, who in 1905 used it to explain a very puzzling phenomenon, known as the **photoelectric effect**. When light of a certain frequency shines on a metal surface, electrons are ejected from it with a range of energies. (This is the basis of today's photocells in door openers, bar code scanners, and hundreds of other applications.) However, the maximum kinetic energy of these electrons does not depend on how bright the source of light is! According to the wave model of light, if we increase the rate of light energy falling on the surface, individual electrons will absorb more energy and will be emitted with larger kinetic energies. This is not what is observed. Increasing the brightness of the light source results in more electrons being emitted, but it doesn't affect their maximum kinetic energy. In addition, there is a minimum frequency that the light must have in order for any electrons to be emitted. This is also contrary to the wave model for light, according to which if we wait long enough, the electrons will be able to absorb enough energy to be emitted, independent of the frequency of the light source.

Einstein's explanation was based on the assumption that light energy striking the metal surface is quantized in small bundles. These bundles behave as if they are a stream of massless particles, that today we call **photons**. The energy of each photon is equal to hf. Convincing evidence that light indeed shows its particle nature when it interacts with matter was provided by the American physicist Arthur Compton in 1922. He measured the change in the wavelength of X-ray photons as they were scattered by free electrons (an observation now called the **Compton effect**) and found it to be exactly as predicted by the theory of a collision between two particles.

What, then, is the nature of light? It seems that light has a **wave-particle duality**. When it propagates through space, light can be described by a wave model (as shown by the double-slit experiment). When it interacts with matter, light can be described by a particle model (as shown by the photoelectric effect and the Compton effect). We cannot say whether light *is* particle or wave. It seems clear that this is not an either/or situation. Light seems to be both particles and waves and so is probably neither.

We have described one atom emitting one photon. In an actual lamp that contains hot hydrogen gas, there are countless atoms gaining energy and countless atoms emitting photons as they lose energy. Different atoms will have different amounts of energy, depending upon the energy they have absorbed (from a collision with another atom, for example). If a particular atom's energy corresponds to the electron being in the third orbit, the atom might release its energy in a single step, as shown in **Figure 4-16a**. On the other hand, it might release energy one step at a time, as shown in Figure 4-16b.

Suppose enough energy is available to cause electrons to move to the fourth orbit. **Figure 4-17** shows all of the possible falls an electron might take to get back to the lowest orbit. Each of these falls corresponds to a certain specific energy and therefore to a certain specific frequency of emitted radiation. The electrons of some atoms will

Figure 4-16
(a) The electron may fall directly from orbit 3 to orbit 1, emitting a single photon. (b) The electron may fall from orbit 3 to orbit 1 in two steps, emitting two photons whose total energy is equal to the photon emitted in part (a).

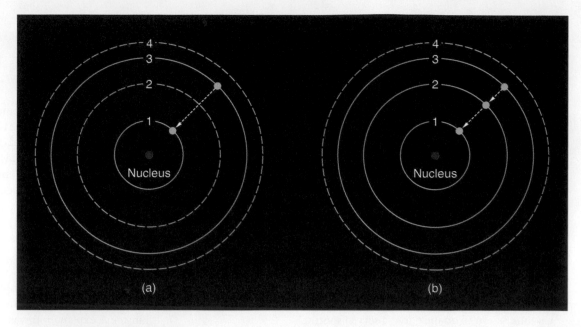

Figure 4-17
The possible paths an electron may take to get from orbit 4 to orbit 1. Each fall corresponds to a different energy change and therefore to a photon of different frequency.

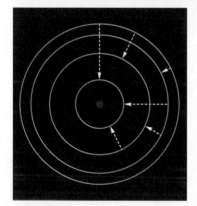

emission spectrum A spectrum made up of discrete frequencies (or wavelengths) rather than a continuous band.

fall by some paths, and the electrons of other atoms will fall by other paths. As a result, radiation of several different frequencies will be emitted from the entire group of atoms. Not all frequencies will be emitted, however—just those that correspond to the electron jumps.

Hence, the spectrum from a heated, low-density gas is not a continuous spectrum. It contains only certain definite frequencies. We call such a spectrum an *emission spectrum*—the bright line spectrum mentioned earlier. An emission spectrum has a few individual bright lines instead of a continuous band of colors.

Refer back to Figure 4-14, which shows the emission spectra of four elements. Each spectrum is different because the allowed energy levels of the atoms are different for each chemical element. No two chemical elements have the same set of energy levels, and thus no two chemical elements have the same emission spectrum. This provides us with a valuable method of identifying elements, since each has a unique spectral "fingerprint." This process has some important applications here on Earth, but since in this book we are more interested in the stars, we'll look at a stellar application.

Absorption Spectra of the Stars

Kirchhoff's laws tell us that dark line spectra result from light with a continuous spectrum passing through a cool gas. Let us examine this in the case of the Sun.

photosphere The region of the Sun from which mostly visible radiation is emitted.

The visible surface (the *photosphere*) of the Sun emits a continuous spectrum. Even though the Sun is, in most ways, more like a gas than a solid, it produces a continuous spectrum rather than an emission spectrum. This is because as atoms are pushed together, their energy levels are broadened (as if slightly different orbits are allowed). As atoms become more and more tightly pressed together, their energy levels begin to overlap, so that a full range of orbital energies is possible. Thus an entire range of photon energies is emitted by the atoms, and instead of separate, distinct spectral lines appearing in the spectrum, an entire range of frequencies appears.

We will discuss the Sun in Chapter 11.

Before the light from the Sun gets to us on Earth, it must pass through the relatively cooler atmosphere of the Sun, as well as through the atmosphere of the Earth.

TOOLS OF ASTRONOMY

The Balmer Series

Notice in Figure 4-14 that there is a pattern to the spacing of bright lines in the hydrogen spectrum—they become progressively closer as they approach the blue end of the spectrum. The mathematical relationship that expresses this pattern was found in 1885 by Johann Jacob Balmer, a Swiss teacher. The lines visible in the figure are therefore called the *Balmer series* of spectral lines. The Balmer series served as the foundation for Bohr's work, and Bohr's model of the atom easily explains the pattern of the lines. In **Figure B4-7**, the spacing of the lines depicting the orbits of the hydrogen atom represents the relative energy of each of the levels. According to Bohr's model, there is less difference between the energy levels as orbits get farther from the nucleus. Thus the levels are spaced closer together toward the top of the figure.

Suppose that an electron is in the lowest energy level, the ground state. This electron will jump to a higher state when a photon of the appropriate energy strikes it. Look at the left side of **Figure B4-8**. Five arrows point upward from the ground state, representing electron jumps to higher orbits. On each arrow is printed the wavelength of the photon corresponding to the jump indicated. Notice that each of these wavelengths is less than 400 nm, the shortest wavelength of visible light. They are in the ultraviolet region of the spectrum.

As a gas becomes hotter, more of its atoms have electrons in energy levels above the ground state. Suppose that hydrogen is at a temperature at which a significant number of its electrons are in energy level 2. The middle of Figure B4-8 shows the wavelengths of photons that would cause electrons at this level to jump to a higher level. Notice that the wavelengths of the photons that cause jumps from the second level are within the visible range—from 400 to 700 nm—and since energy levels are more closely spaced toward the top of the figure, the wavelengths toward the blue end of the spectrum are closer together. These wavelengths correspond to the wavelengths of the Balmer series.

The first set of wavelengths described above, those in the ultraviolet region of the spectrum, form the *Lyman series.* As Figure B4-8 indicates, there is another series that falls in the infrared portion of the spectrum, called the *Paschen series.*

Thus far we have discussed the absorption of photons to produce an absorption spectrum. As we noted in the text, the emission spectrum of hydrogen is produced when electrons fall from higher energies to lower. Thus the spectrum shown in Figure 4-14 is an emission spectrum that resulted from electrons falling from higher energies to lower levels of the atom.

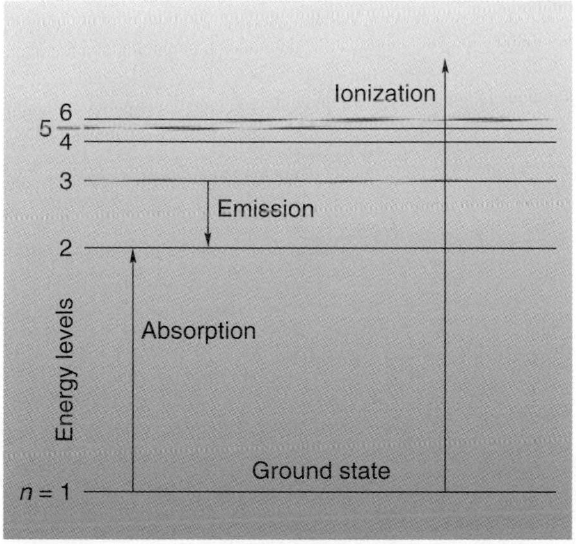

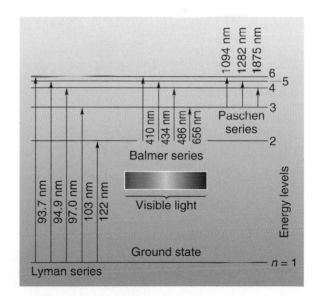

Figure B4-7
The energy levels of the hydrogen atom. The levels are progressively closer in energy as they are farther from the nucleus.

Figure B4-8
Electrons that move from the ground state to higher energy levels do so by absorbing photons of the wavelengths shown. Electron jumps for the first three series of the hydrogen atom's spectrum are represented here.

Figure 4-18
(a) The emission spectrum of an element; (b) the absorption spectrum of the same element. Notice that the absorbed wavelengths in (b) are the same as the emission lines in (a).

(a)

(b)

The Sun does indeed have an atmosphere, and as we will see in a later chapter, its atmosphere is much deeper than Earth's. As the light passes through these gases, atoms of the gases absorb some of it. This absorption of energy raises an atom's energy level, but since only certain specific energy levels are possible, only certain amounts of energy can be absorbed by the atom. This results in the reverse of what we had before: instead of an atom emitting a photon as it releases energy, it absorbs a photon as it absorbs energy. And just as a hot, low-density gas emits photons of certain energies, the same gas when in a cooler atmosphere absorbs photons of the same energies.

If white light (a continuous spectrum) is passed through hydrogen gas that is too cool to emit light, the gas absorbs some frequencies of this white light—the same frequencies it would emit as an emission spectrum. **Figure 4-18a** represents the emission spectrum of some element. Figure 4-18b shows the ***absorption spectrum*** that results when white light is passed through the cool gas of this same element. Notice that the dark lines of the absorption spectrum correspond exactly to the bright lines of the emission spectrum.

You might point out that after the cool gas has absorbed radiation, it must re-emit it. Shouldn't this then cancel out the absorption? No. As **Figure 4-19** shows, the re-emitted light is sent out in all directions. So certain frequencies of the light that was originally

absorption spectrum A spectrum that is continuous except for certain discrete frequencies (or wavelengths).

Figure 4-19
Consider a beam of light leaving a point on the Sun and moving toward the Earth. As it passes through the Sun's atmosphere, some wavelengths are absorbed and then re-emitted in random directions. This results in the solar spectrum being an absorption spectrum. Light from the Sun's atmosphere is an emission spectrum.

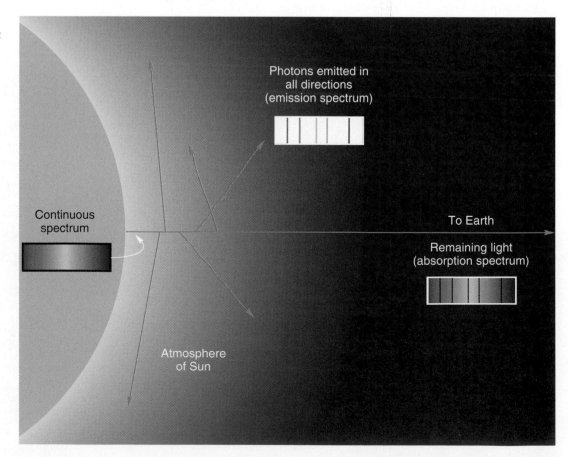

coming toward the Earth are scattered by the atmosphere of the Sun. This results in less light of those frequencies reaching us, and we observe an absorption spectrum.

The spectrum of the light that is re-emitted is an emission spectrum. During a total eclipse of the Sun, light from the main body of the Sun is blocked out, and astronomers can see the light emitted by the Sun's atmosphere and examine its emission spectrum. It was by examining the emission spectrum from the gas near the Sun that astronomers first discovered the element helium (from the Greek *helios*, the Sun).

The Sun, and other stars as well, has various chemical elements in its atmosphere (that is, different kinds of atoms, each with its own pattern of electron orbits and spectral lines). As the white light passes through this gas, many frequencies are absorbed, corresponding to the various chemical elements of the gas. By examining the complicated absorption spectrum that results, we are able to deduce what elements are present in the star's atmosphere. Thus we answer a question that, just a century ago, was thought to be unanswerable. We now know what the surface layers of stars are made of! As you might appreciate from the complexity of the Sun's spectrum (look back at Figure 4-13), the analysis is fairly complicated, but it is now a common one in astronomy.

Absorption spectra are the dark line spectra of Kirchhoff's laws.

The elements in the Earth's atmosphere also absorb radiation, the frequencies of which depend upon what elements are in our atmosphere. But we know what those elements are and can take them into account.

4-7 The Doppler Effect

Have you ever stood near a road and listened to the sound of a car as it sped by you? Recall how the sound changed when the car passed, going from a higher pitch down to a lower pitch. The change is especially noticeable for noisy, fast-moving race cars; perhaps you have heard it on televised auto races. This phenomenon has a parallel in astronomy—a very important one. To understand it, we first consider water waves.

Figure 4-20a is a photo of waves spreading from a disturbance on the surface of water. The waves move away from the source in a regular way and appear the same in all directions from the source. Figure 4-20b was made by moving a vibrating object toward the right as it makes waves on water. Look at the difference between the waves in front of and behind the moving source. Four important points can be made about this case, although only one of them is apparent in the photo.

1. Even though the source of the waves is moving, the waves still travel at the same speed in all directions. The motion of a source does not push or pull the

(a)

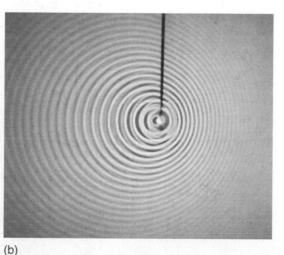

(b)

Figure 4-20

(a) Waves spread evenly from their source. (b) If the source is moving to the right, the waves are compressed in front of the source and stretched behind it.

waves. The source just disturbs the liquid, and the disturbance moves away at a speed that depends only on the characteristics of the liquid—water, in this case.

2. The wavelengths of the waves in front of the moving source are shorter than they would be if the source were stationary, and the wavelengths behind the moving source are longer. Recall that for waves traveling at the same speed, where the wavelength is shorter, the frequency is higher; and where the wavelength is longer, the frequency is lower. This means that if there were corks on the water, the corks in front of the moving source would bounce up and down with a greater frequency than if the source were not moving, and corks behind the moving source would bounce with a lower frequency.

    This change in wavelength is the same effect that causes the sound frequency of a car to change as the car passes. The car emits sound waves just as the vibrating object produces water waves. When you are in front of the car, you are in the region of shorter wavelength and higher frequency. In the case of sound, high frequency means high pitch. After the car has passed, its sound waves are stretched in wavelength, causing you to hear a lower pitch.

    This effect, which we see here both in water waves and in sound waves, is called the ***Doppler effect,*** named after the Austrian physicist Christian Doppler (1803–1853), who first explained it. Before we apply it to light waves and astronomy, let's continue the list of important things to know about it.

3. The sound does indeed get louder as a car approaches (and the water waves get higher as the vibrating object approaches a bobbing cork), but this is *not* what the Doppler effect is about. The Doppler effect refers only to the *change in wavelength* (and therefore frequency) of the wave.

4. The frequency does *not* get higher and higher as the source approaches at a uniform speed. Rather, the frequency is observed to be higher than the source's (but constant in value) as the source approaches and lower than the source's (but constant in value) as it recedes.

The Doppler Effect in Astronomy

We do not receive water waves and sound waves from the stars, but only electromagnetic waves. However, the Doppler effect also occurs for electromagnetic waves. This means that if an object is coming toward us, the light we receive from it will have a shorter than normal wavelength, and from an object moving away we will receive a longer than normal wavelength. The reason you can't observe this for moving objects here on Earth is that the amount of the shortening and lengthening of the wave depends on the speed of the object *compared to the speed of the wave.*

For water waves, which move fairly slowly, you observe the Doppler effect even for a slow-moving wave source. In the case of sound, which has a speed much greater than water waves, you notice the Doppler effect only for fairly fast objects. A car going 70 miles/hour is moving at about 10% of the speed of sound.

In the case of light, you don't perceive the Doppler effect for a car traveling at 70 miles/hour because it is moving at only one ten-millionth the speed of light. To describe the Doppler effect for light, the example is sometimes used of a spaceship with a lamp emitting green light. If the spaceship is moving away from us at a great enough speed, the wavelength of the light we see from the lamp is stretched so that the light appears red. The light is ***redshifted.*** If the spaceship is approaching, the light is ***blueshifted.***

The spaceship example does illustrate the Doppler effect, but it is very misleading in an important respect: Except for very distant galaxies, objects in the heavens do not move with speeds great enough to actually change their colors appreciably. The

---

Doppler effect The observed change in wavelength of waves from a source moving toward or away from an observer.

By *normal* we mean the wavelength of the emitted light as measured by someone traveling with the source.

redshift A change in wavelength toward longer wavelengths.

blueshift A change in wavelength toward shorter wavelengths.

(a)

(b)

Figure 4-21
(a) Emission spectrum of hydrogen when the source is stationary with respect to the observer. (b) Absorption spectrum of hydrogen for a receding source (at 0.7% of the speed of light). Note the shift of all lines toward redder colors.

redshift or blueshift caused by the Doppler effect is very small in most cases. If the spectra of stars were continuous spectra, there would be few cases in which the Doppler effect could be used to detect motion. It is the spectral lines (usually the absorption lines) that make the Doppler effect such a powerful tool.

As an example of how we use absorption lines to detect the Doppler effect for stars, imagine that we record the spectrum of hydrogen gas in the laboratory. **Figure 4-21a** shows the spectrum. Figure 4-21b represents the spectrum of a star having only hydrogen in its atmosphere (which is unrealistic, but this is a simplified example). Notice that the absorption lines in the star's spectrum do not align exactly with the emission lines of the laboratory spectrum. Instead, the absorption lines are shifted slightly toward the red. This indicates that the star is moving away from us.

There are three major differences between our example and a measurement of a real star: First, a real star's spectrum has many more spectral lines. Second, the Doppler shift is almost always much smaller than that indicated in the example. When we later show photos of actual spectra, they will usually be a magnified portion of a small part of the visible spectrum. Third, astronomers do not normally use color film in recording the spectrum. This may seem odd, but recall that color is not easy to describe accurately, whereas wavelength is. The wavelengths of absorption lines can be measured very accurately and compared to the lines from a laboratory spectrum to determine the existence and the precise amount of the Doppler shift.

The Doppler Effect as a Measurement Technique

Thus far we have described the Doppler effect as a method for detecting whether a star is moving toward or away from us, but it is more powerful than this. From measurements of the *amount* of the shifting of the spectral lines, we can determine the **radial velocity** of the star relative to Earth. The radial velocity is the star's velocity toward or away from us, and must be distinguished from its **tangential velocity**, which is its velocity across our line of sight. **Figure 4-22** distinguishes the two velocities. The Doppler effect provides a method to measure radial velocity, making it much easier to detect and measure than tangential velocity. To measure tangential velocity, we must look for motion of the star across our line of sight (like the yellow car in Figure 4-22), and this motion can be detected only for relatively nearby stars.

The following equation allows us to calculate radial velocity from Doppler shift data:

$$\frac{\Delta\lambda}{\lambda_0} = \frac{v}{c}$$

where $\Delta\lambda = \lambda - \lambda_0 =$ wavelength difference

$\lambda =$ observed wavelength

$\lambda_0 =$ wavelength of spectral line from stationary source

$v =$ radial velocity of object

Figure 4-22
The red car has a radial (to-or-fro) velocity with respect to the observer, while the yellow car has a tangential (side-to-side) velocity. The Doppler effect cannot be used to detect tangential velocity.

radial velocity Velocity along the line of sight, toward or away from the observer.

tangential velocity Velocity perpendicular to the line of sight.

This equation applies if the object is moving much slower than the speed of light, as is the case for most galaxies other than the very distant ones. For distant galaxies, a slightly more complex equation is needed, as relativistic effects become important.

c = velocity of light

If we solve this equation for what we are usually calculating—that is, the velocity of the object—we obtain

$$v = c\left(\frac{\Delta\lambda}{\lambda_0}\right)$$

When an object is approaching the observer, the wavelength difference is negative and so is the radial velocity. When an object is moving away from the observer, the wavelength difference is positive and so is the radial velocity.

To illustrate how easy this equation is to apply, let's do an example.

EXAMPLE

The wavelength of one of the most prominent spectral lines of hydrogen is 656.285 nanometers (this is in the red portion of the spectrum). In the spectrum of Regulus (the brightest star in the constellation Leo), the wavelength of this line is observed to appear longer by 0.0077 nanometers. Calculate the speed of Regulus relative to Earth and determine whether it is moving toward or away from us.

SOLUTION

First, notice that the data indicates that the wavelength of the line in the spectrum of Regulus is *longer* by 0.0077 nanometers. This means that the star is moving *away* from us. To determine its speed, we substitute the given values in the equation that relates Doppler shift and speed:

$$v = c\left(\frac{\Delta\lambda}{\lambda_0}\right) = (3.0 \times 10^8 \text{ m/s}) \times \left(\frac{0.0077 \text{ nm}}{656.285 \text{ nm}}\right)$$

$$= 3500 \text{ m/s} = 3.5 \text{ km/s}$$

So Regulus is moving away from Earth at a speed of about 3.5 kilometers/second. This is about 2 miles/second, or 8000 miles/hour, a typical speed for nearby stars. Because the Earth's speed in its orbit around the Sun is about 30 kilometers/second, the Earth's motion would have to be taken into account in making the measurement. Our calculation assumed that the Earth was between the Sun and Regulus, so that the Earth had no motion toward or away from the star.

TRY ONE YOURSELF

The nearest star to the Sun that is visible to the naked eye is Alpha Centauri. With Earth's motion removed, the 656.285 nanometers line of hydrogen has a wavelength of 656.237 nanometers in Alpha Centauri's spectrum. Calculate the radial velocity of this star relative to the Sun and tell whether it is moving toward or away from the Sun.

Other Doppler Effect Measurements

The annual variation in the Doppler shift observed in stellar spectra provides evidence for the Earth's motion around the Sun. The evidence was not available, of course, when the question was controversial.

Police radar measures motorists' speeds by bouncing waves from their cars and detecting the Doppler shift in the reflected waves.

The use of the Doppler effect is not limited to measuring the speeds of stars. Other applications include:

1. Measuring the rotation rate of the Sun. Galileo was the first to observe sunspots on the Sun. He reported that they move across the Sun, thus providing a method to measure the Sun's rotation rate. The Doppler effect gives a second method. The measurement is done by examining the light from opposite sides of the Sun. Light from the side moving toward us is blueshifted and light from the other side is redshifted.

2. In a similar manner, the rotation rates of distant stars, other galaxies, planets, and the rings of Saturn can be measured. The fact that the light in the case of planets has been reflected by the planet (or rings) rather than emitted by it does not matter; the light is still shifted by the Doppler effect. In fact, the rotations of Mercury and Venus were first revealed by reflected radar waves.

3. Many stars are part of a two (or more) star system in which the stars orbit one another. If their orbits happen to be aligned so that each star moves alternately

toward and away from our position in the universe, we can detect this motion by the blueshift and redshift of each star's spectrum. When we discuss such ***binary stars*** in Chapter 12, we'll see that they are very important in our quest to learn more about stars.

> **binary star (system)** A pair of stars gravitationally bound so that they orbit one another.

Relative or Real Speed?

The speed measured by the Doppler effect is the speed of the object *relative to the speed of the Earth*. As mentioned earlier, we must take the Earth's speed into account in any calculation made with the Doppler effect. You may argue that even if we do this, we are only measuring the speed of the object with respect to the Sun. What about the object's *real* speed? There is no such thing.

The reason for this last statement is that *all* speeds are relative to something. When you say that a car is moving at 50 miles/hour, you mean "relative to the surface of the Earth." When you walk up the aisle of a moving plane, you have one speed relative to the seated passengers, another relative to the Earth, another relative to the Sun, and so on.

Just as all motion is relative, all nonmotion is relative, too. When we say that something is at rest, we usually mean that it is at rest relative to the Earth. (Not always, however. You might tell your little brother to sit still in the back seat of your car, even though the car is moving. In this case you are asking him to sit still relative to the car.) It is meaningless to say that something is absolutely at rest.

Thus, there is no loss of meaning to our finding the velocity of an object relative to the Earth and then correcting for the Earth's motion around the Sun. All motion is relative.

> This understanding of the relativity of motion is called *Galilean relativity* (or *Newtonian relativity*). Einstein's theory of relativity goes much further than this.

4-8 The Inverse Square Law of Radiation

Everyone has experienced the fact that the intensity—that is, the brightness—of light decreases when we move farther from the source of the light. The decrease in intensity follows an ***inverse square law*** of radiation, a relationship that states that radiation spreading from a small source decreases in intensity as the inverse square of the distance from the source.

> **inverse square law** Any relationship in which some factor decreases as the square of the distance from its source.

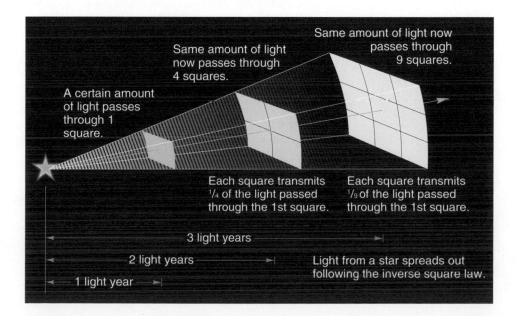

Figure 4-23
As light spreads from a star, its intensity decreases as the inverse square of the distance.

In order to understand this law, consider a small energy source that emits energy equally in all directions. **Figure 4-23** shows light spreading toward the right from a star. The part of the light that we have chosen to show fills the first square, one light year away. The light continues to spread, and when it has gone two light years from the star, it has spread out to cover four squares. Thus less light illuminates each of the four squares, and the star's light is only one-fourth as bright. Finally, when the light has traveled three light years, it covers nine squares and is only one-ninth as intense. We see from this that the intensity of light decreases as the inverse square of the distance from the source.

The inverse square law applies only to small sources of light, but this requirement is easily met in most astronomical cases. Even the Sun and stars qualify as small sources, for compared to our distance from them, their diameters are small.

◆ The force of gravity follows an inverse square law, as does the apparent brightness at some distance from a source of light.

Conclusion

As this chapter has shown, the spectra of stars are used in two very different ways. First, the spectra are treated as continuous spectra and we examine their graphs of intensity versus wavelength. By measuring the wavelength of the maximum intensity of radiation, we can determine the temperature of a star's surface. In doing this analysis, we ignore the absorption lines. These lines, remember, are very narrow; their presence does not change the overall pattern of the blackbody curve.

Second, we examine the absorption lines within the spectrum. These lines allow us to determine the chemical composition of stars. In addition, when we measure the Doppler effect, the shift in the lines allows us to calculate the radial speeds of stars relative to Earth, as well as the speeds of rotation and revolution of celestial objects.

Study Guide

1. The frequency of visible light falls between that of
   A. infrared waves and radio waves.
   B. X-rays and cosmic rays.
   C. ultraviolet waves and X-rays.
   D. short radio waves and long radio waves.
   E. ultraviolet waves and radio waves.

2. Infrared radiation differs from red light in
   A. intensity.
   B. wavelength.
   C. its speed in a vacuum.
   D. [All of the above.]
   E. [None of the above.]

3. The frequency at which a star emits the most light depends upon the star's
   A. distance from us.
   B. brightness.
   C. temperature.
   D. eccentricity.
   E. velocity toward or away from us.

4. In an infrared photo taken on a cool night, your skin will appear brighter than your clothes.
   A. Correct.
   B. Wrong, it depends upon the color of your clothes.
   C. Wrong, your clothes will appear brighter.
   D. Wrong, they will be equally bright.

5. Light waves of greater frequency have
   A. shorter wavelength.
   B. longer wavelength.
   C. [Either of the above; there is no direct connection between frequency and wavelength.]

6. Which of the following choices does not have the same fundamental nature as visible light?
   A. X-rays
   B. Sound waves
   C. Ultraviolet radiation
   D. Infrared waves
   E. Radio waves

7. As an electron of an atom changes from one energy level to a higher energy level by absorbing a photon, the total energy of the atom
   A. increases.
   B. decreases.
   C. remains the same.

8. Which of the following choices is produced when white light is shined through a cool gas?
   A. An absorption spectrum.
   B. A continuous spectrum.
   C. An emission spectrum.
   D. [All of the above.]
   E. [None of the above.]

9. The solar spectrum is which of the following?
   A. An absorption spectrum.
   B. A continuous spectrum.
   C. An emission spectrum.
   D. [All of the above.]
   E. [None of the above.]

10. The spectrum of light from the Sun's atmosphere *seen during a total eclipse* is
    A. an emission spectrum.
    B. an absorption spectrum.
    C. a combination of emission spectrum and absorption spectrum.
    D. a continuous spectrum.
    E. a combination of all three types of spectra.

11. Analysis of a star's spectrum *cannot* determine
    A. the star's radial velocity.
    B. the star's tangential velocity.
    C. the chemical elements present in the star's atmosphere.
    D. [More than one of the above.]

12. According to the Doppler effect,
    A. sound gets louder as its source approaches and softer as it recedes.
    B. sound gets higher and higher in pitch as its source approaches and lower and lower as its source recedes.
    C. sound is of constant higher pitch as its source approaches and of constant lower pitch as its source recedes.
    D. [Both A and B above.]
    E. [Both A and C above.]

13. The Doppler effect causes light from a source moving away to be
    A. shifted to shorter wavelengths.
    B. shifted to longer wavelengths.
    C. changed in velocity.
    D. [Both A and C above.]
    E. [Both B and C above.]

14. We can determine the elements in the atmosphere of a star by examining
    A. its color.
    B. its absorption spectrum.
    C. the frequency at which it emits the most energy.
    D. its temperature.
    E. its motion relative to us.

15. Which list shows the colors of stars from coolest to hottest?
    A. Red, white, blue
    B. White, blue, red
    C. Blue, white, red
    D. Red, blue, white
    E. Blue, red, white

16. The Doppler effect is used to
    A. measure the radial velocity of a star.
    B. detect and study binary stars.
    C. measure the rotation of the Sun.
    D. [Two of the above.]
    E. [All of the above.]

17. Sound waves cannot travel in a vacuum. How, then, do radio waves travel through interstellar space?
    A. They are extra-powerful sound waves.
    B. They are very high frequency sound waves.

C. Radio waves are not sound waves at all.
D. The question is a trick, for radio waves do *not* travel through interstellar space.
E. Interstellar space is not a vacuum.

18. The energy of a photon is directly proportional to the light's
    A. wavelength.
    B. frequency.
    C. velocity.
    D. brightness.

19. In the Bohr model of the atom, light is emitted from an atom when
    A. an electron moves from an inner to an outer orbit.
    B. an atom gains energy.
    C. an electron moves from an outer to an inner orbit.
    D. one element reacts with another.
    E. [Both A and B above.]

20. The intensity/wavelength graph of a "blue-hot" object peaks in the
    A. infrared region.
    B. red region.
    C. yellow region.
    D. ultraviolet region.

21. The emission spectrum produced by the excited atoms of an element contains wavelengths that are
    A. the same for all elements.
    B. characteristic of the particular element.
    C. evenly distributed throughout the entire visible spectrum.
    D. different from the wavelengths in its absorption spectrum.
    E. [Both A and D above.]

22. Each element has its own characteristic spectrum because
    A. the speed of light differs for each element.
    B. some elements are at a higher temperature than others.
    C. atoms combine to form molecules, releasing different wavelengths depending on the elements involved.
    D. electron energy levels are different for different elements.
    E. hot solids, such as tungsten, emit a continuous spectrum.

23. The speed of sound is 335 meters/second. What is the wavelength of a sound that has a frequency of 500 cycles/second?
    A. 0.67 meters
    B. 1.49 meters
    C. 165 meters
    D. 835 meters

24. Suppose the speed of a water wave is 12 inches/second and the wavelength is 4 inches. What is the frequency of the wave?
    A. 48 cycles/second
    B. 18 cycles/second
    C. 8 cycles/second
    D. 3 cycles/second
    E. 1/3 cycles/second

25. Define *wavelength* and *frequency* in the case of a wave.

26. Which has the higher frequency, light of 400 nanometers or light of 450 nanometers?

27. Approximately what are the least and greatest wavelengths of visible light?

28. Name six regions of the electromagnetic spectrum in order from longest to shortest wavelength. (Do not list the colors of visible light as separate regions of the spectrum.)

29. Which parts of the electromagnetic spectrum penetrate the atmosphere?

30. Define the following terms: electron, nucleus, photon.

31. State the three postulates upon which the Bohr model of the atom is based.

32. How is the energy of the photon related to the frequency of the light?

33. Why do the various elements each have different emission spectra?

34. Name the three different types of spectra and explain how each is produced.

35. Explain how we know what elements are in the atmospheres of the stars.

36. Carefully explain what the Doppler effect is. In the case of light waves, does the Doppler effect tell us anything about the intensity of the light in front of and behind a moving light source?

37. Distinguish between the way a spectrum is used to determine the temperature of a star and the way it is used to determine the star's composition and/or motion toward or away from us.

QUESTIONS TO PONDER

1. Why do we have three different temperature scales instead of one? What are the advantages (if any) of each?

2. Two people are having fun in the ocean. One is surfing in on a wave; the other is floating up and down on the waves. The different motions of the two people could be used to illustrate two quantities in the equation $v = \lambda \times f$. Which two?

3. The period of a wave is defined as the amount of time required for the wave to complete one cycle. What, then, is the relationship between the period and the frequency of a wave?

4. It is especially easy to get a sunburn when skiing high in the mountains. How does the high altitude contribute to the danger of sunburn?

5. Why can't we hear radio waves without using the device that we call a "radio"?

6. In today's world, radio waves (both for radio and TV) are striking your body all the time, but you are not allowed to get too many X-rays in one year. Why is this so? After all, both radio waves and X-rays are a part of the electromagnetic spectrum.

7. What does an object do to light to give the object its color? How does this differ from the color of an object that emits its own light?

8. Draw a graph of intensity versus wavelength for a red star and another for a white star. Describe two ways in which the graphs differ.

9. The text explains that cool stars are redder than hot ones. However, couldn't the red color result instead from the Doppler effect if the star is moving away from us? Explain.

10. If absorption lines were to be included in the graph of intensity versus wavelength for a star, how would this change the graph?

11. How would Figure 4-13 be different if it were the spectrum of a red star?

12. How do we know what chemical elements are in the atmosphere of the Sun? Explain.

13. What is the Doppler effect and why is it important to astronomers?

14. Figure 4-22 greatly exaggerates color differences for the two cars. If it showed a car coming toward you, what color would the car be, using the same exaggeration?

CALCULATIONS

1. If the speed of a particular water wave is 12 meters/second and its wavelength is 3 meters, what is its frequency?

2. The speed of sound in air at room temperature is about 344 meters/second. What is the wavelength of a sound wave with a frequency of 256 Hz?

3. Express 500 nanometers in meters.

4. The 656.285 nanometers line of hydrogen is measured to be 656.305 nanometers in the spectrum of a certain star. Is this star approaching or receding from the Earth? Calculate its radial speed relative to the Earth.

5. If a certain star is moving away from Earth at 25 kilometers/second, what will be the measured wavelength of a spectral line that has a wavelength of 500 nanometers for a stationary source?

6. Suppose the Kelvin temperature of blackbody X is three times as great as that of blackbody Y. Compare the energy released by equal areas of the two objects.

7. Use Wien's law to determine the wavelength at which an object at body temperature (98.6°F) emits most of its energy.

(Hint: Make sure you find the corresponding temperature in kelvin.) What kind of radiation is this? Can you see it?

8. Two thermometers, one marked in °F and the other in °C, are placed in the same container. At what temperature will both thermometers read the same?

9. Matter falling toward a black hole gets compressed and heated to about 10^6 K. In what part of the electromagnetic spectrum would you search for black holes?

10. A star is five times as large as the Sun, but its surface temperature is smaller by a factor of three. Does this star emit more energy than the Sun or less? Explain.

11. (a) We said that light takes 8.3 minutes to reach us from the Sun. Show that this is the case. (b) When we are observing an object that is 5000 light years away, do we see it as it is now? Explain.

12. Compare the light we receive from a 100-watt light bulb that is 30 feet away to the light from a similiar bulb that is 10 feet away.

Figure 4-21a shows a hydrogen emission spectrum (in the visible part of the spectrum) when the source is stationary with respect to the observer, while Figure 4-21b shows a corresponding hydrogen absorption spectrum for a receding source. In general, spectra are much more complicated, including many lines from different elements. However, let us imagine that we have isolated the hydrogen lines in this spectrum. In addition, consider the spectrum as a "map," exactly as is. We would like to calculate the speed of the source, using the Doppler effect.

For a map to be useful, it must include the correct scale. The lines in Figure 4.21a correspond to the following wavelengths from right to left: 656.3 nanometers (red), 486.1 nanometers (blue-green), 434.0 nanometers (violet) and 410.1 nanometers (violet). Measure the distances between the lines in millimeters (mm) and calculate the scale with units nm/mm.

Now compare the position of the lines between the two spectra. For example, by how many millimeters has the 656.3-nanometer line shifted to the right (toward longer wavelengths)? Using the scale you found in the previous step, translate this shift (redshift) into nanometers. This is the change between the observed and laboratory-based wavelength for this hydrogen line ($\Delta\lambda$). Using $\lambda_0 = 656.3$ nanometers and the Doppler equation, show that the speed of the receding source is 0.7% of the speed of light.

1. "The Duality in Matter and Light," by B.G. Englert, M. O. Scully, and H. Walther, in *Scientific American* (December, 1994).

2. "The Truth About Star Colors," by P. C. Steffy, in *Sky and Telescope* (September 1992, p. 266).

3. "Unlocking the Chemical Secrets of the Cosmos," by O. Gingerich, in *Sky and Telescope* (July, 1981).

4. "The Electromagnetic Spectrum," by H. Augensen and J. Woodbury, in *Astronomy* (June, 1982).

Quest Ahead to Starlinks: www.jbpub.com/starlinks

Starlinks is this book's online learning center. It features **eLearning,** which contains chapter quizzes and other tools designed to help you study for your class. You can also find **on-line exercises,** view numerous relevant **animations,** follow a guide to **useful astronomy sites** on the web, or even check the latest **astronomy news** updates.

ACTIVITIES

EXPANDING THE QUEST

STARLINKS

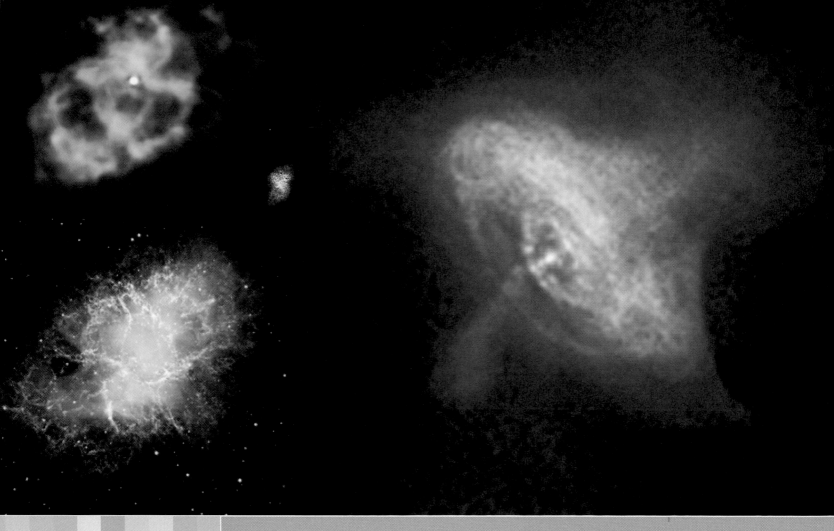

CHAPTER 5

Telescopes: Windows to the Universe

The Crab nebula as seen in ultraviolet (insert) and (clockwise from the right) in X-ray, visible, and radio radiation.

LOOKS CAN BE DECEIVING. DURING THE PAST FEW DECADES, astronomers have come face-to-face with this old saying. Up until the middle of the twentieth century, almost all our knowledge of the sky came from visible light waves. Therefore, photographs of the sky showed what our eyes see, just adding detail because telescopes can detect very dim light and magnify the image. Still, it seemed that "What we see is what we get."

Then along came detectors of radiation in other parts of the spectrum. The chapter-opening photo shows four views of the Crab nebula, which is the leftover debris of a star that exploded almost a thousand years ago. The bottom left image was taken with an optical (visible-light) telescope. The larger image was taken by the *Chandra X-Ray Observatory*, which is in orbit around the Earth. It shows the spinning, top-like structure that may provide the energy for the glowing gases seen in visible light. Which image shows reality? The answer is "Both, and more!" for the same picture taken in radio or ultraviolet radiation looks different still. We must conclude that our eyes don't tell us the whole story. In astronomy, what we see is just part of what we get.

In order to understand the universe, we rely heavily on observations. As with any branch of science, we need observational data to test existing theories or develop new ones. The distances between us—the observers—and the objects of our studies are vast and the cosmos is very dynamic in nature. All we can do is passively collect the radiation sent from these objects and try to decipher the information it carries. The most useful tools for collecting radiation are telescopes.

Galileo Galilei was the first to use a telescope to study the heavens systematically, as we discussed in Chapter 3. Much was learned before Galileo's time by naked-eye observations, but Galileo's telescope changed astronomy—and our outlook on the universe—tremendously. As we make telescopes larger and take them into space, above the distortions of the Earth's atmosphere, we realize that we are just beginning to learn about the mysteries of our universe.

The past century has brought a multitude of telescopic tools, and our view of the skies has expanded to parts of the spectrum well beyond visible light. Telescopes now include instruments used to map the sky in all regions of the electromagnetic spectrum. We begin this chapter by describing some properties of light that are important to the understanding of visible-light telescopes. Then we discuss the use of such telescopes. Finally, the chapter concludes with a look at some of the nonvisible-light telescopes that are so indispensable to modern astronomy.

> O telescope, instrument of much knowledge, more precious than any scepter!
> —*Kepler, in a letter to Galileo*

5-1 Refraction and Image Formation

The discussion of light thus far has concentrated on its wave properties. The effect of these properties will be examined later in this chapter, but for the moment we concentrate not on the nature of light, but on the path that light travels. That path is usually very simple, for light travels in a straight line as long as it remains in the same (uniform) medium. It may, however, change direction upon entering a second medium. **Figure 5-1a** shows a ray of light passing through a wedge of glass. We see that the light travels in a straight line before and after passing through the glass and that it travels in a straight line inside the glass; however, the ray bends when it passes through each surface of the glass. Figure 5-1b shows that when the two sides of the wedge are more nearly parallel, the light bends less. The phenomenon of the bending of a wave as it passes from one medium into another is called ***refraction.***

Two factors determine the amount of refraction that occurs when light crosses from one material into another. The first factor is the relative speeds of light in the two materials. In Chapter 4 we stated that the speed of light in vacuum is about 3×10^8 meters/second. In air, light's speed is just slightly slower. In glass, however, light travels at about 2×10^8 meters/second (more or less, depending upon the type of glass). Because of the change in speed, a ray of light may bend significantly when going from air into glass.

The word *medium* refers, in a generic way, to the "material" through which light travels, such as air, glass, water, or the almost perfect vacuum in the far reaches of the universe.

refraction The bending of light as it crosses the boundary between two materials in which it travels at different speeds.

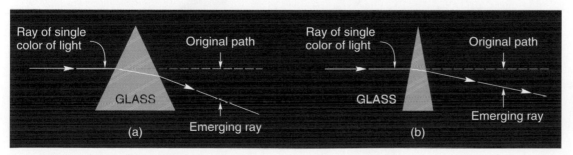

(a)   (b)

Figure 5-1
When light passes through a wedge of glass, it is bent from its straight-line path. The smaller the angle of the wedge, the less the bending. (The "wedge" is really a prism, but we are concentrating here on the bending of the light rather than its separation into colors.) Notice that the bending occurs both when the light enters the glass and when it emerges.

For a simple analogy, consider what happens when you roll a pair of toy wheels connected by an axle from a smooth wooden floor onto a carpet. If the wheels cross the interface between the two surfaces perpendicularly (face-on), then they will slow down but will continue moving in the same direction. If, on the other hand, they cross the interface between the two surfaces non-perpendicularly, the wheel that first hits the carpet will slow down while the other wheel, still on the smooth floor, will continue moving at its original speed (**Figure 5-2a**). As a result, the axle will turn; it will change direction. In a similar way, light changes direction when it crosses the interface between two different media non-perpendicularly. Figure 5-2b shows an everyday example of refraction.

To understand the second factor in refraction, consider **Figure 5-3a**, which shows several rays of light (that came from beyond the left side of the page) passing through a lens-shaped piece of glass. Notice that the rays that strike the surface of the glass (the interface between the two media) at a glancing angle (farther from the perpendicular) bend more than those rays that hit the glass more "head-on." The central ray strikes the glass perpendicularly and does not bend at all. This is a general rule: the larger the angle between the ray of light and the perpendicular line to the interface, the more the light bends upon passing through the surface.

Refer again to Figure 5-3a. The surfaces of the lens have just the right curvature to cause all of the rays of light shown in the figure to pass through the same spot. We could even block some of the incoming light rays (for example, the one labeled *X*) and still get an image of the object, although a bit dimmer. This will be important later in this chapter in our discussion of a class of telescopes.

To see how a lens forms an image, imagine that *each* ray in the figure came from a distant star, which we treat as a point. A point light source emits light rays in all directions; if this source is far from our telescope, only a few of these rays will enter the telescope and they will be essentially parallel to each other. If we put a piece of paper at the point where the light rays come together, all the light from that star that passes through our lens will come to a single point on the piece of paper. In fact, the rays of light coming from other stars will likewise come to a focus on the paper, forming an ***image*** of that area of the sky (Figure 5-3b).

The ***focal point*** of a lens is that point where light from a very distant object comes to a focus. This is the point where the rays converge in Figure 5-3.

The ***focal length*** of the lens is the distance from the center of the lens to the focal point. Depending upon the curvature of their surfaces, different lenses have different focal lengths.

image The visual counterpart of an object, formed by refraction or reflection of light from the object.

focal point (of a converging lens or mirror) The point at which light from a very distant object converges after being refracted or reflected.

focal length The distance from the center of a lens or a mirror to its focal point.

Figure 5-2
(a) As a light beam crosses the interface between two different media (such as from air into water, which is optically denser), it bends due to the change in the speed of propagation (as shown in Figure 5-1). This is similar to the change in direction when a pair of toy wheels travels from a smooth floor onto a rough carpet. However, if we consider two light rays instead of two wheels, both rays change direction at the interface and the distance between them increases by an amount that depends on the type of media. (b) The pencil in a cup of water appears bent due to the refraction of light. As the light rays from the submerged part of the pencil rise toward the observer, they bend at the interface.

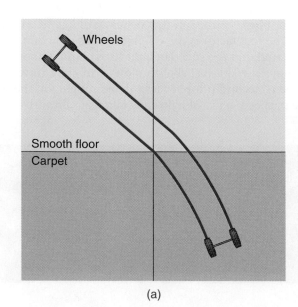

Wheels

Smooth floor
Carpet

(a)

(b)

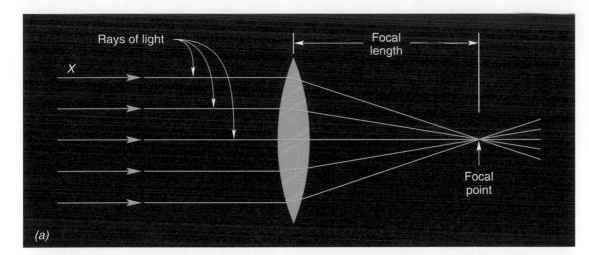

Rays of light

X

Focal length

Focal point

(a)

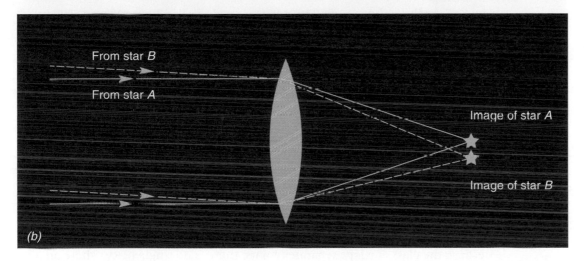

From star B

From star A

Image of star A

Image of star B

(b)

Figure 5-3
(a) A lens bends incoming rays of light toward a single point. When the incoming rays are parallel to the axis (as shown here), they cross at the lens's focal point. (b) Light rays from two different stars are focused into two separate images.

Refraction has important consequences for observations involving the Earth's atmosphere. When light passes from the (almost perfect) vacuum of space into Earth's atmosphere, it continuously refracts as it moves through the air. Each thin layer in the atmosphere is under different conditions and therefore behaves as a different medium. As a result, objects appear higher in the sky than they really are. We must take this into account when measuring positions of celestial objects (**Figure 5-4a**). This phenomenon also explains why the Sun looks "flattened" when it is closer to the horizon. Its lower edge is shifted upward more than its upper edge (Figure 5-4b).

Refraction of light in raindrops is a part of the process that gives us the rainbow.

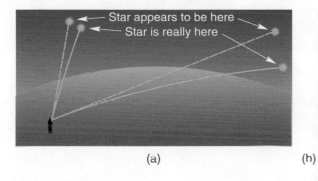

Star appears to be here
Star is really here

(a)

(b)

Figure 5-4
(a) A simplified picture of atmospheric refraction. The position of the observed object seems to be higher in the sky than it really is. (b) The Sun appears flattened at sunset because the lower edge is refracted higher than the upper edge. As shown in part (a), the amount of refraction depends on the altitude above the horizon. As a result, the Sun appears more flattened the closer it is to the horizon.

5-2 The Refracting Telescope

objective lens (or **objective**) The main light-gathering element—lens or mirror—of a telescope. It is also called the primary lens.

Lenses are at the heart of the refracting telescope. (The reflecting telescope will be discussed later in this chapter.) The simplest use of a telescope (in principle, anyway) is as a lens for a camera. **Figure 5-5a** shows a camera mounted on a small telescope. What might not be obvious in the photograph is that the camera's normal lens has been removed. The camera is using the long-focal length lens of the telescope in place of its regular lens. Figure 5-5b shows the arrangement. The telescope lens simply replaces the regular camera lens and brings the image to a focus on the film. As we will see later, telescopes can be mounted so that they can track stars across the sky and allow astronomers to take long time exposure photographs of the heavens. Long time exposures allow us to photograph much fainter objects than can be seen by simply looking through a telescope. Thus the use of a telescope with a camera is much more important to a professional astronomer than its use for direct viewing.

Figure 5-5c shows how a small telescope is used for direct observation. The primary lens—the lens through which the light passes first—is called the *objective lens*, or simply the *objective.* This lens brings the light to a focus at the focal point, which is where the film is placed when the telescope is used with a camera. This is where the image is located. For direct viewing, a second lens, the *eyepiece,* is added just beyond the focal point. This lens simply acts as a magnifier to enlarge the image.

eyepiece The magnifying lens (or combination of lenses) used to view the image formed by the objective of a telescope.

Chromatic Aberration

A prism separates white light into its colors because different wavelengths of light are refracted different amounts. Except for the ray of light that goes straight through the center of a lens, rays go through a lens in much the same way as they go through a prism. And like the light going through the prism, the light passing through a lens separates into colors. This causes the lens to have a slightly different focal length for each wavelength of light. **Figure 5-6a** exaggerates the effect but illustrates the idea.

chromatic aberration The defect of optical systems that results in light of different colors being focused at different places.

Because the glass of a lens separates colors, there is no single place where an image is exactly in focus. If the film of a camera is placed at the point where the red light focuses, the other colors will be out of focus, and the result will be an image with a fuzzy, bluish edge. This phenomenon, called *chromatic aberration,* occurs not just in telescopes, but in regular cameras as well. Fortunately, it can be corrected, at least in part.

(a)

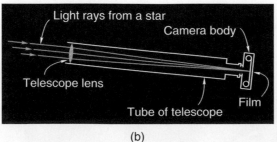

(b)

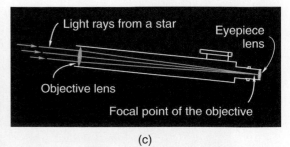

(c)

Figure 5-5
(a) A simple way to use a telescope for photography is to let the telescope serve as the camera's lens. (b) The image is focused directly on the film by the telescope's main lens. (c) When a telescope is used for direct viewing, the eyepiece is used as a magnifier to view the image formed by the objective lens. This image is formed at the focal point of the objective.

The amount of color separation that occurs when light passes through a lens depends not only on the curvature of the glass but also upon the type of glass. Some kinds of glass separate the colors more than other kinds. Telescope and camera manufacturers use this fact to correct for chromatic aberration. In all but the cheapest toy-store telescopes, the objectives of refracting telescopes are made of two lenses instead of one. As Figure 5-6b shows, the second lens has reverse curvature from the first. This curvature is not enough to undo all of the converging effect of the first lens, however, and the light is still brought to a focus. The second lens is made of a different type of glass than the first, and although it does not cancel out the bending of the light, it does cancel out most of the color separation, as shown in Figure 5-6c. Such a combination of lenses is called an *achromatic lens*.

achromatic lens (or **achromat**) An optical element that has been corrected so that it is free of chromatic aberration.

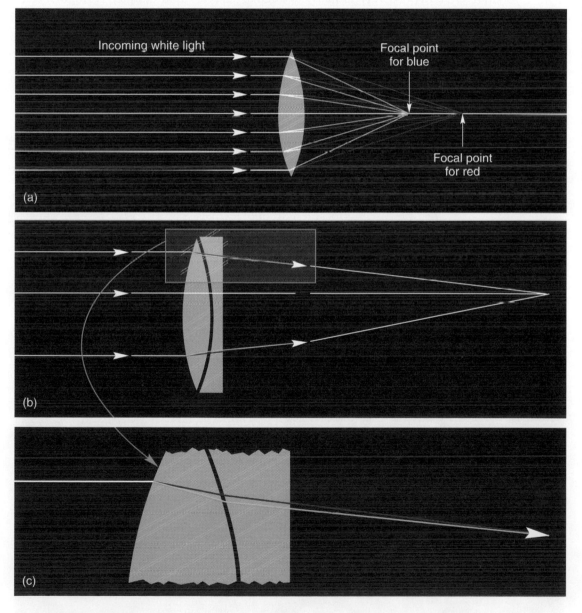

Figure 5-6
(a) Chromatic aberration. A lens exhibits a prism effect, separating white light into its colors. The lens therefore focuses each color at a different place. Only red and blue are shown here. (b) A lens can be corrected for chromatic aberration by the proper combination of lenses of different types of glass. Here the second lens brings the separated colors back together at the image. (c) This shows more detail of the upper portion of the lenses.

Incoming white light

Focal point for blue

Focal point for red

(a)

(b)

(c)

5-3 The Powers of a Telescope

Do telescopes just magnify things?

When most people think of a telescope's power, they think of magnification. Magnification, however, is only one of three major powers of a telescope and is the least important. After all, magnifying a blurry image will not show any more details. Our discussion of the powers of a telescope begins with this least important one and then considers the other two powers: light-gathering power and resolving power. Again, we will describe small telescopes, but remember that the same ideas apply to large telescopes used by professional astronomers. In fact, the reasons for using such large telescopes are related to the powers of telescopes.

Angular Size and Magnifying Power

angular size (of an object) The angle between two lines drawn from the viewer to opposite sides of the object.

The *angular size* of an object is the angle between two lines that start at the observer and go to opposite sides of the object. **Figure 5-7a** shows someone looking at the Moon. The angle between the lines to the sides of the Moon is indicated. (Note that angular size is defined very similarly to angular separation in Chapter 1. Angular size measures the apparent size of one object, while angular separation measures the apparent angular distance between two objects.) The angular size determines, for example, how big the image of the Moon is on the retina of your eye.

magnifying power, or **magnification** (of an instrument) The ratio of the angular size of an object when it is seen through the instrument to its angular size when seen with the naked eye.

For a telescope (as well as binoculars and several other optical instruments), *magnifying power* or *magnification* is defined as the ratio of the angular size of an object when it is seen through the instrument to the object's angular size when seen with the naked eye. Figure 5-7b shows the angular size of the Moon as seen through a particular telescope. You might estimate from the angles in the figure that this telescope has a magnification of about four; that is, the telescope magnifies the object four times.

The magnification of a particular telescope depends upon the focal lengths of both the objective and the eyepiece. The magnification can be calculated using the following formula:

$$\text{magnifying power} = \frac{\text{focal length of objective}}{\text{focal length of eyepiece}} \quad \text{or} \quad M = \frac{f_{obj}}{f_{eye}}$$

Figure 5-7
(a) The Moon seen by the naked eye. (b) In a telescope, the angular size of the Moon is apparently larger. The magnification of the telescope is the ratio of the two angular sizes.

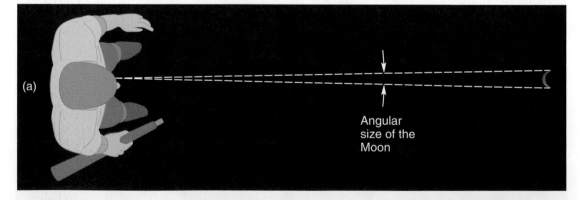

(a)

Angular size of the Moon

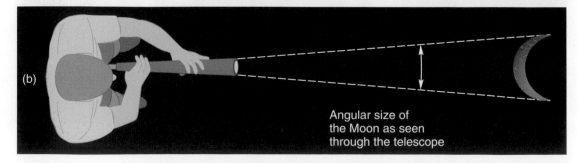

(b)

Angular size of the Moon as seen through the telescope

This means that the greatest magnification can be achieved by having a long-focal length objective and a short-focal length eyepiece.

EXAMPLE

The telescopes used in an introductory astronomy laboratory have objectives with focal lengths of 1250 millimeters. One eyepiece has a focal length of 25 millimeters. What is the magnification produced by this telescope? What is the angular size of the Moon as seen through this telescope, using the 25-millimeter eyepiece? (The naked-eye angular size of the Moon from Earth is about 1/2 degree.)

SOLUTION

To calculate the magnification, we use the equation that relates it to focal lengths:

$$M = \frac{f_{obj}}{f_{eye}} = \frac{1250 \text{ mm}}{25 \text{ mm}} = 50$$

Thus the magnifying power of the telescope is 50. We say that the magnification is 50 times, or 50×.

Now, since the Moon's angular size as seen by the naked eye is 1/2 degree and the telescope magnifies the Moon 50 times, the Moon's angular size in the telescope will be 50 × 1/2 = 25 degrees.

TRY ONE YOURSELF

What is the magnification produced by a telescope having an objective with a focal length of 1.5 meters when it is being used with an eyepiece that has a focal length of 12 millimeters? (Hint: In doing the calculation, you must express the two lengths in the same units: either meters or millimeters.)

Notice that in both the example and the suggested problem, we hinted that you might use an eyepiece of a different focal length. Indeed, it is a minor matter to change the eyepiece of a telescope. An eyepiece costs relatively little, so it is common to have a few different eyepieces for a telescope in order to have different magnifications available. The obvious question is, "Why not always use the greatest magnification?" There are several answers to this question.

The first answer is that as the magnification increases, the *field of view* of the telescope decreases. Field of view refers to how much of the object is seen at one time. **Figure 5-8** shows the decrease in field of view as the magnifying power increases. This is entirely reasonable, for if the object appears larger, not as much of it will be contained within the view of the telescope. When viewing an object—the Moon, for example—you may wish to see the whole object rather than just a portion of it. If so, you need an eyepiece with a long focal length, thus producing less magnification.

field of view The actual angular width of the scene viewed by an optical instrument.

(a) 50x   (b) 100x   (c) 250x

Figure 5-8
Increasing the magnification decreases the field of view and makes the image darker. The entire telescopic view is shown in each case.

The other reasons why we do not always use the greatest magnification relate to the other powers of a telescope.

Light-Gathering Power

There is another difference in the three views in Figure 5-8. Notice that the more magnified the image, the darker it is. To see why this occurs, consider the part of the Moon we see in part (c). The light from this portion of the Moon was concentrated in a small part of the image shown in part (a), but it is now more spread out in the more magnified view. That is, the same light that covers only part of the image in part (a) has to illuminate the entire image in part (c). The image is therefore darker. There are two ways to make the image brighter and still retain this magnification. If we are taking a photograph, a longer time exposure can be used. When we do this, we allow the light's effect to accumulate on the film over a longer time, and the film becomes more exposed. The other way is to capture more light from the Moon in the first place. This can only be done by using a larger objective, which brings us to the second power of a telescope: the *light-gathering power,* which is often the most important power.

light-gathering power A measure of the amount of light collected by an optical instrument.

The light-gathering power of a telescope refers to the amount of light it collects from an object. When bright objects like the Moon and nearby planets are observed, the objects reflect so much light to Earth that lack of light is not a problem. Most objects that are observed with a telescope, however, are very faint, and to obtain an image, we need to capture as much light from them as possible. This is true whether we are using the telescope for photography or for direct viewing.

A telescope does not literally "gather" light. It simply captures the light that hits the objective and brings that light to a focus.

The major way to gather more light is to use a telescope with a larger objective. The amount of light that strikes the objective simply depends upon its area. This is the primary reason it is desirable to have a telescope with a large-diameter objective and why the size of the objective is one of the features specified when discussing a telescope. For example, we might describe a particular telescope as a refractor with a 12-centimeter objective of focal length 140 centimeters. Notice, however, that although we state the *diameter* of the objective, it is the *area* that is important. Since the area of a circle is proportional to the square of the diameter, we must be careful in making comparisons of light-gathering power.

EXAMPLE

How does the light-gathering power of a telescope with a 6-inch objective lens compare to that of a telescope with a 9-inch objective?

SOLUTION

The area of a circle depends upon the square of its diameter, so the squares of the diameters of the telescopes must be compared in order to compare their light-gathering power. We set up a ratio of the squares of the two diameters:

$$\frac{9^2}{6^2} = \frac{81}{36} = 2.25$$

Thus the light-gathering power of the larger telescope is more than twice that of the smaller.

TRY ONE YOURSELF

Compare the light-gathering power of a small telescope, with an objective lens of diameter 10 centimeters, to that of a fully dark-adapted human eye with a pupil of diameter 5 millimeters.

The desire to be able to see fainter and fainter objects in space has led us to make larger and larger telescopes. Before discussing large telescopes, however, another advantage of size, the third and last power, must be considered.

Resolving Power

One property of light that we assume in our everyday life is that it travels in straight lines (unless it reflects from a mirror or is refracted at a surface). If we could not assume this, we would be unsure whether an object we see is in front of us or behind us. Yet there are exceptions to this rule: light does not always travel in straight lines. The exceptions are usually unimportant, but look at **Figure 5-9**. This is a magnified photograph of the shadow of a regular household screw. The shadow was made by holding the screw a few meters from a screen and illuminating it with a small, bright light source. Notice that the edges of the shadow are not distinct; there are light and dark fringes near the edges.

In this case, the light passing near the edge of the object (the side of the screw) has "spread out" slightly. The effect is small and is seldom seen in everyday life because conditions must be right and you must look carefully. We will not discuss the reason why light acts this way except to point out that water waves behave similarly; they bend around corners. Such bending of waves as they pass by the edge of an obstacle is called *diffraction*.

The amount of diffraction that occurs when light passes through an opening depends upon two things: the wavelength of the light and the size of the opening. The longer the wavelength, the more diffraction; and the larger the opening, the less diffraction.

In a telescope, the objective itself forms the opening through which the light passes. The effect is small, but it is there: light that should bend regularly and accurately according to the laws of refraction fans out a slight amount. This results in the image not being exactly clear. The image of a star that should appear as a single point is blurred into a small spot. If we increase the magnification, we simply make the blurred spot bigger and fainter.

Figure 5-10a is a photograph of the Big Dipper and indicates what seems to be a single star in the Dipper's handle. If you have fairly good eyes and look at the Dipper in a clear dark sky, you can see that this star is not one, but two stars (named Mizar and Alcor, as we discussed in Chapter 1). We say that your eyes, and the clarity of the sky, allow you to *resolve* the pair of stars. Someone with poorer eyesight may be unable to resolve the pair. Now suppose you look at this pair of stars with a small tele-

Figure 5-9
If an extremely small light source is used, the shadow of an object will have light and dark fringes at its edge. This is due to diffraction of light. (A laser was used to make this photo, but laser light is not necessary to produce the effect.)

diffraction The spreading of light upon passing the edge of an object.

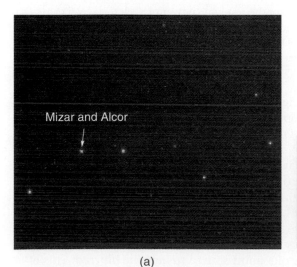

(a)

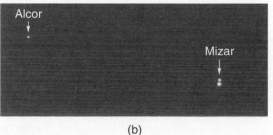

(b)

Figure 5-10
(a) Mizar and Alcor, stars in the handle of the Big Dipper, are seen as a single star by many people, but can be resolved into two stars by those with better eyesight. (b) Even a small telescope resolves Mizar into two stars (lower right). Alcor is at the upper left.

scope. Figure 5-10b shows what you see. The photograph shows three stars, two of them very close together. In fact, the two stars that are close together were seen as one star (Mizar) when viewed with the naked eye; it was the brighter of the naked-eye pair. The fainter of the naked-eye pair, Alcor, is the third star in the photograph, at the opposite side of the field of view. The telescope is able to resolve the group of stars into three.

resolving power (or **resolution**)
The smallest angular separation detectable with an instrument. Thus it is a measure of an instrument's ability to see detail.

The ***resolving power*** of an instrument is the smallest angular separation between two stars that can still be resolved as two by the instrument. Thus resolving power is described in terms of an angle.

What is it about a telescope that determines its resolving power? Naturally, the quality of the lenses is a major factor, but even with perfect optical components, the resolving power of a telescope is limited by diffraction. Since less diffraction occurs with a large objective, the maximum resolving power can be achieved by a telescope with a large-diameter objective.

Assuming that diffraction is the only limitation, a telescope's resolving power can be calculated by using the expression

$$\theta \approx 2.5 \times 10^5 \times \frac{\lambda}{D}$$

where θ = angular resolution in arcseconds

λ = wavelength of the light used, in meters

D = diameter of the telescope objective, in meters

EXAMPLE

What is the resolving power of a human eye?

SOLUTION Assuming that a fully dark-adapted human eye has a diameter of 5 millimeters = 5×10^{-3} meters, for visible light of wavelength λ = 600 nanometers = 6×10^{-7} meters, we get

$$\theta \approx 2.5 \times 10^5 \times \frac{\lambda}{D} = 2.5 \times 10^5 \times \frac{6 \times 10^{-7}}{5 \times 10^{-3}}$$

$$= 30'' = 0.5' = 1/120 \text{ of a degree}$$

TRY ONE YOURSELF

The Hubble Space Telescope has a 2.4-m primary mirror. (The expression for the resolving power of a telescope works for both refracting and reflecting telescopes.) Show that its resolving power for visible light of 600 nanometers is about 0.06" or 1/60,000 of a degree. This is about the equivalent of the angle subtended by a quarter from 80 kilometers (50 miles) away.

Based on size alone, the largest telescopes should have a resolving power far greater than the best backyard telescope, but in fact, the lack of clarity of the Earth's atmosphere becomes a major factor in limiting the resolution of large telescopes. This lack of clarity is caused by two factors: turbulence of the air and air pollution—the latter due to modern civilization or simply to dust. Recall that light is refracted as it passes from one material into another if there is a difference in its speed in the two materials. In fact, light travels at slightly different speeds in air at different temperatures and densities. Our atmosphere always contains some amount of turbulence, and this causes air at various temperatures to move across the line of sight of a telescope. This results in the image moving slightly in nearly random directions, causing the image of a point source to become blurred. This effect places a limit on the resolution of even the largest telescope. The largest telescopes on Earth have a practical resolving power of about 0.5 arcsecond, improving to about 0.25" on the best nights. As will be discussed later, the limit the atmosphere places on resolving power is the primary reason that the Hubble Space Telescope was built and put into orbit.

This atmospheric turbulence is what causes the *twinkling* of the stars when they are viewed with the naked eye. Most planets do not seem to twinkle because their angular size is larger than the scale of the atmospheric turbulence and the distortions are averaged out over the size of the image.

5-4 The Reflecting Telescope

An inwardly curved mirror will bring rays of light to a focus, just like a lens, as **Figure 5-11** illustrates. This allows us to use such a mirror as the objective of a telescope. **Figure 5-12** shows the arrangement devised by Isaac Newton. A small flat mirror is arranged in front of the objective mirror to deflect the light rays out to the eyepiece or camera body.

The largest refractor in existence is the 40-inch diameter telescope at Yerkes Observatory at Williams Bay, Wisconsin (**Figure 5-13**). Reflecting telescopes are made much larger than this, however, and they are less expensive. Astronomers prefer to build and use reflecting telescopes instead of refracting ones for the following reasons.

An inwardly curved mirror is said to be *concave,* so the objective mirror of a telescope is concave. A mirror with an outward curvature—such as the passenger-side mirror on many cars—is said to be *convex.*

1. In order to be achromatic, a refractor requires two lenses. This means that four surfaces of glass have to be shaped correctly. A front-surface mirror, on the other hand, has only one critical surface. Since it is extremely important to obtain perfectly shaped surfaces, limiting the number of surfaces greatly simplifies the construction of the objective. How "perfect" must the surface be? We typically want to keep any deviations from the desired shape of the surface to be less than about $\lambda/20$. For example, if we are observing in the visible part of the spectrum and set $\lambda = 500$ nanometers, any deviations must be kept less than 25 nanometers. Such a small deviation for, say, a 10-meter diameter surface corresponds to requiring that the largest mountain—deviation from a flat surface—on Earth be smaller than about 5 centimeters (2 inches) in height.

2. It is impossible to correct lenses completely for chromatic aberration. When light reflects from a mirror, however, all wavelengths reflect in exactly the same way; thus reflectors automatically eliminate chromatic aberration problems.

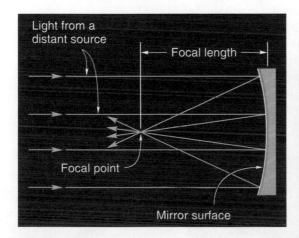

 Figure 5-11
A curved mirror can bring incoming light rays to a focus. Again, the focal point is defined as the point where incoming rays that are parallel to the axis of the mirror converge.

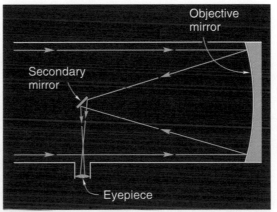

Figure 5-12
The Newtonian focal arrangement places a small flat mirror in the path of the reflected rays so that they are bounced off to the side and into the eyepiece (or camera or other instrument).

Newtonian focus The optical
arrangement of a reflecting telescope in which a plane secondary
mirror is mounted along the axis of
the telescope in order to intercept
the light reflected from the objective mirror and reflect it to the side.

Cassegrain focus The optical
arrangement of a reflecting telescope in which a convex secondary
mirror is mounted so as to intercept the light reflected from the objective mirror and reflect the light
back through a hole in the center
of the primary.

prime focus The point in a telescope where the light from the objective is focused. This is the focal
point of the objective.

Figure 5-14
(a) In the Cassegrain focal arrangement,
the secondary mirror is curved outward
(convex), and the light is reflected back
through a hole in the objective or primary mirror. (b) This is a 5-inch diameter
Cassegrain telescope.

3. Since a reflector's mirror is front-surfaced,
   the light does not pass through the glass of
   the mirror. Thus the glass does not need to
   be as perfect as that used for a refractor. It is
   difficult (and therefore expensive) to make
   large pieces of glass without tiny air bubbles
   or other imperfections. In addition, a fraction of the light going through the lens of a
   refracting telescope gets absorbed, and this is
   especially true for short-wavelength light.
   Reflecting telescopes do not have this problem, since light simply reflects off the mirror.

4. For a refracting telescope, light must pass
   through the objective lens; thus the lens
   must be supported from its edges. However,
   as the size of the lens is increased in order
   to collect more light, its weight increases
   and gravity will deform its shape. In addition, this deformation will change with the
   orientation of the telescope. For a reflecting
   telescope, we can support the mirror's
   weight by a support structure, in a honey-
   comb form, behind the reflecting surface.
   This also allows the use of a computer-
   controlled active system of pressure pads that can slightly change the mirror's
   shape to accommodate for distortions caused by gravity or the atmosphere.

For all of these reasons, and others, a large reflector is much less expensive and
more practical than a large refractor. As a result, all really large telescopes are reflectors. However, even reflectors have some image distortions.

Large Optical Telescopes

A reflecting telescope with an eyepiece arrangement like that in Figure 5-12 is called
a *Newtonian telescope,* or is said to have a **Newtonian focus.** The Newtonian focus is a
common one for small telescopes. **Figure 5-14** shows an arrangement common in large
telescopes, the **Cassegrain focus** (invented by G. Cassegrain, a French optician who
lived at the time of Newton). Notice that the eyepiece or camera body is at the back
of the telescope. In this arrangement, the secondary mirror is not a flat mirror but has
an outward curvature. The effect of the curved secondary mirror is that an objective
that actually has a short focal length can be given a longer effective focal length and
thus can be contained in a short telescope.

In very large telescopes, observing is often done at the **prime focus.** This is the
point where the light from the objective mirror comes to a focus, the focal point.

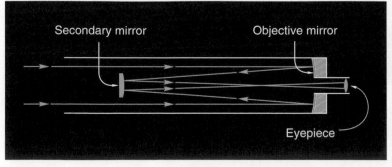

(a)   (b)

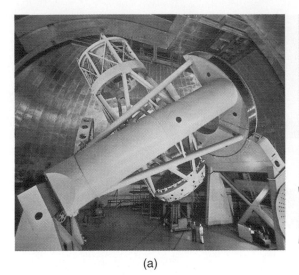

(a)

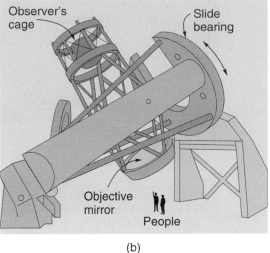

(b)

Observer's cage

Slide bearing

Objective mirror

People

Figure 5-15
The Hale telescope on Mount Palomar. To appreciate the size of the telescope, look for the people at the bottom.

And now our hundred-inch . . . I hardly dare to think what this new muzzle of ours may find. Come up, and spend that night among the stars.

—*British poet Alfred Noyes, in* Watchers of the Skies, *written during the first observations with the 100-inch Mt. Wilson telescope, Nov. 1, 1917.*

Figure 5-15 shows the 200-inch (5-meter) Hale telescope on Mount Palomar. The observer's "cage" is at the prime focus of the instrument, located at the top end of the telescope. The astronomer can sit in the cage and be carried around with the telescope as it moves. The observer's cage does block some light to the objective mirror, but only a small fraction of it.

Losing some light because of reflection from the objective mirror back along the direction of incoming light is a drawback of reflecting telescopes. However, if the secondary mirror or the observer's cage is small enough, only a fraction of the incoming light is lost. Consider the case of a 5-meter objective mirror and a 1.5-meter cage. The amount of light received is proportional to the area of the objective mirror, while we lose an amount of light proportional to the area of the cage. Both areas are proportional to the squares of their diameters and thus we lose about $1.5^2/5^2$ or about 10% of the incoming light.

Figure 5-16 shows another design, the Coudé telescope. This is a useful design for the case when large and heavy equipment, such as a spectroscope, is used for observing through the telescope. The equipment is set at the focal point, which is outside the main telescope tube, and thus does not exert a weight on the telescope that must be compensated for. This design also increases the focal length substantially, which can be very useful for high-resolution observations.

To achieve the best viewing conditions, it is advantageous to locate telescopes high in the mountains in dry, clear climates (**Figure 5-17a**). This greatly reduces the blurring effects of the atmosphere. One technique of partially compensating for atmospheric distortions while observing several objects is to simultaneously observe a star of known size. Looking at how the atmosphere distorts the image of the known star, we can correct the images of the other objects. Another technique involves the use of a powerful laser beam to create an artificial star in the atmosphere. The distortions in this "star's" image by the

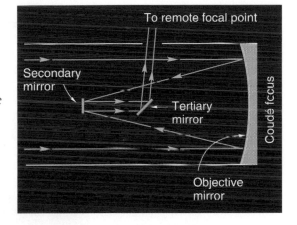

To remote focal point

Secondary mirror

Tertiary mirror

Objective mirror

Coudé focus

Figure 5-16
In the Coudé design, light reflects off three mirrors before it exits the telescope.

(a)

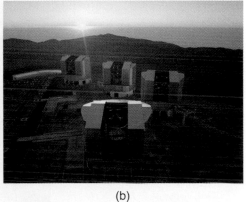

(b)

Figure 5-17
(a) Fifteen major telescopes of the European Southern Observatory cover the summit of this mountain in the Chilean Andes. (b) The four telescopes of the Very Large Telescope, part of the European Southern Observatory in Paranal, Chile.

Figure 5-18
The telescope domes of the Mauna Kea observatory are spread over the summit of Mauna Kea in Hawaii. The large dome in the foreground houses the 10-meter Keck Telescope Facility.

atmosphere can be canceled out by slightly changing the shape of the mirror, as we discuss below.

New telescope technology has produced telescopes larger than the Hale telescope. The problem with making larger mirrors is that a large mirror tends to bend and sag when the telescope moves. The resulting change in the mirror's surface destroys image quality, of course. The Tools of Astronomy box on page 148 describes how a laboratory at the University of Arizona is solving this problem.

A different modern mirror design does not attempt to *prevent* bending, but relies instead on an active optics system that can adjust the mirror's shape. The mirrors of the four 8.2-meter telescopes of the European Southern Observatory's Very Large Telescope (VLT) are designed to be flexible. The image they produce is constantly monitored by a computer. To keep the image sharp under different conditions, the computer controls about 200 actuators that push and pull on the back of each mirror to adjust its shape. This technique is called ***adaptive optics.*** The four telescopes (Figure 5-17b) can work independently or, by 2006, in combined mode, in which case the VLT provides the total light-collecting power of a 16.4-meter single telescope (since $4 \times 8.2^2 = 16.4^2$). The useful wavelength range extends from the near UV up to 25 μm in the infrared.

adaptive optics A technique that relies on an active optics system that monitors and changes the shape of a telescope's objective to produce the best image.

Figure 5-19
The 36 mirrors of the Keck telescope are reflecting the underside of the prime focus cage, where the 1.4-meter secondary mirror is mounted. For scale, note the astronomers on either side of the objective mirror. Each of the 36 hexagonal mirrors is 1.8 m across. The hole at the center of the objective is for the Cassegrain focus.

Meanwhile, a different type of adaptive optics telescope began operating in 1993 on the 4200-meter (13,800-feet) summit of Mauna Kea in Hawaii (**Figure 5-18**). At this remote location, high above the clouds where few lights pollute the atmosphere, the sky is clear, calm, and dry about 300 nights a year. **Figure 5-19** shows the multiple mirrors of the Keck I telescope and **Figure 5-20** is an unusual photo of the telescope inside its dome. The telescope has a 10-meter aperture consisting of 36 hexagonal mirrors, each weighing 880 pounds and each mounted separately and controlled by computer. Keck II, the second of the twin Keck telescopes, was completed in 1996.

Figure 5-20
This "X-ray" of the dome of the Keck telescope was produced by taking a time exposure photo while the open dome rotated. The telescope is pointing straight up, and the objective mirrors are just below the "transparent" dome. The prime focus cage is visible at the top center of the frame.

An innovative modern telescope design is used in the Hobby-Eberly Telescope (HET), put into operation at McDonald Observatory in 1998. Although its primary mirror is one of the largest in the world, using 91 one-meter hexagonal segments to form an array with an effective aperture of 9.2 meters, it does not move the entire mirror to track objects across the sky. **Figure 5-21** illustrates how it works. Another innovative design is used in the Large Binocular Telescope (LBT) at Mount Graham, Arizona (**Figure 5-22**). The telescope consists of two 8.4-meter mirrors on a common mount, making it equivalent in light-gathering power to a single 11.8-meter mirror. Because of its binocular design, the BLT will have a resolving power that corresponds to a 22.8-meter telescope.

Telescope Accessories

Telescopes have many other astronomical uses besides obtaining images of celestial objects. A few of these applications are described here. Each involves accessories that are used with the telescope.

- As described in an earlier section, a camera can be attached to a telescope to take photos. The same basic principle is used in large professional telescopes, although the equipment and methods are more sophisticated. Photographic film is not a very efficient detector of light, however, for only about 5% of the light hitting the film causes the chemi-

Figure 5-21
The 9.2-meter-diameter objective of the Hobby-Eberly Telescope focuses light on a moving collector. (The collector is the device that spans across the hexagonal support at the upper left of the mirror.) Moving only the collector reduces by more than 10 times the amount of telescope mass that must be moved under precise control. The telescope's schedule is computer-controlled to allow many different types of observations to be made in a single night.

TOOLS OF ASTRONOMY

Spinning a Giant Mirror

The 5-meter (200-inch) mirror for the Hale Telescope was cast at the Corning Glass Works in 1934. It was nearly 60 years before a mirror this large was made again in the United States. Then, in 1992, the Steward Observatory Mirror Laboratory transformed 10 tons of glass into a 6.5-meter mirror blank.

The Steward Laboratory is making mirrors in a radically different way than ever before. In the past, mirrors were made by melting glass to form large pieces with flat surfaces. Then, after the glass hardened, the working surface of the glass was carefully ground into a curve to form a concave mirror surface. For the 6.5-meter mirror, that would have meant starting with more than twice the weight of the finished mirror, with 12 tons of glass then to be ground away, taking a year of extra work. Under the stands of the University of Arizona football stadium, technicians of the Steward Laboratory form the mirror surface by spinning the mirror as it cools (**Figure B5-1**).

Rotating the furnace—a technique called spincasting—saves time and money in the production of large telescope mirrors. It shapes a natural curve in the surface of the molten glass in the same way that swirling a liquid in a glass causes a curved surface. The curvature created this way is close to the mirror's final parabolic shape, the rotation speed determining the amount of curvature. To create the 6.5-meter mirror, the furnace was spun at 7.4 revolutions per minute. This may seem slow, but remember that the mirror was 6.5 meters—21 feet—across.

The new mirrors use much thinner glass than in the Hale telescope mirror, making them much lighter. The 6.5-meter mirror has 64% more light-gathering area than the Hale telescope mirror, but weighs half as much. The glass facesheet of the mirror surface averages only a little more than an inch thick; strength is provided by a honey-comb structure cast all in one piece with the facesheet and extending back nearly 30 inches at the edge of the mirror and half that at the center.

Another advantage to spin-casting is a deeply curved, parabolic surface with a focal length much shorter than conventional

Figure B5-1
In the spin-casting technique, molten glass is placed in a rotating furnace to form a mirror with a curved upper surface. Increasing the rotation rate increases the amount of curvature.

mirrors. The shorter focal length means that the mirrors require a much shorter, less expensive enclosure.

The first 6.5-meter Steward Observatory mirrors have been finished and installed at the Multiple Mirror telescope on Mount Hopkins in southern Arizona and at the Magellan Project telescope on Las Campanas, Chile. Two 8.4-meter mirrors have also been cast and are now being polished for the Large Binocular Telescope on Mt. Graham, Arizona.

---

charge-coupled device (CCD) A small semiconductor chip that serves as a light detector by emitting electrons when it is struck by light. A computer uses the pattern of electron emission to form images.

cal reaction that results in an image. For this reason, astronomers often use various electronic light detectors, particularly the *charge-coupled device* (CCD). This device, about the size of a postage stamp (**Figure 5-23a**), is divided into small squares, each capable of detecting the intensity of the light that hits it. Just as a black-and-white newspaper photo (or the screen of an electronic game) is made up of many individual pixels, a CCD may have more than 4 million pixels. Figure 5-23b is a highly magnified CCD image in which individual pixels are visible.

The intensity of light collected in each pixel of a CCD is stored as a number in a computer. The data can then be used to show what the object would "look like" in a regular photograph, or they can be used to produce a false color image that reveals some other aspect of the object (Figure 5-23c). Remember that a false

The devices used in digital cameras are similar to CCDs.

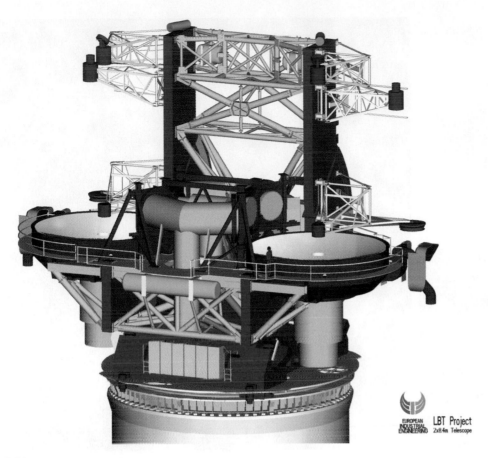

Figure 5-22
This is a drawing of the LBT. Each of the two 8.4-meter primary mirrors weighs about 16 metric tons. First light for the first primary Is expected in 2004 and for the second primary in 2005.

color image does not tell us what the object actually looks like; instead it reveals some other property of the object. For example, a false color image may illustrate the intensity of radiation from the object by showing each brightness level as a different color. Most modern astronomical "photos" are actually CCD images, not film-based photographs.

CCDs have revolutionized the way astronomers count photons. These devices not only can count almost all the incident photons but also can detect a wide range of wavelengths. When viewing objects of very different intensities simultaneously, CCDs can differentiate between them. Finally, CCDs have a linear response; that is, the signal increases linearly with the number of incident photons.

- One important measurement made of celestial objects is the intensity of light that is received at various wavelengths, a procedure called *photometry.* In the past this was done with a device similar to the light meter of a camera, but today a CCD is normally used. The measurement can be made by placing filters in front of the light detector that allow only the wavelength of interest to pass through.

- In Chapter 4 we pointed out that a tremendous amount of information about celestial objects is obtained by spectral analysis—the examination of light that has been separated into its various wavelengths. To obtain data for such an analysis, a *spectrometer* is connected to a telescope. This instrument uses a prism—or, more commonly, a *diffraction grating*—to separate light into its colors. A spectrometer produces either a photograph of the spectrum or numerical data about the intensity of light at various wavelengths. The data can then be converted into a graph of intensity versus wavelength as shown in **Figure 5-24.**

photometry The measurement of light intensity from a source, either the total intensity or the intensity at each of various wavelengths.

spectrometer An instrument that measures the wavelengths present in electromagnetic radiation. (A **spectrograph** is a spectrometer that produces a photograph of the spectrum.)

diffraction grating A device that uses the wave properties of electromagnetic radiation to separate the radiation into its various wavelengths.

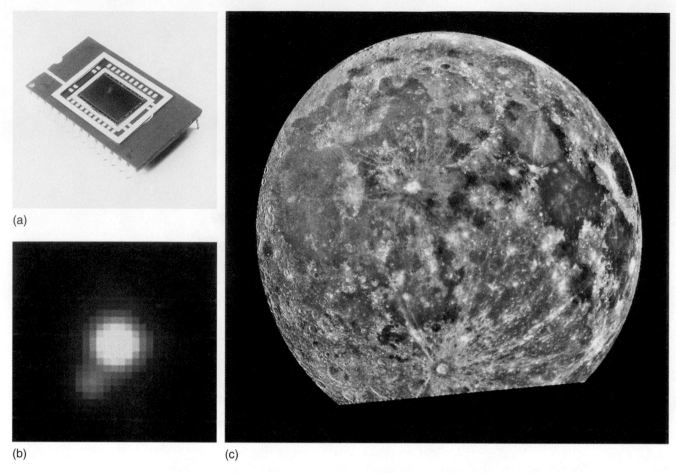

(a)

(b)

(c)

Figure 5-23
(a) A CCD (charge-coupled device) is a rectangle of the semiconductor silicon. This one contains nearly 164,000 electric circuits that detect the intensity of light striking them. A computer analyzes the resulting data and produces an image. (b) This CCD image has been magnified so much that the individual pixels are obvious. (c) This false color image of the Moon not only emphasizes slight natural color differences, but compresses the spectrum from the ultraviolet to the near infrared so that it all shows as visible. The image was made from images from the spacecraft *Galileo*.

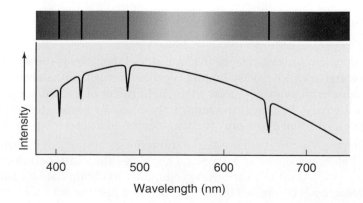

Figure 5-24
This drawing shows two different ways of representing a spectrum. The dark absorption lines in the photograph correspond to the dips in the intensity versus wavelength graph.

5-5 Radio Telescopes

Thus far the discussion has concentrated on optical telescopes—telescopes that gather visible light. Besides visible light, the type of radiation from space that best penetrates the atmosphere is radio waves. In 1931 a scientist working on radio transmission for Bell Laboratories noticed that static received by his antenna originated in the Milky Way. (See the Advancing the Model box on page 153.) When better radio receivers were designed (largely during World War II), astronomers were able to pinpoint the sources of celestial radio waves, and the field of radio astronomy was born.

Two problems arise in examining radio waves from space. First, the intensity of radio waves from a star is much less than the intensity of visible light. Second, since the wavelengths of radio waves are a million times greater than the wavelengths of visible light, there is a corresponding decrease in the resolution of images made with radio waves. (Recall that diffraction is greater with longer wavelengths.)

Both of these problems are solved in the same way—by making radio telescopes extremely large. **Figure 5-25** shows the 140-foot radio telescope at the National Radio Astronomy Observatory at Green Bank, West Virginia. Notice that the telescope does not have a shiny reflecting surface. This is possible because of a feature of waves that goes hand-in-hand with the diffraction effect: Although longer wavelengths diffract more when going through an opening, they do not require as smooth a surface for reflection.

Radio telescopes are similar in principle to the satellite dishes we use to receive television signals from Earth satellites. In each case the reflector simply directs waves to a small detector (antenna) located at the reflector's focal point. You can see the supports for the antenna in the photograph of the radio telescope.

The spherical mirror of the world's largest radio telescope, located in Arecibo, Puerto Rico, is fixed on the ground. However, its various antennas, suspended on a track above the mirror, enable it to receive radio emissions from different parts of the sky. Shown in **Figure 5-26**, it is a telescope constructed by perforated aluminum panels supported by steel cables stretched across a natural bowl between hills. This telescope is 300 meters (1000 feet) in diameter and scans the sky as the telescope moves along with the Earth. Changes in the direction from which it detects radio signals can be achieved by moving its antenna, which hangs from the cables suspended above the bowl.

Figure 5-25
The 140-foot (43-meter) radio telescope at the National Radio Astronomy Observatory near Green Bank, West Virginia. It was closed as a user facility in 1999.

(a)

(b)

Figure 5-26
(a) The radio telescope near Arecibo, Puerto Rico, is the world's largest. Its radio detectors are part of a 900 ton platform suspended above the reflecting surface. (b) This photo was taken under the Arecibo antenna.

Figure 5-27
(a) A typical graph made by one scan of a radio telescope across a source. (b) A contour map can be made from a number of such scans. The strength of the radio signals is greatest in the center. (c) Color has been added to the contour to produce a false color image.

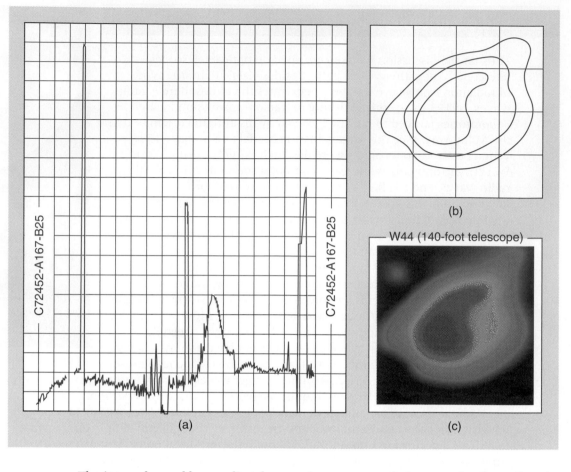

C72452-A167-B25

C72452-A167-B25

(a)

(b)

W44 (140-foot telescope)

(c)

Figure 5-28
(a) The galaxy nearest to the Milky Way is the Andromeda Galaxy, shown here in visible light. (b) A radio image of the Andromeda Galaxy shows that most radio waves are emitted from its spiral arms and from its center.

(a)

The image formed by a radio telescope is not a normal photograph. The radio telescope simply detects the intensity of radio signals from the area of the sky toward which it is pointed. One way to display and examine the data received is to plot a graph of the intensity of the radiation as the telescope moves across a small portion of the sky. **Figure 5-27a** shows what such a plot might look like. A more complete image of the radio-emitting object can be obtained by scanning the radio telescope back and forth across the celestial object and feeding the data into a computer that is programmed to represent the various intensities of the radio waves as different colors.

Parts (b) and (c) of Figure 5-27 were produced in this manner and indicate the intensity of radio waves from a small portion of the sky. **Figure 5-28** illustrates the different appearance of the same object (a galaxy) when seen in visible light and in radio waves.

At Green Bank, West Virginia, the first of a modern generation of radio telescopes is now operational. In a regular radio telescope, the detector and its

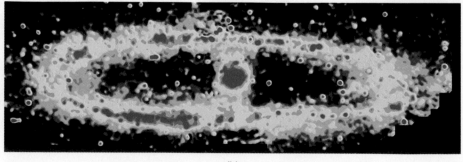

(b)

ADVANCING THE MODEL

Radio Waves from Space

In 1927, Karl Jansky received his B.A. in physics from the University of Wisconsin. After one year of graduate study, he was hired by Bell Laboratories to do research on radio communications. By this time, the first telephone service across the Atlantic had opened, and the telephone signal was sent by radio. (The service cost $75 for three minutes!) Static was a common problem, however, and Jansky was assigned the task of designing and building an antenna to find the source(s) of the static (**Figure B5-2**). He had no specific experience in radio or electrical engineering, but after much study and some dead ends, he built a 100-foot-long antenna with which he was able to detect weak radio signals and to determine the direction from which they came.

In January 1932, Jansky wrote in his monthly report that he detected " . . . a very steady continuous interference—the term 'static' doesn't quite fit it. It goes around the compass in 24 hours. During December this varying direction followed the sun." As the early months of 1932 passed, he found that the direction from which the signals came seemed to move around, getting farther from the Sun. He anticipated that after the summer solstice, the apparent source would move closer to the Sun again, but instead it continued to move around the sky during the year. By August 1932, Jansky decided that the static was coming from a fixed place among the stars; it was some type of "star static."

Jansky's star static was the first detection of radio waves from space. The source of the waves was the center of the Milky Way Galaxy. (The *New Yorker* magazine said that "This is believed to be the longest distance anybody ever went to look for trouble.") Jansky had been slow to recognize the celestial nature of the source in part because he did not know astronomy

well. Astronomers, in turn, were slow to recognize the significance of his work because they were unfamiliar with electronics and radio, and they did not imagine that celestial objects emit radio waves. Radio astronomy did not grow quickly after Jansky went on to other things, but today it is a major branch of astronomy and provides us with information about the heavens that could not be learned in other ways.

Figure B5-2
Jansky's "merry-go-round" antenna could rotate to pinpoint the source of the radio signal.

---

supports not only block out a small portion of the waves, but they cause diffraction effects. The detector of the new Green Bank Telescope (GBT) is located off to the side (**Figure 5-29**). This is done by constructing the reflecting surface in the shape of part of a much larger reflector, as shown in **Figure 5-30**. The surface of the GBT is 100 by 110 meters and consists of 2004 separate panels, each of which is computer controlled. This adaptive capacity, the first use of adaptive optics for radio telescopes, permits the surface to be adjusted as the metal supports flex due to the telescope's motion. The resulting surface accuracy allows the GBT to be useful at shorter wavelengths than would otherwise be possible.

Figure 5-29
The Green Bank Telescope (GBT) is the first radio telescope to take advantage of adaptive optics. A computer controls each of the panels making up the surface of the reflecting dish. The GBT is the world's largest fully steerable radio telescope. The weight of the moving part of the telescope is 16 million pounds (7300 metric tons).

Spark plugs in cars with internal-combustion engines produce radio interference. Therefore only vehicles with diesel engines are used near radio telescopes.

Figure 5-30
(a) The reflecting surface of the Green Bank Telescope can be thought of as part of a much larger reflector. (b) The size of the Green Bank Telescope compared to the Statue of Liberty. The telescope is drawn here in an orientation to receive radio waves from straight above.

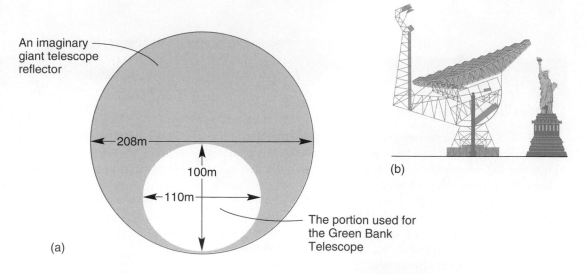

An imaginary giant telescope reflector

208m

100m

110m

The portion used for the Green Bank Telescope

(a)

(b)

5-6 Interferometry

Recall that the resolution formula is

$$\theta \approx 2.5 \times 10^5 \times \frac{\lambda}{D}.$$

Recall that the resolution of a telescope depends upon both the diameter of the telescope and the wavelength of the radiation. The greater the diameter of the telescope, the better the resolution (assuming atmospheric clarity is not a factor), but the greater the wavelength, the poorer the resolution. Even though radio telescopes are very large, radio waves are so long that the best resolution from a single radio telescope is on the order of a number of arcminutes. This means that a radio source the size of a star would still appear in a radio telescope to be a blur as large as half the diameter of the Moon.

The solution to this problem lies in the fact that a giant radio telescope would retain the same resolution if only two portions of its outer surface were being used, as shown in **Figure 5-31a**. Astronomers take advantage of this idea by combining two radio telescopes so that they act as one: in a sense, the two telescopes substitute for two portions of the outer part of a giant telescope, as in Figure 5-31b. In this way, they are able to obtain resolutions equal to that of a single large telescope.

Using two telescopes to act as one is not a simple matter. To understand the problem, refer to **Figure 5-32**. Radio waves from a distant source are shown striking

Figure 5-31
Two small radio dishes (b) can be made to have the same resolution as a large radio telescope (a) that has a diameter equal to the distance between the two small ones. The strength of the signals detected will be much greater in the case of the single large telescope.

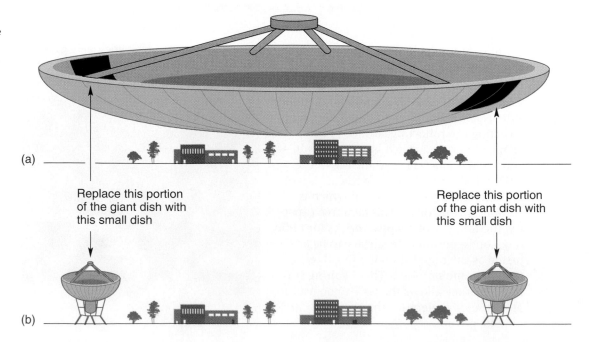

(a)

Replace this portion of the giant dish with this small dish

Replace this portion of the giant dish with this small dish

(b)

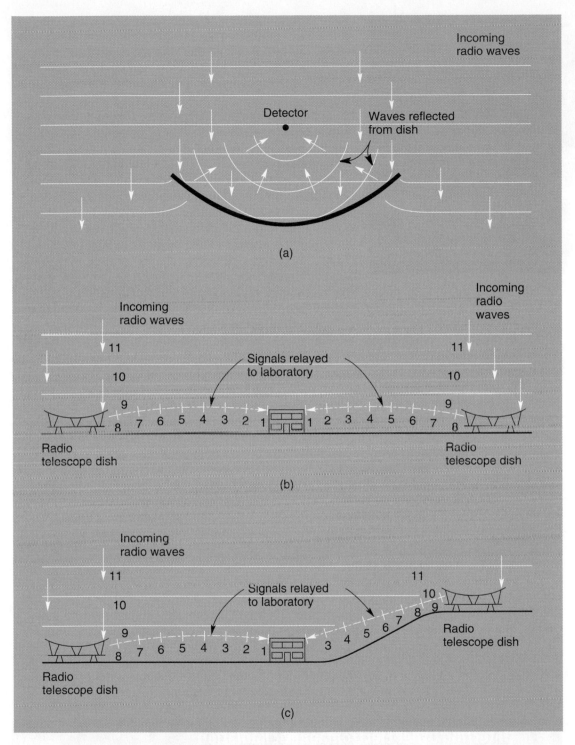

(a)

(b)

(c)

Figure 5-32

(a) All portions of an incoming wave reach the detector of a radio telescope at the same time. (b) If two radio telescope dishes are to function as a single telescope, the waves must likewise reach the detector at the same time, or at least the time difference must be corrected for. (c) This is an obvious (and oversimplified) case of waves whose difference in reception times must be accounted for in the laboratory.

the dish of a radio telescope. Notice that they are reflected so that a single wave gets to the detector (located at the focal point of the dish) at the same time from all areas of the dish. This feature must be retained when two radio telescopes are used as one. Waves from each single dish must be combined in the correct relationship. We say that the waves from the two telescopes "interfere" with one another when they combine; therefore the technique of linking two (or more) telescopes so that they act as one is called ***interferometry.*** Due to today's extremely accurate atomic clocks, interferometry using widely separated radio telescope dishes has become possible.

interferometry A procedure that allows several telescopes to be used as one by taking into account the time at which individual waves from an object strike each telescope.

Figure 5-33
The Very Large Array is spread over 15 miles of the New Mexico desert.

In the New Mexico desert is an array of radio telescopes used for observations by interferometry. **Figure 5-33** is a photograph of part of this Very Large Array. The telescopes ride on a double pair of railroad tracks arranged in a Y shape, so that they can be moved and the arrangements changed.

The farther apart the telescopes, the better the resolution that can be obtained by interferometry. To achieve a longer baseline—distance between telescopes—some telescopes across the United States are being used as part of a single array, the Very Long Baseline Array (VLBA). The VLBA consists of 10 radio antennas, each 25 meters in diameter. The sites include Hawaii, New Hampshire, Washington, and the Virgin Islands, resulting in a baseline of 8600 km. The control center is in Socorro, New Mexico. Very accurate atomic clocks are used to coordinate the signals received by such distant radio telescopes. The VLBA achieves resolutions of fractions of a milliarcsecond, 10,000 times better than earth-bound optical telescopes. An angle of one milliarcsecond is less than the angular diameter of a dime at 1000 miles!

In 1997 the Japanese Space Agency launched a radio satellite as part of the VLBI (Very Long Baseline Interferometry) program. The Japanese *HALCA* satellite is in an elliptical orbit that, at its apogee, provides a baseline three times longer than those achievable on Earth (about 30,000 km). This satellite, as well as others in the series, transmit their signals down to four Earth stations as they orbit.

Interferometry is also being employed in the newest optical telescopes. Recall that the VLT of the European Southern Observatory is actually four separate telescopes, rather than mirrors that focus light at the same point. To use the four telescopes as one, the signals from each must be combined using techniques similar to those originally developed for radio telescopes. Because the wavelength of visible light is much shorter than radio waves, matching waves from different sources is more critical for visible light waves. Therefore, optical interferometry is more difficult than radio interferometry. Using the four 8.2-meter telescopes together with 1.8-meter auxiliary telescopes, the VLT can reach an angular resolution of less than a milliarcsecond in the visible.

The Center for High Angular Resolution Astronomy (CHARA) at Georgia State University has built an optical stellar interferometer, an array of six telescopes of only 1-meter aperture each. They are placed in a Y-shaped array contained within a 400-meter diameter circle on Mount Wilson, California. Because each telescope is relatively small, this system costs much less than most new telescopes. It can produce a limiting resolution of 0.2 milliarcseconds in the visible, better than any other telescope of its kind on Earth's surface.

5-7 Detecting Other Electromagnetic Radiation

Visible light and radio waves pass through our atmosphere, and these two portions of the electromagnetic spectrum were the first that astronomers used in their quest to understand the heavens. (Examine again Figure 4-6, which shows atmospheric absorption at different wavelengths.) But celestial objects emit radiation over the entire range from radio to gamma rays, and modern astronomy has tools that study each region of the spectrum.

In the range of wavelengths from about 1200 nanometers to 40,000 nanometers (called the *near infrared*) are several narrow wavelength regions whose radiation pen-

◆ These wavelengths are called "near infrared" because they are the part of the infrared region that is near the visible portion of the spectrum.

etrates to the surface of the Earth. Since water vapor is the chief absorber of radiation in the infrared, infrared observatories are located on mountains where the air is dry. The extinct volcano Mauna Kea is an ideal location for infrared telescopes, and two major observatories are located there.

Radiation in the far infrared region—wavelengths greater than about 40,000 nanometers—is emitted by cooler celestial objects, such as planets and newly forming stars, and is very valuable in the study of these objects (**Figure 5-34**). Infrared does not penetrate the atmosphere as deeply as shorter wavelengths, and we must therefore locate our instruments higher to detect it. The SOFIA project (Stratospheric Observatory For Infrared Astronomy) involves a modified Boeing 747-SP aircraft carrying a 2.5-meter reflecting telescope at altitudes as high as 12,000 meters (41,000 feet), above 99% of the atmosphere's water vapor.

As we move from infrared to wavelengths shorter than visible—shorter than about 400 nanometers—we come upon ultraviolet, X-rays, and gamma rays. Ozone is the chief absorber of most of this radiation and our atmosphere has a layer of ozone at altitudes between about 20 and 40 kilometers. Therefore, telescopes designed to detect this range of radiation must be located in space.

SOFIA replaces the Kuiper Airborne Observatory, which observed in the infrared from 1974 to 1995. SOFIA is a joint NASA/German Space Agency effort and is expected to begin flying in 2004 and have a 20-year lifetime.

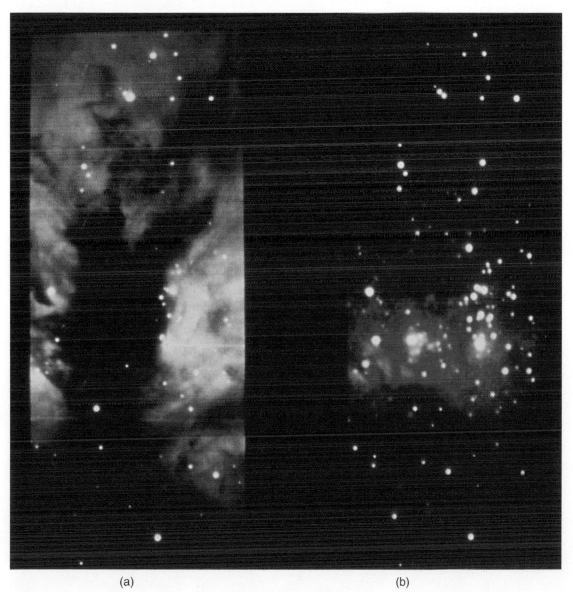

(a)   (b)

Figure 5-34
Two views of a region in the constellation Orion. (a) The left image is what an optical (visible-light) telescope reveals. (b) The photo is an infrared image of the same part of the sky.

The name "Chandra" was the nickname of the Indian physicist S. Chandrasekhar, who shared a Nobel Prize for his theoretical work on the structure and evolution of stars.

The *Chandra X-ray Observatory* was deployed in July 1999 and started operating in August 1999. It is one of the most sophisticated observatories ever built and has an unusual elliptical orbit that takes the spacecraft more than a third of the way to the Moon. Its closest approach to Earth is 10,000 km. *Chandra* is designed to observe X-rays from high-energy regions of the universe, such as hot gas in the remnants of exploded stars. It provides X-ray images that are 25 times sharper than previously obtained. The chapter-opening photograph is an example of how X-ray images provide different information than visible-light images of the same object.

The shortest wavelength in the electromagnetic spectrum is gamma radiation, which was the target of the *Compton Observatory*. Since its 1991 launch, and up to its deorbiting in 2000, the *Compton* project has provided high-quality data on subjects ranging from solar flares to supernova explosions to accreting black holes of stellar mass. The *Compton Observatory* was designed to map the gamma ray sky in order to solve some of the outstanding questions that earlier missions had posed and—perhaps most importantly—to watch for the unexpected. One of its most important findings was that gamma-ray burst sources are distributed in a manner consistent with a cosmological population.

The *Integral* satellite (launched in 2002) and the *XMM-Newton* (launched in 1999) are being operated by the European Space Agency and cover the gamma-ray and X-ray parts of the spectrum, respectively.

5-8 The Hubble Space Telescope

The *Hubble Space Telescope (HST)* is mainly an optical telescope but can observe across the spectrum from the near infrared to the near ultraviolet regions (115–2500 nm). The *HST* (**Figure 5-35**) has a modular design so that on subsequent shuttle missions astronauts can replace faulty or obsolete parts with new and/or improved instruments. This was fortunate, because soon after its 1990 launch, astronomers discovered that the objective mirror of the *HST* was slightly misshaped. The telescope has performed flawlessly after a repair mission in late 1993, during which corrective optics packages were installed. Another mission in late 1999 replaced, among other things, all six gyroscopes.

The *HST* has a 2.4-meter primary mirror and is roughly cylindrical in shape, 13.1 meter end-to-end and 4.3 meters in diameter at its widest point. Radiation enters the telescope through an opening below the open door at upper left in Figure 5-34. The following instruments are aboard the spacecraft:

- The Wide Field/Planetary Camera 2 is actually four cameras. The "heart" of WF/PC2 consists of an L-shaped trio of wide-field sensors and a smaller, high-resolution ("planetary") camera tucked in the square's remaining corner. This camera is the instrument that produces many of the beautiful astronomical images that you see in newspapers and magazines, as well as in this book. The "2" in the name of this camera is there because this instrument replaced an earlier version in the *HST*. The Advanced Camera for Surveys (ACS) is an upgrade to the WF/PC2. It was installed in the *HST* during a March 2002 servicing mission. It doubles the telescope's field of view and collects data ten times faster than before.

- The Space Telescope Imaging Spectrograph (STIS) can study celestial objects across a spectral range from the UV (115 nanometers) to the near-IR (1000 nanometers).

Figure 5-35
The crew of the Shuttle mission STS-82 took this photo of the *Hubble Space Telescope* after they had released it.

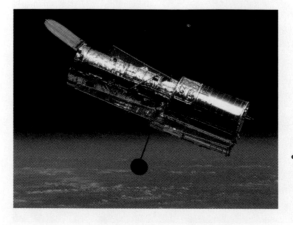

The main advance in STIS is its capability to record the spectrum of many locations in a galaxy at the same time, rather than observing one location at a time. As a result, STIS is much more efficient at obtaining scientific data than the earlier *HST* spectrographs.

- The Near Infrared Camera and Multi-Object Spectrometer (NICMOS) consists of three cameras designed for simultaneous operation. Since infrared radiation is heat radiation, NICMOS's surroundings have to be cooled to very low temperatures (as do all infrared telescopes). NICMOS, installed in the *HST* during a 1997 mission, keeps its detectors cold inside a thermally insulated container that contains frozen nitrogen ice. A new cooling system was installed during the 2002 mission to replace the block of nitrogen ice that was depleted three years earlier.

- The Faint Object Camera, built by the European Space Agency, is extremely sensitive to dim light. Consider the faintest star that the naked eye can see. For this camera to observe a star one-millionth as bright as that, the light must be dimmed by a filter system to avoid saturating the camera's detectors.

It takes about 95 minutes for the *HST* to complete one orbit around the Earth. Only part of this time is spent observing, with the remainder spent on "housekeeping" functions. These functions include receiving command loads and sending data to Earth, turning the telescope to find a new target or avoid the Sun or Moon, and similar activities. To keep the telescope operating efficiently, commands are sent to the *HST* several times a day.

The successor to the *HST* is the *James Webb Space Telescope (JWST)*. It is scheduled to launch in 2011 for a 5- to 10-year mission. Its range is from visible green light through the invisible mid-infrared (0.6 – 28 μm) and has a 6.5-m primary mirror. The *JWST* will see objects 400 times fainter than those currently observed by large infrared telescopes. Its goal is to observe the first stars and galaxies in the universe and to give us a better understanding of the dark matter problem.

The *HST* is a unique instrument, a cooperative program of the European Space Agency and NASA, and an invaluable resource for astronomers worldwide. The following are but a few of its contributions: It has uncovered evidence for the existence of super-massive black holes in space and evidence in support of the Big Bang theory, uncovered details of the processes involved in the formation of planets, stars, and galaxies, and is helping astronomers calculate the age of the universe.

Conclusion

In this chapter we showed how refraction and reflection of light allow us to gather radiation from dim stellar objects and focus it to form an image. We saw that the powers of a telescope include not only magnification, but also light-gathering power and resolving power. This analysis showed the importance of large telescopes and led to a discussion of reflecting telescopes, which can be made much larger than refractors.

Non-optical telescopes are becoming more and more important in our quest to understand the universe by permitting us to observe objects that are invisible to the eye yet emit vast quantities of electromagnetic energy. New and different telescopes—including space telescopes—have provided us with information that was impossible to obtain by other means. **Figure 5-36** clearly shows the different appearances an object—in this case our entire galaxy—can have when observed at different wavelengths. These differences provide us with an enormous amount of information that was not available from ground-based observations centered on visible light. As later chapters will show, entirely new celestial objects have been discovered in recent years by the new generation of telescopes. Undoubtedly, telescopes of the future will continue to bring us new and unexpected results and open whole new areas of exploration in astronomy.

Galileo's telescope began a revolution in astronomy nearly 400 years ago. Today the *Hubble Space Telescope* and other new telescopes are producing a comparable revolution. We do indeed live in exciting times.

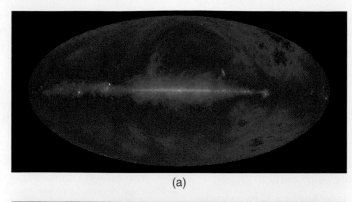
(a)

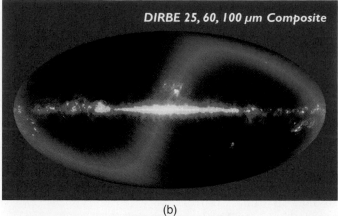
(b)

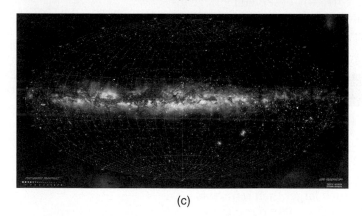
(c)

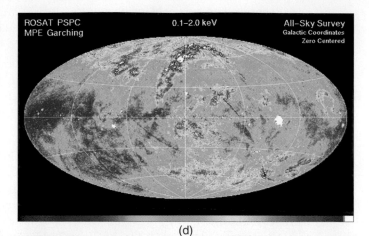
(d)

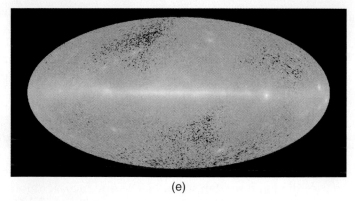
(e)

Figure 5-36
Our Galaxy as observed in (a) radio, (b) infrared, (c) visible, (d) X-rays, and (e) gamma rays.

Study Guide

1. The best site for an optical telescope is a place where the air is
   A. thin and dry.
   B. thick and dry.
   C. thin and moist.
   D. thick and moist.

2. Which of the following features determines the light-gathering power of a telescope?
   A. The diameter of the objective.
   B. The focal length of the objective.
   C. The focal length of the eyepiece.
   D. [Two of the above.]

3. Which of the following features determines the resolving power of a telescope?
   A. The diameter of the objective.
   B. The focal length of the objective.
   C. The focal length of the eyepiece.
   D. [Two of the above.]

4. Which of the following features determines the magnifying power of a telescope?
   A. The diameter of the objective.
   B. The focal length of the objective.
   C. The focal length of the eyepiece.
   D. [Two of the above.]

5. When the magnification of a telescope is increased by changing eyepieces,
   A. the apparent angular size of the object is increased.
   B. the field of view is decreased.
   C. the brightness of the object is decreased.
   D. [All of the above.]
   E. [None of the above.]

6. Which of the following telescopes has the greatest light-gathering power?

Telescope	Focal Length of Eyepiece	Focal Length of Objective	Diameter of Objective
A	24 mm	150 cm	12 cm
B	6 mm	100 cm	8 cm
C	18 mm	125 cm	20 cm
D	12 mm	90 cm	6 cm
E	12 mm	100 cm	10 cm

7. Which of the telescopes in question 6 has the greatest magnification?

8. The objective of most radio telescopes is similar to the objective mirror of a reflecting optical telescope
   A. in being concave in shape.
   B. in its approximate diameter.
   C. in being made of Pyrex glass.

9. The field of view of a telescope is
   A. the range of distance from the telescope over which it is in focus.
   B. the particular object being viewed by the telescope.
   C. the range of practical magnifying powers for the telescope.
   D. the range of wavelengths that can be detected by a particular telescope.
   E. the actual angular width of the scene viewed by the telescope.

10. Why are achromatic lenses used in optical telescopes?
    A. They reduce diffraction.
    B. They reduce color fringing.
    C. They produce greater magnification.
    D. They allow more light-gathering power.

11. Which of the following choices puts a limit to the useful magnification of a given telescope?
    A. Diffraction of light.
    B. Redshift of distant objects.
    C. The limit to how well lenses can be made.
    D. Reflection of light from parts of the telescope.

12. The resolving power of a telescope is a measure of its
    A. magnification under good conditions.
    B. overall quality.
    C. ability to distinguish details in an object.
    D. [All of the above.]

13. The light-gathering power of a telescope is determined by
    A. the telescope's magnification.
    B. the diameter of the objective of the telescope.
    C. the clarity of the sky.
    D. [All of the above.]
    E. [None of the above.]

14. Radio telescopes need not have finely polished surfaces because
    A. we are not interested in detail in the radio image.
    B. the speed of radio waves is less than that of light.
    C. the speed of radio waves is greater than that of light.
    D. radio telescopes can be used during the day.
    E. radio waves are much longer than light waves.

15. The primary purpose of a typical radio astronomer's work is to
    A. look for signals from other beings.
    B. send out radio waves to other beings.
    C. send out radio waves to be reflected back from stars and galaxies.
    D. receive radio waves sent out by radio sources.
    E. [All of the above.]

16. The eyepiece of a telescope is primarily used
    A. to collect as much light as possible.
    B. as a magnifier.
    C. as a prism to break light into its component colors.

17. A 40-inch telescope has _____ times the light-gathering power of a 10-inch telescope.
    A. 4
    B. 8
    C. 16
    D. 40
    E. [Either A, B, or C, above, depending upon the eyepiece used.]

18. The Keck telescope is
    A. a large Earth-bound optical telescope.
    B. an orbiting telescope.
    C. a single radio telescope.
    D. an array of radio telescopes.

19. The Hubble telescope is
    A. a large Earth-bound optical telescope.
    B. an orbiting optical telescope.
    C. a single radio telescope.
    D. an array of radio telescopes.
    E. an orbiting infrared telescope.

20. The Arecibo telescope is
    A. a large Earth-bound optical telescope.
    B. an orbiting optical telescope.
    C. a single radio telescope.
    D. an array of radio telescopes.
    E. an orbiting infrared telescope.

21. Interferometry is used to increase
    A. magnifying power.
    B. resolving power.
    C. light-gathering power.
    D. [All of the above.]
    E. [None of the above.]

22. Which of the following effects is reduced by using a larger telescope?
    A. diffraction
    B. refraction
    C. reflection
    D. chromatic aberration
    E. [All of the above.]

23. Define refraction, chromatic aberration, and diffraction.

24. How is an achromatic lens made? Describe the defect it corrects.

25. Define magnifying power. Why do we not always use the highest magnification available?

26. What is meant by light-gathering power, and what is it about a telescope that determines its light-gathering power?

27. Explain what is meant when we say that a certain telescope can "resolve" a particular pair of stars.

28. Sketch a Newtonian reflecting telescope, showing the relative positions of the primary mirror, the secondary mirror, the eyepiece, and the image produced by the objective.

29. About how large is the largest refracting (visible light) telescope? The largest reflecting (visible light) telescope? Explain why one type can be made larger than the other.

30. Why do we not have to correct the objectives of reflectors for chromatic aberration?

31. Why must radio telescopes be made so large? About how large is the largest?

32. Why are most research telescopes located on mountains?

33. What is interferometry, and what is its advantage?

34. What is the primary advantage of locating a telescope in space?

35. What is spectroscopy?

QUESTIONS TO PONDER

1. Define and distinguish between reflection and refraction, and explain how each phenomenon is used in telescopes.

2. In the drawings showing light rays that come from very distant objects, the rays are represented as being parallel. If two rays come from a single point, how can they ever be parallel? A drawing may be helpful.

3. What determines the resolving power of a telescope? The magnification? The light-gathering power?

4. Consider the following telescopes:

Telescope	Focal Length of Eyepiece	Focal Length of Objective	Diameter of Objective
A	24 mm	150 cm	12 cm
B	6 mm	100 cm	8 cm
C	18 mm	125 cm	20 cm
D	12 mm	100 cm	10 cm

Which telescope has the greatest light-gathering power? Which one has the greatest magnification? What is the magnification of that telescope?

5. If the magnification of a telescope can be changed by changing eyepieces, why don't we use an eyepiece that will magnify objects thousands of times?

6. If radio telescopes use the principle of reflection, why do they not require a shiny reflecting surface?

7. Describe some special features of two recently constructed telescopes.

8. Why are the telescopes of the Very Large Array arranged so far apart? It would seem more convenient to have them in a tight cluster.

9. Write a report on the future plans that NASA has for space observatories. (You might begin with articles in *Sky and Telescope* or *Astronomy*. Also, visit NASA's web page.)

10. Why are all large telescopes reflectors?

11. Some of the largest telescopes in the world are located on Mauna Kea in Hawaii, at an altitude of 4.2 km. What are some of the factors that astronomers consider when deciding the location for a telescope?

12. Suppose you and your neighbor each have a satellite dish to receive television signals from a satellite, but the signals from each dish are weak. Suppose that you decide to solve your problem by combining the signals from the two dishes to produce a signal twice as strong. Why won't this work well?

CALCULATIONS

1. Suppose you have lenses of the following focal lengths: 30, 10, and 3 centimeters. If you wish to construct a telescope of maximum magnification, which two lenses would you use? Which would be the objective and which the eyepiece? What magnification would this telescope produce?

2. The pupil of your eye is the opening through which light enters. The maximum diameter of the pupil of a human eye is about 0.5 centimeter. How does the light-gathering power of two eyes compare to that of a telescope with a 10-centimeter objective?

3. The telescope at Mount Pastukhov in Russia has a 6-meter objective. Compare its light-gathering power to that of the 5-meter Hale telescope on Mount Palomar.

4. In order to have four similar mirrors with the same light-gathering power as one 16-meter circular mirror, what must be the diameter of each of the four?

5. The *HST* has a 2.4-meter objective mirror. If we would like to use it in order to observe Pluto in the visible, could we distinguish any of Pluto's features? (Diameter of Pluto = 2390 kilometers. Closest distance of Pluto from Earth ≈ 38 AU, at which distance Pluto's angular size is about 0.08 arcseconds.)

6. In the text we discussed the concept of using a space radio telescope (such as the 8-meter Japanese *HALCA* telescope) in conjunction with ground-based radio dishes in order to increase the baseline. The *HALCA* satellite is at a distance of about 21,000 kilometers at apogee and 560 km at perigee. Observations are made at 18, 6, and 1.3 centimeters. What is the best possible angular resolution that we can achieve?

1. Making a Telescope

In order to build your own simple telescope you will need two cardboard tubes that fit tightly together. You may even use kitchen paper tubes, as long as one tube fits tightly inside the other and you can roll the small tube in and out of the larger one. You will also need two converging lenses: the first, the objective, should be the size of the opening of the large tube and the second, the eyepiece, should be about half the size of the opening of the smaller tube. Both lenses should have their flat sides facing out of the tubes and their curved sides facing in. The objective can be held in place with a piece of tape wrapped around the tube. To keep the eyepiece in place, you may use a foam ring (a piece of foam that fits inside the smaller tube and has its central region cut so that it plays the role of the eyepiece holder). By moving the smaller tube you can focus on a specific object. Find the magnification of the refracting telescope you just built and use it to observe objects near and far.

2. Making a Spectroscope

In order to make a simple spectroscope you will need a small piece of diffraction grating, some black construction paper, tape, scissors, and a box. (Plastic diffraction grating is available for a few dollars from Edmund Scientific Co., http://www.scientificsonline.com.) The simplest box that works for this purpose is a toothpaste box. A toothpaste box has at each end a "tongue" that folds in and a pair of "ears" that cover the opening. Cut off the ends of the ears so that there is an opening about half an inch wide between them. Cut a piece of the black construction paper so that after rolling it and inserting it in the box, it fits loosely inside. Also, cover the inside face of each tongue of the box with this paper by using some tape. The purpose of the paper is to create a dark interior, reducing extra reflected light. Punch a hole at the center of one tongue and cover this hole with the small piece of diffraction grating by taping the edges of the grating on the inside face of the tongue. On the other tongue you will create a thin slit by making two parallel and closely spaced cuts along the tongue (from the tip of the tongue to the body of the box). Close the box, tape around the edges of the tongues, and your spectroscope is ready. Use it to look at any light source by putting the end with the grating close to your eye. Compare the spectra you see when looking at an ordinary incandescent light bulb, a fluorescent light source, a mercury vapor streetlight and the sky. **(Warning: Never look directly at the Sun.)**

3. The Moon Illusion

When the Moon is low on the horizon, it sometimes appears to be much larger than when it is high in the sky. This phenomenon does not involve refraction, since refraction would tend to "flat-ten" the Moon. If anything, the Moon's angular size is a bit smaller closer to the horizon, since in this case you are a bit farther from it than when it is at your zenith. This is a purely "psychological" illusion.

One explanation involves the inverted "railroad track" illusion. (In this illusion, a set of "parallel" tracks is drawn on a piece of paper so that they converge in the distance, due to perspective. If we put two equal-size blocks between the tracks, one "farther down" the tracks than the other, then the farther block will appear larger because the "background" influences our perception of the "foreground" objects—the two blocks.) In this scenario, the illusion depends on our perception that the "horizon sky" is farther away than the "zenith sky"; when we view the horizon over a typical landscape, the familiar objects we see (mountains, trees, etc.) form distance-cue patterns that signal "very far" for objects at the horizon, while such patterns are absent when we view the "zenith sky." Thus, when we see the Moon "against" an apparently more distant sky, it will appear larger than when we see it "against" an apparently closer one.

To verify this explanation, use a tube to observe the Moon close to the horizon and high in the sky. The tube blocks light from other objects. Do you see the Moon at its "normal" size? Another way is to turn your back to the Moon when it is close to the horizon and look at it from between your legs. An argument against this explanation is based on the fact that for some people the horizon Moon looks larger but also closer than, or at the same distance as, the zenith Moon. Thus, a better understanding of this illusion will come from considering together the ideas of angular size, linear size and distance.

A second explanation is based on how our eyes focus on an object. For the Moon, which has a fixed linear size, located at approximately a fixed distance from us, its approximately constant angular size will look slightly smaller when our eyes are focused and converged to a closer distance than the Moon's (the zenith Moon case), while it will look slightly larger when our eyes are adjusted to a greater distance (the horizon Moon case). The reason for the change in focus and convergence of the eyes is that, as in the first explanation, there are different distance cues because of the background. To test this second explanation, while observing the horizon Moon, deliberately "cross" your eyes, thus focusing at a closer point. This, of course, will create blurring and double vision, but do you notice a decrease in the Moon's angular size? Do you also see a change in either the Moon's apparent linear size or distance? Please contact us and let us know what you think about this illusion.

---

1. Four articles on observational astronomy are included in *Physics Today,* volume 44 (April, 1991).

2. The article "Untwinkling the Stars" by R. Fugate and W. Wild in *Sky & Telescope* (May, 1994, p.24 and June, 1994, p.20) discusses adaptive optics.

3. The book *Astronomical Centers of the World* by K. Krisciunas (1988, Cambridge Univ. Press) includes a history of the major observatories.

4. If you are interested in buying a telescope, you may want to read the article "Buying the Best Telescope" by A. Dyer in *Sky & Telescope* (December, 1997).

CHAPTER 6

The Earth-Moon System

The Earth–Moon system as seen by the *Mariner 10* spacecraft in November, 1973. This composite photo shows the Earth and Moon in their correct relative sizes.

COLUMBUS'S FOURTH VOYAGE TO THE NEW WORLD was a particularly difficult one. He lost two of his four ships on the voyage, and the remaining two were infested with shipworms, forcing him to land in Jamaica. By February, 1504, he had been trapped there for more than six months, and his attempts to barter with the natives for food were not going well. He learned from his almanac that an eclipse of the rising Moon was to occur in a few days, so he warned the natives that God was angry at them for their laxity in supplying him with food and would punish them with famine and pestilence. Furthermore, God would demonstrate his intent by causing the Moon to become inflamed. Some natives were frightened, but others scoffed at Columbus's threats. When the Moon rose on the evening of February 29, 1504, the eclipse had already begun, so the Moon appeared to have a piece missing. As the eclipse grew, the Moon became darker and took on a red glow (see Figure 6-18). On seeing this, the natives came running to Columbus with supplies and begged him to ask God to forgive them. Columbus knew that the eclipse would remain total for about an hour, so he told them that he would speak with God. When the eclipse reached maximum, he returned to the natives and told them that God had agreed to forgive them and would soon remove the inflammation from the Moon. From that night, Columbus was supplied with whatever he needed.

Lunar eclipses also served a much more practical purpose for sailors of that time. Although a ship's latitude could be determined easily by measuring the altitude of Polaris, determination of longitude depended upon knowing the difference in local time between the ship's location and some location at a known longitude. Columbus attempted to use the eclipse to make this determination, for his almanac listed the local time of the eclipse at locations in Europe. Unfortunately, he made an error in his calculations, and the error confirmed his belief that he had sailed to Asia. He went to his death not knowing that he had traveled to a world that was unknown in the old country.

In previous chapters we laid the groundwork for our study of the cosmos. We presented the historical development of some fundamental ideas in astronomy and described some of the tools we will use in our efforts to understand celestial phenomena. We are now ready to begin our voyage. We must not forget that our explorations are based on the very important assumption that the physical laws derived from and supported by our experiments and observations are valid throughout the universe. This is a bold assumption, but seems to be supported by the data.

If we are to understand our universe, we must be able to understand first our immediate environment. The knowledge we have about the Earth–Moon system serves as a basis for our studies of the entire solar system. On the other hand, information we obtain about other planets and satellites helps us better understand our own past and future. So we begin our quest of the universe by first examining our home planet and its satellite.

6-1 Measuring the Moon's Distance and Size

Astronomers have learned a great deal about the Earth and Moon since the advent of the space program in the late 1960s, but simple naked-eye observations had allowed people to calculate the size of the Earth long before we even journeyed around it. In Chapter 1 we described how Eratosthenes measured the size of the Earth more than two thousand years ago. We begin this chapter by describing how the Moon's distance and size were measured.

The Distance to the Moon

Recall the discussion of parallax in Chapter 1. Parallax is the phenomenon we observe when we hold up a thumb and look at it first with one eye and then with the other. When we do this, the thumb's position changes with reference to the background. Similarly, the Moon exhibits parallax when seen from different positions on the Earth. For example, if a person in Chicago and another person in Paris, France, happen to be looking at the Moon at the same time, they will observe it in slightly different positions against the background of stars (**Figure 6-1**).

In reality, the Moon is far enough away that its parallactic shift among the stars is very small. But the shift can be observed, and the observation allows us to measure the distance from the Earth to the Moon. Using parallax, Ptolemy determined that the distance from the Earth to the Moon is 27.3 Earth diameters—very close to today's value of 30.13 for the average distance to the Moon. Ptolemy's measurement is significant not only because it is so close to the correct value, but also because it shows how close the ancient Greeks were in having a realistic map of the solar system about 2000 years ago. If Aristarchus's heliocentric model had been accepted,

Figure 6-1

When viewed from two different places on Earth (*A* and *B* in the figure), the Moon seems to be at two different places among the stars. The effect is greatly exaggerated in the drawing.

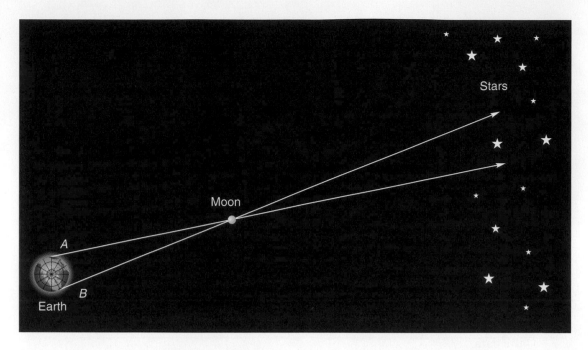

then Ptolemy's determination of the Earth–Moon distance would have provided the final clue in having realistic sizes for the Earth-Moon-Sun system. Recall from Chapter 1 that Aristarchus had calculated the relative sizes of these three objects and their relative distances, thus providing a map, albeit one without a scale. Eratosthenes's and Ptolemy's measurements provide the missing scale and thus a realistic map of the solar system.

Today we know that Earth's diameter is about 12,800 kilometers. A distance of thirty Earth diameters puts the Moon about 380,000 kilometers (or 240,000 miles) from Earth.

The Size of the Moon

The size of the Moon cannot be measured directly, of course. To calculate it, we need to know the distance to the Moon. **Figure 6-2** illustrates the idea.

When we look up at the Moon from Earth, we have no way of judging its actual size. If you ask children how big the Moon looks, one might say that it is about the size of her play ball. Another might claim that it is gumball size. Just how do we judge the size of something we see? For example, suppose you see two objects like those in **Figure 6-3a**. Could you tell how large they are? One looks larger than the other, but in fact they are the same size and one looks smaller simply because it is farther from the camera. When we view an object at a distance, we estimate its size in several ways.

Figure 6-2

Both characters are right. To determine size, one must know the distance, and the word "looney" comes from "lunar."

(a)

(b)

Figure 6-3
(a) It is difficult to judge the real size of the two balls here. They are actually the same size. (b) Here we can use the surroundings to judge the size of the balls.

One way is by comparing it to objects of known size that are near it. That makes the judgment of the size of the two balls in Figure 6-3b easier.

When we look at the sky, there are no familiar objects near the things we see. In this case, we can estimate the size of an object only if we know how far away it is. By combining knowledge of its angular size and its distance, we can determine the object's size.

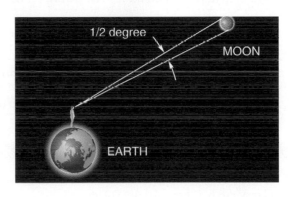

Figure 6-4
The Moon's angular diameter is about 1/2 degree.

Recall that the angular size of an object is the angle between two lines that start at the observer and go to opposite sides of the object. **Figure 6-4** indicates the angular size of the Moon. Refer again to Figure 6-3a. As you look at the two balls, the angular size of the ball on the left is less than that of the ball on the right. So, not knowing its distance, you would say that it looks smaller.

The angular size of the Moon seen from Earth is very close to 1/2 degree. This is much larger than the angular size of a star, but is still not very big. For example, since there are 360 degrees in a complete circle, this means that about 720 Moons could fit in a circle around the Earth (**Figure 6-5**).

The distance to the Moon and the angular size of the Moon (as seen from Earth) can be used to calculate the Moon's diameter. This can be done with a simple equation, called the small-angle formula, which relates the angular size, the distance, and the width of an object. (The term *width* is used here instead of diameter so that the equation will be general for all objects. In the case of the Moon, the width is about equal to its diameter.)

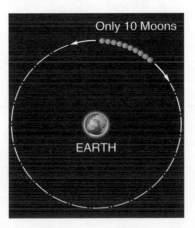

Figure 6-5
A total of about 720 Moons could be drawn around the Earth if they were drawn at the correct distance. They wouldn't fit here; the 10 Moons shown in this figure are drawn much too close to Earth (about 8 times closer than the correct distance).

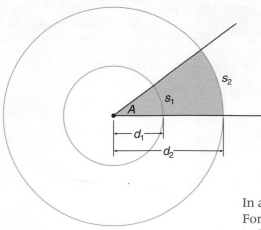

Figure 6-6
The angular size and diameter of an object are related to its distance from the observer.

The Small-Angle Formula

If we have two circles (**Figure 6-6**) with the same center and radii d_1 and d_2, and we draw an angle A from the center, then the angle "cuts" arcs s_1 and s_2 from the two circles. Clearly, the larger the radius of the circle, the larger the arc that subtends the angle A.

Since 360 degrees correspond to the circumference of a circle of radius d (that is, to a length $2\pi \cdot d$), then an angle A corresponds to an arc of length s. In mathematical terms this implies the following relationship:

$$\frac{360°}{A} = \frac{2\pi \cdot d}{s} \quad \Leftrightarrow \quad s = \frac{2\pi \cdot A}{360°} \cdot d$$

In astronomy, however, the angles subtended by an object are, in general, very small. For example, for the Sun and Moon the angle is about 0.5°. A more convenient unit is the arcminute (where $1° = 60'$) and the arcsecond (where $1' = 60''$). If we measure our angles in arcseconds (and write A'' as a reminder), then 360 degrees correspond to about 1.3×10^6 arcseconds and thus $360°/2\pi$ is about $206{,}265''$. Also, the length s of the arc is approximately equal to the diameter (or width) D of the object we are observing. Therefore, the previous equation now takes the simpler form

$$D \approx \frac{A''}{206{,}265} \times d \quad \text{(the small-angle formula)}$$

where D = diameter (width) of an object

d = distance from observer

A'' = angular size of object (in arcseconds)

This "small-angle formula" is very convenient and will be used again later in the book.

This equation cannot be used accurately for large angles (more than about 5 degrees), but in most cases in astronomy, angular sizes are much less than this. Before using the equation for the Moon, let's use it in a terrestrial example.

EXAMPLE

Suppose you look at a photograph on the wall across a large room. You happen to have a device with you that allows you to measure the angular width of the photo, and you determine it to be 1.5 degrees. Suppose further that you know that the photo is 5 meters away from you. How wide is the photograph?

SOLUTION

Substituting into the small-angle formula:

$$D \approx \frac{A''}{206{,}265} \times d = \frac{1.5 \times 60 \times 60}{206{,}265} \times 5 \text{ m} \approx 0.13 \text{ m or 13 cm}$$

TRY ONE YOURSELF

Use the small-angle formula to calculate the diameter of the Moon, using the fact that the Moon's angular size is 0.52 degree and its distance is 384,000 kilometers (more accurate values than given above). Why does your answer differ from the known diameter of 3476 kilometers?

If we use the most accurate data, an exact calculation yields 3476 kilometers (2160 miles) for the Moon's diameter. This is very close to one-fourth of the diameter of the Earth. **Figure 6-7** shows two balls, one four times the diameter of the other. Form a mental image of these two balls. You would probably not say that the large one is four times the size of the small one. The word *size* is an indefinite term; you can't tell whether it refers to diameter, area, or volume. *Size* will not be used in a quantitative sense here; that is, an object that has four times the diameter of another will not be described by saying that its *size* is four times greater.

To appreciate the scale of the Earth–Moon system, imagine the Earth to be a large grapefruit (about 5 inches in diameter). On this scale, the Moon would be a Ping-Pong ball 12 feet away. Between the Earth and Moon, there would be nothing but empty space.

Figure 6-7
The larger ball has a diameter four times the smaller. Would you say that its size is four times larger? Its volume is actually 64 times greater.

Summary: Two Measuring Techniques

As our study of astronomy proceeds, we will describe some important relationships that allow us to measure features of objects in the heavens. Two have been introduced so far. First, we used parallax to measure the distance to the Moon. This method is often called *triangulation* because it involves using a triangle to find a distance. As we will explain in a later chapter, this method is also used to measure distances to nearby stars. Parallax is an important phenomenon in astronomy. Notice that using it involves a relationship among three quantities: the size of the baseline, the angle of parallax, and the distance to the object. Knowing any two of the three, you can calculate the third.

triangulation The use of parallax to determine the distance to an object.

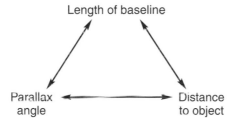

A second relationship (the small-angle formula) involves angular size, actual diameter, and distance. Again, if you know any two of the quantities, you can calculate the third. This is an important relationship, and in later chapters we will explain its use in measuring the diameters of planets.

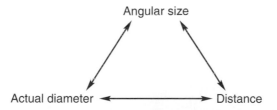

Figure 6-8
These are two photos of the Moon put side by side; one was taken when the Moon is closest to Earth, the other when it is farthest.

The Moon's Changing Size

Figure 6-8 shows two photographs of the full Moon taken at two different times by the same camera. The photographs make it obvious that the apparent diameter of the Moon changes. But apparent diameter depends on distance and actual diameter.

The actual diameter of the Moon doesn't change, of course, so our distance from the Moon must change. This occurs because the Moon's orbit is an ellipse, just like the orbits of all orbiting objects. And the Moon's orbit is fairly eccentric; that is, it is not very circular. The larger apparent diameter occurs when the

perigee/apogee The point in the orbit of an Earth satellite where it is closest/farthest from Earth.

Moon is at its ***perigee,*** or closest distance to the Earth, about 363,300 kilometers. The smaller apparent diameter occurs at the Moon's maximum distance from Earth, 405,500 kilometers—when the Moon is at its ***apogee.*** (Note the similarity to the terms "perihelion" and "aphelion," which refer to the distances closest to and farthest from the Sun, as discussed in Section 2-5.)

6-2 The Moon's Phases

rotation The spinning of an object about an axis that passes through it.

revolution The orbiting of one object around another.

phases (of the Moon) The changing appearance of the Moon during its revolution around the Earth, caused by the relative positions of the Earth, Moon, and Sun.

The words "month" and "Moon" have the same root, as you might guess by their similarity.

Terms related to the phases of the Moon are defined based on their elongation as shown in Table 6-1.

The Moon orbits the Earth in such a way that its same face points toward Earth at all times. "The Man in the Moon" always faces us. At first thought, you might be tempted to say that the Moon does not ***rotate,*** but this is not so. Consider **Figure 6-9**. The Moon is shown with a spot on its surface. In (a), you see that if the Moon did not rotate, this spot would always face the same way in space. In (b), that spot continues to face the Earth as the Moon goes around its orbit. But this means that the Moon must rotate once for every ***revolution*** around the Earth. The fact that the rotation period and revolution period of the Moon are exactly equal seems remarkable and cannot be attributed to mere coincidence. In fact, this is another phenomenon that can be explained by the law of universal gravitation, as will be described later in this chapter.

The photographs in **Figure 6-10** show the ***phases*** of the Moon at various times during a period of about a month. The cause for the Moon's phases is fairly straightforward and has been known since antiquity. To explain the Moon's phases, we need only consider three objects: the Earth, the Sun, and the Moon.

The Moon circles the Earth, completing one orbit in slightly less than a month. This causes its position in the sky, relative to the Sun, to change with time. **Figure 6-11** shows various positions of the Moon in its orbit around the Earth. The Sun is out of the picture, far to the left. The drawings of the Earth and the Moon are dark on the side away from the Sun, because sunlight does not reach that side of them. Consider the Moon in position *A*. Most of the side that faces the Earth is dark, and only a small portion of that side is lit by the Sun. Figure 6-10a shows how the Moon appears from Earth when it is in position *A*. We call such a Moon a ***waxing crescent*** Moon.

When the Moon is at position *B* on its monthly trip around the Earth, we call it a ***first-quarter*** Moon, for if we start the cycle when the Moon is at position *H*, it is now one-quarter of the way around.

Figure 6-9

The Moon rotates as it revolves. If it did not rotate, a dot on its surface would always point in the same direction in space, as in (a). Instead, the Moon always keeps the same face pointed toward Earth (b).

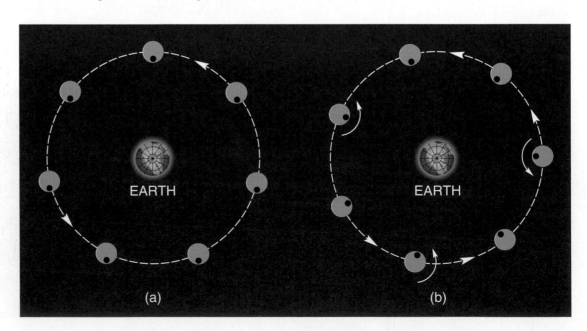

(a)   (b)

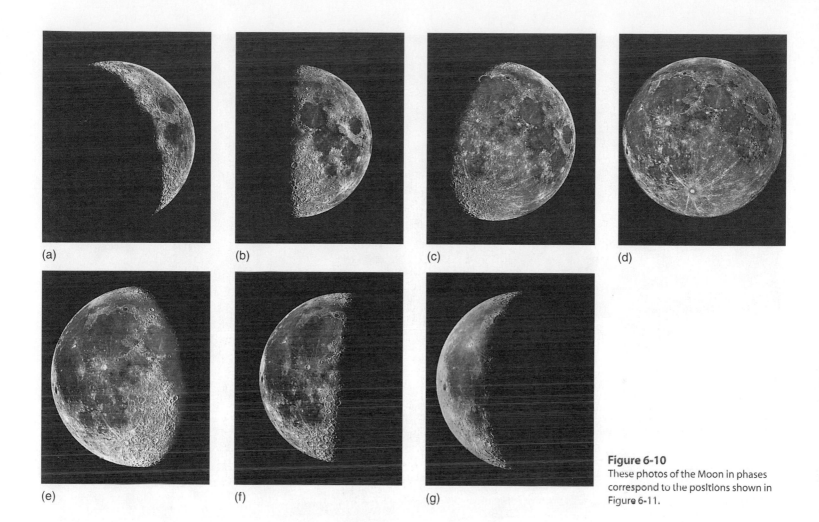

(a) (b) (c) (d)

(e) (f) (g)

Figure 6-10
These photos of the Moon in phases correspond to the positions shown in Figure 6-11.

Position *C* will appear from Earth like the photograph of Figure 6-10c and is called a ***waxing gibbous*** Moon. The word "waxing" is derived from an old German term that means "growing." Between points *H* and *D*, the Moon is waxing; the visible portion is growing nightly from a thin crescent near *H* toward the ***full Moon*** (when the Moon is at position *D*). When we see the Moon in a gibbous phase, most of its sunlit side is facing the Earth.

Observe the photograph of the Moon when it is in position *E*. It is again in a gibbous phase, but here the gibbous phase is called ***waning gibbous*** because from night to night the lit portion that we observe is decreasing (waning) in size.

At position *F* the Moon is again in a quarter phase, the ***third quarter,*** or ***last quarter.*** Then around position *G* we have the ***waning crescent*** phase, and finally the Moon is back to where we start the cycle, at position *H*. Because we (arbitrarily) begin the cycle here, we call this a ***new Moon.*** In this position, the Moon is not visible in our sky because no sunlight strikes the

Table 6-1

Terms Relating to Moon Phases

Phase	Elongation (in degrees)
Waxing	0–180° east
Waning	0–180° west
Crescent	0–90° east or west
Gibbous	90°–180° east or west
New	0
First quarter	90° east
Full	180°
Third quarter	90° west

Questions to Ponder #14 explains the term "Once in a blue Moon."

The phases of the Moon are the subject of two Activities at the end of this chapter.

elongation The angle of the Moon (or a planet) from the Sun in the sky.

The angle of elongation is illustrated in Figure 6-11.

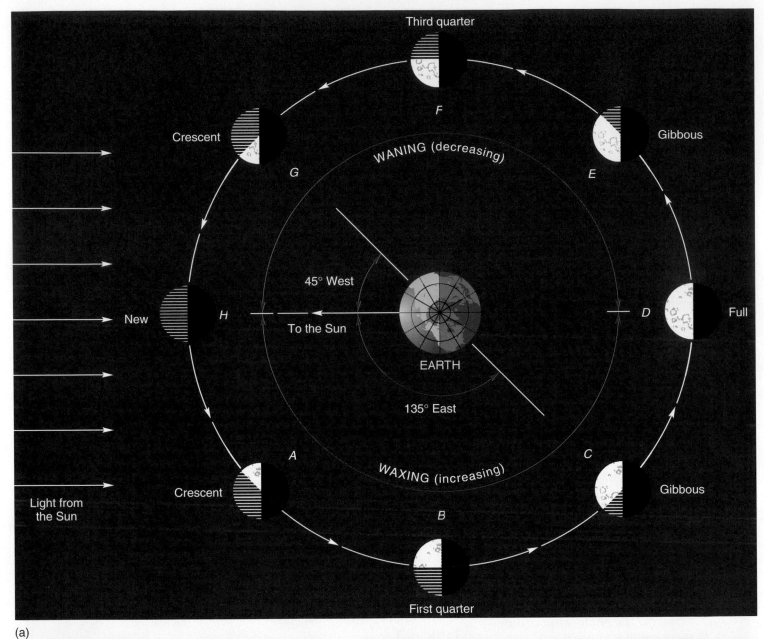

(a)

(b)

Figure 6-11

The Moon in various phases. (a) Seen from above the Earth's North Pole, no light reaches the half of the Moon shown as black. The half of the Moon's surface facing the Sun is always lit; the portion shown in white can be seen by an observer on Earth, while the lined portion cannot be seen. The elongation of the Moon, stated as an angle either east or west of the Sun, is indicated for the waxing gibbous phase and for the waning crescent phase. We are viewing the system from above the north, so east is counterclockwise. (b) Seen from Earth, the Moon appears in phases. (The Moon-Earth system is not drawn to scale.)

side facing Earth. Only at this phase can an eclipse of the Sun occur. Since it takes about a month for the Moon to revolve around the Earth, you may expect to see a solar eclipse approximately once a month. This does not occur, as you may know from personal experience. In the next section we will see why such an eclipse does not occur every time there is a new Moon. Before you read the next section, try to think of a reason why this may be so.

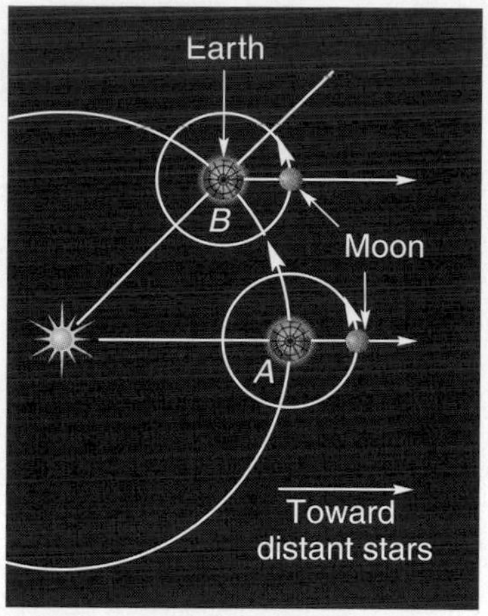

Because the Earth moves in its orbit while the Moon revolves around it, we must distinguish between two revolution periods of the Moon. Refer to **Figure 6-12**. The drawing shows the Earth in two positions, labeled *A* and *B*. At position *A*, the Moon is full. About 27-1/3 days later, when the Earth reaches position *B*, the Moon has completed one *sidereal* revolution, one revolution with respect to the distant stars. Notice, though, that the Moon is not full at this time. Slightly more than two additional days are required for the Moon to reach the full phase again. At this point the Earth will have moved a bit beyond position B, the Moon will have completed one *synodic* revolution, and a *lunar month* will have passed. The lunar month is about 29-1/2 days. (A Calculation at the end of the chapter shows you how to calculate the difference of about 2 days between the sidereal and synodic revolutions.)

Figure 6-12
This drawing shows the difference between the Moon's sidereal period and its synodic period. When the Earth is at point *A*, the Moon is full. At point *B*, the Moon has completed one sidereal period, but about two more days are required for it to reach full phase again, at which time one synodic period will have been completed.

sidereal period The amount of time required for one revolution (or rotation) of a celestial object with respect to the distant stars.

synodic period The time interval between successive similar alignments of a celestial object with respect to the Sun.

lunar month The Moon's synodic period, or the time between successive similar phases.

6-3 Lunar Eclipses

During its orbit of the Earth, the Moon sometimes enters the Earth's shadow; when this occurs, sunlight is blocked from reaching the Moon. This phenomenon is known as a *lunar eclipse.* You might wonder why this doesn't happen at the time of every full Moon. There are several reasons. First, as we have seen, the Moon and Earth are very small compared to their distance apart: the Moon is 30 Earth diameters away. Thus it is unlikely that they will align so accurately that one eclipses the other. Think of trying to align a grapefruit, a Ping-Pong ball 12 feet away, and a distant object.

Another reason a lunar eclipse does not occur during each lunar orbit is that the Earth's shadow is smaller than the Earth. This is because the source of light—the Sun—is so much larger than the Earth. Look at **Figure 6-13**. Although distances are not nearly to scale, the principle illustrated is correct. Notice that there is a cone of darkness behind the Earth. When the Moon is in this area (at point *A*, for example), it is in the full shadow of the Earth. This full shadow of the Earth—called the *umbra*—tapers down to a point. At the distance of the Moon, the width of the umbra is only three-fourths of the diameter of the Earth. So the Moon is less likely to pass through the shadow than if the shadow were the size of the Earth.

If the Moon is at point *B* of Figure 6-13, on the other hand, it is only in partial shadow, for light from the left part of the Sun (as seen from the Moon) is hitting it. When the Moon is here, in the *penumbra,* it will not receive the full light from the Sun and will appear dim to Moon-watchers on Earth. The penumbral shadow increases in size at greater distances from Earth, but it is not equally dark across its width. Right next to the umbra, the shadow is very dark, but it gets brighter and brighter out toward its edge. When the Moon passes through the outer penumbra, we don't even notice the darkening.

However, there is a more important factor in explaining why a lunar eclipse does not occur at each full Moon. The Moon's plane of revolution is tilted relative to

The lunar month is $29^d 12^h 44^m 2^s.9$.

Why don't we have a lunar eclipse every month?

lunar eclipse An eclipse in which the Moon passes into the shadow of the Earth.

umbra The portion of a shadow that receives no direct light from the light source.

penumbra The portion of a shadow that receives direct light from only part of the light source.

Figure 6-13

(a) Point *A* is in the umbra of the Earth's shadow. Point *B* is in the penumbra, where light from part of the Sun hits it. Distances are not to scale in the drawing. (b) This observer is looking at the Earth and Sun from point *A*, in the umbra of the Earth's shadow. (c) This observer is looking at the Earth and Sun from point *B*, in the penumbra of the Earth's shadow.

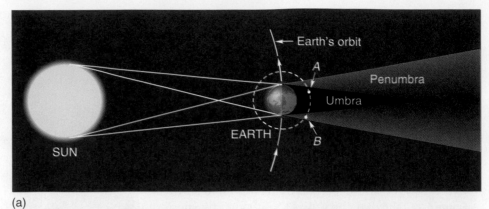

(a)

(b) This observer is looking at the Earth and Sun from point *A*, in the umbra of the Earth's shadow.

(c) This observer is looking at the Earth and Sun from point *B*, in the penumbra of the Earth's shadow.

Eclipse seasons occur when the Earth is at points *A* and *C* in Figure 6-14(a). Eclipse seasons are slightly less than six months apart because the orientation of the Moon's orbit changes slightly as time passes.

eclipse season A time of the year during which a solar or lunar eclipse is possible.

penumbral lunar eclipse An eclipse of the Moon in which the Moon passes through the Earth's penumbra but not through its umbra.

total lunar eclipse An eclipse of the Moon in which the Moon is completely in the umbra of the Earth's shadow.

the Earth's plane of revolution around the Sun. Consider **Figure 6-14a**, which shows both the Earth's and the Moon's orbits. The tilt of the Moon's orbit with respect to the Earth's is actually only 5 degrees, but we have exaggerated it in the drawing. Notice that when the Earth is at positions *B* and *D*, its shadow cannot hit the Moon. In fact, only when the Earth is near points *A* and *C* can the Moon pass through its shadow. These points represent the two *eclipse seasons* that occur each year. Thus in most cases of a full Moon, the Moon will not be in the Earth's shadow but will be either north or south of it. The plane of the Moon's orbit changes relatively little as the year progresses, so eclipses can only occur about twice a year.

Types of Lunar Eclipses

The Earth's umbra gets smaller as the distance from Earth increases. At the Moon's average distance, the umbra has a diameter of about 9200 kilometers. Since the diameter of the Moon is less than 3500 kilometers, the Moon can easily be covered by the umbra; however, the Moon might not pass right through it. **Figure 6-15** shows three possible paths of the Moon through the shadow of the Earth. If the Moon moves along path *A*, it only passes through the penumbra, producing a *penumbral lunar eclipse.* In this case, it darkens slightly as it does so, but such an effect is not obvious from Earth and is only noticeable if the Moon passes into the darkest part of the penumbra, near the umbra.

If the Moon follows path *B*, it slowly darkens as it moves toward the umbra. **Figure 6-16** is a triple exposure of the Moon during an eclipse. The Moon is moving along a path such as path *B*, and the exposure at the right side of the photo shows the Moon while only part of it is in the Earth's umbral shadow. As the Moon continues to move into the umbra, the shadow slowly moves across its surface until the Moon appears as it does in the center exposure, where we see a *total lunar eclipse.* Depending upon the Moon's distance from Earth, it may take an hour from the time of first con-

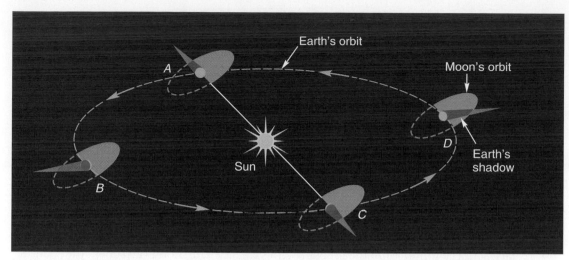

(a)

(b)

 Figure 6-14
(a) When the Earth is very near either *A* or *C*, lunar eclipses can occur, but when it is at other points in its orbit, the Moon does not pass through its shadow. (b) This imaginative drawing shows a flying fish (the Moon) "orbiting" a duck (the Earth) while the duck swims around a beachball (the Sun). The Moon-fish at the left never comes between the Sun and the Earth, so no eclipse occurs, but the orbit of the Moon-fish at the right is such that an eclipse might occur.

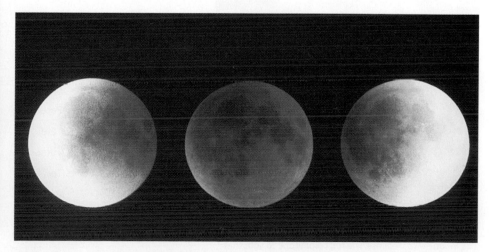

Figure 6-15
Three of the possible paths of the Moon through the Earth's shadow. Path *A* produces only a penumbral eclipse. Path *B* produces a total eclipse and *C* produces a partial eclipse.

Figure 6-16
This is a triple-exposure photo of the Moon taken before, during, and after the total eclipse of August 16, 1989. (The photo was taken with a 4-inch refractor with 30-second exposures.)

Figure 6-17
Though the Moon is in total eclipse, some light is refracted toward the Moon by the Earth's atmosphere. As the light passes through the atmosphere, however, the blue end of the spectrum is scattered away more than the red is, so the light that makes it to the Moon is more reddish. After reflection from the Moon's surface, this light must again pass through the Earth's atmosphere and thus the light seen by an observer on the ground is mostly red. (This selective scattering also explains why the Sun looks red at sunsets and sunrises and why the sky is blue.)

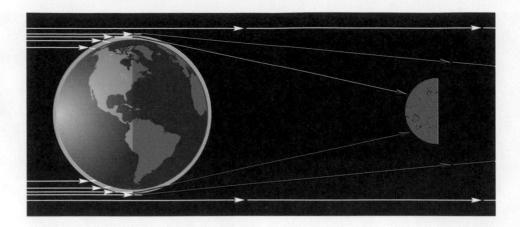

partial lunar eclipse An eclipse of the Moon in which only part of the Moon passes through the umbra of the Earth's shadow.

tact with the umbra until the eclipse reaches totality. The Moon can stay in the shadow for up to about 1-1/2 hours. On the left side of Figure 6-16, we see the Moon leaving the Earth's umbra again.

If the Moon follows path *C* of Figure 6-15, it is never entirely covered by the umbra, and we see only a ***partial lunar eclipse.*** The dark shadow creeps across the Moon, covering (in the case shown) only the top part of the Moon.

An eclipse of the Moon, especially a total eclipse, is a beautiful sight. The totally eclipsed Moon is not completely dark, however. Even when the Moon is completely in the umbra, some sunlight strikes the Moon. This light has been refracted by the Earth's atmosphere, as shown in **Figure 6-17**. As the light passes through the atmosphere, however, the blue end of the spectrum is scattered away more than the red is, so the light that makes it to the Moon is more reddish. After reflection from the Moon's surface, this light must again pass through the Earth's atmosphere; thus the light seen by an observer on the ground is mostly red. For this reason the eclipsed Moon appears a dark red color (**Figure 6-18**).

Table 6-2 shows the dates of coming lunar eclipses. You cannot be sure that you will be able to see any particular lunar eclipse, however, for two reasons. First, in order to be able to see it, you must be on the dark side of the Earth when the eclipse

Figure 6-18
The Moon at total eclipse. The red color is due to light that has passed through the Earth's atmosphere before striking, and after being reflected from, the Moon.

Table 6-2
Dates of Total/Partial Lunar Eclipses, 2004-2010

Date	Type of Eclipse	Visible from the Following Regions
May 4, 2004	Total	S. America, Europe, Africa, Asia, Australia
October 28, 2004	Total	Americas, Europe, Africa, Central Asia
October 17, 2005	Partial	Asia, Australia, Pacific, N. America
September 7, 2006	Partial	Europe, Africa, Asia, Australia
March 3, 2007	Total	Americas, Europe, Africa, Asia
August 28, 2007	Total	East Asia, Australia, Pacific, Americas
February 21, 2008	Total	Central Pacific, Americas, Europe, Africa
August 16, 2008	Partial	S. America, Europe, Africa, Asia, Australia
December 31, 2009	Partial	Europe, Africa, Asia, Australia
June 26, 2010	Partial	East Asia, Australia, Pacific, West Americas
December 21, 2010	Total	East Asia, Australia, Pacific, Americas, Europe

occurs. So, on the average, only half the people on Earth have a chance to see a given lunar eclipse. For this reason, the last column of the table indicates where each eclipse will be visible. Second, there is the weather factor. A cloudy night can ruin a long-planned eclipse-viewing party.

6-4 Solar Eclipses

We have seen that a lunar eclipse occurs when the shadow of the Earth falls on the full Moon. An eclipse of the Sun—a *solar eclipse*—occurs when the Moon, in its new phase, passes directly between the Sun and the Earth so that the Moon's shadow falls on the Earth. There is a major difference between these events, however. The Earth's size is such that its umbral shadow reaches back into space nearly a million miles, and at the distance of the Moon, it is easily large enough to cover the entire Moon. The umbral shadow of the Moon, however, reaches only about one-fourth that far, or about 377,000 kilometers (234,000 miles). Recall that the average distance from the Earth to the Moon is about 384,000 kilometers (239,000 miles). Thus if the Moon stayed at this distance, its umbra would never reach the Earth. A dark shadow of the Moon would never fall on the Earth.

The Moon, though, follows an eccentric orbit, coming as close as 363,300 kilometers to Earth. So it does get close enough that its umbra can reach the Earth. When this occurs, we can experience one of the most spectacular of natural phenomena, a *total solar eclipse.* **Figure 6-19** shows two cases of the Sun, Moon, and Earth being aligned when the Moon is close enough for its umbra to reach the Earth. Even when the Moon is at its closest, the width of its umbral shadow at the Earth's distance is only about 130 kilometers. If the shadow hits the Earth as in the lower case in Figure 6-19, its width is between 130 kilometers and about 270 kilometers. If the shadow happens to strike the Earth as in the upper case, the shadow on the surface may measure wider than 270 kilometers, but it seldom exceeds 400 kilometers.

This explains why relatively few people ever experience a total solar eclipse. Only people in that small area of Earth covered by the umbra see the Sun entirely hidden by the Moon. As the Moon moves along, its shadow swings in an arc across the surface of the Earth (**Figure 6-20**). Thus there is a strip across the Earth's surface in which the total eclipse may be seen. This strip may be many thousands of kilometers long, but its width can reach a maximum of only about 400 kilometers. You have to be within this strip at the exact moment of totality, in clear weather, to see the total eclipse of the Sun.

solar eclipse (or **eclipse of the Sun**) An eclipse in which light from the Sun is blocked by the Moon.

total solar eclipse An eclipse in which light from the normally visible portion of the Sun (the photosphere) is completely blocked by the Moon.

130 kilometers ≈ 80 miles, 270 kilometers ≈ 170 miles, and 400 kilometers ≈ 250 miles.

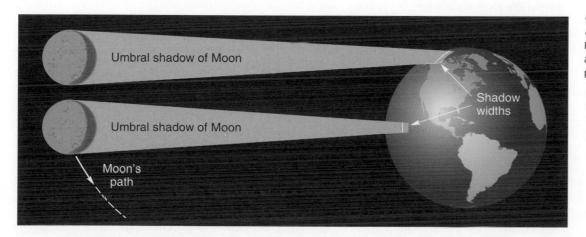

Figure 6-19
If the Moon's umbra strikes the Earth at an angle, a wider area on Earth will experience a total eclipse.

Figure 6-20
The Moon's motion causes the path of a total solar eclipse to sweep across the Earth. The eclipse shown moves primarily across land and therefore would be seen by many people.

Umbral shadow of Moon

Eclipse path

Moon's path

At totality, the sky is dark enough that planets and the brightest stars can be seen in the sky. The appearance of the Sun is shown in **Figure 6-21**. What you see around the dark disk where the Moon blocks out the Sun is the glowing outer atmosphere of the Sun, called the *corona*. This is a layer of gas that extends for millions of miles above what normally appears to be the surface of the Sun. The gas glows because of its high temperature, but the glow is so much dimmer than the light we receive from the main body of the Sun that it is observed only during an eclipse. The opportunity to observe the corona is one of the scientific values of an eclipse, although today we are able to block out the Sun by artificial means in order to observe its outer layers. We will discuss the Sun further in Chapter 11.

A total solar eclipse is truly an awesome experience. As **Table 6-3** indicates, it will be some time before one is visible in the United States. If you get a chance to travel to the path of totality of a solar eclipse, don't pass it up. It is one of nature's grandest spectacles.

corona The outer atmosphere of the Sun. (We will discuss the corona in Chapter 11.)

The Partial Solar Eclipse

Figure 6-22 shows not only the Moon's umbra, but also its penumbra. People on Earth who are in the umbra see a total eclipse, but anyone within the penumbra sees a *partial solar eclipse.* The penumbra covers a much greater portion of the Earth's surface, stretching about 3000 kilometers (2000 miles) from the central path of totality, so most of us have the opportunity to see a few partial solar eclipses during our lifetimes.

partial solar eclipse An eclipse in which only part of the Sun's disk is covered by the Moon.

Figure 6-21
During a total solar eclipse, the glowing light of the Sun's outer atmosphere— the corona—is visible. This photo of the June 30, 1991 eclipse was taken from the window of a DC-10 41,000 feet above the ground.

Table 6-3
Dates of Total/Annular Solar Eclipses, 2004–2010

Date	Type of Eclipse	Visible from the Following Regions
October 3, 2005	Annular	Portugal, Spain, Libya, Sudan, Kenya
March 29, 2006	Total	Central Africa, Turkey, Russia
September 22, 2006	Annular	Guyana, Suriname, French Guiana, Southern Atlantic
February 7, 2008	Annular	Antarctica
August 1, 2008	Total	Northern Canada, Greenland, Siberia, Mongolia, China
January 26, 2009	Annular	South India, Sumatra, Borneo
July 22, 2009	Total	India, Nepal, China, Central Pacific
January 15, 2010	Annular	Central Africa, India, Myanmar, China
July 11, 2010	Total	South Pacific, Easter Island, Chile, Argentina

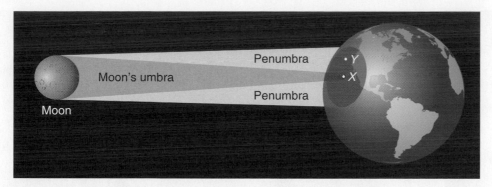

(a)

(b)

Figure 6-22
(a) A person at point *X* sees a total solar eclipse, while a person at *Y* sees a partial solar eclipse, with only the southern part of the Sun blocked by the Moon. The umbra and penumbra in the Moon's shadow can be seen in the image on page 201. (b) This series of photos shows the progression of a partial solar eclipse as would be seen by a person standing at point *Y*.

In a partial solar eclipse, the dark disk of the Moon moves across the Sun, but its path is not perfectly aligned with the Sun and it does not move across the center of the Sun. The closer you are to the path of totality, the more of the Sun is blocked out by the Moon.

The Annular Eclipse

Remember that a total solar eclipse can occur only when two conditions are present: the Moon is directly between the Sun and the Earth, and the Moon is close enough to the Earth that its umbral shadow reaches the Earth. When the Moon is at its average distance from the Earth, it is a little too far away to cause a total eclipse on Earth. Therefore somewhat fewer than half the solar eclipses that occur are total. **Figure 6-23** illustrates what happens when the Moon is too far from Earth to allow a total eclipse. A person at point *P* on the Earth sees a partial eclipse where the disk of the Moon crosses that of the Sun.

Think of what an observer at point *A* on Earth would see in this case. The Moon is so far away that its disk is not large enough to cover the Sun, even when it is directly centered on the Sun. The photograph in **Figure 6-24** shows what the observer would see at eclipse maximum. The Latin word *annulus* means *ring,* and from the figure you can see why such an eclipse is called an ***annular eclipse.*** Note the spelling; this is not an annual eclipse. Slightly over half of the solar eclipses are annular. Table 6-3 indicates which of the coming eclipses will be total and which annular. **Figure 6-25** shows the paths of total solar eclipses through 2024. Notice that the next total solar eclipse to

There are about twice as many total or annular solar eclipses as total lunar eclipses, but you are much less likely to see a solar eclipse because of the narrow path of the shadow.

annular eclipse An eclipse in which the Moon is too far from Earth for its disk to cover that of the Sun completely, so the outer edge of the Sun is seen as a ring.

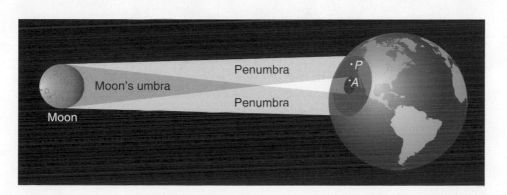

Figure 6-23
When the Moon is far away during a solar eclipse, the eclipse will be annular. The person at point *A* sees the annular eclipse, while the person at *P* sees a partial eclipse.

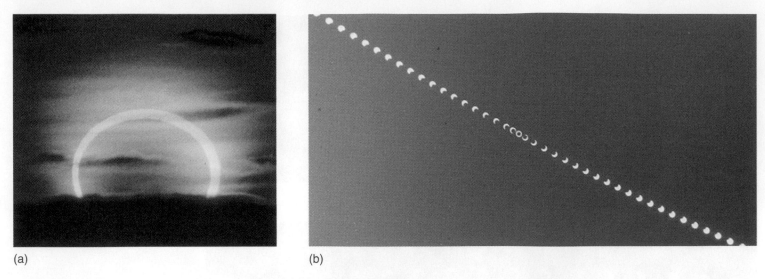

(a)   (b)

Figure 6-24
(a) During an annular eclipse, we can see the entire ring—annulus—of the Sun around the Moon. This photo shows the annular eclipse of May 30, 1984. The irregularity of the ring is due to mountains and valleys on the Moon's surface.
(b) This series of photos shows the progression of the Moon's motion across the Sun during an annular eclipse.

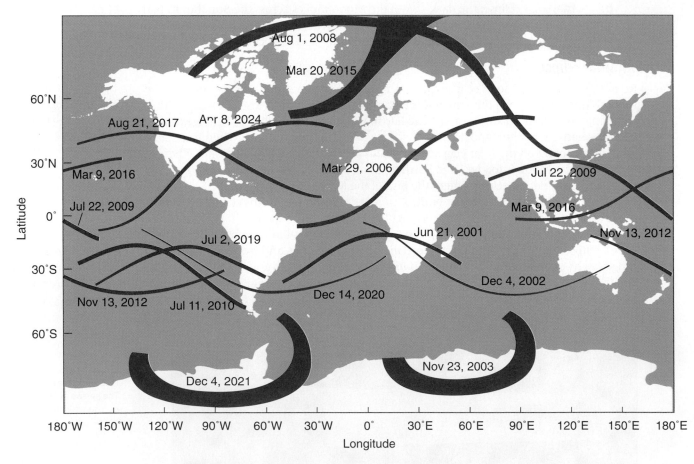

Figure 6-25
The map shows the path of the Moon's shadow during total solar eclipses for the period 2001–2025. On each path, the position of greatest eclipse is marked.

cross the continental United States will occur on August 21, 2017 (and the next after that will be on April 8, 2024).

6-5 Earth

The next few chapters will focus on the details of the objects that make up the solar system. Before moving outward from the Earth and Moon, however, a close examination of these two objects is in order to provide a basis for comparison.

The Interior of the Earth

How do we know what is inside the Earth? One of the first properties of the interior that would aid our quest is the *density* of the Earth. The density of an object is defined as the ratio of the object's mass to its volume. To do the calculation, we need to know the Earth's mass and volume. As we have seen, the diameter of the Earth has been known for centuries, and, of course, the space program has greatly increased the accuracy of our measurements of the Earth's size and shape.

The mass of the Earth was first determined by applying Kepler's third law (as revised by Newton) to the period and radius of the Moon's orbit. We saw in Chapter 3 that the period of revolution of an orbiting object depends only upon the size of the orbit and total mass of the objects. Today we can calculate the gravitational pull on orbiting satellites and use this to further refine our values for the Earth's mass. Values of the mass and diameter of the Earth are given in Appendix C.

Using the mass and volume of the Earth, we calculate its density to be 5.52 grams per cubic centimeter (g/cm^3) or 5520 kilograms per cubic meter (kg/m^3). For comparison, the density of water is 1 gram/centimeter^3, of aluminum is 2.7 grams/centimeter^3, and of iron is 7.8 grams/centimeter^3.

The ability to calculate the density of a celestial object by using its diameter and mass allows us to make a reasonable guess about its composition. For example, assume that you discover a planet whose average density is 5.5 times that of water. The surface of this planet seems to be rocky and is mostly covered by water. Based on this information, what can you say about the interior of this planet? Since the density of water is 1 gram/centimeter^3 and that of rock about three times larger, the interior must have elements whose densities are larger than the density of rock. This is necessary for the *average* density to be 5.5 grams/centimeter^3. You may reasonably guess that the interior must have metals—iron and nickel, for example. Even though we cannot accurately predict the ratios at which the different elements exist in the interior, the simple knowledge of the average density of the planet allows us to say something reasonable about its interior. A comparison closer to the situation of the Earth is a snowball with a rocky-metallic interior. Like the Earth, the snowball is made up of different materials at different levels of its interior.

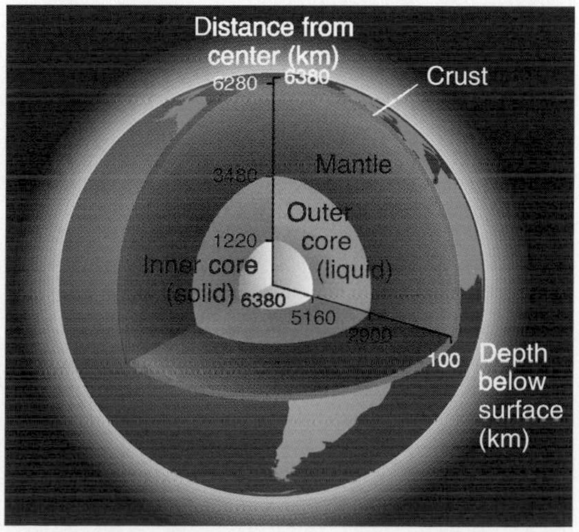

The interior of the Earth is made up of three layers, as shown in **Figure 6-26**. The *crust,* which is the outer layer, extends to a depth of less than 100 kilometers and is made up of the common rocks with which we are familiar here on the surface. The density of the crust is about 2.5 to 3 grams/centimeter^3.

density The ratio of an object's mass to its volume.

For example, if you take a cube of metal 1 centimeter on each edge (1 centimeter^3 of metal) and determine that its mass is 5 grams, then its density is 5 grams/centimeter^3.

The applicable equation is

$$\frac{a^3}{P^2} = \frac{m_1 + m_2}{M_{Sun}}$$

where a is the orbital semimajor axis in AU, P is the orbital period in years, and m_1 and m_2 are the masses of the objects—the Moon and Earth in this case. See Chapter 3 for how we know what fraction of the total mass of the Earth–Moon system is the Earth's.

Figure 6-26
The interior of the Earth, showing its primary layers.

Distance from center (km)
6280 6380
Crust
3480
Mantle
1220
Outer core (liquid)
Inner core (solid) 6380
5160
2900
100 Depth below surface (km)

crust (of the Earth) The thin, outermost layer of the Earth.

mantle (of the Earth) The thick, solid layer between the crust and the core of the Earth.

core (of the Earth) The central part of the Earth, consisting of a solid inner core surrounded by a liquid outer core.

chemical differentiation The sinking of denser materials toward the center of planets or other objects.

P waves Seismic waves analogous to the waves produced by pushing a spring back and forth.

S waves Seismic waves analogous to the waves produced by shaking a rope, attached to a wall, up and down.

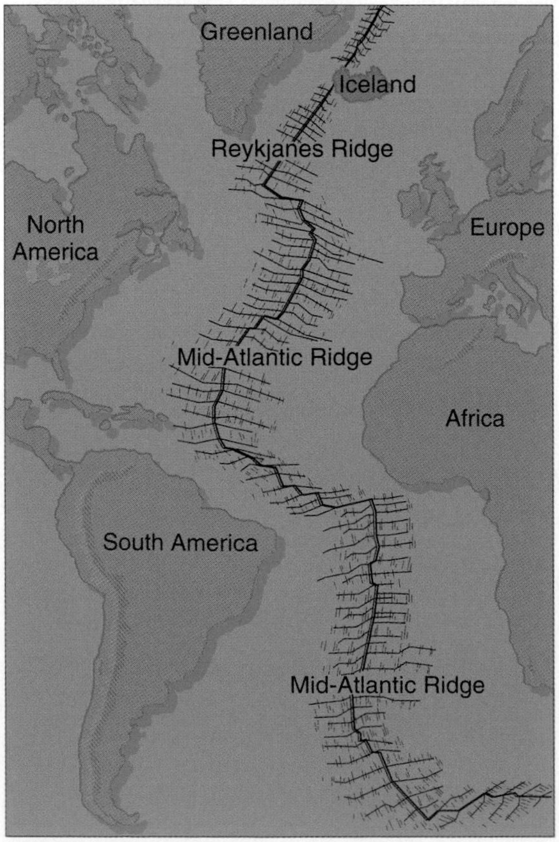

Figure 6-27
If the Atlantic Ocean were drained, we could see the Mid-Atlantic Range of mountains and the rift down its center. As lava is forced out of the rift, the plates are pushed aside, causing continental drift.

continental drift The gradual motion of the continents relative to one another.

rift zone A place where tectonic plates are being pushed apart, normally by molten material being forced up out of the mantle.

The *mantle* extends nearly halfway to the center of the Earth, about 2900 kilometers below the surface. This layer, although solid, is able to flow very slowly when steady pressures are exerted on it; however, it can crack and move suddenly under extreme sudden pressures. The mantle is denser (3 to 9 grams/centimeter^3) than the crust, and therefore the crust floats on top of the mantle.

The *core* of the Earth seems to be divided into two parts: a liquid outer core and a solid inner core. The core is even denser than the mantle, ranging from 9 to 13 grams/centimeter^3. Because of its high density, the core of the Earth is thought to be made up primarily of iron and nickel, the most common heavy elements.

The pattern of increasing density of materials within the Earth tells us something of the Earth's past, for such *chemical differentiation* could only have come about when the Earth was in a molten state, when the heavier elements would have sunk through the less dense layers. The molten state must have resulted from heating during the formation of the planet and from energy released by radioactive elements early in the Earth's history.

Looking at Figure 6-26 naturally brings up the question of how we know the internal structure of our planet to such a degree. After all, the deepest wells only go down a few kilometers from the surface and the information we get from volcanic eruptions takes us only a bit deeper. We learn about the makeup of the Earth's interior primarily by detecting two types of waves that result from earthquakes. These waves travel through the Earth, and by analyzing their times of travel from distant earthquakes, geologists can deduce some properties of the materials that lie deep within the Earth. The first type of seismic waves, called primary or **P waves,** are analogous to the waves produced by pushing a spring back and forth. The second type, called secondary or **S waves,** are analogous to the waves produced by shaking a rope, attached to a wall, up and down. Just as light refracts when traveling from one medium into a different one, seismic waves refract when they travel from one region inside the Earth into another region of different composition. Using seismographs at many locations around the Earth, we can measure at a specific location whether the observed wave has P or S characteristics and also determine its travel time from the center of the earthquake. Taking into account that S waves cannot travel far through liquids, the absence of S waves at certain regions and the strength of P waves in general allows us to draw the internal structure of the Earth as shown in Figure 6-26.

Greenland
Iceland
Reykjanes Ridge
North America
Europe
Mid-Atlantic Ridge
Africa
South America
Mid-Atlantic Ridge

Plate Tectonics

You may have noticed while looking at a map of the Earth that there seems to be a rough fit between the eastern edge of the American continents and the western edge of Europe and Africa. Early in this century it was proposed that the continents were once in contact. No acceptable mechanism for continents moving relative to one another (*continental drift*) was obvious, however, and the idea was put aside. (Alfred Wegener, a German meteorologist, is credited with first developing the idea of continental drift.)

Later, geologists discovered that there is a line near the center of the Atlantic Ocean where lava flows upward, called the *rift zone,* forming an extensive range of underwater mountains (**Figure 6-27**). They discovered further that as the lava solidified,

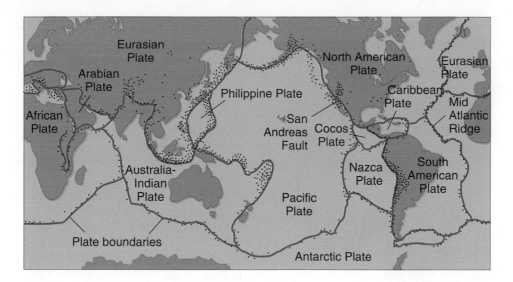

Figure 6-28
The major tectonic plates of the Earth. The dots represent earthquake locations. Note that earthquakes tend to occur near plate boundaries.

magnetic material in it oriented in specific directions, depending upon the direction of the Earth's magnetic field at the time. (We will see later in this section that the magnetic field of the Earth has changed through the ages.) From the magnetic properties of the material, they were able to determine roughly when it solidified. Then, by examining the sea floor at various distances from the center of the rift zone in the mid-Atlantic, they learned that the sea floor has gradually spread from the rift. Finally, laser light was bounced from satellites to detect any relative motion between continents, and it was discovered that Europe and North America are moving apart at the rate of 2-4 centimeters/year. From this evidence, our present theory of ***plate tectonics*** developed.

The Earth has about a dozen tectonic plates, which are sections of the Earth's crust (and perhaps upper mantle) that extend about 50 to 100 kilometers deep. **Figure 6-28** shows the major plates.

If the plates are spreading from places where lava flows from beneath the crust, they must be jamming together at some other places. **Figure 6-29** shows how one plate might be pushed below another at such a location. South America is on a plate that is moving westward from the rift zone in the mid-Atlantic. At the western boundary of South America, this plate is forced against the Nazca Plate (Figure 6-28), which is moving eastward. Consequently, the Nazca Plate gets pushed under the South American continent. Two major effects have resulted from these movements: (1) A deep ocean trench has opened off the western coast of South America. (2) As the material of the Nazca Plate is pushed downward, it melts and low-density rock is forced upward, erupting from the Earth's surface as volcanic lava. Over millions of years, this volcanic

It is just as if we were to refit the torn pieces of a newspaper by matching their edges and then check whether the lines of print run smoothly across.
-*Wegener*

plate tectonics The motion of sections of the Earth's crust (plates) across the underlying mantle.

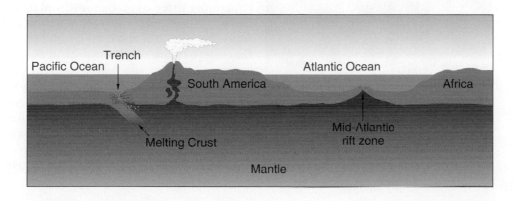

Figure 6-29
As the South American Plate is pushed westward (to the left here) from the mid-Atlantic rift, it is pushed against the Nazca Plate under the Pacific Ocean. This forces that plate down into the mantle, resulting in the volcanoes of the Andes Mountains.

Figure 6-30
(a) The Earth today and (b) about 200 million years ago. The motion of the continents in (b) is indicated by arrows.

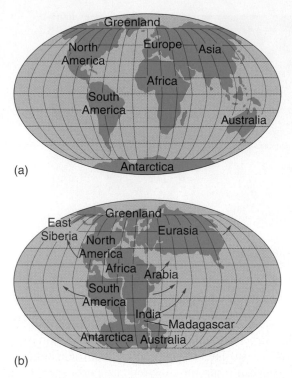

(a)

(b)

action has raised the Andes mountain range that runs the length of the western side of the continent. The volcanoes of the west coast of North America and those of Japan were formed in the same manner.

In some places, plates are slipping by one another. This is the primary motion that occurs between the Pacific Plate and the North American Plate. The slippage is not smooth and continuous, however. Instead, forces build up over a number of years until motion occurs very suddenly, causing an earthquake.

By projecting the motion of the plates backward in time, we can conclude that some 200 million years ago, a map of the Earth's continents looked somewhat like that in **Figure 6-30b**. Today's continents started separating shortly after that and continue to move apart today. The great mountains, the canyons, and indeed the continents, that may appear to us as permanent monuments, are instead transitional stages of an ever-changing planet.

Earth's Atmosphere

Compared to the size of the Earth, the atmosphere reaches a very small distance above the surface. The atmosphere gets thinner and thinner farther from Earth, so it is impossible to put a definite boundary on it, but at a distance of 100 or 150 kilometers above the surface, the atmosphere is essentially nonexistent.

Our atmosphere consists of about 78% nitrogen and 21% oxygen by volume. Other constituents, such as water vapor, carbon dioxide, and ozone, make up a very small percentage of the atmosphere, even though they are very important to life on Earth. We will discuss the atmospheric importance of carbon dioxide when we discuss the planet Venus in Chapter 8.

Most of the mass of the atmosphere—about 75%—lies within 11 kilometers (7 miles) of the surface. This portion of the atmosphere, called the *troposphere,* is where all of our weather occurs. The troposphere receives most of its heat from infrared radiation emitted by the ground, so the troposphere is cooler as one gets higher. **Figure 6-31** illustrates the temperature of the atmosphere at different heights.

Centered at about 50 kilometers above the surface is the ozone layer. Ozone is a molecule containing three oxygen atoms, and it is an efficient absorber of ultraviolet radiation (UV) from the Sun. This absorption is the reason that temperature reaches a peak at the ozone layer.

The ozone layer is extremely important to life on Earth, for most life has developed while being sheltered from all but a little UV. Ultraviolet radiation breaks apart molecules that make up living tissue, as you have experienced if you have ever been sunburned, a

◆ Commercial planes fly at the top of the troposphere in order to avoid the turbulence of weather in that layer.

troposphere The lowest level of the Earth's (and some other planets') atmosphere.

Figure 6-31
The temperature of the atmosphere varies with altitude because of the way solar energy is absorbed by the different layers. The temperature at altitudes of 200-300 kilometers (where satellites and the space shuttle orbit) is much larger than on the surface. This poses no threat because atmospheric density at these altitudes is extremely small (although there is still enough air resistance in the lower thermosphere to affect a satellite's motion and eventually cause it to reenter and burn up in Earth's lower atmosphere).

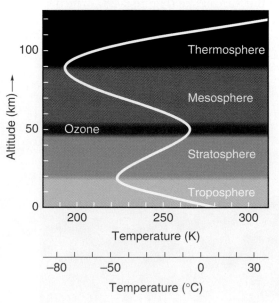

Data Page

The Earth

Earth

Moon

Earth Data

Earth	Value
Equatorial diameter	12,756 km
Oblateness*	0.00335
Mass	5.97×10^{24} kg
Density	5.52 g/cm³
Surface gravity	9.78 m/s²
Escape velocity	11.19 km/s
Sidereal rotation period	23.9345 hours
Solar day	24.0 hours
Albedo**	0.306
Tilt of equator to orbital plane	23.45°
Orbit	
Semimajor axis	1.496×10^{8} km
Eccentricity	0.0167
Sidereal period	365.26 days
Moons	1

*Recall that oblateness tells us how "flattened" an object is. The Earth's polar diameter is 12,714 km.
**The albedo of a solar system object is the fraction of incident sunlight that the object reflects without absorption.

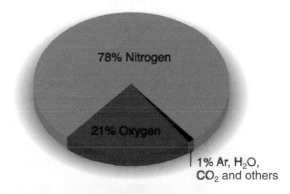

Mantle

Outer core (liquid)

Inner core (solid)

Interior

78% Nitrogen

21% Oxygen

1% Ar, H_2O, CO_2 and others

Atmospheric Gases

QUICK FACTS

- *Third planet from the Sun. Very circular orbit. Except for Pluto, has largest satellite compared to its size. Only planet with liquid water on surface. Active interior.

*This Quick Facts Box may be trivial for Earth, but for other planets (Chapters 8 and 9) it should help you review.

Figure 6-32
(a) An edge-on view of iron filings being sprinkled onto a piece of paper that covers a bar magnet. (b) The resulting magnetic field pattern made by the iron filings. Notice in particular the pattern the filings form beyond the magnet.

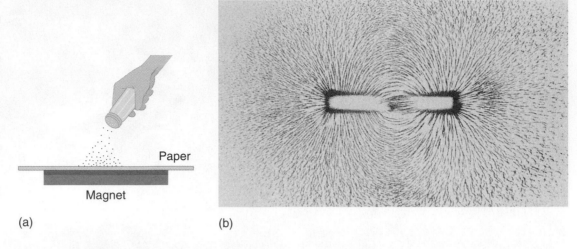

Paper

Magnet

(a)

(b)

condition resulting from a fairly small amount of UV. As we are frequently warned, too much exposure to UV can cause skin cancer. Modern civilization releases into the atmosphere chemicals that are reducing the amount of ozone, primarily chlorofluorocarbons from aerosol cans and chlorine from various sources. The United Nations is seeking international agreements to limit the amount of damaging chemicals that are released, but even if we suddenly and drastically reduce the release of these gases, the problem will not be solved, for chlorofluorocarbons remain in the atmosphere for over a hundred years, continuing to damage the ozone layer.

Earth's Magnetic Field

magnetic field A magnetic field exists in a region of space if magnetic forces can be detected in that region.

If a piece of paper is laid over a magnet and iron filings are sprinkled on the paper, a pattern such as that shown in **Figure 6-32b** appears. We say that a *magnetic field* exists in a region of space if we can detect magnetic forces in that region; for example, by observing the behavior of a magnetic compass. In a magnetic field, a small magnet aligns with the field.

A magnetic compass is simply a magnet that is free to rotate so that it can align with a magnetic field.

When a magnet is suspended so that it is free to rotate near the Earth, it always aligns so that its opposite ends point in specific directions, indicating that the magnet is aligning with a magnetic field related to the Earth. By plotting the direction of the magnetic field at various places on and off the Earth's surface, we can determine that the Earth's magnetic field has a shape similar to that of a bar magnet, as shown in **Figure 6-33**.

Notice that the poles of the Earth's magnetic field—those points toward which the magnetic field converges—are located near (but not exactly at) the poles of the Earth's rotation axis. As shown in **Figure 6-34**, the Earth's magnetic north not only does not coincide with the Earth's geographic north (fixed by our planet's spin axis), but can actually wander hundreds of miles. In addition, we have a terminology problem since the Earth's magnetic north is actually a "south" pole (and the Earth's magnetic south is actually a "north" pole). This is why a magnetic compass points north, since the north-seeking pole of its needle is attracted to the Earth's

Figure 6-33
The magnetic field of the Earth is similar in shape to that of a bar magnet.

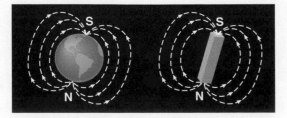

magnetic south pole (which is located north)! Actually a magnetic compass rarely points exactly north; in New York it points about 15° west of north, in Los Angeles it points about 15° to the east of north, while in Chicago it points almost due north.

Even though you might picture the Earth as containing within it a bar magnet that is not quite aligned with its rotation axis, this is not the case. The origin of the field is not completely understood. Magnetic fields (even those near a magnet) are caused by moving electric charges, and we are confident that the cause of the Earth's magnetic field depends upon the existence of Earth's molten iron core. However, the Earth's core is too hot to remain magnetic, which means that we are not dealing with a solid magnetic region. Apparently, circulation within the liquid regions of the Earth's core causes electric currents that result in the magnetic field. We know that the locations of the Earth's magnetic poles wandered in the past (Figure 6-34), and the magnetic field has even undergone complete reversals (about 300 times in the past 170 million years, perhaps as recently as just 30,000 years ago). At least in part, these changes are due to movements of the Earth's crust with respect to the interior. It has been hypothesized that a major reason for the reversals was a change in the Earth's rotation rate that resulted from bombardment with **meteoroids** and from volcanic eruptions. Many questions remain before hypotheses concerning pole wandering can be confirmed or disproved.

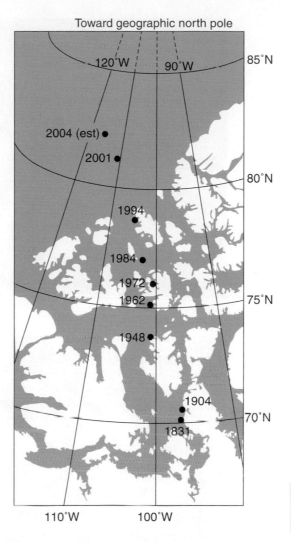

Figure 6-34
The Earth's magnetic poles are not located at its poles of rotation. The location of the magnetic north (magnetic "south" pole) is shown here; note also that its location changes with time.

meteoroid An interplanetary chunk of matter smaller than an asteroid.

One effect of the Earth's magnetic field became the first scientific discovery of the space age. Early spacecraft discovered the existence of electrically charged particles swarming in doughnut-shaped regions high above the Earth's surface (**Figure 6-35a**). The reason for these "Van Allen belts" is well understood. Just as an electric current (moving electric charges) causes a magnetic field, a magnetic field exerts a force on charged particles. When charged particles enter a magnetic field, they are forced to move in a spiral around the lines of the field, as shown in Figure 6-35b. In the case of the Van Allen belts, the charged particles (protons and electrons) come primarily from the Sun (the "solar wind") and have been captured by the magnetic field of the Earth.

You can produce a magnetic field by generating a current (moving electric charge) in a coiled wire.

We see the effect of the particles trapped within the Earth's magnetic field in the **auroras** that are often visible in the skies near the North and South Poles of the Earth (**Figure 6-36**). These beautiful displays of light are the byproducts of violent events on the Sun's surface, as a result of which streams of very energetic protons and electrons are sent out into space. Some of these particles manage to penetrate the Earth's magnetic field (our "shield"), but are forced by the field to spiral toward the magnetic north and south poles. These energetic particles collide with atoms and molecules in our atmosphere, exciting them to high energy levels. The atoms and molecules then release this energy as visible light. The displays of northern or southern lights can be breath-taking. (In Chapter 11 we will discuss the reason for the violent events on the Sun's surface.)

aurora Light radiated in the upper atmosphere due to impacts from charged particles.

The model that explains the Earth's (and other planets') magnetic fields as due to currents within a molten iron core is called the *dynamo model*.

Figure 6-35
(a) The Van Allen belts are regions where the magnetic field of the Earth traps charged particles from the Sun (the solar wind). The belts surround the Earth except near the poles. (b) Charged particles move in spirals around the lines of a magnetic field. This causes them to become trapped in the Earth's field.

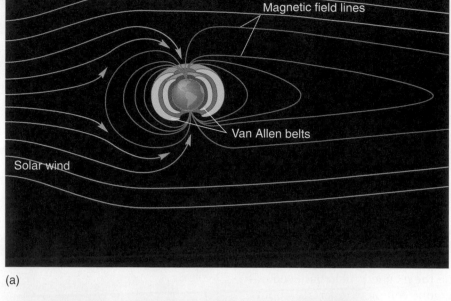

(a)

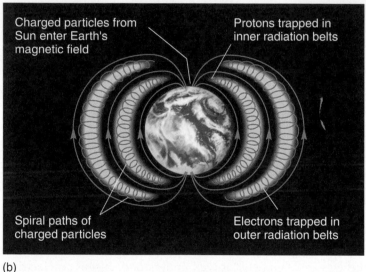

(b)

Figure 6-36
An aurora is caused by charged particles trapped in the Earth's magnetic field striking atoms in the upper atmosphere.

6-6   The Moon's Surface

The Earth is a planet of rich diversity, with interior heat and motion, moving crustal plates, a rich atmosphere, and the only known life forms in the universe. The Moon is a very different place. It is cold and lifeless, with little interior heat and no plate tectonics. The Moon has an extremely tenuous and "transient" atmosphere; it is continuously being produced by evaporation of surface materials, which are then lost by escape or impact back onto the surface. But there is much to learn about the solar system from studying the rocks and surface of the Moon.

The surface of the Moon can be divided into the ***maria*** (singular ***mare***) and the mountainous, cratered regions. From Earth, the maria appear darker than the other regions. The far side of the Moon, first seen when a Soviet spacecraft photographed it in 1959, is covered almost completely with craters.

Until the middle of this century, it was assumed that lunar craters were formed the same way almost all earthly craters are, by volcanic action, instead of by impacts of ***meteorites*** from space. Astronomers had two primary reasons for thinking that the

mare (plural **maria**) Any of the lowlands of the Moon or Mars that resemble a sea when viewed from Earth.

meteorite An interplanetary chunk of matter that has struck a planet or moon.

TOOLS OF ASTRONOMY

The Earth from Space

We Earthlings have learned much about neighboring planets by sending spacecraft past them, putting craft into orbit around them, and sending landers to their surfaces. Let us consider what could be learned about Earth if we lived on another planet and used similar technology to study the Earth.

The cloud layer of the Earth's atmosphere hinders detailed viewing of the surface from afar. Inhabitants of nearby planets, however, could detect differences between our continents and oceans, and they could see the Earth's white polar caps (as we see those on Mars). In order to determine what makes up the continents, oceans, and polar caps, our neighbors could analyze the spectra of light reflected from these areas. They would be able to determine that most of our planet is covered by water, and that our polar caps are frozen water. In addition, they would see our changing cloud cover and determine that it, too, consists of water.

Could they detect signs of life? Probably not. Although we think of ourselves as important, we have not changed our planet in a way that would be obvious to an observer on another planet or moon. The only sign of life they are likely to detect from afar is our electromagnetic noise; that is, radio and television broadcasts.

As an alien spacecraft approached Earth and went into orbit around it, it would obtain views such as those in **Figure B6-1**. Now its occupants might be able to see signs of life. Perhaps their first visible evidence of life would come when they viewed the

Figure B6-1
These photos of Earth show how beautiful our planet is. In the photo at the left, we can clearly see Egypt's Sinai Peninsula. (Note that the Red Sea and the Persian Gulf form part of a rift zone between the African and Arabian crustal plates.) At the right we see a storm system over the ocean. The windowsill of the spacecraft is at the bottom and the Earth's horizon is at the top.

dark side of our planet, for urban areas would be visible at night because of the wasted light that escapes from them into space (**Figure B6-2**).

If landers were sent to Earth to look for life, their ease in finding it would obviously depend upon where they landed. But if they were equipped like our *Viking* missions that landed on Mars to search for life, they would not only be capable of photographing our large animals, but they would be able to detect organic molecules in the soil (or ice) no matter where they landed.

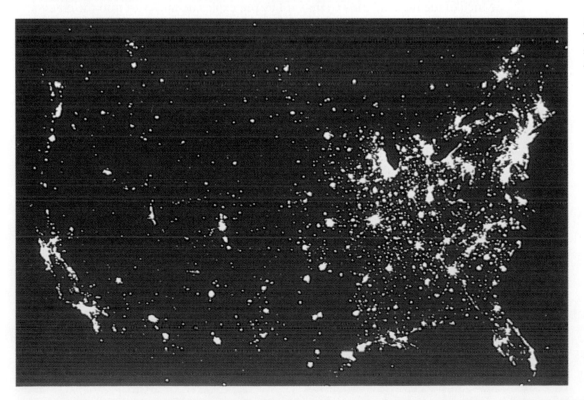

Figure B6-2
This is a composite of several photos showing the United States from space at night.

craters were volcanic: (1) Since almost all craters on Earth were thought to have formed that way, it was reasonable to assume that lunar craters were similar. (2) Lunar craters are very circular, but if you form a crater by throwing a rock into sand, the crater will only be circular if the rock is thrown straight down. Otherwise the crater will be elongated. Since meteorites would be likely to hit the Moon's surface at various angles, one would expect many impact craters to be somewhat elliptical in shape if they were formed by meteorite impact.

One observation, on the other hand, provided evidence against a volcanic origin for the craters: The floors of lunar craters are lower than the surrounding surface (as we can tell by measuring the shadows cast by the crater walls). Volcanic craters typically occur at the top of volcanic mountains, and their floors are higher in elevation than the surrounding territory.

We now realize that almost all (and perhaps all) lunar craters are the result of impacts. The arguments against an impact origin are answered as follows:

1. Earth has few impact craters because its atmosphere prevents any but the largest meteoroids from reaching its surface. Small meteoroids burn up in the air or are slowed enough by the air that they do not produce craters. In addition, craters produced far back in Earth's history have been eroded so that they are no longer noticeable. Lunar craters, on the other hand, do not suffer erosion on the airless Moon. Now that we have observed the Earth from space, we have found many more impact craters on Earth.

2. The argument that we would expect some impact craters to be elongated does not apply to lunar craters because they are not formed simply from material being "splashed" away by the impact. Instead, an approaching meteoroid is pulled toward the Moon by the force of gravity and strikes with such great speed that it penetrates below the surface, compressing and heating the Moon's material until an explosion occurs. Such an explosion is similar to a nuclear bomb ignited below the surface of the ground; it results in a circular crater in spite of the angle of the meteoroid's fall.

Much of the material thrown upward by the explosion falls back into the crater and forms its floor. Other material is thrown away from the crater to form the *rays* that are seen radiating from some craters (**Figure 6-37a**). The peaks found in the center of many

◆ Until about 1940, Meteor Crater in Arizona (Chapter 10) was thought to have a volcanic origin.

lunar ray A bright streak on the Moon caused by material ejected from a crater.

Figure 6-37
(a) Light-colored rays can be seen radiating from the prominent crater (named Tycho) near the bottom of this photo of the full Moon. They were formed by material ejected from the crater. (b) A close up of Tycho shows that it has a prominent central peak, the result of its surface rebounding after the impact of the meteorite that caused it.

(a)   (b)

craters (Figure 6-37b) are caused by a rebound of the surface after the explosion. Both the low floor levels and the central peaks are in accord with the idea that the craters are impact craters.

Figure 6-38
This bootprint on the Moon's surface was made by an *Apollo 11* astronaut, one of the first two humans to visit our nearest neighbor in space. The astronaut's prints will remain on the Moon for millions of years, for the only erosion that occurs there is due to tiny meteorites and is extremely slow.

Volcanic action did occur in the Moon's past, however, and it resulted in the Moon's maria. They were caused by the flow of dark lava onto lowland areas of the Moon. Most of the maria are roughly circular, leading us to believe that they were originally the floors of very large craters. We know, however, that the dark lava that fills them was not produced by the impacts that formed the craters, for within the maria are old, smaller craters that have themselves been mostly covered by lava. Thus the lava flowed from beneath the surface after the small craters had been formed inside the giant ones.

The top few centimeters of the Moon's surface are made up of loose powdery lava, small rocks, and mostly spherical pieces of glass, the result of bombardment by countless meteorites through the ages. **Figure 6-38** is a photograph of a footprint made in the dust of the Sea of Tranquility during the first visit to the Moon by a human, on July 20, 1969.

Many of the glassy spheres resulted from the melting and solidification of rock upon ejection from a crater.

The crust of the Moon (**Figure 6-39**) ranges in depth from about 60 to 100 kilometers and is thinner on the side facing the Earth than on the other side. When the maria were formed, molten lava released from within the Moon flowed toward the side with the thinner crust. It is no accident that the side of the Moon with more maria faces the Earth, for lava has a greater density than the rocks of the highland areas. Recall from Chapter 3 that tidal forces are what caused the Moon to slow its rotation. These same forces acted on the Moon's uneven distribution of mass to cause the denser side to face the Earth.

If you have studied geology, you know that earthly mountains were formed by the motion of plates within the Earth and by volcanic action. The lunar mountains were formed differently. They are simply the results of millions of ancient craters, one on top of another. Mountain ranges that border maria are the walls of giant craters whose floors are now covered with lava.

The density of the Moon is 3.35 grams per cubic centimeter. Since this is close to the average density of rocky material, we conclude that if the Moon has an iron core, that core must be small. In addition, the magnetic field of the Moon is less than one ten-thousandth of the Earth's, so this also indicates that the Moon cannot have a large molten iron core. It is thought that any iron core must be less than about 700–800 kilometers in diameter. Questions remain, however: First, we are not certain that planetary magnetic fields are due to molten iron cores, and second, some rocks brought back by *Apollo* astronauts were magnetized more than we would have expected in such a weak magnetic field. Perhaps the Moon's field was stronger in the past, or perhaps these rocks were magnetized by the Earth's or the Sun's magnetic fields.

Even though we still do not have answers to these questions, we have

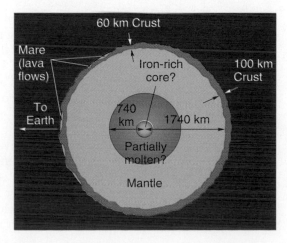

Figure 6-39
Studies of the Moon's interior have been made by striking the surface to produce "moonquakes" and then investigating the vibrations that result at other points on the surface. Cracks under the mare allowed lava to reach the floor of the ancient craters.

better data to work with thanks to the *Lunar Prospector.* This small spacecraft (drum-shaped and about 300 kilograms when full of fuel) was launched on January 6, 1998 for an 18-month polar-orbiting mission to the Moon. Its mission ended with a planned dive into a shadowed crater near the Moon's south pole on July 31, 1999. The science objectives of the mission were (a) to study the lunar atmosphere and crust for resources, such as minerals, certain gases, and water ice; (b) to learn more about the Moon's core, its size and content; (c) to map the Moon's gravitational and magnetic fields.

Tentative results from data collected by *Lunar Prospector's* instruments suggest that there is water ice in some polar craters, thus answering a question first raised in the early seventies and later by a 1994 mission. It is estimated that a few billion metric tons of water ice exist at each pole. However, this is indirect evidence, since the instrument detects hydrogen levels from which we infer water ice. The planned crash of the spacecraft was meant to liberate water vapor and dust, which could be observed by instruments on Earth. No such signature was seen, but then the chances of positive detection were less than 10%. Thus the question is still open as to whether ancient impacts by comets on the Moon deposited water ice that still exists in the permanently shadowed polar regions.

We now have a precise gravity map of the entire lunar surface, which indicates the existence of a lunar core, probably iron, with a diameter of more than about 600 km. Also, even though it was believed that the Moon had no global magnetic field such as Earth's, we now know that strong magnetic fields exist locally and that they are diametrically opposite large young impact basins on the surface. Could it be that an object such as the Moon can acquire magnetic characteristics from impacts, such as with asteroids and comets? Finally, we now have the first global map of abundances of ten key elements, several of which give us clues to the Moon's formation and evolution. Our knowledge of our satellite is changing as the data collected by *Lunar Prospector* is analyzed.

Another source of information about the Moon are the sensors left on the Moon by the *Apollo* missions to measure moonquakes. Some of these quakes were artificially produced by striking the Moon at various places, but we have detected about 3000 natural moonquakes a year, far fewer than the hundreds of thousands detected on Earth every year. These moonquakes were very weak, much weaker than our earthquakes, but they tell us that the interior of the Moon is essentially dead, with no major changes taking place. There is no evidence for plate tectonics on the Moon's surface and the cause of these quakes is the tidal interactions between Earth and Moon.

The sensors on the Moon that measure the intensity of quakes are called *seismographs.*

6-7   Theories of the Origin of the Moon

Evidence indicates that the Moon formed about 4.6 billion years ago. (The next section and its accompanying Advancing the Model box explain some of the evidence.) Until recently, there have been three theories of the origin of the Moon, called the *double planet (or co-creation),* the *fission,* and the *capture* theories. We will briefly describe each of these theories and look at the evidence to see which best fits the data. Then we will introduce a modern theory that seems to work better than any of the other three.

double planet (or co-creation) theory A theory that holds that the Moon was formed at the same time as the Earth.

- The ***double planet theory,*** which was suggested in the early 1800s, is the oldest. It holds that as the Earth formed from a spinning disk of material, not all of that material coalesced to form the Earth. A small part of it was left orbiting the Earth and formed into the Moon. In Chapter 7 we will discuss theories of the origin of the solar system and will see that this idea is entirely consistent with those theories.

ADVANCING THE MODEL

The Far Side of the Moon

Figure B6-3 is an image that includes most of the far side of the Moon, the side that is never seen from Earth. Although the image is in false color, you can see immediately that this side is different from the side that faces Earth (see Figure 5–23). The most obvious difference is that the far side has fewer maria. (The large green area is not really a mare. If you look closely, you can see that it is filled with craters. The colors are explained below.) To understand the reason for the lack of maria, refer to Figure 6-39, which shows a cross section of the Moon. The crust is thicker on the side opposite Earth. Maria are thought to have formed as the result of volcanism between 4 billion and 2.5 billion years ago. Lava is more likely to be forced to the surface where the crust is thinner, and since the side of the Moon that faces Earth has a thinner crust, that is where most of the volcanism occurred.

The image of Figure B6-3 is a composite of images made by *Mariner 10* in 1973 (as it passed the Moon on the way to Venus and Mercury) and the Jupiter-bound *Galileo* spacecraft in December, 1990. The image was made from separate photos that were taken with different filters on the cameras. This technique allows astronomers to determine the mineral content of the surface. For example, green and yellow indicate a large amount of iron and magnesium. Blue indicates a high abundance of titanium oxide, a mineral that is associated with maria. Knowing the mineral composition of each location helps astronomers in their quest to understand the Moon's past.

Figure B6-3
The far side of the Moon is shown here in false color. The prominent bull's-eye crater toward the left is called Mare Orientale. Different colors represent different mineral abundances.

The Moon

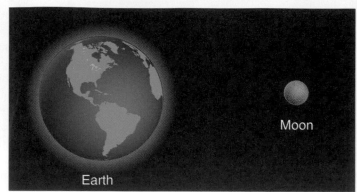

Earth

Moon

Moon Data

Moon	Value	Compared to Earth
Equatorial diameter	3476 km	0.27
Oblateness	0.0012	0.36
Mass	7.35×10^{22} kg	0.0123
Density	3.35 g/cm³	0.61
Surface gravity	1.62 m/s²	0.166
Escape velocity	2.4 km/s	0.21
Revolution period	27.322 days	
Sidereal rotation period	27.322 days	
Synodic period (phases)	29.531 days	
Surface temperature	−170° C to 130° C	
Albedo	0.11	0.36
Tilt of equator to orbital plane	6.68°	
Orbit		
Average distance from Earth	384,400 km (center-to-center)	
Closest distance	363,300 km	
Farthest distance	405,500 km	

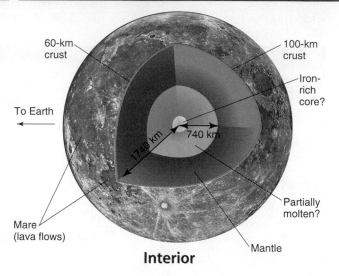

60-km crust

100-km crust

Iron-rich core?

To Earth

1740 km

740 km

Mare (lava flows)

Partially molten?

Mantle

Interior

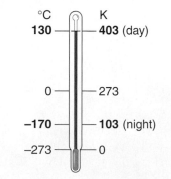

°C		K
130	—	**403** (day)
0	—	273
−170	—	**103** (night)
−273	—	0

Surface Temperature Extremes

QUICK FACTS

- One-fourth of Earth's diameter. Surface craters were caused by meteorite impacts. Maria were caused by lava flow. Extremely tenuous and transient atmosphere. Weak magnetic field. Likely to have formed as a result of a collision between Earth and a large object early in Earth's history.

A simple comparison of densities seems to rule out the double planet theory, for if the Moon formed along with the Earth, the two bodies should have about the same density. The Earth's density, however, is 5.52 grams/centimeter^3, much greater than the Moon's 3.35 grams/centimeter^3.

- In 1878, the astronomer Sir George Howard Darwin, son of the biologist Charles Darwin, proposed that the Moon was once part of the Earth and broke (or *fissioned*) from it due to forces caused by a fast rotation and solar tides. This ***fission hypothesis*** proposed that the large basin of the Pacific Ocean was the place from which the Moon was ejected.

fission theory A theory that holds that the Moon formed when material was spun off from the Earth.

The difference in density between the Earth and the Moon might at first glance seem to rule out the fission theory along with the double planet theory, but the crust of the Earth does have a density close to that of the Moon. If the Moon formed from material from the Earth's crust, we would expect its density to be just as we find it.

There is a problem with the fission theory, however. Astronomers have difficulty explaining how an object as massive as the Moon might have been pulled out of—or thrown off from—the Earth. No satisfactory mechanism for this event has been proposed. In addition, the Moon does not orbit in the plane of the Earth's equator, as it should if it were ejected from a spinning Earth.

- Early in the twentieth century, another theory was proposed. It holds that the Moon was originally a separate astronomical object that happened to come near the Earth and was captured by the Earth's gravitational field, so that it settled into orbit as the Moon. This is the ***capture theory.***

capture theory A theory that holds that the Moon was originally solar system debris that was captured by Earth.

There are also problems with the capture theory. If one astronomical object comes close to another, each of their paths will be changed by the gravitational force between them (**Figure 6-40**). However, one will not capture the other unless there is contact between the two or unless a third object is involved, so that the interaction of the three objects results in one of them being slowed down to an orbital speed. Such a near-collision between three objects seems highly unlikely.

Although we have known the density of the Moon for a long time, its chemical composition was not well known until the *Apollo* astronauts brought back soil and rock samples. These new data posed new problems for the three theories. In many ways, the chemical composition of the Moon is similar to that of the Earth's crust, for both have about the same proportions of some of the major elements: silicon, magnesium, iron, and manganese. The Moon, however, has smaller proportions of easily vaporized (***volatile***) substances (such as potassium and sodium) than does the Earth. The Moon has higher proportions of nonvolatiles (such as aluminum and titanium), which require a very high temperature to vaporize, than does the Earth's crust. The differences between terrestrial and lunar rocks suggest that the Moon was formed out of material that was at higher temperature than the material from which the Earth formed. This offers an additional argument against the fission theory.

volatile Capable of being vaporized at a relatively low temperature.

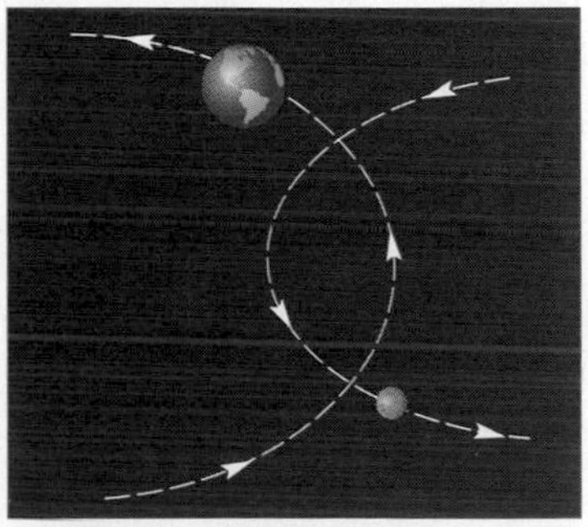

The differences between the rocks could support the capture theory if the Moon were formed elsewhere, such as closer to the Sun. In addition, the plane of the Moon's orbit is closer to the ecliptic than to the plane of the Earth's equator, supporting the idea that the Moon was orbiting the Sun before the Earth captured it. The conditions for such capture, however, are very improbable.

In order for the double planet theory to explain the differences between the rocks, we must assume that the Moon was formed by rocky debris orbiting the Sun in the plane of the ecliptic. This debris lost its volatile elements because of the solar heat and later coalesced to form the Moon, a very unlikely scenario.

Figure 6-40
If the Earth and another object (in this figure, a smaller one) had a near-collision, each object's path would be altered, but they would not begin to orbit one another.

Thus none of these three theories for the origin of the Moon gave a satisfactory answer to the question "How did the Moon get there?"

The Large Impact Theory

In the 1970s, A. G. W. Cameron and William Ward of Harvard proposed a new theory of Moon formation. They proposed that early in the Earth's history, our planet was struck at a glancing angle by a large object. The impact resulted in a fusion of the two objects, and material was thrown off from the combined object to form the present Moon (**Figure 6-41**). Computer simulations of such collisions show that if the impacting object has a mass nearly as great as Mars, heat resulting from the collision would vaporize material and eject enough of it into orbit to account for the mass of the Moon, once the material coalesced in its orbit around the Earth. The ***large impact theory* (or *collisional ejection theory)*,** as it is called, is able to explain both the similarities and the differences in the compositions of the Earth and the Moon. Since the impact would have vaporized the rocks, they should be depleted in water and volatile elements, in agreement with observations. If the collision occurred after the Earth was chemically differentiated, then the resulting Moon would have proportionally much less iron than the Earth, which is the case, as has been recently confirmed by the *Lunar Prospector.* If the plane of the impacting object were close to the ecliptic, so would be the plane of the resulting debris and thus the Moon, again in agreement with observations. The impact would most likely tip the Earth's rotation axis in the process.

Since the mid-1980s, a consensus has been building among astronomers that the large impact theory fits the data better than the other three theories. Recent theoretical work on the formation of the planets indicates that without large impacts, the Earth would rotate once every 200 hours instead of every 24 hours. A glancing impact by a large object would explain its present rotation rate. Like all new theories, the large impact theory needs to be tested against both existing data and new data as the years pass. Although it will probably have to undergo modification, it appears that astronomers may finally have found the answer to the age-old question of the origin of the Moon.

large impact theory A theory that holds that the Moon formed as the result of an impact between a large object and the Earth.

(a)   (b)   (c)

Figure 6-41

(a) According to the large impact theory, the young Earth was hit by a large object, perhaps as big as Mars.
(b) The two objects fused together, but much material was thrown off and went into orbit around the new Earth.
(c) Eventually, the material coalesced into the Moon.

6-8 The History of the Moon

The history of the Moon can be pieced together in several ways. For example, when we observe that one crater overlaps another (**Figure 6-42**), we know that the overlapping crater was formed after the other. Likewise, we know that the crater Tycho was formed relatively recently because its rays overlap the craters around it. Lunar rays darken with time, providing more information on the order of lunar events. However, such examination of the surface only provides information about the order of events; it does not allow us to determine how long ago the events happened. Reliable assessments of time scales had to wait for the *Apollo* missions and the 840 pounds of lunar material they brought back to Earth. The accompanying Advancing the Model box describes a procedure that uses radioactivity of rocks to determine their age. Such ***radioactive dating*** techniques have been indispensable in forming a model of the Moon's history.

The darkening is caused by sunlight and bombardment by tiny meteorites.

radioactive dating A procedure that examines the radioactivity of a substance to determine its age.

The Moon formed about 4.6 billion years ago. (The oldest rocks we found there are 4.42 billion years old.) We know that its surface was molten a few million years after the Moon was formed and conclude that the surface was probably molten from its formation, in agreement with the large impact theory. As the Moon cooled and solidified, cratering marked its surface. Most craters were formed between 4.2 and 3.9 billion years ago. Giant impacts that occurred near the end of the cratering period produced the areas where we see maria today. After most of the cratering ended, the interior of the Moon became hot enough (probably due to radioactivity) that molten lava flowed from beneath the surface and gathered in the floors of the giant craters. Rocks that astronauts gathered from the maria are between 3.1 and 3.8 billion years old, leading us to believe that this volcanic stage ended about 3.1 billion years ago.

The volcanic action described here consisted of lava seeping through cracks in the surface, so volcanic mountains did not form.

Cratering continues today, but at a much reduced rate from the Moon's early history. Three rocks have been found on Earth that are thought to have been ejected from the Moon by impacts within the past few million years. Our region of the solar system has been swept clear of most large chunks of matter, however, so meteorites large enough to produce noticeable craters on the Moon are now infrequent. We will see in Chapter 10 that the Earth is constantly being struck by debris from space, but much of it burns up in the atmosphere and never reaches the surface. The Moon has no atmosphere, however, so meteorites large enough to produce small craters must still strike it occasionally, although no new crater has ever been observed. In addition, ***micrometeorites*** strike the Moon's surface, further pulverizing its soil. Except for these impacts—and some recent visits by humans and their machines—the surface of the Moon changes very little. This is fortunate for astronomers, for the surface becomes a book in which we can read the Moon's distant history.

In Chapter 7 we will discuss the formation of the solar system and the sweeping up of its original matter into planets and moons.

micrometeorite A tiny meteorite.

Conclusion

To determine the diameter of the Moon, we must first know its distance, and to determine this distance, we must know the diameter of the Earth. By measuring the position of the Sun as seen from different locations on Earth, Eratosthenes in the third century B.C. was able to measure the Earth's diameter (see Chapter 1). A few centuries later, Ptolemy successfully used the method of parallax to measure the distance to the Moon, and he was therefore able to calculate its diameter.

The familiar phases of the Moon are caused by the relative positions of the Sun, Earth, and Moon and have been understood for more than two thousand years. Likewise, as the Moon orbits the Earth, alignments of the two bodies with the Sun produce both solar

Figure 6-42
In this photo, a small crater overlaps a larger one (named Gassendi), so that we know that it was formed after Gassendi was. Note that Gassendi has a central peak.

ADVANCING THE MODEL

Measuring the Ages of the Earth and Moon

From rocks brought back from the Moon by *Apollo* astronauts (**Figure B6-4**), we know that the Moon is at least 4.5 billion years old. We have found rocks on Earth that are 3.9 billion years old. How do we determine the age of these rocks—how do we "date" them? The method used is called *radioactive dating*, and to understand it, we must first look at the makeup of the nuclei of atoms.

Figure B6-4
Astronaut Jack Schmitt collects rock samples during the *Apollo 17* trip to the Moon. The astronauts brought back 243 pounds of rock on this mission alone.

There are two primary particles within the nucleus: protons and neutrons. The number of protons in a nucleus determines what element the atom is. For example, any atom with only one proton in its nucleus is necessarily hydrogen; two protons, helium; and so forth, up to uranium, which has 92 protons, more than any other naturally occurring element. Although different atoms of the same element all have the same number of protons, the number of neutrons in their nuclei may differ. Such different forms of an element are called *isotopes*. To specify which isotope we are talking about, we state the name of the element and the total number of protons and neutrons the nucleus contains. Thus uranium-238 has a total of 238 protons and neutrons in its nucleus (and it has 92 protons because every uranium nucleus has 92 protons).

The nuclei of some isotopes are unstable, in that their nuclei change spontaneously. When this happens, some isotopes emit a gamma ray and others emit a nuclear particle. The gamma ray or particle emitted is the radiation that comes from a radioactive material. We say that the isotope *decays*, but the term *decay* here does not refer to deterioration, such as wood undergoes when it decays. Rather it simply means that the isotope spontaneously emits radiation and in doing so changes to another isotope. For example, when rubidium-87 undergoes radioactive decay, it changes into strontium-87. Each isotope has its own characteristic rate of decay, which is called the *half-life* of the isotope—the amount of time needed for *half* of the isotope to emit radiation and change to another isotope. Half-lives range from tiny fractions of a second to billions of years. Isotopes with long half-lives are the ones that are useful in dating geological samples. However, we are making an important assumption here, one that

does seem to be supported by the observations. We accept that an atom that is thousands of years old is identical to an atom of the same species that is one second old. We accept that an unstable nucleus decays spontaneously, that there is no way to predict when it will decay, and that identical atoms behave, by themselves, in non-identical ways that follow rules of statistics.

The basic idea of radioactive dating is quite simple. Consider uranium-238, which has a half-life of 4.5 billion years. If we begin with a pure sample of uranium-238, we know that in 4.5 billion years half of it will have decayed to another isotope (and then, by a series of quicker decays, to lead-206). Thus by comparing the percentage of uranium-238 and lead-206 in a sample that was once pure uranium-238, we can tell how much time has passed since the sample was pure. The problem, of course, is that we must know that the sample was pure at the beginning.

There are several radioactive dating techniques, all of which depend upon an assumption about the original condition of the matter. One technique uses the fact that when molten material solidifies, the crystals that are formed have certain specific chemical elements in them. If uranium is present when crystallization takes place, certain types of newly formed crystals will contain uranium but no lead. If we examine rocks containing such crystals and find lead-206, we can be confident that this resulted from radioactive decay of uranium-238. The relative percentages of the two isotopes then allow us to calculate the time elapsed since the crystals formed.

Fortunately, the dating of a sample of rock does not depend upon just one isotope. For example, the radioactive isotope rubidium-87 (whose half-life is known) is found in some crystals, along with its decay product strontium-87. By comparing the concentrations of these isotopes to that of strontium-86, which is stable and is not created by radioactive decay, a second value can be found for the age of the sample. This value does not depend on whether or not the sample was initially pure, and thus provides a way to check the value found by uranium dating. Other techniques depend upon other isotopes; for example, potassium-40 decays into argon-40 with a half-life of 1.3 billion years.

Note that radioactive dating does not actually tell us the age of the Moon or the Earth, but only the minimum age, for the Moon and Earth existed before the rocks solidified. Knowledge about the solidification of matter and theories of planetary formation are used to determine how much time elapsed between the formation of the planetary body and the solidification of its surface rocks.

and lunar eclipses, exciting events that were mysterious to primitive people but are easily explained by simple geometry.

Understanding the Moon and the Earth allows us to better understand our immediate neighborhood, the solar system, and in return better understand our place in it. With the aid of what we learned from the *Apollo* missions, we are able to explain the features on the surface of the Moon and are becoming more and more confident that it originated in a tremendous impact between some object and the Earth. We are also piecing together an outline of the Moon's history from the information we have obtained from telescopic observations, from visits during the 1960s and 1970s, and more recently during the *Lunar Prospector* mission.

Study Guide

1. Knowledge of which of the following quantities will allow us to calculate the diameter of the Moon?
   A. The Moon's distance and speed.
   B. The Moon's angular size and speed.
   C. The Moon's angular size and distance.
   D. All three—distance, speed, and angular size—must be known.
   E. [None of the above.]

2. Suppose some object is known to be 6 kilometers away and is observed to have an angular size of 0.25 degree. What is its actual size?
   A. 26 kilometers
   B. 57.3 kilometers
   C. 60 kilometers
   D. 1.5 kilometers
   E. [None of the above.]

3. The phase of the Moon on a particular night is determined by
   A. where the Earth's shadow hits the Moon.
   B. the distance from the Earth to the Moon.
   C. the season of the year.
   D. the relative positions of the Sun, Earth, and Moon.
   E. the speed of the Moon in its orbit.

4. If the Moon was new last Saturday, what phase will it be this Saturday?
   A. Waning crescent.
   B. Waxing gibbous.
   C. At or very near first quarter.
   D. At or very near full.
   E. [Any of the above, depending upon other factors.]

5. If the Moon revolved around the Earth in the opposite direction, but with the same period,
   A. it would not exhibit phases.
   B. it would exhibit crescent phases but not gibbous phases.
   C. it would exhibit gibbous phases but not crescent phases.
   D. we would see it rise in the west and set in the east.
   E. [None of the above.]

6. If you observe the Moon rising in the east as the Sun is setting in the west, then you know that the phase of the Moon must be
   A. new.
   B. first quarter.
   C. full.
   D. third quarter.
   E. [Any of the above, depending upon other factors.]

7. Paris is about one-fourth of the way around Earth from Chicago. On a night when people in Chicago see a first-quarter Moon, people in Paris see
   A. a new Moon.
   B. a first-quarter Moon.
   C. a full Moon.
   D. a third-quarter Moon.
   E. [Any of the above, depending upon the time of night.]

8. Suppose that astronauts land somewhere on the Moon when it is new for Earth-bound observers. Which of the following statements would be true?
   A. Earth would appear full to the astronauts (assuming that they could see it).
   B. Around the landing site there might be bright sunlight.
   C. The landing site might be dark.
   D. [All of the above.]
   E. [None of the above.]

9. If the Moon always orbited directly above the equator, we would
   A. have eclipses every month.
   B. never have eclipses.
   C. have eclipses only at solstice.
   D. have eclipses only at equinox.

10. A lunar eclipse can occur
    A. only around sunset.
    B. only near midnight.
    C. only near sunrise.
    D. at any time of day or night.

11. The Moon is _____ during an annular eclipse as/than during a total solar eclipse.
    A. farther from Earth
    B. closer to Earth
    C. the same distance from Earth
    D. [No general statement can be made.]

12. The Sun is _____ during an annular eclipse as/than during a total solar eclipse.
    A. farther from Earth
    B. closer to Earth
    C. the same distance from Earth
    D. [No general statement can be made.]

13. Earth doesn't experience an eclipse of the Sun every month because
    A. sometimes the Moon is too far away.
    B. the Moon always keeps its same side toward the Earth.
    C. the Moon's orbit is not in the same plane as the Earth's orbit.
    D. you have to be in the right place to see a solar eclipse.

14. You are likely to see more of which type of eclipse during your lifetime?
    A. Solar eclipse.
    B. Lunar eclipse.
    C. [No general statement can be made.]

15. During a solar eclipse, the smallest possible width of the Moon's umbra is _____, and its largest possible width is _____.
    A. zero . . . a few kilometers.
    B. zero . . . a few hundred kilometers.
    C. zero . . . across the entire earth.
    D. about 100 kilometers . . . about a thousand kilometers.
    E. about 100 kilometers . . . across the entire Earth.

16. The density of an object is defined as
    A. its thickness.
    B. how much solid material the object contains.
    C. its mass.
    D. its volume.
    E. the ratio of its mass to its volume.

17. The Earth's atmosphere is made up of about
    A. 80% oxygen and 20% nitrogen.
    B. 50% oxygen and 50% nitrogen.
    C. 20% oxygen and 80% nitrogen.
    D. equal amounts of oxygen, nitrogen, and carbon dioxide.
    E. equal amounts of oxygen, hydrogen, and carbon dioxide.

18. Auroras result from
    A. the Earth's magnetic field and its rotation.
    B. the Earth's magnetic field and its revolution around the Sun.
    C. the Earth's magnetic field and the solar wind.
    D. the solar wind and the Sun's rotation.
    E. the motion of the Moon around the Earth.

19. Maria are
    A. lunar mountains.
    B. lunar highlands.
    C. flat plains.

D. near the lunar poles, and nowhere else.
E. [More than one of the above.]

20. Which of the following theories of the Moon's origin seems to fit the data best?
    A. The capture theory.
    B. The fission theory.
    C. The double planet theory.
    D. The large impact theory.
    E. [Either A or C above.]

21. The fact that the density of the Moon is somewhat different from that of the Earth is an argument against which theory of the origin of the Moon?
    A. The capture theory.
    B. The fission theory.
    C. The double planet theory.
    D. The large impact theory.
    E. [Both B and C above.]

22. When we see an unfamiliar object at a distance, how can we judge its size?

23. What is the angular size of the Moon? How does this compare to the angular size of the Sun?

24. Name eight different phases of the Moon in the order in which they occur.

25. Define umbra and penumbra.

26. Total (and annular) solar eclipses occur more frequently than do total lunar eclipses. Why, then, will you probably observe many more lunar eclipses than solar eclipses during your lifetime?

27. Describe the appearance of a totally eclipsed Moon. Why is it not completely hidden from view?

28. Why are penumbral lunar eclipses not easily detectable?

29. Why might a partial solar eclipse go unnoticed by most people?

30. Why are some solar eclipses annular rather than total?

31. Show on a sketch the relative positions and sizes of the Earth's core, mantle, and crust.

32. What is a magnetic field, and how can one be detected?

33. What are the Van Allen belts and what causes them?

34. List some evidence for the theory of plate tectonics.

35. Name and describe four theories for the origin of the Moon. Which best fits present data?

36. What caused the craters and rays on the Moon?

37. Explain how we can determine the relative order in which events occurred in the formation of the Moon's surface.

QUESTIONS TO PONDER

1. Since Ptolemy lived long before instant distant communication was possible, he was not able to coordinate his observations of the Moon with someone far around the Earth. Propose a method by which he might have been able to observe parallax of the Moon.

2. If you observe the Moon with first one eye and then the other, do you detect a parallax shift against the stars? Why or why not?

3. Suppose that you see an object in the sky that you have never seen before. You estimate that it is 100 feet long. Another person sees the same object and estimates that it is 20 feet long. Explain how this can happen.

(Hint: Each of you is making a different assumption about some other factor.)

4. If the Moon were twice as big as it is, but four times as far away, how much smaller or larger would its angular size be?

5. If you look at the Moon tonight and then again tomorrow night at the same time, will it be farther east or farther west on the second night? If you observe it at 8 P.M. tonight and then again at 9 P.M., will it appear farther east or farther west at the later time? Explain the discrepancy.

6. At about what time does the first quarter Moon rise? Set?

7. Will the Moon appear crescent or gibbous when it is at position *X* in **Figure 6-43**? At position *Y*? At about what time will the Moon rise if it is at position *Y*?

11. Discuss the danger involved in viewing a solar eclipse and describe four ways to view such an eclipse safely. (Hint: See the Activity "Observing a Solar Eclipse.")

12. We say that the Moon's maria were formed by volcanic action, but that the craters of the Moon are not volcanic in origin. Explain this apparent contradiction.

13. **Figure 6-44** is a multiple exposure of the Moon and Venus as they set one night over Tulsa. Explain why the Moon's distance from Venus changes as time passes.

14. The expression "Once in a blue Moon" is commonly used to describe a rare phenomenon. It was thought that the origin of this expression had to do with the rare occurrence of two full Moons in the same month, the second one being called "Blue Moon." However, this is not the case. A literature search suggests that the term "Blue Moon" was assigned to the third full Moon in a season that had four full Moons. During which month(s) can we not have a "Blue Moon" if we adopt the first (erroneous) explanation? During which month(s) do you expect to get "Blue Moons" if we adopt the second explanation? (Hint: You may want to consult *Sky & Telescope*; the magazine was actually the source of the erroneous modern convention for the meaning of the expression.)

15. Describe some features of the Earth's surface that are direct consequences of the motion of tectonic plates.

16. Craters have been formed on the Earth and on the Moon by meteorite impact. The Earth has a much stronger gravitational field than does the Moon, and yet we find more craters on the Moon. Explain this apparent contradiction.

17. What leads us to conclude that the Moon does not have a large iron core?

18. Name and describe three different theories of the formation of the Moon. Which of these theories is considered most likely correct? Describe the evidence that leads us to that conclusion.

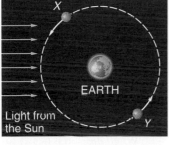

Figure 6-43 (Question 7)

Figure 6-44 (Question 13)

8. If you see the Moon high overhead shortly before sunrise, about what phase is it in?

9. At the same time that people in Chicago see a total solar eclipse, what type of eclipse is seen by people in Evansville, Indiana? (Evansville is about 250 miles south of Chicago.) Answer the same question for a total lunar eclipse.

10. Even at the midpoint of most total lunar eclipses, the Moon is not uniformly illuminated. Instead, one edge of the Moon is usually much brighter than the rest. Why is this? What would be necessary in order for the red color to appear uniform across the Moon?

1. Compare the distances reached by the umbral shadows of the Moon and the Earth. Compare the diameters of the two bodies. Discuss any relationship you see.

2. In order to show that the Moon's synodic period is about 2 days larger than its sidereal period, refer to Figure 6-12 and consider the following steps. (a) During one sidereal period (27.322 days), how many degrees has the Earth covered on its orbit around the Sun, from point A to point B? (Hint: The Earth covers 360° in about 365 days.) (b) Let us assume that we need an additional x days for a synodic period to be completed. During this period, both the Earth and the Moon move on their respective orbits. How many degrees does the Earth cover on its orbit around the Sun in this period? (c) To find x, set up a ratio by taking into account that the Moon covers 360° in a sidereal period.

3. You just bought a telescope that allows you to see clearly two separate stars if the angular distance between them is at least 1 arcsecond. What is the diameter of the smallest crater you can see on the Moon?

4. At what distance could we see clearly an asteroid of diameter 200 kilometers if the best telescope on Earth could give us a resolution of 0.25"?

CALCULATIONS

1. Do-It-Yourself Phases

This is an important activity that is worth the trouble if you want to understand the phases of the Moon. We will simulate the Sun/Earth/Moon system as it is shown in Figure 6-11. For the Moon you will need something like an orange, a grapefruit, or a softball. The softball would be best because it is rounder than the other choices. For the Sun, you can use a bright light across an otherwise dark room. (You could go outside and use the real Sun, but you must be careful not to look directly at it. This method would work best when the Sun is low in the sky.) Your head will be the Earth, and one of your eyes will be you.

Hold the ball (your Moon) out at arm's length so that it is nearly between you and your Sun. Now observe it as you move it around to the left until it is at 90 degrees to the Sun, the first-quarter position, as shown in **Figure 6-45**. Did you see its growing crescent as you were moving it? Continue to move it around your head and observe it as it changes phase. (When it gets directly behind you from the Sun, you will eclipse it.)

ACTIVITIES

Figure 6-45
The person is holding a ball that represents the Moon at first quarter. When doing this, you should put the light bulb farther away than indicated here, perhaps 10 to 15 feet away.

Now fix the Moon at the first-quarter position by laying it down on something in that position. Turn your head (and body) slowly around and around toward the left to simulate the Earth rotating. When the Sun is directly in front of you, it is noontime. As you lose sight of the Sun, it is sunset, about 6:00 P.M. It is midnight when the back of your head is toward the Sun. When the Moon is at first quarter and it is sunset on Earth, where do you see the Moon?

Put the moon in various other positions and observe it as you rotate your head to simulate different times of day. Answer the following questions:

1. If the Moon is at third quarter, at about what time will it rise? At about what time will it set?

2. At about what time will a full Moon appear highest overhead?

3. If you see the Moon in the sky in mid-afternoon, about what phase will it be?

2. Observing the Moon's Phases

This exercise will take several nights. Find a calendar or daily newspaper that lists Moon phases, or ask your instructor for this information, so that you can begin your observations three or four nights after the new Moon. Observations should start at sunset or shortly thereafter. At least four observations should be made, continuing for at least three more clear nights during the two weeks following your first night.

It is important that the observations be made from exactly the same place and at exactly the same time of night. Stand at the same place, not just in the same parking lot, for example. You might even go so far as to mark your location with chalk.

1. On each night of your observations, after you have arrived at your observing location, use a full sheet of paper to make a sketch of the position of the Moon relative to buildings, trees, and the like on the horizon. This sketch should show how things look to you and should not be a map. Label buildings, such as "Student Union." Also include prominent stars you see near the Moon. Draw the Moon in the shape it appears, and with the correct apparent size relative to objects on the horizon. Finally, write the date and time on your paper.

2. On one of the nights, repeat the observation after waiting about an hour. Use the same sketch and show the new position of the Moon.

3. When you have finished your four (or more) observations, make a general statement about how the Moon changed position and phase from one night to another. Look for a pattern in the Moon's behavior and explain that pattern based on the Moon's motion around the Earth.

3. Observing a Solar Eclipse

There is a misconception that the Sun emits especially harmful rays of some kind during a solar eclipse. This is a particularly anthropomorphic idea, for it would mean that somehow the Sun knows when Earth's Moon is about to block sunlight from the Earth. Naturally, the radiation emitted by the Sun during an eclipse is no different from that emitted at any other time.

Like most misconceptions, however, there is an element of truth in this idea. In fact, the Sun continuously emits radiation that is harmful to our eyes: *infrared* radiation.* If you were to look at the Sun anytime, this radiation would harm your eyes, but normally you are not able to look at the Sun. If you attempt it, your eye will quickly close because of the intense light. When the Sun is nearly totally eclipsed, however, its light is dim enough that you are able to look at it. So during an eclipse it would be possible for a person to stare directly at the Sun for some time, all the while unknowingly absorbing the harmful infrared rays in his or her eyes.

There are a few safe methods of observing the Sun during an eclipse. First, we might use a telescope with a solar filter attached. This is a filter that blocks out some 99.99% of the Sun's light, allowing just enough through for us to see the Sun.

A more convenient way to use a telescope to observe an eclipse is illustrated in **Figure 6-46**, which shows one of the authors using a telescope to project the partially eclipsed Sun onto a screen. In this case, a reflecting device was mounted on the telescope to cause the image to appear off to the side. This photo was

* The Sun also emits ultraviolet radiation that can harm the eyes. However, the primary danger during solar eclipses is infrared radiation.

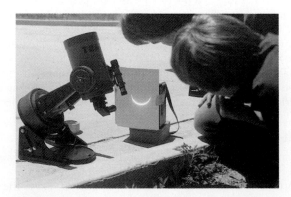

Figure 6-46
The telescope has a mirror (called a *star diagonal*) attached to it so that it projects the image of the partially eclipsed Sun onto the screen at the side.

taken when the Moon had progressed much of the way across the solar disk. It was near noon at the time, but the sky had still not darkened noticeably. Even a little of the Sun's light is enough to give us a bright day on Earth.

A third method of safely observing a solar eclipse—one requiring little equipment—is by pinhole projection. To use this method, all you need is a piece of cardboard with a hole in it and a piece of paper to use as a screen. **Figure 6-47** illustrates the method. Try holes of different sizes, from one millimeter to one made with a paper punch. (A hole made by a pin will probably be too small.) To see the image better, you might use large pieces of cardboard to shield your screen from reflected light or work in a dark room that has an opening facing the Sun. Block all of the opening except for your pinhole.

A fourth method of observing an eclipse is to obtain a safe filter through which you can view the Sun directly. Extreme care must be exercised here, however. If your filter does not block the Sun's infrared radiation, it may damage your eyes. (Smoked glass is definitely not recommended.) To avoid the possibility of injury, it is recommended that you be very sure of your filter or that you use one of the other methods.

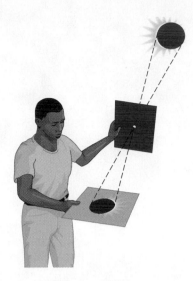

Figure 6-47
A pinhole projector can be used to view a solar eclipse, but you must shield your screen better from scattered light than is done here.

The web sites for the magazines *Sky & Telescope* and *Astronomy* (and the magazines themselves) have information about upcoming eclipses. An authoritative site for eclipses is maintained by Fred Espenak at NASA's Goddard Space Flight Center. At the same Center you can also find information about past, current and future NASA missions (including *Lunar Prospector*) and pages devoted to our planet and the Moon.

1. "Solar Eclipses That Changed the World," by B. E. Schaeffer, in *Sky & Telescope* (May, 1994).

2. "The Evolution of Continental Crust," by S. R. Taylor and S. M. McLennan, in *Scientific American* (January, 1996).

3. "The Dynamic Aurorae," by S.-I. Akasofu, in *Scientific American* (May, 1989).

4. "The Scientific Legacy of Apollo," by G. Jeffrey Taylor, in *Scientific American* (July, 1994).

EXPANDING THE QUEST

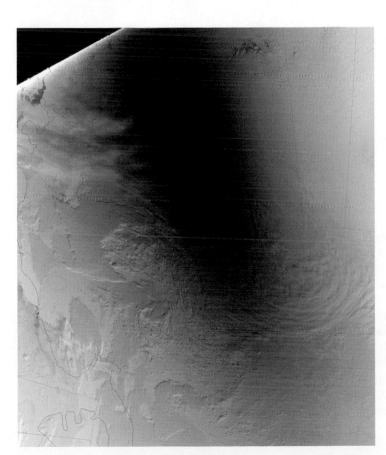

The umbra and penumbra in the Moon's shadow can clearly be seen in this true-color image of the November 13, 2003 total solar eclipse over Antarctica. The tip of the roughly 500-km long shadow is pointing toward Africa. The Sun was about 15° above the horizon, just rising over Antarctica, when this image was taken from space.

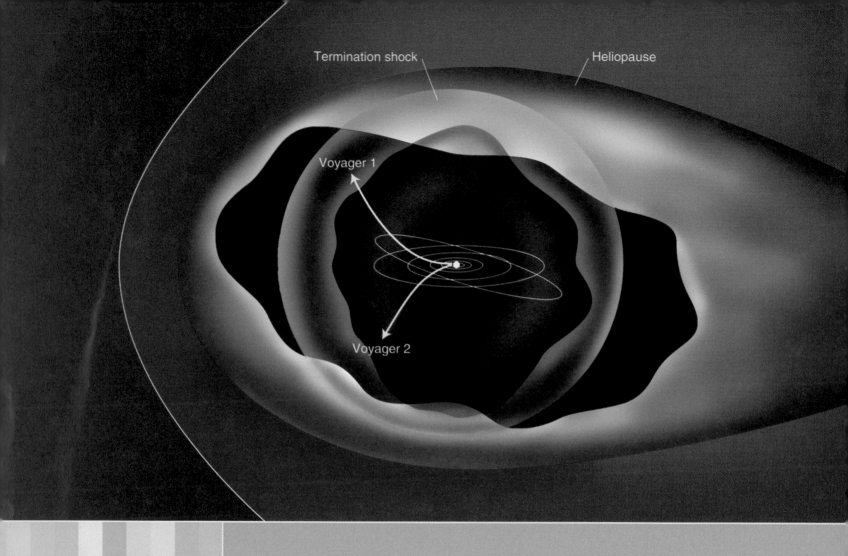

Termination shock   Heliopause

Voyager 1

Voyager 2

A Planetary Overview

The *Voyager 1* and *Voyager 2* spacecraft started their Grand Tour from Earth to the outer planets in 1977.

IN THE PERIOD 1976–78, AN ASTRONOMICAL EVENT TOOK PLACE that only occurs about once every 177 years. During this time the large outer planets Jupiter, Saturn, Uranus, and Neptune were bunched closely together looking out from Earth, as they traveled their orbits around the Sun. This had not happened since the time of Napoleon. Starting in 1972, NASA scientists and engineers planned to take advantage of this situation by sending out two space probes to explore these planets. *Voyager 1* was launched on September 5, 1977, on a faster trajectory than *Voyager 2*, which was launched 16 days earlier. Over the next few years they revolutionized our understanding of the solar system. Much of what we know about the outer planets, and most of the photos of them in this book, are due to these missions.

Keeping in touch with the two tiny spacecraft turned out to be a trial in overcoming adversity and avoiding disaster. *Voyager 2*'s onboard computers often detected emergencies when none existed. For example, during the initial launch of the rocket from Earth, the computers interpreted the rapid acceleration of the spacecraft as outside normal operation and tried to reprogram the thrusters to slow it down. Later, the craft's radio receiver blew a fuse and the craft could receive only a limited range of signals from Earth. NASA engineers overcame these problems, and the two space probes continued on their way.

Voyager 1 arrived at Jupiter in March of 1979, about four months earlier than *Voyager 2*, and the two craft sent back unprecedented views of the planet during the spring and summer. They discovered the rings of Jupiter, which had never before been seen, and then went on to explore the large moons Io, Ganymede, Callisto, and Europa.

Voyager 1 began making discoveries about Saturn in October, 1980, when it was still 30 million miles away from the planet. *Voyager 2* followed in August of 1981 and continued its string of calamities; it was struck by a stray piece of one of Saturn's rings, jamming the camera platform. But between them, the spacecraft sent back images of the complex structure of Saturn's rings and the violent storms in Saturn's atmosphere, and detected an atmosphere on Titan, Saturn's largest moon.

Four and a half years passed before *Voyager 2* reached Uranus, making its closest approach in January, 1986. NASA engineers moved the craft itself to position its camera, which was still jammed on its platform. Considering the lack of light from the far-distant Sun and the speed of the spacecraft, this was an incredibly difficult task, but it worked. Photos came back of the Uranian rings and its major moons, before the craft was reprogrammed for the trip to Neptune. *Voyager 2* passed within 3100 miles of Neptune in August, 1989, and then went on into deep space.

The spacecraft are now on an extended mission, searching for the outer limits of the Sun's magnetic field and outward flow of the solar wind (the heliopause boundary). Once the spacecraft cross this boundary, they will be able to take measurements of the interstellar fields, particles, and waves without being influenced by the solar wind. In July 2003, *Voyager 1* (*Voyager 2*) was at a distance of 88 AU (70 AU) from the Sun, escaping the solar system at a speed of about 3.6 AU/year (3.3 AU/year), 35 degrees out of the ecliptic plane to the north (48 degrees, south).

Before we begin to examine the individual planets and satellites in our solar system, we need to develop a common framework in order to understand similarities and differences among these objects. We cannot possibly expect to understand the universe by simply describing in detail all known characteristics of all its objects. The information we collect must be used to develop an understanding (a model) of how these characteristics developed, and why certain objects (or groups of objects) seem to have similar or different characteristics from others in the same system. Such knowledge will allow us to make reasonable generalizations about similar systems everywhere in the universe.

In previous chapters we have pointed out some patterns among the planets of our solar system. For example, Kepler's third law tells us of the relationship between a planet's distance from the Sun and its period of revolution. Before turning to the individual planets, an examination of other patterns of similarities and differences among the planets will be helpful. In this chapter we present our current understanding of our own planetary system. In the next three chapters we will study in detail the major objects in it.

You might think 3100 miles is a large distance, but remember that Neptune is 30,800 miles across and almost 2,700 million miles from Earth. Sending a space probe that close to Neptune is like using a rifle to shoot a penny 2 miles away—and hitting it!

7-1 Sizes and Distances in the Solar System

To say that the Sun is the largest object in the solar system is a gross understatement. It contains almost all the mass (about 99.85%) of the solar system and is about ten times larger in diameter than the largest planet (Jupiter). **Figure 7-1** shows the planets drawn to scale. At the bottom of the drawing, you see the partial disk of the Sun. The Sun is so large that if it had been drawn as a complete circle fitting the page, many of the planets would have been too small to see. The Sun's diameter is about 1,390,000 kilometers, while the Earth's diameter is about

Figure 7-1
This shows the Sun, planets, and a few of the large moons drawn to scale.

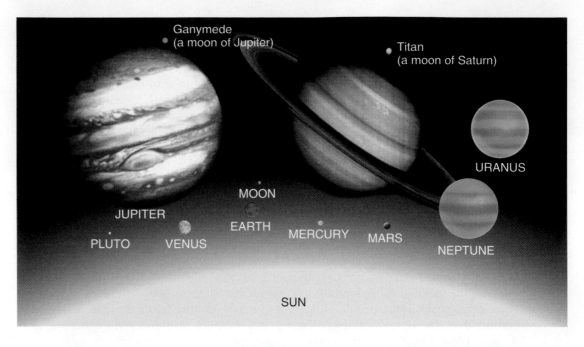

Ganymede
(a moon of Jupiter)

Titan
(a moon of Saturn)

URANUS

MOON

JUPITER

EARTH   MERCURY   MARS

PLUTO   VENUS

NEPTUNE

SUN

Figure 7-1 shows the planets as disks, but they are actually spheres. In trying to imagine their comparative sizes, think of them that way.

13,000 kilometers. Thus the diameter of the Sun is more than 100 times that of Earth. To better picture this, think of the Sun as an object the size of a basketball, a sphere about one foot in diameter. On this scale the Earth would be a BB, about the size of the head of a pin, an eighth of an inch in diameter (**Figure 7-2**).

Jupiter, the largest planet, is much larger than the Earth. Its diameter, in fact, is about 11 times that of the Earth. On our scale, with the Sun as a basketball and the Earth as the head of a pin, Jupiter would have a diameter of about 1-1/4 inches, which is about the size of a Ping-Pong ball. It's still not much compared to the Sun.

Pluto is the smallest planet, with a diameter about one-fifth that of Earth. In our scale model it would be a grain of sand, less than 1/32 inch across! Appendix C lists the actual sizes of the planets, along with their sizes compared to the Sun and the Earth.

Now let us consider the distances between the planets. **Table 7-1** shows the average distance of each planet from the Sun in astronomical units and according to our

As we discuss the sizes of solar system objects and the distances between them, try to form a mental picture of the relative distances rather than just memorizing the values.

Figure 7-2
If the Sun were the size of a basketball, the Earth would be the size of the head of a shirtpin.

Table 7-1

Average Distances of the Planets from the Sun

Object	Distance from the Sun (AU)	On Our Scale	
		Mean Diameter	Distance from Sun
Sun	—	1 foot (30.48 cm)	—
Mercury	0.39	0.04 inch (0.11 cm)	42 feet (13 m)
Venus	0.72	0.10 inch (0.27 cm)	78 feet (24 m)
Earth	1.0	0.11 inch (0.28 cm)	107 feet (33 m)
Mars	1.52	0.06 inch (0.15 cm)	164 feet (50 m)
Jupiter	5.20	1.21 inch (3.06 cm)	559 feet (170 m)
Saturn	9.58	1.00 inch (2.55 cm)	1030 feet (0.2 mile) (314 m)
Uranus	19.20	0.44 inch (1.11 cm)	2064 feet (0.4 mile) (629 m)
Neptune	30.05	0.42 inch (1.08 cm)	3229 feet (0.6 mile) (984 m)
Pluto	39.24	0.02 inch (0.05 cm)	4217 feet (0.8 mile) (1285 m)

model. To continue the model in which the Sun is a basketball, we might put the basketball at one end of a tennis court. A BB at the opposite end of the tennis court would be the Earth. Two football fields away is the Ping-Pong ball Jupiter. Pluto would be a grain of sand almost a mile away! Between these objects we put nothing—or almost nothing. There are only the other planets, all smaller than Jupiter, and some even smaller objects.

More than 4000 *asteroids* that are too small to include in our scale model have been discovered in the solar system. The largest of these asteroids has a diameter of about 1000 kilometers, or 600 miles. Perhaps another 100,000 much smaller asteroids orbit the Sun, most of them in the asteroid belt between Mars and Jupiter. In addition, trillions of *comets* orbit the Sun in a huge shell at a distance of about 40,000–100,000 AU. We will discuss these objects in more detail in Chapter 10.

asteroid Any of the thousands of minor planets (small, mostly rocky objects) that orbit the Sun.

comet A small object, mostly ice and dust, in orbit around the Sun.

Measuring Distances in the Solar System

In Chapter 2 we discussed how Copernicus used geometry to calculate the relative distances to the planets. That is, he was able to calculate that Mars is 1.5 AU from the Sun, although he could not determine the value of an astronomical unit. Today we can measure the distances to planets using radar. We send radar signals to a planet and measure the time required for the signal to reach the planet and bounce back. Then, knowing that radar signals travel at the speed of light (3×10^8 meters/second, or 3×10^5 kilometers/second), we can calculate the distance to the planet.

EXAMPLE

Suppose we bounce a radar signal off of Mars. The signal returns to Earth 22 minutes after being transmitted. How far away is Mars?

SOLUTION First, we realize that 22 minutes is the time the signal takes to reach Mars and return to Earth. So a one-way trip requires 11 minutes. Now let's change 11 minutes to seconds (since our signal speed is given in kilometers/second).

$$11 \text{ min} \times \frac{60 \text{ s}}{1 \text{ min}} = 660 \text{ s}$$

Now,

$$\text{velocity} = \frac{\text{distance}}{\text{time}}$$

so

$$\text{distance} = \text{velocity} \times \text{time} = (3.0 \times 10^5 \text{ km/s}) \times (660 \text{ s}) = 2.0 \times 10^8 \text{ km}$$

To check that this is a reasonable answer, recall that one astronomical unit is 1.5×10^8 kilometers. So our calculated distance is 1 1/3 AU. Since the orbit of Mars is 1.5 AU from the Sun, the distance from Earth to Mars varies from about 0.5 to 2.5 AU. Therefore, at some point in its orbit, it is possible for Mars to be at our calculated distance from Earth.

TRY ONE YOURSELF

When Venus is at its closest distance to Earth, it requires about 4.7 minutes for a radar signal to travel to Venus and back. What is the distance to Venus? Convert the answer to astronomical units and check it with the correct distance given in Table 7-1.

ADVANCING THE MODEL

The Titius-Bode Law

As we pointed out in Chapter 2, the relative distances to the planets were known in Copernicus's time. We saw that Kepler used this data to formulate his third law. From the time of Copernicus, people have wondered why the planets are at the distances they are. Is there any pattern to these distances?

In 1766, a German astronomer named Johann Titius found a mathematical relationship for the distances from the Sun to the various planets. The rule was publicized by Johann Bode, the director of the Berlin Observatory, in 1772, and is known today as the Titius-Bode law or simply Bode's law. **Table B7-1** illustrates how the law works. Column 1 shows a series of numbers starting with zero, jumping to three, and then doubling in value thereafter. Column 2 was obtained by adding 4 to each of those values. Finally, to get column 3, we divide each of the column 2 values by 10. Now compare these figures to the measured distances (in AU) of each of the planets from the Sun.

The table shows that the Titius-Bode law fits fairly well, except that there is a gap: No planet is found at 2.8 AU from the Sun. The law seems to indicate that there should be a planet between Mars and Jupiter and, further, that the planet should be 2.8 AU from the Sun. In addition, the law predicts that if other planets were found beyond Saturn, the next one would be about 19.6 AU from the Sun.

Bode published the law in 1772. Nine years later, in 1781, the planet Uranus was discovered by William Herschel in England. Its distance from the Sun was 19.2 AU. The Titius-Bode law predicted 19.6 AU. A close fit!

With this confirmation of the validity of the Titius-Bode law, a group of German astronomers (who called themselves the Celestial Police) divided the zodiac into regions, planning to assign a specific region to each of a number of astronomers who would systematically search for the missing planet at 2.8 AU. The searchers did not find it, however. Instead, a monk who was working on a different project discovered the largest of the asteroids at the distance predicted for Bode's missing planet. (See the Advancing the Model box later in this chapter.) Although it was first thought that this was the missing planet, the discovery within a few years of other objects at about the same distance made it obvious that things were not this simple. The Titius-Bode law could not account for the large number of "planets" between Mars and Jupiter. (However, it is possible that tidal interactions from Jupiter did not allow all these objects to coalesce and form a planet at this distance from the Sun.)

Two more planets have been discovered since the discovery of the asteroids. How well do they fit the Titius-Bode law? **Table B7-2** shows the complete comparison. Notice that Neptune does not fit

Table B7-1

Planetary Distances According to the Titius-Bode Law Compared to Today's Values

A Series of Numbers	Add 4	Divide by 10	Today's Measured Distance (AU)	Planet
0	4	0.4	0.39	Mercury
3	7	0.7	0.72	Venus
6	10	1.0	1.0	Earth
12	16	1.6	1.52	Mars
24	28	2.8		
48	52	5.2	5.20	Jupiter
96	100	10.0	9.58	Saturn

Table B7-2

How Celestial Objects Fit the Titius-Bode Law

Bode's Law Prediction	Today's Measured Distance (AU)	Object
0.4	0.39	Mercury
0.7	0.72	Venus
1.0	1.00	Earth
1.6	1.52	Mars
2.8	2.77	Ceres
5.2	5.20	Jupiter
10.0	9.58	Saturn
19.6	19.20	Uranus
	30.05	Neptune
38.8	39.24	Pluto

ADVANCING THE MODEL

The Titius-Bode Law (Cont'd)

the prediction at all, but that Pluto comes fairly close. So here we see that the law no longer fits all the data. (Since Pluto's characteristics are more like those of a satellite than a Jovian planet, it has been suggested that early in the history of the solar system, collisions and close encounters between the outer planets and Pluto-like objects could have changed Pluto from being a satellite to being a planet, while knocking Uranus on its side and moving Neptune closer to the Sun. However, recent work has all but ruled out the idea that Pluto was originally a satellite of Neptune. The suggested explanations for the rotation of Uranus, and for that matter Venus, are still at the speculative stage.)

Today the Titius-Bode law is considered little more than a curiosity rather than a scientific law.* There are several reasons for this: The law is not accurate even for the planets it fits; it does not fit Neptune at all; and it is not internally consistent. (Note that the num-

ber in column 1 of Table B7-1 is not doubled in one case.) Surely, though, it is not just coincidence that the law fits even in its limited way. In this chapter we show that theories proposed for the formation of the solar system account for the fact that the more distant a planet is from the Sun, the farther it is from other planets. Still, the fact that the Titius-Bode law fits as well as it does must be considered a coincidence. Astronomers may be able to judge its significance better when, at some future date, they are able to observe planetary spacing around other stars.

*Relationships such as the Titius-Bode law are said to be *empirical*. This means that they are found to work, but they are not related to any theoretical framework; we don't know why they work. The Titius-Bode law isn't a particularly good empirical law, however, for the reasons noted above.

When the nearest planet, Venus, is closest to Earth, a radar signal still requires nearly 5 minutes to get there and back. The great distances in the solar system become clearer when we realize that if such a signal could be emitted in New York City and reflected from something in Washington, D.C., only 0.002 second would be required for the round trip.

The radar signal is typically a burst of 400 kilowatts of power, but the returning signal is only 10^{-21} watt, which is so weak that it is very difficult to detect.

7-2 Measuring Mass and Average Density

How do we know the masses of the Sun and planets? To answer this we must return to Kepler's third law, which relates each planet's distance from the Sun to its period of revolution. We saw in Chapter 3 that Newton's formulation of Kepler's third law was more complete than the original statement by Kepler. The equation for Newton's version of the third law is

$$\frac{a^3}{P^2} = K \times (m_1 + m_2)$$

where a = semimajor axis of the orbit

P = period of the orbit

m_1, m_2 = the masses of the two objects

K = a constant whose value depends on the units of the preceding quantities

Using standard units (meters for distance, seconds for period, and kilograms for mass), the value of K is given by $G/4\pi^2$, where G is the gravitational constant. Thus, $K = 1.69 \times 10^{-12}$ m^3/(s$^2 \cdot$ kg). However, in some cases it is easier to measure the semimajor axis a in AU and the period P in years; then K is equal to the inverse of the mass of the Sun ($K = 1/M_{\text{Sun}}$).

Let us now consider the case of a planet orbiting the Sun. Since the mass of even the largest planet, Jupiter, is less than 0.001 times the mass of the Sun, the sum of the two masses ($m_1 + m_2 = M_{\text{planet}} + M_{\text{Sun}}$) is essentially equal to the mass of the Sun and is therefore the same for each planet. Thus for objects in orbit around the Sun, we can write the preceding equation as

$$\frac{a^3}{P^2} \approx K \times M_{\text{Sun}}$$

Recall that the semimajor axis of a planet's elliptical orbit is essentially its average distance from the Sun.

where a = average distance to the Sun

P = orbital period

M_{Sun} = mass of the Sun

K = a constant

This is an example of the correspondence principle, which states that a new theory must make the same predictions as the old one in applications where the old one worked. (See Section 3-11.)

Notice that Newton's statement does not actually conflict with Kepler's unless great accuracy is demanded. Since the Sun's mass is constant, the value on the right side of Newton's equation is very nearly the same for each of the planets, just as Kepler said. Newton's statement reduces to Kepler's when data are used that are no more precise than Kepler had. Newton's statement of the law, however, allows us to calculate something else—the mass of the Sun. All we need to know in order to do this is the semimajor axis of one planet's elliptical orbit and that planet's period of revolution around the Sun.

There is even more value to the equation. Kepler's third law, as completed by Newton, applies to *any* system of orbiting objects. Recall that Galileo had compared Jupiter's system of moons to the solar system. Here the equation lets us calculate the mass of Jupiter, which is the central object in this case. As we will see, every planet except Mercury and Venus has at least one natural satellite. Thus, to calculate the mass of one of these planets, we need only know the distance and period of revolution of at least one of its satellites.

EXAMPLE

The average distance of the Earth from the Sun is $a = 1$ AU $\approx 1.5 \times 10^{11}$ m and the Earth's orbital period is $P = 1$ year $\approx 365 \cdot 24 \cdot 60 \cdot 60$ seconds $\approx 3.15 \times 10^7$ s. Using Kepler's third law as modified by Newton, find the mass of the Sun.

SOLUTION

We use the equation

$$\frac{a^3}{P^2} \approx K \times M_{\text{Sun}}$$

and thus

$$\frac{(1.5 \times 10^{11})^3}{(3.15 \times 10^7)^2} \approx 1.69 \times 10^{-12} \times M_{\text{Sun}}$$

Therefore $M_{\text{Sun}} \approx 2 \times 10^{30}$ kg

TRY ONE YOURSELF

The Moon orbits the Earth with a period of 27.32 days (which is 2.36×10^6 seconds) and its semimajor axis is 3.844×10^8 meters. Assuming that the Moon's mass is negligible compared to Earth's, use this data to calculate the mass of the Earth. When you check your answer in Appendix C, remember that the Moon's mass is not really negligible compared to Earth's.

What about Mercury and Venus, which have no moons? Their masses have been calculated on a few occasions by observing their effects on the orbits of passing asteroids and comets. No asteroid or comet has passed close enough to provide highly accurate data, however, and thus the accuracy of the calculations was limited until space probes flew by these planets. If a space probe is put into orbit around a planet, the equation above applies to it and allows us to calculate the mass of the planet. In practice, the space probe does not actually have to be put into orbit. By analyzing how the gravitational force of the planet changes the direction and speed of a probe during a flyby, we can calculate the planet's mass, although by a more complicated method than the equation we have used.

When we consider the masses of the objects that make up the solar system, we should be impressed by the fact that the Sun makes up almost the entire system. **Table 7-2** shows the masses by percentages of the total; the Sun's mass is almost 99.9 percent of the total. Jupiter makes up most of the rest, having more than twice as much mass as the remainder of the planets combined.

Table 7-2
Percentages of the Total Mass of the Solar System

Object	Percent of Solar System's Mass
Sun	99.85
Jupiter	0.095
Other planets	0.039
Satellites of planets	0.00005
Comets	0.01 (?)
Asteroids, etc.	0.0000005 (?)

Calculating Average Density

The density of an object is defined as the ratio of the object's mass to its volume. In the previous section we discussed how we could calculate the mass of a solar system object by using Kepler's third law or other more complicated methods. We have also discussed (in Chapter 6) how to use the small-angle formula to find the diameters of objects if we know their distances from Earth and their angular size. Unless the object is too small, its angular size can be measured with observations. The object's distance from Earth can be obtained by using the methods described in Section 7-1.

Knowing the mass (m) and radius (R) of an object, its average density is given by

$$\text{average density} = \frac{\text{mass}}{\text{volume}} = \frac{m}{\frac{4\pi}{3} \times R^3} = \frac{3}{4\pi} \times \frac{m}{R^3}$$

where we assumed that the object is approximately spherical.

The formula for the volume of a sphere of radius R is $V = \frac{4}{3}\pi R^3$.

EXAMPLE

When Venus is close to Earth, we measure its angular size to be about 54 arcseconds. Its distance from Earth at that point is 44.8 million kilometers. Using the small-angle formula (Section 6-1), we find that Venus's diameter is 12,100 kilometers and therefore its radius is about 6050 kilometers. Kepler's third law allows us to find Venus's mass, about 4.87×10^{24} kilograms. Now we are ready to calculate Venus's average density.

$$\text{average density} = \frac{3}{4\pi} \times \frac{4.87 \times 10^{24} \text{ kg}}{(6{,}050{,}000 \text{ m})^3} = 5250 \text{ kg/m}^3$$

This is equal to 5.25 grams/centimeter^3, in good agreement with the value of 5.24 grams/centimeter^3 found in Appendix C.

TRY ONE YOURSELF

The angular size of the Moon is 0.52 degrees and its distance from Earth is 384,000 kilometers. The Moon's mass is 7.35×10^{22} kilograms. Using all this data, find the average density of the Moon and compare it to the value given in Appendix D.

As we mentioned in Chapter 6, once we know the average density of an object we can compare it to the densities of well-known materials such as water, rock, and iron. This allows us to make a reasonable guess about its composition. For example, the average density of Jupiter is 1.33 grams/centimeter^3, barely greater than that of water (1 gram/centimeter^3) and less than silicate rock (about 3 g/cm^3). We can infer that Jupiter consists mostly of low-density materials (gas, liquids) with a small (compared to its overall size) core of iron, rock, and water. However, knowing the value of an object's average density does not mean we can accurately predict its exact composition. Different combinations of materials can result in the same average density and an object's gravity can affect the density of certain materials. However, even though there are limitations, we can gain reasonable insights into the makeup of an object by calculating its average density.

7-3 Planetary Motions

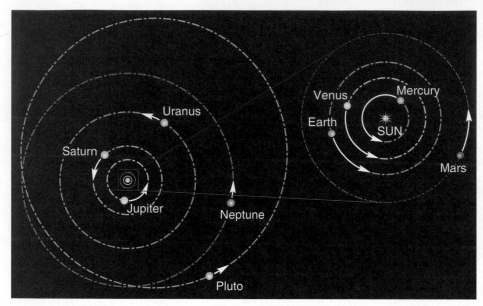

Figure 7-3
The orbits of the planets are ellipses, according to Kepler's first law, but most are very nearly circular. The obvious exception is Pluto, which for about 20 years of its 248-year period is closer to the Sun than Neptune is.

inclination (of a planet's orbit) The angle between the plane of a planet's orbit and the ecliptic plane.

Figure 7-3 illustrates the orbits of the planets, drawn to scale. They are all ellipses, as Kepler had written. Most are very nearly circular, but Pluto's orbit is eccentric enough that it overlaps the orbit of Neptune. In 1979 Pluto moved to a location in its orbit where it was inside Neptune's orbit. Until 1999 it remained closer to the Sun than Neptune was. For the next 220 years we can safely list Pluto as the planet most distant from the Sun.

All of the planets revolve around the Sun in a counterclockwise direction as viewed from far above the Earth's North Pole. Their paths are very nearly in the same plane. This means that we can draw them on the same piece of paper without having to change their paths to view them face-on. **Figure 7-4** illustrates the angles between the planes of the various planets' orbits and the plane of the Earth's orbit (this angle is known as the planet's *inclination*). Notice that Pluto's orbit is the most "out of line." We will see that Pluto is unusual in several other ways.

In Chapter 2 we showed that the eccentricity of an elliptical orbit is a measure of how much the orbit is less than perfectly circular. The eccentricities of the planets' (and Ceres's) orbits are given in **Table 7-3**. Notice how much Pluto and Mercury differ from the other planets in eccentricity.

All of the planets except Mercury and Venus have natural satellites revolving around them, just as the Earth does. The direction of revolution of most of these satellites is also counterclockwise, although there are some exceptions. Finally, as we will see in the next section, most of the planets also rotate counterclockwise about an axis.

The fact that the planets have their orbits in basically the same plane, that they all orbit in the same direction, that most of them rotate in that same direction, and that most of their satellites revolve in that direction cannot be coincidence. We will recall these similarities when we discuss theories concerning the formation of the solar system; we must be sure that the theories explain these properties.

Figure 7-4
The orbits of most of the planets are in the same plane as the Earth's (the ecliptic), but Mercury's plane is inclined at 7 degrees and Pluto's at 17 degrees.

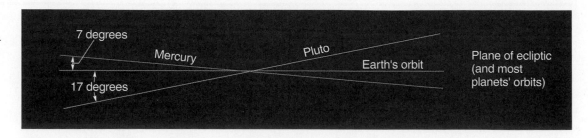

7-4 Classifying the Planets

When we examine the properties of individual planets in the following chapters, it will be clear that they divide easily into two groups. It is convenient to classify the four innermost planets—Mercury, Venus, Earth, and Mars—in one group, which we call the terrestrial planets because of their similarity to Earth (in Latin, "earth" is "terra"). The next four planets—Jupiter, Saturn, Uranus, and Neptune—are called the Jovian planets because of their similarity to Jupiter. As already noted, Pluto is unusual in several ways. Indeed, it does not fit well into either of the two categories of planets. Closer examination of some of the properties of the various planets reveals even more cases in which Pluto fits into neither category. As a result, there have been efforts through the years to declassify Pluto as a planet.

Size, Mass, and Density

Figure 7-5 is a bar graph of the diameters of the planets, in kilometers and as ratios to Earth's diameter. Notice that although the four terrestrial planets differ quite a bit from one another, they are all much smaller than the Jovian planets. Although Pluto is out beyond the Jovian planets, it has a size more like the terrestrials. In fact, it is the smallest of the planets.

The masses of the planets present even bigger differences between the terrestrial and Jovian planets. Look at the planetary masses shown in **Figure 7-6**.

Many people have a tendency to skip over tables and graphs. You are not expected to memorize the values given, but a few minutes looking at patterns and thinking of their meaning will yield much knowledge about the solar system. Study

Table 7-3

Eccentricities of the Orbits of the Planets and Ceres

Planet	Eccentricity of Orbit
Mercury	0.206
Venus	0.007
Earth	0.017
Mars	0.094
(Ceres)*	0.079
Jupiter	0.049
Saturn	0.057
Uranus	0.046
Neptune	0.011
Pluto	0.244

*Ceres is the largest of the asteroids, which we will discuss later. (Also see the Advancing the Model boxes in this chapter.)

Jupiter was named after the most powerful of the Roman gods. "Jove" is another form of the name "Jupiter" and "Jovian" is an adjective form of the name.

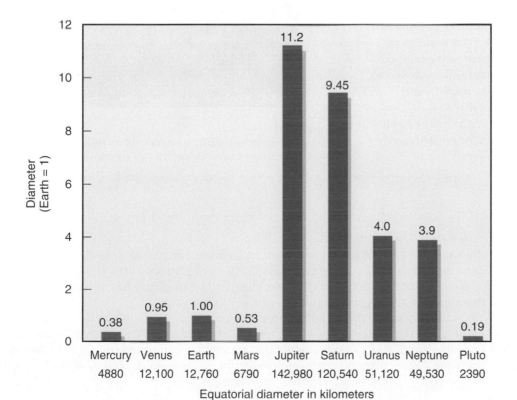

Figure 7-5
A plot of planetary equatorial diameters (Earth's diameter equals one here) makes obvious the distinction between the terrestrial and Jovian planets. The diameter of each planet is shown in kilometers below the planet's name. The values have been rounded (see Appendix C for more details).

Equatorial diameter in kilometers

	Mercury	Venus	Earth	Mars	Jupiter	Saturn	Uranus	Neptune	Pluto
Diameter (Earth = 1)	0.38	0.95	1.00	0.53	11.2	9.45	4.0	3.9	0.19
km	4880	12,100	12,760	6790	142,980	120,540	51,120	49,530	2390

ADVANCING THE MODEL

The Discovery of the Asteroids

Johannes Kepler once proposed that there might be an undiscovered planet between Mars and Jupiter, because the large distance between their orbits does not follow the pattern of other orbits. The Titius-Bode law also seemed to predict such a planet. This led Francis von Zach, a German baron, to plan a systematic search for the planet. Giuseppe Piazzi, a Sicilian astronomer and monk, was one of the astronomers who had been chosen to search in one of the sectors into which von Zach had divided the sky. Before he was notified where he was to search, however, Piazzi discovered what he first thought was an uncharted star in Taurus. The object he discovered (on January 1, 1801) was very dim, far too dim to see with the naked eye. He named it Ceres after the goddess of the harvest and of Sicily. Continuing to observe it, he saw that it moved among the stars, and by January 24 he decided that he had discovered a comet. He wrote two other astronomers (including Bode) of his discovery, but on February 11 he became sick and was unable to continue his observations. By the time Bode and the other astronomer received their letters (in late March), the object was too near the Sun to be observed.

Bode was convinced that the hypothesized new planet had been discovered, but he also realized that it would not be visible again until fall. By that time, it would have moved so much that astronomers would have a difficult time finding it again. This was because relatively few observations had been made of the object's position, not enough for the mathematicians of the time to calculate its orbit. Fortunately, a young mathematician named Carl Friedrich Gauss, one of the greatest mathematicians ever, had recently worked out a new method of calculating orbits. He worked on Bode's project for months and was able to predict some December positions for the object. On December 31, 1801, the last day of the year in which Piazzi had seen it on the first day, von Zach—the man who had sought to organize the original search—rediscovered the object.

The elation over finding the predicted planet did not last long, however, for another "planet" was found in nearly the same orbit about a year later. Its discoverer, Heinrich Olbers, was looking for Ceres when he discovered another object that moved. He sent the results of a few nights' observations to Gauss, and the mathematician calculated its orbit. The object was given the name Pallas, and a new classification of celestial objects had been found: the new objects were called *asteroids*.

By 1890, about three hundred asteroids had been found using the tedious method of searching the skies and comparing the observations to star charts, looking for uncharted objects. In 1891 a new method was introduced: a time exposure photograph of a small portion of the sky was taken, and the photograph was searched for any tiny streaks. The streaks (**Figure B7-1**) would be caused by objects that did not move along with the stars. These objects were then watched very closely and their orbits determined. Using such methods, well over 4000 asteroids are now known and named, and it is predicted that some 100,000 asteroids are visible in our largest telescopes.

Figure B7-1
The two streaks (arrows) on the time exposure photo are caused by the motion of two asteroids as the camera follows the stars' apparent motions across the sky.

the values of the masses of the planets in terms of Earth's mass. Notice the tremendous difference between the two classifications of planets. Earth is the most massive of the terrestrial planets, but the least massive Jovian planet has more than 14 times the mass of Earth. Again, Jupiter stands out as the giant.

A density graph (**Figure 7-7**) shows another difference between the terrestrial planets and the Jovian planets. The terrestrials are denser. This is because they are primarily solid, rocky objects, while the Jovians are composed primarily of liquid. At one time Jovian planets were commonly called "gas planets," but now we know that they actually contain much more liquid than gas.

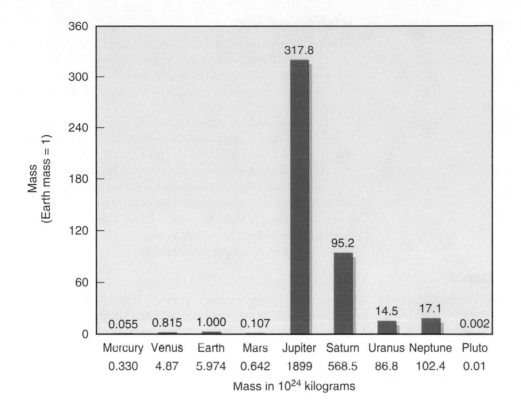

Figure 7-6
When we plot the masses of the planets, we see that most of their total mass is in Jupiter. On this graph we have made Earth's mass equal to one. The masses are given in 10^{24} kilograms below each planet's name. To find the mass in kilograms, multiply the value given by 10^{24}.

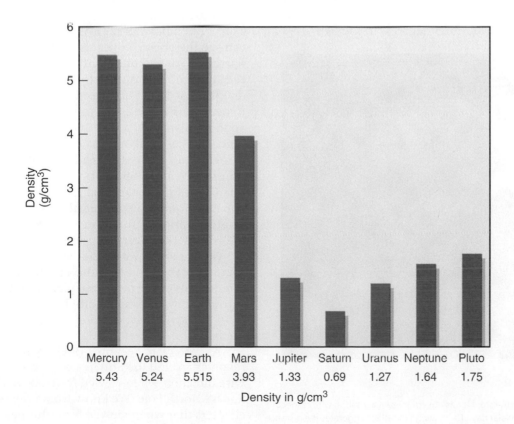

Figure 7-7
The average density of the terrestrial planets is significantly greater than that of the Jovian planets. Density values are given in grams per cubic centimeter below the names of the planets.

Table 7-4

The Number of Known Planetary Satellites

Planet	Known Satellites	Planet	Known Satellites*
Mercury	0	Jupiter	61 + r
Venus	0	Saturn	31 + r
Earth	1	Uranus	27 + r
Mars	2	Neptune	11 + r
		Pluto	1

*An "r" indicates that the planet has a ring system

Satellites and Rings

Table 7-4 shows the number of natural satellites of each planet. Notice that although there is no obvious pattern, the Jovian planets have more satellites. More details of the planetary satellites are found in Appendix D and in discussions of the planets in future chapters.

The table also indicates that all Jovian planets have rings. A planetary ring is simply planet-orbiting debris, ranging in size from a fraction of a centimeter to several meters. Motions of particles within the rings are extremely complex, due largely to gravitational interactions with nearby planetary satellites. In a sense, each particle of a ring may be considered a satellite of the planet, but if we do so, counting satellites becomes meaningless. So we will continue to speak only of the larger "moons" as being planetary satellites.

Rotations

solar day The amount of time that elapses between successive passages of the Sun across the meridian.

meridian An imaginary line that runs from north to south, passing through the observer's zenith.

sidereal day The amount of time that passes between successive passages of a given star across the meridian.

In discussing the rotation periods of the planets we must distinguish between a solar day and a sidereal day. The **solar day** is defined as the time between successive passages of the Sun across the **meridian,** so that the length of a solar day on Earth is 24 hours. However, this is not the same amount of time that the Earth takes to complete one rotation. With respect to the stars, the Earth rotates once in 23 hours, 56 minutes.

Refer to **Figure 7-8** to see why there is a difference between the solar day and the **sidereal day,** which is the amount of time required for an object to complete one rotation with respect to the stars. In that figure we have designated a particular location on the Earth as A. When the Earth is at location X in the figure, point A faces the Sun directly. When the Earth has turned so that point A again faces the Sun directly, one solar day will have passed. However, the Earth moves around the Sun as it rotates, moving about 1/365 of the way around the Sun in one day; that is, about 360/365 degrees. In the figure we have exaggerated how far it moves so that you can see the effect. Notice that when the Earth has rotated once with respect to the stars, moving from X to Y, point A does not face the Sun directly. For this to occur again, the Earth must rotate an additional angle of about 360/365 degrees. Since it takes us about 24 hours to rotate through 360 degrees, this additional rotation requires 24/365 hours or 4 minutes. The Earth's sidereal period is 23 hours and 56 minutes, but the Earth has to rotate for another 4 minutes to complete a solar day (location Z).

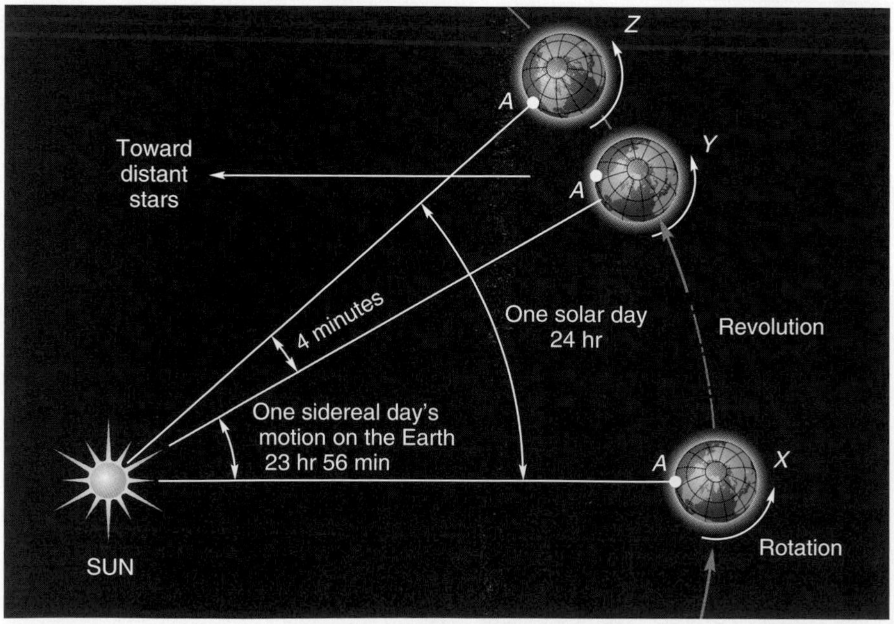

Figure 7-9a shows the sidereal periods of rotation of the planets. Although the rotation periods of the terrestrial planets differ considerably from one another, notice that all of the Jovian planets rotate faster than the fastest of the terrestrials.

As we saw in Section 7-3, every planet revolves around the Sun in a counterclockwise direction when viewed from above the Earth's North Pole. We know from previous chapters that when viewed from this perspective, the Earth rotates on its axis in this

Figure 7-8

When the Earth has rotated once with respect to the stars, it has not completed a rotation with respect to the Sun. Thus the sidereal day is shorter than the solar day. (The drawing greatly exaggerates the distance the Earth moves in one day.)

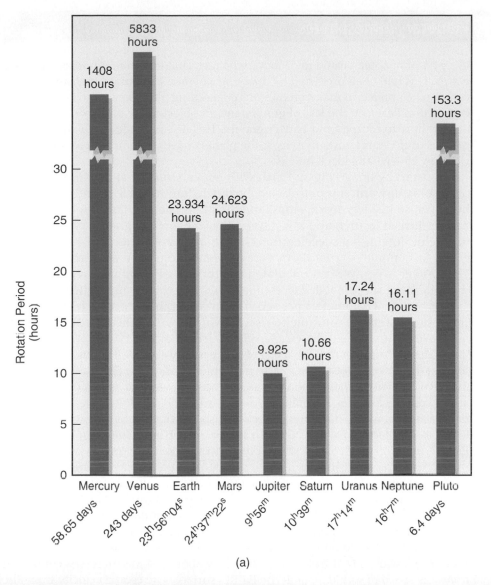

(a)

(b)

Figure 7-9
(a) The rotation periods of Mercury, Venus, and Pluto are so much greater than those of the other planets that their bars go far off the graph. (b) The tilt between a planet's axis of rotation and its orbital plane varies among the planets in our solar system. Venus, Uranus, and Pluto rotate in a direction opposite that of Earth's as seen from far above the Earth's North Pole. The arrows point in the direction of the planets' north poles, and the direction of rotation is also shown. If you grab a planet with your right hand in a manner such that your fingers point in the direction of its rotation, then your thumb will point to the planet's north pole. This "right-hand rule" is used in many applications in Physics and Astronomy.

same counterclockwise direction and that the Moon also orbits the Earth in a counterclockwise direction. We might ask if this pattern holds elsewhere in the solar system. The answer is yes, in most cases. As shown in Figure 7-9b, all of the planets except Venus, Uranus, and Pluto rotate in a counterclockwise direction as seen from far above the Sun's North Pole.

7-5 Planetary Atmospheres

Long before people visited the Moon, we knew that it contained no air and no liquid water. This had been predicted by applying Newton's law of gravity, and the same law can also be applied to make predictions concerning planetary atmospheres. To see the connection between the law of gravity and an object's lack of atmosphere, we first discuss how to escape from Earth's gravity. This discussion leads to an idea that will help us understand not only why some planets have no atmosphere, but also—in Chapter 15—what a black hole is.

We start by imagining an Earth with no air. On such an Earth, if we throw something upward, it is not slowed by air friction. It still feels the effect of gravity, however, so it slows down, stops, and then falls back to Earth. So we throw it harder. It rises farther, and as it gets higher, the force of gravity on it is less. Thus its rate of slowing—its deceleration—is less at greater heights. Could we throw the object fast enough so that Earth's gravity could not stop it and bring it back down? The answer is yes. We can calculate from the laws of motion and gravitation that the minimum speed needed to escape Earth's gravity, assuming we start at the surface, is about 11 kilometers per second. An object fired upward from Earth at this speed or greater will continue to rise, slowing down all the time, but never stopping. We call this speed the *escape velocity* from Earth.

When we consider that the speed of sound is only about 0.3 kilometer per second, we see that the escape velocity from Earth is a tremendous speed. The reason that we imagined an Earth without air friction is that, in practice, if we fired an object from the surface at 11 kilometers/second, it would be slowed—and probably destroyed—by air friction. In the space program we have fired objects into space with a velocity exceeding escape velocity; every space probe that has been sent to other planets has completely escaped the Earth. The probes were not destroyed by air friction because rockets carried them above the atmosphere before increasing their speed to escape velocity (**Figure 7-10**).

The escape velocity of a projectile launched from an astronomical object depends on the gravitational force at the object's surface (or from whatever height we are launching the projectile). Recall that the gravitational force at the surface of the Moon is only one-sixth of that at the surface of Earth. A 120-pound astronaut weighs only about 20 pounds on the Moon. The escape velocity from the Moon is therefore less than that from Earth. It is only about 2.4 kilometers/second—eight times the speed of sound in air at the Earth's surface. The escape velocity from Phobos, a Martian moon, is only 50 kilometers/hour (30 miles/hour). If you have a good arm, you could throw a ball from Phobos so that it would never return.

To see what escape velocity has to do with the question of the atmosphere of an astronomical object, we must briefly discuss the nature of a gas.

Gases and Escape Velocity

There are three states of matter in our normal experience: solid, liquid, and gas. Some understanding of the gaseous phase is necessary to understand planetary atmospheres. To envision a gas, picture a swarm of bees buzzing around a hive (**Figure 7-11**). The bees represent individual molecules of a gas. However, this analogy is faulty in a few

escape velocity The minimum velocity an object must have to escape the gravitational attraction of another object, such as a planet or star.

Even though we use the terms "speed" and "velocity" as if they are the same, they are not. An object's velocity tells us not only how fast it is moving (its speed), but also the direction of its motion.

1 km/s = 3600 km/hr = 2237 mi/hr

Figure 7-10
Sixty seconds after takeoff, the main engines on the space shuttle cut back to 65% thrust to avoid stress on the wings and tail from the Earth's atmosphere. When the shuttle reaches thinner air at higher altitudes, the engines resume full power until the shuttle reaches orbiting speed.

ways. First, compared to their size, the molecules of a gas are much, much farther apart than are the bees. Second, compared to the volume occupied by the gas, the volume of a molecule is much smaller. Third, molecules move in straight lines until they collide, either with one another or with the walls of their container. Then they bounce off and move in straight lines again. Finally, remember that molecules are moving through completely empty space. There is nothing between them—nothing.

There are three additional things we must keep in mind about the molecules of a gas:

1. As gas molecules bounce around, at any given time, different molecules have different speeds. Some will be moving fast and some moving slow. In this sense, they are like the bees.

2. The average speed of the molecules depends on the temperature of the gas. Gases at higher temperature have faster-moving molecules.

3. At the same temperature, less massive molecules have greater speed. For example, since a molecule of oxygen has less mass than a molecule of carbon dioxide, in a mixture of oxygen and carbon dioxide gases, the oxygen molecules will, on the average, be moving faster.

Bees around a hive…

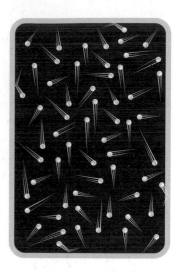

…are like molecules in a gas.

Figure 7-11
Molecules in a gas, like bees around a hive, move at different speeds and in random patterns. But molecules are very tiny; their average speed depends on temperature; and less massive molecules have higher speeds.

In fact, the temperature of a gas is defined as a measure of the average energy of motion of the gas molecules: average kinetic energy = $3/2 \cdot k_B \cdot T$, where T is the temperature in kelvin and k_B is Boltzmann's constant.

Now let's consider the atmosphere of the Earth. Our atmosphere is held near the Earth by gravitational forces. Consider a molecule in the upper part of our atmosphere. At great heights above Earth, the atmosphere has a low density; we say that it is "thin." This means that the molecules are much farther apart than down here at the surface. Suppose that at some instant a particular molecule up there happens to be moving away from Earth. There are very few other molecules around, so a collision is unlikely and our molecule acts just like any other object moving away from Earth. The force of gravity slows it down. Whether the molecule returns to Earth or escapes depends upon how the speed of the molecule compares to the Earth's escape velocity. If the molecule's speed is greater than escape speed, the molecule is gone, never to return to Earth.

Obviously, since an atmosphere exists on Earth, the velocities reached by molecules of the air do not exceed escape velocity. Recall, however, that molecules of lower mass have greater speeds. Hydrogen molecules have less mass than those of any other element. It is therefore no coincidence that there is little hydrogen in the Earth's atmosphere: the temperature of the upper atmosphere is high enough for hydrogen molecules to escape. Any pure hydrogen that is released into the Earth's atmosphere is eventually lost. The chemical element hydrogen does not exist alone in our atmosphere, but only as part of more massive molecules. A molecule of water vapor, for example, contains hydrogen.

As noted earlier, the escape velocity from the surface of the Moon is about 2.4 kilometers/second. At the temperatures reached on the sunlit side of the Moon, all but the most massive gases attain speeds greater than this, and therefore the Moon has essentially no atmosphere. The *Apollo* astronauts were not surprised to find no air to breathe when they landed on the Moon.

Does the Earth keep all of its atmosphere forever?

Figure 7-12
This graph shows how the speeds of various gases depend upon their temperatures. The dashed lines represent 10 times the gases' average speeds. All the planets and some planetary satellites are indicated at their corresponding temperatures and escape velocities.

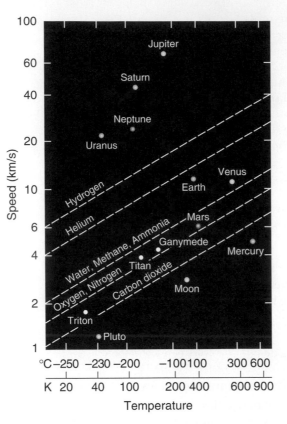

The Atmospheres of the Planets

The average speed of a particular type of molecule depends upon the temperature of the gas, but at any given time some molecules are traveling faster than average. This means that although the average speed of a particular gas may be less than the escape velocity from a planet, the gas may still gradually escape because the speed of a small fraction of its molecules exceeds the escape velocity. Because of this, we must use a multiple of the average speed in considering whether or not a gas will escape from a planet. Theory shows that rather than the average speed, the value we should use in determining whether the planet will retain the gas for billions of years is 10 times the average speed of the molecules of that gas.

Figure 7-12 is a graph of the average speeds of various molecules versus their temperatures. The dashed lines represent 10 times the average molecular speed. All the planets, as well as some planetary satellites, are plotted on the graph at their respective temperatures and escape velocities. A planet can retain a gas if the planet lies *above* the dashed line for that gas. Find the Moon on the graph and notice that every gas has an average molecular speed that allows it to escape. The planet Mercury is similar. On the other hand, only hydrogen and helium escape from Earth and Venus. The four Jovian planets retain all of their gases, including hydrogen and helium.

Even though a graph such as the one in Figure 7-12 is a powerful tool, we can only accurately find the chemical composition of an object's atmosphere by using spectroscopy. This is necessary because we have directly examined samples of only four worlds: Earth, Moon, Venus and Mars. Spectroscopy, the analysis of reflected sunlight from a distant object, allows us to find the chemical composition of the object's atmosphere.

As light from the Sun penetrates a planet's or satellite's atmosphere before it gets reflected back into space, some of its wavelengths are absorbed. The spectrum we observe from the reflected light has absorption lines. Since every element has its own unique spectral "fingerprint," by studying the specific absorption lines we can infer the chemical composition of the atmosphere. (Of course, we must take into account that some of these lines result from absorption that occurs in the atmospheres of the Sun and Earth.)

If an object does not have an atmosphere, we can still use spectroscopy to get useful information about the chemical composition of the object's surface. In this case the observed absorption lines are broad, instead of the sharp lines produced by molecules in a gas.

Table 7-5 summarizes the differences we have discussed between the terrestrial and Jovian planets. You should keep these differences in mind as we discuss theories of the

Table 7-5

Characteristics of the Jovian and Terrestrial Planets

Terrestrials	Jovians
Near the Sun	Far from the Sun
Small	Large
Mostly solid	Mostly liquid and gas
Low mass	Large mass
Slow rotation	Fast rotation
No rings	Rings
High density	Low density
Thin atmosphere	Dense atmosphere
Few moons	Many moons

origin of the solar system, since any successful theory must be able to explain these differences.

7-6 The Formation of the Solar System

More than five billion years ago, the atoms and molecules that now make up the planets—and our own bodies—were dispersed in a gigantic cloud of dust and gas. In a later chapter we will examine how such an interstellar cloud condenses to form a star, but with what we have learned about the patterns within the solar system, we can discuss the formation of the system. A study of the beginnings of the solar system is interesting as an example of the way science in general (and astronomy in particular) progresses, because the theory is still in its early development and many gaps remain. The search for answers here resembles a mystery story where there are many clues; new ones appear all the time and some of the clues seem to contradict others.

There are two main categories of competing theories to explain the origin of the solar system: evolutionary theories and catastrophe theories. This section examines the evidence for each and shows why one is gaining favor among astronomers. First, however, it will be helpful to review the clues, many of which were discussed earlier.

Note that we are concerned here only with the formation of the solar system, and not the formation of the Galaxy or the origin of the universe, both of which occurred much earlier. These questions are considered in later chapters.

Evidential Clues from the Data

The members of the solar system exhibit several patterns that must be explained by any successful theory of the system's origin. In addition, a theory should be able to account for exceptions to the patterns. Here is a list of significant data that must be explained.

First of all there came Chaos . . . From Chaos was born Erebos, the dark, and black night, And from Night again Aither and Hemera, the day, were begotten . . .
Hesiod (Theogony), about 800 BC

1. All of the planets revolve around the Sun in the same direction (which is the direction the Sun rotates), and all planetary orbits are nearly circular except that of Pluto (and, to a lesser degree, Mercury).

2. All of the planets lie in nearly the same plane of revolution.

3. Most of the planets rotate in the same direction as they orbit the Sun, the exceptions being Venus, Uranus, and Pluto.

4. The majority of planetary satellites revolve around their parent planet in the same direction as the planets revolve around the Sun (and as the planets rotate). In addition, most satellites' orbits are in the equatorial plane of their planet.

5. There is a pattern in the spacing of the planets as one moves out from the Sun, with each planet being about twice as far from the Sun as the previous planet.

The pattern of the spacing of planets was discussed in an Advancing the Model box, The Titius-Bode Law, near the beginning of this chapter.

6. The chemical compositions of the planets have similarities, but a pattern of differences also exists, in that the outer planets contain more volatile elements and are less dense than the inner planets.

7. All of the planets and moons that have a solid surface show evidence of craters, similar to those on our Moon.

8. All of the Jovian planets have ring systems.

9. Asteroids, comets, and meteoroids populate the solar system along with the planets, and each category of object has its own pattern of motion and location in the system.

10. The planets have more total **angular momentum** (to be described later in this chapter) than does the Sun, even though the Sun has most of the mass.

11. Recent evidence indicates that planetary systems in various stages of development exist around other stars.

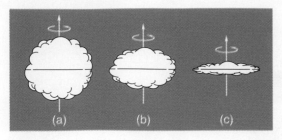

Figure 7-13
As a cloud contracts, its rotational motion causes it to form a disk.

The French mathematician and philosopher René Descartes (1596–1650) is best known for developing the most common coordinate system used in geometry. However, his ideas about motion and the nature of the universe were influential in the seventeenth century, particularly for Isaac Newton.

angular momentum An intrinsic property of matter. A measure of the tendency of a rotating or revolving object to continue its motion.

conservation of angular momentum
A law that states that the angular momentum of a system does not change unless there is a net external influence acting on the system, producing a twist around some axis.

Figure 7-14
If an ice skater begins a spin with his arms extended (a), he will spin faster and faster as he draws his arms in (b). This effect is explained by the law of conservation of angular momentum.

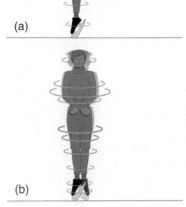

(a)

(b)

As astronomers try to solve the mystery of the origin of the Sun and its companions, they must be sure their theories explain these clues. In addition, any successful theory of the origin of the solar system should be consistent with the theory of star formation.

Evolutionary Theories

There is no single evolutionary theory for the solar system's origin, but there are several theories that have in common the idea that the solar system came about as part of a natural sequence of events. These theories have their beginning with one proposed by René Descartes in 1644. He suggested that the solar system formed out of a gigantic whirlpool, or vortex, in some type of universal fluid and that the planets formed out of small eddies in the fluid. This theory was rather elementary and contained no specifics as to the nature of the universal fluid. It did, however, explain the observation that the planets all revolve in the same plane, the plane of the vortex.

After Isaac Newton showed that Descartes's theory would not obey the rules of Newtonian mechanics, Immanuel Kant (in 1755) used Newtonian mechanics to show that a rotating gas cloud would form into a disk as it contracts under gravitational forces (**Figure 7-13**). Thus to explain the disk aspect of the solar system (our clue #2), Kant changed the philosophical "universal fluid" of Descartes into a real gas subject to the natural laws of mechanics. Later, in 1796, the French mathematician Pierre Simon de Laplace found that such a rotating disk would break up into rings similar to the rings of Saturn. He suggested that perhaps these rings could form into the individual planets while the Sun was coalescing from material in the center.

However, application of Newtonian mechanics to such a contracting gas cloud caused another problem. To understand this problem, consider what happens when a spinning ice skater pulls in his arms. His rotation speed increases greatly (**Figure 7-14**). This increase in speed is predicted by Newton's laws and is a result of the law of ***conservation of angular momentum.*** We will not define angular momentum mathematically, but simply state that the angular momentum of a rotating (or revolving) object is a direct measure of how fast the object rotates (or revolves) and how far it is from its axis of rotation (or revolution). In other words, an object's angular momentum is greater if the object is rotating (or revolving) faster or if the object is farther from the axis of its rotation (or revolution). As the skater pulls in his arms, he decreases their distance from the axis of rotation; in the process, he decreases their angular momentum. To make up for this, his entire body increases its rotation speed, keeping the total angular momentum approximately constant ("conserved"). (The total angular momentum would remain perfectly constant if it were not for air resistance and small friction forces with the ice.)

The law of conservation of angular momentum must also apply to a contracting, rotating cloud of gas. Like the ice skater, the cloud speeds up its rotation as its parts come closer to the center of rotation. When calculations are made for a cloud contracting to form the Sun and planets, we find that the Sun should rotate much faster than it does; it should spin around in a few hours. Chapter 11 will provide evidence that the Sun's period of rotation is close to a month. This means that the angular momentum possessed by the Sun is much less than the theory predicts. In fact, the total angular momentum of the planets (because of their greater distance from the center) is observed to be much greater than the angular momentum of the Sun. This should not occur, according to Newton's laws.

The contradiction of these well-established laws caused the evolutionary theories to lose favor early in the 20th century. The alternative theory was a catastrophe theory.

Catastrophe Theories

Contrary to what the name may imply, a ***catastrophe theory*** does not refer to a disaster, but rather to an unusual event—in this case, the formation of the solar system by an unusual incident. In 1745 Georges Louis de Buffon proposed such an event: the passage of a comet close to the Sun. Buffon suggested that the comet pulled material out of the Sun to form the planets. In Buffon's time, comets were thought to be quite massive, but in the 20th century we learned that a comet's mass is not great enough to cause this breakup of the Sun. However, his basic idea—that a massive object exerted gravitational forces on the Sun, pulling material out and causing it to sweep around the Sun until it eventually coalesced to form the planets—still seemed a reasonable hypothesis. Such an event as the passage of a massive object so near the Sun would be very unusual, but not impossible.

More recently, it was suggested that the Sun was once part of a triple-star system, with the three stars revolving around one another. As we will see, such star systems are common, so this in itself is not a far-fetched idea. This particular catastrophe theory holds that the configuration was unstable and that one of the stars came close enough to cause a tidal disruption of the Sun, producing the planets. The close approach of this star also caused the Sun to be flung away from the other two stars.

Starting around the 1930s, astronomers began to find major problems with catastrophe theories. First, calculations showed that material pulled from the Sun would be so hot that it would dissipate rather than condense to form planets. A second problem involved deuterium, an ***isotope*** of hydrogen. Even the outer portions of the Sun are too hot for deuterium to be stable, so not much deuterium exists in the Sun. However, much more deuterium is found on the planets than in the Sun, indicating that the material of the planets could not have been part of the Sun.

Finally, as we will discuss later, we now know that other nearby stars have planetary systems around them. A catastrophe theory would predict that such systems are rare, since they are produced by unusual events. If we find planetary systems elsewhere, there is probably some common process that forms them.

At the same time as these problems were becoming apparent, a solution appeared for the angular momentum problem of the evolutionary theories; as a result, catastrophe theories have been nearly abandoned in favor of modern evolutionary theories.

Present Evolutionary Theories

In the 1940s the German physicist Carl von Weizsäcker showed that a gas rotating in a disk around the Sun would rotate differentially (the inner portion moving faster than the outer). This would result in the formation of eddies, as shown in **Figure 7-15**. As the figure shows, the eddies would be larger at greater distances from the Sun. According to his view, these eddies are the beginnings of planet formation, and the eddies therefore explain the pattern of distances between the planets.

catastrophe theory A theory of the formation of the solar system that involves an unusual incident, such as the collision of the Sun with another star.

The evolution of the world may be compared to a display of fireworks that has just ended: some few red wisps, ashes and smoke. Standing on a cooled cinder, we see the slow fading of the suns, and we try to recall the vanished brilliance of the origin of the worlds.
G. Lemaître (1894–1966),
Astronomer and Catholic priest

isotope One of two (or more) atoms whose nuclei have the same number of protons but different numbers of neutrons.

Figure 7-15
Carl von Weizsäcker showed that eddies would form in a rotating gas cloud and that the eddies nearer the center would be smaller.

Figure 7-16
(a) The bright bluish stars visible in the Rosette nebula are apparently hot, young ones forming from dense dust clouds. The nebula is about 3000 light years away. (b) Hodge 301, at lower right, is a cluster of massive, brilliant stars in the Tarantula nebula, about 160,000 light years away. Many of the stars in the cluster have exploded as supernovae, blasting material into the nebula at high speed. These explosions compress the gas into filaments, seen at upper left. Near the center of the image are small, dense gas globules and dust columns where new stars are being formed today.

(a)

(b)

A real breakthrough occurred when it was realized that a mechanism exists to account for why the Sun does not rotate faster than it does. Before turning to this mechanism, let's consider the beginning of the scenario that today's theory envisions.

Chapter 13 will explain that new stars form from enormous interstellar clouds of gas and dust. **Figure 7-16** shows the Rosette nebula and the Tarantula nebula, which are stellar nurseries where newly formed stars can be seen.

Recall that when an interstellar dust cloud collapses, any slight rotation that it had at the beginning, before the collapse, results in a greatly increased speed of the central portion (explained by the conservation of angular momentum). The material in the center becomes a star—the Sun in the case of our solar system. During the few million years that this is occurring, the matter surrounding the newly forming Sun is condensing into a disk.

As the gases in the disk cool, they begin to condense to liquids and solids, just as water vapor condenses on the cool side of an iced drinking glass. Nonvolatile elements such as iron and silicon condense first, forming small chunks of matter, or dust grains. Each of these grains has its own elliptical orbit about the center, and as time passes, more matter condenses onto its surface. The orbits of these tiny objects are elliptical, so that they intersect one another. The resulting collisions between particles have two effects: (1) particles involved in gentle collisions (as if they were rubbing shoulders with one another as they orbit) occasionally stick together and form larger particles, and (2) particles are forced into orbits that are more nearly circular.

As the matter sticks together, small chunks grow into larger chunks. Their increased mass causes nearby particles and molecules of gas to feel a greater gravitational force toward them. Since this force is still very small, the coalescing is a very slow process, but, over a few hundred thousand years, larger particles—now called *planetesimals*—sweep up smaller ones. Some planetesimals, resembling miniature solar systems, have dust and gas orbiting them—material that eventually condenses to become the moons we know today.

As the force of gravity shrinks a celestial object, gravitational energy causes it to heat up. A simple case of gravitational energy being converted to thermal energy occurs whenever you drop something. The object hits the floor and heats up slightly. (The heating is very slight, and to experience it, you should probably cheat and *throw* an object, such as modeling clay, down to the floor a few times.) Perhaps you can visualize a release of heat when an object falls from the heavens onto a planet, but the same effect occurs when gravitational forces cause the collapse of a cloud of gas and dust. The material heats up as it falls toward the center.

This heating effect occurs with our solar-system-in-formation. The material that falls inward to form the Sun gets hot, and the high temperatures near the new Sun do not allow for condensation of the more volatile elements. This means that the planets that form in the inner solar system are made primarily of nonvolatile, dense material.

Farther out, matter orbiting the new Sun (the *protosun*) is moving at a more leisurely pace, and the swirling eddies around protoplanets are more prominent. The situation at this time is illustrated in **Figure 7-17**.

A particularly large outer planet, Jupiter, gravitationally stirs the nearby planetesimals of the inner system so that the weak gravitational forces between them cannot pull them together. Today's asteroids are the remaining planetesimals.

planetesimal One of the small objects that formed from the original material of the solar system and from which a planet developed.

In Chapter 9 we will present evidence that Jupiter has not yet lost all of the excess thermal energy that resulted from its formation.

Figure 7-17
At the stage of development shown here, planetesimals have formed in the inner solar system, and large eddies of gas and dust remain at greater distances from the protosun.

ion An electrically charged atom or molecule.

A charged particle moving in a magnetic field experiences a force unless it is moving parallel to the magnetic field lines.

stellar wind The flow of particles from a star.

For information on the distant shell of comets, refer to the Oort cloud in Chapter 10.

While planet formation is taking place, the Sun continues to heat up. It heats the gas in the inner solar system and causes electrons there to leave their atoms, forming charged atoms (***ions***) and electrons. A magnetic field does not exert a force on an uncharged object, but if a magnetic field line sweeps by a charged object, a force is exerted on that object. This is what must have slowed the Sun's rotation; the magnetic field of the rapidly rotating Sun exerted a force on the ions in the inner solar system, tending to sweep them around with it. However, Newton's third law tells us that if the Sun's magnetic field exerts a force to increase the rotational speed of these particles, they must exert a force back on the Sun to decrease its rotational speed. So it is the magnetic field of the Sun, discovered rather recently, that provides the explanation for why the Sun rotates so slowly—a fact that was once a stumbling block for evolutionary theories.

The solar system of our story is getting close to what we see today. However, gas and dust were still more plentiful between the planets than in today's solar system, and the inner solar system of our story contained much more hydrogen and other volatile gases than exist there today. To help answer the question of how these gases were moved to the outer solar system and how, in general, the system was "cleaned up," we can again look into space at interstellar dust clouds.

In these clouds we see stars at various stages of formation. There is evidence that many newly forming stars go through a period of instability during which their ***stellar wind*** increases in intensity. The stellar wind, called the solar wind in the case of our Sun, consists of an outflow of particles from the star. It continues throughout a star's lifetime, as we explain in Chapter 14. If the instabilities we observe in other stars occurred during the formation of the Sun, the pulses of solar wind would sweep the volatile gases from the inner solar system. Even without this increased activity, it is expected that the solar wind would gradually move this material outward, but if the Sun did go through this active period, there is certainly no difficulty explaining why hydrogen and helium exist on the outer planets but not the inner. Once in the outer system, this material would gradually be swept up by the giant planets there.

Explaining Other Clues

As millions of years passed, remaining planetesimals crashed onto the planets and moons, resulting in the craters we see on these objects today.

Comets are thought to be material that coalesced in the outer solar system, the remnants of small eddies. These objects would feel the gravitational forces of Jupiter and Saturn, and many would fall onto those planets. (Recall the discussion at the start of the Prologue of Comet Shoemaker-Levy 9 crashing onto Jupiter.) Small objects that formed beyond the giant planets' orbits, however, would be accelerated by Jupiter and Saturn as those planets passed nearby and would be pushed outward. As we will explain in Chapter 10, there is reason to think that great numbers of comets exist in a region far beyond the most distant planet.

Notice that the evolutionary theories explain that nonvolatile elements would condense in the inner solar system, but that volatiles would be swept outward by the solar wind. This accounts for the differences in the planets' chemical composition. In fact, astronomers find that when compression forces are taken into account in calculating density, planets closest to the Sun contain the most dense and least volatile material, as would be expected from the theories.

Further confirmation of evolutionary theories is found in Jupiter's Galilean satellites. As we point out in Chapter 9, these satellites also decrease in density and increase in volatile elements as we move outward from Jupiter. The formation of Jupiter and its moons must have resembled the formation of the solar system; thus we see the same density pattern in Jupiter's system.

The next two chapters will examine each planet in turn and will point out some exceptions to the patterns described here. Some of the exceptions are easy to explain using evolutionary theories. Others cannot be explained by these theories and require a hypothesis of collisions—"catastrophes"—within the early solar system.

Catastrophes may well have played a part in the formation of the solar system, as in the formation of our Moon (Section 6-7). However, the evidence indicates that it was a fairly minor part, involving relatively few objects, and that the overall formation of the system in which we live was evolutionary in nature. Nonetheless, the origin of the solar system is poorly understood. Pieces continue to fall into place, but we still have much to learn.

7-7 Planetary Systems Around Other Stars

Is the existence of our planetary system unusual, or is it common for stars to have planets? Until recently, it has been very difficult to answer this question. Direct observation of extrasolar planets (or exoplanets) is necessarily based on light reflected by these objects (or weak infrared light emitted by their cool surfaces). But planets are very small and the light emitted from their companion star overwhelms their light. For example, our Sun is ten billion times brighter than Earth at visible wavelengths. In most cases we have to use indirect methods to search for exoplanets. Different categories of evidence can help answer the question we asked. We'll examine each category in turn.

- *Infrared companion.* The star T Tauri has a companion (**Figure 7-18a**) that emits significant radiation only in the infrared region. The companion has too little mass for it to become a star itself, yet its infrared radiation indicates that it has a high temperature. A possible explanation for this high temperature is that it is a giant

Figure 7-18
(a) A false-color image of the T Tauri system, a prototype laboratory for the formation of low-mass stars like our Sun. The binary system consists of the star T Tauri North and its companion T Tauri South. (b) Gliese 229B, a brown dwarf of 20–50 times the mass of Jupiter, orbits the cool red star Gliese 229A, located in the constellation Lepus, about 18 light years from Earth. The image was made at infrared wavelengths, but false colors are used to represent different levels of brightness. The diagonal spike is caused by the telescope optics.

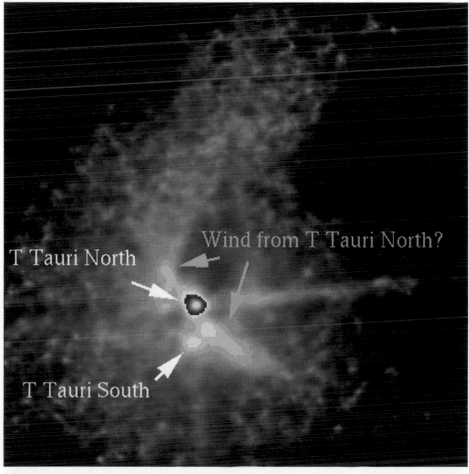

T Tauri North

Wind from T Tauri North?

T Tauri South

(a)

(b)

planet in the process of formation, with dust and gas still falling into it. If so, this scenario lends support for evolutionary theories of planet formation—at least, for large Jupiter-size planets. Even though directly detecting a planet orbiting a star is difficult, we have found planet-like objects around other stars. For example, the *Hubble Space Telescope* captured an image of a **brown dwarf** (Gliese 229B, about 20–50 times as massive as Jupiter) orbiting the cool red star Gliese 229A (Figure 7-18b).

What makes direct observations of exoplanets difficult, of course, is the fact that they are very faint and very close to the star(s) they orbit. The reflected starlight of planets similar to the ones in our solar system is about one-billionth as bright as the star and about one-millionth in the infrared. Other factors that make such direct observations difficult include changes in our atmosphere and scattering of light by telescope optics. Adaptive optics, however, should eventually make it possible to directly image and study planets around nearby stars.

- *Dust disks.* Disks of dust and gas about the size of the solar system have been detected around several stars. The most obvious case is that of β (the Greek letter beta) Pictoris. **Figure 7-19a** shows a false color photo of an infrared image of β Pictoris. The bright central star was blocked out so that the radiation from the dimmer disk that surrounds it could be observed. Studies of the bright band that extends from upper left to lower right show that it contains fine dust with an icy consistency, which fits with our theory of the development of the solar system. The image shows the disk to a distance of 24 AU from the star and it clearly shows a warp in the inner part of the disk. The presence of this warp indicates the existence of a planet the size of Jupiter orbiting at a distance of less than 20 AU from the star, with an inclined orbit of 3°. The intensity of the light we receive from this star varies, which may be the result of a partial eclipse by an orbiting planet.

brown dwarf A star-like object that has insufficient mass to start nuclear reactions in its core and thus become self-luminous.

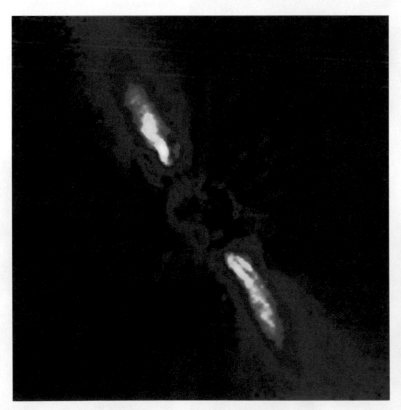

(a)

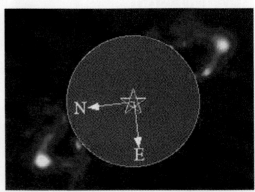

(b)

Figure 7-19

(a) This is an infrared photo of β Pictoris, obtained with the ESO ADONIS adaptive optics system (Grenoble Observatory). The star itself has been blocked out, and the disk of particles around it is visible. This disk is dynamically "ringing like a bell," probably due to the gravitational influence of a star that passed near β Pictoris some 100,000 years ago. Such close encounters with nearby stars can disrupt the evolution and appearance of thin disks, where planetary systems are formed. (b) This infrared image was taken by the *Hubble's* NICMOS camera and shows a disk around the star HR 4796A.

The β Pictoris system is an exceptional case; in general, we cannot see disks around stars in visible light, since the disk's visible light is hidden by the brightness of the star it orbits. However, in the infrared, the brightness of the star is reduced while the brightness of a planet peaks, allowing us to detect a planet more easily in the infrared. Figure 7-19b is an infrared image of a disk around the star HR 4796A. Since the star is about 1000 times brighter than the disk, the star's light has been blocked in order to see the disk better. Other evidence comes from the *Hubble Space Telescope*, which has found numerous cases of dust disks around new stars, making it appear that at least half of the Sun-like stars in the Galaxy have planetary systems.

- *Pulsar companion.* In 1992 astronomers reported that they had found variations in the rate of the signals from the ***pulsar*** PSR 1257+12 (about 2630 light-years away). Pulsars are stars that result from supernova explosions and emit beams of radio waves. When these beams sweep past the Earth, we observe them as radio pulses that normally are very constant in their frequency of pulsation. The variations in the frequency of a pulsar can be explained if it has one or more companion objects, such as a planet. We now have a confirmed detection of at least three planets around this pulsar. Two of these exoplanets have masses about three times that of Earth, a third is about the size of the Moon (but might be an observational artifact), while a fourth might be 100 times the mass of Earth. Since 1992, we have another confirmed observation of a planet around pulsar PSR B1620−26 (about 12,400 light-years away). Other observations, which have not been confirmed yet, suggest that planets may be orbiting two other pulsars. As additional pulsars are examined for similar evidence, and if any of these observations are confirmed, we will have more pieces to add to the puzzle of how common are planetary systems.

- *Binary systems and visual wobble.* As we showed in Chapter 3, gravitationally bound objects revolve around their common center of mass. The case we discussed there was the Earth–Moon system. Among the stars, we observe many cases of ***binary star systems,*** in which two stars revolve around one another in this manner. In some cases, only one star of a binary system is visible, but we can deduce the existence of the dimmer star from the motion of the visible one. If we see a star that appears to wiggle in its position or that exhibits an elliptical motion, we can conclude that it is in orbit with another object. This provides us with a possible method of detecting the presence of a large planet in orbit around another star. If we hope to find a planet by such means, we must look at nearby stars, whose motion is easier to observe.

A star named Barnard's star is the second-closest star to the Sun. In the first half of the 20th century, a back-and-forth motion was reported for this star. However, the motion is very slight and detecting it involved comparing photographs taken over long periods of time. Most astronomers considered the data very suspect, for the photographs were taken under different conditions with instruments that had been changed over the course of the observations. However, measurements of changes in the *radial* velocity of Barnard's star agreed with the original conclusions and indicated that the star may indeed have at least one planet of 1.5 times the mass of Jupiter revolving around it at a distance of 4 AU. (Large planets, of course, would cause the star to move more than small ones would.) Measurements on this object are continuing, since even recent observations with the *Hubble Space Telescope* have not yet confirmed the existence of any planets around this star.

Detecting the wobble of a star by carefully measuring its position in the sky relative to other stars is not easy (**Figure 7-20a**). For example, if we were to look at the wobble of our Sun (mainly due to Jupiter) from 30 light years away, the small circle made by the Sun would appear to be about the size of a quarter seen from a distance of 6000 miles. Measurements made using this direct method must be extremely accurate and must be made over long periods of time to capture the orbital period of

pulsar A celestial object of small size (typical diameter of 30 kilometers) that emits beams of radio waves, which, when sweeping past the Earth, are observed as pulses of radio waves with a regular period between a millisecond to a few seconds.

Pulsars will be described in Chapter 15.

binary star system A pair of stars that are gravitationally linked so that they orbit one another.

Recall that radial velocities are measured by the Doppler effect and that radial motion can therefore be measured with greater sensitivity than can tangential motion.

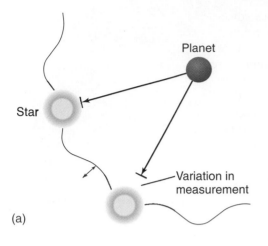

(a)

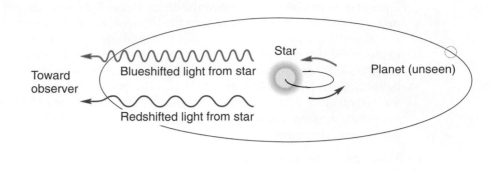

(b)

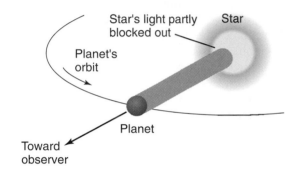

(c)

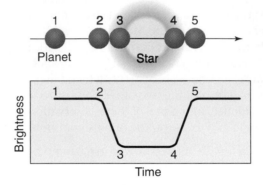

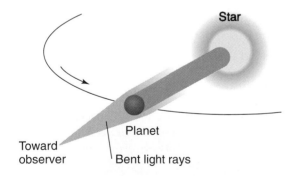

(d)

Figure 7-20

Different methods of detecting exoplanets. (a) The astrometric method. A planet causes a slight movement in the star, a to-and-from motion that can be detected. (b) The Doppler (radial-velocity) method. The presence of a planet around a star results in the star alternately wobbling toward and away from the Earth. Astronomers measure the Doppler shift in the star's spectrum and use the shift to determine the star's orbit. This in turn allows us to determine the orbit of the unseen companion (planet-like object). (c) The transit photometry method. When a planet passes directly between its star and an observer's line of sight, it blocks out a portion of the star's light. The shape of the light curve allows us to find the sizes of the orbit and of the planet. (d) The gravitational microlensing method. Light rays from a star bend when they pass through space that is warped by the mass of the planet. The planet's gravity acts as a lens briefly increasing the star's apparent brightness.

astrometry The branch of astronomy that deals with the measurement of the position and motion of celestial objects.

the star's motion. The advantage of this astrometric technique is that it allows us to find the mass of the planet because the star's motion is detectable in two dimensions.

- *Binary systems and Doppler wobble.* Another method for detecting the wobble of a star due to the presence of planets around it involves measuring the Doppler shift of the star's spectrum as it alternately wobbles toward and away from the Earth (Figure 7-20b). Such shifts are tiny because the star's motion is very slow in its orbit: For example, Jupiter causes the Sun's speed to vary with an amplitude of 12.5 meters/second (or 28 miles/hour); Saturn's effect is the next largest, with an amplitude of 2.7 meters/second (or 6 miles/hour). The current precision achieved by researchers using this radial-velocity technique is 3 meters/second, allowing them to find Jupiter-like planets orbiting Sun-like stars. The wobble motion of a star can provide a wealth of information about its planetary companion, such as its mass, distance, and orbital period. This method has proven to be very successful. In 1995, astronomers used this method to discover the first planet orbiting a normal, Sun-like star, 51 Pegasi. Since then, many groups of researchers have discovered several exoplanets. As of March 2003, about one hundred normal stars have been discovered to have planet-like objects (including some stars with orbiting brown dwarfs). Many of these planets seem quite different from those in our solar system (**Figure 7-21**). Many have masses larger than Jupiter's, yet they orbit much closer to their parent star. Also, many have large eccentricities. This is very unlike the planetary orbits in our solar system. The first two Saturn-sized planets were discovered in March 2000, suggesting that as time goes on we should be able to detect even smaller planets; this discovery supports our current theory that planets form in a disk of dust and gas by a snowball effect of growth.

Even though this radial-velocity (Doppler) method has proven to be very successful, precise measurements require a large number of spectral lines; since the hottest stars have far fewer spectral features than cooler Sun-like stars, this method cannot be used for the hottest stars. It also does not allow us to find the exact mass of the planet but only a lower limit to it. Finally, finding Earth-like planets in Earth-like orbits with this method is beyond our current capabilities.

Figure 7-21

Some of the exoplanets around normal stars. Most of these objects are closer to their parent star and more massive than the planets in our solar system. The figure includes the name of the parent star, the mass of the planet (in units of the mass of Jupiter), and the orbital semimajor axis (in AU). Most of the exoplanets known to date were discovered using the Doppler method.

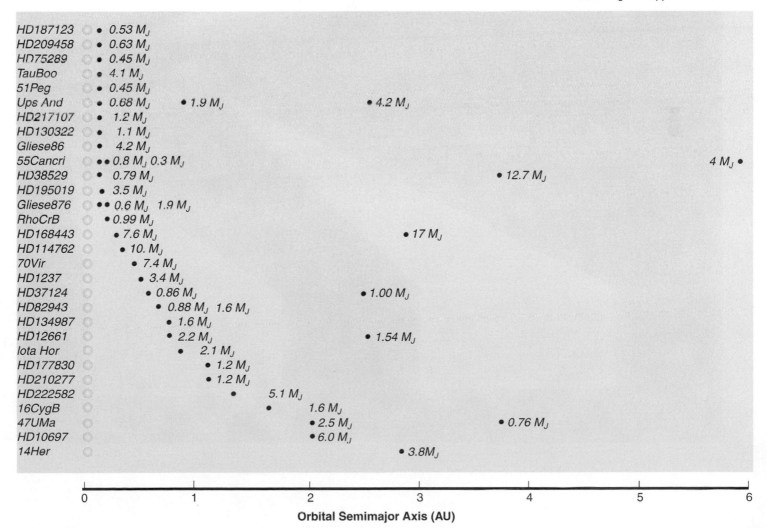

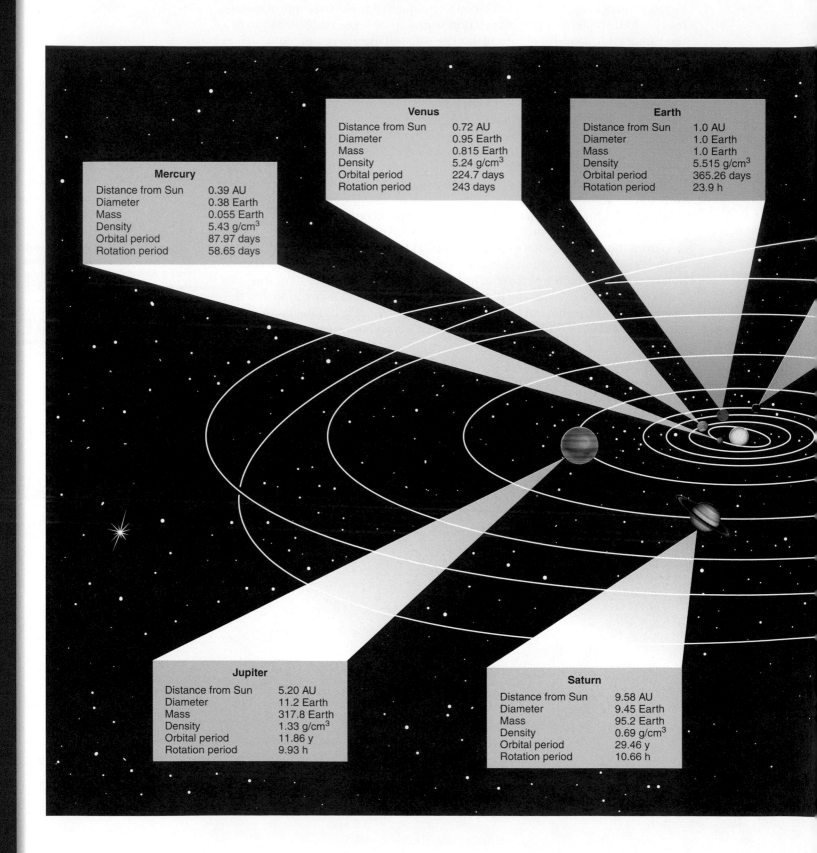

Mercury

Distance from Sun	0.39 AU
Diameter	0.38 Earth
Mass	0.055 Earth
Density	5.43 g/cm^3
Orbital period	87.97 days
Rotation period	58.65 days

Venus

Distance from Sun	0.72 AU
Diameter	0.95 Earth
Mass	0.815 Earth
Density	5.24 g/cm^3
Orbital period	224.7 days
Rotation period	243 days

Earth

Distance from Sun	1.0 AU
Diameter	1.0 Earth
Mass	1.0 Earth
Density	5.515 g/cm^3
Orbital period	365.26 days
Rotation period	23.9 h

Jupiter

Distance from Sun	5.20 AU
Diameter	11.2 Earth
Mass	317.8 Earth
Density	1.33 g/cm^3
Orbital period	11.86 y
Rotation period	9.93 h

Saturn

Distance from Sun	9.58 AU
Diameter	9.45 Earth
Mass	95.2 Earth
Density	0.69 g/cm^3
Orbital period	29.46 y
Rotation period	10.66 h

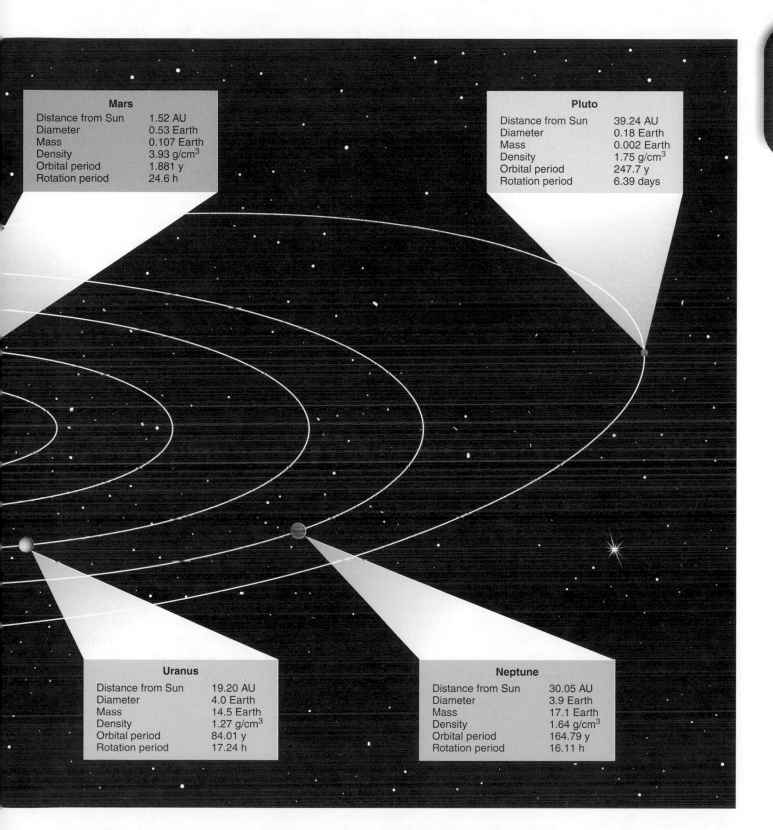

Mars

Distance from Sun	1.52 AU
Diameter	0.53 Earth
Mass	0.107 Earth
Density	3.93 g/cm^3
Orbital period	1.881 y
Rotation period	24.6 h

Pluto

Distance from Sun	39.24 AU
Diameter	0.18 Earth
Mass	0.002 Earth
Density	1.75 g/cm^3
Orbital period	247.7 y
Rotation period	6.39 days

Uranus

Distance from Sun	19.20 AU
Diameter	4.0 Earth
Mass	14.5 Earth
Density	1.27 g/cm^3
Orbital period	84.01 y
Rotation period	17.24 h

Neptune

Distance from Sun	30.05 AU
Diameter	3.9 Earth
Mass	17.1 Earth
Density	1.64 g/cm^3
Orbital period	164.79 y
Rotation period	16.11 h

Planetary Data Comparison

◁ NASA's *Kepler Mission,* scheduled for
launch in 2006, will use a space-
borne telescope to search for Earth-
like (or even smaller) exoplanets
using the transit method.

◁ Since we cannot directly see the
planet, we do not know the inclination of its orbital plane to our line of
sight. As a result, we can only calculate a lower limit to the planet's
mass.

◁ In Chapter 13 we will study how the
material in interstellar space forms
stars and planetary systems.

- *Stellar occultation*. When one celestial object passes in front of another, we say that an "occultation" occurs. In late 1999 astronomers observed the dimming of star HD209458, caused by its planet passing in front of it (such a passage is called a transit). This confirmed the earlier discovery, made using the Doppler method, that this star has a Jupiter-size planet. The star's intensity dims by only 1.7% during occultation. This method is very useful in confirming previously discovered planets. An advantage of this method is that it is not sensitive to the planet's mass; we are looking for the amount of starlight the planet obscures and not the planet's gravitational pull on its star (Figure 7-20c). It is clear that this method can be used only in cases where the Earth lies in or near the orbital plane of an exoplanet. However, this method is very efficient because within the field of view of a telescope there are thousands of stars that can be surveyed. In January 2003, astronomers announced the first detection using the transit method of a Jupiter-sized planet (OGLE-TR-56b) orbiting a *normal* star about 5,000 light-years away, in the Sagittarius arm of our Galaxy.

- *Microlensing*. When a massive object passes between the observer and a distant source of light (say a star), the gravity of the object will bend the light according to the general theory of relativity. The massive object plays the role of a lens, causing the source to appear to slowly brighten to a few times its usual intensity over a period of time. For the case of a star and its planet(s), when a planet passes in front of the star along the observer's line of sight, the planet's gravity acts like a lens and produces a brief enhancement to the star's brightness (Figure 7-20d). It is possible to detect Earth-like planets using this method; however, there are many parameters that influence microlensed light.

All the methods previously mentioned are currently being used from the ground and in the future they will also be used from space. Astronomers are also exploring additional methods that could be used to detect exoplanets. These include the following: detailed observations of the structure of the disks around stars, showing clumps of dust and gas that could be the seeds of planet formation; gaps in the disks surrounding young stars, possibly caused by the gravity of a newly formed planet; and spectroscopic detection of elements expected to be present in the atmospheres of planets such as water vapor and sodium (atomic sodium has been detected by the *HST* in the case of exoplanet HD2094586).

The properties of the exoplanets found so far have defied expectations and are forcing us to reexamine our theories of how planets form. It is possible that the mechanism that formed some of these planets is different from that which operated in our solar system. On the other hand, maybe these planets did form at large distances from their parent star but migrated inward. Since a newly formed planet separates the dust disk around the star into two regions, the protoplanet and the inner region of the disk will lose energy and angular momentum to the outer region of the disk. As a result, the protoplanet will migrate inward toward the star. The protoplanet may finally reach a stable orbit close to the star as a result of tidal interactions with it. Another possibility for the differences is that some of these objects may not be planets at all. The masses shown in Figure 7-21 are lower limits and in reality they may be much larger, so these objects may be brown dwarfs that follow a different evolutionary path.

The recent discoveries of planetary systems around other stars suggest that our solar system formed by an evolutionary rather than a catastrophic process. The reasoning is as follows. If catastrophic processes are necessary for the formation of planetary systems, relatively few such systems would be expected, since the special circumstances needed are, by their nature, rare. On the other hand, if stars form by the process described by evolutionary theories, we would expect planetary systems to be common. Notice that although the formation of planetary systems by catastrophic events is unlikely, this in itself is not an argument against such a scenario in the case of the solar system. If there is any possibility at all that a catastrophic event can cause a planetary system, it could well have happened here. However, there is now clear evidence that evolutionary development of planetary systems is common, and it lends support for hypotheses that postulate that this is what occurred in our own system.

It is too early for us to draw any conclusions based on the newly discovered exoplanets. After all, the recent discoveries seem to suggest two contradictory ideas. On one hand, we now know that planets are more plentiful than we once thought, and therefore it is more likely that Earth-like planets exist where life could develop. On the other hand, life on Earth could be unique, since all planetary systems discovered so far are not suitable for life. We do not yet have the ability to detect Earth-like objects, and the few discovered objects may be exceptions to the general rules of planetary formation. There is no doubt that in the next few decades we will discover many more exoplanets as our telescopes and instruments allow us to make even more precise observations. A planned space-borne interferometer might be able to detect other Earths as early as the next decade and it should be able to use spectroscopy to determine the chemical composition of their atmospheres or surfaces, looking for any signatures of life on them.

During the 4th century B.C., Aristotle argued in favor of the uniqueness of our planet, since it was at the center of the universe. Another great philosopher at the time, Epicurus, argued that the universe must be infinite and therefore must contain an infinite number of other worlds. For about 2400 years we have been trying to answer the question of whether our solar system is truly unique. We are getting very close.

The European Space Agency has two planet-search missions in development. *Gaia* (around 2010) is an astrometry mission that will conduct a survey of a billion stars in our Galaxy. *Darwin* (after 2014) is a flotilla of eight spacecraft that will use interferometry to survey 1000 of the closest stars looking for the most likely places for life to develop.

Conclusion

In this chapter we showed that our solar system contains a myriad of objects, vastly different from one another. Similarities are also present, however, and enable the planets to be divided into two categories: terrestrial and Jovian. We also described some of the characteristics that prohibit us from listing the (normally) most distant planet, Pluto, in either category.

One of the most important topics discussed during our overview of the solar system was how we measure some of the properties of the objects circling the Sun, including distances between them, their sizes, and their masses.

The patterns we observe in the solar system, along with what we know about the formation of stars, allow us to put together a reasonably detailed story of the origin of the solar system. Although many questions remain, observations of newly forming stars are confirming our theories of the development of our system. Recent observations of exoplanets lend support to an evolutionary development of planetary systems, while bringing forward many new questions about the process.

In the next three chapters we will examine each of the planets in more detail and will look at the lesser objects within the solar system—comets, meteoroids, and asteroids.

In some worlds there is no Sun and Moon, in others they are larger than in our world, and in others more numerous. In some parts there are more worlds, in others fewer . . . ; in some parts they are arising, in others failing. There are some worlds devoid of living creatures or plants or moisture.

Democritus (~460–370 B.C.)

Study Guide

1. Which planet is most massive?
   A. Mercury
   B. Mars
   C. Earth
   D. Jupiter
   E. Saturn

2. The only planet whose orbit is more eccentric than Mercury's is
   A. Saturn.
   B. Earth.
   C. Pluto.
   D. Venus.
   E. Neptune.

RECALL QUESTIONS

3. Whether a planet or moon has an atmosphere depends upon the planet's (or moon's)
   A. orbital speed.
   B. temperature.
   C. escape velocity.
   D. [Both A and C above.]
   E. [Both B and C above.]

4. Which planet has its plane of rotation tilted most with respect to its plane of revolution?
   A. Uranus
   B. Earth
   C. Venus
   D. Mars
   E. Mercury

5. Venus might be called Earth's sister planet because it is similar to the Earth in
   A. size.
   B. mass.
   C. rotation period.
   D. [Both A and B above.]
   E. [Both A and C above.]

6. Saturn is one of the _____ planets.
   A. Jovian
   B. inner
   C. inferior
   D. minor

7. Which of the following statements is true of all of the planets?
   A. They rotate on their axes and revolve around the Sun.
   B. They rotate in the same direction.
   C. They have at least one moon.
   D. Their axes point toward Polaris.
   E. [More than one of the above is true of all of the planets.]

8. Saturn's density is
   A. less than that of Jupiter.
   B. more than that of Jupiter.
   C. similar to the Earth's.
   D. greater than that of Earth.
   E. [Two of the above.]

9. Which of the following choices lists the four planets from smallest to largest?
   A. Mars, Mercury, Earth, Uranus
   B. Mercury, Uranus, Mars, Earth
   C. Uranus, Mercury, Mars, Earth
   D. Mars, Mercury, Earth, Uranus
   E. [None of the above.]

10. Compared to Jovian planets, terrestrial planets have a
    A. more rocky composition.
    B. lower density.
    C. more rapid rotation.
    D. larger size.
    E. [More than one of the above.]

11. Which of the following statements is true of Jovian planets?
    A. They have low average densities compared to terrestrial planets.
    B. Their orbits are closer to the Sun than the asteroids' orbits.
    C. They have craters in old surfaces.
    D. They have smaller diameters than terrestrial planets do.
    E. They have fewer satellites than terrestrial planets do.

12. Most asteroids orbit the Sun
    A. between Earth and Mars.
    B. between Mars and Jupiter.
    C. between Jupiter and Saturn.
    D. beyond the orbit of Saturn.
    E. [None of the above. No general statement can be made.]

13. Distances to the planets are measured today by the use of
    A. geometry.
    B. calculus.
    C. spacecraft flybys.
    D. analysis of the motion of their moons.
    E. radar.

14. The mass of Jupiter was first calculated
    A. using its distance from the Sun and its revolution period.
    B. using its angular size and distance from the Earth.
    C. using data from spacecraft flybys.
    D. by analysis of the motion of its moons.
    E. [Two of the above.]

15. Which is a longer time on Earth?
    A. A sidereal day.
    B. A solar day.
    C. [Either of the above, depending upon the time of year.]
    D. [Neither of the above, for they are the same.]

16. At a greater distance from the surface of a planet, the escape velocity from that planet
    A. becomes less.
    B. remains the same.
    C. becomes greater.
    D. [Neither of the above, for the behavior of different planets is different in this regard.]

17. At the same temperature, the average speed of hydrogen molecules is _____ that of oxygen molecules.
    A. less than
    B. the same as
    C. greater than
    D. [No general statement can be made.]

18. The escape velocity from the top of Earth's atmosphere is _____ the escape velocity from the surface of the Moon.
    A. less than
    B. the same as
    C. greater than
    D. [No general statement can be made, for the escape velocity depends upon temperature.]

19. Which planet (of those listed) gets closest in distance to the Earth?
    A. Jupiter
    B. Mercury
    C. Venus
    D. Saturn
    E. Mars

20. If planetary systems are caused as proposed by the catastrophe theories, there should be
    A. many planetary systems besides ours.
    B. few planetary systems besides ours.
    C. [Neither of the above; the theories would make no predictions in this regard.]

21. Evolutionary theories now account for the slow rotation rate of the Sun by pointing to
    A. the slowing effect on the Sun of the solar wind.
    B. friction within the gases involved, which would prevent the Sun from rotating fast.
    C. the effect of the inner planets on the Sun.
    D. the effect of the large planets—particularly Jupiter—on the Sun.
    E. the conservation of angular momentum, which predicts a slowly rotating Sun when it formed.

22. Which of the following observations *cannot* be accounted for by evolutionary theories of solar system formation?
    A. All of the planets revolve around the Sun in the same direction that it rotates.
    B. All of the planets revolve in nearly the same place.
    C. Planets farther from the Sun are farther apart.
    D. The outer planets contain more volatile elements than the inner planets do.
    E. [All of the above are accounted for by evolutionary theories.]

23. After the evolutionary theory of the formation of the solar system was proposed, it was almost dismissed because it seemingly could not explain
    A. planetary masses.
    B. planetary distances from the Sun.
    C. the existence of comets.
    D. why some planets—particularly Jupiter—have a strong magnetic field.
    E. the observed rotation rate of the Sun.

24. According to the evolutionary theories of solar system formation, the outer planets contain much more hydrogen and helium than the inner planets because these elements
    A. never fell in near the Sun.
    B. condensed quickly to liquids and solids and remained far from the Sun.
    C. were blown away from the inner solar system by the solar wind.
    D. [Both A and B above.]
    E. [All of the above.]

25. Astronomers are now reasonably confident that the planets of the solar system
    A. formed when a comet pulled material from the Sun.
    B. formed when another star passed very close to the Sun.
    C. evolved from a rotating disk when the Sun was forming.

D. [None of the above. There is currently no satisfactory explanation for the origin of the planets.]

26. In a previous chapter we saw that Kepler's third law relates a planet's period to its distance from the Sun. This law was expanded by Isaac Newton to include what other quantity?

27. How does the Sun's mass compare to the total mass of all other objects in the solar system?

28. What is the largest planet, and how does it compare in size and mass to the Earth?

29. What was the first celestial object discovered after Bode's law was proposed? Did it fit predictions made by the law? (Hint: See the Advancing the Model box.)

30. How do we know the masses of the planets?

31. Distinguish between a sidereal day and a solar day. Which is longer on Earth?

32. The planets' directions of rotation and revolution have certain features in common. Describe these features. Which planets have an unusual direction of rotation?

33. Name the terrestrial planets and the Jovian planets. Why are the latter called Jovian?

34. In what ways are the terrestrial planets similar to one another but different from the Jovians?

35. How does the eccentricity of the Earth's orbit compare to that of other planets?

36. How does the Earth compare to the other planets in density?

37. What is meant when we say that the escape velocity from the Earth is 11 kilometers per second?

38. What two factors determine the speed of the molecules of a gas?

39. Explain why hydrogen escapes from the Earth's atmosphere but carbon dioxide does not.

40. How do evolutionary theories explain the pattern of chemical abundance among the planets?

41. We now have definite evidence that planetary systems exist around other stars. Does this lend support to either the catastrophe or evolutionary theories? If so, which?

42. How do evolutionary theories account for asteroids? The Oort cloud?

43. Describe the evidence that planetary systems exist around other stars.

---

1. Copernicus did not know the distance from the Earth to the Sun. Yet the text states that he calculated the relative distances to the planets. Explain.

2. How do we measure the distances to the planets today?

3. Is density the same thing as "hardness"? Why do we bother calculating the density of an object when we could just state its mass? (That is, what additional information is gained by calculating its density?)

4. Name the planet that
   a. has the most eccentric orbit.
   b. is largest.
   c. has the least density.
   d. has the most mass.
   e. has the most moons.
   f. has the least atmosphere.

5. If a planet rotates slowly in a retrograde direction (clockwise as seen from above the Sun's north pole), which will be longer, its sidereal day or its solar day? Explain.

6. Is the escape velocity from Earth the same whether we are considering a point just above the atmosphere or a point higher up? Explain.

7. How do evolutionary theories explain observations 1, 2, and 4 as listed in the section "Evidential Clues from the Data"? How do catastrophe theories explain these observations?

8. Explain the problem the law of conservation of angular momentum presents for evolutionary theories and describe how today's theory accounts for the observations.

9. Explain in your own words why there is a pattern of changes in the composition of planets as we move from those close to the Sun to those far away.

10. What techniques are used to detect planets that orbit other stars? Name some limitations to these techniques.

11. If we wish to find planets near other stars, why not just use telescopes to look for the planets directly?

CALCULATIONS

1. The eccentricity of Pluto's orbit is 0.24. Use a drawing of an ellipse to explain what this means quantitatively.

2. If another planet were found beyond Pluto, how far would Bode's law predict it to be from the Sun (assuming it is the next planet in his scheme)?

3. At a time when Jupiter is 9.0×10^8 kilometers from Earth, its angular size is 33 seconds of arc. Use this data to calculate the diameter of Jupiter and then check your answer in Appendix C.

4. Suppose that we bounce a radar signal off of Jupiter. It returns to Earth 100 minutes after being sent. How far away is Jupiter at the time of the measurement?

5. The escape speed of an object from a planet or star of mass m and radius R is proportional to $\sqrt{m/R}$. The escape speed from

Earth's surface is about 11 kilometers/second. Using data from Appendices B and C, find the escape speed from the Sun's surface and compare your result to that given in Appendix B.

6. Phobos is a small satellite orbiting Mars. The average distance between the two objects (from center to center) is about 9400 kilometers and the orbital period is 0.319 days (see Appendix D). What is the mass of Mars? What assumption did you make to get this answer? Since Mars' radius is 3400 kilometers, what is its average density? What can you say about its chemical composition? You can find information about current explorations of our solar system at NASA's home page (http://www.nasa.gov).

EXPANDING THE QUEST

1. "Other Suns, Other Planets?" by D. C. Black, in *Sky & Telescope* (August, 1996).

2. "Searching for Life on Other Planets," by J. R. P. Angel and N. J. Woolf, in *Scientific American* (April, 1996).

3. "The Diversity of Planetary Systems," by G. Marcy and R. P. Butler, in *Sky & Telescope* (March, 1998).

4. "The Formation of the Earth from Planetesimals," by G. W. Wetherill, in *Scientific American* (June, 1981).

5. Peter Ward and Donald Brownlee, *Rare Earth: Why Complex Life Is Uncommon in the Universe* (Copernicus Books, 2000).

STARLINKS

Quest Ahead to Starlinks:
www.jbpub.com/starlinks

Starlinks is this book's online learning center. It features **eLearning,** which contains chapter quizzes and other tools designed to help you study for your class. You can also find **on-line exercises,** view numerous relevant **animations,** follow a guide to **useful astronomy sites** on the web, or even check the latest **astronomy news** updates.

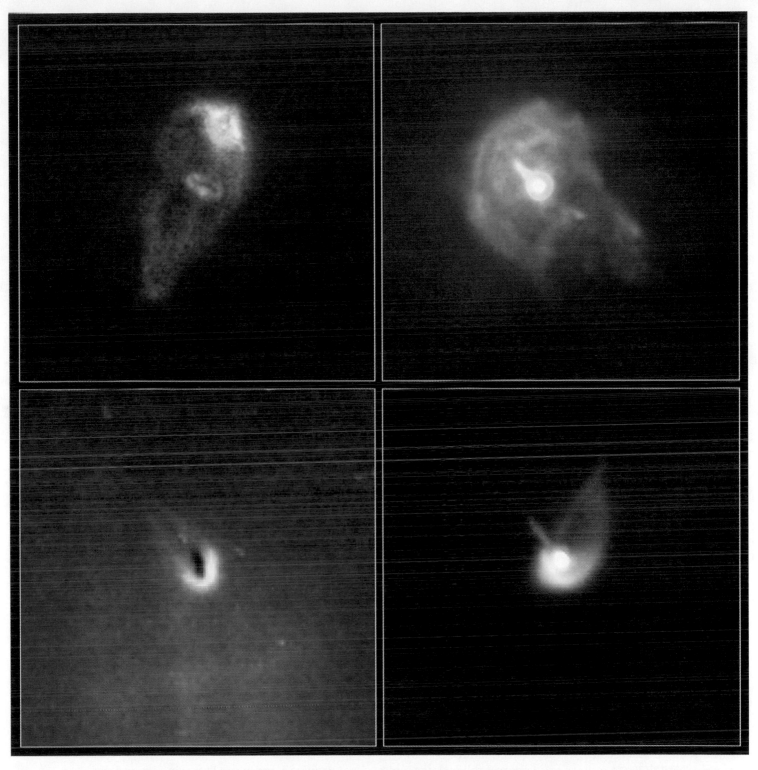

These *Hubble* images taken in 1998 and 1999 show dust disks around very young stars in the Orion nebula (about 1,500 light-years away). Ultraviolet radiation from the nebula's hottest star (Theta 1 Orionis C) is interacting with these protoplanetary disks, which include the seeds of planets. It is possible that these seeds will not grow into full-fledged planets as a result of the effects of the UV radiation from the star. In the image at top left, the green-colored oval near the center represents the disk. The cocoons of gas surrounding the disks were formed from material that evaporates from the surface of the disks.

CHAPTER 8

The Terrestrial Planets

This color composite image of Mars was taken by the *HST* on August 27, 2003 when Mars was at its closet distance to Earth in about 60,000 years. It reveals the southern polar ice cap, the largest volcano in the solar system (Olympus Mons, the oval-shaped feature above center), and a system of canyons (Valles Marineris, right of center). When this image was taken, Mars' Northern (Southern) hemisphere was in the midst of winter (summer). Clouds cover the region around the northern polar cap, while the orange streaks show dust activity above the southern polar cap.

THE MARS *PATHFINDER* MISSION TOOK OFF FROM EARTH on December 4, 1996, and landed on Mars July 4, 1997. Cameras on the lander took pictures of the Martian landscape for almost three months, until contact was lost. The photos showed a bone-dry landscape that might once have been a flood plain, though when floods of water existed on Mars is still not known. But the star of the *Pathfinder* show was its robot rover, named *Sojourner*.

Sojourner rolled down a ramp from *Pathfinder* and started crawling around the Martian surface at a rate of 2 feet/minute, guided by cameras and laser range-finders. At a size of 2 feet long and 1 foot wide, *Sojourner* looked like a child's remote-control car, bumping along from rock to rock (see Figure 8-34). *Pathfinder* sent back pictures of *Sojourner* that headlined newspapers and television broadcasts around the world. NASA engineers made up whimsical names for each rock that *Sojourner* stopped to study, including Barnacle Bill, Yogi, and Scooby Doo. When *Sojourner* bumped into the rock called Yogi, headlines reported the first "fender bender" on Mars. Weather reports noted morning temperatures on Mars of 76 degrees below zero Fahrenheit, with a forecast of 10 degrees above zero for an afternoon high.

The X-ray analysis of rocks and soil completed by *Sojourner* supported theories of geological convulsions through much of Mars's history. It's no accident that the largest volcano and the largest canyon known in the solar system are both on Mars. Yet the rocks also appeared to be more like those of Earth than are the lunar rocks brought back by the *Apollo* missions. Geologists reported that the landing site was a true "rock festival," but not really a nice place to visit.

Our trip to the planets begins with the planet closest to the Sun and then proceeds outward. As we discuss each planet in turn, you should compare that planet with Earth and with planets previously discussed. Concentrate on remembering patterns and comparisons between planets rather than memorizing numerical facts.

8-1 Mercury

In legend, Mercury was the god of commerce and travel, the Roman counterpart of the Greek god Hermes, the messenger of the gods. He wore winged sandals and delivered his messages with godlike speed. It is for this speedster that the fastest planet of the solar system is named.

Mercury as Seen from Earth

Although Mercury was one of the planets known to the ancients, it is the planet least seen by people on Earth without a telescope. Even Copernicus supposedly lamented near the end of his life that he had not seen it. Mercury is hard to see because it is so close to the Sun that it can be seen with the naked eye only either shortly after sunset in the western sky or shortly before sunrise in the eastern sky (**Figure 8-1**).

Looking at Mercury through a telescope, we are able to see that it exhibits phases like Venus and the Moon. Surface detail is difficult to discern, however, primarily because when Mercury is near the horizon, its light is passing through so much of the Earth's atmosphere. **Figure 8-2** illustrates this situation. The best telescopic views of Mercury actually are made when it is high overhead during the day. However, even then, surface features are not well defined, and the telescope shows only that Mercury's surface contains bright and dark areas. Early in this century some reports indicated a cratered surface, but this idea was not widely accepted. Details of Mercury's surface features were not seen clearly until the planet was visited by space probes.

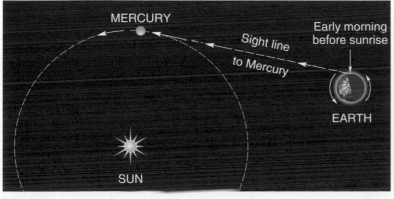

(a)

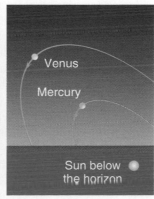

(b)

Figure 8-1
(a) Mercury can only be seen from Earth either shortly before sunrise (shown) or shortly after sunset. Mercury's maximum elongation (its maximum angle from the Sun) is about 28 degrees. (b) Possible positions of Mercury and Venus in the evening relative to the horizon. Mercury never appears in a really dark sky.

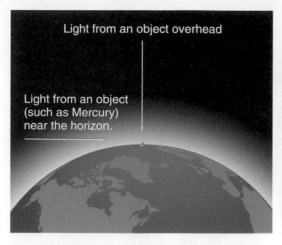

Figure 8-2
The thickness of Earth's atmosphere is exaggerated here, but notice that light from an object near the horizon must pass through more atmosphere than light from an overhead object. The dot on the Earth's surface is the location of a person looking at the two objects.

Figure 8-3
A mosaic of photos of Mercury taken by *Mariner 10*.

Mercury via *Mariner*—Comparison with the Moon

In November 1973, *Mariner 10* left Earth on a mission to Mercury. After passing near Venus, it swept by Mercury in March 1974, taking numerous photographs. Its orbit was then adjusted so that after passing Mercury, it orbited the Sun with a period just twice that of Mercury's, coming back near Mercury once during each of its orbits. *Mariner's* systems lasted long enough to enable it to take photographs on two passes of Mercury after the first, providing us with more than 4000 photographs.

Figure 8-3 is a mosaic of *Mariner's* views of Mercury. Notice its similarity to our Moon; both are covered with impact craters produced by debris from space. We find, however, that the walls of the craters of Mercury are less steep than those of the Moon, so that Mercury's craters are less prominent than the Moon's (**Figure 8-4**). This was expected, because Mercury's surface gravity is about twice that of the Moon and loose material does not stack as steeply under the greater gravitational force.

When you compare photographs of Mercury and our Moon, notice that ray patterns of material ejected from craters are less extensive on Mercury than on the Moon. When craters were formed by meteorite impacts, the greater gravitational force on the surface of Mercury kept the material from being thrown as far from the craters.

Notice also that Mercury lacks the large maria we see on the Moon. Instead, Mercury's craters are fairly evenly spread across the surface, separated by smooth plains. This difference between the two surfaces is related to differences in the rate at which the objects cooled after formation. As we discussed in Chapter 7, the terrestrial planets and the Moon began as balls of molten rock. As they cooled, a solid crust formed on their surfaces. Debris falling from space struck the crust with enough energy to penetrate it and let lava flow up through the break. The results of these lava flows are obvious on the Moon, where large maria were formed from the lava that welled up from beneath the surface.

However, a larger object cools more slowly than a small one, so Mercury's crust formed more slowly than the Moon's. To understand this, consider the amount of energy included in the volume of a cooling object. This energy is released into space through the object's surface. The larger the surface area, the easier it is for the object to release energy into space and cool down. However, the larger the volume of the object, the more energy it has to release. In other words, the time it takes for an ob-

Like the craters on the Moon, most (or all) of Mercury's craters were formed by impacts with infalling objects—meteorites.

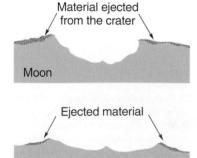

Figure 8-4
Craters on the Moon have steeper walls than those on Mercury, and because of the greater surface gravity on Mercury, material ejected from the crater does not travel as far.

ject to cool is directly related to the ratio of the object's volume to its surface area. For a spherical object, this ratio is simply proportional to the object's radius:

$$\text{time to cool} \propto \frac{\text{volume}}{\text{surface area}} = \frac{(4\pi/3) \cdot R^3}{4\pi \cdot R^2} \propto R$$

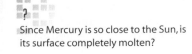

?
Since Mercury is so close to the Sun, is its surface completely molten?

EXAMPLE

The radius of Mercury is 1.4 times the radius of the Moon. Considering only this difference between the two objects, which would take longer to cool and what would be the ratio of their cooling times?

SOLUTION The equation above tells us that the cooling time of a sphere depends directly on its radius. Since Mercury is larger, it will take longer to cool. In fact, its radius is 1.4 times as great, so its cooling time will be 1.4 times as long.

TRY ONE YOURSELF

Jupiter and Earth were both hot when they were formed, but although Earth has cooled as much as it will, Jupiter is still cooling. Considering only their sizes, what would be the ratio of their cooling times? (Hint: See Appendix C for planetary sizes.)

Thus, the larger the object is, the longer it takes for the object to cool. As a result, Mercury's crust formed more slowly than the Moon's. For a long time after the crust started forming, meteoritic debris was still able to penetrate it, allowing lava to flow out and obliterate older craters. This resulted in the smooth plains we see between the evenly spread craters.

Another difference between Mercury's surface and the Moon's is the great number of long cliffs (called *scarps*) that are found on Mercury (**Figure 8-5**). These are found at many locations around the planet and suggest that after its crust hardened during its formation, Mercury shrank a little, causing the cliffs to be formed.

Mariner photographs revealed a large impact crater, named the Caloris Basin, on Mercury's surface (**Figure 8-6**). It consists of several rings somewhat like those of Mare Orientale on the Moon (**Figure 8-7**), although the Mercurian feature is larger. Both "bull's eyes" were caused by large objects striking the surfaces; the impacts resulted in shock waves that caused the rings of cliffs. On the opposite side of Mercury from Caloris Basin is a jumbled, wavy area, strange enough that it has been dubbed "weird terrain." This was probably the outcome of the shock wave

scarps Cliffs in a line. They are found on Mercury, Earth, Mars, and the Moon.

The Latin word *calor* means *heat*. A unit for measuring heat is the *calorie*.

(a)

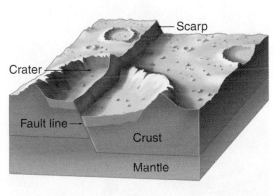

(b)

Figure 8-5
(a) Near the upper right center of this photo, you can see a black line of cliffs called scarps. These are much more extensive on Mercury than on the Moon. Some are hundreds of kilometers in length and up to 3 kilometers high.
(b) It is estimated that Mercury's radius decreased by about 1 kilometer after its crust hardened, causing the cliffs to be formed along fault lines.

Figure 8-6
The center of the Caloris Basin lies just out of the photo toward the left. Notice the ring pattern that was formed by the impact of a large object (now buried under the surface). The basin is about 1400 kilometers (870 miles) in diameter, the size of Texas.

4880 kilometers ≅ 3033 miles.

Figure 8-7
Mare Orientale, on the Moon, is similar to the Caloris Basin, but is only about 1000 kilometers (620 miles) in diameter. Mare Orientale is clearly visible in Figure B6-3 on page 191.

that was sent around the planet as a result of the Caloris Basin impact. When the waves met on the other side of the planet, they caused a permanent disruption of the surface there. We will see another consequence of the Caloris Basin impact when we discuss Mercury's motions.

Data from *Mariner* confirmed that Mercury has negligible atmosphere, as was expected considering the escape velocity from Mercury's surface and the high daytime temperatures caused by its proximity to the Sun. The thin atmosphere is created by the ejection of atoms from the planet's surface, a process that also occurs on our Moon. Refer to Figure 7-12 and notice that although the escape velocity from the surface of Mercury is greater than from the Moon, Mercury's temperature is higher; therefore we would expect none of the gases shown to be found on Mercury.

Size, Mass, and Density

Mercury is the second smallest planet, exceeding only Pluto in size. In fact, there are two natural satellites in the solar system that are larger than tiny Mercury. (**Figure 8-8** compares Mercury with some other solar system objects.) With a diameter of about 4880 kilometers, Mercury's total surface area is only slightly greater than that of the Atlantic Ocean.

The mass of Mercury, accurately determined from its gravitational pull on the *Mariner 10* probe, is only 0.055 of the mass of the Earth. Its average density is 5.43 times the density of water, slightly less than the Earth's 5.52.

The density of an astronomical object is determined not only by the material of which it is composed, but also by how much that material is compressed by the gravitational field of the object. For example, an object on the surface of Mercury feels less gravitational pull downward than it would on Earth. Thus, if Mercury were made up of the same elements as the Earth, matter below the surface would be less compressed by material above, and Mercury would have a density considerably less than the Earth's. The fact that Mercury's average density is only slightly less than Earth's means that Mercury has a higher concentration of heavy elements than the Earth does. However, evidence from *Mariner* indicates that the rocks on Mercury's surface are similar to Earth rocks, so this higher-density material must be below the surface. The answer must be that Mercury has a very large iron core, one that accounts for as much as

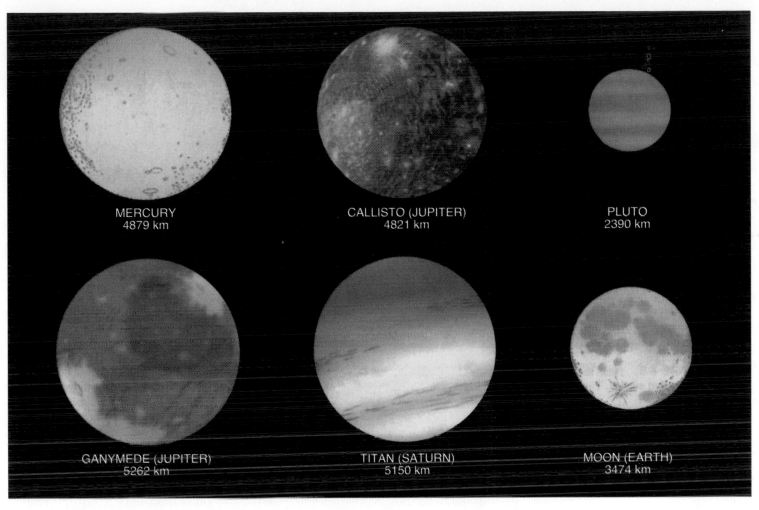

Figure 8-8
The diameter of Mercury compared to the mean diameters of some other solar system objects. Pluto is smaller than Mercury, but one of Jupiter's and one of Saturn's moons are larger. (In the case of moons, the parent planet is shown in parentheses.)

65 to 70% of the planet's mass. **Figure 8-9** shows Mercury's theorized interior and compares it to Earth's. Astronomers speculate that early in Mercury's history a catastrophic collision with a large asteroid blasted away much of Mercury's rocky mantle, leaving behind a planet with a greater-than-normal percentage of iron.

Before the *Mariner 10* mission, Mercury was thought to have no magnetic field. To see why, recall that the Earth's field is caused by the rotation of the planet and its liquid metallic core. The Earth's rotation period is about 24 hours while Mercury's is about 59 days. As a result, Mercury's rotation rate was thought to be too slow to produce a magnetic field.

Mariner 10 did detect a magnetic field on Mercury, although it is not very strong—about 1% as strong as the Earth's field. The field appears to be shaped like the Earth's, with magnetic poles nearly aligned with the planet's spin axis. The presence of the magnetic field has been a puzzle since it was detected in 1974. First, it led astronomers to hypothesize that Mercury's metallic core is molten, because the ***dynamo effect*** (Section 6-5)—the predominant explanation for planets' magnetic fields—requires molten magnetic material within the planet. More recent measurements indicate that Mercury does not release more heat than it receives from the Sun. Therefore the planet's interior must no longer retain heat from its formation,

dynamo effect The generation of magnetic fields due to circulating electric charges, such as in an electric generator.

Figure 8-9
Mercury's core occupies a greater portion of the planet's volume than does Earth's core.

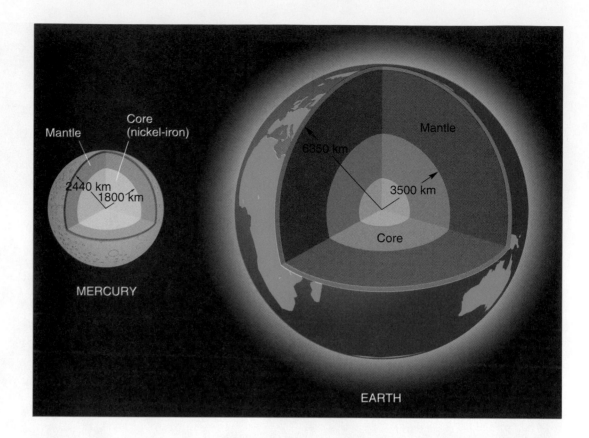

making it difficult to imagine a molten core. Many astronomers still think that some form of the dynamo effect is responsible for the magnetic field, but the question remains open. It could be that the core is partly molten as a result of impurities such as sulfur, which would lower the temperature at which the metal would solidify.

Mercury's Motions

Being the closest planet to the Sun, Mercury circles the Sun in less time (about 88 days) than any other planet and moves faster in its orbit than any other (average speed: 48 kilometers/second). Again, except for Pluto, Mercury's orbit is the most eccentric of the planets. Its distance from the Sun varies from 46 to 70 million kilometers. Careful observations made in the 19th century showed that Mercury's orbital path could not be exactly explained using Newton's laws. To accommodate for the small differences, some astronomers proposed the existence of another planet (tentatively named Vulcan) in an orbit near Mercury's. Einstein's General Theory of Relativity finally gave the correct explanation of Mercury's motions (see Section 3-11).

Because of Mercury's proximity to the Sun and its elongated orbit, it exhibits an interesting effect: One solar *day* on Mercury lasts two Mercurian *years*. To explain this, recall that because of tidal effects, the Moon keeps the same face toward the Earth. Astronomers once thought that Mercury likewise points the same face toward the Sun at all times. However, radar observations indicate that this is not the case. They show that Mercury rotates on its axis once every 58.65 Earth days, which is precisely two-thirds of its orbital period of 87.97 days. This means that the planet rotates exactly 1-1/2 times for every time it goes around the Sun.

Why would this pattern occur? The answer is that Mercury is not perfectly balanced; one side is more massive than the other. Because of this, the Sun exerts a torque on the planet, especially when it is closest to the Sun (at *perihelion*).

48 km/s is about 107,000 miles/hour.

perihelion (aphelion) The point in its orbit when a planet (or other object) is closest (farthest) to the Sun.

Mercury

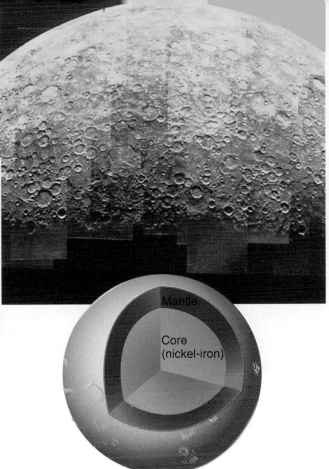

Earth

Mercury

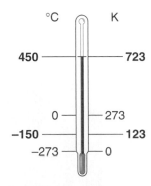

Mantle

Core
(nickel-iron)

Interior

Mercury Data

Mercury	Value	Compared To Earth
Equatorial diameter	4879 km	0.38
Oblateness	0	
Mass	3.302×10^{23} kg	0.055
Density	5.43 g/cm³	0.98
Surface gravity	3.7 m/s²	0.38
Escape velocity	4.3 km/s	0.38
Sidereal rotation period	58.65 days	58.79
Solar day	176 days	176
Surface temperature	−150°C to +450°C	
Albedo	0.12	0.39
Tilt of equator to orbital plane	0.01°	
Orbit		
Semimajor axis	5.79×10^7 km	0.387
Eccentricity	0.2056	12.3
Inclination to ecliptic	7.0°	
Sidereal period	87.97 days	0.24
Moons	0	

°C	K
450	723
0	273
−150	123
−273	0

Surface Temperature Extremes

QUICK FACTS

- Of all the planets, only Pluto is smaller (Mercury is one-third Earth's diameter). Closest planet to the Sun. Most eccentric orbit (except Pluto). Greatest difference between maximum and minimum temperatures. Difficult to see from Earth. Almost no atmosphere. Surface cratered somewhat like Moon. Large iron core. Period of revolution 1-1/2 times period of rotation. Magnetic field not fully explained.

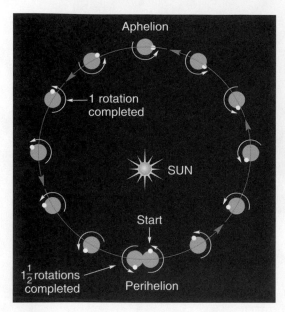

Aphelion

1 rotation completed

SUN

Start

1½ rotations completed

Perihelion

Figure 8-10
Mercury's rotation period is such that a point on its surface that is under the Sun at one perihelion position is opposite the Sun on the next.

450°C is about 840°F.

albedo The fraction of incident sunlight that an object reflects.

Figure 8-11
Radar reflections from Mercury show strong reflectivity from the planet's north pole. The reflections (using 3.5-centimeter wavelength radar) can be explained if ice exists at the pole.

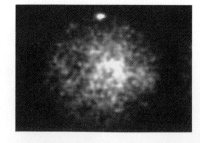

After countless revolutions, this has resulted in the rotational period of the planet being coupled with its orbital period. **Figure 8-10** shows this situation; note the light spot on Mercury's surface, which has been added to illustrate that opposite sides of the planet face the Sun at each perihelion passage.

The point marked on Mercury's surface in Figure 8-10 could represent Caloris Basin, for that Mercurian feature falls directly under the Sun at every other perihelion position. This is unlikely to be a coincidence. Astronomers have hypothesized that the object that hit Mercury's surface and created the Caloris Basin must have been made of dense material and that the object remains under the planet's surface, causing Mercury to be lopsided and therefore to have its periods of rotation and revolution coupled as they are.

Mercury's solar day is far different from its sidereal day. Refer again to Figure 8-10 to see why. Notice that after two-thirds of an orbit, Mercury has completed one rotation with respect to the distant stars. This does not mean that it has completed one solar day, however. Mercury does not complete one rotation with respect to the Sun until 176 Earth days have passed. Notice that a person standing at the light spot when the planet is at the position marked "start" would see the Sun at its highest point in the Mercurian sky. As time passes, the Sun would get lower, finally setting when the planet has completed half a revolution. When Mercury reaches "start" again, it is midnight for that person. It takes 88 days for Mercury to complete its orbit and that constitutes only half of a day for the person on the planet. Thus, a solar *day* on Mercury lasts 176 Earth days (or two Mercurian *years*) as mentioned earlier!

Imagine the Earth with almost no atmosphere to shield us from the Sun and with a daylight period lasting 88 of our days. This is the situation on Mercury. Daytime temperatures there reach as high as about 450°C. The element lead melts at about 330°C, so lead would be molten at midday on Mercury.

On the other hand, the nighttime side of Mercury gets very cold, since there is no atmosphere to hold in the heat during its long nights. It therefore gets as cold as −150°C (which is −240°F). Mercury's temperature variations are much greater than any other planet's.

Mercury's poles are thought to remain in constant cold, because the planet's equator aligns almost exactly (within a few arcminutes) with its orbit. This means that Mercury has no seasons, as we define seasons on Earth. Therefore even if the polar surface were perfectly smooth, only a very small amount of sunlight would strike it. But the surface isn't smooth, and the bottoms of some craters never receive sunlight. There are indications that the surface temperature inside these polar craters hovers forever at around −150°C, about the same temperature reached by the nighttime side of the planet.

Recent radar results from Mercury show a very high **albedo** for the polar regions (**Figure 8-11**). The most plausible explanation for the reflections is that Mercury has ice at its poles. However, even cold ice should evaporate over the ages, and the question of why it still exists remains a mystery at the time of this writing.

8-2 Venus

Venus is never seen farther than 46 degrees from the Sun, so, like Mercury, it is visible only in the evening sky after sunset or in the morning sky before sunrise. Except for the Sun and Moon, however, Venus can be the brightest object in our sky. Seeing such a bright object above the horizon has fooled many people into thinking they were seeing something else. During World War II, pilots of fighter planes sometimes shot at Venus, thinking that it was an enemy plane. Today, reports of UFOs often increase when Venus is at its brightest in the sky.

Venus is such a beautiful sight that it is little wonder that the ancients named it after the goddess of beauty and love. However, we'll see that as a possible home for a human colony, it is not such a beautiful planet after all.

Size, Mass, and Density

Historically, Venus has been known as the Earth's sister planet. It is similar to Earth in many ways: its diameter is 95% of Earth's, its mass is 82% of Earth's, and of all the planets, its orbit is located closest to us. From the values for its diameter and mass, we can calculate that Venus has a density 5.24 times that of water—not much different from Earth's 5.52. It would have a slightly higher density if its material were compressed as much as Earth material. Soviet spacecraft that landed on Venus have shown that its surface rocks have about the same composition as Earth rocks, and we therefore conclude that to have a density as high as it does, Venus must have a very dense interior and is probably differentiated, with a metallic core.

No magnetic field has been detected on Venus, and based on the sensitivity of the instruments that have searched for it, if one does exist, its strength must be less than 1/10,000 of Earth's. Because of Venus's slow rotation (about 243 days), we would not expect it to have a strong magnetic field. However, according to present theories for the origins of planetary magnetic fields, Venus's field should be strong enough for us to detect. Recall that Earth's field is known to have reversed its direction in the geologic past. Perhaps the magnetic field of Venus is now in the process of reversing, which would explain why it is so weak.

Venus's Motions

Venus orbits the Sun in an almost circular orbit, more nearly circular than any other planet. Its period is about 225 days, resulting in a very nearly constant orbital speed of about 35 kilometers/second.

The surface of Venus is covered by dense clouds, so it is not visible from Earth. However, astronomers learn about nearby planets by bouncing radar waves off them. Radar waves can penetrate Venus's cloud cover, and since 1961 we have been bouncing radar signals off its surface. This is how we first learned about its rotation rate and its surface features.

The use of radar to study Venus told us that its rotation is very unusual. The planet rotates *backward* compared to the direction of rotation of most other objects in the solar system. That is, if we view the solar system from far above the Sun's North Pole, we see that all planets circle the Sun counterclockwise and that all the planets except Venus, Uranus, and Pluto rotate in this same direction. Venus's rotation is very slow, with a sidereal rotation period of 243 days. Its period of revolution around the Sun, however, is only 225 days. These rotation and revolution rates result in the solar day on Venus being about 117 Earth days.

Tables showing the tilt of Venus's equator relative to its orbital plane list an angle of 177 degrees. Since a tilt of 180 degrees would turn a planet completely over, wouldn't a tilt of 177 degrees be the same as a 3-degree tilt? Not really. To show the difference, we first need to define what we mean by the north pole of a planet. After all, only the Earth has its axis oriented so that Polaris, the North Star, is nearly above its North Pole. To decide which pole of a planet is to be called north, imagine standing on the planet facing one of its poles. When we do this, if the rotation of the planet carries us around to our right, we are facing the north pole. The north pole of Venus, as defined in this manner, lies on the other side of the plane of the ecliptic from the Earth's North Pole. **Figure 8-12** illustrates this difference. Thus, we usually list the tilt of Venus's axis as 177 degrees rather than 3 degrees. The fact that the angle is greater than 90 degrees tells us that Venus has a backward rotation compared to most objects in the solar system.

Adjectives for Venus include "Venusian," "Venerian," and "Cytherean," the latter from the Greek island Cytherea that was associated with the goddess Aphrodite.

On page 275, the drawing of Venus's magnetic field shows a shock structure and a magnetotail around Venus. These features result from the interaction of the solar wind with Venus's upper atmosphere (ionosphere).

35 kilometers/second is about 78,310 miles/hour.

Another way to define the north pole of a planet is to imagine grabbing the planet with your right hand in a manner such that your fingers point in the direction of its rotation. Your thumb will then point to the north pole.

Figure 8-12
An arrow on Earth's surface indicates its rotation. Notice that Venus's rotation arrow points in the same direction around its pole as does Earth's arrow (the "right-hand" rule).

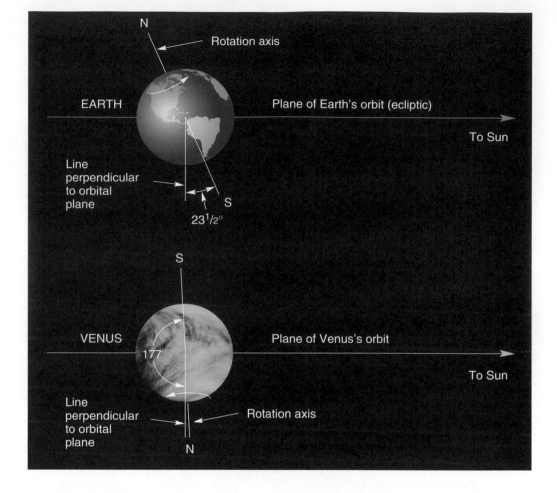

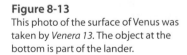 Through a telescope, Venus looks as smooth as a big white cue ball. No markings are visible. Images like the one in Figure 8-12 greatly exaggerate slight color variations.

The Surface of Venus

Since 1962, many spacecraft have visited Venus, none of them with humans aboard. From 1962 to 1975, three *Mariner* spacecraft from the United States flew by the planet, obtaining data as they did so. The former Soviet Union landed 11 spacecraft on Venus, and some of them produced close-up photos of the surface (**Figure 8-13**). The photos indicate that the surface is rock-strewn, at least at the landing sites. They showed that rocks in some locations are more weathered than in others, indicating that they have been on the surface for a longer period of time. The sharp-edged rocks were a surprise to scientists studying Venus until they realized that the wind is fairly calm at the surface. If there had been great winds at the surface, the dense atmosphere (see the next section) would have caused considerable dust movement and erosion of the rocks.

Figure 8-13
This photo of the surface of Venus was taken by *Venera 13*. The object at the bottom is part of the lander.

ADVANCING THE MODEL

Our Changing View of Venus

Galileo was the first person to observe the phases of Venus. As discussed in Chapter 3, he argued that the fact that Venus exhibits a gibbous phase proves that it orbits the Sun and not the Earth. Although its phases are not visible to the naked eye, they have a definite effect on how bright Venus appears to us. When Venus is between the Earth and the Sun, it is closer to us than any other planet gets. However, it is then in a new phase and cannot be seen from Earth. Just before and just after the new phase, its crescent is so small that the planet appears dim to the naked eye. Also, when it is nearly full and we see almost its entire disk, it is on the other side of the Sun from us and is so far away that it appears dim again. Venus appears brightest at an intermediate position, when it shows less than a full face but is fairly close to Earth (**Figure B8-1**). This position occurs when Venus is 39 degrees from the Sun, about

36 days before and after its new phase. Table A8-2 in the Activity at the end of the chapter shows when Venus will be at its brightest over the next few years.

In a telescope, the angular diameter of Venus varies from 10 arcseconds when it is most distant to 66 arcseconds when it is closest. The photographs of the planet in Figure 3-4 show the great change in its angular diameter as seen from Earth.

It was during Venus's solar transits in 1761 and 1769 (when Venus passed in front of the Sun) that combined observations from different sites on Earth allowed us to measure the actual distance between Earth and Venus using the method of parallax. As a result, we finally had the actual value of the astronomical unit and thus the true dimensions of our solar system, compared to only knowing the relative sizes of the planetary orbits.

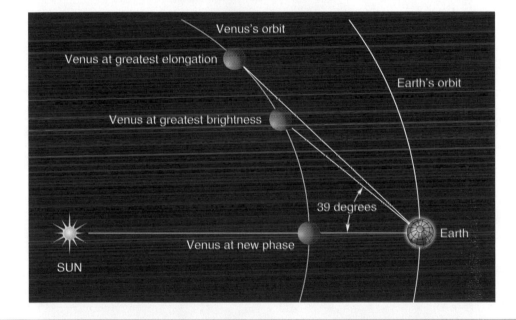

Figure B8-1
Venus is most brilliant when its elongation is 39 degrees, which occurs about 36 days before or after (shown here) its new phase. (Recall that a planet's elongation is the angular distance between the planet and the Sun as viewed from Earth.)

The surface of Venus is not visible from above its atmosphere, but before we ever visited Venus we had produced Venusian maps using radar from Earth. Radio waves penetrate the atmosphere, and from the characteristics of the returned signals, we are able to make maps of the surface. *Pioneer Venus 1* was put in orbit around Venus in 1978 and continued sending radar data until 1992. This allowed us to make higher resolution maps of the surface. Just as radar from airport control towers is used to determine the distances to airplanes, radar from Venus orbiters can determine the distances from the orbiter to points on the surface of the planet. From 1990 to 1994 the *Magellan* orbiter returned images of much higher resolution (**Figure 8-14**), so that objects smaller than 100 meters could be detected. *Magellan* mapped 98% of the surface of Venus.

About two-thirds of the surface of Venus is covered by rolling hills, with craters here and there. Highlands occupy less than 10% of the surface and lower-lying areas make up the rest. Remember that maps of the surface (**Figure 8-15**) are made from

Figure 8-14

(a) The Venusian crater Golubkina is shown as a mosaic of two images. The image on the left side of the photo was produced by the Soviet *Venera 15/16* craft that discovered the crater, and the one on the right was produced from *Magellan* data. The crater, named after Anna Golubkina (1864–1927), a Soviet sculptor, is about 34 kilometers across. Its features, including the central peak, terraced inner walls, and surrounding ejected material, are characteristic of impact craters on the Earth, Moon, and Mars. (b) This three-dimensional image was made from *Magellan* data and enhances the height of structural features of Golubkina Crater.

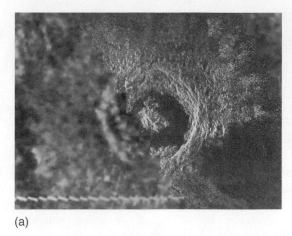

(a)

(b)

Figure 8-15

(a) A color-coded map of Venus's surface shows the highest elevations in brown/red and the lowest in blue/violet. (b) All but one Venusian feature are named for women, as is the planet. Aphrodite Terra, the second largest "continent," is named for the Greek equivalent of Venus.

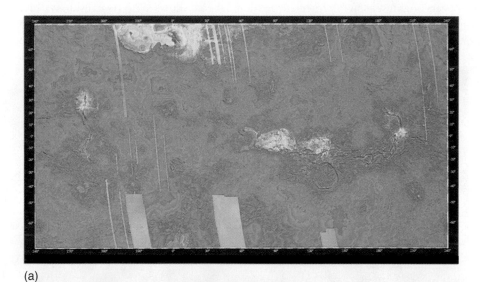

(a)

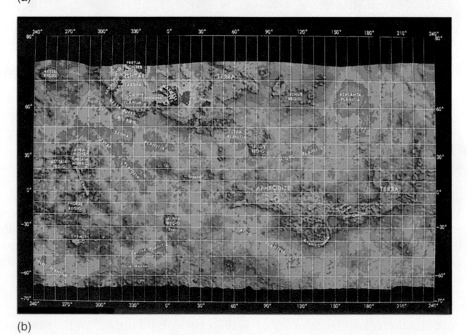

(b)

radar data, so their color is not the color that would appear in a photo. Instead, colors are used to indicate altitude. In Figure 8-15, brown/red indicates the highest altitude and blue/violet the lowest.

Do not be fooled by the blue in the map of Venus. All of Venus, including the rolling plains that make up most of its surface, is dryer than the driest desert on Earth. Earth has rolling water; Venus, rolling desert.

Venus has about a thousand craters (**Figure 8-16**) that are larger than a few kilometers in diameter. This is many more than are found on Earth, but far fewer than are found on the Moon. We know that the terrestrial planets experienced more frequent cratering early in the history of the solar system than they do now, for there was much more interplanetary debris. The cratering rate decreased as this debris was swept from space. This allows us to determine the age of a planet's surface, for volcanic action and motion of tectonic plates remove old craters. Venus has no craters older than about 800 million years, and we conclude that the average age of the planet's surface is no more than about 500 million years. This is about twice as old as Earth's, but much younger than the Moon's.

On Venus's surface we find evidence of past volcanic and tectonic activity, including mountains, volcanoes, and large lava flows (**Figure 8-17**). Assuming that Venus has been struck by impacts at about the same rate as other objects in its vicinity (such as the Moon or Mercury), we can compare the number of impact craters found on Venus with that on the Moon. The relatively low number of craters on Venus suggests that large-scale lava flows occurred about 500 million years ago. Also, as a surface becomes smoother due to weathering, it becomes a poorer reflector of radio waves, allowing us to compare the age of lava flows over the planet. Some *Magellan* radar data suggest that there are regions of very young lava flows on Venus, less than 10 million years old.

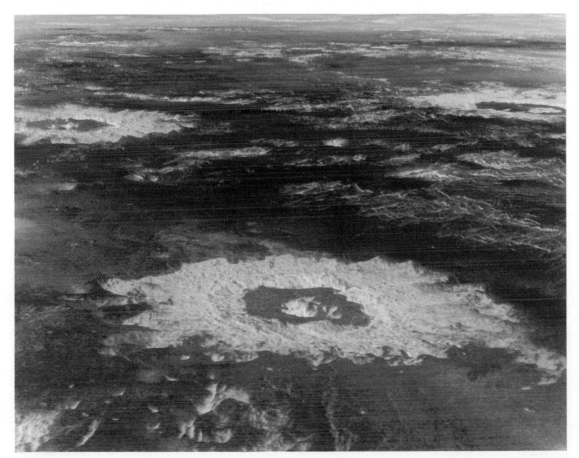

Figure 8-16
The thick atmosphere of Venus shields the surface from small interplanetary debris, but impacts of larger objects result in craters such as these. The foreground crater (named Howe) is 37 kilometers wide. Most of the floors of the craters are flat because lava has flooded them.

Figure 8-17
The lava near Maat Mons is the youngest crustal material yet found on Venus. The mountain is 8 kilometers high, exaggerated 10 times in this image made from *Magellan* radar data.

Probes landed by the USSR contained instruments to measure the composition of the Venusian soil and rocks.

?

Since Venus is covered with clouds, does that mean it's a wet planet?

350 kilometers/hour is about 218 miles/hour.

Computer models suggest that dragging from Venus' heavy atmosphere changed the planet's rotation from its original Earth-like mode to its current slow backward rotation.

The density of the gas in Venus's atmosphere is so high that if we could survive its hazards, we could strap on wings and fly.

Additional evidence for recent volcanic activity is provided by the decrease in the amount of sulfur dioxide over the past two decades, and the detection of radio bursts (thought to be lightning). Since sulfur dioxide is converted to sulfuric acid by ultraviolet radiation in the upper Venusian atmosphere, the decrease in sulfur dioxide suggests that a major eruption may have occurred in the 1970s. The presence of volcanism suggests that Venus has a molten interior. In the case of Earth, the molten interior gives rise to plate tectonics. For Venus, however, we have no evidence to suggest the existence of plate tectonics.

Notice how the shapes of some rocks in Figure 8-13 indicate that they fit together like pieces of a puzzle. Apparently, these are pieces of lava that fractured as it cooled and solidified. Most of the surface of Venus is covered with lava rock. Venus has no large tectonic plates. If it did, we would see mountain ranges as we find on Earth. One explanation is that Venus's wrinkled surface is the result of a thin crust that moves and flexes over most of its area. This thin crust is not strong enough to form large tectonic plates. Another explanation is that the crust is relatively thick, but Venus's molten interior is not moving fast enough, so that convection stresses are relieved in many relatively small surface areas instead of plate boundaries as on Earth.

The Atmosphere of Venus

The length of a day on Venus and the lack of water on its surface are not the only features that make the planet a poor candidate for the title of Earth's twin sister. The Soviet *Venera* landers confirmed what we had already begun to learn about the unusual atmosphere of Venus.

The Earth's atmosphere consists of nearly 80% nitrogen and 20% oxygen, with small amounts of water, carbon dioxide, and ozone. Venus, on the other hand, has an atmosphere made up of about 96% carbon dioxide, 3.5% nitrogen, and small amounts of water and sulfuric acid—the same acid used in car batteries. In fact, the clouds we see on Venus are made up in large part of sulfuric acid droplets. Chemical reactions between the sulfuric acid in the atmosphere and fluorides and chlorides in surface rocks result in some very corrosive substances that can dissolve even lead. Venus is indeed inhospitable.

The upper atmosphere of Venus is very windy. The winds there reach speeds up to 350 kilometers/hour. The wind blows in the direction of the planet's rotation, but remember that the planet is rotating very slowly. As one moves lower in the atmosphere, the wind velocity decreases until it is nearly zero at the surface. We do not fully understand the causes of Venus's wind patterns. Undoubtedly, the slow motion of the Sun across Venus's surface (because of its slow rotation rate) is the basic cause, but the details are not known. Further study of the weather on Venus would not only help us understand that planet better, but would increase our understanding of weather patterns on Earth.

As the space probes descended through the atmosphere of Venus, they encountered hazards from more than the acidic atmosphere and the great winds. The space probes were also subjected to tremendous pressures. The atmospheric pressure on the surface of Venus is about 90 times that on the surface of Earth, about the same as the pressure at a depth of 1 kilometer in Earth's oceans. As if the acidic atmosphere and high pressure were not enough, Venus is also inhospitable because of its high temperatures: about 464°C (867°F) near the surface.

The clouds of Venus form a layer between altitudes of about 50 and 70 kilometers (**Figure 8-18**). A layer of haze extends from the cloud layer down to about 30 kilometers. From there to the ground, the Venusian atmosphere is surprisingly clear. However, the light that filters through the clouds has a distinct yellow/orange hue, caused primarily by the sulfur dust in the clouds. Refer back to Figure 8-13 and note the color of the rocks at the base of the spacecraft. This color is not the natural color of the objects, but is the same result you would get if you turned out all the white lights in a room and lit only a yellow light—everything would have a yellow tint. If the surface of Venus were lit with white light, it would appear gray.

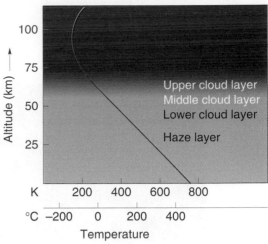

A Hypothesis Explaining Venus/Earth Differences

As noted earlier, both Earth and Venus have small amounts of water vapor in their atmospheres. However, Venus's atmospheric water is about all the water that exists on the planet. Why is this planet, so close to the size of the Earth, so different from Earth?

Terrestrial planets get much of their atmospheres from gases released from their interiors, primarily through volcanic action. These gases were presumably trapped (accreted) during planetary formation. Volcanoes on Earth release large amounts of both water vapor and carbon dioxide, and it is likely that this is also the case for volcanoes on Venus. Both planets are theorized to have once had water on their surfaces and substantial amounts of carbon dioxide in their atmospheres. Most of the carbon dioxide on Earth is now dissolved in the oceans and in rocks such as limestone, which is found in the oceans. Venus, however, would have had a higher surface temperature due to its position nearer the Sun, and this is thought to have prevented water from condensing to liquid form. Thus, carbon dioxide remained in the atmosphere. The clouds of water vapor, along with large amounts of carbon dioxide, resulted in an effect we experience only to a mild degree here on Earth: the greenhouse effect.

Did you ever notice that our coldest nights occur when the sky is clear? When we have cloudy skies at night, the difference between day and night temperatures is minimized. The reason for this is that in order for the Earth's surface to cool down at night, it must radiate into space a portion of the energy it gained during the day. The primary energy loss occurs in the form of infrared radiation. However, infrared radiation does not readily pass through water, so the clouds reflect the infrared waves back to Earth. Thus clouds form a sort of blanket for the Earth.

Early in Earth's and Venus's history, both planets experienced this blanketing effect, but because of Venus's higher temperature, it had more water vapor than Earth did. In addition, Venus had large amounts of carbon dioxide in its atmosphere. This gas is also opaque to infrared radiation. Visible light, on the other hand, is not blocked by carbon dioxide. Thus, although most of the visible light from the Sun was reflected by the clouds of Venus, some of it passed through the atmosphere and was absorbed by the surface, causing the surface to heat up. Meanwhile, high in Venus's atmosphere, ultraviolet radiation from the Sun was breaking water molecules into their constituents: hydrogen and oxygen. The hydrogen escaped from the planet, and the oxygen combined with other elements in the atmosphere. This left the atmosphere with large amounts of carbon dioxide, which continued to trap infrared radiation. In contrast, on Earth, carbon dioxide dissolves in the oceans and is chemically bound into carbonate rocks such as limestone and marble.

The hotter Venus's surface got, the more infrared radiation it emitted. Because of the carbon dioxide in the atmosphere, though, much of this radiation could not escape. Thus the planet continued to heat up. The high surface temperature on Venus baked more carbon dioxide out of the surface rocks. So a chain reaction resulted: the

Figure 8-18
At the top of Venus's atmosphere, the temperature is about 170 K ($-100°C$). It increases smoothly to about 737 K ($464°C$) at the surface. The cloud layer causes the surface to appear orange.

If oceans once formed on Venus, they quickly evaporated in the heat.

Venus

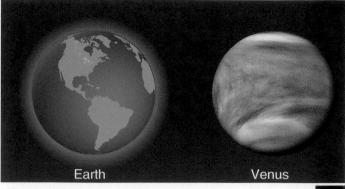

Earth   Venus

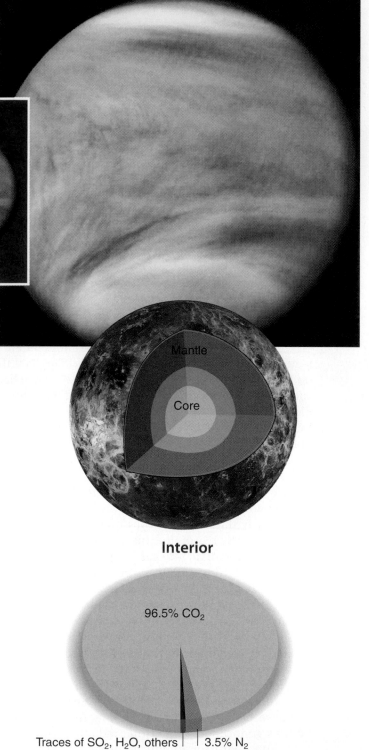

Interior

Venus Data

Venus	Value	Compared to Earth
Equatorial diameter	12,104 km	0.95
Oblateness	0	
Mass	4.869×10^{24} kg	0.82
Density	5.24 g/cm^3	0.95
Surface gravity	8.87 m/s^2	0.91
Escape velocity	10.36 km/s	0.93
Sidereal rotation period	243 days	243.7
Solar day	117 days	117
Surface temperature	+464°C	
Albedo	0.75	2.5
Tilt of equator to orbital plane	177.4°	7.6
Orbit		
Semimajor axis	1.082×10^8 km	0.723
Eccentricity	0.0067	0.40
Inclination to ecliptic	3.39°	
Sidereal period	224.7 days	0.62
Moons	0	

96.5% CO_2

Traces of SO_2, H_2O, others   3.5% N_2

Atmospheric Gases

QUICK FACTS

- Visible as morning and evening "star"; Earth's twin in size, mass, and density. Surface is hidden by cloud layer, mapped by radar. Slow retrograde rotation. Dry, hot surface. No large tectonic plates. Surface primarily solidified lava. Surface older than Earth's, younger than Moon's. Dense carbon dioxide atmosphere. Greenhouse effect causes high atmospheric temperature.

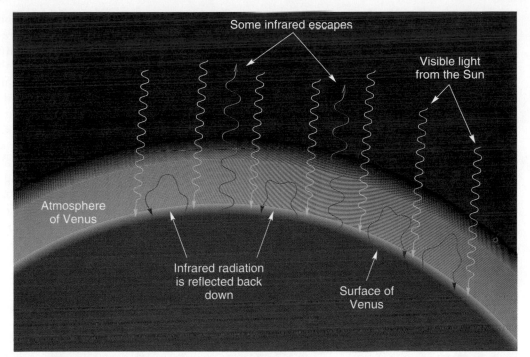

Figure 8-19
Some visible light penetrates the atmosphere of Venus and reaches the surface, heating it. Most of the infrared radiation emitted by the surface, however, is reflected back by the atmosphere, causing Venus to be very hot.

The effect illustrated in Figure 8-19—the greenhouse effect—is a well-established theory. However, the explanation for why Venus developed an atmosphere so different from Earth's is not fully accepted. Even though it seems likely that at least a portion of a terrestrial planet's atmosphere results from volcanic activity, it could be that comets and meteorites have delivered significant fractions of these atmospheres.

more carbon dioxide released, the hotter the surface became and the more carbon dioxide was released.

The planet did not continue heating forever, of course. The atmosphere was not *completely* opaque to infrared. A small fraction of the infrared radiation that entered the atmosphere was reemitted into space and that energy was lost to the planet. The amount of solar energy being absorbed by Venus was constant, while the amount being released depended upon the temperature of the planet. Thus Venus continued to heat up until the amount of radiant energy leaving it was equal to the amount striking it. At this point an equilibrium condition was reached. **Figure 8-19** illustrates the situation.

Venus continues to have high temperatures due to trapped infrared radiation. Such heating is called the ***greenhouse effect*** because it is somewhat similar to what happens in greenhouses on Earth. Sunlight penetrates the glass of a greenhouse's roof and walls and is absorbed by the plants and soil within. These surfaces then emit infrared radiation. Glass, like water vapor and carbon dioxide, is a poor transmitter of infrared radiation, so the air in a greenhouse is kept warmer than the outside air. However, this is not the entire story in the case of a greenhouse. A very important factor in the heating of greenhouses is that the walls prohibit the warm air inside from mixing with the cooler outside air. Since this trapping of the air is not a factor on a planet, the heating of Venus and of a greenhouse are not truly similar. A similar phenomenon to a greenhouse is what happens inside a car whose windows are closed during a hot summer day.

The greenhouse effect is also at work in Earth's atmosphere. If we calculate what Earth's temperature should be due to the amount of solar radiation that strikes it and the energy it emits into space due to its temperature, we obtain a temperature 35 degrees Celsius colder than it actually is. The greenhouse effect, primarily caused by water and carbon dioxide, makes up the difference. As humans burn more and more fossil fuel, we add more carbon dioxide to the atmosphere. The carbon dioxide (and other greenhouse-causing chemicals) in Earth's atmosphere is increasing at a rate of about 0.5% per year, and many scientists fear the effect this may have in the future. Computer simulations of the effect of an increase in carbon dioxide depend greatly on what assumptions are made, but various studies indicate that if we double the amount of carbon dioxide in the atmosphere, we will cause an increase in Earth's temperature of somewhere between 2.8 and 5.2 degrees Celsius.

greenhouse effect The effect by which infrared radiation is trapped within a planet's atmosphere through the action of particles, such as carbon dioxide molecules, within that atmosphere.

The Stefan-Boltzmann law (Section 4-4) allows us to calculate the energy radiated from Earth due to its temperature.

◆ Plant growth has the effect of decreasing atmospheric carbon dioxide. However, humans are destroying tropical forests at the rate of 1 acre per second!

◆ So much solar radiation is reflected from the clouds of Venus that if Venus had no greenhouse effect, its atmosphere would be cooler than Earth's.

If we don't change the direction in which we are going, we are likely to end up where we are headed.

Chinese proverb

The great disparity between conditions on Venus and on Earth appears to be due to a fairly slight difference in temperature far back in their histories. This difference was magnified by the greenhouse effect so that now Venus's atmosphere might be compared to an inferno. We know that an increase of just a few degrees in Earth's temperature would cause major disruptions in our way of living because of changes in growing conditions. However, the greater fear is that a change of just a few degrees may cause a runaway greenhouse effect, destroying Earth as we know it. More study is needed before we know at what point a runaway greenhouse effect would begin, but can we afford to take chances?

During the last decade we have experienced, worldwide, some of the warmest years in the last century. Working with scientists from other fields, including astronomy, climatologists are trying to determine whether this represents a chance fluctuation or whether it is due to the greenhouse effect. If the greenhouse effect is responsible, significant changes in our lifestyles may be necessary to reverse the process.

As we leave Venus and move outward from the Sun, we next encounter Earth. This planet, our home, is so important that we discussed it separately in Chapter 6. That makes Mars next.

8-3 Mars

The ancient Greeks called the fourth planet from the Sun "Ares," the name of their mythical god of war. However, our names for the planets come from their Roman names, and indeed, in Roman mythology the god of war is Mars. Why is this planet associated with warfare? Perhaps because of the association of its red color with blood. Mars does indeed look red. Earlier in this text we mentioned red stars, but the red stars that are visible to the naked eye are not very red at all compared to Mars.

Mars as Seen from Earth

Mars is the only planet with surface features that can be seen from Earth (aside from Earth itself, of course). However, the amount of surface detail that is visible depends on several factors. First, the surface of Mars is often obscured by dust storms on that planet, which will be discussed later. Second, surface visibility varies greatly depending on Mars's distance from Earth.

As you might expect, Mars is best seen when it is directly opposite the Sun in the sky. When a planet is in such a position, 180 degrees from the Sun as seen by an observer on Earth, it is closest to Earth and is said to be in *opposition.* This happens about every 2.2 years for Mars. But Mars's orbit is eccentric enough that the distance from Earth to Mars at opposition might be as little as 55 million kilometers or as great as 100 million kilometers.

Figure 8-20 shows three views of Mars, the one on the right taken by the Hubble Space Telescope. The red color of the light areas is obvious. Before the advent of color photos such as these, some observers claimed that the darker areas appear green. These markings were observed as early as 1660, and the rotation rate of Mars was determined from their motion. In addition, you can see a white cap at the pole of the planet. Observing this cap as Mars orbits the Sun, we can watch it diminish in size as that pole faces the Sun and then grow again when that pole faces away from the Sun. This effect is similar to the Arctic and Antarctic areas on Earth. The tilt of Mars's axis is very similar to Earth's, and we are observing these poles as they experience the Martian summer and winter.

Other seasonal changes can be observed on Mars. Large parts of the planet change periodically from a dark to a light color and back, depending on the position

opposition The configuration of a planet when it is opposite the Sun in our sky. That is, the objects are aligned as Sun-Earth-planet.

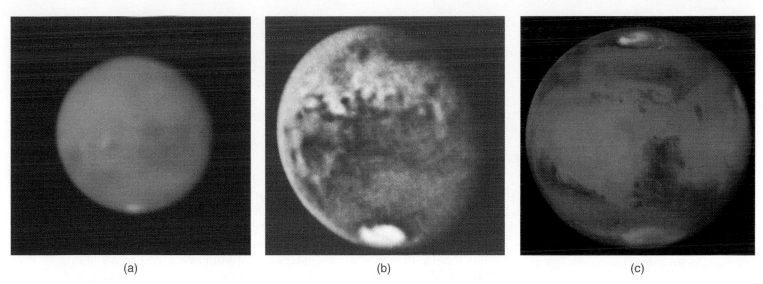

(a) (b) (c)

Figure 8-20
(a) This 1971 photo of Mars shows its rusty color. The small polar cap indicates that the photo was taken during the Martian summer in that hemisphere. (b) This 1988 image of the same side was made with CCD technology, which can produce an image of higher resolution. (c) This 1997 image taken by the *Hubble Space Telescope* shows the northern hemisphere of Mars and clearly shows how much better resolution the *HST* provides. The colors, however, have been exaggerated and are false.

of the planet in its orbit. Since the darker areas appear green (to some observers, at least), this color change led to speculation that there is vegetation on the planet and that it changes color in response to seasonal growth. If there is vegetation, could there be animal life on Mars? What about intelligent life, then? Before exploring the possibility of life on Mars, we will consider the planet itself.

The reason for this reported green color is probably because green is the complementary color to red. Stare at something red for a little while and then close your eyes. You'll see green.

Size, Mass, and Density

Mars is a small planet, closer in size to our Moon than to the Earth. **Figure 8-21** compares Mars to Mercury, Venus, Earth, and the Moon. Mars falls between Mercury and Venus in size, having a diameter about half of Earth's. This makes the surface area of Mars about one-fourth that of Earth. The volume of Mars, then, is about one-eighth, $(1/2)^3$, of Earth's. Now recall that both Mercury and Venus have densities comparable to Earth's. If Mars were to follow this pattern, its mass should have been about one-eighth that of Earth's. The mass of Mars, however, is only one-tenth of Earth's, and its density is only 3.93 times that of water, about 0.7 of Earth's density.

Area depends on the square of the diameter and $(1/2)^2 = 1/4$. Volume depends on the cube of the diameter and $(1/2)^3 = 1/8$. Also, mass = density × volume.

Everything we know about the interior of Mars comes from data from its surface, its overall characteristics, and its tidal interaction with orbiting spacecraft. Surface rocks are rich in iron and silicon. The planet rotates almost as fast as the Earth, but has no global magnetic field. From this data and the planet's relatively low density, we conclude that a significant fraction of Mars's iron is

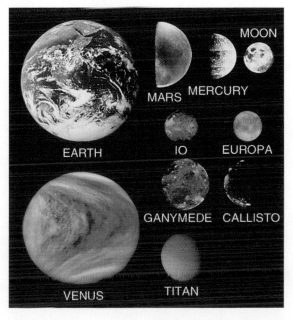

Figure 8-21
This mosaic of photos shows the objects with their correct relative sizes.

probably distributed throughout the planet's body instead of being concentrated in a dense core. Measurements of the tidal interactions between Mars and the *Mars Global Surveyor (MGS)* spacecraft currently orbiting the planet suggest that the planet's interior may be more like Earth's than previously thought. It seems that Mars has not cooled to a completely solid iron core but that there is at least a layer of liquid outer core surrounding the solid inner core. Also, the core has a significant fraction of a lighter element, such as sulfur, and its overall size is about half the size of the planet.

Mars's Motions

Mars orbits the Sun at an average distance of 1.524 AU (about 228 million kilometers). Its orbit, however, is fairly eccentric, and its distance from the Sun varies from about 210 to 250 million kilometers. From Kepler's third law we can calculate that Mars requires 1.88 years to complete its orbit around the Sun.

We saw that a solar day on Mercury was 176 Earth days, and on Venus 117 Earth days. In this sense, it is Mars that should be called Earth's sibling, for it has a sidereal rotation period of 24 hours and 37 minutes, while its day is 24 hours and 40 minutes.

The equator of Mars is tilted 25.2 degrees with respect to its orbital plane. This is very close to Earth's 23.4 degrees. Since the tilt of a planet's axis causes opposite seasons in the planet's two hemispheres, we expect such seasons on Mars and, indeed, this is the case. However, another feature of Mars's motion affects its seasons. The eccentricity of Mars's orbit causes it to be much closer to the Sun at some times of its year than at others. It turns out that Mars is 19% closer to the Sun during the northern hemisphere's winter than during its summer (**Figure 8-22**). Being closer to the Sun in winter and farther away in summer means that seasonal temperature variations are moderated in the northern hemisphere.

In the southern hemisphere, the reverse is true. Mars is closer to the Sun during summertime and farther from the Sun in winter. Thus the southern hemisphere experiences greater seasonal temperature shifts than the northern hemisphere.

The same effect occurs for Earth, which is closest to the Sun in January (wintertime in the Northern Hemisphere). The Earth's orbit is so nearly circular, though, that we are only about 3.3% closer to the Sun at one time than the other (rather than 19%).

Life on Mars?

Speculation about life on other planets—particularly Mars—is nothing new. The famous mathematician Carl Friedrich Gauss was so convinced that intelligent life existed on Mars that in 1802 he proposed that a huge sign be marked in the snows of Siberia to signal the Martians.

During the favorable (close) opposition of Mars in 1863, Father Angelo Secchi, an Italian astronomer, observed Mars and drew a colored map of the planet's surface. The map showed some lines that Father Secchi called *canali*, an Italian word best translated as *channels*. Then in 1877, the director of the Rome Observatory, Father Giovanni Schiaparelli, observed Mars and drew a more elaborate map showing many *canali*. In translation to English, however, Schiaparelli's *canali* became *canals*, rather than channels, giving the implication of artificial waterways. This struck a responsive chord with the public, and Schiaparelli's findings were taken as confirmation of an intelligent race of beings on Mars.

In 1894, American astronomer Percival Lowell (1855–1916) built an observatory on a hill near Flagstaff, Arizona, to concentrate on the study of Mars and its intelligent life. Lowell's drawings of Mars include as many as 500 canals (**Figure 8-23**). Recall that white polar caps are easily visible on Mars and that they change in size according to the season. Lowell believed that the canals were built to transport water from the polar caps to the drier parts of the planet.

The eccentricity of Mars's orbit is 0.093, while Earth's is 0.017.

You should be able to figure out why the two periods are nearly the same for Mars.

Figure 8-22
Mars is closer to the Sun during summer in its southern hemisphere. In this drawing both the tilt of Mars's axis and the change in its distance from the Sun are exaggerated.

Lowell's observatory is on Mars Hill, now within the city of Flagstaff. The observatory is still engaged in active research, financed largely by the Lowell family. It is open at times to the public.

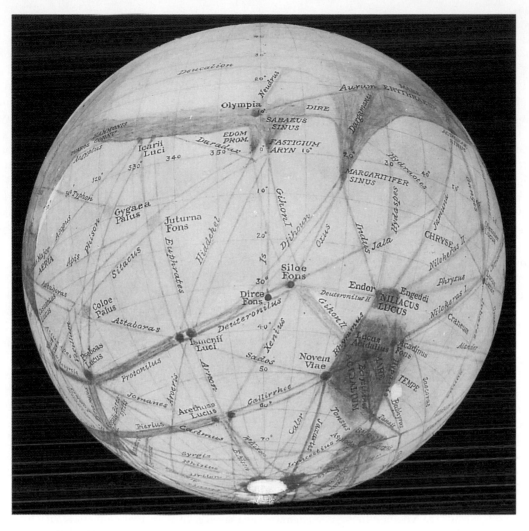

Figure 8-23
This map of Mars was drawn by Percival Lowell in 1907.

Other astronomers failed to see canals, however, even though some of them were using telescopes much larger than Lowell's and Schiaparelli's. In 1894, as Lowell was completing his observatory, the astronomer Edward Barnard said, "to save my soul I can't believe in the canals as Schiaparelli draws them." Barnard was probably the keenest telescopic observer of his day, and he reported to friends that he saw craters on Mars. For fear of ridicule, he did not publish his drawings. Lowell's assistant, Vesto Slipher, also believed that he saw Martian craters, but he also did not publish his views.

In 1898, H. G. Wells wrote *The War of the Worlds,* describing a fictional invasion of Earth by Martians. When Orson Welles dramatized this novel in a very realistic radio broadcast in 1938, he made it appear that a radio music show was being interrupted frequently to report on the invasion as it occurred. Many people did not hear (or forgot about) the announcement at the beginning of the show that it was a dramatization, and there was widespread panic, especially in New Jersey, where the invaders were supposed to be landing.

Invasion and Its Results

A series of Martian invasions did start in the late 1960s and is still in progress. The invasion, however, is proceeding in the opposite direction than H. G. Wells envisioned, and it is a peaceful endeavor, although competition between major political powers has certainly been a factor. In 1965 *Mariner 4* passed by Mars, sending back 22 pictures of the surface. These pictures ended speculation about canals, for none was observed. The planet was seen to be covered by deserts and craters. However, other

In December, 1907, the *Wall Street Journal* said that one of the most important events of the year was the proof that there was intelligent life on Mars.

A headline in the November 9, 1913, *New York Times* read "THEORY THAT MARTIANS EXIST STRONGLY CORROBORATED."

A mind of no mean order would seem to have presided over the system [of canals] we see. Certainly what we see hints at the existence of beings who are in advance of, not behind, us in the journey of life.

—*Percival Lowell*

Viking's *Search for Life*

In the late summer of 1975, *Viking 1* and *Viking 2* were launched from Cape Kennedy (now Cape Canaveral) toward Mars. After traveling to Mars, each spacecraft separated into two parts, one part staying in orbit and another descending to a soft landing on the planet. *Viking 1* lander touched down on Mars on July 20, 1976, seven years after our first step on the Moon. Then on August 7, *Viking 2* lander touched down nearly on the other side of the planet.

One of the primary purposes of the *Viking* mission was to look for signs of life on our neighboring planet. Various tests were conducted toward this end. The first was made by a television camera that scanned the area for any signs of plant or animal life or even for footprints. Scientists did not really expect that such large life-forms existed on Mars, so the negative results were not particularly disappointing. The other tests, however, were more sophisticated and were designed to detect less obvious life-forms.

To perform these experiments, each lander contained an arm with a scoop at its end (**Figure B8-2**) so that it could retrieve some soil from the surface. The first of these experiments was called the "labeled release experiment." It was performed by taking about a teaspoon of soil and dampening it with a rich nutrient that should have been absorbed by any organism in the soil. Some of the carbon atoms in this nutrient were carbon-14, a radioactive form of carbon. It was thought that after an organism absorbed nutrients, it should release carbon into the air, including some of the radioactive carbon. Such radioactivity can be easily detected.

To try to ensure that a positive result from this experiment would be due to biological processes rather than chemical processes, other soil samples were heated to 300°F before being fed the nutrient. This was done to sterilize the soil and kill any living matter that might be in it. In this way, the investigators could see whether the unheated soil acted differently from soil that had been heated to kill life-forms. (In a procedure like this, the sample that was sterilized is called the *control*. Such a procedure allows a comparison to be made and is common in all branches of science.)

When the labeled release experiment was tried on unsterilized soil by both *Viking 1* and *2,* the radioactivity of the air increased quickly. In the case of the control, the change in radioactivity was

much less. Thus the experiment seemed to indicate that life was present. But there was a problem: the radioactivity showed up *too* quickly, faster than should occur if it had been caused by an organism absorbing and releasing the carbon. And when more nutrient was added, there was no further increase in the radioactivity of the air. This argued *against* the presence of life in the soil.

Although the initial results of the experiment caused some excitement among the researchers monitoring the results from Earth, they finally concluded that the release of carbon was caused by simple chemical reactions with elements in the soil and did not involve any biological processes. The surface of Mars is thought to have plentiful oxygen, enough to rust iron. A hydrogen peroxide molecule is a water molecule with an extra oxygen atom. It seems likely that hydrogen peroxide is present in the soil and that this chemical reacted with the carbon to produce carbon dioxide, which was then detected by the radiation monitors. Once the reaction used up the hydrogen peroxide from the Martian soil, the reaction stopped and no more carbon was released. The reason the control sample did not show the activity was that the heat broke down the hydrogen peroxide before it was exposed to the nutrient containing carbon.

Another experiment aboard the *Viking* was a test for respiration. Here, the soil sample was put in a container with inert gases (gases that don't react chemically), and a nutrient was added. Finally, the gas was examined for changes in its chemical composition. If the living organism released any gases, they would be detected. When the experiment was performed, no more new gases appeared than would be expected from normal chemical reactions.

A common procedure in laboratories on Earth is to use a *mass spectrometer* to search for very small quantities of given chemicals in a sample. (A mass spectrometer is a device used to measure the masses of isotopes.) The *Viking* landers contained mass spectrometers capable of finding organic molecules even if they made up only a few billionths of the sample. All life on Earth contains these very large molecules that use carbon as their foundation. No such molecules were found.

We have learned a great deal about the surface of Mars since the original *Vikings* landed there. This knowledge could be used to improve the experiments on a future mission and avoid the confusion created by the hydrogen peroxide reactions. Is there life on Mars? Before we landed, most scientists doubted it. However, the experiment was judged worth trying, for if we had found life, the knowledge we could have gained by examining a life-form different from our own would have been tremendous.

Since the *Viking* missions, scientists simulated the environment on Mars's surface in the laboratory and found that the combination of ultraviolet light, dry conditions, atmospheric oxygen, and mineral grain surfaces produce oxidants, which can destroy organic molecules. The failure of the *Viking* landers to find any organic molecules does not absolutely show that life does not exist on Mars. Life might still exist beneath the planet's surface or in places protected from oxidants. The latest observations of Mars clearly suggest that there are vast amounts of water ice under the planet's surface. And wherever there is water and energy, there is a good chance that life exists.

Figure B8-2
Viking 1's sampler scoop is at the end of the arm extending from right.

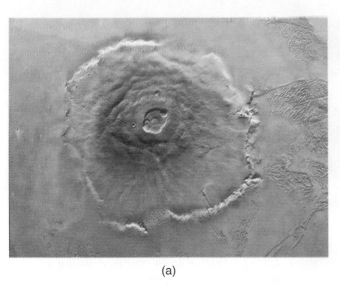

(a)

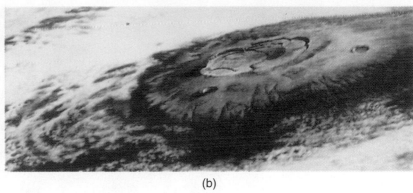

(b)

Figure 8-24
(a) A computer-enhanced photo of Olympus Mons at latitude 18° north. (b) The peak of Olympus Mons extends above the white clouds of frozen carbon dioxide.

questions were raised, for the pictures confirmed that major dust storms are common on Mars, yet the surface contains numerous craters—craters that should have been worn down by the constant pounding of wind and dust. We know now that erosion does not occur very quickly because the atmosphere of Mars is extremely thin, so that the pressure at the surface is about 1/160 of the air pressure at Earth's surface. Dust stirred up by this thin atmosphere must be extremely fine and not at all like a sandstorm in one of our deserts.

Mariner 4 was followed in 1969 by Mariners 6 and 7 and in 1971 by Mariner 9. The latter spacecraft went into orbit around Mars and provided us with a wealth of information about the planet's surface, particularly about the spectacular canyons and volcanoes. On July 4, 1997, the Pathfinder spacecraft landed on Mars. The Mars Global Surveyor (MGS) began orbiting Mars in September of 1997 and started mapping the planet's surface in March 1999. Mars Odyssey started orbiting Mars in October 2001 and began studying the radiation environment and mapping the amount and distribution of chemical elements and minerals on the planet's surface. The twin Mars Exploration Rovers (Spirit and Opportunity) are expected to land on Mars in January 2004 and will study the surface for about three months. The European Space Agency launched Mars Express, which will arrive at Mars in late December 2003; the spacecraft will carry remote observations and a lander, Beagle 2, which is dedicated to looking for life on the planet. More missions are in the planning stages, showing our immense interest in understanding Mars.

The Surface of Mars

The largest of the Martian volcanoes, and the largest mountain known in the solar system, is Olympus Mons. **Figure 8-24a** shows the gigantic cliffs around its base, and **Figure 8-25** compares it to two of the largest mountains on Earth. Its height of 15 miles is twice that of our largest mountain, and its base would cover much

The base of Olympus Mons has a diameter of 400 miles and the collapsed depression at its top is 50 miles across!

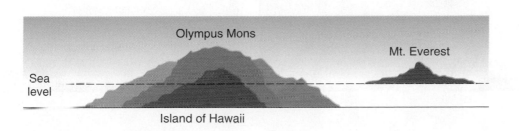

Figure 8-25
Olympus Mons compared to Mount Everest and the volcanic island of Hawaii.

Figure 8-26
A composite photo of Olympus Mons and three companion volcanoes has been overlaid on the western United States to show comparative sizes.

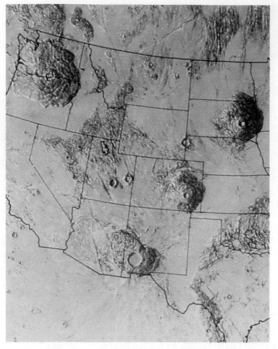

of the area of Washington and Oregon. In **Figure 8-26** Olympus Mons and its three companion volcanoes are prominently superimposed on a map of the western United States.

Venus also has at least one mountain (Maat Mons) larger than any mountain on Earth. Why do these planets have such tall volcanoes? On all three planets, volcanoes are formed above hot spots that lie deep within the planets. From time to time, material wells up from these hot spots, spilling out lava and building up the volcanoes. The reason that Mars and Venus have larger mountains than Earth has to do with the motions of their crusts. On Earth, the crust is divided into plates that move across the underlying material. This means that the crust moves across a given hot spot and the volcanic action slowly shifts along the surface. This is why we often find volcanoes on Earth in a straight line, with only the volcano on the end being active. The crusts of Venus and Mars do not contain large tectonic plates, so the same part of the crust stays above a particular hot spot. This causes the volcano to grow higher and higher. Mars is a small planet and therefore its interior cooled quickly. Its crust may now be thick enough that it is too strong to be broken into individual pieces.

Stretching away from the area of Olympus Mons is a canyon that more than matches the volcano. If placed on Earth, *Valles Marineris* (named for the *Mariner* spacecraft) would stretch all the way across the United States, as shown in **Figure 8-27a**. Part (b) of the figure shows its size relative to the Grand Canyon.

On the seventh anniversary of the first human landing on the Moon, *Viking 1* landed on a cratered plain of Mars and took the first close-up photographs of the surface. **Figure 8-28** shows the landscape near *Viking 1* in approximately true color. The red color we see from Earth is real!

We'll see the reason for the red color later.

Figure 8-27
(a) Valles Marineris would stretch nearly across the United States. (b) In width and depth, Valles Marineris dwarfs the Grand Canyon. Note that it is the product of stresses in the Martian crust, not of erosion by flowing waters, as is a canyon.

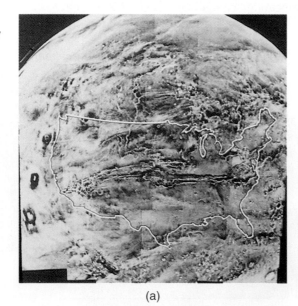

(a)

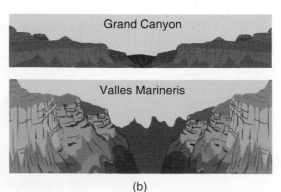

(b)

The *Viking* orbiting modules confirmed something that had been seen by *Mariner 9:* It appears that running water was once common on Mars. **Figure 8-29** shows dry riverbeds that look very similar to the arroyos found in the southwestern U.S. deserts. Arroyos are formed on Earth by infrequent flows of large amounts of water from rainstorms. At present, however, there can be no rainfall on Mars because atmospheric pressure on Mars is much too low for liquid water to exist. The Martian arroyos are so similar to those found on Earth that we conclude that at one time there must have been liquid water flowing on Mars. Where would the water have gone? The answer lies, at least partly, in those white polar caps.

Figure 8-28
The horizon is about 2 miles away in this photo from *Viking 1.* Color photos were made by combining three photos, each made with a different color filter on the camera.

The polar caps of Mars consist of two parts: a water ice base that is covered during the winter by frozen carbon dioxide. In summer the carbon dioxide evaporates, leaving behind the ice. The temperature on Mars does not rise high enough to melt that ice. If it did, there would still be only a small fraction of the water found on Earth. This "freezing out" of carbon dioxide from the atmosphere in the form of dry ice frost and snow is the most important seasonal change on Mars and involves almost a quarter of the atmosphere.

Other water seems to be hidden in permafrost below the surface of Mars. We have no way at present of knowing exactly how much water is there and whether this water and the water in the polar caps are enough to have provided the moisture for a warmer, more hospitable Mars at some time in its history. An alternative hypothesis is that the riverbeds were formed in brief, cataclysmic periods when ice was melted by meteorite impacts or volcanic heating. Look at **Figure 8-30**. The channel stretching along the photo is 800 feet deep. Some astronomers hypothesize that this channel—and others like it—were formed by the motions of glaciers during an ice age on Mars.

The amount of water vapor in Mars's atmosphere would barely fill a small pond. However, the water volume in the ice on Mars's north polar cap is about 4% of the Earth's south polar ice sheet.

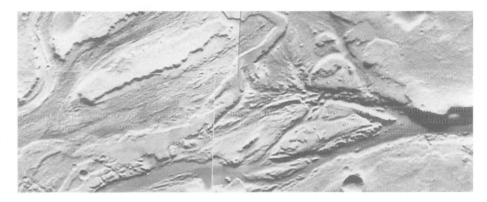

Figure 8-29
What appear to be dry riverbeds on Mars lead us to believe that water once flowed on the surface.

About 28 meteorites found on Earth have been identified by astronomers as having come from Mars (based on their composition and on gases trapped inside). The meteorites are causing controversy among scientists. Some who study the meteorites claim to find evidence that some of them were under water while they were on Mars. Further, they claim that the meteorites contain the remains of single-cell organisms, showing that life once existed on Mars. Other scientists, however, claim that the evidence can be understood by simply considering inorganic reactions in Mars's atmosphere and contamination, not biological processes. **Figure 8-31** shows one of the meteorites to cause excitement about the possibility of life on Mars.

Figure 8-30
The Martian channel shown here is 800 feet deep. It is thought to have been formed by the motion of a glacier sometime in the last 300 million years.

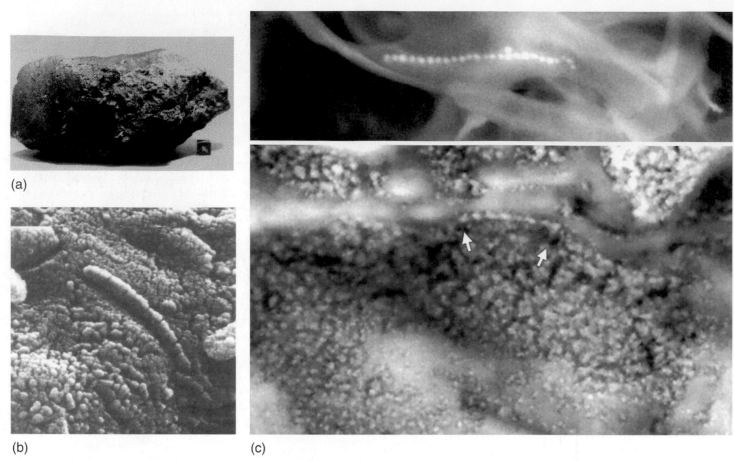

(a)

(b)

(c)

Figure 8-31

(a) This meteorite (ALH84001) is by far the oldest Martian meteorite, with a crystallization age of 4.5 billion years. It is a sample of the early Martian crust. It is hypothesized to have stayed more or less undisturbed on Mars until a huge asteroid or comet smacked into that planet 16 million years ago. The rock wandered in space until about 13,000 years ago, when it fell on the Antarctic ice sheet, where it was found in 1984. (b) A thin slice from the same meteorite. Some scientists claim that it shows evidence of ancient life inside. The tubular structure measures about 100 nanometers long. (c) The top figure shows a chain of magnetite crystals of biological origin. The bottom figure shows magnetite crystals and chains of such crystals (one conspicuous chain is indicated by arrows) in the ALH84001 meteorite. The diameter of a single crystal is approximately one-millionth of an inch. A comparison between the two figures suggests, but does not prove, that the magnetite crystals in ALH84001 might be of biological origin.

Atmospheric and Surface Conditions

◆ 30°C corresponds to 86°F and –135°C corresponds to –210 °F.

Near the Martian equator, noontime temperatures reach as high as 30°C, a comfortable temperature for humans. At night, however, the temperature at the same location might drop to –135 °C. The thin atmosphere is the reason for this extreme difference in temperature. As we saw in the discussion of Venus and Earth, a planet's atmosphere shields it from the Sun during the day and serves as a blanket at night by reflecting back some infrared radiation toward the surface. The amount of shielding and reflection is determined by the amount and type of atmosphere. The atmosphere of Mars is 95% carbon dioxide, and one might suppose that Mars would have a greenhouse effect as Venus does. However, the low atmospheric pressure at the surface of Mars means that there is simply too little atmosphere of any type to significantly moderate the temperature.

The escape velocity from Mars is 5 kilometers/second, less than half of Earth's escape velocity. Even though Mars is colder than Earth, almost all of the water vapor, methane, and ammonia have escaped the atmosphere, along with the less massive gases. (Refer again to Figure 7-12.)

If we assume that the primitive atmospheres of Mars and Earth were similar, their current differences are due to their different evolutionary paths (**Figure 8-32**). As we discussed for Earth and Venus, the atmosphere's carbon dioxide gets locked up in carbonate rocks after being washed out by rainfall. However, since Mars did not exhibit a continuous plate tectonic activity, the recycling of carbon dioxide through volcanic activity stopped. This resulted in a permanent depletion of carbon dioxide from the atmosphere and the destruction of the greenhouse effect. As Mars got colder, water could not exist in liquid form.

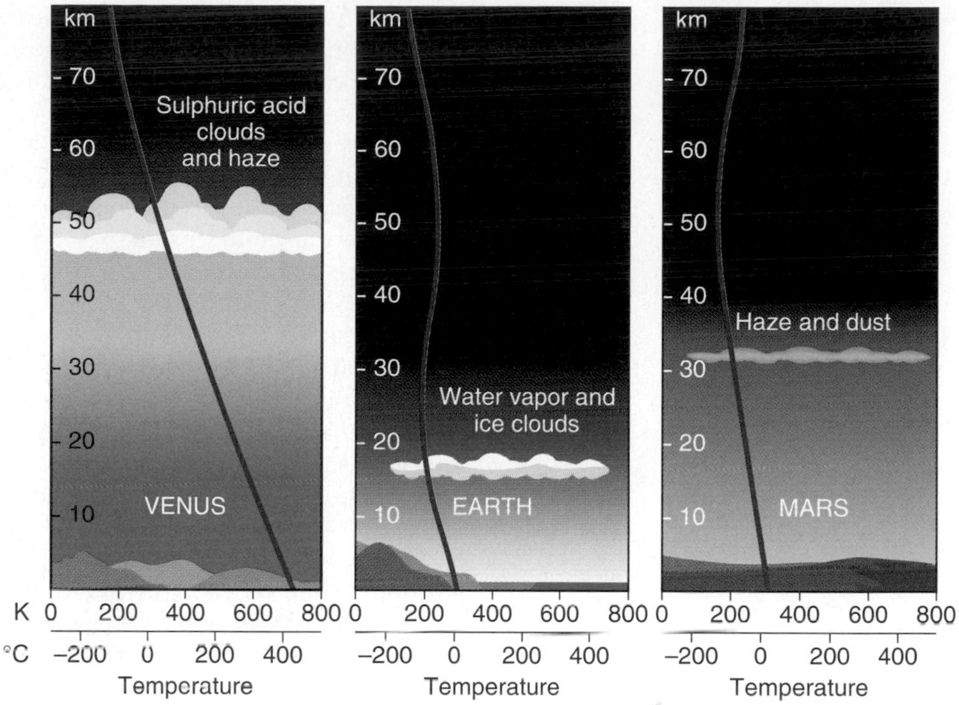

Another difference between Mars and Earth is that Earth has a layer of ozone that prevents most of the Sun's ultraviolet radiation from reaching the surface. Mars has little ozone, so the Sun's intense ultraviolet radiation passes through the atmosphere and breaks up water vapor into hydrogen and oxygen. The hydrogen escapes and the oxygen enters into chemical reactions with other elements.

One of the elements with which the oxygen combines is iron. The surface of Mars is rich with iron, and when oxygen and iron react, we get a compound that is usually a nuisance to us on Earth—rust. It is this rust that gives Mars its characteristic red color. The surface is rusty! The views of Mars in **Figure 8-33** were taken by the *Hubble Space Telescope* when Mars was in opposition in 1997, at a distance of 60 million miles. The photos show different views of Mars and make the red, rusty color obvious.

Figure 8-32

A comparison of the atmospheres of three of the terrestrial planets. Mercury is omitted because it has almost no atmosphere.

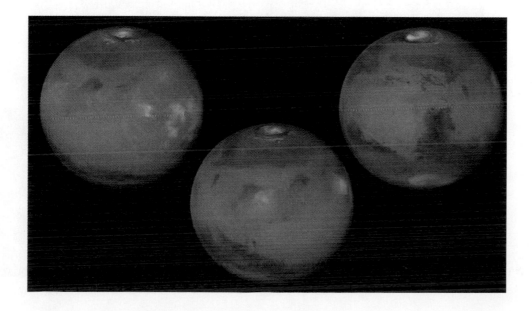

Figure 8-33
The three views of Mars (taken by the *Hubble Space Telescope*) show its northern polar cap during the transition from spring to summer in the Martian northern hemisphere.

From the *Viking* landers, we also learned what causes the seasonal color variations. The fine grains that make up the dust storms are a lighter color than the underlying surface. In the springtime, this dust is stripped away by the wind, exposing the darker surface. Global dust storms on Mars, larger than any seen on Earth, have been observed by the *HST* and *MGS* to last for months. Such events provide a laboratory for observing global warming in real time. The dust traps sunlight and warms the upper atmosphere (by as much as 80°F), while at the same time the planet's surface cools under the cloud shroud.

So there is no vegetation on Mars. In fact, the *Viking* landers found no signs of any vegetation at all. Nor did they find any animal life or even life of microscopic size. The *Viking* landers scooped up some of the Martian soil to analyze its chemical composition and to look for signs of microbial life. They found no organic chemicals whatsoever in the soil, meaning that they found no evidence for life ever existing on Mars. It is important to understand that this is not the same as saying they found evidence that there was never life on Mars. No evidence for *current* life was found. Perhaps our experiments were too limited in what they were searching for, or perhaps we simply need to look in other locations.

The *Pathfinder* spacecraft landed on Mars on July 4, 1997. It immediately sent back photos that showed the rock-strewn area where it landed. The photos further confirmed that Mars once had a great amount of water on its surface, and NASA scientists think that the region where *Pathfinder* landed may have once been covered with water a few meters deep.

The next day, *Pathfinder*'s roving robot, named *Sojourner,* began a slow journey to investigate the composition of some of the many rocks in the area (**Figure 8-34**). The major surprise from this mission has been that Mars is more similar to Earth than had been thought. The first rock investigated turned out to be rich in silica, the quartz material found in sand. Such a rock may have been brought to the surface by volcanic action or by meteorite impact.

Since 1999, the *MGS* has produced detailed 3D maps of Mars's surface that show differences between the planet's hemispheres (**Figure 8-35**). The northern hemisphere seems to consist of young plains, low in elevation, while the southern hemisphere is mainly ancient cratered highlands. Scientists were able to probe beneath the planet's surface by combining high-resolution topographic maps of Mars (like the ones shown in Figure 8-35) with measurements of Mars's gravity. They found that the crust thins progressively from the south pole toward the north pole (**Figure 8-36**). Beneath the

Figure 8-34
This photo was taken soon after the *Sojourner* rover moved onto the Martian surface at the Carl Sagan Memorial Site. The rock that *Sojourner* is examining was dubbed Yogi.

 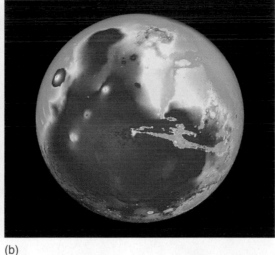

(a)   (b)

Figure 8-35
These maps show features on the surface of Mars. The false colors in the images show elevation, from white (highest elevation) to red to green to blue (lowest elevation). The images were made possible by bouncing a laser off the surface of the planet and calculating the distance traveled by the beam of light. (a) Olympus Mons is shown on the upper right side of the map. (b) Olympus Mons is shown on the upper left side of the map and Valles Marineris on the right, with the Tharsis rise between the two.

southern highlands the crust is about 45 miles thick, while the northern lowlands have a crust of uniform thickness, about 22 miles thick. The measurements also show evidence for large channels (buried under tons of sediment) that could have formed from the flow of large quantities of water from the southern highlands to the northern lowlands, which could have created early oceans on Mars. These subsurface channels, about 125 miles wide and over 1000 miles long, continue the path begun by the Valley Marineris canyon that is visible on the surface (**Figure 8-37**). The differences between the two hemispheres on Mars are very similar to the differences between the oceans and continents on Earth. These measurements also provide information on how Mars cooled over time, which in turn is related to the climate and history of water on the planet. It seems that early in Martian history there was high heat loss from the planet's interior through the northern lowlands; this was probably the result of a period of plate recycling and strong convection inside Mars. During this time period, large amounts of gases and water or ice trapped in the interior could have been released in the atmosphere, the planet could have had a warmer climate, with liquid water flowing on the surface and a strong global magnetic field protecting the planet's surface.

Figure 8-36
This visualization shows a cross-section of the planet's skin, with surface features and crustal thickness displayed in relative (but exaggerated) sizes to each other. The crust on the right side of the cross-section (the Martian south pole) appears significantly thicker than the crust on the left (the Martian north pole). This could be the result of subcrustal erosion by a large-scale mantle convection cell or one large impact (or multiple impacts) in the lowlands.

Differences also exist between the two polar regions on Mars (**Figure 8-38**). The north polar cap has a flat and pitted surface, resembling cottage cheese, and the residual cap (the part that survives the summer) is made mostly of water ice. The south polar cap has larger pits and troughs, giving it a holey swiss-cheese appearance, and the residual cap is made mostly of dry ice. This supports the idea that a

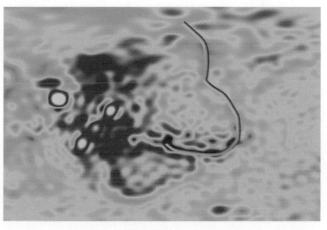

Figure 8-37
This is a flat topographic map, showing Olympus Mons as a white area at upper left and Valles Marineris in the middle of the map. The image shows a channel draining from Valles Marineris into the wide, flat area of the north.

Mars

Earth

Mars

Mars Data

Mars	Value	Compared to Earth
Equatorial diameter	6794 km	0.53
Oblateness	0.0065	1.93
Mass	6.419×10^{23} kg	0.11
Density	3.93 g/cm^3	0.71
Surface gravity	3.69 m/s^2	0.377
Escape velocity	5.03 km/s	0.45
Sidereal rotation period	24.623 hours	1.03
Solar day	24.66 hours	1.03
Surface temperature	−135°C to +30°C	
Albedo	0.25	0.82
Tilt of equator to orbital plane	25.19°	1.07
Orbit		
Semimajor axis	2.28×10^8 km	1.524
Eccentricity	0.0935	5.6
Inclination to ecliptic	1.85°	
Sidereal period	1.881 years	1.881
Moons	2	

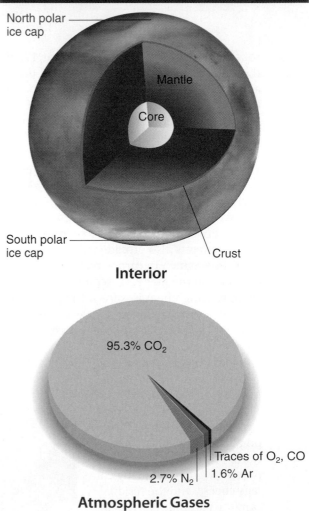

North polar ice cap

Mantle

Core

South polar ice cap

Crust

Interior

95.3% CO$_2$

Traces of O$_2$, CO

1.6% Ar

2.7% N$_2$

Atmospheric Gases

QUICK FACTS

- Known as the red planet. Half Earth's diameter. Least dense of terrestrial planets. Rotation rate and equatorial tilt very similar to Earth's. Has seasons. Water-carbon dioxide polar caps. Has largest mountain (Olympus Mons) and canyon (Valles Marineris) known. Low-pressure carbon dioxide atmosphere. Evidence of warmer, wetter past.

large ocean once covered most of the northern part of the planet. The latest observations by the *MGS* and *Mars Odyssey* show that the pitted layer is dry ice but that water ice makes up the bulk of both polar caps. The main difference between the two poles is the thickness of their dry ice cover; the north polar cap has a 1-meter covering of dry ice, while the south pole dry ice cover is about 8 meters deep and does not disappear completely during summertime. These new findings present new mysteries. Planetary scientists have assumed that Earth, Venus, and Mars were formed with similar total amounts of carbon dioxide. Earth's carbon dioxide is mostly in the form

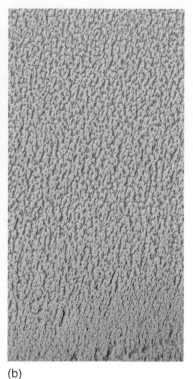

(a)   (b)

of marine carbonates, while Venus's is in the atmosphere causing the greenhouse effect. But the latest observations suggest that the total amount of carbon dioxide on Mars is a small fraction of that found on Earth and Venus.

Observations from the *MGS* also show features that suggest there may be current sources of liquid water at or near Mars's surface (**Figure 8-39**). These features look like gullies formed by flowing water and the deposits of soil and rock transported by these flows. The gullies appear to have been formed in the recent past and they are seen at some of the coldest places on Mars. Almost all of them are in latitudes between 30° and 70°, mostly in the southern hemisphere, and usually on slopes that get the least amount of sunlight. Since there is a very low surface atmospheric pressure on Mars, if liquid water were to get exposed at the planet's surface, it would begin to boil in a violent and explosive way. How then can these gullies form? One explanation assumes the existence of a ground water supply, similar to an aquifer, located near the surface. The location of the gullies indicates that ice plays a role in protecting the liq-

Figure 8-38
(a) An *MGS* picture, covering an area of 3 × 9 kilometers, of the residual (summertime) south polar cap. The upper layer of the cap looks like pieces of sliced and broken swiss cheese. The cap has been eroded, leaving flat-topped regions (mesas) standing as high as 4 meters, into which there are circular depressions. (b) An *MGS* picture of the residual (summertime) north polar residual, covering an area of 1.5 × 3 kilometers. The surface of the cap shows a cottage cheese-like texture, and it is covered by pits (estimated to be less than 2 meters deep), cracks, small bumps, and knobs. The residual north cap is thought to contain mostly water ice; the summertime temperature at the north cap is usually close to the freezing point of water, and water vapor has been observed by the orbiters.

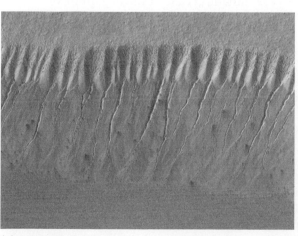
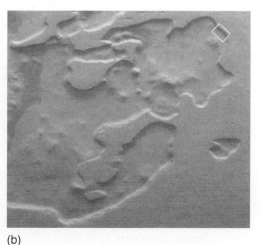

(a)   (b)

Figure 8-39
(a) An *MGS* picture (July 1999) showing gullies with sharp, deep, v-shaped channels on the pit walls at latitudes 70° S. The area above the channels is layered and has been eroded by dry avalanching of debris. The picture covers an area of about 3 kilometers wide by 2 kilometers high. (b) A *Viking 2* picture (1977). The small white box shows the location of the high resolution *MGS* view in (a).

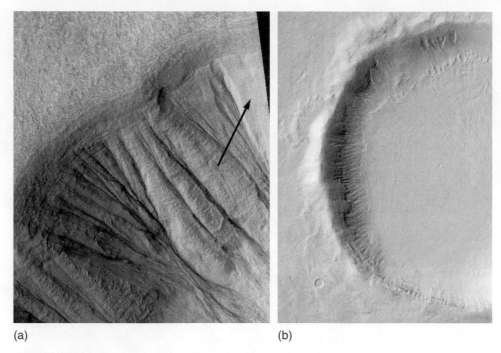

(a)   (b)

Figure 8-40

(a) This *MGS* image shows gullies on Martian crater walls (33.3°S, 92.9°E). The arrow shows the remnant of the snow pack proposed to be the source of the water that eroded the gullies. The image covers an area of 2.8 kilometers by 4.5 kilometers. (b) This visible-light *Mars Odyssey* image shows gullies and snow packs on the walls of another Martian crater (43°S, 214°E). The gullies at the top upper-right of the image appear to be emerging from under a snow pack that is gradually disappearing. This is the cold, pole-facing area of the crater wall. The area of the crater wall on the left is warmer and the snow cover has completely disappeared, exposing the gullies. The image covers an area of 14.8 kilometers by 21.6 kilometers.

uid water from evaporating until enough pressure builds for it to be released catastrophically down a slope. In an alternative explanation for the gullies, liquid carbon dioxide vaporizes as it comes out from the ground, expands and cools quickly, and forms carbon dioxide snow. This snow, together with rock debris, forms slurry that acts like a liquid. No explanation to date has been universally accepted and new possibilities are being examined. One such possibility is supported by observations made by the *Mars Odyssey* spacecraft. **Figure 8-40** shows gullies on crater walls that may be carved by water melting from remnant snow packs. As the snow melts and evaporates in the thin atmosphere of Mars, the gullies are being exposed. According to this scenario, Mars goes through 100,000-year climatic cycles. At the beginning of a cycle, ice packs form as a result of snowfall. Sunlight warms the snow, which begins to melt from below, creating liquid water that flows downhill forming the gullies over a 5000-year period. A layer of snow above the liquid water protects the melt from evaporating immediately.

Layers of sedimentary rocks discovered on Mars (**Figure 8-41**) reveal clues about the planet's early history, when impact craters were forming more frequently than today. This history may have included warm, wet climates that lasted for millions of years, with many crater lakes filled with liquid water. On Earth, layered rock structures are commonly found in lakes; over long periods of time, sediments settle to the lake bottom and form sheets of rock, like a stack of pancakes. However, it is also possible that these rocks are providing evidence for a drier and much colder Martian surface, which went through many climate changes and thick deposits of airborne dust. This view of a cold and dry Mars through its geologic history is supported by the abundant presence on the planet's surface of the mineral olivine, an iron-magnesium silicate that weathers easily by water. It is also possible that Mars is cold, dry, and quiet for hundreds of millions of years and then it becomes actively warmer and wetter during brief episodes, triggered by internal planetary heat, that lasts thousands of years. In such a scenario, water and volatiles remain frozen under the planet's surface due to the extreme cold temperatures. This frozen permafrost acts like a cap on a soda bottle. A burst of internal planetary heat can trigger a dramatic release of gas and water under the permafrost, just as in the case of a capped soda bottle when heated. During such episodes, an ocean can form over the northern hemisphere, while the released carbon dioxide promotes the greenhouse effect that keeps liquid water stable near the surface. Since Mars lacks a soil layer like Earth's, when it rains, water removes carbon dioxide from the atmosphere and filters underground. As a result, Mars cools to the point that permafrost forms again and enters another long-lasting cold and dry period.

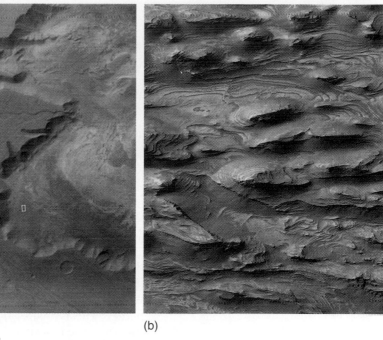

(a) (b)

Figure 8-41
(a) A context view of the West Candor Chasma. The small white box shows the location of the high-resolution *MGS* view in (b). The picture in (b) covers an area of 1.5 kilometers by 2.9 kilometers and shows a large number of layers on the chasma's floor. Each layer has about the same thickness, estimated at about 10 meters.

The channels, gullies, and layers of sedimentary rock discovered on Mars clearly suggest that the planet was very different 3.5 billion years ago than it is today. Liquid water seems to have played an important role in shaping the planet's surface, but it is likely that it was not the only force at play. Other factors such as volcanism, tectonic shifts, ice, and strong winds almost certainly contributed to the features we see today.

Mars does not show any evidence for a global magnetic field but it probably had at some point an interior dynamo like Earth's. Observations suggest that Mars probably lost its magnetosphere about 4 billion years ago, and since then the planet's atmosphere has been feeling the full force of the Sun's solar wind. As a result, most of Mars's atmosphere has since disappeared, even though it seems very likely that Mars had at one time a dense enough atmosphere to allow liquid water to flow on its surface. However, *MGS* has detected a few magnetized regions in the planet's crust. These regions may be relics of an era when Mars had a global magnetic field and a very active interior. These crustal fields (some of which are ten times stronger than Earth's field) could have played an important role in the evolution of the planet's atmosphere by shielding parts of it from being swept away by the solar wind.

Another important *MGS* discovery is the lack of carbonates, at least on the planet's surface. This is important because atmospheric carbon dioxide is supposed to be locked up in carbon-rich sediments, after being dissolved by rainfall early in the planet's history. This is one of the two ways that carbonate rocks are formed on Earth. The second is by marine organisms that produce carbonates for shells or other hard parts; when these organisms die, the accumulated material eventually forms a carbonate deposit at the bottom of the water. (An example of such carbonate is blackboard chalk.) So far we have not discovered similar deposits on Mars's surface and that is a discouraging observation about the possibility of life on Mars. However, carbonate patches have been found on Martian meteorites like the one shown in Figure 8-31.

Observations made by *Mars Odyssey* suggest that there is enough water (in the form of ice) just a few feet under the surface of Mars to cover the entire planet ankle-deep. These observations were based on the detection of hydrogen on the surface of

MGS did discover the presence of carbonate in Martian dust in quantities between 2% and 5%. These trace amounts probably did not come from marine deposits but are instead the result of the interaction between tiny amounts of water in Mars's atmosphere with dust. This finding is pointing to a cold and icy Mars through its history (because otherwise the carbonate rock layers formed in oceans should be observable). To understand this result we may have to wait for the discoveries to be made by future missions.

Figure 8-42
The images are equatorial projections of the Martian surface and show epithermal neutron flux, which is sensitive to the amount of hydrogen present. (a) A view during late summer in the south. The magenta region in the south implies large amounts of water ice (60% by volume) buried less than 1 meter beneath the surface. At this time a thick cap of dry ice covers the north. (b) A view during early summer in the north. The thick dry ice frost has disappeared and the water ice regions are now visible. The water ice in the north is greater than in the south. At this time the water ice regions in the south are beginning to be covered by dry ice.

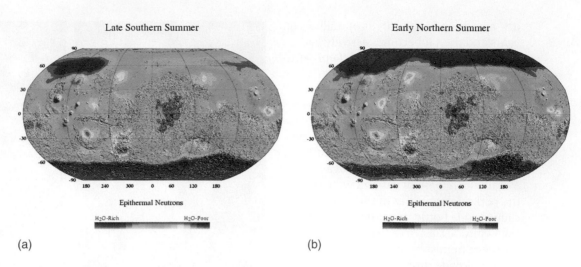

(a)   (b)

the planet (**Figure 8-42**). The hydrogen-rich areas are also very cold, and water ice should be stable in such regions. It is then likely that this hydrogen is in the form of water ice. But how did water get into the soil and rocks under the surface of the planet? This is an area of active research, and we will simply mention two possible theories. The first theory is based on the results of numerical modeling of the changes in the tilt of Mars's equator to its orbital plane; these results show that about a million years or so ago Mars's axis was tilted about 35°. This could have caused the polar icecaps to evaporate and for a short time create enough water in the atmosphere to make ice stable over the entire planet. The soil and rocks then absorbed this thick layer of frost. According to the second theory, the source of the water is the vast water icecaps at the poles. Their thickness is such that they could bottle up enough geothermal energy from below, resulting in melting of their bottom layers, which could then supply a global water table.

In late 2002, scientists confirmed for the first time the presence of water ice at the surface of the southern polar regions of Mars using thermal emission images made by *Odyssey* and temperature data from the *MGS*. The implications of the latest findings are not just important for understanding the water cycle of Mars but also for any future exploration of the planet.

The history of Mars may be more complicated than our current models allow. All this new data makes comparisons between Earth and Mars even more interesting. *Pathfinder, Mars Global Surveyor, Mars Odyssey*, and other explorations underway will continue to change the way that we think of our neighbor, the red planet.

The Moons of Mars

Phobos's dimensions are 17 × 14 × 11 miles. Deimos's dimensions are 9 × 8 × 6 miles.

Mars has two natural satellites, named Phobos and Deimos. Both are small and irregularly shaped. Phobos, for example, is 27 kilometers across its longest dimension, but it is 22 kilometers and 18 kilometers across its other dimensions. Thus it is shaped like a potato. Deimos is even smaller and is similarly shaped.

The surfaces of both Martian moons are very dark, similar to those of many asteroids. Their densities are 1.9 and 1.8 times that of water (for Phobos and Deimos, respectively), also similar to the densities of the rocky asteroids. Such similarities lead us to believe that these satellites are captured asteroids rather than objects that were formed in orbit around Mars during the formation of the solar system. A capture like this could have taken place as the asteroid passed very close to Mars and was either slowed by friction with the Martian atmosphere, by collision with a smaller asteroid, or by gravitational pull from another asteroid. Such an event may not seem likely, but given the number of asteroids that have passed by Mars over billions of years, it is certainly a possibility.

ADVANCING THE MODEL

The Discovery of the Martian Moons

Although Johannes Kepler made important contributions to science, he was very nonscientific in some ways. For example, he practiced numerology, the study of occult meanings of numbers. One of the patterns he saw in the heavens concerned the numbers of satellites of the various planets. Mercury and Venus have none, Earth one, and Jupiter four (known at that time). Between Earth and Jupiter was Mars, and Kepler proposed that Mars must have two moons if it is to fit the pattern. Scientists don't use logic like this today, but that was Kepler's way. From the time of Kepler, popular thought held that Mars must have two moons circling it. The idea appears a few times in literature, most notably in Jonathan Swift's *Gulliver's Travels,* where Swift makes up such details as the moons' size (small) and orbital periods (10 and 21.5 hours).

There was no evidence for the existence of any satellites of Mars until 1877 when Asaph Hall, the astronomer of the newly constructed U.S. Naval Observatory near Washington, D.C., decided to search very near the planet for moons. To do so, he used a disk that blocked the planet's brightness from his view. There is a controversy about how quickly he found the moons, but Carl Sagan reports that Hall was unsuccessful for the first few nights and was about to quit when his wife encouraged him to continue searching. He did find the two moons and named them Phobos (Fear) and Deimos (Terror), the names of the horses that pulled the chariot of the Greek god of war.

In his book, *The Cosmic Connection,* published before *Viking* sent back photographs of the moons, Sagan states that a feature on one of the moons should be named after Mrs. Hall. Look at **Figure B8-3**, a photograph of Phobos. The large crater is about 10 kilometers across and is named Stickney, Angelina Hall's

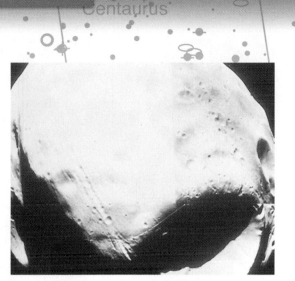

Figure B8-3
The large, prominent crater on Phobos is named Stickney, the maiden name of Mrs. Hall, the wife of the discoverer of Mars's moons.

maiden name, and the name proposed by Sagan. Another large crater on Phobos is named for her husband.

The actual orbital periods of the two moons are 7.7 and 30.3 hours. Although Swift had predicted 10 and 21.5 hours, he hit close enough to the actual times that some people believe that he had special knowledge of some sort. Actually, it was reasonable for him to assume that the moons, if they existed, must be close to the planet and must be small. Otherwise they would already have been discovered. And if they are close to the planet, Kepler's laws tell us that their periods of revolution must be short. There is no reason to hypothesize strange explanations for Swift's guesses. He just knew about Kepler's work.

The masses of Phobos and Deimos are so small, you would weigh almost nothing standing on their surfaces. A 120-pound person on Earth would weigh about 1 ounce at the lowest point on Deimos. Escape velocity from Deimos is about 13 miles/hour, so you could easily throw a ball completely off this moon, never to return. If you threw a ball at a lower speed, it would go into orbit, circling the moon in about three hours. A baseball game on Deimos would be quite strange.

The Martian moons are not only small, but they orbit very close to the planet and have short periods of revolution, about 0.3 days for Phobos and 1.3 days for Deimos. Both revolve in the same direction as most other solar system objects, counterclockwise as seen from a point above the solar system, north of Earth.

The counterclockwise motion of our Moon results in it moving slowly eastward among the stars as viewed from Earth. However, as we watch the Moon during a

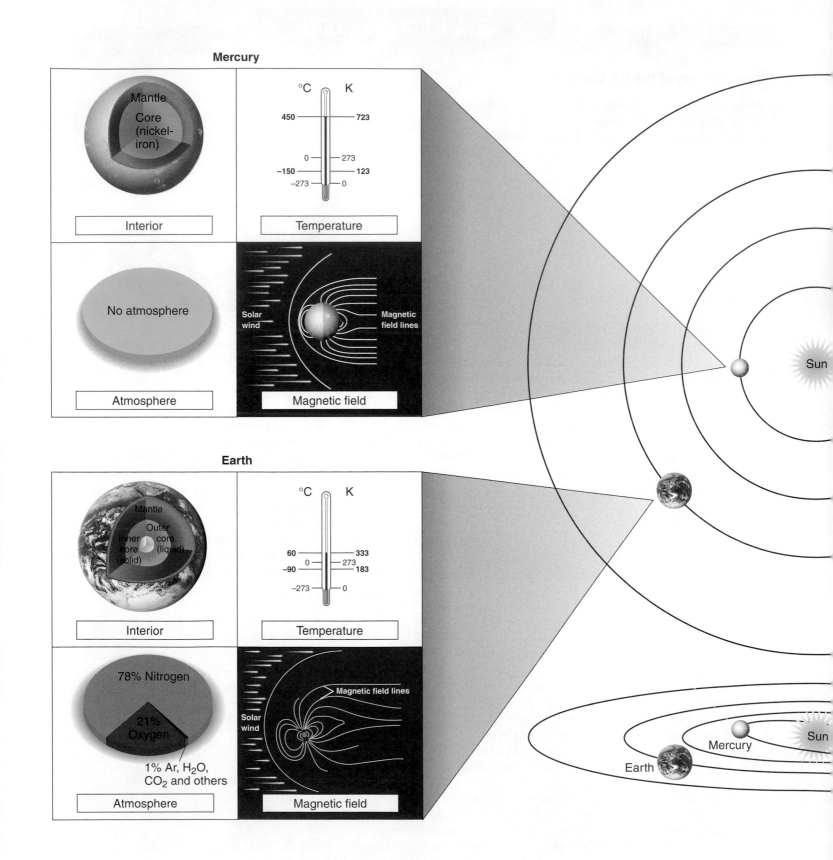

Mercury

Interior

Mantle

Core (nickel-iron)

Temperature

°C   K

450   723

0   273

−150   123

−273   0

Atmosphere

No atmosphere

Magnetic field

Solar wind

Magnetic field lines

Earth

Interior

Mantle

Inner core (solid)   Outer core (liquid)

Temperature

°C   K

60   333

0   273

−90   183

−273   0

Atmosphere

78% Nitrogen

21% Oxygen

1% Ar, H_2O, CO_2 and others

Magnetic field

Solar wind

Magnetic field lines

Sun

Mercury

Earth

Sun

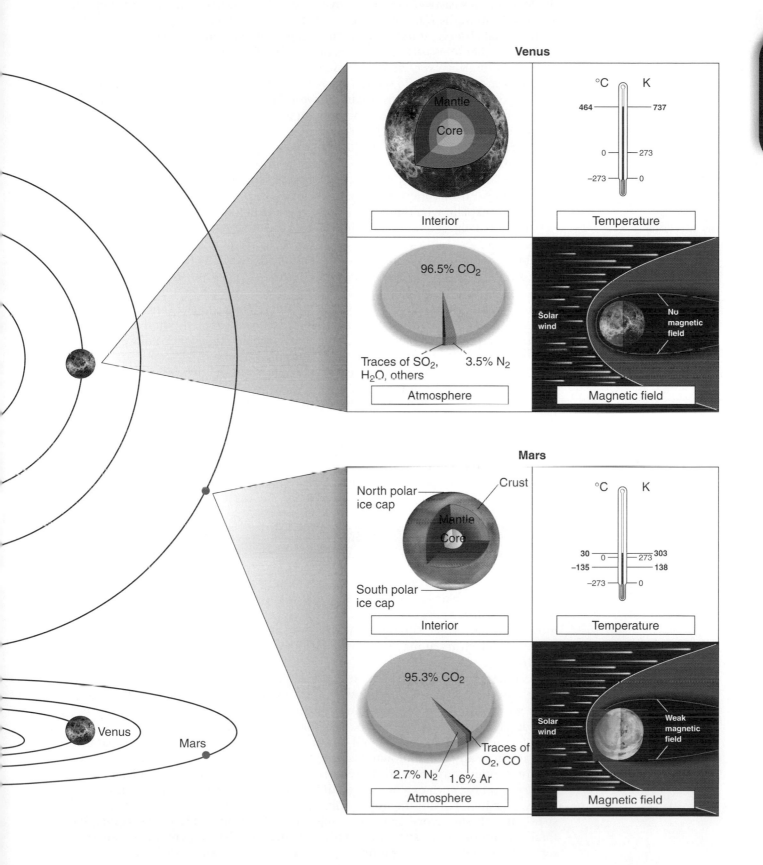

Venus

Mantle
Core

Interior

°C   K

464 ——   —— 737

0 ——   —— 273

−273 ——   —— 0

Temperature

96.5% CO$_2$

Traces of SO$_2$,   3.5% N$_2$
H$_2$O, others

Atmosphere

Solar wind   No magnetic field

Magnetic field

Mars

North polar ice cap   Crust
Mantle
Core
South polar ice cap

Interior

°C   K

30 —— 0   273 —— 303
−135 ——   —— 138
−273 ——   —— 0

Temperature

95.3% CO$_2$

Traces of O$_2$, CO
2.7% N$_2$ 1.6% Ar

Atmosphere

Solar wind   Weak magnetic field

Magnetic field

Venus   Mars

Planetary Data Comparison

single night, this eastward motion is overwhelmed by its apparent westward motion that results from the rotation of the Earth. The eastward motion is far from obvious to the casual observer during one night. Things would be quite different for a Martian. Recall that Mars rotates with a period of about 24-1/2 hours in a counterclockwise direction. However, Phobos takes only about 8 hours to circle Mars. As a result, an observer on Mars would see stars move across the sky just as we do on Earth, but that person would see Phobos rise in the west and move across the sky toward the east! Deimos would hover overhead for long periods of time, moving only very slowly toward the west.

Mars has an oblateness of 0.006, which means that its polar diameter is 0.6% smaller than its equatorial diameter. Mars is the most oblate of the terrestrial planets, so these planets are all very nearly perfectly spherical. We have seen that Phobos and Deimos are far from spherical. Before explaining why they are not spheres, we should first explain why one might expect celestial objects to be spherical.

A celestial object tends to be spherical because the force of gravity pulls its parts toward its center. As a result, the object takes the shape that will give it the smallest possible surface area. This shape is a sphere. Consider what would happen if the Earth were a cube. In this case, a rock at a corner of the cube could get closer to Earth's center by moving to the center of one face of the cube. In that event, the Earth would not remain a cube. Earthly mountains, which are formed by geological phenomena within the Earth (such as volcanic action), are forever being disturbed by wind and rain so that their parts can be pulled closer to the center, making the planet a more perfect sphere.

The same tendency exists in the case of Phobos and Deimos, but the gravitational force toward the center is too small to form these small objects into a spherical shape. The strength of the rock of which the moon is made resists the weak gravitational force that tends to reshape it.

Phobos and Deimos are more interesting to us than their size might seem to indicate. In fact, we may visit them some day. When people go to Mars, some plans call for using Phobos as a landing area before proceeding to the surface of the planet. The moon's weak gravitational field will make landing and taking off easy.

> We will explain why the equatorial diameter of a planet might be greater than its polar diameter when we discuss Jupiter, a noticeably out-of-round planet.

> Most asteroids are likewise non-spherical.

8-4 Why Explore?

The Martian invasion continues. In spite of failures of some recent (1999) missions to Mars, ambitious plans are being made for its further exploration. In 1989, President Bush established a goal of a human expedition to Mars. This venture will probably be international in scope, and the earliest it could occur would be the first decade of the twenty-first century. It may be preceded by an unmanned mission that returns samples from the planet.

We have learned much about our planetary neighbors by observing them from Earth, but there is a limit to what we can learn without sending spacecraft to visit these worlds. Some people question the value of using our resources for these endeavors. There are two answers to their questions.

First, seeking knowledge about our universe is one of the things that separates humans from any other known creature. It is our nature to explore, just as it is our nature to enjoy music and art. If we never achieve one practical benefit from planetary exploration, such exploration would be valuable simply because seeking new knowledge is one of our highest aspirations. The amount of money that should be spent on such endeavors might be a subject for discussion, but when we compare the total amount of money spent on the basic sciences to what is spent on warfare or entertainment, we see that the money spent on space exploration is trivial.

Second, planetary exploration offers many practical benefits. Few would argue that knowledge of Earth is without practical benefit, but many people do not realize that a study of other planets is a valuable source of knowledge about our Earth. If we did not study other planets, we would be severely limited in what we would know about our own planet. For example, by learning more about the greenhouse effect on Venus and the lack of such an effect on Mars, we will find out how serious the threat of a runaway greenhouse effect is here on Earth. We cannot afford to experiment with our planet, but an examination of other planets provides us with just such an experiment. By studying weather systems on other planets, we learn about our own. Was there once life on Mars? If so, why does it not seem to be there now? Could the same thing happen on Earth? Surely such knowledge is of practical benefit.

Conclusion

Throughout history, people have wondered about the planets. With the advent of telescopes, particularly the larger ones, we began to learn something about their surfaces. However, it was the planetary visits of the space age that made us think of these planets as *places,* rather than as celestial objects. We now have detailed maps of their surfaces, and landers have invaded Venus and Mars to begin exploration.

As we learn more about the other terrestrial planets, we are finding that in many ways they are similar, for they share a common history of formation; but we also find major differences. Some of the differences exist because the planets differ in size and mass. These factors determine whether a planet will retain an atmosphere, and the atmosphere of a planet determines such properties as the planet's range of temperature. Other differences occur simply because of the planets' different distances from the Sun.

In the next chapter, we move out to the Jovian planets. Each of these giants forms another piece of the puzzle of how the solar system—including the inhabitants of its third planet—came to be.

Study Guide

1. Mercury's atmosphere is
   A. mostly hydrogen.
   B. mostly hydrogen and helium.
   C. mostly nitrogen and oxygen.
   D. virtually nonexistent.
2. Mercury's rotation and revolution are linked so that Caloris Basin is
   A. directly under the Sun at every perihelion passage.
   B. directly under the Sun at alternate perihelion passages.
   C. never directly under the Sun.
3. Mercury's surface is difficult to map from Earth because
   A. Mercury is such a small planet.
   B. Mercury rotates so slowly.
   C. the surface is hidden below a thick cloud cover.
   D. Mercury is always close to the Sun.
   E. [All of the above.]
4. Mercury's diameter is about
   A. one-fifth of Earth's.
   B. one-third of Earth's.

   C. the same as Earth's.
   D. twice that of Earth's.
   E. more than twice that of Earth's.
5. The property of Mercury that makes its temperature variations greater than those of any other planet is primarily
   A. its lack of an atmosphere.
   B. its proximity to the Sun.
   C. its small size.
   D. the carbon dioxide in its atmosphere.
   E. [The statement is false; Mercury is so close to the Sun that it is always hot.]
6. *Mariner* spacecraft found that Mercury has _____, and this leads us to conclude that the planet _____.
   A. no magnetic field . . . rotates very slowly
   B. no magnetic field . . . has little or no iron in it
   C. a magnetic field . . . rotates faster than we had thought
   D. a magnetic field . . . has a metallic core
   E. [Both A and B above.]

7. The greenhouse effect heats a planet because
   A. more sunlight strikes the planet's surface than normal.
   B. the surface of the planet is darker than normal.
   C. infrared radiation is trapped by the planet's atmosphere.
   D. cloud cover prevents the atmosphere from escaping.
   E. cloud cover prevents visible light from striking the surface.

8. In which of the following ways is Venus most like the Earth?
   A. Its period of rotation.
   B. Its average surface temperature.
   C. Its average density.
   D. The length of its day.
   E. The composition of its clouds.

9. When we see Venus in the sky, the light we receive is sunlight reflecting from
   A. its oceans.
   B. its solid surface.
   C. the top of its cloud layer.
   D. [Both A and B above.]
   E. [All of the above.]

10. A planet that has a high albedo
    A. is necessarily a large planet.
    B. is necessarily a small planet.
    C. has water or ice on its surface.
    D. has clouds in its atmosphere.
    E. reflects a relatively high percentage of the light that hits it.

11. The primary constituent of Venus's atmosphere is
    A. oxygen.
    B. water vapor.
    C. nitrogen.
    D. sulfuric acid.
    E. carbon dioxide.

12. Mars is least similar to Earth in
    A. the tilt of its equator to its orbital plane.
    B. its period of rotation.
    C. that Mars's surface is not hard.
    D. its atmosphere.

13. The two moons of Mars are named
    A. Ceres and Pallas.
    B. Io and Europa.
    C. Galileo and Copernicus.
    D. Helios and Juno.
    E. Phobos and Deimos.

14. The seasonal color changes on Mars probably result from
    A. vegetation.
    B. rain.
    C. dust movement.
    D. dry ice.
    E. [Both A and B above.]

15. The Martian polar caps
    A. are all water ice.
    B. are all frozen carbon dioxide ("dry ice").
    C. are a combination of carbon dioxide and water ice.
    D. completely disappear during the Martian summer.
    E. [Two of the above.]

16. Which of the following statements about the Martian satellites is true?
    A. Both are large satellites (nearly the size of Earth's Moon).
    B. Both are small satellites compared to Earth's.

C. One is very large and the other is small.
D. Mars has only one satellite.

17. The main source of erosion on Mars today is
    A. flowing liquids in the channels.
    B. ice floes that cover the entire planet in winter.
    C. microscopic organisms in a layer just below the surface.
    D. giant dust storms.
    E. [The statement is false; there is no erosion on Mars.]

18. The channels of Mars (observed in the nineteenth century) were found to be
    A. irrigation ditches dug by now-extinct Martians.
    B. faults in the Martian crust.
    C. straight mountain ranges.
    D. optical illusions.

19. Mars's moons are not spherical like Earth's Moon because
    A. of cratering.
    B. their gravitational force is not strong enough.
    C. they are much older than our Moon.
    D. they are much younger than our Moon.
    E. [The statement is false; they are spherical.]

20. Olympus Mons has a base _____ miles in diameter and a height of _____ miles.
    A. 5000 . . . 40
    B. 400 . . . 15
    C. 20 . . . 2
    D. 10 . . . 1

21. Why do Venus and Mars have larger volcanoes than Earth?
    A. Differences in atmospheres result in volcanoes being larger.
    B. Movement of the Earth's crust has prevented our volcanoes from growing as large as those on the other planets.
    C. Erosion on Earth is more prominent than on either of the other planets.
    D. [The statement is false; Earth's volcanoes are larger.]

22. Water is thought to be present on Mars
    A. in its polar caps.
    B. in permafrost below its surface.
    C. in liquid form on its surface.
    D. [Both A and B above.]
    E. [All of the above.]

23. If we list the terrestrial planets in order of increasing atmospheric pressure at the surface, the list should read
    A. Mercury, Venus, Earth, Mars.
    B. Mercury, Earth, Venus, Mars.
    C. Mercury, Earth, Mars, Venus.
    D. Mercury, Mars, Earth, Venus.
    E. Mars, Mercury, Earth, Venus.

24. The escape velocity from the surface is least for which of the following planets?
    A. Mercury
    B. Venus
    C. Earth
    D. Mars

25. Which is the smallest of the terrestrial planets?

26. Which planet is most similar to the Earth in size? In rotation period?

27. Why is the surface of Mercury difficult to see from Earth?

28. What is unusual about the revolution and rotation rates of Mercury and why has this occurred?

29. How was the most accurate value for Mercury's mass obtained?

30. Explain why Mercury's surface experiences such great extremes of temperature.

31. In what ways is Venus the Earth's sister planet? In what ways is it different?

32. What is the greenhouse effect? Why is the effect not an exact analogy to a greenhouse on Earth?

33. Describe the surfaces of Mercury, Venus, and Mars.

34. How does the tilt of Mars's axis compare to Earth's?

35. What is meant by a planet being in opposition? In inferior conjunction?

36. Of what are Mars's polar caps composed?

37. Why can't liquid water exist on Mars? What is the evidence that liquid water once flowed on Mars?

38. How do the daily temperature changes on Mars compare to those on Earth? Why does this difference exist?

39. Explain why large celestial objects tend to be spherical and how some small ones avoid that fate.

QUESTIONS TO PONDER

1. What is unusual about Mercury's period of rotation? What is the hypothesized explanation for this regularity of motion?

2. What causes Mercury's temperature variations to be so great?

3. Why do we not expect Venus to have a strong magnetic field?

4. Why is Venus's orbital tilt usually shown as 177 degrees, since this would seem to be the same as saying it is tilted 3 degrees?

5. Describe the evidence that Venus is—or recently has been—volcanically active.

6. We see hot-spot volcanoes on Venus but not in "linear chains" like the Hawaiian chain on Earth. What does this mean about the differences in tectonic activity between the two planets?

7. What would be the effect on a planet's temperature if its atmosphere reflected most of the visible light striking it but let infrared radiation pass through?

8. The difference between the length of the solar day and the sidereal day is much greater for Earth than for Mars. Why?

9. Mars has no ozone layer protecting its surface from ultraviolet radiation. Our industrial society may be destroying the ozone layer here on Earth. Report on the latest research concerning depletion of the Earth's ozone layer.

10. Why is it reasonable to assume that Venus, Earth, and Mars started with roughly the same primordial atmospheres? Why are the atmospheres of the three planets so different today?

11. In discussing our search for life beyond Earth, researchers point out that "absence of evidence is not evidence of absence." Explain this statement.

1. If a planet has a diameter one-third of Earth's, how does its surface area compare to Earth's? Its volume?

2. Jupiter's diameter is 11 times Earth's diameter. Compare the volume of Jupiter to the volume of Earth.

3. If planet X has a diameter 4 times that of planet Y, and both start at the same temperature, which will cool faster? By what factor?

4. You are observing Mercury using a telescope that can resolve features as small as 1 arcsecond. Observing conditions are good and the planet is at about 1 AU away from Earth. Using the small-angle formula, find the size of the smallest surface features you could see on Mercury. Compare your answer to the size of Caloris Basin, shown in Figure 8-6.

5. Using Figure 8-1, show that Mercury never rises (or sets) more than about 2 hours before (or after) the Sun.

6. Mars's semimajor axis is 1.524 AU. Using Kepler's third law, show that the planet's sidereal orbital period is 1.88 years.

CALCULATIONS

Viewing Mercury, Venus, and Mars

Table A8-1 shows the dates from 2004 through 2008 when Mercury will be situated for best viewing. Between the pairs of dates listed, Mercury and the Sun are separated by at least 12.5 degrees in altitude. The greater the separation in altitude between the two objects, the easier it is to find Mercury in the sky. The best chance of seeing Mercury is about midway between each pair of dates, and the longer the time between the dates, the better the viewing at the midpoint. For example, in 2006 the May/June period provides a better viewing opportunity than the February/March period, and the best viewing occurs around June 15.

Within about 45 minutes of sunrise and sunset, the sky is so bright that Mercury is particularly difficult to see with the naked eye. Thus, if you are viewing in the morning, you should begin your search at least an hour before sunrise. In the evenings, there is no need to begin searching until about 45 minutes after sunset. Mercury is never easy to find for the first time, and you must have a very clear sky and a low horizon.

Table A8-2 shows favorable dates for viewing Venus in the morning and evening sky. Between each pair of dates, Venus rises at least 40 minutes before the Sun or sets at least 40 minutes after. Between the dates indicated, Venus should be easy to find any time the sky is clear. Venus is bright enough that you can find it as close as 20 minutes to sunrise or sunset.

Table A8-3 shows some positions of Mars. Suppose you wish to know where to look for Mars on May 15, 2007. Since the planet rose at about 5 A.M. the previous March 15 and will rise at about 1 A.M. on July 15, you can figure that it will rise at about 3:30 A.M. on May 15 (since May 15 is about midway between the two). Thus in the early morning it will be somewhere in the southeastern sky.

If a telescope is available to you, use it to observe the planets. You should be able to see the phases of Venus (and perhaps Mercury).

ACTIVITIES

Helpful advice is available for the amateur astronomer in several monthly magazines, particularly *Astronomy* and *Sky and Telescope*. These include diagrams of planetary locations and hints on viewing.

Table A8-1

Favorable Dates for Viewing Mercury

Year	Evening Viewings	Morning Viewings
2004	Mar. 17 – Apr. 8 July 5 – July 28	Sep. 2 – Sep. 20 Dec. 17 – Jan. 8
2005	Mar. 2 – Mar. 20 June 17 – July 15	Aug. 16 – Sep. 9 Dec. 1 – Dec. 24
2006	Feb. 14 – Mar. 4 May 30 – June 30	July 31 – Aug. 18 Nov. 15 – Dec. 9
2007	Jan. 29 – Feb. 15 May 14 – June 14	July 14 – Aug. 2 Oct. 31 – Nov. 22
2008	Jan. 14 – Jan. 30 Apr. 27 – May 26 Dec. 30 – Jan. 12	June 27 – July 15 Oct. 14 – Nov. 3

Table A8-2

Favorable Dates for Viewing Venus

Year	Range of Dates	When Visible
2003–2004	Nov. 10 – May 28	Evening
2004–2005	June 20 – Jan. 17	Morning
2005–2006	May 13 – Jan. 7	Evening
2006	Jan. 19 – Sept. 22	Morning
2006–2007	Dec. 27 – July 29	Evening
2007–2008	Aug. 27 – Mar. 3	Morning
2008–2009	Aug. 13 – Mar. 22	Evening

Table A8-3

Approximate Rising (R) and Setting (S) Times* for Mars

Year	March 15	July 15	November 15
2004	S: 11 P.M.	Not visible	R: 5 A.M.
2005	R: 4 A.M.	R: midnight	R: 4 P.M., S: 6 A.M.
2006	S: 1 A.M.	S: 10 P.M.	Not visible
2007	R: 5 A.M.	R: 1 A.M.	R: 7 P.M.
2008	S: 3 A.M.	S: 11 P.M.	Not visible

* Times given include daylight saving time for North America.

You can find information about current explorations of our solar system at NASA's home page (http://www.nasa.gov).

1. "Mercury: The Forgotten Planet," by R. Strom, in *Sky & Telescope* (September, 1990).

2. "The Significance of Planet Mercury," by W. Hartmann, in *Sky & Telescope* (May, 1976).

3. "The New Face of Venus," by E. R. Stofan, in *Sky & Telescope* (August, 1993).

4. "What Makes Venus Go?" by R. Burnham, in *Astronomy* (January, 1993).

5. "Global Climate Change on Venus," by M. A. Bullock and D. H. Grinspoon, in *Scientific American* (March, 1999).

6. "Welcome to Mars!" by C. C. Petersen, in *Sky & Telescope* (October, 1997).

7. "*Mars Global Surveyor:* You Ain't Seen Nothin' Yet," by S. Parker, in *Sky & Telescope* (January, 1998).

8. "Global Climatic Change on Mars," by J.S. Kargel and R. G. Strom, in *Scientific American* (November, 1996).

EXPANDING THE QUEST

STARLINKS

Quest Ahead to Starlinks: www.jbpub.com/starlinks

Starlinks is this book's online learning center. It features **eLearning**, which contains chapter quizzes and other tools designed to help you study for your class. You can also find **on-line exercises,** view numerous relevant **animations,** follow a guide to **useful astronomy sites** on the web, or even check the latest **astronomy news** updates.

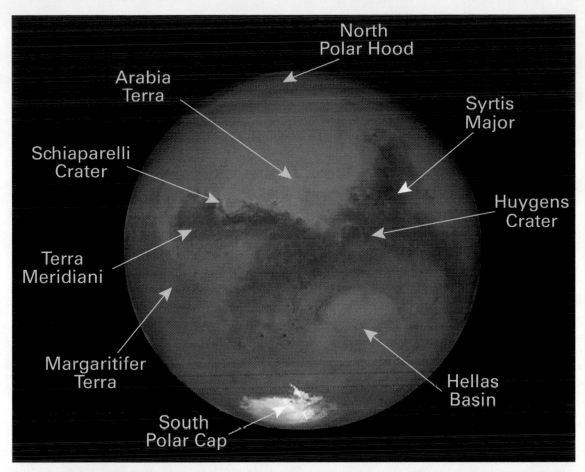

North
Polar Hood

Arabia
Terra

Syrtis
Major

Schiaparelli
Crater

Huygens
Crater

Terra
Meridiani

Margaritifer
Terra

Hellas
Basin

South
Polar Cap

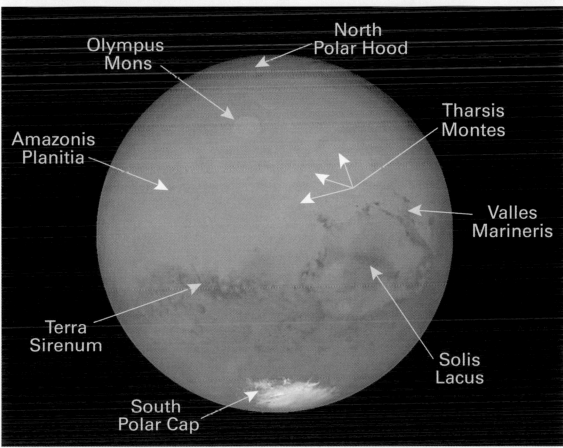

North
Polar Hood

Olympus
Mons

Tharsis
Montes

Amazonis
Planitia

Valles
Marineris

Terra
Sirenum

Solis
Lacus

South
Polar Cap

These color composite images of Mars were taken by the *HST* in August 2003 when Mars was at it closest distance to Earth in about 60,000 years. They reveal the southern polar ice cap, the largest volcano in the solar system (Olympus Mons, the oval-shaped feature above center on the bottom image), and a system of canyons (Valles Marineris, right of center on the bottom image). When these images were taken, Mars' Northern (Southern) hemisphere was in the midst of winter (summer). Clouds cover the region around the northern polar cap, while the orange streaks show dust activity above the southern polar cap.

CHAPTER 9

The Jovian Planets

This false-color image of Saturn, taken by the *Hubble Space Telescope*'s infrared camera, shows two violent storms near the planet's equator. In this photo, blue indicates clear atmosphere or ammonia ice; green and yellow indicate atmospheric haze; and red and orange indicate clouds high in Saturn's atmosphere.

In October 1997, the spacecraft *Cassini* was launched toward Saturn. (It was named for astronomer Giovanni Cassini, who in 1675 observed that Saturn's rings were separated by a gap, now called the Cassini Division.) Its trajectory will get the spacecraft to Saturn in July 2004. *Cassini* (with a mass of about 5650 kilograms, the mass of a small school bus) was launched toward Venus rather than Saturn in order to take advantage of a "gravitational assist," which uses the gravitational field of a planet to increase the speed of a passing spacecraft. *Cassini* got two assists from Venus, one from Earth, and one from Jupiter before finally heading toward Saturn. Without such assists, *Cassini* would have taken decades to get to Saturn. (Gravitational assists come with a price. When the *Voyager* spacecraft got an assist from Jupiter, *Voyager* gained 36,000 miles/hour at a cost of slowing down Jupiter by 1 foot every trillion years!)

Nineteen days before arrival at Saturn, *Cassini* will pass close to Phoebe, Saturn's most distant satellite. Of all the moons in the solar system, Phoebe is one of the most curious. Its inclined, retrograde orbit suggests that the moon may be a captured object, perhaps an old comet or asteroid that wandered close to Saturn in ages long past.

When *Cassini* reaches Saturn, it will approach to an altitude only one-sixth the diameter of Saturn itself. The spacecraft will cross Saturn's rings (in a "gap," of course) and then ignite its rockets in order to slow down enough to be captured by Saturn's gravity in a 5-month-long orbit. During the rocket burn, *Cassini* will decrease its speed by more than 600 meters/second (1300 miles/hour).

The tour of Saturn will continue for four years, as *Cassini* makes about 60 orbits with various orientations, from as close as three Saturn radii to more than seven. It will collect large amounts of data about the planet, its rings, and its moons, and it will make more than 30 flybys of Saturn's moon Titan—the second largest moon in the solar system. In November 2004 *Cassini* will drop the *Huygens* probe (supplied by the European Space Agency) to Titan; the probe will explore Titan's clouds, atmosphere, and surface.

The *Cassini-Huygens* venture includes the *Cassini* orbiter and the *Huygens* probe. It is a collaboration of three space agencies (NASA, ESA, ASI) and scientists and engineers from 15 countries on both sides of the Atlantic.

In this Chapter, we turn our attention to the planets of the outer solar system—the Jovian planets. (Pluto will be reserved for the next chapter.) Although Mercury, Venus, and Mars are certainly strange worlds for us Earthlings, in this chapter we will show that the outer planets are even stranger. We will consider these planets in order as we continue our journey out from the Sun. There is still a lot that astronomers do not know about our neighboring worlds, and you will see that as we get farther and farther from our home base, less and less is known about the objects there.

9-1 Jupiter

The size of a planet is not apparent to the naked eye, but long before the invention of the telescope, Copernicus had deduced that Jupiter was larger than Venus, even though Venus at its brightest is brighter than Jupiter. From the two planets' relative brightnesses and distances, he concluded that since Jupiter is so much farther away, if it shines only by reflected light it would have to be much larger to appear so bright. Galileo observed the planets' angular sizes with a telescope and was able to determine that Jupiter was indeed larger, for he could use their angular sizes and relative distances to calculate their relative sizes. Today we know that Jupiter is the largest object in the solar system besides the Sun. It is appropriate, then, that it is named after the king of the Roman gods.

Copernicus simply assumed that every planet reflects the same percentage of the light that hits it.

Jupiter as Seen from Earth

We can calculate Jupiter's mass by observing the radii of the orbits of its moons and their periods of revolution, as explained in Chapter 7. And Jupiter is massive! It has more than twice the combined mass of all the other planets, their moons, and the asteroids. When we compare Jupiter to the Earth, we find that Jupiter is 318 times more massive than our little planet.

In size, Jupiter is even more remarkable. Look back at Figure 7-1 to see how it compares in size to the other planets. Jupiter's diameter is about 11 times that of the Earth, making its volume about 1400 times that of Earth.

Since Jupiter's volume relative to Earth is so much greater than its mass relative to Earth, you might deduce that its density must be less than that of Earth. That's exactly right; Jupiter's density is only about one-fourth of Earth's. This means that Jupiter must be composed of a higher percentage of light elements than is the Earth.

While the Earth's density is about twice as great as common rock, Jupiter's density is 1.3 times that of water.

You might have a tendency to picture Jupiter as a lumbering, slow-moving giant. Indeed, it is 5.2 AU from the Sun and takes nearly 12 years to circle the Sun. However, its rotation rate is not slow. This gigantic planet spins on its axis once every 9 hours and 55 minutes. This is the first Jovian planet we have studied, and we will find that fast rotation is common for Jovians. Their rotational speeds are much greater than those of the terrestrial planets.

Things are not quite this simple, however. Through even a small telescope we can see that Jupiter has dark- and light-colored bands parallel to its equator. The pho-

Figure 9-1
The bands around Jupiter are easily visible in this photo taken from Earth. The red spot is called—imaginatively—"the Great Red Spot." The circle has been drawn to show how oblate Jupiter is.

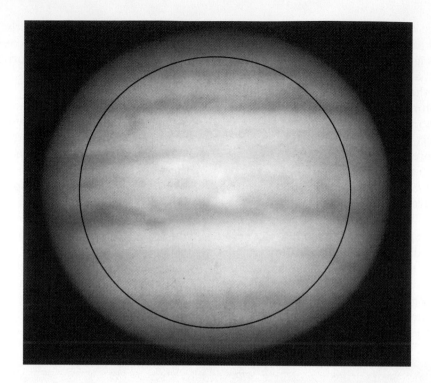

tograph in **Figure 9-1**, which was taken with a large Earth-based telescope, shows obvious parallel bands around the planet. Observations show that the bands near the equator rotate slightly faster than those nearer the poles. The band at Jupiter's equator has a rotation period of 9 hours, 50 minutes. Bands closest to the poles complete one rotation in 9 hours, 56 minutes. This "spreading" of the rotation, or **differential rotation,** indicates that the visible surface of Jupiter is not solid. Instead, it must be at least partially fluid. Like a pot of liquid when stirred, different regions of Jupiter take different amounts of time to rotate around the planet's axis.

differential rotation Rotation of an object in which different parts have different periods of rotation.

oblate Flattened at the poles.

Another thing about Jupiter that can be observed fairly easily is that it is slightly flattened at the poles (*oblate*). A circle has been drawn on the photo of Figure 9-1 to make it obvious that Jupiter's equatorial diameter is definitely greater than its polar diameter. In fact, it is 6.5% greater. Recall that both Earth and Mars are oblate, but not nearly as much as Jupiter. The cause for the greater oblateness of Jupiter is its great rotation rate. (Picture a spinning ball of Jello; its spin would cause it to flatten out.)

Jupiter as Seen from Space

Because color filters were used when the black-and-white photos were taken, color images could be reconstructed.

In December, 1973, *Pioneer 10* flew to within 130,000 kilometers of the surface of Jupiter. Then in December, 1974, its twin, *Pioneer 11*, came within 50,000 kilometers of the surface. These two craft sent back great amounts of data, from which a computer produced black-and-white photographs of the surface. In addition, the craft contained instruments to detect charged particles, radiation from the planet, and Jupiter's magnetic field. After passing Jupiter, *Pioneer 11* continued on to Saturn, also giving us our first close view of that planet. Both of these spacecraft are now beyond the solar system, gliding endlessly through space.

Knowledge gained from the *Pioneer* missions helped in the design of *Voyager 1* and *2*, which visited Jupiter in 1979. They sent back 33,000 images, including that of **Figure 9-2**. The spacecraft *Galileo* orbited Jupiter and its moons during the period December 1995 to September 2003, before it finally plunged into Jupiter's atmosphere. Upon reaching Jupiter, *Galileo* dropped a probe into the planet's atmosphere, and some of the information that follows has been learned from *Galileo* and its probe.

As we pointed out in Chapter 8, if we knew more about weather on Venus, we would know more about weather on our own planet. Jupiter may provide an even better study of weather systems than does Venus, for weather patterns in the upper atmosphere of Jupiter are far removed from any solid surface Jupiter may have and therefore are almost unaffected by complications produced by surface irregularities. Jupiter's weather system should therefore be simpler in nature and allow us to learn what happens when surface features are not a major factor. (The practice of examining a simple system to form a working hypothesis about a complicated one is common in science.)

The fact that surface features have little effect on Jupiter's upper atmosphere is probably what allows Jupiter's weather patterns to last for such long periods. The prime example of this is the giant red spot on Jupiter (**Figure 9-3**). This spot was seen as early as the mid-1600s, so it has lasted for more than 300 years. The spot is about 40,000

Figure 9-2
This photo of Jupiter was taken by *Voyager 1* at a distance of 40 million kilometers from the planet.

The spot is actually more gray than red and is difficult to see in a small telescope.

Figure 9-3
Jupiter's Great Red Spot (at upper right) is seen with the fierce turbulence around it. Wind speeds in this system reach 500 kilometers/hour. Earth is superimposed at the same scale to help us appreciate the size of the spot. Another persistent feature in Jupiter's atmosphere is seen near the Great Red Spot: It is a white oval region where winds also flow counterclockwise. This is probably an area of high-altitude, cold clouds.

kilometers in length and nearly 15,000 kilometers across. When we realize that the Earth is about 13,000 kilometers in diameter, we can appreciate the immensity of this feature. From time to time over the centuries, the red spot has faded in intensity, but it has never disappeared completely.

Data from Jupiter indicate that the red spot is a storm system of rising high-pressure gas whose cloud tops are colder and about 8 kilometers higher than the surrounding regions. It is similar to the much smaller Earth systems reported by your local weather forecaster. High-pressure systems in Earth's Northern Hemisphere rotate in a clockwise direction, and in the Southern Hemisphere they rotate in the opposite direction. The red spot is in Jupiter's southern hemisphere, and indeed it rotates counterclockwise, with a period of about 6 days. As shown in Figure 9-2, the winds to the north and south of the Great Red Spot move in opposite directions. The motion of the gas in this region resembles that of a wheel spinning between two surfaces moving in opposite directions. Figure 9-3 is a close-up view of part of the spot; we can see the swirling currents at its edges.

The light-colored bands around the planet mark the tops of low-pressure, low-temperature, high-altitude regions. These clouds formed by gas that, after being warmed by heat from the planet's interior, rose from inside Jupiter and cooled. The dark-colored bands mark the tops of high-pressure, high-temperature, low-altitude regions. These clouds formed by gas that gets warmer as it descends (**Figure 9-4**). Jupiter's differential rotation moves the regions around the planet so that it has a banded appearance. This has been the standard interpretation of atmospheric motion in the bands and is based on our experience with Earth's atmosphere, in which many clouds form where air is rising. However, pictures from the *Cassini* spacecraft show that almost without exception individual storm cells of upward-moving bright-white clouds exist in the dark-colored bands. This suggests that the dark-colored bands are the regions of net upward-moving gas motion in Jupiter's atmosphere, with the implication that the light-colored bands correspond to a net downward-moving gas motion. This is exactly the opposite of our current interpretation, shown in Figure 9-4. We still have a lot to learn about the atmospheres of the Jovian planets. In any case,

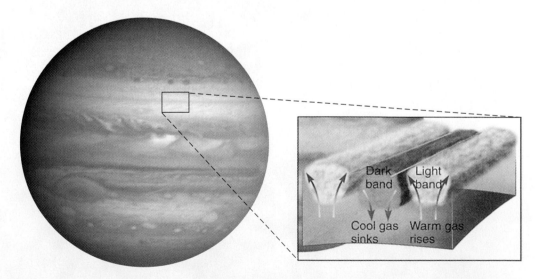

Figure 9-4
The light and dark bands on Jupiter are each thousands of kilometers wide, blowing around the planet at 300 to 500 kilometers/hour. The dark bands are sinking cool gas; the light bands are rising warm gas. The white patches are ammonia ice. This true-color view of Jupiter is the result of four images taken by the *Cassini* spacecraft during its flyby in December 2000. The shadow on the planet is due to its moon Europa. Keep in mind that the interpretation of atmospheric motion in the bands may have to be revised, as we explain in the text.

measurements made by *Voyager* show that at the boundaries of each band, the wind velocities are different and in opposite directions. As a result of these different speeds, there is considerable swirling at the boundaries between bands. The stormy nature of these swirls is obvious in the view of the bands in Figures 9-2 and 9-3.

Data obtained by *Galileo* support the idea that lightning storms beneath the upper cloud cover are the energy source for Jupiter's colorful weather patterns. Observed at depths of 80 to 100 kilometers, lightning indicates the presence of water, since nothing else can condense at those depths. As is the case on Earth, lightning may occur in water clouds where positive and negative charges separate as a result of turbulent upward and downward motions of partially frozen ice droplets. Lightning then points to areas where swirling currents occur, which in turn get pulled apart by shear flow and give up their energy to the large-scale features observed on Jupiter. Scientists have also observed the merging of giant storms that had lasted for decades to a single, bigger storm. It is possible that a similar merger took place centuries ago giving us the Great Red Spot.

The Composition of Jupiter

The terrestrial planets consist mostly of hard, rocky material. Jupiter is different. The compositions of Jupiter and the other Jovian planets are more similar to the Sun than to the terrestrial planets. The *Galileo* probe measured Jupiter's atmospheric composition to be about 90% hydrogen and 10% helium, with small amounts of methane, ammonia, and water vapor. This is similar to the composition of the original solar nebula. *Galileo* detected small amounts of heavier elements (carbon, nitrogen, and sulfur) at concentrations about 3 times higher than the solar composition, suggesting that meteorites and other small bodies have contributed to the planet's composition. Few complex organic compounds were evident, so the likelihood of finding life as we know it here on Earth is extremely remote.

The *Galileo* probe did not detect the thick, dense clouds that were expected, despite the fact that data received from the probe go down to about 150 kilometers below the top of the atmosphere. Infrared images from Earth-based telescopes and by *Galileo* indicated that the site where the probe descended is not typical.

The probe found that Jupiter has the same helium content as the outer layers of our Sun, but only a tenth of the Sun's neon concentration. Since the outer layers of our Sun lose helium, these observations suggest that there must be a mechanism on Jupiter that removes helium and neon from its upper atmosphere. It is possible that there is helium rain in Jupiter's atmosphere, with neon dissolving in the helium raindrops under certain conditions. Also, the concentration of deuterium—one of the heavy forms of hydrogen—was found to be similar to that of the Sun, but very different from that of comets or of Earth's oceans. This finding minimizes the possible effect that comets may have had on the composition of Jupiter's atmosphere.

The *Galileo* probe also found that the concentrations of the noble gases argon, krypton, and xenon are 2 to 3 times higher than the solar composition. The only way for Jupiter to get such quantities of these gases is to trap them by condensation or freezing, which requires very cold temperatures, colder than what we find on the surface of Pluto. This challenges our current theory of planetary formation, as described in Chapter 7. According to this theory, Jupiter and the other giant planets were built up with material found in the original solar nebula. Planetesimals clumped together to form a protoplanet that, after growing to a critical size, started sweeping up gas directly from the nebula. Thus, Jupiter's original composition was similar to the early Sun's. Since then, the planet has been shaped by continuing collisions with space debris and by its internal differentiation. However, the data suggest that the material that makes up Jupiter must have originated at a much colder place than Jupiter occupies today. It is possible that the original solar nebula was much colder than we currently think, or that planetesimals could have started forming even before the original cloud of dust and gas collapsed to form our solar nebula. It is also

Figure 9-5
The *Hubble Space Telescope* took this ultraviolet image of Jupiter on July 21, 1994, and shows the blotches formed by impact from fragments of the Shoemaker-Levy 9 comet. Many dark patches are seen from different impacts. The reason the spots are dark in the ultraviolet is because sunlight gets absorbed by the large quantity of dust deposited high in Jupiter's atmosphere as a result of the fireballs.

This amount of energy is hundreds of times larger than the energy that would be released if we detonated all the nuclear weapons on Earth at the same time.

possible that Jupiter formed farther from the Sun and then drifted inward. The latter is a very tempting possibility, since we have now discovered many planetary systems in which large planets are very close to their stars, as we discussed in Chapter 7. As is usually the case, new observations provide new insights while creating new and exciting problems.

The colors seen in Jupiter's upper atmosphere are thought to be the result of chemical reactions induced by sunlight and/or by lightning in its atmosphere, but this is still an open question. Another possibility is that impurities (possibly in the form of sulfur or phosphorus) in the cloud droplets of water, ammonia, and ammonia sulfides, result in the distinctive colors.

In addition to direct observations and the data we received from *Galileo* and its probe, we had an opportunity to learn a great deal about Jupiter's atmosphere by observing a spectacular show in July, 1994. During the week of July 16–22, 23 fragments of the comet Shoemaker-Levy 9 crashed into Jupiter's atmosphere, as we talked about at the beginning of the Prologue. The strong gravitational tidal forces from Jupiter tore the comet into fragments smaller than 1 kilometer in size. Jupiter's gravity accelerated these fragments to speeds of about 60 kilometers/second (130,000 miles/hour) and the largest of the fragments produced energy upon collision equivalent to 6 million megatons of TNT. As each fragment entered the atmosphere, the resulting shock wave vaporized it, resulting in a huge, hot fireball that sent millions of tons of material up to 3000 kilometers (1900 miles) above Jupiter's clouds. As the material cooled and fell back to the atmosphere, it left a signature of dark-colored blotches, some of which were the size of the Earth (**Figure 9-5**). Unfortunately, there is no consensus as to the makeup of the fragments or the depth to which they penetrated the atmosphere. As a result, the analysis of the fireball spectra has not provided conclusive information about the chemical composition of Jupiter's atmosphere.

Jupiter's Interior

The gaseous atmosphere of Jupiter is fairly thin, only a few thousand miles in depth. Naturally, the lower you go in the atmosphere, the greater the pressure is (because the gas at lower elevation supports the gas above). On Jupiter, the pressure soon becomes so great that the hydrogen no longer acts as a gas, but starts acting like a liquid. The molecules are forced so close together that the hydrogen acquires a liquid nature; it becomes liquid hydrogen. On Earth we have a gaseous atmosphere above a liquid ocean. On Jupiter, however, there is no distinct boundary between the liquid and the gas. At the top of the atmosphere, we would classify the hydrogen as a gas. Much lower, we would call it a liquid. In between, the gas becomes more and more dense as we move deeper and deeper into the atmosphere. If we could travel inward from the gaseous region, we would find ourselves in thicker and thicker gas until we finally decided that we were no longer in a gas but were in a liquid.

Figure 9-6
The interior of Jupiter consists mostly of liquid metallic hydrogen, possibly with a core made up of heavy elements.

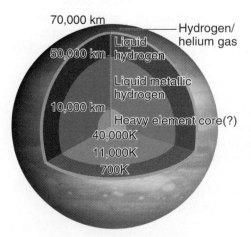

70,000 km
— Hydrogen/ helium gas
50,000 km — Liquid hydrogen
Liquid metallic hydrogen
10,000 km — Heavy element core(?)
40,000K
11,000K
700K

At perhaps 15,000 kilometers below the cloud tops, another change in Jupiter's composition occurs, caused by the increasingly greater pressures and temperatures. At these depths, electrons move easily from one atom to another, making the hydrogen a good electrical conductor. Because it conducts electricity like a metal, we call it *liquid metallic hydrogen*. Hydrogen in this form cannot be produced on Earth because we cannot create such a combination of high temperature and high pressure, but it is predicted by atomic theories. As **Figure 9-6** indicates, most of the planet is made up of this state of matter.

Figure 9-6 shows a core of heavy elements at the center of Jupiter. We have no direct evidence that such a core exists, but it is hypothesized on the

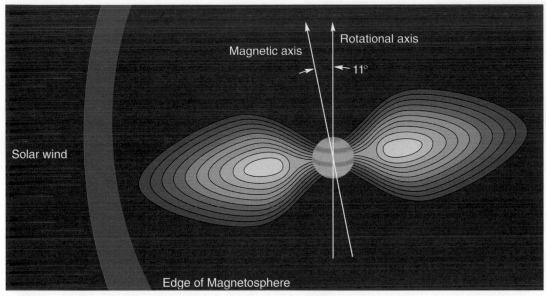

(a)

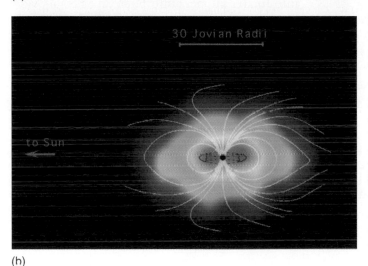

(b)

Figure 9-7

(a) Jupiter's radiation belts are flattened by the planet's rapid rotation. Compare them to the Van Allen radiation belts shown in Chapter 6. The magnetosphere of Jupiter is so large that if you could see it from Earth it would cover an area in the sky about 3 times larger than the full Moon. (b) This image of Jupiter's vast magnetosphere was taken by the *Cassini* spacecraft shortly after its closest approach to Jupiter, in December 2000, about 10 million kilometers from the planet. Three features are sketched in to provide context: a black circle showing the size of Jupiter, lines showing Jupiter's magnetic field, and a cross-section of a doughnut-shaped ring of charged particles circling Jupiter at about Io's orbit. The magnetosphere is normally invisible. However, fast-moving ions within the magnetosphere can pick up electrons to become neutral atoms and thus escape Jupiter's magnetic field. *Cassini*'s ion and neutral camera detects these atoms and thus can derive information about their source.

basis that there must have been heavy elements in the original material of which Jupiter is made. These heavy elements exist on the moons of Jupiter, and there is good reason to think that they also exist on the planet. If so, they would have sunk to the center of Jupiter. The size of this portion of the planet is unknown, but it is thought to have a diameter perhaps as small as Earth's or as much as a few times Earth's diameter. In any case, it makes up a very small portion, perhaps 1%, of the entire planet.

When the *Pioneer* spacecraft passed Jupiter, we learned of that planet's very strong magnetic field. *Voyager* provided more data, and then in February 1992, the *Ulysses* spacecraft passed by the planet and found that Jupiter's magnetic field is even larger and stronger than had previously been thought. Jupiter's magnetic field is nearly 20,000 times stronger than Earth's!

Recall from Chapters 6 and 8 that the dynamo theory of planetary magnetic fields holds that in order for a planet to have a strong magnetic field, two conditions are necessary: rotation and electrically conductive material within the planet. The terrestrial planets appear to have iron (or nickel-iron) cores that are good electrical conductors. Jupiter does not depend upon its core to produce its magnetic field, for the liquid metallic hydrogen that makes up much of the planet is responsible—along with Jupiter's fast rotation—for its strong magnetic field.

Like the Earth's magnetic field, Jupiter's field deflects the solar wind around the planet, as well as trapping some of the charged particles of the wind in belts around it. Volcanic eruptions on Jupiter's moon Io also add large quantities of particles (such as hydrogen, oxygen, sulfur, and sulfur dioxide) into Jupiter's *magnetosphere*. The magnetosphere extends to perhaps 15 million kilometers from the planet, enveloping most of Jupiter's satellites. (The field has presented problems to the electronic circuits of spacecraft that passed through it.) **Figure 9-7** illustrates the radiation belts and magnetosphere of Jupiter.

The *Ulysses* spacecraft was built by the European Space Agency. In 1992, it used a gravitational assist from Jupiter to go into a polar orbit around the Sun, where it arrived in 1994. After gathering data, it began its voyage back to Jupiter in 1995, where it looped around and returned for its second orbit around the Sun in September 2000.

magnetosphere The volume of space in which the motion of charged particles is controlled by the magnetic field of the planet, rather than by the solar wind.

Figure 9-8
A radio image of Jupiter at 21 centimeters. The bright features beyond the disk are due to synchrotron radiation, which is mainly concentrated along Jupiter's magnetic equator. The radiation is also seen at higher magnetic latitudes as indicated by the two "horns."

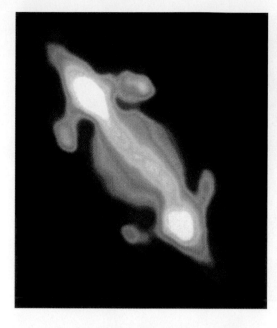

Is the center of our Sun the "hottest" place in our solar system?

Jupiter's magnetic field accelerates the charged particles in the magnetosphere to high speeds. As a result, the temperature of this hot, gas-like mixture of charged particles can reach 400 million K. This is about 25 times larger than the temperature at the center of our Sun! However, you should not expect nuclear reactions to occur in Jupiter's magnetosphere, since the average density of this gas is very low. In order to have nuclear reactions, we need both high densities and high temperatures. However, the fast-moving charged particles trapped in Jupiter's magnetic field do emit synchrotron radiation, which is observed at radio wavelengths (**Figure 9-8**).

As you might expect, powerful auroras form close to Jupiter's poles, a thousand times more powerful than on Earth (**Figure 9-9**). These auroras are formed when energetic charged particles spiral along the magnetic field lines and crash into Jupiter's upper atmosphere. As a result, violent winds blow in these regions of the auroras, producing lots of energy (due to friction) as they interact with the rest of Jupiter's atmosphere. This might explain why the temperature at the top of Jupiter's atmosphere is about 1000 K, much hotter than expected for a planet so far away from the Sun.

Figure 9-10 summarizes what we currently know about Jupiter's space environment.

Energy from Jupiter

Several puzzles concerning Jupiter remain besides the composition of its interior. One such puzzle results from the fact that Jupiter emits more energy than it absorbs from the Sun. This was known before the *Pioneer* and *Voyager* missions, and those missions only confirmed the observations. Let's explain why this is a problem.

We would expect the radiation coming from a planet to be simply the sum of the solar radiation that is reflected from the planet and the infrared radiation that is emit-

Figure 9-9
This *HST* ultraviolet image was taken in November 1998. The auroral lights are the bright emissions above the dark blue background.

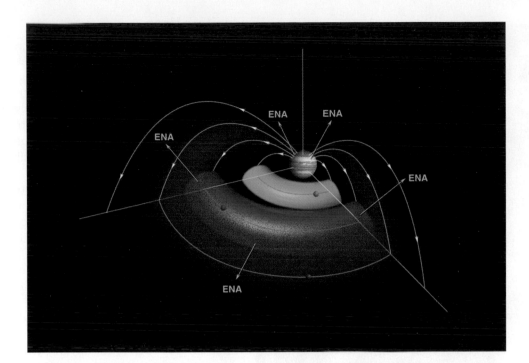

Figure 9-10
A cut-away schematic of Jupiter's space environment. Magnetically trapped radiation ions are shown in red, the neutral gas torus of the volcanic moon Io is shown in green, and the newly discovered neutral gas torus of the moon Europa is shown in blue. The latter is the result of severe bombardment of radiation from Jupiter on Europa's surface, kicking up and pulling apart water-ice molecules. Magnetic field lines are represented in white. "ENA" stands for energetic neutral atoms; they are emitted from the Europa torus regions because of the interaction between the trapped ions and the neutral gases. They also come from Jupiter when radiation ions collide with particles in Jupiter's upper atmosphere. ENAs were discovered in images by *Cassini* in early 2001 during its flyby of Jupiter.

ted as the planet re-emits the absorbed solar radiation. (When we speak of solar radiation, we include—along with visible light—ultraviolet, infrared, and all other types of radiation.) We don't expect a planet to have an energy source of its own. It is easy to calculate the amount of solar energy that strikes any planet. In Chapter 4 we discussed the inverse square law of radiation—the fact that as light emitted by the Sun spreads out in a spherical volume, the amount of energy collected per unit area decreases as the square of its distance from the Sun. Therefore, we can use our knowledge of how much solar energy strikes a square mile of Earth's upper atmosphere, along with the inverse square law, to calculate the energy that strikes a square mile of another planet's surface (or atmosphere). Then we multiply this value by the cross-sectional area of the planet to get the total solar energy that strikes that planet.

When the energy that strikes a planet is compared to the energy reflected and emitted from the planet, we expect to find that the energy coming from the planet is the same as the energy absorbed. If it isn't, the planet must either be heating up as it gains energy, or cooling as it loses energy.

Jupiter, though, does not behave as expected; it emits about twice as much energy as it absorbs. There are three possible ways for it to do this. One possibility is that it may have an internal energy source, such as chemical reactions or a source of radioactivity, within it. There is no reason to believe that much energy is produced in Jupiter by either of these methods. It was once hypothesized that Jupiter acts like a miniature star, with significant nuclear fusion reactions going on within it. (Nuclear fusion reactions provide the energy of the stars.) However, further calculations showed that Jupiter is not massive enough and does not have sufficient internal pressures or temperatures to support fusion reactions.

A second possibility: recall from Chapter 7 that as an object shrinks due to gravitational force, it heats up. Calculations show that Jupiter may still be shrinking and producing heat. However, if this were the case, the amount of shrinking would still not be enough to explain all of the extra energy from Jupiter.

It is now thought that the excess energy from Jupiter is energy left over from its formation. By now, the smaller planets have lost their excess energy, but Jupiter's immense size has served to insulate its interior so that it cools slowly, and the cooling continues even today. Computer models show a temperature of about 40,000 K for Jupiter's center, and a density about 20 times larger than the density of water.

Recall that in discussing Venus, we said that the greenhouse effect results in the planet being so hot that it emits as much energy as it absorbs.

Jupiter would have to be nearly 80 times more massive to support nuclear fusion.

Do not get the idea that Jupiter gets significantly cooler from year to year. The planet is so large that the extra energy emitted from it corresponds to an extremely small temperature change.

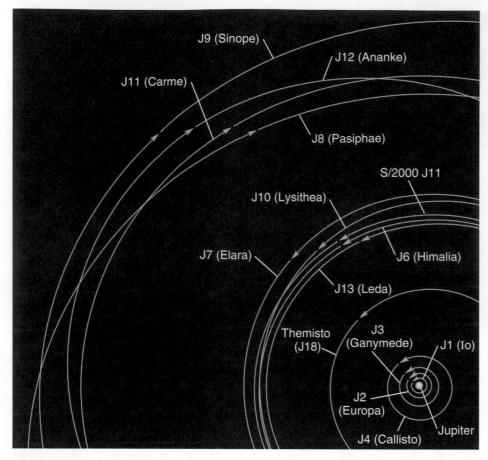

Figure 9-11
The orbits of 14 of Jupiter's satellites. (Satellites are numbered in the order of their discovery.) Data on planetary satellites are given in Appendix D.

Jupiter's Moons

Jupiter's family of 61 moons (as of May 2003) can be divided mainly into three groups. The first group includes four moons that are not shown in **Figure 9-11**, for they are very close to Jupiter, inside the smallest orbit shown. These satellites are generally referred to as fragmented moonlets. The second group includes the four Galilean satellites (Io, Europa, Ganymede, and Callisto) shown orbiting in nearly circular orbits. The third group includes the remaining 53 moons. The majority of the outermost of these moons (four of which are shown) orbit in a clockwise direction, different from the rest of Jupiter's moons (and most objects in the solar system), and their orbits are fairly eccentric. Astronomers hypothesize that the moons in the third group are captured asteroids. That is, they once orbited the Sun, but came close enough to the Jovian system that they were captured by the planet's gravity. In its early history, Jupiter probably had a more extensive atmosphere, which might have been sufficient to slow down the asteroids so they could be captured. The moons have black surfaces, as do many asteroids, providing further support for the hypothesis. It has been suggested that these outer moons are divided into two groups orbiting in opposite directions because only two asteroids were captured, each of which broke into pieces when it hit Jupiter's atmosphere.

Among the Galilean satellites, Europa, the smallest, is 7000 times more massive than the largest of the non-Galilean moons. **Figure 9-12a** shows the relative sizes of the four Galilean moons, along with the four terrestrial planets, Earth's Moon, and

Figure 9-12
(a) Compare the sizes of the Galilean moons to other objects we have studied. (b) This is a "family portrait" that includes Jupiter (with its Great Red Spot) and its largest moons. From top to bottom the Galilean moons shown are Io, Europa, Ganymede, and Callisto.

(a)

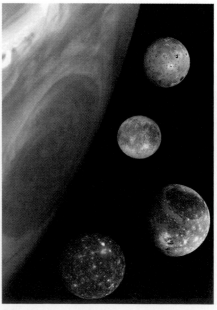

(b)

Figure 9-13
Io, the innermost of the Galilean satellites, has several active volcanoes on its surface. This *Galileo* image is color enhanced with data taken in the near-infrared, green and violet filters.

Saturn's largest moon (Titan). Due to the large gravitational forces exerted by Jupiter, the Galilean moons keep the same face toward Jupiter as they move in their orbits. In addition, the three moons closest to Jupiter influence each other's orbital period, with Io orbiting Jupiter twice for each orbit of Europa, and Europa orbiting Jupiter twice for each orbit of Ganymede.

1. Io. The Galilean moon nearest Jupiter is Io. Io is a strange sight; **Figure 9-13** shows its mottled surface. Its yellow-orange color is caused by the element sulfur, which covers its surface. During the *Voyager* mission, the biggest surprise to astronomers came when Linda Moribito of NASA, while examining *Voyager* photographs, discovered an active volcano on Io. Further examination showed that Io has many active volcanoes or geysers. *Galileo* and the two *Voyager* craft photographed many more active volcanoes (**Figure 9-14**). The circles and dark spots you see in the photograph of Io are not impact craters; they are the results of volcanoes. In fact, the volcanoes explain the lack of impact craters, for the surface of Io is constantly changing by the release of hot sulfur and sulfur dioxide from below the surface. (Sulfur dioxide is a gas on Earth, but on Io it can be either a solid or a gas at the surface, or a subsurface liquid.) Sulfur can take many different colors if heated and then suddenly cooled, and this probably explains the many colors on Io. Originally, *Voyager* images suggested that the lava flows on Io's surface were mostly molten sulfur. However, sulfur vaporizes at about 700 K, while *Galileo* has observed lava flows at temperatures as high as 1800 K—about 1/3 the temperature of the Sun's surface. Such temperatures imply that the lava flows consist of rock formed by a large amount of melting of Io's mantle. *Galileo*'s images give us an opportunity to see what our planet might have been like early in its evolution. About 2 billion years ago, volcanoes on Earth were as hot as Io's and about 15 million years ago, they were as big. Although the vents of Io's volcanoes are very hot, the temperature of most of Io's surface is very low, reaching as low as about −180°C (−290°F), and Io's thin atmosphere is space-cold. Gases from volcanic eruptions quickly freeze as they rise in the air, sulfur-dioxide snow forms in the plumes, and frost collects on the surface.

What is the source of energy for these volcanoes? In the case of geysers and volcanoes on Earth, the energy source is primarily internal heat retained from the Earth's formation and energy released by radioactive materials within the Earth. But a small object cannot retain heat nearly as long as a large one, and Io would have lost any heat that resulted from its formation. Nor can radioactivity account for the tremendous volcanic activity. Instead, the heat within Io is produced by tidal forces. These tides can be 100 meters high and result from the eccentricity of Io's orbit, so that Io is sometimes

Recall that Galileo discovered the four moons named for him.

The term *geyser* is more appropriate than *volcano* because the eruptions do not come from mountain tops. The propulsive agent here is sulfur dioxide, which changes from liquid into high-pressure gas in Io's interior.

An interesting example of the confirmation of a scientific prediction occurred here. Just three days prior to the discovery of Io's volcanoes, S. Peale published a paper predicting that tidal heating of Io would result in volcanism.

Figure 9-14
This *Galileo* image shows two volcanic plumes on Io. One plume (at the edge of the moon and at upper right inset) is 140 km high. The second plume (at center and at lower right inset) is 75 km high and may have been continuously active for more than 18 years. We now recognize at least 120 different active volcanoes on Io.

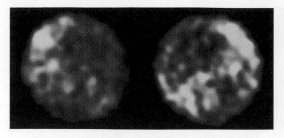

Figure 9-15
The image of Io at the left was made in visible light, while the one at the right is by ultraviolet light. Both show the same surface of Io. Areas that are bright in visible light but dark in ultraviolet are probably covered with sulfur dioxide frost.

This is an example of how we can deduce characteristics of a surface from its different reflectivities in different spectral regions.

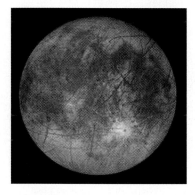

Figure 9-16
There are few craters on Europa, indicating a young and active surface. This is a *Galileo* false-color composite of violet, green and infrared images. Dark brown areas represent rocky materials. Long, dark lines (some are more than 3000 km long) are fractures in the icy crust.

Is Earth the only place in the Solar System where liquid water can exist?

slightly closer to Jupiter than at other times. Even a small difference in distance from massive Jupiter results in powerful, varying tidal pulls on Io. The resulting "kneading" of Io adds energy to it until the molten interior bursts through cracks in the crust—hence, volcanoes. The energy added into Io is estimated to be equivalent to 25 tons of TNT exploding every second. As a result, a volcano on Io ejects about 10,000 tons of material every second, sending it to heights of up to 500 km from the surface.

Figure 9-15 shows two images of Io taken by the *Hubble Space Telescope*. The one on the left was taken in visible light, while the one on the right shows the same surface of Io in ultraviolet radiation. Some areas are bright in visible light but dark in ultraviolet. Sulfur dioxide frost absorbs ultraviolet but reflects visible light quite well, and this is thought to cause the differences between the two images.

Io is surrounded by a halo of sodium atoms, some of which are swept away from the moon by Jupiter's magnetic field, causing a faint ring of sodium near Io's orbit. Other elements observed escaping from Io's atmosphere are sulfur, oxygen, potassium, and chlorine. There seems to be enough salt in Io's volcanic atmosphere to supply the observed amounts of sodium atoms in the neutral clouds and chlorine in the plasma torus.

Io is the most geologically active object in our solar system, with mountains up to twice as tall as Mt. Everest. But these mountains are not themselves volcanoes. They are instead the result of heating, melting, and tilting of giant blocks of Io's crust. Io does not generate its own magnetic field due to lack of convection inside its molten iron core. However, electrical currents that connect Io to Jupiter result in beautiful auroral shows.

The mass of Io has been calculated based on its gravitational attraction on passing spacecraft, and its density is calculated to be about 3.5 times that of water. This, along with data obtained by *Galileo,* leads us to conclude that the satellite has an iron core but is composed mostly of rock, with a relatively shallow layer of sulfur on its surface. No water is found on Io; Io is a strange, volcanic desert.

2. Europa. Europa, 1.5 times as far from Jupiter as Io, presents a far different picture. Even though sulfate has been observed on Europa, its surface is something more familiar to us: ice. **Figure 9-16** shows its cracked-billiard-ball appearance. Europa's density is slightly less than Io's, and again we conclude that it has an iron core but consists mostly of rock covered by an ocean of frozen (and possibly liquid) water. *Galileo* data suggest that Europa has its own magnetic field that reverses every 5.5 hours. This is consistent with changes that could be produced by an underground layer of conductive liquid, such as salt water. This observation raises the possibility that beneath the ice surface is a layer of liquid water and indirectly supports the case for primitive organisms surviving in this environment, as they do deep in Earth's oceans near volcanic vents.

There are few craters on Europa. We know of only about ten impact craters with diameters greater than 20 kilometers. Again, this does not indicate that meteoroids have not struck Europa, but rather that its surface is young and active so that craters do not remain visible for long. Although it is farther from Jupiter than Io is, Europa also experiences tidal heating. There appears to be at least one volcano on Europa, although what looks like a volcano may simply be water spraying out through a crack in the moon's crust, forced out by pressure resulting from tidal flexing. The cracks in Europa's surface are hypothesized to be another result of this flexing.

3. Ganymede. Ganymede is the largest moon in the solar system; it is larger than the planet Mercury. On its surface (**Figure 9-17**) we see ice, but notice how different Ganymede is from Europa. There are craters, indicating that its surface is not as active. Its ice is also darker than Europa's. This darker color is due to dust from mete-

Jupiter

Earth   Jupiter

Jupiter Data

Jupiter	Value	Compared to Earth
Equatorial diameter*	142,984 km	11.2
Oblateness	0.065	19.36
Mass	1.899×10^{27} kg	317.8
Density	1.326 g/cm³	0.24
Surface gravity*	23.12 m/s²	2.36
Escape velocity	59.5 km/s	5.32
Sidereal rotation period	9.925 hours	0.415
Albedo	0.34	1.12
Tilt of equator to orbital plane	3.13°	0.13
Orbit		
Semimajor axis	7.79×10^8 km	5.204
Eccentricity	0.0489	2.93
Inclination to ecliptic	1.3°	
Sidereal period	11.86 years	11.86
Moons	61	

* Measured where the pressure is one atmosphere, Earth's
  pressure at sea level.

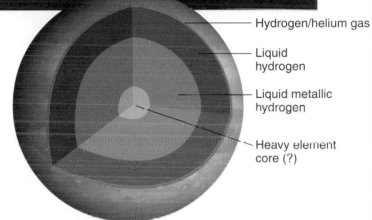

Hydrogen/helium gas

Liquid hydrogen

Liquid metallic hydrogen

Heavy element core (?)

Interior

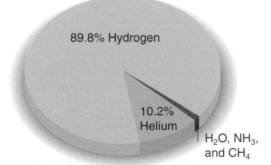

89.8% Hydrogen

10.2% Helium

H_2O, NH_3, and CH_4

Atmospheric Gases

QUICK FACTS

- The largest planet. Contains more than twice the mass of all other planets combined. Fast rotation. Atmosphere shows bands and the Great Red Spot and is mostly hydrogen and helium. Giant magnetic field, due to liquid metallic hydrogen that makes up much of the planet. Emits more energy than it receives. Has a large family of moons and a faint ring system.

Figure 9-17
Ganymede is larger than Mercury. The craters on the surface appear bright, possibly due to clean ice at lower depths.

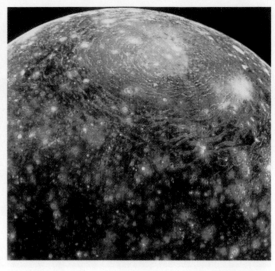

Figure 9-18
Callisto, the Galilean satellite that is farthest from Jupiter, has the least active surface.

orites that spreads across the surface of the satellite after impact. On Europa, the surface is constantly being refreshed with water from below, causing the meteorite dust to be spread through a much deeper layer of the satellite. The less active surface of Ganymede, on the other hand, leaves the dust on the surface. Notice, though, that there are light-colored streaks on Ganymede, whose origins are clearly tectonic. These are where cracks have formed and icy slush from below has welled up to fill the cracks. Data from *Galileo* suggest that Ganymede has a small iron or iron/sulfur core, surrounded by a rocky mantle and a shell of ice at the surface. *Galileo*'s data also suggest that Ganymede generates its own magnetic field, most likely due to a thick layer of liquid, salty water under its crust.

4. Callisto. The outermost of the Galilean moons is Callisto (**Figure 9-18**). Now we see a more familiar cratered surface. At its greater distance from Jupiter, Callisto experiences little tidal heating and has a very inactive surface. It actually has the most cratered surface of any observed object in the solar system. The newer craters are the whitest, showing where a meteorite impact has brought clean ice to the surface. The large white crater near the top of the photograph is named Valhalla and is the largest impact crater known in the solar system. Callisto seems to have a relatively uniform mixture of ice (40%) and rock (60%), with the percentage of rock increasing toward the center. Observations of the region exactly opposite to the Valhalla crater on the moon's surface do not show any signs of such an impact. Such signs are seen on similar-sized worlds, like Mercury and our Moon. It is then possible that a liquid layer exists under Callisto's crust acting as a shock absorber.

Galileo's magnetometer data support the idea that Callisto may have a liquid layer under its crust.

Summary: The Galilean Moons

As one moves from the innermost Galilean moon to the outermost, patterns of change are obvious. The farther the moon is from Jupiter,

1. the less active the surface,
2. the lower the average density of the moon, and
3. the greater the proportion of water.

Points 2 and 3 listed here are related, for since water has a lower density than most other substances on the moons, the more water on a moon, the lower its density.

According to the standard model describing the formation of the Galilean moons, the moons formed from a disk orbiting Jupiter. The combined mass of the moons suggests that the mass of this disk was 2% of Jupiter's mass. The model assumes that the disk formed first and then the moons formed within it. However, the patterns of change we mentioned previously are not consistent with such a model. The temperatures in such a massive and gas-rich disk would be too high for ice to exist for long enough periods for Ganymede and Callisto to form. A new model is emerging in which the material necessary to form the moons is delivered to the disk over a long period of time as Jupiter itself is growing by accreting gas and particles from the original solar nebula. According to this model, the Galilean moons formed slowly, over 100,000 to 1 million years, in a disk where the temperatures remained low enough for ice to exist naturally.

Figure 9-19
The ring system of Jupiter was imaged by the *Galileo* spacecraft on November 9, 1996. The rings clearly show a structure that had only been hinted at in the Voyager images.

Jupiter's Rings

A surprising discovery by *Voyager 1* was that Jupiter has rings. **Figure 9-19** shows the rings as photographed by *Galileo*. The photograph was made when the camera was in Jupiter's shadow. The light that reached the camera from the rings was therefore scattered forward by the material of the rings; since large chunks of matter cannot scatter light in this manner, the rings are known to be made of very tiny particles (as small as particles of cigarette smoke). The rings are fairly close to Jupiter, extending to about 0.8 planetary radius from the planet's surface.

Calculations indicate that the particles of Jupiter's rings could not have been there since the formation of the solar system because radiation pressure from the Sun, as well as forces from Jupiter's strong magnetic field, would gradually send some particles down into the planet and others away into space. Therefore, the rings are thought to be continually replenished, probably by meteoroid impacts on small moonlets within it or near it. Two of these moonlets, Adrastea and Metis, which are not seen in Figure 9-19, orbit through the outer portion of the rings. We will discuss planetary rings further when we discuss Saturn and its very prominent system of rings.

9-2 Saturn

In Roman mythology, Saturn was the god of agriculture and the father of Jupiter. The planet Saturn is probably the most impressive object visible with a small telescope. Galileo, who first observed it in 1610, called Saturn "the planet with ears" and was unable to explain what appeared to be bumps on opposite sides of it. Some 50 years later, Christian Huygens recognized that the "ears" were due to rings around the planet.

Size, Mass, and Density

Saturn has an equatorial diameter of 120,000 kilometers, not much smaller than Jupiter's and more than nine times that of the Earth (**Figure 9-20**). Except for its obvious

Figure 9-20
Here photos of Saturn, Earth, and Jupiter are reproduced to scale. Jupiter's diameter is 11.2 times Earth's, while Saturn's diameter is 9.5 times Earth's.

Huygens (1629–1695) was a Dutch physicist and astronomer who made major advances in the field of optics. He also discovered Saturn's moon Titan, and the probe to be dropped on Titan from the *Cassini* spacecraft has been named *Huygens* in his honor.

rings, Saturn is in many ways similar in appearance to Jupiter. If the planets are indeed similar, we would suspect that Saturn has a density like Jupiter's; but in fact, Saturn's density is less, only 0.7 as much as water. (Jupiter's density is nearly twice that of Saturn.) As we will see later, the low density of Saturn is probably due to a less dense core and a lower percentage of liquid metallic hydrogen.

The composition of Saturn's atmosphere is about the same as Jupiter's: approximately 96% hydrogen and 3% helium, with only about 1% of heavier materials. This is about the same as the composition of the Sun. The similarity is not a coincidence, for all of these objects formed from the same material—the interstellar cloud that collapsed to form the solar system. Recall from Chapter 7 that the reason the terrestrial planets have less hydrogen and helium is that these volatile gases were swept away from the hot, inner portions of the system.

◆ These percentages correspond to the number of particles of each element in the atmosphere.

Saturn's Motions

Saturn orbits the Sun at an average distance of 9.6 AU; its distance from the Earth varies from about 8.5 AU to 10.5 AU. Its appearance from Earth varies greatly, but not because of changes in distance. **Figure 9-21** shows several views of Saturn taken at various times during half of its 29.5-year period of revolution. To see the reason for

Figure 9-21
These photos of Saturn were taken at various times during half of Saturn's 29-year period. They were taken at the Lowell Observatory in Flagstaff, Arizona.

the change in appearance of its rings, think of Saturn revolving at its great distance from the Earth and Sun, as shown in **Figure 9-22**. The rings of Saturn are in the plane of the planet's equator, and the planet is tilted 27 degrees with respect to its orbital plane. Saturn keeps this same tilt while moving around the Sun (and the Earth), so when it is viewed from the Earth, we see the rings at different orientations at different times.

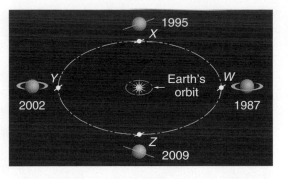

Figure 9-22
Saturn is shown at four points in its orbit. The outer drawings show its appearance from Earth with respect to the background stars when it is at each of those locations. At points X and Z the rings are edge-on to Earth. At position W we see the "top" of the rings, while at position Y we see the "underside" of the rings. The cycle repeats every 29.5 years.

When we consider the rotation of Saturn, we find two more similarities to Jupiter. The various belts of Saturn's atmosphere rotate at different rates, just as in the case of Jupiter, and its equatorial rotation rate of 10 hours and 39 minutes is close to Jupiter's 9 hours and 50 minutes.

Recall that Jupiter's oblateness was explained by its high rotation rate. Saturn has a similar rotation rate, but is even more oblate. This is because the gravitational field at its surface is weaker than at Jupiter's; since there is less gravitational force tending to keep the planet in a spherical shape, Saturn is more oblate than any other planet.

Saturn's oblateness is about 0.1, meaning that its equatorial diameter is 10% greater than its polar diameter.

Pioneer, Voyager, and Cassini

As you might expect, we have learned much about Saturn from space probes. *Pioneer 11* passed Saturn in September 1979, *Voyager 1* about a year later, and *Voyager 2* a year after that. (*Pioneer 10* swung around Jupiter in such a manner that its path did not take it to Saturn.) Knowledge gained from each of these probes was used to guide scientists in decisions concerning the following ones.

The atmosphere of Saturn is similar to that of Jupiter except that Saturn has a slightly lower percentage of helium. Since Saturn is smaller, one must descend deeper into its interior to reach a pressure great enough for hydrogen to form a liquid metallic state. Whereas liquid metallic hydrogen extends out about two-thirds of the way to Jupiter's cloud tops, it extends only about halfway to Saturn's clouds. As a result, Saturn has a weaker magnetic field than Jupiter does. **Figure 9-23** shows a cross section of Saturn. **Figure 9-24** shows auroras at Saturn's north and south poles.

Saturn's magnetic field is only 5% as strong as Jupiter's, which makes it only 1000 times stronger than Earth's!

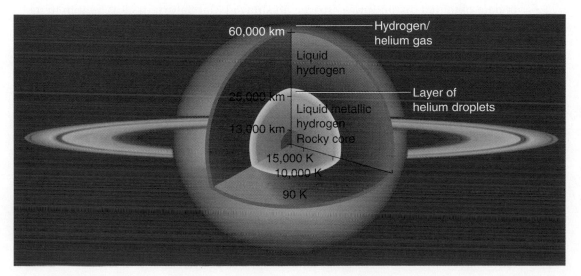

60,000 km — Hydrogen/helium gas
Liquid hydrogen
25,000 km — Layer of helium droplets
Liquid metallic hydrogen
13,000 km — Rocky core
15,000 K
10,000 K
90 K

Figure 9-23
Saturn's interior resembles Jupiter's, but Saturn contains a smaller percentage of liquid metallic hydrogen. The rocky core contains silicates, minerals, and ices under tremendous temperature and pressure. As for the case of Jupiter, Saturn's interior structure is inferred from models and extrapolation of data from the outer layers.

Figure 9-24
This photo in the extreme ultraviolet range shows glowing auroras at the north and south poles of Saturn, reaching as high as 1200 miles above the top clouds. (This photo could not be taken from an Earth-based telescope because the Earth's atmosphere would absorb the ultraviolet radiation.)

Figure 9-25
The *Hubble Space Telescope* took this photo of Saturn in 1994. The white storm system near Saturn's equator appeared that year and lasted several months.

Except for its rings, Saturn appears much blander than Jupiter for two reasons. First, because of its greater distance from the Sun, Saturn is much colder than Jupiter. The extreme cold inhibits the chemical reactions that give Jupiter's atmosphere its varied colors. Second, a layer of methane haze above Saturn's cloud tops blurs out color differences. However, *Voyager* photos show that Saturn has atmospheric features similar to Jupiter, and that its winds reach speeds three to four times faster than Jupiter's. **Figure 9-25** is a color-enhanced photo taken by the *Hubble Space Telescope*, showing a major storm (white) that occurred on Saturn in 1994.

The *Cassini* spacecraft will arrive at Saturn in July 2004 and drop its *Huygens* probe into the atmosphere of Saturn's largest moon, Titan, in November 2004. As was the case of the *Galileo* spacecraft around Jupiter, *Cassini* and *Huygens* will surely provide new surprises and expand our understanding of Saturn and its moons.

Saturn's Excess Energy

Like Jupiter, Saturn radiates more energy than it absorbs. We once thought that the source of the energy is the same as Jupiter's—leftover energy from the planet's formation—but calculations show that because of its smaller size, Saturn would have cooled quickly enough that it would not still be radiating excess energy.

Saturn also has another unusual feature that astronomers must explain: analysis of data obtained by the *Voyager* spacecraft showed that Saturn has less helium in its upper atmosphere than Jupiter has. By mass, Saturn's upper atmosphere contains 11% helium, much less than the nearly 20% on Jupiter. Astronomers had expected Saturn to have about the same composition as the original material of the solar system—the composition of Jupiter and the Sun.

An interesting hypothesis explains both of these anomalies. Some astronomers propose that the cooling of Saturn's atmosphere causes helium to condense to liquid form and rain downward. This would explain the low percentage of helium in the upper atmosphere. In addition, as the helium droplets fall, they lose gravitational energy, and this energy is converted to thermal energy. This thermal energy, the hypothesis holds, results in Saturn emitting more energy than it absorbs.

◀ Saturn's cloud tops have a temperature of −180°C.

Titan

Saturn has 31 moons, most of which consist of dirty ice such as we see on some of Jupiter's moons. Although each of Saturn's moons (**Figures 9-26** and **9-27**) is unique, we will discuss only Titan, the largest. Refer to Appendix D for details concerning the moons of Saturn, as well as other moons of the solar system.

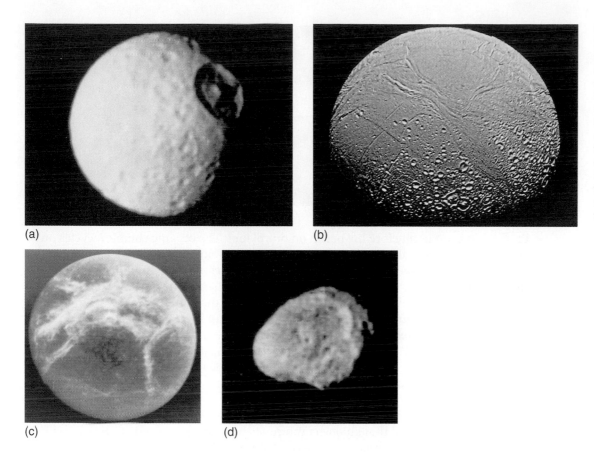

Titan (Figure 9-27) may turn out to be the most interesting moon in the solar system, but our knowledge of its surface is limited by the haze that covers it. The atmosphere of Titan is mostly nitrogen, with a few percent methane and argon. There are also traces of water and organic compounds such as ethane and carbon dioxide. When sunlight breaks down the methane in Titan's upper atmosphere, organic molecules are formed; they then slowly drift down to the surface. This is similar to a thicker version of the smog found over large cities. Radar bounced from its surface varies in intensity as time passes, indicating that the surface is not uniform and that therefore the moon is probably not completely covered by an ocean of organic soup. Perhaps it has lakes or large seas of organic material. This raises the question of whether life might have formed from these organic molecules. Titan is probably the best analogue to the

Radar observations made in 2001 and 2002 revealed mirror-like reflections from Titan's surface with properties that are consistent with liquid hydrocarbon surfaces. However, such reflections could also be from very smooth, solid surfaces.

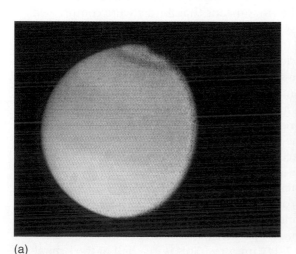

Figure 9-28
Seen edge-on, Saturn's rings are very thin. In this photo, Titan is above the rings on the left, casting a shadow on Saturn. On the right are the moons (left to right) Mimas, Tethys, Janus, and Enceladus.

Titan's diameter is 5150 km (3200 miles), making it the second largest moon in the solar system, after Ganymede.

In proportion to their diameter, Saturn's rings are much flatter than a compact disk.

?

Are Saturn's rings solid?

Earth's environment before life began on our planet. The organic chemical cycles on Titan may be a natural laboratory for understanding some of the steps leading to life.

Astronomers have recently discovered methane clouds and rapidly evolving storms, which suggest that an important feature of Titan's surface is erosion by methane rainfall. They also confirmed previous data that show a bright continent-sized feature on the moon's surface, which could be an icy highland surrounded by ethane seas. All the existing evidence suggests that Titan uniquely resembles Earth with clouds, rain, and seas. If all goes well with the *Cassini-Huygens* mission, we should be able to look under Titan's dense smog and learn not only about this interesting moon but also about how we came to be.

Titan is only slightly larger than Mercury, which has no appreciable atmosphere. Titan, however, has a denser and ten times more massive atmosphere than Earth's. How can this be? Recall that an object can retain an atmosphere only if the escape speed from its surface is greater than about ten times the speed of gaseous molecules there. The escape speed from Titan is somewhat less than from Mercury (Figure 7-12). However, the difference in temperature between the two objects allows Titan to retain an atmosphere. On Mercury's Sun side, we find temperatures up to 450°C, but the temperature of Titan's surface is only about −180°C. Thus the speed of the gas molecules on Titan is less than the escape speed from Mercury, and it has retained its atmosphere.

Planetary Rings

Notice in Figure 9-21 that the rings of Saturn are not visible when they are edge-on to us. This is because they are very thin; they are only a few tens of meters across (**Figure 9-28**). The rings, however, are not solid sheets but are made up of small particles of water ice, or perhaps rocky particles coated with ice. The great amount of empty space between the particles in the rings means that if they were compressed, their thickness would be reduced to less than a meter.

Each of the particles that make up the rings of Saturn revolves around the planet according to Kepler's laws. Thus, particles nearer the planet move faster than those farther out. Each of the particles is, in a sense, a separate satellite of Saturn.

Three distinct bands are visible from Earth, and long before the advent of space flight they were named (from outer to inner) rings A, B, and C. Photographs from space indicate that there are many more rings than these, as is evident in **Figure 9-29**. Several reasons have been given for the existence of the spaces between the rings, and it is now evident that different spaces have different causes. Some explanations are quite complicated, but others are much simpler.

One of the spaces that is easily explained is the one between rings A and B, known as *Cassini's division* after the astronomer G. D. Cassini. To explain this space between rings, we must compare the motion of a particle in the Cassini division to the motion of Mimas, one of Saturn's moons. A particle in Cassini's division would orbit the planet with a period of 11 hours, 17-1/2 minutes. This is just half of the orbital

Saturn

Earth   Saturn

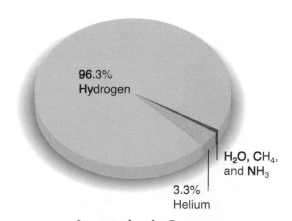

Interior

Saturn Data

Saturn	Value	Compared to Earth
Equatorial diameter*	120,536 km	9.45
Oblateness	0.098	29.24
Mass	5.685×10^{26} kg	95.16
Density	0.69 g/cm³	0.125
Surface gravity*	8.96 m/s²	0.916
Escape velocity	35.5 km/s	3.2
Sidereal rotation period	10.656 hours	0.445
Albedo	0.34	1.12
Tilt of equator to orbital plane	26.73°	1.14
Orbit		
Semimajor axis	1.434×10^9 km	9.58
Eccentricity	0.0565	3.38
Inclination to ecliptic	2.49°	
Sidereal period	29.46 years	29.46
Moons	31	

* Measured where the pressure is one atmosphere, Earth's pressure at sea level.

96.3% Hydrogen

H_2O, CH_4, and NH_3

3.3% Helium

Atmospheric Gases

QUICK FACTS

- The ringed planet. Diameter is 84% of Jupiter's, but density is only half. Primarily hydrogen and helium, like Jupiter. Magnetic field 1/20 as strong as Jupiter's. Emits more energy than it receives—not fully explained. Large family of moons. Rings have detailed features and, seen from Earth, change greatly during its year.

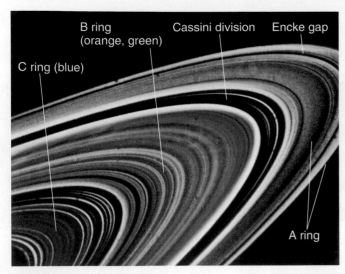

Figure 9-29
This computer-enhanced *Voyager 2* image of Saturn's rings shows clear color variations across the rings. These variations may be the result of differences in chemical composition from one part of the rings to another.

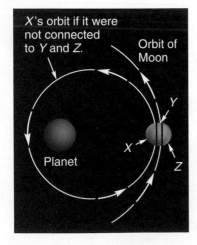

Figure 9-30
The moon is divided into three parts for analysis. Part *X* is closer to the planet than the rest of the moon and would have to move faster if it were a separate satellite. If there were no attractive force toward the rest of the planet, at *X*'s present speed, it would fall inward and begin an eccentric orbit.

period of Mimas. Thus any particle found in the division will be at the same place in its orbit each time Mimas passes near it. Each time this happens, Mimas exerts a slight gravitational tug on the particle, deflecting it slightly from its path. If Mimas did this at random times during the orbiting of the particle, the gravitational effect would be random and the particle's path would not be severely affected. However, as a result of the synchronous relationship between the periods, the tug is repeated regularly, and the overall effect is to pull the particle out of its orbit. This is what happens to particles that drift into Cassini's division, and it is the reason the division exists.

If such synchronous gravitational tugs were the only mechanisms at work determining the structure of Saturn's rings, Cassini's division might be completely clear of particles, but in fact it is not. Many other forces are at work, including gravitational forces from other moons and particles as well as electromagnetic forces from Saturn's magnetic field.

Other features of the rings are explained by the existence of small moons orbiting near the rings. These satellites, called *shepherd moons,* cause some of the rings to keep their shape. Shepherd moons are discussed in an Advancing the Model box in this chapter. Still other features of the rings seem to be similar to spiral wave patterns seen in galaxies. Various hypotheses have been used to explain the various features of the rings, but no comprehensive theory has yet been proposed.

The Origin of Rings

The origin of the rings of Saturn is not well understood. The most likely scenario is that an icy moon (or moons) once orbited near Saturn and was shattered in a collision with a passing asteroid or comet. Another possibility is that an object from the outer solar system came too close to Saturn and was torn apart by the planet's gravity. The particles of the shattered moon eventually dispersed in a ring around the planet. This event may seem unlikely, but in fact we know that small objects in orbit around a planet do arrange themselves in a flat ring. Whether or not this is the true origin of the rings, we do understand why the rings have not formed into moons. To see why, we must review the effect of tides.

Recall how the gravitational force between the Earth and the Moon causes tides on both the Earth and the Moon. However, it is not simply the existence of a gravitational force that causes the tides, but the fact that there is a *difference* in the amount of gravitational force exerted on masses that are at different places on the Earth (or the Moon). We considered tides in Chapter 3, but we now look at them in another way, focusing here on Kepler's laws.

According to Kepler's third law—expanded to apply to planets and moons—if two objects are orbiting a planet, the object closer to the planet must move faster in its orbit if it is to stay in that orbit. **Figure 9-30** illustrates a moon in a circular orbit around a planet, but in the figure we have divided the moon into three parts, each at a different distance from the planet. Kepler's law tells us that if the three parts of our moon were not connected, part *X* would have to move faster. If it were to move at the same speed as part *Y,* it would fall inward toward the planet. Likewise, part *Z* should travel more slowly than *Y,* and if it were to travel at the same speed as *Y,* it would fly out of *Y*'s orbit. This, in fact, is just another way to look at the cause of tides, for each part of the moon tends to do what we have just described.

Why doesn't a moon fly apart because of this tidal effect? The answer is that the moon has its own forces holding it together. A rock on its surface is pulled by the force of gravity toward the center of the moon. Artificial satellites, on the other hand, are held together by the strength of the materials of which they are made.

It is important to see that it is the *relative difference* in distance from the planet that causes the effect. If a moon is far from its planet, the outside edge may be only a fraction of a percent farther from the planet than the center of the moon is. If the same moon is located close to its planet, the difference in distance between the moon's outer edge and its center will be a greater fraction of the moon's distance from the planet. This causes the effect of tides to be greater if a moon is nearer a planet. There is a critical distance, called the ***Roche limit,*** inside which the tidal force on a moon will be greater than the moon's own gravitational force. Inside this distance, a large moon will be unable to hold itself together.

Now we come to Saturn. Until 1859 it was thought that Saturn's rings might be solid sheets of material, but calculations of Saturn's Roche limit showed that they are inside that limit and therefore the gravitational force between particles within the rings is less than the tidal force that tends to pull them apart. That they are made up of separate particles was experimentally confirmed in 1895 by means of the Doppler effect, which showed that different parts of the rings have different speeds. Because the rings are inside the Roche limit, no moons have formed from the particles.

The particles of the rings vary in size from small grains to irregularly shaped pieces more than a meter across (**Figure 9-31**). It is interesting to note that if all of the material of the rings of Saturn were formed into a single moon, the moon would be about the mass of Janus (one of the smallest of Saturn's moons) and only 1/20,000 the mass of the Earth's Moon.

Roche limit The minimum radius at which a satellite (held together by gravitational forces) may orbit without being broken apart by tidal forces.

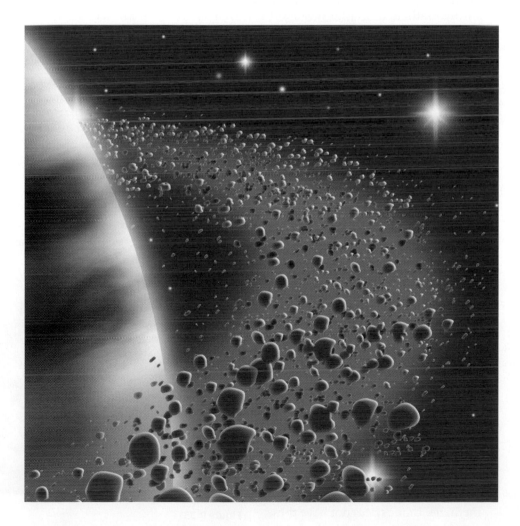

Figure 9-31
Saturn's rings consist of thousands of fragments, probably of water ice covered with rocky dust. Most of these chunks are the size of small stones and pebbles, but some may be as big as a house.

9-3 Uranus

In Greek myth, Uranus (YOOR-uh-nuss) was the earliest supreme god, the father of Cronus (Saturn). The planet was unknown to the ancients even though they undoubtedly saw it, for it is barely visible to the naked eye under perfect viewing conditions. It was plotted on star charts made by telescope as early as 1690. However, in anything but large, high-resolution telescopes, Uranus appears only as a speck of light rather than a disk. This, coupled with the fact that it moves very slowly, caused it to go unnoticed as a planet until 1781, when the English astronomer William Herschel (1738–1822) noticed that this particular "star" seemed to have a size. He thought at first that it was a comet, but after calculations showed that this object's orbit was nearly circular, Herschel realized that he had discovered a new planet.

The size of Uranus is difficult to determine from the Earth because its disk is so indistinct in a telescope (**Figure 9-32**). Recall that the size of an astronomical object can be determined from its distance and angular size. If the image is indistinct, the angular size cannot be determined accurately, of course. The first reliable value for Uranus's diameter came from a telescope aboard a high-altitude balloon operated by Princeton University. The balloon lifted this telescope to an altitude of 25 kilometers, above most of the distortion caused by the Earth's atmosphere.

An improved determination of the diameter of Uranus was made in 1977 when Uranus was observed passing in front of a star. Such an event, when a moon or planet eclipses a star, is called an **occultation** and provides us with a valuable measuring method. Since the speed of Uranus in its orbit was known, all that had to be measured during the occultation in order to calculate Uranus's diameter was the time during which the star was occulted.

The diameter of Uranus is 51,000 kilometers, about half the diameter of Saturn and one-third that of Jupiter. Long before *Voyager 2* passed Uranus in 1986, the mass of the planet had been calculated by applying Kepler's third law to its five major satellites known at the time. When its density is then calculated, we find that Uranus has a density of 1.27 times that of water; greater than Saturn but slightly less than Jupiter. Taking into account the surface gravity on Uranus (which is less than on Saturn or Jupiter), we can calculate what Uranus's density would be if the planet were made up entirely of hydrogen and helium. Astronomers once thought that a dense rocky core about the size of the Earth was necessary to produce Uranus's measured density, but recently researchers have subjected material that matches the composition of Uranus to a pressure two million times that of Earth's atmosphere and found that it becomes dense enough to account for the planet's density. Thus it may be that Uranus has a very small rocky core or no core at all.

Figure 9-33 shows a theoretical model of Uranus's interior with a small rocky core.

The atmosphere of Uranus is similar to those of Jupiter and Saturn, consisting primarily of hydrogen and helium, with some methane. **Figure 9-34** is a *Voyager* photo of Uranus. The reason for the planet's blue coloring is that the methane absorbs red light and reflects the rest of the spectrum. The atmospheres of Jupiter and Saturn also contain methane, but they have high cloud layers that reflect sunlight before it reaches the higher concentrations of methane. Uranus does not have these cloud layers.

Herschel proposed that the new planet be named "Georgium Sidus" (the Georgian star) after King George III, but the name didn't take. Johann Bode (of Bode's law fame) suggested "Uranus," after the mythical Greek god of the sky.

Seeing is in some respect an art, which must be learnt.
William Herschel

occultation The passing of one astronomical object in front of another.

51,000 km is about 32,000 mi. This is four times Earth's diameter.

Figure 9-32
Uranus as seen from Earth, with three of its moons.

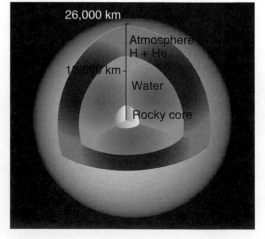
26,000 km
13,000 km
Atmosphere H + He
Water
Rocky core

Figure 9-33
A model of Uranus's interior.

ADVANCING THE MODEL

William Herschel, Musician/Astronomer

William Herschel, who has been called the greatest observational astronomer ever and the Father of Stellar Astronomy, was a musician. At least that was his training and how he earned his living for a great part of his life. His father Isaac was an oboist in the band of the Hanoverian Foot Guards and raised all his children as musicians. (Hanover was a province in the former state of Prussia in Germany.) Friederich Wilhelm Herschel, who adopted the name William when he moved to England, learned to play the violin as soon as he could hold the small one his father obtained for him. When he was 14, he joined his father and older brother in the military band. The semimilitary life was not for him, however, so he and his brother requested a discharge, and just after William turned 19, they moved to England together. William served for a brief time as instructor to another military band, but spent most of his young adult life teaching, performing, and composing music.

One of the interests that William inherited from his father was the study of astronomy. He began to build his own telescopes, and after teaching as many as eight music lessons a day, he would spend the evenings observing the heavens. By 1778 he had built an excellent reflecting telescope about six inches in diameter. This was no easy task, for construction of the mirror involved metal casting. Herschel's life changed after March 13, 1781. On that night he saw a hazy patch in the sky and upon observing it over a few nights, he saw that it moved. What he thought was a comet turned out to be the first planet to be discovered in recorded history.

William Herschel was not immediately accepted by the scientific community. His telescopes were of such quality that he was able to see things in the heavens that other astronomers couldn't, and many did not believe his claims about the details of what he was seeing. There seems to have been some professional jealousy toward this musician who, in his writings on astronomy, did not use the jargon accepted by the profession. However, Herschel's skills as a telescope maker and as an observer won out, and his discovery of Uranus could not be denied. He was soon recognized and celebrated by scientific societies. Within a year after he discovered Uranus, he received a pension from King George III so that he could devote full time to astronomy, his former hobby. In 1787 he discovered two moons of Uranus and soon after found two smaller moons of Saturn.

While he was still devoting most of his time to music, Herschel had brought his sister Caroline to live with him. (She served as a vocalist at some of his performances.) Now that he was an astronomer, she became a very important assistant and an accomplished astronomer herself. Together they made the first thorough study of stars and nebulae (faint, hazy objects not understood at the time). One important discovery they made was the existence of double stars—stars in orbit about one another. After Herschel's death, the orbits of these stars were analyzed in enough detail to determine that the stars obey Newton's laws of motion and gravitation, an indication that these laws are valid not only here in our solar system but even among the distant stars.

Herschel continued serious observing until he was nearly 70. After his death at the age of 84, Caroline compiled a large catalog of his observations and received many honors for her own astronomical knowledge. She discovered eight comets from her own observations and was one of the first two women elected to honorary membership in the Royal Astronomical Society.

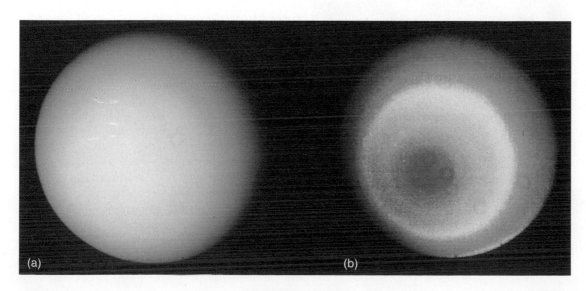

(a)   (b)

Figure 9-34
(a) Images of Uranus taken by *Voyager 2* under ordinary light show a featureless planet. (b) When color is enhanced by computer processing techniques, Uranus is seen to have zonal flow patterns in its atmosphere. One of Uranus's poles is at the center of the red spot.

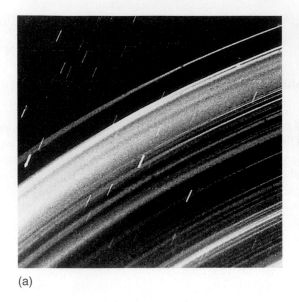
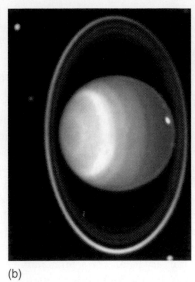

(a)   (b)

The purpose of the Uranus observations during the occultation of 1977 was to determine the planet's size more accurately. However, a surprise was in store for the observers. A short while before Uranus was expected to occult the star, the light from the star blinked five times. Then, after the occultation, similarly timed blinks were observed, but in the opposite order. From this, astronomers concluded that Uranus has at least five rings that are invisible from the Earth but are dense enough to obscure the light from a star behind them. By comparing the results obtained by several observatories, details of the rings' orientations and sizes could be determined. Use of the same method during an occultation in 1978 revealed a total of nine rings. **Figure 9-35a**, taken by *Voyager,* indicates the complexity of the ring system. A more recent, false-color photo from the *Hubble Space Telescope* (Figure 9-35b) shows the complete ring system, along with several moons.

The rings of Uranus cannot be seen from the Earth because they reflect only about 5% of the sunlight that hits them. The rings must be made of material as dark as soot. By comparison, Saturn's rings reflect 80% of the light that hits them.

Figure 9-36
Uranus's axis is tilted 98°, which tells us that the planet has retrograde rotation.

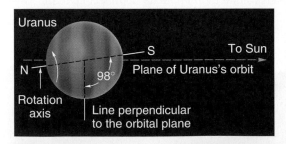

Figure 9-37
Uranus's axis is tilted so that its poles point nearly directly at the Sun at times.

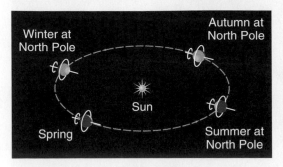

Uranus's Orientation and Motion

Recall that Venus is unusual in that it rotates in a backward direction: clockwise as observed from north of the Sun. Except for Venus, every other planet we have studied so far rotates normally: counterclockwise as observed from the north. In addition, all planets we have studied, including Venus, have their equatorial plane tilted less than 30 degrees to their plane of revolution around the Sun. Here is where Uranus is unique. Its equatorial plane is tilted nearly 90 degrees to its plane of revolution. (The tilt of Uranus's orbit is listed as 98°, indicating its retrograde rotation. See **Figure 9-36**.) This means that during its 84-year orbit, its north pole at one time points almost directly to the Sun and at another time faces nearly away from the Sun. **Figure 9-37** illustrates this. Remember that the tilt of a planet's axis is what causes the planet to have seasons. One might think that such a large tilt of Uranus's axis would have a great effect on its weather. However, data from *Voyager* reveal that Uranus has a fairly uniform temperature over its entire surface (about −200°C), indicating that the atmosphere is continually stirred up, with winds moving from one hemisphere to another.

Features of Uranus's surface are not visible from the Earth, so until *Voyager* approached Uranus, there was uncertainty about its period of rotation. *Voyager* photographs told us that, as on Jupiter and Saturn, there are various bands of clouds on Uranus and that these bands rotate with different periods—from 16 hours at the equator to nearly 28 hours at the poles. Uranus has about the same chemical makeup as Jupiter, about 83% hydrogen with most of the remainder helium.

Uranus's magnetic field is comparable to Saturn's. Probably it originates in electric currents within the planet's layer of water. The magnetic field is unusual in that its axis is tilted 59 degrees with respect to the planet's rotation axis. No other planet has such a large angle between the two axes, although Neptune's—at 47 degrees—is close.

Five moons of Uranus (**Figure 9-38**) were known before *Voyager*, and now we know of 27 (as of November 2003). All are low-density, icy worlds. The innermost, Miranda, is perhaps the strangest looking

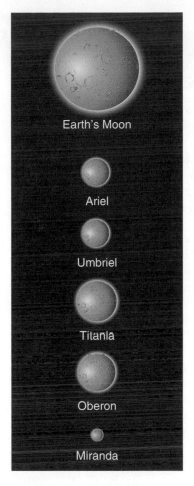

Figure 9-38
Some of Uranus's moons compared in size to Earth's.

Earth's Moon

Ariel

Umbriel

Titania

Oberon

Miranda

On Earth, there is little air exchange between the Northern and Southern Hemispheres.

The corresponding angle between the rotation axis and the magnetic dipole axis for Earth is currently 11.5°.

? Do all planets rotate so that the equator lies in the plane of revolution?

ADVANCING THE MODEL

Shepherd Moons

The 1977 occultation that resulted in the discovery of Uranus's rings showed that they are very narrow and that their edges are well defined. Soon after this discovery, two astronomers—Peter Goldreich and Scott Tremaine—proposed a mechanism to explain why a ring might stay in a narrow band rather than spread out over space.

They hypothesized that there are a pair of moons, one orbiting just inside and another just outside the narrow ring (**Figure B9-1**). Kepler's third law tells us that the moon closer to the planet moves faster than the particles of the ring. As that moon catches and passes the particles on the inside portion of the ring, gravitational pull from it increases the particles' energy just a little, causing them to move outward in their orbit. (This energy comes from the moon, and therefore the moon loses energy and moves inward a bit. However, the moon is so much more massive than the particles that the change in its path is almost insignificant.)

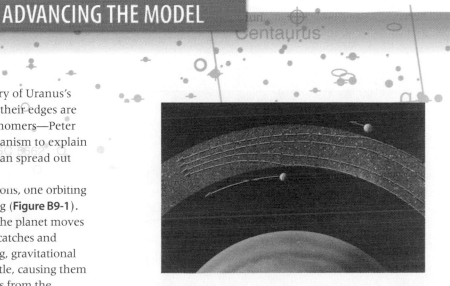

Figure B9-1
As the inner moon passes the inner portion of the ring, it forces ring particles outward. The outer moon forces particles inward. The effect is greatly exaggerated here.

(Continued)

ADVANCING THE MODEL

Shepherd Moons, *Continued*

In an opposite manner, as particles on the outside edge of the ring pass the more distant moon, they lose energy and move inward. The effect on each particle is very small, much less than is indicated by the figure. Nevertheless, Goldreich and Tremaine showed that the effect is significant enough that as the moons continue to orbit near the rings, they force the particles together; in other words, the moons act as shepherds for the flock of particles in the ring.

The scientific community awaited *Voyager*'s passing by the rings of Uranus to see if it found such moons. The answer came earlier than expected. *Voyager 2* at that time had not yet passed Saturn. When it did, it photographed Saturn's rings, and on each side of Saturn's narrow F ring was a moon (**Figure B9-2**). Calculations showed that the two moons near the F ring would indeed perform the function of shepherds for that ring.

In January 1986, *Voyager 2* passed Uranus and returned a remarkable photograph (**Figure B9-3**) of the epsilon ring of Uranus along with two shepherds. Other rings around Uranus exist without the aid of shepherd moons, so although Goldreich and Tremaine were right in their prediction of the action of shepherd moons, this mechanism is only a part of what is happening in planetary ring systems.

Inside the Roche Limit?

The shepherd moons of Saturn's F ring and those orbiting Uranus are within the Roche limit of each planet. How can they exist there without being pulled apart by gravitational forces? There are several answers to this question. First, it is an oversimplification to refer to a single Roche limit. For a smaller moon, the limit is closer to the planet than it is for a larger moon. The shepherds are small moons. Second, the position of the limit depends upon the density of the moon. But most important, the Roche limit concerns only moons that are held together by gravitational forces. A single large rock can exist inside the Roche limit, for it is held together by cohesive forces within the material. Each shepherd moon may well be a single large rock rather than being made of smaller particles held together by gravity.

Chaotic Motions

In the mid-1990s, observations of the two shepherd moons in Figure B9-2 found them far from where they should have been based on *Voyager* data in the early 1980s. It seems that the best explanation for this discrepancy is based on chaotic interactions between the moons and the rings. The movements of the moons are not predictable because a very small difference in starting conditions can result in a big difference in later positions. Pandora pushes the particles in the ring inward toward Saturn, while Prometheus pushes them outward; by the principle of action and reaction (Newton's third law), the two moons are slowly pushed away from the ring. (The influence of Saturn's A ring on the moons is even more important in pushing them outward.) This is the first observation of chaotic orbital motions in our solar system. (However, it has been known since 1984 that Saturn's moon Hyperion exhibits a chaotic rotational motion.)

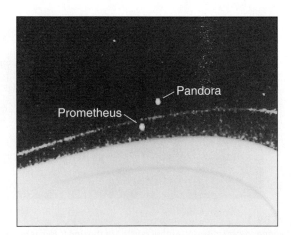

Figure B9-2
Voyager photographed these shepherd moons near the F ring of Saturn.

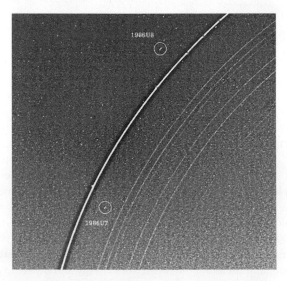

Figure B9-3
Moons 1986U7 and 1986U8 were photographed shepherding a ring of Uranus.

Uranus

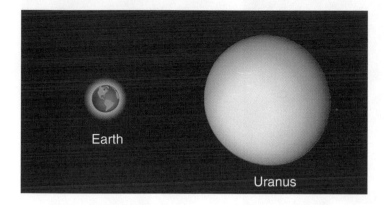

Earth

Uranus

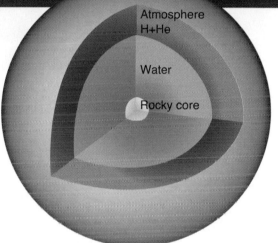

Atmosphere
H+He

Water

Rocky core

Interior

Uranus Data

Uranus	Value	Compared to Earth
Equatorial diameter	51,118 km	4.0
Oblateness	0.023	6.84
Mass	8.683×10^{25} kg	14.5
Density	1.27 g/cm³	0.23
Surface gravity*	8.69 m/s²	0.89
Escape velocity	21.3 km/s	1.9
Sidereal rotation period	−17.24 hours	0.72
Albedo	0.30	0.98
Tilt of equator to orbital plane	97.77°	4.17
Orbit		
Semimajor axis	2.872×10^{9} km	19.20
Eccentricity	0.0457	2.7
Inclination to ecliptic	0.77°	
Sidereal period	84.01 years	84.01
Moons	27	

*Measured where the pressure is one atmosphere, Earth's pressure at sea level.

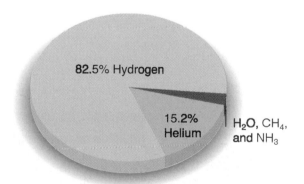

82.5% Hydrogen

15.2% Helium

H_2O, CH_4, and NH_3

Atmospheric Gases

QUICK FACTS

- Four times Earth's diameter, but only one-third of Jupiter's. Composition like other Jovian planets. Almost featureless atmosphere. Far greater tilt of axis than any other planet. Retrograde rotation. Uniform temperature. Many small moons, no large one.

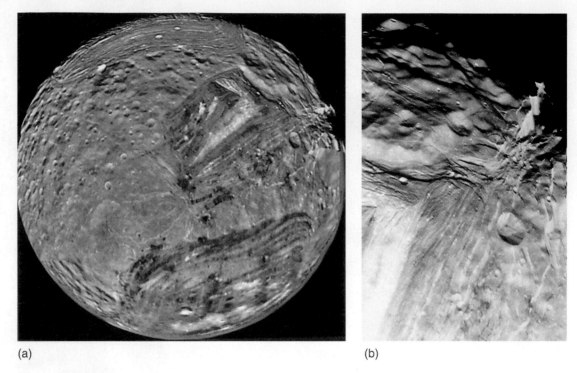

(a)   (b)

Figure 9-39
Miranda. It is hypothesized that an impact once tore Miranda apart. (a) The V-shaped feature is called the Chevron. (b) Detail of the upper right portion of part (a) shows a cliff that is 5 kilometers high.

object in the solar system. **Figure 9-39** shows its varied terrain. You may wonder who has been farming it. Because of its racetrack-like grooves, one feature has been named Circus Maximus after the arena of ancient Rome. One possible explanation for the strange features of Miranda is that it experienced a tremendous collision with another object and was broken into pieces that later fell back together.

Two of Uranus's moons are shepherd moons (see the Advancing the Model box), keeping one of the rings in formation. However, material in the rings is very sparse; so much so that there is more material in Cassini's division of Saturn's system than in all of the rings of Uranus! Nine rings had been observed (by occultations) from the Earth; only one more was found by *Voyager*. The 10 rings, however, contain hundreds of smaller ringlets, as seen in Figure 9-35.

9-4 Neptune

Uranus and Neptune are similar in many ways. The first similarity is size; they are very nearly the same. Uranus appears as only a small, indistinct disk in a telescope, and when we consider that Neptune is twice as far away from Earth, it is no surprise that Neptune appears even more indistinct as seen from the Earth, even in the largest telescope. Most of what we know about Neptune was learned when *Voyager 2* passed it in August, 1989, and sent back more than 9000 images, as well as other data.

The composition of Neptune matches that of Uranus, and its blue color (**Figure 9-40**) has the same cause—methane in its upper atmosphere. However, light penetrates deeper into Neptune's atmosphere, resulting in a deeper blue than Uranus. The other major difference in appearance is that while Uranus is almost featureless, Neptune exhibits weather patterns. It has parallel bands around it, and its *Dark Spot* (**Figure 9-41**) is similar in appearance to Jupiter's Red Spot. Like the Red Spot, it is in the planet's southern hemisphere and rotates counterclockwise. The Great Dark Spot is about the same size relative to Neptune as the Great Red Spot is relative to Jupiter.

Neptune radiates more internal energy than Uranus, although the cause of the energy is not understood. This energy, however, is what drives the weather on Neptune, stirring the liquids and gases of the planet so that winds reach speeds of 700 miles/hour.

The wispy white clouds seen on Neptune are thought to be crystals of methane. **Figure 9-42** shows that the clouds are higher than the surrounding atmosphere, for they cast shadows. One of the clouds (Figure 9-41) seems to be permanently associated with the Great Dark Spot and is probably caused by methane rising with other gases in the spot until it freezes, becoming visible as a white cloud.

The rotation rate of Neptune was uncertain until *Voyager* measured it. Like the other gas planets, Neptune exhibits differential rotation: near the equator it rotates in 18 hours, but near the poles its rotation period is only about 12 hours. However, these great differences are confined to the upper few percent of the atmosphere. Neptune's magnetic field rotates with a period of 16 hours, 7 minutes. Since the magnetic field is thought to result from electric currents in liquid layers deeper within the planet, this is taken as Neptune's basic rotation rate.

The temperature of Neptune's surface is remarkably uniform at the poles and the equator, −216°C. This is another similarity to Uranus and was a surprise to astronomers, for Neptune's axis is tilted less than

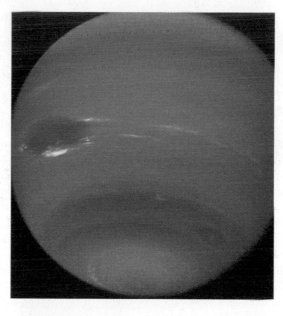

Figure 9-40
Neptune was photographed by *Voyager 2* in August, 1989. The photographs revealed the previously undiscovered cloud features, including the Great Dark Spot at left center.

Neptune's dark spot is approximately Earth-size.

Neptune receives only 1/900 as much solar energy per square meter as Earth does.

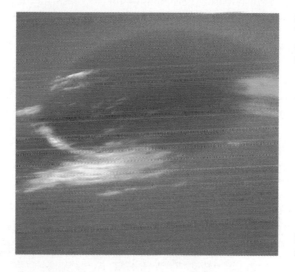

Figure 9-41
The Great Dark Spot of Neptune. The narrow clouds are made up of crystals of methane.

Some features on Neptune have periods as long as 18 hours.

Figure 9-42
The methane clouds are at a higher altitude than the surrounding atmosphere, as indicated by the shadows cast on the other side from the Sun.

ADVANCING THE MODEL

The Discovery of Neptune

The discovery of Neptune is especially interesting because it reveals some very human aspects of science.

After Uranus was discovered in 1781, astronomers examined old star charts and found that Uranus had been plotted on charts as far back as 1690. Those who had plotted it had mistaken it for just another star, but the positions they marked allowed later astronomers to calculate the orbit of the planet. As time went on, however, it was clear that Uranus was not following the orbit predicted by Newton's laws, even with the gravitational effects of Jupiter and Saturn taken into account. Most astronomers thought that an unknown planet was disturbing Uranus's orbit, but others thought that perhaps the law of gravity acts differently at great distances from the Sun.

John C. Adams, a 26-year-old British mathematician, took up the problem of calculating the position of the hypothesized planet. In late September 1845, after two years of calculations, he took his results to Sir George Airy, the British Astronomer Royal, telling him that the planet would be found near a certain position in the constellation Aquarius. Adams first had problems contacting Airy. He must have felt like a student trying to find his adviser, for the first time he tried to deliver his results in person Airy was out of the country. Another time, Airy was out of his office. A third time, he was at dinner and his butler would not let him be disturbed. Adams left messages but they never reached Airy.

Airy was one of the astronomers who doubted the existence of the new planet, thinking that Newton's laws were inexact at the distance of Uranus. As a result, when Airy finally got one of Adams's messages, he refused to use valuable time on the large telescope for what he thought would be a fruitless search.

Adams's youth and inexperience also seem to have contributed to Airy's skepticism.

During the same period, the French mathematician Urbain Le Verrier was making similar calculations of Uranus's orbit. In November, 1845, Le Verrier published the first part of his calculations in a scientific journal. Then in June, 1846, he published his final results. Although neither knew of the other's work, Adams's and Le Verrier's predicted positions for the planet were within one degree of one another.

Similar to Adams's experience with his own countrymen, Le Verrier was unable to get French astronomers to search for the planet. Finally, on September 23, 1846—one year after Adams had first requested a search—Le Verrier asked Johann Galle in Berlin to look for the planet. Galle quickly convinced the director of the observatory where he worked of the worthiness of the project, and the new planet was discovered on the very first night of searching.

Credit for scientific discovery usually goes to the first person who documents his or her results. (Today such documentation normally takes the form of publication in a scientific journal.) There is little doubt, however, that if Airy had used Adams's calculations to find Neptune, the prediction would be credited to Adams. For years, controversy swirled about the priority of prediction. In this case, not only was the reputation of the two mathematicians involved, but also national prestige. (Similar disagreements have arisen in this century concerning scientific discoveries made in the United States and the former Soviet Union.) Today the two men are given joint credit, and we find the story of the discovery interesting enough to retell it in nearly every astronomy text.

Figure 9-43

Neptune's interior is probably similar to that of Uranus (see Figure 9-33), although Neptune may have a larger core.

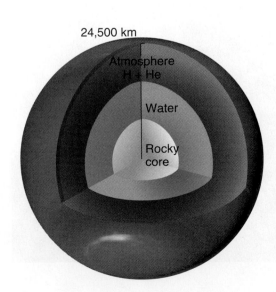

24,500 km

Atmosphere
H + He

Water

Rocky
core

Jupiter's Galilean moons, our Moon, and Saturn's Titan are larger than Triton. All seven of these moons are larger than Pluto.

30 degrees to its orbit, rather than the 98-degree tilt of Uranus. The reasons for many of the similarities and differences between these two planets remain unknown.

Neptune is presumed to have an interior much like that of Uranus, but its central rocky core is probably larger, thus resulting in its greater average density (see table on Neptune data page). **Figure 9-43** illustrates one model of Neptune's interior.

Neptune's Moons and Rings

Before *Voyager*'s visit, Neptune was known to have two moons, both unusual. Triton, its largest moon (and the seventh largest in the solar system), revolves around the planet in a clockwise direction as seen from the north.

Neptune

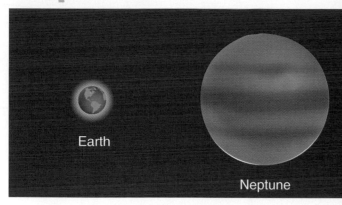

Earth

Neptune

Neptune Data

Neptune	Value	Compared to Earth
Equatorial diameter	49,528 km	3.9
Oblateness	0.017	5.1
Mass	1.02×10^{26} kg	17.1
Density	1.64 g/cm³	0.30
Surface gravity*	11.0	1.12
Escape velocity	23.5 km/s	2.1
Sidereal rotation period	16.11 hours	0.67
Albedo	0.29	0.95
Tilt of equator to orbital plane	28.3°	1.2
Orbit		
Semimajor axis	4.495×10^{9} km	30.05
Eccentricity	0.0113	0.68
Inclination to ecliptic	1.77°	
Sidereal period	164.79 years	164.79
Moons	11	

*Measured where the pressure is one atmosphere, Earth's pressure at sea level.

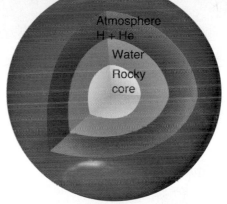

Interior

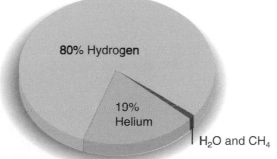

Atmospheric Gases

QUICK FACTS

- Blue surface due to methane. Jupiter-like atmospheric features, including Great Dark Spot. Composition is thought to be similar to other Jovian planets. Unexplained internal heat source. Extreme differential rotation. Largest moon, Triton, has retrograde revolution. The other major moon, Nereid, has the most eccentric orbit of any moon in the solar system. Lumpy rings.

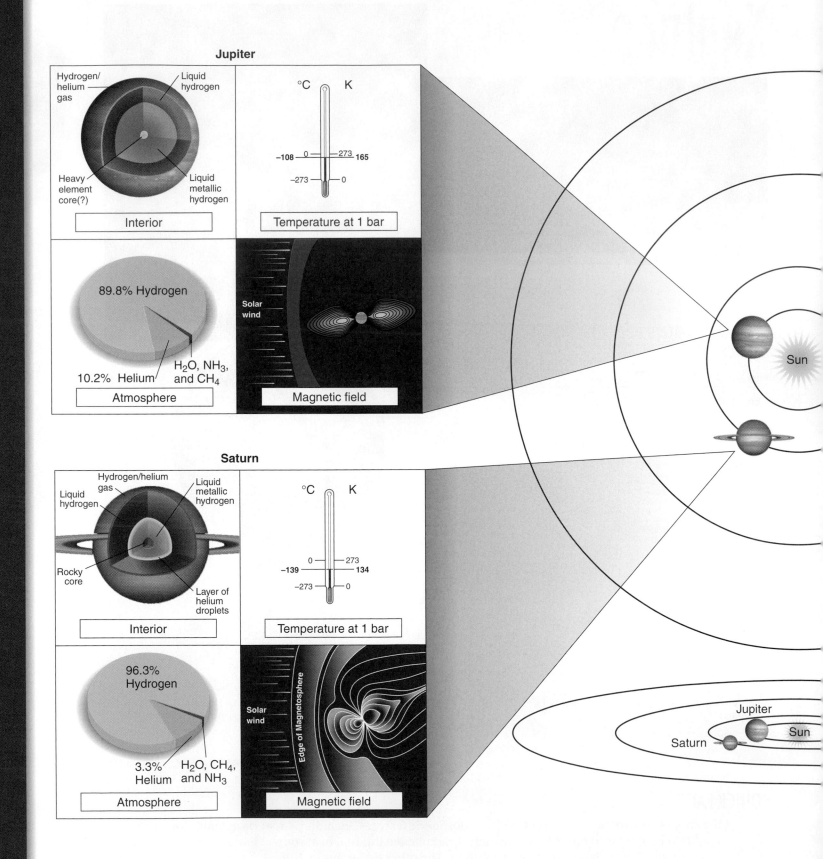

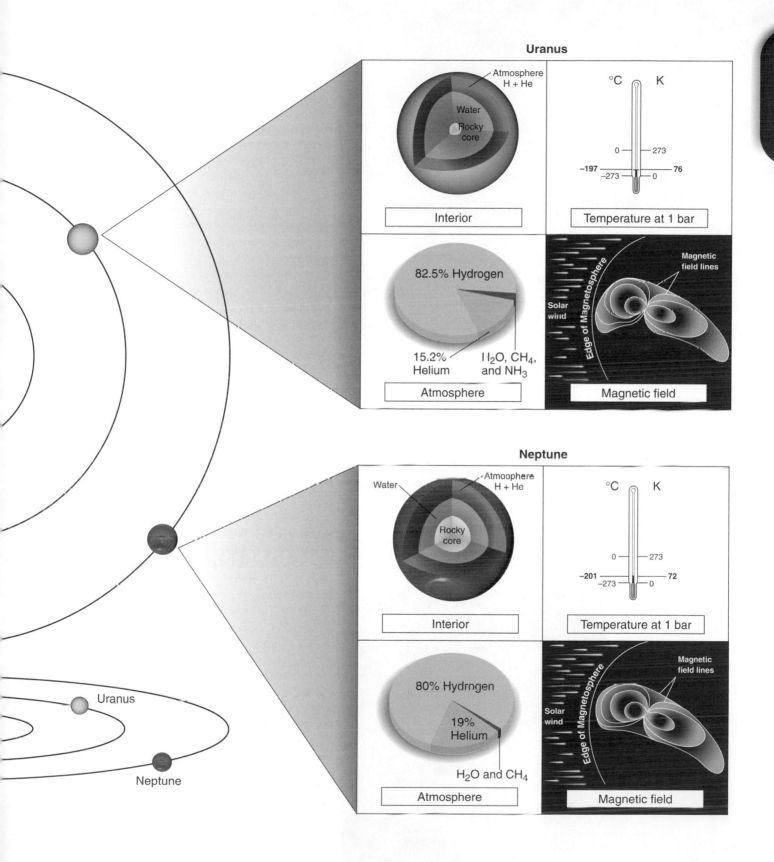

Uranus

Atmosphere H + He

Water

Rocky core

Interior

°C   K

0 — 273

−197 76
−273 0

Temperature at 1 bar

82.5% Hydrogen

15.2% Helium   H₂O, CH₄, and NH₃

Atmosphere

Solar wind

Edge of Magnetosphere

Magnetic field lines

Magnetic field

Neptune

Water

Atmosphere H + He

Rocky core

Interior

°C   K

0 — 273

−201 72
−273 0

Temperature at 1 bar

80% Hydrogen

19% Helium

H₂O and CH₄

Atmosphere

Solar wind

Edge of Magnetosphere

Magnetic field lines

Magnetic field

Uranus

Neptune

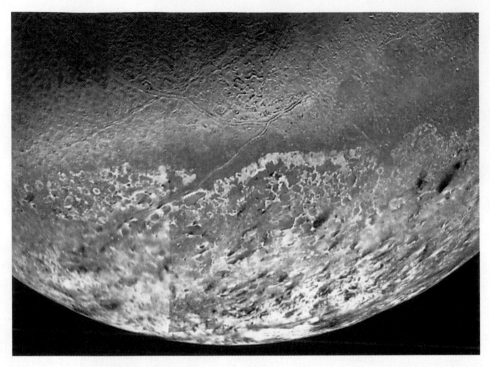

Figure 9-44
The southern part of Triton has been exposed to sunlight for some 30 years, and erupting volcanoes (not obvious here) were photographed by *Voyager*.

This makes Triton the third object, after Earth and Io, that we know to be volcanically active.

Figure 9-45
Neptune's rings are seen clearly in this photo. Note their "lumpiness."

Data on Neptune's moons can be found in Appendix D.

The *Voyager* spacecraft were originally designed to operate for only five years. *Voyager 2*, launched in 1979, is now expected to communicate with us until 2010.

Although some smaller moons of the Jovian planets move in retrograde orbits, Triton is the only major moon to do so. The other moon known prior to *Voyager*, Nereid, orbits in the "right" direction, but it has the most eccentric orbit of any known moon in the solar system, eccentric enough that its distance from the planet is five times greater at some times than at others. Its 360-day orbital period is the longest of any moon of its size (340-kilometer diameter) in the solar system.

Voyager showed us that the surface of Triton is as unusual as its orbit. **Figure 9-44** shows the region surrounding its south pole, which is near the bottom of the photograph in an area of very irregular terrain. Other parts of the moon are much smoother, as seen in the top part of the photograph. The surface consists primarily of water ice, with some nitrogen and methane frost. Three observations lead us to believe that the surface is very young: (1) It is very light in color. When methane is exposed to sunlight for a long time, darker organic compounds are formed. (2) There are very few craters, indicating that they have been obliterated by surface activity. (3) Active volcanoes have been seen on the moon.

A current hypothesis explains many of the unusual properties of Triton and Nereid. The mass of Triton was determined by its gravitational influence on *Voyager,* and we find that its density (about 2 grams/centimeters^3) is about the same as the density of Pluto. It could be that both Triton and Nereid are objects that were captured by Neptune after the initial formation of the solar system. If so, Triton probably once had an eccentric orbit like Nereid. If its orbit came very close to Neptune, tidal forces would eventually cause it to settle into a circular orbit such as the one in which we find it. In addition, the tidal forces would heat the interior of Triton by continually distorting its shape. This heat would result in the volcanic action that has covered old craters. Much more remains to be learned about Neptune's unusual moons before we can be confident of this hypothesis.

Voyager discovered six more moons orbiting Neptune, one of them larger than Nereid. It had not been seen from Earth because it is very close to the planet. The other five moons are very small.

Stellar occultations observed in 1984 had revealed that Neptune has rings, but the rings were thought to be incomplete, not circling the entire planet. *Voyager* photographs revealed that they are indeed complete rings (**Figure 9-45**), but that they are "lumpy," perhaps as a result of undiscovered moons orbiting with them. Thus we find that each of the Jovian planets has a ring system.

Conclusion

The Jovian planets certainly are different from the worlds of the inner solar system. Compared to the terrestrial planets, the Jovians are larger, more massive, more fluid (rather than solid), less dense, and rotate faster. Their many moons are vastly different from one another; the differences are caused at least partly by their distances from the Sun and from their planet.

As we learn more about the Jovian planets, we are even more amazed at these beautiful giants. We see great similarities among them, but we also see great differences. Jupiter, Saturn, Uranus, and Neptune are similar enough that we classify them into a single group, but each is an individual world with its own history.

One more planet is yet to be discussed: Pluto. This planet is such an enigma that it has been placed in the next chapter, along with various small objects that are part of our solar system.

There are many mysteries remaining in the solar system. These mysteries are what make its study so interesting. If we were to run out of mysteries, science would cease to exist, and we would have lost one of the pursuits that makes life so exciting.

Study Guide

1. The most abundant constituent of Jupiter's atmosphere is
   A. helium.
   B. carbon dioxide.
   C. ammonia.
   D. hydrogen.
   E. sulfuric acid.

2. The Great Red Spot is
   A. a continent.
   B. a storm.
   C. an optical illusion.
   D. a shadow of one of Jupiter's moons on its surface.
   E. a mountain protruding above Jupiter's atmosphere.

3. The chemical composition of Jupiter is most similar to which of the following?
   A. The Earth.
   B. The Sun.
   C. Mars.
   D. The Moon.
   E. Venus.

4. A planet is said to have differential rotation if
   A. it changes speed when orbiting the Sun.
   B. it changes its rate of spin from time to time.
   C. different parts of it have different rotation periods.
   D. its rotation rate is far different from the planets closest to it.
   E. most of its moons revolve around it in a different direction than it rotates.

5. Jupiter's weather patterns should be simpler than Earth's because
   A. Jupiter is so large.
   B. Jupiter is so massive.
   C. Jupiter is made up primarily of hydrogen.
   D. Jupiter has a small ring.
   E. Jupiter's core is a small fraction of the total planet.

6. The bands around Jupiter are due to the fact that the planet
   A. rotates differentially.
   B. has a fairly low rotation rate.
   C. has so many moons.
   D. is made up primarily of hydrogen.
   E. has no solid core.

7. Most of the extra energy that Jupiter emits is thought to be energy
   A. left over from its formation.
   B. that results from the fact that it is still shrinking.
   C. that results from radioactive materials within it.
   D. that results from the greenhouse effect.
   E. [All of the above are considered about equally responsible.]

8. The energy for the volcanic activity on Io results from
   A. radioactive substances within it.
   B. chemical reactions.
   C. tidal action.
   D. energy radiated from Jupiter and the Sun.
   E. [Both A and B above.]

9. Saturn's rings are
   A. composed of ice particles.
   B. in the plane of the planet's equator.
   C. within the planet's Roche limit.
   D. [All of the above.]

10. Titan is able to retain an atmosphere even though it is just slightly larger than Mercury because it is
   A. close to Saturn.
   B. very dense.
   C. far from Saturn.
   D. far from the Sun.
   E. [Three of the above.]

11. Saturn has rings instead of additional moons because of
    A. magnetic fields.
    B. tidal forces.
    C. electrical forces.
    D. radiation.
    E. the proximity of Jupiter.

12. Saturn is unique in that it
    A. is the only planet with rings.
    B. is the brightest planet in our sky.
    C. is the only Jovian planet with more than four moons.
    D. has the least average density of the planets.
    E. [Three of the above.]

13. Liquid metallic hydrogen
    A. can exist only at great pressures.
    B. exists both in Jupiter and Saturn.
    C. is a conductor of electricity.
    D. [Two of the above.]
    E. [All of the above.]

14. Uranus is unique in that
    A. it has the least average density of the planets.
    B. its poles point nearly toward the Sun at some times.
    C. it is the most distant of the Jovian planets.
    D. its north pole is south of its orbital plane.
    E. it has a more eccentric orbit than any other planet.

15. The fact that Uranus has a fairly uniform temperature over its surface indicates that
    A. it has a great angle between its equatorial plane and its orbital plane.
    B. it has no solid core.
    C. it has a large solid core.
    D. its winds move from one hemisphere to the other.
    E. [Both A and B above.]

16. The oblateness of Jupiter and Saturn is caused primarily by
    A. expansion forces from within.
    B. rotation.
    C. revolution around the Sun.
    D. tidal forces from their moons.
    E. storms in their atmospheres.

17. Observation of the occultation of a star by a planet allows us to determine
    A. a planet's diameter.
    B. whether a planet has rings.
    C. the chemical composition of a planet.
    D. [Both A and B above.]
    E. [Both B and C above.]

18. The existence of Neptune was proposed
    A. on the basis of photographs.
    B. because of its effect on Pluto's orbit.
    C. because of its effect on Uranus's orbit.
    D. on the basis of Bode's law.
    E. [None of the above; it was discovered by accident.]

19. Which of the following is a spacecraft that is analyzing Jupiter's atmosphere?
    A. *Viking*
    B. *Galileo*
    C. *Skylab*
    D. *Mariner*
    E. *Mir*

20. We say that Jupiter has a rotation rate of 9 hours, 50 minutes. What is it on the planet that has this rotation rate?

21. What is the *Galileo* probe?

22. Starting at the cloud tops and descending downward, describe Jupiter's interior. What leads us to think that Jupiter has a core made up of heavy elements?

23. What causes the swirling between the bands we see in Jupiter's atmosphere?

24. Explain why, as the years go by, Saturn appears so different when viewed from the Earth.

25. Describe the rings of Saturn and explain the Roche limit.

26. Describe two ways the diameter of a planet can be measured from the Earth.

27. Why were the rings of Uranus not observed directly from the Earth? How were they observed?

28. What is unusual about the orientation of Uranus?

29. How do the densities of the Jovian planets compare to the densities of the terrestrial planets?

30. What causes the heating of the inner Galilean moons that results in volcanism? On which moons is this most prevalent?

31. In what ways are Triton and Nereid unusual?

1. What causes a planet to be spherical (rather than some other shape, like a pancake, for instance)? What causes Jupiter to be a "flattened" sphere?

2. Analysis of data from the *Galileo* probe has produced some surprising results regarding the composition of Jupiter. Explain why these results are causing astronomers to re-evaluate today's theories of the formation of the solar system.

3. It would seem that a planet should absorb and release the same amount of energy. Jupiter doesn't. Describe the situation and list possible explanations.

4. Among the terrestrial planets and their moons there are several cases of gravitational linkage between rotation and revolution. In this chapter we have encountered another. Describe at least three such linkages among terrestrial planets/moons and one that involves a Jovian planet and its moon(s).

5. What is the source of energy for the geysers on Io?

6. Do you think it is possible for Europa to harbor some sort of life? What kind of life would you expect and what energy source would it use?

7. Why does Saturn appear so different year-to-year when viewed with a telescope?

8. Report on the status of the *Cassini* mission. (The internet and the magazines *Sky and Telescope* and *Astronomy* are suggested sources.)

9. How can it be that Titan has an atmosphere, while Mercury (only slightly smaller) has almost none?

10. Describe a theory that explains why some planets have rings and others do not.

11. How can we determine the sizes of Uranus and Neptune?

12. Name each Jovian planet and in each case describe at least one property of the planet that makes it unique from other planets. For example: The biggest planet, the hottest planet, etc.

13. Does one pole of every planet point to Polaris? What is meant by the direction *north?* What defines which pole of a planet is its north pole?

14. Each of the Jovian planets has differential rotation. Kepler's third law might lead you to expect that the rotation period at a planet's equator would be greater than near a pole, but the opposite is true. Explain why the rotation of a gas planet is fundamentally different from a situation that is covered by Kepler's third law.

---

1. If an object has a diameter equal to four times the Earth's diameter, how would its volume compare to the Earth's?

2. If an object has a mass equal to three Earth masses and a volume five times that of the Earth, how would its density compare to the Earth's?

3. Suppose a planet has an angular diameter of 11 arcseconds when it is 7.5 AU from the Earth. Use the small-angle formula to calculate its diameter.

4. Estimate the wind speeds in the Great Red Spot, assuming that it is approximately circular with radius 10,000 kilometers.

5. Choose relevant data from Appendix D for a Jovian satellite and calculate Jupiter's mass.

6. Io is losing material into Jupiter's magnetosphere at a rate of 1000 kilograms/second. How long would it take to lose 1% of its mass? Is your answer larger or smaller than the age of the solar system (about 5 billion years)?

CALCULATIONS

ACTIVITIES

Observing Jupiter and Saturn

Jupiter and Saturn are beautiful objects to observe if you have access to a telescope. Even with binoculars you may be able to see the four Galilean moons of Jupiter and the rings of Saturn. Make a careful sketch of the location of Jupiter's moons and then view them a few hours later or the next night. You will be able to see that they have moved. In addition, see if you can tell that Jupiter and Saturn are not perfectly round. Look for dark bands across the disk of Jupiter.

Table A9-1 shows selected positions of Jupiter. For dates between those listed, you can find Jupiter as shown in the following

example: Suppose you wish to find Jupiter in early March, 2008. The table shows the planet to be in *conjunction* with the Sun in late December of 2007. (An astronomical object is said to be in *conjunction* with another if the two are in approximately the same direction in the sky.) When a planet is in conjunction with the Sun, it can't be seen. Thus in late December of 2007 Jupiter rises with the Sun, and in early May of 2008 the chart indicates that it rises at midnight. Since early March is about half-way between those two dates, we can figure that it rises half-way between midnight and sunrise.

Table A9-2 shows some rising and setting times for Saturn. Note that Saturn can't be seen between the dates when it sets in evening twilight and rises in morning twilight. In each table, midnight refers to standard time rather than daylight saving time.

Although these tables should be of help to you in finding the planets, a better plan would be to pick up an issue of either *Astronomy* or *Sky and Telescope,* two monthly magazines (or *SkyNews,* a bimonthly Canadian magazine) that contain maps and hints for viewing the planets and other celestial objects.

Table A9-1

Selected Positions of Jupiter, 2004-2008

2004	Early Mar.	Rises at 6 P.M.
	Early June	Sets at midnight
	Late Sep.	Conjunction with Sun
2005	Mid Jan.	Rises at midnight
	Early April	Rises at 6 P.M.
	Late June	Sets at midnight
	Late Oct.	Conjunction with Sun
2006	Late Feb.	Rises at midnight
	Late May	Rises at 6 P.M.
	Mid July	Sets at midnight
	Late Nov.	Conjunction with Sun
2007	Early April	Rises at midnight
	Early July	Rises at 6 P.M.
	Early Aug.	Sets at midnight
	Late Dec.	Conjunction with Sun
2008	Early May	Rises at midnight
	Mid August	Rises at 6 P.M.
	Mid Sep	Sets at midnight

Notice that as time passes, Jupiter rises and sets earlier each day.

Table A9-2

Rising and Setting Times for Saturn, 2006

Mid Jan.	Rises at 6 P.M.
Early May	Sets at midnight
Early Aug.	Conjunction with Sun
Early Nov.	Rises at midnight

The chart can be used for three to four years before and after 2006 by subtracting a week for each year before 2006 and adding a week for each year after. For example, in 2008 Saturn will rise in morning twilight in late November.

You can find information about current explorations of our solar system at NASA's home page (http//www.nasa.gov).

1. "*Galileo* in Retrospect," by S. J. Goldman, in *Sky & Telescope* (December, 1997).

2. "Europa: Distant Ocean, Hidden Life?" by M. Carroll, in *Sky & Telescope* (December, 1997).

3. "The Hidden Ocean of Europa," by R.T. Pappalardo, J.W. Head, and R. Greeley, in *Scientific American* (October, 1999).

4. "Big Blue: The Twin Worlds of Uranus and Neptune," by T. Cowling, in *Astronomy* (October, 1990).

5. "Uranus," by A.P. Ingersoll, in *Scientific American* (January, 1987); "The Rings of Uranus," by J.N. Cuzzi and L.W. Esposito, in *Scientific American* (July, 1987).

6. "Into the Giant," by J. Beatty, in *Sky & Telescope* (April, 1996).

7. "Voyage of the Century," by B. Smith, in *National Geographic* (August, 1990).

8. "Bound for the Ringed Planet," by J. Rogan, in *Astronomy* (November, 1997).

Quest Ahead to Starlinks: www.jbpub.com/starlinks

Starlinks is this book's online learning center. It features **eLearning,** which contains chapter quizzes and other tools designed to help you study for your class. You can also find **on-line exercises,** view numerous relevant **animations,** follow a guide to **useful astronomy sites** on the web, or even check the latest **astronomy news** updates.

Saturn
March 7, 2003

Ultraviolet

Visible

Infrared

Saturn as seen by the *HST* in ultraviolet, visible, and infrared. In these images Saturn's rings were at a maximum tilt of 27° toward the Earth.

CHAPTER 10

Pluto and Solar System Debris

Comet Hyakutake was first spotted in January, 1996. Its nucleus is 5 to 10 miles in diameter, but its tail is 62,000 miles long.

IN THEIR NOVEL, *LUCIFER'S HAMMER*, LARRY NIVEN AND JERRY POURNELLE write of two characters who do not know one another, but who co-discover a comet. One, Tim Hammer, is rich and owns the most expensive astronomical equipment. The other, Gavin Brown, works with a home-made telescope.

In real life, on July 22, 1995, two astronomers co-discovered Comet 1995O1, better known as Comet Hale-Bopp. Alan Hale is a professional astronomer who observes known comets about once a week, recording and reporting their changing brightness to the Central Bureau for Astronomical Telegrams in Cambridge, Massachusetts. At the time the comet was discovered, Tom Bopp worked for a construction materials company, and although he is a serious amateur astronomer, he did not own a telescope when the discovery was made. He was at a star party with friends and was using a friend's telescope.

Each astronomer, the professional and the amateur, found a faint, fuzzy object in the sky, an object that was new to him and that did not appear on any sky chart he had. After observing it for a few hours and finding that it was moving among the stars, each man contacted the

Central Bureau to report his findings. In fiction, the comet was named Hammer-Brown; in real life it is Hale-Bopp. Strange, but true.

We devoted a chapter to the four terrestrial planets and another to the four Jovian planets. That leaves only Pluto. This planet was not included among the others because it is so different from them that it fits in neither category. We will discuss it here before moving on to the debris of the solar system: the asteroids, comets, and meteoroids.

10-1 Pluto

Pluto is so far away from the Earth that less is known about it than any other planet. In fact, it is the only planet that was not even discovered until the twentieth century.

The Discovery of Pluto

The fact that astronomers were able to predict the existence and location of Neptune, based on irregularities in the orbit of Uranus, led them to try the method once more. (See Advancing the Model: The Discovery of Neptune, in Chapter 9.) Analysis of the orbital data of Uranus led to the conclusion that although the gravitational pull from Neptune accounted for about 98% of the variation from Uranus's expected orbit, there still were unexplained irregularities. Some astronomers used the data to predict a ninth planet. The one whose work led to success was Percival Lowell, a successful businessman-turned-astronomer. Lowell had built an observatory near Flagstaff, Arizona, in 1894, and in 1905 he made his prediction of the existence of the new planet. He used the Lowell Observatory telescope to search for the disk of "Planet X" until he died in 1916, but he had no success.

This is the same Percival Lowell who thought he had found evidence for life on Mars.

In the 1920s a new photographic telescope, donated by Percival Lowell's brother, was installed at Lowell Observatory. On April 1, 1929, Clyde W. Tombaugh started a new search for the predicted planet. Tombaugh, however, used a different method of searching. Instead of looking for a small, faint disk, he concentrated on looking for the motion a planet must exhibit.

Tombaugh searched for the moving planet with the aid of an instrument called a *blink comparator*. This instrument is essentially a microscope containing a mirror that can be flipped quickly, allowing the observer to look alternately at two different photographs of the sky. The astronomer takes photographs of the same area of the sky a few days apart. The two photographs are then arranged in the comparator so that the stars in each photograph appear at the same place as he or she shifts from viewing one to viewing the other. If a moving object such as a planet is in the photographs, it will appear to jump from one spot to another as the astronomer changes views.

Searching for an object in this manner is very tedious work because the comparator must be scanned slowly over one pair of photographs after another. Tombaugh's search was especially difficult because the predicted position of Planet X was in the constellation Gemini, which is near the Milky Way and therefore in a region with a great many faint stars. As a result, each photograph contained some 300,000 star images. Nevertheless, Tombaugh was successful.

After the discovery of Pluto was confirmed by other astronomers, young Tombaugh went to college, enrolling as a freshman at the University of Kansas.

On February 18, 1930, more than 10 months after he began, Tombaugh detected a difference, indicated by the arrows in the two photographs in **Figure 10-1**. Consider that in searching for the planet, even with the help of the comparator, Tombaugh had to examine not just this pair of photographs but numerous pairs. Tombaugh announced the discovery of the new planet on March 13—the 75th anniversary of Lowell's birth and the 149th anniversary of the discovery of Uranus. It was named Pluto after the mythical Greek god of the underworld.

Pluto was discovered nearly 6 degrees from where Lowell had predicted, and it soon became clear that Pluto's size was so small that its mass must also be small—too small to cause the irregularities that had been seen in Uranus's orbit. More recent

326 CHAPTER 10 **Pluto and Solar System Debris**

Figure 10-1
These are the two photos on which Tombaugh discovered Pluto. Notice how difficult it is to detect the change in Pluto's position, which is indicated here by the arrows.

Neptune's discovery was not an accident, of course, but a triumph of the ability of Newton's theories to make predictions.

Recall that from 1979 to 1999 Pluto was inside Neptune's orbit. Refer back to Figure 7-3.

Recall that *aphelion* refers to the point in a planet's orbit when the planet is farthest from the Sun.

Is Pluto always the farthest planet from the Sun?

analysis of the orbital data used by Lowell and his contemporaries indicates that the irregularities they perceived were not caused by another planet at all, but were simply variations due to the limited accuracy of the data. This is an example of the importance of knowing the amount of uncertainty involved in a measurement.

Thus we must conclude that Pluto was discovered by accident. However, if Lowell's calculations and predictions had not been made, the search would not have been carried on so diligently, and the planet probably would not have been discovered until much later.

Pluto as Seen from Earth

Pluto was as close to Earth in 1989 as it has been for 248 years, but its image in a telescope was very small and fuzzy even then. It has a very eccentric orbit, more eccentric than any other planet, so that although its average distance from the Sun is about 40 AU, its distance ranges from 30 AU to about 50 AU. Recall from Chapter 7 that Pluto's orbit has another unusual aspect: it is tilted 17 degrees to the ecliptic, whereas no other planet's orbit is tilted more than 7 degrees.

Stellar occultations revealed that Pluto has an atmosphere, and spectroscopic examination of the light from Pluto indicates the presence of nitrogen with traces of methane and carbon monoxide. It is probable, however, that this atmosphere is not present during Pluto's complete orbit, for when Pluto is at aphelion it receives so little energy from the Sun that its surface is probably below the temperature at which methane freezes. Thus its atmosphere is probably a temporary phenomenon that occurs only when it comes within a certain distance of the Sun. Comparing observational data from successive Pluto occultations, astronomers discovered that its atmosphere is undergoing global cooling, while its surface is getting slightly warmer. Only a part of this change can be explained by the motion of Pluto on its eccentric orbit around the Sun. It seems that Pluto is a more complex system than expected. Also, the planet's surface temperature is not uniform. The coldest regions have a temperature of about −235°C, while the warmest may reach −210°C.

In 1956, astronomers observed that Pluto's brightness changes slightly every 6.4 days, leading to the conclusion that it has a dark area on its surface and that its period of rotation is 6.4 days. Observations of an occultation in 1965 had placed an upper limit on the planet's size, but not until the discovery of Pluto's moon did we have an opportunity to learn much about Pluto's mass and size.

Pluto and Charon

In 1978, James W. Christy of the U.S. Naval Observatory was analyzing data and noticed that Pluto seemed to have a bump on one side. **Figure 10-2a** shows this bump and indicates the limited resolution of photographs of this distant, tiny planet. Continued

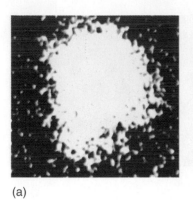
(a)

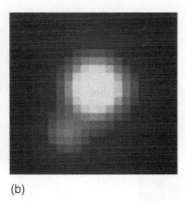
(b)

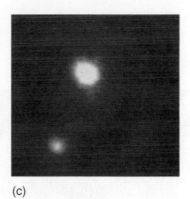
(c)

Figure 10-2
(a) James Christy concluded from a photo such as this that Pluto has a moon, although Pluto and Charon cannot be seen as separate objects. Charon causes the "bump" at the lower left. (b) In this Earth-based CCD photo, two objects can definitely be seen. (c) This image was made by the *Hubble Space Telescope*.

observation showed that the bump moved from one side to another, and it was concluded that Pluto has a moon. The moon was named Charon (pronounced KEHR-on), after the mythical boatman who ferried souls to the underworld to be judged by Pluto, and also after Christy's wife, Charlene. Figure 10-2b is a more recent ground-based photo of Pluto and Charon. In this photo, taken with a CCD camera, two separate objects are obvious. Finally, in 1991 the *Hubble Space Telescope* provided the image shown in Figure 10-2c. Here we see two obviously distinct objects.

Charon's orbit is tilted at 61 degrees to Pluto's orbit around the Sun. Its orbital period around Pluto is 6.4 days, the same as both Charon's and Pluto's rotational period. Thus, these two objects rotate synchronously with their orbital motion and always keep the same face toward each other. As a result, each one of these two objects seems to hover in the sky, as seen from the other, never rising or setting. **Figure 10-3a** shows the orbits of Charon and Pluto in 1978. As Pluto continued to move to the left in its orbit, our view of the system shifted so that between 1985 and 1990, Charon occulted Pluto during each revolution (Figure 10-3b). The duration of each occultation allowed astronomers to calculate the sizes of the two objects with greater accuracy. Still, a lack of knowledge of the nature of Pluto's atmosphere limits the accuracy of measurements of the planet's diameter. Its diameter may be as little as 2300 kilometers (if the atmosphere is clear and we are seeing through it) or as great as 2340 kilometers (if the atmosphere is hazy).

Pluto's mass is about 12 times that of Charon, but only a fifth of our Moon's mass. Its density (between 1.8 and 2.1 grams/centimeters³) leads us to conclude that its interior is probably a mixture of ice (about 30%) and rock (about 70%), similar to Triton. Pluto and Charon are drawn to scale in Figure 10-3 according to our best data.

The two photos of Pluto in **Figure 10-4** were taken by the *HST* and are the first photos to reveal surface features on the planet. Some of these variations in surface reflectivity may be due to topographic features such as basins or fresh impact craters. However, most of these surface features, including the prominent northern polar cap, are likely produced by the complex

Our Moon was the first case where we saw synchronous rotation and revolution.

This series of occultations occurs only twice during each of Pluto's orbits of the Sun. (Each orbit takes 248 years.)

Charon's diameter is about 1200 kilometers.

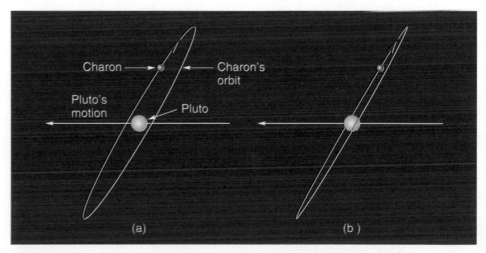

Figure 10-3
The plane of Charon's orbit is tilted at a great angle with respect to Pluto's path around the Sun. (a) This represents the view from Earth when Charon was discovered. (b) A decade later Pluto had moved such that Charon passed in front of and behind Pluto.

Data Page

Pluto

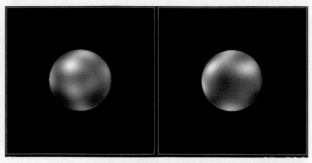

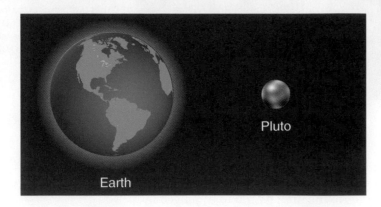

Pluto

Earth

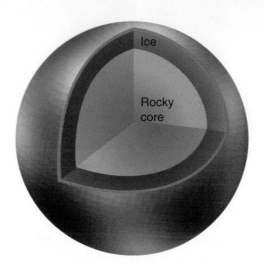

Ice

Rocky core

Interior

Pluto Data

Pluto	Value	Compared to Earth
Equatorial diameter	2300–2340 km	0.2
Mass	1.3×10^{22} kg	0.002
Density	2.00 g/cm³	0.32
Surface gravity	0.58 m/s²	0.06
Escape velocity	1.1 km/s	0.1
Sidereal rotation period	–6.39 days	6.4
Surface temperature	–223°C	
Albedo	0.15	0.47
Tilt of equator to orbital plane	122.5°	
Orbit		
Semimajor axis	5.870×10^9 km	39.24
Eccentricity	0.244	14.6
Inclination to ecliptic	17.2°	
Sidereal period	247.7 years	247.7
Moons	1	

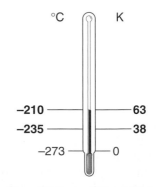

°C		K
–210		**63**
–235		**38**
–273		0

Surface Temperature Extremes

QUICK FACTS

- The most distant planet. Most eccentric orbit, loops inside Neptune's orbit. Orbit has by far the greatest angle to the ecliptic of any planet. Receives 1/1600 the intensity of sunlight that Earth receives. Occultations by its moon, Charon, provided a method of determining its size and mass.

distribution of frost that migrates across Pluto's surface with its orbital and seasonal cycles. The frost is a byproduct deposited out of Pluto's nitrogen-methane atmosphere.

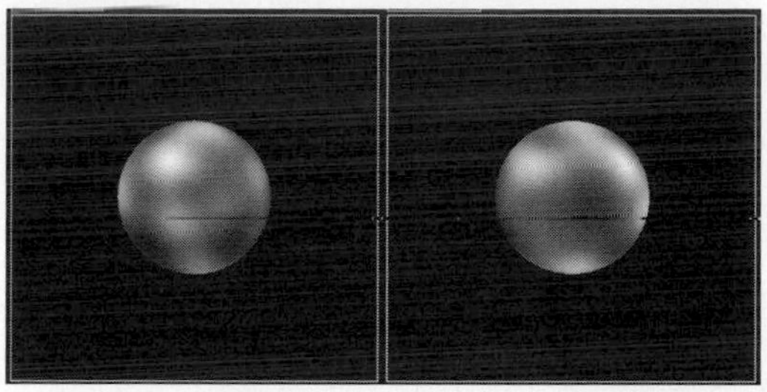

Figure 10-4
These images of Pluto, taken with the *Hubble Space Telescope* during the course of a solar day on Pluto, indicate polar ice caps and bright features near the equator.

A Former Moon of Neptune?

For several reasons, including Pluto's very eccentric orbit, the similarity in its bulk properties with Triton, and the fact that it is so much smaller than any Jovian planet, some astronomers have proposed that Pluto was once a moon of Neptune. Jupiter, Saturn, Neptune, and the Earth all have moons that are larger than Pluto. Perhaps a collision or near passage of some unknown celestial object caused Pluto to be ejected from the Neptunian system. Such a collision would be highly unlikely, but not impossible.

The discovery of Charon made it seem less likely that Pluto was once a moon of Neptune, for how could Pluto itself have acquired a moon? However, the density of Charon (1.2 to 1.3 grams/centimeters^3) is a bit less than that of Pluto, indicating that the two may not have formed out of the same material, but that Charon was indeed captured by Pluto after the planet already existed.

An alternative hypothesis is based on the overall similarities between Charon and Pluto. In contrast to other planet–moon systems in our solar system, Pluto and Charon have comparable sizes and masses. It is possible that this system was formed when Pluto collided with a similar object, leaving behind enough mass to form Charon, which was then captured by Pluto's gravity. Such a collision must have created a large number of fragments that may still exist today.

Some astronomers have suggested that Pluto and Charon should be classified as merely two of the largest examples of a group of objects orbiting in a swathe of the outer solar system known as the Kuiper belt. We will study this disk-shaped region past the orbit of Neptune (roughly 30 to 500 AU from the Sun) later in this chapter. This suggestion indicates how arbitrary our classification is of celestial objects. Nature is unified; we divide objects within it into discrete groups only to help our understanding. Nevertheless, Pluto's official designation as a planet will most likely not change for historical reasons.

Although the orbits of Neptune and Pluto intersect on a two-dimensional drawing, the two objects don't actually cross paths.

Recall the "double planet" and "capture" theories for the origin of our Moon, discussed in Section 6-7.

10-2 Solar System Debris

One dictionary defines debris as "an accumulation of fragments of rock," which is similar to what we will be describing in the rest of this chapter. It could be said that the solar system is made up of one large object (the Sun), some medium-sized objects (the planets), and a lot of debris. The medium-sized objects (as well as some smaller moons) were discussed in the preceding two chapters and the large object will be discussed in the next chapter. Now we look at the debris.

The material of which the debris is composed is not basically different from that which makes up the planets. Rather it is material that did not become part of the Sun or the planets when the solar system was formed. The debris comes in several forms, including asteroids, meteoroids, comets, and dust.

10-3 Asteroids

Since the first night of the nineteenth century, when Father Giuseppe Piazzi discovered Ceres (see Advancing the Model: The Discovery of the Asteroids, in Chapter 7), we have accurately determined the orbits of some 5000 asteroids. Approximate orbits

Ceres was named for the patron goddess of Sicily by its discoverer, Giuseppe Piazzi. German astronomers wanted to call it Juno or Hera, while some French astronomers wanted to call it Piazzi. Piazzi the man won out over Piazzi the name.

Figure 10-5
The telescope was following the stars during this time exposure. The two streaks (identified with arrows) are asteroids.

are known for thousands more. The photograph in **Figure 10-5** shows the motion of two asteroids across the background of stars; it is from photographs like this that most asteroids are discovered and their orbits analyzed. Astronomers estimate that about 100,000 different asteroids appear on today's photographs of the heavens. These objects traditionally have been called *minor planets,* indicating their stature as objects that orbit the Sun.

Only when the orbit of an asteroid is determined well enough that its location can be predicted on succeeding oppositions do we give it a name. Since calculations of asteroid orbits are no trivial task and usually are of no great value to astronomical knowledge, most asteroids go nameless.

Ceres is by far the largest asteroid, with a diameter of about 950 kilometers (590 miles). In fact, Ceres makes up about a third of the entire mass of all asteroids. Two others (Pallas and Vesta) have diameters greater than 500 kilometers, about 30 more are between 200 and 300 kilometers in diameter, and about 100 are between 100 and 200 kilometers. All the rest are smaller.

The masses of the three largest asteroids have been calculated by observing the perturbations they cause on smaller asteroids passing nearby. Masses of other asteroids have been estimated based on their sizes and expected densities.

For some time, astronomers observing from Earth have seen that the brightness of most asteroids changes with periods that are typically a few hours. They conclude that asteroids are not spherical, but are irregularly shaped, which causes them to reflect different amounts of light toward Earth as they rotate. In October, 1990, the *Galileo* spacecraft passed 16,000 kilometers from the asteroid Gaspra and took the photo shown in **Figure 10-6a**. In all, 16 images were made, and they show that this $19 \times 12 \times 11$-kilometer asteroid rotates with a period of about 7 hours. It is appar-

Figure 10-6
(a) The asteroid Gaspra was photographed by *Galileo* on the spacecraft's trip to Jupiter. Gaspra is about 19 kilometers (12 miles) end-to-end. (b) *Galileo* also photographed the asteroid Ida, which is orbited by a tiny satellite named Dactyl. Ida is 56 kilometers (33 miles) in its longest dimension; Dactyl is 1.4 kilometers across. (c) This picture of Eros was taken by the *NEAR* spacecraft. (d) The size and shape of Eros are close to that of Manhattan Island, but would you really want to live there? (e) A mosaic of four images of Eros taken by *NEAR Shoemaker* on September 21, 2000 from about 100 kilometers (62 miles) above the asteroid.

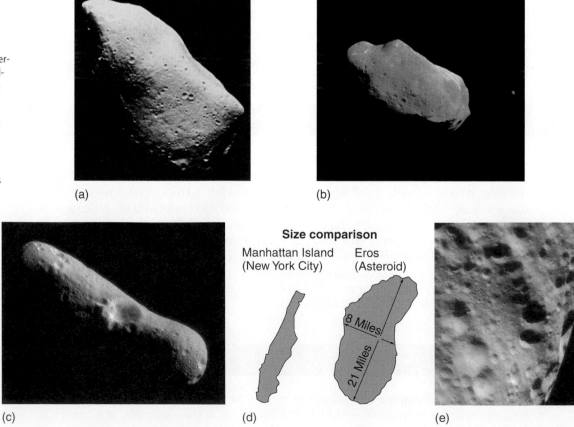

(a)

(b)

(c)

Size comparison

Manhattan Island (New York City)   Eros (Asteroid)

8 Miles

21 Miles

(d)

(e)

The Mission to Eros

The *Near Earth Asteroid Rendezvous* (*NEAR*) mission was the first to orbit an asteroid and the first that involved a controlled descent to and elegant landing on the surface of an asteroid (even though the spacecraft was not built as a lander). *NEAR* was launched in February 1996, and after a 5-year, 2-billion-mile journey it began a yearlong orbit of Eros in February 2000. Renamed *NEAR Shoemaker*, the spacecraft provided the most detailed profile yet of a small celestial object.

The mission provided astronomers with about 160,000 images covering the asteroid's entire surface, which includes craters, boulders, and dust. Data from millions of laser pulses that were bounced on the asteroid's surface resulted in a detailed shape model of Eros. The mission also provided infrared, X-ray, and gamma-ray readings on Eros's composition and spectral properties, along with data on its weak gravity and solid but cracked interior.

We now know that Eros has a composition similar to the most primitive rocks in the solar system, the chondritic meteorites (the building blocks of the terrestrial planets). It is a homogeneous mixture of heavy and light materials, with aluminum, magnesium, and silicon having the same relative abundances that they do in the Sun. This suggests that Eros was never exposed to intense heating and thus never subjected to melting. It is a relic from the dawn of our solar system, a planetesimal that was not captured by a growing protoplanet.

On the surface of Eros there are many craters smaller than 1 kilometer in diameter, while the largest one is 5.5 kilometers wide. There are also many rocks and boulders, 30 to 100 meters across, created by impacts. There are fewer young craters on Eros than expected from our experience with our Moon, and more boulders than expected. Some small craters have flat, smooth floors, as if a fine-grain material is covering the surface. It is possible that dust grains levitate downhill into the bottom of craters due to static electricity, which results from the bombardment of the dust grains by the solar wind.

The asteroid lacks a measurable magnetic field and its density is similar to that of Earth's crust. The gravity on Eros is very weak but enough to hold a spacecraft. A 90-kilogram person (200 pounds) on Earth would weigh about 2 ounces on Eros, making it possible for a person on the surface to escape into space. Eros is about 21 by 8 by 8 miles (33 by 13 by 13 kilometers) in size, orbits the Sun once every 1.76 Earth years, and spins on its shortest axis once every 5.27 hours. The rotation axis appears to remain steady as the asteroid moves in space, as if Eros were a football thrown in a perfectly tight spiral pass. Eros is a "near Earth asteroid," its next close approach to Earth being in 2012 at a distance of only 0.18 AU, but there is no chance that it will collide with our planet.

---

ently covered with about a meter of dusty, rocky soil. Gaspra is one example of a type of asteroid that is bright and has a reddish tint. These asteroids are more reflective than our Moon. Others are dark like a lump of coal. Figure 10-6b is an image of asteroid Ida, which has a tiny satellite of its own, called Dactyl. Figure 10-6c shows asteroid Eros, which was the target of a spacecraft exploration that began in February 2000 and lasted for a year (see Advancing the Model: The Mission to Eros). Some idea of the size of these asteroids is shown in Figure 10-6d.

The bright reddish asteroids are called S-type. They are metallic nickel-iron, mixed with iron- and magnesium-silicates. Bright nonreddish asteroids are M-type, which are pure nickel-iron. Dark ones are C-type, with a chemical composition similar to the proportions of metals within the Sun.

The Orbits of Asteroids

The direction of motion of all of the asteroids, like that of the planets, is counterclockwise as viewed from north of the solar system. Most of their planes of revolution are near the plane of the ecliptic, although we know of one with a plane of revolution 64 degrees from Earth's.

Most of the asteroids orbit the Sun at distances from 2.2 to 3.3 AU (in what is called the **asteroid belt**). This corresponds, by Kepler's third law, to orbital periods of about 3.3 to 6 years. Most of the asteroids in the asteroid belt have fairly circular orbits, although not as circular as the orbits of the major planets.

If the 100,000 asteroids that appear in photographs were in exactly the same plane and were spread evenly across the asteroid belt, neighboring asteroids would be separated by 2 million kilometers, or more than a million miles. In reality, the asteroids are irregularly spaced in the asteroid belt and are not all in the same plane. So

asteroid belt The region between Mars and Jupiter where most asteroids orbit.

TOOLS OF ASTRONOMY

You Can Name an Asteroid

So you want to name an asteroid? That sounds like a magazine advertisement. But the fact is that only about 3000 of the perhaps 100,000 asteroids that appear on photographs (and thus have a diameter of about 1 kilometer or greater) have been named. Why? In order to name an asteroid, you must know its orbit accurately. You determine its orbit by taking photographs of it and perhaps searching photographs taken by other astronomers who were studying other objects. After you have determined the orbit, you must wait until the asteroid has completed at least one more cycle around the Sun in order to check your predicted orbit.

If you have indeed calculated a reliable orbit, you will be given the privilege of naming the asteroid. The official name of the asteroid will include not only the name you give, but a number indicating the order of discovery. For example, 1 Ceres and 2 Pallas. These first asteroids were named after gods of Roman and Greek mythology. Names from mythology quickly ran out, and today almost any name is acceptable. Asteroid number 1001 is Gaussia, named after the mathematician who discovered the method of calculating orbits that made possible the determination of the orbit of Ceres. As you can verify in the internationally accepted catalog of asteroids, the *Minor Planets Ephemeris,* asteroid number 1814 is named for Bach and 1815 for Beethoven. Others are named Debussy, McCartney, and Clapton. There is no Elvis yet. Some of the more recently named asteroids honor the seven astronauts lost in the *Challenger* explosion in January, 1986.

Is there a major risk of collision when a spacecraft passes through the asteroid belt?

Apollo asteroids Asteroids that cross the Earth's orbit and have semi-major axes larger than Earth's.

2 million kilometers is a low estimate for the average distance between the 100,000 asteroids large enough to appear in photographs. There are many more asteroids smaller than this, but even so, the asteroids present no major hazard to spacecraft. Compared to the popular picture of a crowded asteroid belt, the belt is almost empty. To illustrate this, imagine a scale model in which the largest asteroid—Ceres—is the size of a grain of sand. In this case, a normal-size asteroid would be one meter away and would be too small to be seen.

As the orbits of more and more asteroids were calculated during the nineteenth century, it became clear that they are not evenly distributed across the asteroid belt. At certain distances from the Sun, gaps occur in the belt. **Figure 10-7** illustrates the situation. One very prominent gap occurs at 2.50 AU and another at 3.28 AU. Daniel Kirkwood, an American astronomer, first explained these gaps in 1866. He noticed that 3.28 AU corresponds to a period of revolution of 5.93 years, which is just half of Jupiter's period of 11.86 years. An asteroid at 2.50 AU would orbit the Sun in 3.95 years, which is one-third of Jupiter's period. We saw a similar situation in Chapter 9 in the explanation for Cassini's division in Saturn's rings. In both cases, synchronous tugs from a large object on smaller particles gradually move those particles out of their orbits and result in gaps.

Jupiter has cleared out other regions of the asteroid belt where it exerts a regular gravitational pull on objects that may once have been there. Besides the gaps at locations corresponding to one-third and one-half of Jupiter's period, there are also major gaps corresponding to two-fifths and three-fifths of the period of the giant planet.

Not all asteroids orbit the Sun between Mars and Jupiter. About 500 are currently known to have orbits eccentric enough that they cross the Earth's orbit; these are known as ***Apollo asteroids.*** About 70 of these are potentially hazardous to Earth, in the sense that they are larger than 1 kilometer in diameter and their orbits bring them closer than 0.05 AU to Earth. Some of these have passed

Figure 10-7
Kirkwood gaps appear where asteroids would have periods of one-third and one-half of Jupiter's.

fairly close to the Earth in recent history. On May 19, 1996, an asteroid passed within about 450,000 kilometers of the Earth, which is slightly more than the distance of the Moon from Earth. The predicted closest encounter by a potentially hazardous asteroid through the 21st century will occur on August 7, 2027, at a distance of about 400,000 kilometers. When we discuss meteors, we will describe what happens when a large asteroid hits the Earth.

The Origin of Asteroids

Are asteroids pieces of a destroyed planet?

Astronomers once thought that asteroids are the remains of the explosion of a planet. This theory has been abandoned today for two reasons. First, there is no known mechanism by which a planet could explode. Second, if all of the asteroids were combined into one object, that object would be only about 1500 kilometers in diameter—much less than the 3500-kilometer diameter of our Moon. Such a small object does not fit the pattern of planetary sizes.

It is considered much more probable that the asteroids are simply primordial material that never formed into a planet. There is a good reason that the material in the asteroid belt did not form into a planet; the reason is Jupiter. The gravitational pull from that planet causes a continual stirring effect on the objects in the asteroid belt. This prevents the small gravitational forces that exist between the asteroids from pulling them together to form a larger object.

Even today, there is evidence that Jupiter causes collisions between asteroids that result in their fragmentation. In fact, some of the asteroids that have orbits outside the main belt are thought to have resulted from such collisions. Analyses of the orbits of such asteroids have identified groups ("families") whose orbits, when traced backward, indicate that they were once together. They apparently broke apart as the result of a collision with another asteroid, caused by the gravitational force of Jupiter.

Asteroids will be discussed again later in the chapter when we take up the question of the origin of meteoroids. Now we turn to another category of solar system debris: comets.

10-4 Comets

One of the most spectacular astronomical sights available to the naked eye is a comet. **Figure 10-8** is a photograph of Comet West, which made its appearance in 1976, and **Figure 10-9** is a 1997 photograph of comet Hale-Bopp. People who have not seen a comet commonly assume from photographs such as Figures 10-8 and 10-9 that the comet streaks across the sky in the direction opposite to its tail. Perhaps from seeing streaks behind fast-moving characters in cartoons, we assume that the comets in the photographs are moving at great

Figure 10-8
Comet West, discovered by Richard West of the European Southern Observatory in 1976.

Does a comet move visibly across the sky?

Figure 10-9
In this March 29, 1997, photo of Comet Hale-Bopp, the blue ion tail is obvious. It is straight and streams directly away from the Sun. The white dust tail is curved.

Wherefore if according to what we have already said [the comet] should return again about the year 1758, candid posterity will not refuse to acknowledge that this was first discovered by an Englishman.

Edmond Halley

coma (KOH-mah) The part of a comet's head made up of diffuse gas and dust.

nucleus (of comet) The solid chunk of a comet, located in the head.

tail (of comet) The gas and/or dust swept away from a comet's head.

speed. In fact, if you see a comet in the sky, you observe no rapid motion at all. Unless you observe carefully, the comet appears to stay in the same place among the stars, having only the motion across our sky caused by the rotation of the Earth. (It does move among the stars, of course, and its motion can easily be seen over a few days.) Don't confuse a comet with a meteor, which *does* streak across the sky, as we will discuss later.

Cometary Orbits—Isaac Newton and Edmund Halley

Isaac Newton proposed that comets orbit the Sun according to his laws of universal gravitation and motion, just as the planets do. He concluded that since comets are visible from Earth for only short periods of time (typically a few months), their orbits are very eccentric; that is, very elongated.

Edmund Halley was a friend of Newton. In fact, he talked Newton into publishing the *Principia,* personally financed the book, and was the only person for whom Newton expressed appreciation in the book. Halley used Newton's methods, his own observations, and descriptions of previous comet sightings to calculate the orbits of several comets. Halley noticed that the orbits of the comets of 1531, 1607, and 1682 were very similar and suspected that they might be the same comet. However, he was at first confused by the fact that the time that elapsed between one appearance of the comet and the next was not always the same. When he realized that the comet's path would be changed slightly by the gravitational pull of a planet— particularly Jupiter or Saturn—when the comet passed nearby, he hypothesized that these three comets were in fact three appearances of the same comet. In 1705, Halley predicted that the comet would reappear in 1758.

On Christmas night of 1758, the comet was sighted. In honor of its predictor, it was named Halley's comet. The brilliant Edmund Halley, however, had died 16 years earlier at the age of 85 and was unable to see his hypothesis verified. By investigating reports of comets in literature, we can now trace Halley's comet (or *Comet Halley*) back as far as the year 239 B.C.

Today we discover about a dozen comets each year. Of the nearly 1000 whose orbits have been calculated, about 100 have a period of revolution around the Sun of less than 200 years. Most have extremely long periods of up to a million years.

The planes of revolution of comets are not limited to the plane of the planets, but are randomly oriented, so comets sweep past the Sun from all directions. **Figure 10-10** is a diagram of the orbit of Comet Hale-Bopp, which reached perihelion in 1997. Many of us will remember it as the naked-eye comet of a lifetime. The gray surface in the figure is the Earth's orbit, and the blue spheres show the location of the Earth from January 1 to May 1, 1997. The red dot represents the comet and shows its location on the same dates. You can see the importance of the relative positions of the Earth and a comet in determining the comet's visibility.

The Nature of Comets

We normally think of a comet as having three parts (**Figure 10-11a**). What is commonly called the head of a comet consists of the *coma* and the *nucleus.* Sweeping away from the coma is the comet's *tail,* which varies greatly in size and appearance from comet to comet. The coma of a comet may be as large as a million kilometers in diameter—almost as large as the Sun. Some comet tails have been as long as an astronomical unit.

In 1950, Harvard astronomer Fred L. Whipple proposed what remains today as the basic model of a comet. He proposed that the nucleus resembles a dirty snowball made up of water ice, frozen carbon dioxide, a few other frozen substances,

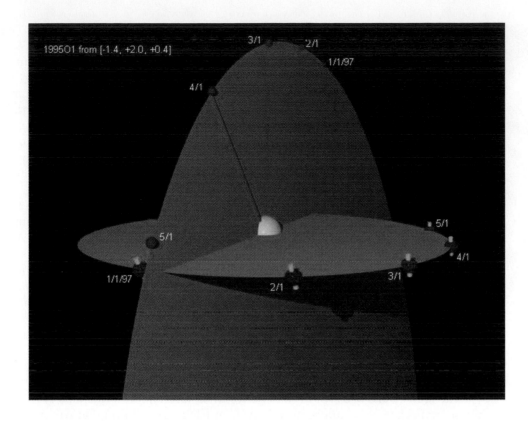

Figure 10-10
The orbits of Comet 1995O1, commonly known as Comet Hale-Bopp, and the Earth.

and small solid grains—the "dirt." Observations of comets since that time have confirmed his model, but have caused it to change to include a crusty layer on the surface of the nucleus, with the ices inside. Figure 10-11b shows a cross-sectional view of the nucleus of a comet. When the comet is far from the Sun, the nucleus (which is only a few miles in diameter) is all there is. As the comet approaches the Sun, it becomes warmer and the ices inside melt and vaporize. These materials then break through the crust, forming a jet. These gases, along with dust particles that have been torn away, hover near the nucleus, held there by the small gravitational field of the nucleus. This is what makes up the coma.

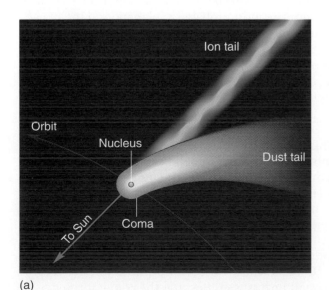

(a)

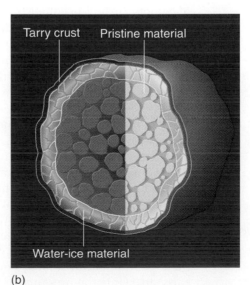

(b)

Figure 10-11
(a) The head of a comet consists of its nucleus and coma. There are usually two tails: a straight ion tail and a curved dust tail. (b) The nucleus of Comet Halley (and presumably other comets) is made up of a mixture of ice and dust surrounded by an ice mantle and covered with a dark crust. Material from inside is ejected through the crust as the comet is heated by the Sun.

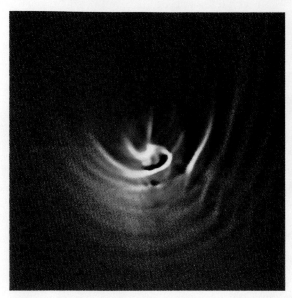

(a)

(b)

Figure 10-12

(a) The layers below the nucleus of Comet Hale-Bopp are caused by the slow spinning of the nucleus as material jets from its surface. The red color was added to enhance the image. (b) The nucleus of Comet Halley is about 5 kilometers wide and shaped like a potato. In this photo the Sun is illuminating the left side, causing the release of bright jets of dust.

 Comets Halley and Hale-Bopp are now moving through the outer portions of the solar system. Comet Haley will go almost to Pluto's orbit and then return in 2061. Hale-Bopp is in a much larger orbit; it won't return until about 2380 years from now. (It was 4200 years since its last appearance, but its orbit has been influenced by the gravity of the Jovian planets.)

The spacecraft *Giotto* was named after the Italian artist who included Comet Halley as the star of Bethlehem in his fresco of the nativity scene. He had just seen a return of the comet a few years earlier, in 1301.

Figure 10-12a is a close up of the coma of Comet Hale-Bopp. This comet has a fairly large nucleus (25 miles across) and the nucleus spins around about once every 12 hours. As it spins, parts of it shoot away in a few geysers. As the ejected material spirals away from the nucleus, it causes the appearance of rings around the nucleus. These are the "layers" seen in the photo. They were visible even in small telescopes in March and April 1997.

The coma and tail of a comet are mostly empty space. When Comet Halley came through the inner solar system in the late 1980s, the European Space Agency launched the spacecraft *Giotto* to intercept it. *Giotto* passed within 600 kilometers (350 miles) of the nucleus (Figure 10-12b), going right through the coma. *Giotto* found that the coma is billions of times less dense than the atmosphere of Earth at sea level. It observed the abundances of chemical elements and confirmed that the comet was composed of material as primitive as the original solar nebula. *Giotto* also showed that there were two major groups of particles. The first group was rich in mineral-forming elements such as silicon, calcium, and magnesium. The second group was composed mainly by organic grains made of carbon, hydrogen, oxygen, and nitrogen.

The entire mass of a typical comet is less than one-billionth the mass of the Earth, and more than 99.99% of this mass is in the nucleus. The reason that we can see the coma at all is that its molecules and dust particles reflect sunlight to us. In addition, ultraviolet light from the Sun causes the molecules to fluoresce, in much the same manner that a "black light" poster glows in ultraviolet light. There are very few molecules and dust particles in any particular volume of the coma, but the coma is extremely large—hundreds of thousands of kilometers in diameter—so there is a lot of material along any particular line of sight.

Astronomers have observed not only ultraviolet radiation from comets but also intense X-ray emission from the gas surrounding the nucleus of a comet. It was an unexpected and puzzling result when energetic X-rays (100 times more intense than anyone had predicted) were first observed in 1996 coming from the vicinity of comet Hyakutake. Since then, satellites have detected such radiation from a number of comets, including comet Hale-Bopp. The most probable explanation to the phenomenon of X-rays from cosmic snowballs is the following. Highly ionized elements in the solar wind (such as oxygen, nitrogen, and carbon) interact with the gas being released by the comet. The strong positive charge of the ions attracts electrons from the atoms and molecules from the comet. The electrons that leap from the neutral atoms to the ions emit X-rays as they move from high-energy to low-energy orbits of the ions.

The high brightness of comet Hale-Bopp allowed astronomers to detect the presence of a noble gas, argon. This is important because noble gases do not interact chemically with other elements and are easily lost from comets at low temperatures. The high abundance of argon in comet Hale-Bopp suggests that it has always been cold and, therefore, was most likely formed in the zone around Uranus and Neptune or possibly farther out. Thus, the presence or absence of such gases provides a tool for measuring the thermal history of comets. Another tool is provided by the structure (amorphous versus crystalline) of the dust that comets carry. Astronomers now believe that comets could have formed at different times during the evolution of our solar system. Comets whose dust is amorphous formed very early on, while the presence of crystalline dust signifies that the comet formed later, as the dust clouds were heated by the forming Sun.

ADVANCING THE MODEL

Astronomer Maria Mitchell, A Nineteenth-Century Feminist

Maria (pronounced ma-RYE-a) Mitchell (1818–1889) grew up on Nantucket Island, Massachusetts, and learned astronomy both from her schoolteacher father and from her readings while she worked as a librarian. In 1847, while observing the sky from her rooftop, she discovered a comet. The discovery resulted in her becoming the first woman elected to the American Academy of Arts and Sciences (although the all-male membership refused to name her a "fellow," calling her instead an "honorary member").

At that time, the king of Denmark awarded a gold medal to each discoverer of a comet, and receiving the medal brought Mitchell some fame. Her astronomical work remained on the amateur level, however, until her mother died, when finally, at the age of 40, she sought a professional career. Vassar College had just been founded, and Maria Mitchell was hired to teach.

Mitchell was an early believer in women's rights and saw astronomy and science as an avenue for liberation and a way for women to break from domestic tradition. Her outspoken nature was revealed on one occasion when her wish to observe the Moon as it occulted a star conflicted with something her college president wanted her to do. She wrote the following note to him*:

My good natured President, I want to hear you preach to-morrow, and I also want to see the moon pass over Aldebaran. Can't you let me do both? Will you stop at eleventhly or twelfthly? Or, why need you show us *all* sides of a subject? The moon never turns to us other than the one side we see, and did you ever know a finer moon? If I could stop the moon and do no more harm than Joshua did, I wouldn't ask such a favor of you, knowing, as I do, what a difficult thing it is for you to pause, when you are once started, and knowing also, that I never want you to do so—*except this once*. Yours with all regret, even if it doesn't appear—M. Mitchell. (Mitchell's italics)

*This story is from E. A. Daniels, "Maria Mitchell, the Star of Vassar College," *Star Date*, November/December, 1989. To subscribe to this delightful magazine, write Star Date, RLM 15.308, University of Texas at Austin, Austin, Texas 78712.

Comet Tails

As indicated earlier, the tail of a comet does not necessarily follow the head through space. Rather, a comet's tail always points away from the Sun. Thus after a comet has passed the Sun, its tail actually *leads* it through space (**Figure 10-13**). This takes some explaining.

Comets usually have two tails, although one or both may be very small and they may change greatly as time passes. **Figure 10-14** shows four photographs of Comet Mrkos, taken in 1957. Notice that one tail is very straight, while the other curves away from the comet. The straight tail consists of charged molecules (*ions*) being swept away from the comet by the solar wind—charged particles that are always being emitted by the Sun. These molecules move away at a great speed and form a straight tail.

The curved tail is caused by grains of dust in the coma being pushed away by the weak pressure of solar radiation. This radiation pressure is not detectable by us in everyday life, but light does indeed exert a tiny force on objects it strikes. In the weak gravita-

Direction of comet's motion

Sun

Figure 10-13

In general, a comet's tails point away from the Sun. This led Ludwig Biermann to predict the existence of the solar wind, a decade before it was discovered in 1962 by *Mariner 2*. Notice that the tails are largest shortly after the comet has passed the Sun.

ion A charged atom or molecule resulting from its loss or gain of at least one **electron.**

Figure 10-14
The two tails of Comet Mrkos are obvious in these photos, taken on different dates during its appearance in 1957. See also Figure 10-9 for the two tails of Comet Hale-Bopp.

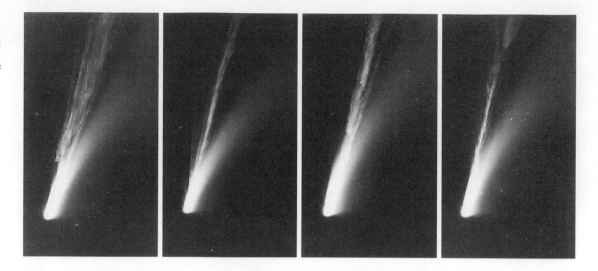

tional field of the comet's nucleus, the force is great enough to move dust particles away from the head. However, they move much more slowly than the molecules that form the ion tail, and **Figure 10-15** shows why this results in the dust tail being curved. Dust particles that left the comet when it was at location X have reached X', those that left at Y are at Y', and so on.

Sometimes only one tail of a comet is visible because one is behind the other as seen from Earth. In addition, some comets are not as dusty as others and do not contain a prominent dust tail.

In general, the tails of comets point away from the Sun regardless of the motion of the comet nucleus because the agents responsible for forming the tails (the solar wind and light) are emitted radially from the Sun. Comet tails are typically 10^7 to 10^8 kilometers long and may be as long as 1 AU.

Since a comet loses material as its gas and dust are pushed off to form its tails, comets have limited lifetimes. Comet Halley spews about 25 to 30 tons of its mass through its jets each second when it is close to the Sun. This sounds like a great deal, but it means that the nucleus loses less than 1% of its mass on each pass of the Sun. Comet Halley is now on its way toward the outer portion of its elliptical orbit, and since its jets are now inactive, it is no longer losing mass.

Figure 10-15
Dust that was blown off the comet when it was at point X is now at X'. Dust blown off at Y is at Y'. This causes the dust tail to be curved.

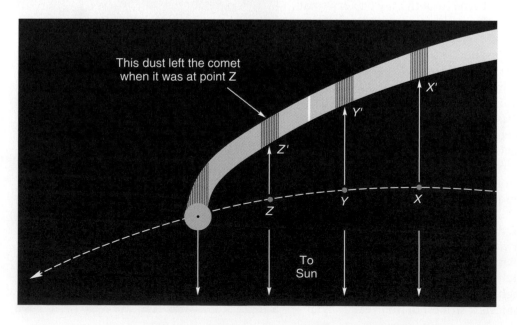

Comets do gradually lose their ices, however. It is thought that some comets finally evaporate away all of their nuclei so that they just fizzle out. Other comets, after they lose all of their volatile materials, become simple chunks of rock and probably would be classified as asteroids. A third way that comets die is to fall into the Sun. A comet that comes close to the Sun is slowed by the Sun's atmosphere, and after several passes, it becomes slowed enough that it falls into the Sun.

Is it dangerous for Earth to pass through a comet's tail?

10-5 The Oort Cloud and the Kuiper Belt

Thus far we have been considering only short-period comets—those with periods of less than a few hundred years. Most comets have much longer periods and approach us from far beyond the orbits of the planets. About a dozen comets are discovered each year; almost all are long-period comets.

In 1950, the Dutch astronomer Jan Oort proposed that great numbers of comets, all of which orbit the Sun, exist in a region of space that lies from 10,000 to 100,000 AU from the Sun (**Figure 10-16**). This hypothesis was based on three observations. First, no comet has been observed that is not gravitationally bound to the Sun. Second, the orbits of long-period comets show a strong tendency to have aphelia at distances around 50,000 AU. Finally, as new long-period comets continually join the inner solar system, they show they come from no preferential direction. This shell of comets surrounding the solar system has come to be known as the *Oort cloud.* Naturally, the Oort cloud is too far away for its comets—which are simply nuclei at that distance—to be visible from Earth.

Many of the comets we observe are in elliptical orbits with extremely long periods. Since they obey Kepler's second law, they move at extremely low speeds when they are far from the Sun. As a result, they spend the overwhelming majority of the time well beyond planetary distances. For example, a comet with a period of 500,000 years would spend about 499,998 of those years beyond the orbits of the planets. Thus, many of the comets in the Oort cloud are there simply in accordance with Kepler's laws.

In addition, there must be many comets in the Oort cloud whose orbits never approach the inner solar system. This would not necessarily mean circular orbits, for a comet could vary from 10,000 to 100,000 AU from the Sun and still remain in the Oort cloud. From time to time, one of the comets of this category must pass near another comet, causing it to change its path. This may result in a more circular path, or it may result in the comet being moved into an orbit that takes it toward the inner solar system. In addition, although the outer Oort cloud stretches only about one-third of the way to what is now the nearest star, every few million years a star passes closer to the Sun than this, and gravitational forces from the star

Oort cloud The theorized spherical shell, lying between 10,000 and 100,000 AU from the Sun, containing billions of comet nuclei.

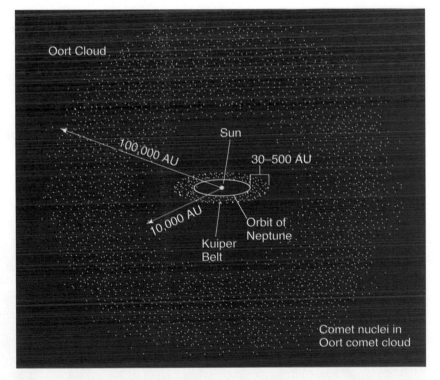

Figure 10-16
Long-period comets are believed to originate in the Oort cloud, a *spherical shell* between 10,000 and 100,000 AU surrounding the Sun. The Kuiper belt is a *disk-shaped* region, 30 to 500 AU from the Sun (past the orbit of Neptune), containing many small icy objects. It is believed to be the source of the short-period comets. Distances are not to scale.

ADVANCING THE MODEL

Jan H. Oort, 1900–1992

On November 5, 1992, Jan Oort died at the age of 92 in his home country, the Netherlands. Although he is best known for the cloud of comets that bears his name, he considered his work with distant comets a sidelight to his other astronomical work. He first drew international attention in 1927, when he and Bertil Lindblad of Sweden discovered that the Milky Way Galaxy rotates (see Chapter 16). Later he established that the Sun is about 30,000 light years from the center of the galaxy, moving in an orbit around the center with a period of about 225 million years.

Oort's accomplishments are numerous. He was among the first to realize the value of radio astronomy and was the driving force behind the Westerbork Radio Telescope. His work resulted in establishing the link between the Crab nebula that we observe and the supernova observed by the Chinese in 1054 (see Chapter 15). Oort served as director of the Leiden Observatory from 1945 until his retirement in 1970. His work continued, however, and he published an article in *Mercury* magazine as recently as the March/April, 1992, issue.

---

cause changes in the orbits of comets. Astronomers believe some comets are deflected into the inner solar system by this method. These comets from the Oort cloud become long-period comets. Statistics imply that the Oort cloud may contain up to a trillion comets, with a total mass of more than the mass of Jupiter.

The Oort cloud does not explain all comets, however, and in 1951, Gerard Kuiper proposed that a second, smaller band of comets must exist inside the Oort cloud. This disk-shaped region is beyond the orbit of Neptune, from about 30 to 500 AU from the Sun. It is estimated that this region contains about 100,000 small icy worlds greater than 100 kilometers in diameter and a billion short-period comets. The first object in this *Kuiper belt* was discovered in 1992 (**Figure 10-17**). Since then, more than 550 objects have been discovered, mostly confined within a thick band around the ecliptic. The field of study of these objects is evolving rapidly. Astronomers are trying to understand the origin of the observed differences between groups of Kuiper belt objects; these differences include orbital inclinations, physical properties, and probable regions of formation.

Kuiper belt A disk-shaped region beyond Neptune's orbit, 30–500 AU from the Sun, closer to the solar system than the Oort cloud and presumed to be the source of short-period comets.

The Origin of Short-Period Comets

The age of the solar system is about 5 billion years. Why have those 100 comets with periods of less than 200 years not evaporated away their ices so that they are no longer observed? The answer must be that these short-period comets are relative newcomers to the inner solar system. We must ask where they come from, and an

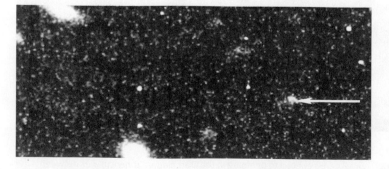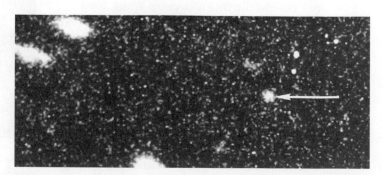

Figure 10-17

The arrows point to an object that moved during a 2-hour period on August 30, 1992. The object was given the name 1992 *QB1* and is thought to be the first object observed in the Kuiper belt.

obvious hypothesis is that long-period comets sometimes become short-period comets. How might this occur?

If the path of a comet depended solely on its gravitational attraction toward the Sun, there would be no way in which its orbit could be changed to make it a short-period comet. However, other objects (particularly the massive Jupiter) affect a comet's orbit. The combined effects of the gravitational forces from a Jovian planet and the Sun can cause one of these comets to change its orbit so that it becomes a short-period comet. **Figure 10-18** illustrates one way this could occur. All of the short-period comets were captured in the inner solar system by such a mechanism. On the other hand, the Sun and a planet can have the effect of changing the orbit of a comet so that it leaves the solar system and the Oort cloud entirely.

The word *cloud* in reference to comets in orbit around the Sun can be misleading. When you think of a cloud on Earth, you may think of a volume fairly crowded with water droplets. But the Oort cloud and Kuiper belt are far from crowded. We pointed out that the asteroid belt is mostly empty space, but it is crowded compared to the Oort cloud. If the Oort cloud contains a trillion comets—a high estimate—there would still be an average distance of 16 AU (1.5 billion miles) between comets. A future interstellar traveler stranded in the middle of the Oort cloud with a telescope probably would not be able to find a comet nucleus.

It seems that objects in the Oort cloud formed closer to the Sun than objects in the Kuiper belt. Objects that formed at the distance of Jupiter and Saturn were either accreted by these massive planets or ejected from the solar system as a result of a gravitational slingshot. Objects that formed farther out, at the distance of Uranus and Neptune, were deflected outward when they passed too close to these planets; they now make up the Oort cloud. Finally, objects that formed beyond the orbit of Neptune never combined to form a planet; they now make up the Kuiper belt. It is almost certain that these objects are "leftover" material from the original cloud that formed our solar system. Understanding how these objects are distributed and what they are made of will put important constraints on models describing the formation and early evolution of our solar system.

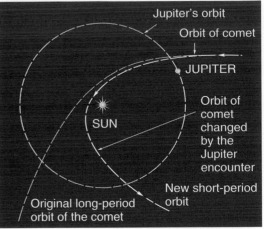

Figure 10-18
Gravitational attraction to a planet—Jupiter, here—can cause a comet to alter its orbit, perhaps changing a long-period comet to a short-period one.

The large range of estimates illustrates how little we know of the Oort cloud.

10-6 Meteors and Meteor Showers

Almost everyone has seen the flash of light in the sky that is sometimes called a *falling star* or *shooting star*. This phenomenon is better termed a ***meteor*** because obviously it is not a star. (Recall the size of our Sun—a typical star—compared to the Earth.) The streak of light in the photograph in **Figure 10-19** is a meteor. The nature of these sudden flashes of light across the sky must have been of great concern to people since the beginning of time. The idea that they are caused by rocks falling from the heavens can be found in ancient writings of the Greeks, the Romans, the Chinese, and in the Old Testament, but some people found the idea hard to accept.

The first confirmation by modern science that rocks do indeed fall from the heavens occurred on April 26, 1803. Citizens of the small town of L'Aigle, France, saw an exceedingly bright meteor that exploded and formed a shower of 2000 to 3000 fragments that fell to Earth. Reportedly, some fragments were still warm when found. The French Academy of Science sent the respected physicist J. B. Biot to investigate the incident, and his report confirmed the ancient writings.

Before describing the stones and the flashes of light in any more detail, some definitions are in order. A meteor is the phenomenon of the flash of light; it is not the object itself. The object out in space that causes the meteor in our atmosphere is

meteor The phenomenon of a streak in the sky caused by partially the burning of a rock or dust particle as it falls, but also by the air molecules in the particle's path that, after being excited, then give off light as they de-excite.

Figure 10-19
The streak of light in this photograph is a Perseid meteor.

Suddenly a blazing stream of fire pierced the sky, lighting the landscape as though Nature had pressed a giant electric switch. The blade of light vanished with equal suddenness, leaving a darkness seeming thicker than before.

Meteorite collector Harvey Nininger, upon seeing a fireball November 9, 1923.

meteoroid An interplanetary chunk of matter smaller than an asteroid.

meteorite An interplanetary chunk of matter after it has hit a planet or moon.

fireball An extremely bright meteor. (More than 10,000 fireballs can be seen each day over the Earth.)

These speeds correspond to 20,000 miles/hour and 150,000 miles/hour.

called a ***meteoroid*** and is in orbit around the Sun. Finally, if the object survives its fall through the atmosphere and lands on Earth, it is called a ***meteorite.***

Meteors

Most meteors are very dim. You might see one out of the corner of your eye and wonder if you saw something or not. Others, however, known as ***fireballs,*** are very bright and might cause a long streak across the sky. In fact, the brightest are brighter than a full Moon. The light is the result of the meteoroid burning itself up as it passes through our atmosphere. As the stone enters the atmosphere, it experiences friction caused by the molecules of air. This heats up the object as well as the air it passes through. At first glance, this may seem odd to you if you have held your hand out of a car window and experienced the *cooling* effect of the air striking your hand. The difference is in the speeds of the objects—your hand and the meteoroid. A meteoroid's speed is typically 50 kilometers/second (100,000 miles/hour). Thus the air molecules are striking the object at tremendous speeds. This causes the surface of the object to heat up until it vaporizes, streaming its atoms in its wake. These atoms, along with the similarly heated air, glow like the gas in a fluorescent lamp and present us with the phenomenon known as a meteor.

It is difficult for an individual observer to estimate the height or speed of a meteor, but if two observers at different locations each record the same meteor on a photograph, they can use triangulation to calculate these quantities. By means of such measurements, it has been determined that the typical meteor begins to glow at a height of about 130 kilometers (80 miles) and burns out at about 80 kilometers (50 miles). The speed of a meteoroid as it enters the Earth's atmosphere might be anywhere from about 10 kilometers/second to about 70 kilometers/second.

Meteoroids

Most meteoroids vaporize completely in the atmosphere and never reach the Earth's surface. The energy source that produces the light we see is simply the motion of the meteoroid. By calculating how much light is emitted from the meteoroid as it burns up and by estimating how much of its original energy of motion changes to light (rather than thermal energy), we can calculate the mass of the original meteoroid. Most meteors are produced by meteoroids with masses ranging from a few milligrams (a grain of sand) to a few grams (a marble-size rock).

The "energy of motion" referred to here is called the *kinetic energy* of the meteoroid.

Under ideal viewing conditions, a person looking for meteors can see an average of five to eight meteors per hour. Since a meteor can be seen only if it is within 150 to 200 kilometers of the viewer, we can calculate that over the entire Earth there must be about 25,000,000 meteors a day bright enough to be seen by the naked eye. The number visible in telescopes would be hundreds of billions. Although the average meteor may have a mass of only a fraction of a gram, it is estimated that 1000 tons of meteoritic material hit the Earth each day.

It has been estimated that only 1 in 1 million meteoroids that hit the atmosphere survives to reach the surface.

If a meteor trail is recorded from more than one location, its path can be determined accurately enough to calculate the path of the original meteoroid before it was slowed by the atmosphere. In this way we find that, as expected, most meteoroids are simply tiny particles orbiting the Sun. Meteoroids differ from asteroids in that most asteroids orbit the Sun close to the plane of the ecliptic, while small meteoroids do not suffer this limitation; their orbit around the Sun may be in any orientation. A majority have very eccentric orbits rather than the nearly circular orbits of most asteroids.

However, this does not rule out the asteroid belt as the origin of these meteoroids. It is thought that many small meteoroids are debris from collisions between asteroids. Such collisions would break the asteroids into smaller pieces, including pieces as small as sand grains. These tiny pieces would emerge from the collision in all directions and move in elliptical orbits at all orientations with the ecliptic.

Many meteors, however, are from a source other than the asteroids—they are due to material evaporated from a comet's nucleus and then blown off the comet by the action of the Sun. This leads us to a discussion of meteor showers.

Figure 10-20
This photo shows the Leonid meteor shower of November 1966. The streaks radiating from one area of the sky are all part of the shower.

Meteor Showers

On some nights, we see many more meteors than normal, and if we observe closely, we see that there is a pattern to the directions of the meteors. **Figure 10-20** is a long-time exposure photograph showing this phenomenon. The streaks are meteors. Notice that they seem to point to (or rather, originate *from*) one point in the sky. This phenomenon is called a ***meteor shower*** and is caused by the Earth passing through a swarm of small meteoroids, as illustrated in **Figure 10-21**. A cluster of tiny particles is shown in the Earth's path. These tiny meteoroids cause the shower.

meteor shower The phenomenon of a large group of meteors seeming to come from a particular area of the celestial sphere.

To see why the meteor seems to originate in a certain constellation, think of the Earth moving through space. When it encounters the swarm of meteoroids, the Earth is moving in a particular direction toward some constellation in the sky. The meteor trail seems to originate in that constellation or close to it. The shower may not appear to come exactly from the constellation toward which the Earth is moving because the swarm itself has a speed and direction of motion. Therefore its direction of hitting the Earth is determined by a combination of the Earth's and the swarm's direction of motion.

Meteor showers are named after the constellation from which they

Figure 10-21
The Earth here is moving toward the constellation Leo and is about to encounter a swarm of meteoritic particles.

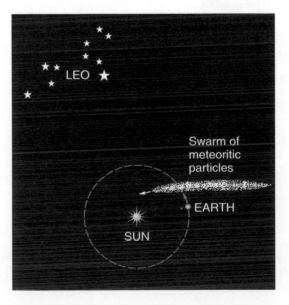

seem to originate. Notice that the meteors in Figure 10-20 seem to radiate from a single point off to the upper right of the photograph (this point is known as the ***radiant***). They appear to come from within the constellation Leo, and the shower is therefore known as the Leonid meteor shower. The meteor shower about to occur in Figure 10-21 will likewise be a Leonid shower.

The particles that cause a meteor shower strike the Earth along nearly parallel paths, yet they appear to diverge from a point. The reason for this apparent contradiction is the same as the reason that the parallel sides of a long, straight highway seem to diverge as they near a viewer standing in its center (**Figure 10-22**). The particles forming the shower are coming into the Earth's atmosphere along parallel paths, but as they near us, they seem to be spreading out.

For centuries we have known that some meteor showers repeat regularly each year, but the origin of the showers was not known. Then, in 1866, it was shown that the particles that cause the Perseid meteor shower (which occurs around August 12 and appears to originate in the constellation Perseus) have almost exactly the same orbital path that an 1862 comet had. Astronomers concluded that the meteoroids of the shower were simply particles that had long ago come from the comet and formed its tail. Those particles were in orbit about the Sun when their paths intersected the Earth's. Today we are able to associate most of the major annual meteor showers with some comet. **Table 10-1** lists the major showers and their associated comets.

The table also indicates the average number of meteors expected to be seen during the maximum of the shower. However, the intensity of some showers changes greatly from year to year. **Figure 10-23** shows why this occurs. In part (a), the particles that had been blown from the comet have become spread fairly evenly around the comet's orbit, so each time the Earth passes through this orbit, about the same number of particles strike our planet. In part (b) of the figure, however, the particles are clumped in one region of the comet's orbit. If the Earth and this swarm happen to meet, we see much greater meteor activity.

As an example, the Leonid showers are irregular in intensity from year to year. In 1833, 1866, and again in 1899, the Earth passed through a major swarm of particles in the comet's orbit, and nearly 100,000 meteors could be seen in an hour. The max-

Figure 10-22

(a) Meteors in a meteor shower seem to come from a single point in the sky. (b) The divergence of the meteors is the same phenomenon seen by a person standing on a long, straight highway. The road seems to diverge as it comes toward the observer.

Table 10-1

Some Major Meteor Showers

Shower	Date of Maximum*	Associated Comet	Expected Hourly Rate
Quadrantids	Jan. 3	??	40
Lyrids	April 21	Comet Thatcher	15
Eta Aquarids	May 4	Comet Halley	20
Delta Aquarids	July 30	??	20
Perseids	Aug. 12	Comet Swift-Tuttle	50
Draconids	Oct. 9	Comet Giacobini-Zinner	15
Orionids	Oct. 21	Comet Halley	20
Leonids	Nov. 17	Comet Tempel-Tuttle	15
Geminids	Dec. 13	Asteroid Phaethon**	50

*The dates given are for the approximate date of maximum and may vary slightly from year to year.

**This asteroid is in the same orbit as the meteoroids and is thought to be the remains of the comet's nucleus.

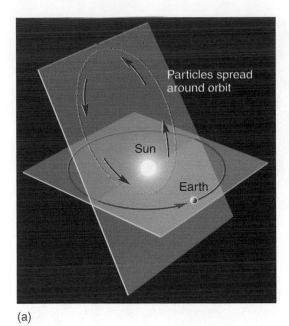

Particles spread around orbit

Sun

Earth

(a)

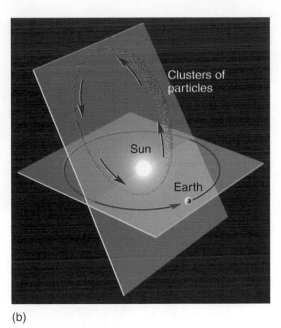

Clusters of particles

Sun

Earth

(b)

Figure 10-23
In part (a) the particles are spread around the entire orbit, but in (b) they are clustered in one area. The plane of the original comet's orbit is inclined to the plane of the Earth's orbit.

ima of the Leonid showers occur about every 33 years, which is just the period of the comet from which they originated. As the years go by, the particles spread out more and more along their orbit, but the display was spectacular in November, 1966, when Figure 10-20 was taken. Another beautiful display of the Leonids was seen in 1999, when observers in Europe saw up to 2500 meteors/hour. The Leonids can be very active for 3-4 years after the Comet's passage and model predictions for even bigger storms in 2001 and 2002 proved accurate.

Many meteor showers occur every year; Table 10-1 lists only the major ones. The best time to observe a meteor shower (assuming a clear, moonless night and an observing site far from city lights) is in the early morning hours. As the Activity at the end of the chapter explains, this results from Earth's rotation and its motion through the swarm of particles left behind by the comet.

Stars can be seen equally well at any time during the night; are meteors the same in this regard?

10-7 Meteorites and Craters

Meteorites are classified in three categories: iron meteorites (or *irons*), which are made up of 80 to 90% iron (most of the rest being nickel); stony meteorites (*stones*), which are just what their name implies, but often contain flakes of iron and/or nickel; and *stony irons*, about half iron and half rock. **Figure 10-24** shows three different meteorites.

About 95% of all meteorites are stones, but if you find a meteorite, it will likely be an iron. The reason for this apparent paradox is that stony meteorites are very similar to regular earthly stones, especially after the meteorites have weathered for a few years. Thus they are difficult to recognize. Irons, however, can be found by using metal detectors and are different in appearance from regular earthly rocks.

Some iron meteorites are found by people who realize that the rock they just kicked was "too heavy."

How do we know, then, that most meteorites are stones? Meteorites can be found in the Antarctic, where they become exposed when snow is blown away from them. Here, where there are no natural rocks to cause confusion, 95% of the meteorites we find are stones.

The largest meteorite ever found, the Hoba meteorite, weighs about 65 tons and is in Namibia, in southwestern Africa. Iron meteorites such as this are much

Figure 10-24

(a) A part of an iron meteorite that fell in Canyon Diablo, Arizona. (b) The sawed face of the stony Martian meteorite ALH84001,0. (Figure 8-31a shows this meteorite before processing.) (c) This 14-ton iron is the largest meteorite ever found in the United States. The farmer and his son who found it in 1902 moved it three-fourths of a mile and then charged people 25 cents to view it.

(a)

(b)

(c)

The crater is 600 feet deep and rises 150 feet above the surrounding ground.

stronger than stones and therefore do not break up in the atmosphere as stones do. The Hoba meteorite (so called after the name of the farm on which it lies) barely disturbed the surrounding surface of the Earth, apparently because it entered the atmosphere at a very small angle and was greatly slowed down before reaching the surface.

The second largest meteorite is on display in New York City. This 34-ton giant was hauled from Greenland in the 1890s by the explorer Robert Peary and is now in the Hall of Meteorites of the American Museum of Natural History.

What happens when a meteorite strikes the Earth? Quite naturally, it makes a hole. You can confirm that such a hole will be much larger than the meteorite by throwing a rock into mud to simulate a meteorite strike. You will see that the impact produces a crater that is much larger than the rock. **Figure 10-25** is a photograph of one of the most prominent impact craters on Earth, Meteor Crater near Winslow, Arizona. The crater is in the desert about 40 miles east of Flagstaff. The crater, nearly a mile across, is 180 meters deep and has a rim rising 45 meters above the surrounding desert ground. To appreciate the size of the crater, look at the left center of Figure 10-25 and find the parking lot and guest building.

Figure 10-25
This view of Meteor Crater shows the guesthouse and entrance road at the upper right edge of the crater.

Some 25 tons of iron meteorite fragments have been found around the crater, some as far as 4 miles away. The meteorite that formed Meteor Crater is estimated to have had a total mass of 300 million kilograms (300,000 tons) and to have been about 45 meters across. It struck about 25,000 years ago at a speed of about 11 kilometers/second (25,000 miles/hour), releasing an energy equivalent to a 25-megaton bomb.

One megaton corresponds to about 40 times the size of the bomb that destroyed Hiroshima in 1945.

About 120 large meteorite craters have been found so far on Earth, but most are far less impressive than Meteor Crater because weather has worn them down or they are under the sea. Meteor Crater is not only the most recently formed major crater, but it is also well preserved because of its location in the dry Arizona desert.

It has been estimated that a meteorite larger than 1 kilometer in diameter strikes the Earth once every few hundred thousand years, on the average. A hit by a meteorite 1 kilometer in diameter would be equivalent to a 5000-megaton bomb and would produce a crater 10 kilometers in diameter. Such an explosion would have effects far beyond the area of impact because it could send enough dust into the atmosphere to block out a significant amount of sunlight. A hypothesis put forward in 1980 suggests that such a meteorite strike was responsible for the extinction of the dinosaurs. (See the Advancing the Model box on page 349.)

The crater shown in **Figure 10-26** was formed on May 17, 1990, in a freshly planted wheat field in Russia. The 10-meter diameter crater was caused by an iron meteorite about 1 meter across. The impact energy is calculated to have been equivalent to 1 1/2 tons of TNT.

Figure 10-26
This 10-meter-wide crater was formed in Russia in 1990 by a 1500-kilogram iron meteorite.

ADVANCING THE MODEL

Hit by a Meteorite?

Something spectacular fell in Tunguska, Siberia, on June 30, 1908. The fireball was bright enough to be seen in daylight, and the sound that resulted from its explosion some 10 kilometers above the Earth was heard as far away as 1000 kilometers (600 miles). The explosion, which was equivalent to about a 50-megaton bomb, knocked down trees as far as 30 kilometers away (**Figure B10-1**), and a man 80 kilometers away was knocked down by the shock wave from the explosion.

Recent work on the fate of asteroids as they plunge through the atmosphere indicates that the object that caused the Tunguska event was an asteroid. Astronomers Jack Hills and M. Patrick Goda calculate that an asteroid at least 80 meters across traveling at 22 kilometers/second would have broken up violently in the atmosphere and caused the disruption at Tunguska in 1908.

Human bodies take up a small portion of the Earth's surface, so it is unlikely that a person will be hit by a meteorite or even see one land. There is an unconfirmed report of a monk being killed by a meteorite in 1650 in Milan, but no good evidence that a meteorite has ever killed anyone. In 1954, a meteorite came through the roof of a house in Sylacauga, Alabama, and hit a woman on the bounce, severely bruising her hip. In 1971, a house in Wethersfield, Connecticut, was hit by a meteorite, and in an extremely unusual coincidence, a house about a mile away was hit just 11 years later. These meteorites were a few inches across and did no major damage to the houses (if puncturing the roof and ceiling is not major damage).

Figure B10-1
These trees were broken like matchsticks by an explosion in 1908 in the Tunguska region of Siberia.

On August 31, 1991, in Noblesville, Indiana, two boys were standing by the sidewalk talking when they heard a whistle followed by a thud. Twelve feet from them they found a meteorite in a crater about 9 centimeters wide and 4 centimeters deep. They reported that the rock felt slightly warm when they picked it up.

Another close call occurred when a meteorite plunged through the rear of a car about 40 miles north of New York City on October 9, 1992 (**Figure B10-2**). The football-sized meteorite had a mass of 27 kilograms (about 60 pounds). The fireball caused by the meteorite (or from the larger one from which it broke) had been seen from as far away as Frankfort, Kentucky.

On September 23, 2003 a large meteorite (about 20 kg) fell in downtown New Orleans. It went through the roof of a house, penetrated the upstairs room floor, and then punctured a hole through the bottom floor of the house before shattering into many pieces on the ground.

Figure B10-2
Michelle Knapp of Peekskill, New York, examines the damage done by a stony meteorite in October 1992. She has been offered $69,000 for the stone.

10-8 The Importance of the Solar System Debris

Several spacecraft have been launched—and plans are being made for a small fleet to be launched in the next few years—to investigate asteroids and comets. *NEAR,* a NASA spacecraft that was launched in 1996, orbited asteroid Eros for a year, beginning in February, 2000. This was the first time a satellite was put in orbit around an asteroid. Eros is one of the largest and best-observed asteroids whose orbit crosses Earth's. *Stardust,* launched in 1999, will reach comet Wild-2 in January 2004 and, for the first time ever, collect dust from both the comet's coma and from interplanetary space and bring these samples back to Earth for study in January 2006. *Hayabusa,* a joint venture between Japan and the U.S., launched in May 2003 for a rendezvous

Hayabusa is the Japanese word for falcon.

ADVANCING THE MODEL

Meteorites and the Death of the Dinosaurs

Sixty-five million years ago, dinosaurs became extinct. In fact, nearly 75% of all species on Earth became extinct during the same short time period (as geological time is measured). The reason for this mass extinction has been debated by scientists for some time, and several different catastrophes have been hypothesized as the cause for the extinctions.

In 1980, astronomy entered the picture. In that year, geologists Walter Alvarez, Frank Asaro, Helen Michel, and Luis Alvarez (Walter's physicist father) proposed a solution for which they had real evidence. At Gubbio, Northern Italy, they found a layer of clay that contained the elements iridium, platinum, and osmium in much greater abundance than normally found on Earth. (Similar layers were later discovered at several other sites around the Earth.) How could this layer have been deposited? These elements are much more abundant in meteorites than in the crust of the Earth. Perhaps a giant meteorite—an asteroid—struck the Earth, exploded, and sent its debris high into the atmosphere to fall to Earth and form the layer. To account for the thickness of the layer found around the world, the meteorite must have been about 10 kilometers in diameter—not an unusual size for an asteroid. The debris, which also would have included earthly material pulverized by the impact, would spread as a cloud around the Earth and then gradually settle to form the layer found by the Alvarez team.

The energy released by such an impact would correspond to a few billion megaton bombs. But could this explosion have wiped out entire species all around the Earth? No, not directly. But consider what effect a giant cloud of dust, soot from global fires, and sulfate aerosols produced from impacted rocks would have on vegetation on Earth. Such a cloud could remain in the atmosphere for more than a year, darkening the Earth below. Much vegetation would die, along with many animal species that depended on vegetation for food. Dinosaurs are included in this category, as well as many other species. However, small creatures such as rodents that could forage for seeds and nuts could survive. And, indeed, the fossil record shows that such creatures did survive this period.

One test of this asteroid hypothesis is to find out how long ago the clay layer was formed; that is, to see if it is 65 million years old. If you have studied geology, you know that such dating of past events is a common procedure. The result? The clay was deposited about 63 million years ago, remarkably close to the value given for the mass extinction as shown by the fossil record. Another test is to find the location of such an impact. Observations suggest that the 180-kilometer crater buried below the jungle near Chicxulub, in the Yucatan Peninsula, is the likely impact site, based on its size, age, and the deposits found in its surrounding area (**Figure B10-3**).

Since 1980, when the Alvarez team presented their data, their hypothesis has become more and more accepted. (In fact, it is usually referred to as a *theory*, in accordance with the practice of calling an idea a theory only after it has gained some acceptance, as was discussed in the first few chapters.) However, a number of scientists believe that this mass extinction was not the result of a single event. Instead, they suggest that an intensive period of volcanic eruptions in addition to a number of asteroid impacts stressed the Earth's ecosystem to the breaking point. They also suggest that the Chicxulub impact was not as powerful as originally believed and probably occurred 300,000 years before the mass extinction.

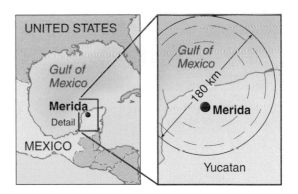

Figure B10-3
The Chicxulub crater off the Yucatan coast of Mexico measures 180-km across, making it one of the largest craters formed in the inner solar system during the last 4 billion years. It is not completely visible from above, but was detected by measuring slight variations in Earth's gravitational pull—measurements made originally for oil exploration.

with an asteroid in June 2005. It will survey the asteroid's surface, collect samples and return them to Earth for analysis in June 2007. The European Space Agency plans to launch *Rosetta* in February 2004, which will rendezvous with, orbit around, and put a lander on the surface of a comet to study its composition and structure. Finally, NASA's *Deep Impact* mission will launch in December 2004 for an impact with the surface of comet Tempel 1 in July 2005. The impact will create a crater with depth larger than a seven-story building and area larger than a football field. The purpose of the mission is to study the inner structure of the comet and the formation process of the crater.

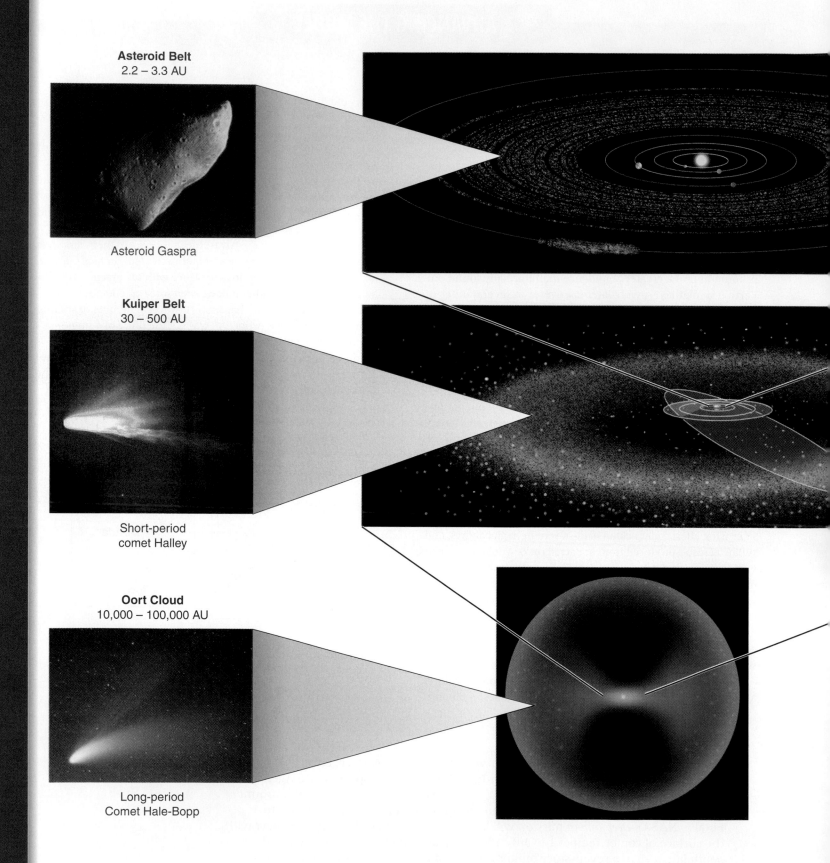

Asteroid Belt
2.2 – 3.3 AU

Asteroid Gaspra

Kuiper Belt
30 – 500 AU

Short-period
comet Halley

Oort Cloud
10,000 – 100,000 AU

Long-period
Comet Hale-Bopp

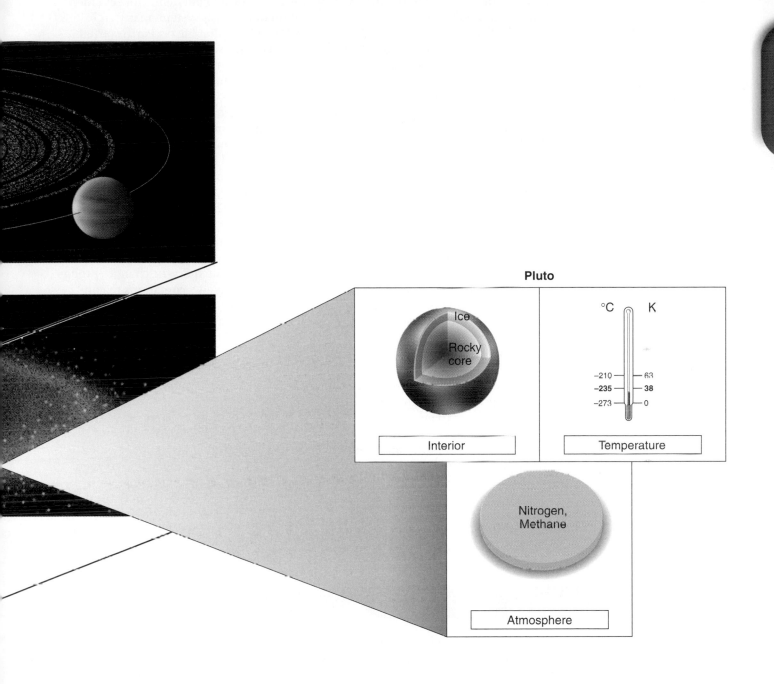

Pluto

Ice

Rocky core

Interior

°C   K

−210 —— 63

−235 —— **38**

−273 —— 0

Temperature

Nitrogen, Methane

Atmosphere

Astronomers are interested in what seems to be nothing but debris for several reasons. During the early period of the formation of the solar system, many collisions occurred between the planets and this debris. We believe that life on Earth started at the end of this period, about 3.8 billion years ago. This seems plausible because carbon-based molecules could not have survived the intense heat resulting from the collisions. There is evidence for biological activity at the end of the heavy bombardment period and known fossils on Earth date as far back as 3.5 billion years ago. This very short, on a geological scale, time period during which life formed on our planet begs the question of how it could have started during a time when most of the water in Earth's oceans was still vapor and there were few carbon-based molecules around. It has been suggested that the building blocks of life could have been delivered to Earth by asteroid and comet impacts. These collisions supplied the Earth with large supplies of water, volatiles, and carbon-based molecules. Understanding the chemical makeup of comets opens a window to our understanding of the composition and conditions from which our planet formed 4.6 billion years ago.

It is ironic that asteroid and comet impacts may have brought the building blocks of life to Earth, while similar collisions later on may have wiped out many developing species and significantly modified our planet's biosphere. Such drastic changes probably allowed only species capable of quickly adapting to the new conditions to continue evolving and possibly opened the way for mammals to dominate our planet.

There is another strong reason for interest in solar system debris. Most of the meteorites reaching Earth's surface originate as fragments of collisions between asteroids and are small in size. The larger the size of the asteroid, the less frequent a possible collision. However, even though the probability of a catastrophic collision is small, the consequences are such that it becomes imperative that we understand their structure, composition, and orbits in order to better protect our planet. Catastrophic collisions happened in the past and they will happen again.

Finally, asteroids and comets will most likely play an important role in any future exploration and colonization of the solar system. Comets can supply water and carbon-based molecules necessary to sustain life. Liquid hydrogen and oxygen obtained from the water can be used to generate rocket fuel. Asteroids are rich resources of raw materials that we can use to build space structures. This may sound like science fiction now, but what was science fiction yesterday is the reality of today. Colonization of the inner solar system will most likely occur in the 21st century and solar system debris may well play an integral part in our efforts.

Conclusion

The solar system is indeed a collection of diverse objects. As our space probes venture out to study it in more detail, we are continually surprised by the beautiful diversity we find. As the last few chapters have shown, great differences exist not only among planets, asteroids, comets, meteors, and satellites, but even between objects within these various categories. On the other hand, we cannot help but be impressed by the similarities we see—even between objects we place in different categories.

As we explained in Chapter 7, the objects in our solar system formed at about the same time, and their formation and differentiation were determined by the conditions that existed at the time. We are beginning to understand why they are different and why they are the same, but we have a long way to go. That is what makes the study of the solar system exciting.

Study Guide

1. The connection between comets and meteors is demonstrated when one sees
   A. sporadic meteors on almost any night.
   B. a comet in one part of the sky and meteors in another part of the sky on the same night.
   C. a predictable shower of meteors.
   D. [None of the above; there is no connection.]

2. Which of the following statements about Pluto is true?
   A. It has two moons (that we know of).
   B. Its orbit is within 2 degrees of the plane of the Earth's orbit.
   C. Its orbit is as circular as the Earth's orbit.
   D. We are confident that it was once a moon of Neptune.
   E. [None of the above is true.]

3. The search for Pluto was undertaken based on
   A. Bode's law.
   B. calculations concerning the orbit of Uranus.
   C. perturbations in the orbits of asteroids.
   D. [None of the above, for it was discovered accidentally.]

4. Pluto was discovered by
   A. William Herschel in England in the last century.
   B. Percival Lowell in Arizona in 1905.
   C. Clyde Tombaugh in Arizona in 1930.
   D. James Christy in Arizona in 1978.
   E. Asaph Hall in California in 1921.

5. Pluto's mass has been calculated from
   A. its effect on passing spacecraft.
   B. its effect on passing asteroids.
   C. the motion of Charon.
   D. [All of the above.]
   E. [None of the above.]

6. How was Ceres discovered?
   A. It was discovered by accident, during a search for comets.
   B. It was observed when it fell to Earth.
   C. It has been known since antiquity, so the manner of its discovery is unknown.
   D. It was found during a search for a "missing planet."
   E. It was found accidentally as a streak on a stellar photo.

7. The discovery of asteroids depends on the fact that, compared to the stars, the asteroids
   A. look bigger.
   B. look brighter.
   C. move.
   D. vary in brightness.
   E. are a different color.

8. The largest asteroid is closest to _____ in diameter.
   A. 6 feet
   B. 600 feet
   C. 6 miles (10 kilometers)
   D. 600 miles (1000 kilometers)
   E. 60,000 miles (100,000 kilometers)

9. The total radiation reflected and radiated by an asteroid depends on
   A. its size.
   B. the total radiation it receives from the Sun.
   C. its speed in orbit.
   D. [Both A and B above.]
   E. [All of the above.]

10. The Kirkwood gaps result from
    A. radiation pressure from the Sun.
    B. the previous passage of a comet through the asteroid belt.
    C. regular gravitational pull from the largest of Jupiter's moons.
    D. regular gravitational pulls from Jupiter.
    E. previous explosions of asteroids at points in the asteroid belt.

11. The masses of the largest asteroids are measured by observing
    A. their motions as they come closest to Jupiter.
    B. perturbations they cause on smaller asteroids.
    C. the motion of spacecraft passing nearby.
    D. the total radiation received from them.
    E. their orbital speed and distance from the Sun.

12. Which of the following statements about the *Voyager* spacecraft that passed through the asteroid belt is true?
    A. They were guided from Earth so that they did not collide with asteroids.
    B. They had their own guidance system to prevent collisions with asteroids.
    C. They passed through the asteroid belt with no collision-avoidance system.

13. Which of the following statements best describes cometary orbits?
    A. They are circular, lying in the plane of the ecliptic.
    B. They lie between the orbits of Mars and Jupiter.
    C. They are elongated ellipses, lying within one astronomical unit of the Sun.
    D. They are very elongated ellipses tens to hundreds of astronomical units across.
    E. They are within the orbit of Mercury.

14. When a comet is visible in the night sky, it appears to the naked eye to
    A. move so rapidly across the sky that it is easily missed if one is not looking in the right direction.
    B. stay in the sky for a week or more, with only slight shifting among the stars each night.
    C. [Either of the above, depending upon the particular comet's motion.]

15. The nucleus of most comets is
    A. much smaller than the Earth.
    B. slightly smaller than the Earth.
    C. slightly larger than the Earth.
    D. much larger than the Earth.

16. A comet's tail results when
    A. part of the comet drifts in the direction opposite to the motion of the comet.
    B. the solar wind carries some of the gas of the coma away from the Sun.
    C. gravitational force pulls loosely held material from the Sun.
    D. tidal forces tear the comet apart.

17. The best-known comet was named after Halley because
    A. he was the first to see it.
    B. he first calculated its orbit.
    C. although he had nothing to do with the comet, he was a famous astronomer and it was named to honor him.
    D. Galileo discovered it and named it after him.
    E. [Both C and D above.]

18. A comet's tail
   A. precedes its head through space.
   B. follows its head through space.
   C. is farther from the Sun than its head is.
   D. is closer to the Sun than its head is.
   E. [None of the above.]

19. The Oort cloud is located
   A. between the orbits of Mars and Jupiter.
   B. just beyond Pluto's orbit.
   C. far beyond the orbit of Pluto.
   D. at about the distance of the nearest star.

20. Why do many comets appear brighter after passing the Sun?
   A. They are more massive after the passage.
   B. The Sun starts nuclear reactions in them.
   C. They are moving faster after passing the Sun.
   D. Their head and tail are larger after passing the Sun.

21. Short-period comets are hypothesized to
   A. be permanent parts of the inner solar system similar to asteroids.
   B. be formed when the orbits of long-period comets are changed.
   C. be formed in the inner solar system and then ejected, becoming long-period comets.

22. Why do most meteoroids not reach the surface of the Earth?
   A. They bounce off the atmosphere and go back into space.
   B. They burn up in the air.
   C. They are light enough that they remain suspended in the air.
   D. They land in the ocean (since oceans cover most of the Earth).
   E. [The statement is false; almost all reach the Earth, but they are not found.]

23. Meteor showers are caused by
   A. the Earth crossing the asteroid belt.
   B. a comet's nucleus striking the atmosphere.
   C. the eccentricity of the Earth's orbit.
   D. meteorite impacts on the Earth's surface.
   E. the Earth crossing a meteoroid stream.

24. Meteors are most easily seen after midnight because
   A. the sky is darker then.
   B. the Sun is closer to rising.
   C. you are then on the "leading" side of the Earth.
   D. meteor showers occur then.
   E. [The statement is false; they are seen equally well anytime.]

25. List two ways in which Pluto's orbit is unusual compared to the other planets.

26. We cannot determine the size of Pluto from the size of its image in a telescope. How, then, can we know its size?

27. What did the discovery of Charon allow us to calculate about Pluto?

28. Why have we not named all of the asteroids that have been observed?

29. What causes the Kirkwood gaps?

30. Why have the asteroids not formed into a planet between Mars and Jupiter?

31. Why do we think that it is unlikely that the asteroids are the remains of a planet that exploded?

32. Describe the three main parts of a comet. Include approximate sizes of each part for a typical comet.

33. Describe the modified "dirty snowball" model of a comet's nucleus.

34. What causes a comet to have two tails? Why are they different shapes?

35. What is the Oort cloud?

36. We cannot hope to see the Oort cloud. What makes us think that it exists?

37. Explain why the passing air heats a meteoroid, while a wind cools a person.

38. What causes a meteor shower and why do the meteors appear to come from just one part of the sky?

39. Name and describe the three main types of meteorites. Which is most common? Which is easiest to find? Why?

QUESTIONS TO PONDER

1. Charon occulted Pluto during the time that Pluto was closest to the Sun. Astronomers (and police detectives) do not like coincidences. Propose a reason that Pluto and Charon act this way.

2. List reasons why Pluto should be considered a planet and other reasons why it should not.

3. *NEAR* (Near Earth Asteroid Rendezvous), a NASA spacecraft that was launched in 1996, began orbiting asteroid Eros in 2000. Use *Astronomy* and/or *Sky & Telescope* magazines to write a report on the latest results from this mission.

4. Objects in the Oort cloud cannot be seen, even with the best telescopes. Why, then, do we believe it exists?

5. If a comet is seen in the west shortly after sunset, in what direction will its tail point—toward or away from the horizon? In what direction will it point if it is seen in the east in the morning sky?

6. What effect on the motion of a comet would you expect to result from the jetting of material from its nucleus?

7. Compare the size and shape of Mars's moon Phobos to that of the nucleus of Halley's comet.

8. Comets are generally brighter during the few weeks after they pass perihelion than during the few weeks before perihelion. Why? (Hint: How well does water ice retain heat?)

9. Use *Astronomy* and/or *Sky & Telescope* magazines and information from NASA's web site on "Asteroid and Comet Impacts Hazards" to write a report on the latest efforts to counter a possible catastrophic impact. Compare an impact with a 10-kilometer asteroid to the destruction resulting from a global thermonuclear war.

10. Write a report about a meteor crater in your part of the country or the world. The geology department of your college is a suggested source of information.

11. Find out more about the "theory" that a catastrophic impact led to the extinction of the dinosaurs. Do you think that there is enough observational support for it? Are there any competing theories? If so, describe them.

12. Assume that the dinosaur-extinction-by-meteorite theory is correct. If the impact had not occurred 65 million years ago, would mammals still have risen to the point of dominating our planet? Discuss.

13. Based on what you know about the makeup of the solar system debris and their constant bombardment of the planets, what effect do you think the debris had on the evolution of life on our planet?

CALCULATIONS

1. Assume that 1000 metric tons (one million kilograms) of meteoritic material strikes the Earth each day. Calculate the area of the Earth and determine how much material strikes each square meter each day.

2. Use the answer to calculation 1 to determine how long it would take to build up a layer of meteoritic material weighing 1 kilogram on a desktop that has an area of 2 square meters.

3. Assume that the period of Halley's comet is 76 years. What is, approximately, the farthest distance that this comet gets from the Sun? (Hint: Use Kepler's third law and ignore the comet's distance from the Sun at perihelion.)

4. Imagine that we have just observed an asteroid on an orbit that will bring it somewhere within 400,000 kilometers from Earth. What is the probability that this asteroid will actually hit Earth? (Hint: Look at this problem from the point of view of the asteroid. You have a dartboard of radius 400,000 kilometers and the bull's-eye has a radius of 6400 kilometers.)

5. Consider two comets that get very close to the Sun at perihelion, while their aphelion distances are 100 AU and 100,000 AU, respectively. Assume that the comets can survive only 100 passages around the Sun. What is the lifetime of each comet? How does it compare to the age of our solar system?

6. Assume that a 1-kilometer asteroid is on a direct collision course with Earth, moving at about 10 kilometers/second relative to the Earth. We first observe this object at a distance of 10 million kilometers. How much warning time do we have? How long would it take for this object to pass through our atmosphere (about 100 kilometers thick)?

7. Assume we have the technology to mine asteroids. Consider a small asteroid of radius 100 meters, which is made of mostly nickel-iron of density 8,000 kilograms/meter^3. If the cost for one kilogram of this alloy is $1, how much money can we make by mining this asteroid?

ACTIVITIES

Observing Meteors

To observe meteors, you should have a clear sky away from city light, so your first step is to get out into the country. Second, you must avoid a night with a bright Moon that will light up the sky so that you have difficulty viewing. Finally, the best time for observing is after midnight, preferably from around 2 A.M. until the beginning of morning twilight.

The reason for choosing a time after midnight can be seen by referring to **Figure 10-27**. In this figure, the Sun (not shown) is to the left and the Earth is moving toward the top of the figure. Thus the half of the Earth lined in blue is on the leading edge as the Earth moves through space. Now consider a car driving through a rain shower and you see why most meteoroids hit the Earth on this leading side. The analogy is not quite exact because different meteoroids have very different motions through space, while raindrops all fall in about the same direction. Meteoroids are able to catch up with the Earth from behind and hit the trailing half of the Earth. Still, many more strike the Earth's leading edge as the Earth sweeps them up.

Referring again to the figure, you see that it is midnight for a person standing at point A and that the Sun is rising for a person at B. Meteors, then, are better observed in the early morning hours.

Once you have found your observing location, you must wait for your eyes to adapt to the darkness. Unless you have chosen a night of a heavy meteor shower, you must be patient. Lie back and relax. (Did you bring a lawn chair?) Patience should reward you with a streak of a meteor across the sky. It is likely that you will see the meteor in some direction other than where you are looking—out of the corner of your eye. There is a good reason for this: the part of your retina that sees things out of your direct line of sight is more sensitive to motion and to changing light. An experienced observer takes advantage of this effect when looking, for example, for the motion of a satellite in the sky. The observer looks to the side of where the object is expected to be seen.

The best advice to the person looking for meteors is to pick a night when a meteor shower is expected. Table 10-1 lists the dates of the most prominent annual meteor showers.

Figure 10-27
The Earth is moving upward in the drawing and strikes most meteoroids on its leading edge. Those striking it from behind must catch up. It is midnight at point A and sunrise at point B.

You can find information about current explorations of our solar system at NASA's home page (http//www.nasa.gov).

1. "Killer Crater in the Yucatan?" by J. K. Beatty, in *Sky & Telescope* (July, 1991).

2. "The Origins of the Asteroids," by R. P. Binzel, M. A. Barucci, and M. Fulchignomi, in *Scientific American* (October, 1991).

3. "The Oort Cloud," by P. Weismann, in *Scientific American* (September, 1998).

4. "The Kuiper Belt," by J. X. Luu and D. C. Jewitt, in *Scientific American* (May, 1996).

Quest Ahead to Starlinks:
www.jbpub.com/starlinks

Starlinks is this book's online learning center. It features **eLearning,** which contains chapter quizzes and other tools designed to help you study for your class. You can also find **on-line exercises,** view numerous relevant **animations,** follow a guide to **useful astronomy sites** on the web, or even check the latest **astronomy news** updates.

EXPANDING THE QUEST

STARLINKS

This *Hubble* image of comet LINEAR (C/1999 S4), taken on August 5, 2000, shows that the comet's nucleus has been reduced to a shower of small fragments. The distance to the comet was 102 million kilometers.

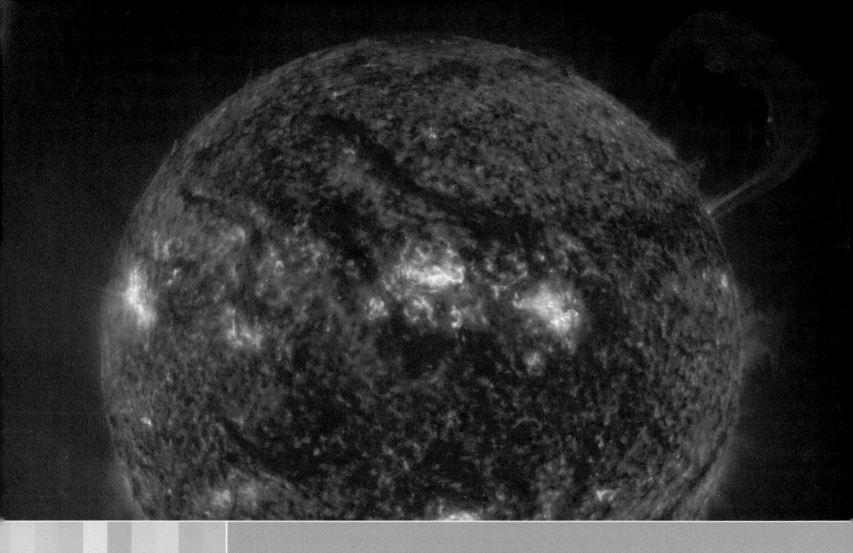

The Sun

Prominences on the Sun are huge eruptions of gas that often form arches along magnetic field lines. The entire Earth could easily fit beneath one of these arches.

THE CHAPTER-OPENING PHOTOGRAPH, WHICH WAS TAKEN FROM SPACE, shows prominences on the Sun's surface, giving us a hint of the tremendous energy within. Over the years, many well-known physicists tried to explain where all this energy comes from. In 1854, Hermann von Helmholtz hit on the idea of gravitational contraction as the source of the Sun's heat. He calculated that the Sun could have existed for 25 million years this way, which was not long enough to match the findings of geologists of the time concerning the age of the Earth. In 1862, Lord Kelvin, one of the most imposing figures in science, refined the calculations to say that the Sun could have been shining for 500 million years, but no more. By this time, Darwin's theory of evolution was being debated all over Europe, and for evolution to be correct, the Earth (and Sun) had to be much older than this. But Kelvin obtained the same result in several versions of the same calculation, always basing his work on gravitational contraction. His only concession to the possibility of error was to say "I do not say there may not be laws which we have not discovered."

There is such a law: radioactivity. Henri Becquerel announced his discovery of radioactivity in 1896, touching off a flood of research activity. One result was the work of young Ernest Rutherford, who showed that radioactive materials could produce large amounts of energy. This energy could provide additional heat in the Earth and the Sun that would mean these objects are far older than Kelvin's results; billions of years older, in fact. Here is Rutherford's description of the address he gave on this topic at a meeting of England's Royal Institution:

Figure 11-1
The Sun is the ultimate source of oil (including that pumped by this rig) and of all energy on Earth except nuclear and geothermal energy.

> I came into the room, which was half dark, and presently spotted Lord Kelvin in the audience and realized that I was in for trouble at the last part of my speech dealing with the age of the earth, where my views conflicted with his. To my relief, Kelvin fell fast asleep, but as I came to the important point, I saw the old bird sit up, open an eye and cock a baleful glance at me! Then a sudden inspiration came, and I said Lord Kelvin had limited the age of the earth, provided no new source [of energy] was discovered. That prophetic utterance refers to what we are now considering tonight, radium! Behold! the old boy beamed upon me. (From A.S. Eve, *Rutherford,* Cambridge University Press, 1939, p. 107.)

In fact, we now believe that the source of energy in the Sun is not due to the energy of radioactive decay, but comes from reactions of nuclei in the hot, dense core. We will discuss these reactions in this chapter.

In our quest to understand our universe, we started by first examining the basic tools we have for deciphering the information we get from celestial objects. Then we used these tools to understand the formation and evolution of our own planet and that of the remaining objects in our neighborhood—the solar system. We now study the Sun, the most important member of our solar system. Without it, life on Earth would not be possible. It is also the closest laboratory for studying how other stars evolve, and thus it is a logical next step in our quest to understand the physical universe.

Although the Sun is the celestial object of most importance to life on Earth (**Figure 11-1**), it is just an ordinary star. The cosmic importance of the Sun is limited to the fact that it is the central object of the planetary system in which we live. Many other stars are bigger and brighter, many are more interesting and unusual, and most will far outlive the Sun.

This chapter begins with a brief overview of the major properties of the Sun. Then we examine the source of the Sun's energy, along with various theories for the production of that tremendous energy. Finally, we describe the Sun in more detail, starting at its center, where energy production takes place, and proceeding outward.

11-1 Solar Properties

As viewed from Earth, the Sun has an average angular diameter of 31'59", just barely less than 32 minutes of arc. By taking 1.50×10^8 kilometers as the average distance from Earth to Sun and by using the small-angle formula (the relationship between angular size, distance, and actual size discussed in Section 6-1), we can calculate that the diameter of the Sun is 1.39×10^6 kilometers. Thus the Sun's diameter is about 110 times Earth's diameter and about 10 times Jupiter's. **Figure 11-2** illustrates the

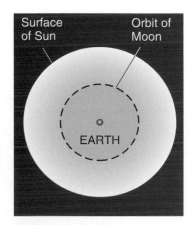

Figure 11-2
If the Earth were at the center of the Sun, the Moon would orbit about halfway to the Sun's surface.

The Sun

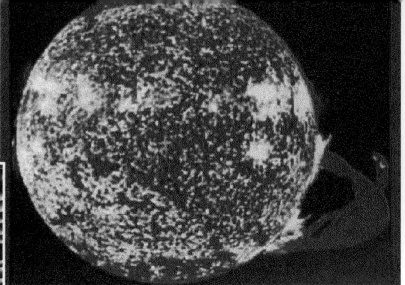

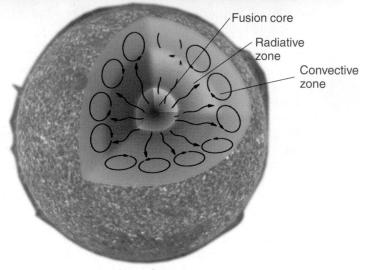

Solar wind shapes the magnetic fields of all planets in the solar system, as Earth's field shown.

Interior

Fusion core

Radiative zone

Convective zone

Sun Data

Sun	Value	Compared to Earth
Diameter	1,392,000 km	109.2
Mass	1.989×10^{30} kg	332,980
Density	1.41 g/cm³	0.255
Surface Gravity	274 m/s²	28.0
Escape velocity	618 km/s	55.2
Effective temperature	5778 K	
Luminosity	3.85×10^{26} watts	
Tilt of equator to ecliptic	7.25°	
Sidereal rotation period		
Equator	25.05 days	
40° latitude	27.40 days	
80° latitude	33.83 days	

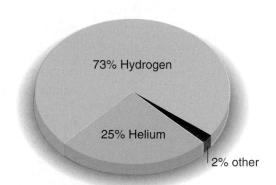

73% Hydrogen

25% Helium

2% other

Atmospheric Gases

QUICK FACTS

- Typical star, consisting mostly (98%) of hydrogen and helium gas. Contains 99.85% of all mass in solar system. Gravitational contraction causes nuclear fusion in the core, where four hydrogen nuclei fuse into a helium nucleus (proton-proton chain). Resulting energy transported to surface of Sun by radiation and convection. Complex magnetic field causes eruptions of gas as spicules, flares, and prominences. Solar wind of high-energy particles affects magnetic fields of all planets in the solar system. Sunspot activity follows 11-year cycle of activity.

TOOLS OF ASTRONOMY

The Distance to the Sun

How do we measure the distance to the Sun? Knowledge of this distance is key to calculating the size of the Sun, and the size must be known before we can determine other solar properties, such as density and total luminosity.

As we pointed out when the question of the heliocentric-versus-geocentric system was discussed in Chapter 2, the relative distances to the planets can be calculated by applying simple geometry to the heliocentric system. Thus, Copernicus was able to calculate that Mars is 1.5 times as far from the Sun as the Earth is. We say that Mars is 1.5 astronomical units from the Sun. But how large is an astronomical unit? The situation is similar to the following scenario: You have a map of some unknown country. The map is drawn to scale, so you can tell from the map that the nation's capital is one-third of the way from the mountains to the ocean. However, unless you know the scale of the map, you have no way of knowing the distance from the mountains to the capi-

tal. You know only relative distances. If a scale tells you that one inch on the map corresponds to 120 miles, you can make measurements on the map and calculate all distances. In the case of the solar system, Copernicus lacked the scale factor—the number of kilometers in an astronomical unit.

Today, radar allows us to determine the scale factor. We can bounce radio waves off of other objects and use the time of travel of the waves to determine the distance to the object. We cannot get good results bouncing radio waves from the Sun, both because the Sun does not reflect them well and because the Sun emits its own radio waves. We can, however, reflect such waves from other planets. Suppose we bounce radio waves off of Mars when Mars is 180 degrees from the Sun in the sky. Since Earth's distance to Mars at that time is 0.5 astronomical units, we can determine the number of kilometers in 0.5 astronomical units and therefore determine the scale of the solar system.

---

great size of the Sun. (Activity 1 at the end of the chapter shows how you can measure the diameter of the Sun.)

From Kepler's third law (as revised by Newton), we calculate that the mass of the Sun is 1.99×10^{30} kilograms. This is about 333,000 times the mass of the Earth! The Sun's average density, then, is 1.41 grams/centimeters^3, about the same as the density of Jupiter. This is just the first of several similarities we will see between the Sun and some of the planets in our solar system.

Nearly 400 years ago, when Galileo first used a telescope to view the Sun, he observed dark spots moving across its face (**Figure 11-3a**). Figure 11-3b is a series of photographs of the motion of such

(a)

Figure 11-3
(a) Sunspots appear as dark spots on the Sun's surface. (b) This photo series shows the motion of sunspots as time passes.

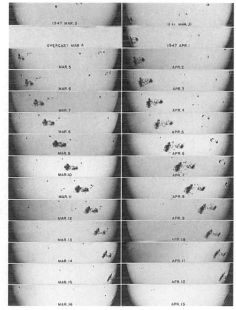

(b)

sunspots. Galileo concluded from sunspot motions that the Sun rotates with a period of more than a month. Today we know that the Sun exhibits differential rotation; it rotates with a period of 25 days at its equator and nearly 35 days near its poles. Recall that the equatorial regions of the Jovian planets also rotate faster than their polar regions.

sunspot A region of the Sun's surface that is temporarily cool and dark compared to surrounding regions.

11-2 Solar Energy

luminosity The rate at which electromagnetic energy is emitted.

watt A unit of power that corresponds to a specific amount of
energy each second. (Imagine how
dim a one-watt light bulb would be.)

power The amount of energy
exchanged per unit time.

The Sun emits energy in all portions of the electromagnetic spectrum. A valuable
piece of information about the Sun is the rate at which it emits its energy, or total
power output. Fortunately, this is not too difficult to determine.

We start by measuring the rate at which solar energy strikes the Earth's atmosphere. This determination was made long ago by measuring the amount of energy
from the Sun that strikes an area on the Earth's surface and then correcting for the
energy absorbed by the atmosphere. Today it is done most accurately from satellites
above the atmosphere. Measurements show that solar energy strikes the upper atmosphere of the Earth at the rate of about 1370 watts per square meter. This value is
used in the following example to show that the Sun's *luminosity*—the energy the
Sun radiates into space per second—is 3.85×10^{26} *watts.* (If the Sun were a light
bulb, this would be its "wattage.") This is an awesome amount of *power.* The energy
that the Sun releases *in one second* is about the same as the energy released by the simultaneous explosion of 4 trillion atomic bombs! The solar energy received by the
surface of the Earth, assuming we could collect and harness it efficiently, is enough
to cover the energy needs of the entire world population 10,000 times over.

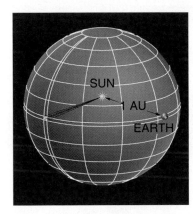

Figure 11-4
An imaginary sphere is shown drawn
around the Sun at the distance of the
Earth. It is a simple matter to calculate
the area of such a sphere and to determine the solar energy that strikes each
square meter of it.

EXAMPLE

Calculate the energy output of the Sun in watts, given that solar energy strikes the
Earth at the rate of 1370 watts/meter^2 and that the Sun is 1.496×10^8 kilometers
from Earth.

SOLUTION

First, since we have expressed the area on Earth in square meters, we should also
express the Earth–Sun distance in meters:

$$1.496 \times 10^8 \text{ km} \times \frac{1000 \text{ m}}{1 \text{ km}} = 1.496 \times 10^{11} \text{ m}$$

Imagine a sphere around the Sun at the Earth's distance. We must calculate how
many square meters are on that surface (see **Figure 11-4**). To do this, use the equation
for the area of a sphere, with 1.496×10^{11} meters being the radius of the sphere in
the calculation:

$$\text{area of sphere} = 4\pi r^2 \quad (r = \text{radius of the sphere})$$

$$= 4(3.14)(1.496 \times 10^{11} \text{ m})^2 = 2.81 \times 10^{23} \text{ m}^2$$

Each of these square meters receives 1370 watts of solar power. Therefore,

$$\text{total solar power} = (1370 \text{ watts}/\text{m}^2) \times (2.81 \times 10^{23} \text{ m}^2)$$

$$= 3.85 \times 10^{26} \text{ watts}$$

Given the power striking the Earth's
surface, you could use the inverse
square law of radiation, Section 4–8,
to calculate this.

TRY ONE YOURSELF

How many watts of power strike one square meter of Mars's surface? (Mars is
2.3×10^{11} meters from the Sun.) Solve this by using the total solar power calculated
in the example and the area of a sphere at Mars's distance.

The example also illustrates nicely why the inverse square law applies to radiation from the Sun (or any distant object). The surface area of a sphere that is centered
on the Sun depends upon the square of the sphere's radius. Thus a sphere twice the

TOOLS OF ASTRONOMY

Observing the Sun

Observing the Sun through a telescope poses unique problems. You know that you can't look at the Sun directly because the intense light can damage your eyes. But the energy of sunlight can also cause air to move around in the telescope tube, blurring the image. Astronomers have developed several specialized instruments to deal with these problems. Most of these observatories are at high altitudes, like most regular observatories, to minimize scattering of light by dust and water vapor.

The largest solar telescope in the world is the McMath Solar Telescope on Kitt Peak in Arizona (**Figure B11-1**). The 492-foot optical tunnel is oriented parallel to the Earth's axis of rotation to make it easier to follow the Sun across the sky. An 80-inch rotating mirror collects the sunlight and directs it down the tunnel to the 60-inch primary mirror. More than half the tunnel is buried in Kitt Peak to help keep the telescope's components at a uniform temperature. The tunnel is also surrounded by coolant pipes to stabilize the air within the instrument.

The Pic du Midi Observatory is located in the Pyrenees Mountains in France. The main instrument is a coronagraph, which contains an "occulting disk" to block out the main body of the Sun and allow only light from the Sun's corona to pass into the main optical system. Astronomers using the instrument often have to bundle up against the cold; it gets chilly at the top of a 9450-foot mountain.

Figure B11-1
The McMath Solar Telescope contains a rotating mirror on top of a 100-foot tower that minimizes air turbulence at ground level. The light is directed down a sloping 492-foot tunnel, most of which is buried in the ground, to the primary mirror. Visitors are allowed in the tunnel when it is not operating.

Perhaps the most exotic locale for an observatory is the site of the solar oscillation telescope at Amundsen-Scott Station, in Antarctica. Near the South Pole, the Sun doesn't set for half the year, so the telescope can image the Sun continuously from October to February in the crystal-clear air. It is designed to identify patterns in the Sun's surface oscillations, which can be done more accurately here than anywhere else on Earth.

distance from the Sun has a surface area four times as great, and only one-fourth as much energy strikes a square meter on the surface of this more distant sphere.

The Source of the Sun's Energy

It has been estimated that a 1 percent change in solar luminosity would result in a temperature change on Earth of 1 or 2 degrees Celsius (about 2 to 4 degrees Fahrenheit). When we consider that the last major ice age on Earth resulted from a temperature decrease that averaged only 5 degrees Celsius across the planet, we see how critical it is that the Sun maintain a uniform rate of energy production. In reality, this rate is not uniform; the amount of solar energy that strikes the Earth varies with time, the variations over the centuries being of the order of fractions of a percent. However, the more we study the interaction between the Sun's energy output and our planet, the more it seems that even these small variations have consequences for life on Earth.

What is the source of this energy, that must have remained approximately constant far into the past?

Prior to the 20th century, several hypotheses had been suggested to explain the source of the Sun's energy. All have now been rejected, based on additional data. For example, it was proposed that chemical reactions (such as the burning of a fuel) are the source of the Sun's energy. We now know that this cannot be the case, simply because if the Sun were made of a fuel such as coal or oil, it would burn out in a few centuries at the rate that it is releasing energy.

In the mid-nineteenth century, Hermann von Helmholtz and Lord Kelvin proposed that the source of the Sun's energy is a very slow gravitational contraction. Such contraction compresses the gases inside the Sun, raising their temperature. This is similar to the air in a bicycle tire getting warmer when you compress it with a pump. When the Sun's gases got hot enough, they started radiating energy out into space. The calculations of Helmholtz and Kelvin showed that gravitational contraction could have produced the Sun's energy output with a reduction in the Sun's diameter of only a few tens of meters per year—so slight that it would not have been enough to notice in recorded history. However, assuming that the Sun was formed from a large diffuse cloud, they calculated that gravitational contraction could not have started more than a few hundred million years ago. This time period seemed long enough in the 19th century, since the Earth was thought to be much younger. Their theory seemed to be a good one; it fit the available data.

Then, in the 20th century, geologists discovered that the Earth's age is not a few hundred million years, but rather a few *billion* years—ten times longer. The contraction theory had to be abandoned, and the source of the Sun's energy was again an open question.

In the first decade of the 20th century, as a result of Einstein's special theory of relativity, we started considering mass and energy as interconvertible. That is, one can be transformed into the other. In the late 1920s it was hypothesized that this process could be the source of energy in the Sun. Then during the 1930s, physicists worked out the theory that today explains how the Sun has produced its tremendous power for the past 4 to 5 billion years and how it will continue this production for another similar period of time.

Solar Nuclear Reactions

Recall from Chapter 4 that the Bohr model of the atom proposed that the atom consists of a nucleus surrounded by orbiting electrons (**Figure 11-5**). That nucleus will be our focus now. An atom's nucleus makes up about 99.98% of the mass of the atom and consists of two kinds of particles: ***protons*** and ***neutrons.*** Protons have a positive electrical charge and neutrons have no electrical charge. The number of protons in the nucleus determines what element the atom is. For example, if the nucleus of an atom contains 1 proton, that atom is necessarily an atom of hydrogen. If it contains 2 protons, it is helium; if 6, carbon; if 92, uranium. A nucleus of hydrogen, on the other hand, is not limited to a specific number of neutrons. Although most hydrogen nuclei contain no neutrons, some have one neutron and a few have two.

It is important to distinguish nuclear reactions from chemical reactions. Chemical reactions, which we encounter in everyday life, involve atoms changing the ways in which

---

Calculation 3 takes you through the steps for proving this statement.

The relationship between mass and energy is what the equation $E = mc^2$ is about. In this equation, E stands for the amount of energy that can be created from a certain amount of mass, m. The conversion factor c is the speed of light.

proton The positively charged particle in the nucleus of an atom.

neutron The nuclear particle with no electrical charge.

Figure 11-5
The Bohr atom, with electrons (which have a negative charge) circling the positive nucleus. If the atom were this size, the nucleus would still be too small to see. Drawings such as this one are used only as visualization tools. The orbits shown for the electrons correspond to the most probable distances from the nucleus at which electrons are observed.

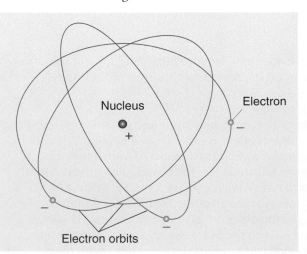
Nucleus

Electron

Electron orbits

they are combined with other atoms. When we burn paper, for example, the paper's carbon atoms combine with oxygen atoms from the air, producing carbon dioxide molecules. **Figure 11-6** illustrates this reaction and emphasizes that this is a *chemical bond,* formed by the sharing of electrons between the carbon atom and each of the oxygen atoms. The forces involved here are electromagnetic in nature. Forces responsible for the structure of the nuclei are not involved in chemical reactions.

Nuclear reactions, on the other hand, involve forces between nuclear particles; orbiting electrons are not part of these reactions. There are many types of nuclear reactions, but only one will be of interest to us: the ***fusion*** reaction. In a nuclear fusion reaction, two nuclei combine to form a larger nucleus. They "fuse."

The core of the Sun is too hot to allow for complete atoms to exist. Instead, nuclei and electrons are separate from one another and bounce around at great speeds. The primary source of energy in the Sun (and in all stars during most of their lifetimes) is a series of nuclear fusion reactions in which four hydrogen nuclei are fused to form one helium nucleus. In the process, a small fraction of the mass of the nuclei is changed into energy. This is where Einstein's theory comes into play. Let's look at the process.

Most hydrogen nuclei consist simply of one proton. Prior to the fusion reaction, we have four hydrogen nuclei, and after the reaction, there is one helium nucleus. Let us subtract the mass of one helium nucleus (6.6447×10^{-27} kilogram) from the mass of four hydrogen nuclei (each having a mass of 1.6726×10^{-27} kilogram):

$$\text{mass of 4 hydrogen nuclei} = 6.6905 \times 10^{-27} \text{ kg}$$
$$-\ \text{mass of 1 helium nucleus} = 6.6447 \times 10^{-27} \text{ kg}$$
$$\text{difference} = 0.0458 \times 10^{-27} \text{ kg}$$

The difference between the mass of the original matter and the resulting matter is very small, not only in terms of the actual amount (less than 10^{-28} kilogram per reaction), but also in that the "lost" mass (which has been completely converted to energy) is only seven-tenths of 1% of the original mass of four hydrogen nuclei. Not even 1% of the mass is changed into energy. In fact, the energy produced by a trillion such fusion reactions is only enough to lift up this book by about a foot.

If the energy produced per fusion reaction is so small, how does the Sun produce an output of 3.85×10^{26} watts? The answer lies in the huge number of fusion reactions occurring in the Sun's core every second—about 10^{38} reactions per second. This implies that nearly 5 million metric tons of matter must be completely converted into energy each second. This involves the transformation of some 626 billion kilograms of hydrogen to about 622 billion kilograms of helium every second. Although this is a tremendous amount of matter by human standards, it is almost insignificant when compared to the Sun's total mass. If the Sun were originally pure hydrogen, it would take about 100 billion years for the Sun to convert its entire mass to helium at the present rate of consumption. (As we will discuss later in the chapter, only the inner portion of the Sun—about 30% of its mass—is involved in the reaction and is converted to helium.)

In practice, the process by which hydrogen is converted into helium incorporates three steps, called the ***proton-proton*** chain. (This chain is the main fusion process in the Sun, responsible for 98.5% of the energy production. The remaining 1.5% is produced by a different process, the carbon cycle, which we will study in Chapter 14.) The reactions start with a fusion of two protons (hydrogen nuclei) and end with the production of a helium nucleus containing two protons and two neutrons. During the process, two other smaller nuclear parti-

Do not confuse the terms *proton* and *photon;* they are very different. Refer to Section 4-6 for a description of photons.

fusion (nuclear) The combining of two nuclei to form a different nucleus.

proton-proton (p p) chain The series of nuclear reactions that begins with four protons and ends with a helium nucleus.

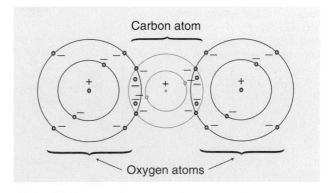

Figure 11-6
A carbon dioxide molecule (like all molecules) is held together by chemical bonds, in this case involving the sharing of electrons between two oxygen atoms and one carbon atom. Nuclear forces do not come into play in the bonding. Drawings such as this one are used only as visualization tools.

Table 11-1

The Proton-Proton Chain

Reaction	Explanation
$^1_1H + ^1_1H \rightarrow ^2_1H + e^+ + \nu_e$	Two protons combine to produce a deuterium nucleus (^2_1H; the 2 indicates the total number of protons and neutrons in the nucleus); a positron (e^+, which is a positively charged electron); and an electron neutrino (ν_e).
$^2_1H + ^1_1H \rightarrow ^3_2He + \gamma$	A deuterium nucleus joins with another proton to produce a helium nucleus (^3_2He, containing two protons and one neutron) and a gamma ray (γ).
$^3_2He + ^3_2He \rightarrow ^4_2He + 2 \cdot ^1_1H$	Two helium nuclei fuse to form the common type of helium (^4_2He) and two protons.

cles are produced, as well as a gamma ray. The solar fusion reactions are presented in more detail in **Table 11-1** and **Figure 11-7**.

Look at the first reaction in the table. Here, two hydrogen nuclei fuse to form the nucleus of another type of hydrogen, called ***deuterium,*** which has a neutron in its nucleus along with the proton. In addition, two other particles are formed, and these two particles fly away from the reaction at great speeds. One of them, the positive electron, or ***positron,*** combines with an electron and completely annihilates into gamma-ray photons. The other particle, the electron ***neutrino,*** escapes from the Sun and does not cause significant heating within the Sun. This particle is discussed later in this chapter.

In reality, only about 85% of the Sun's energy is produced by the chain described above. The remaining 15% is produced by two other, similar chains, where the third step involves the temporary formation of the element beryllium before finally producing a helium nucleus. These additional two chains are important because they produce electron neutrinos with larger energies than the chain shown in Figure 11-7. In the next section we will discuss the importance of these neutrinos in our understanding of the interior of the Sun.

deuterium A hydrogen nucleus that contains one neutron and one proton.

positron A positively charged electron emitted from the nucleus in some nuclear reactions.

neutrino An elementary particle that has little rest mass and no charge but carries energy from a nuclear reaction.

Figure 11-7
The proton-proton chain begins with four protons and ends with a helium nucleus. The four protons are shown at left combining in separate reactions to produce two deuterium nuclei (each with a proton and a neutron).

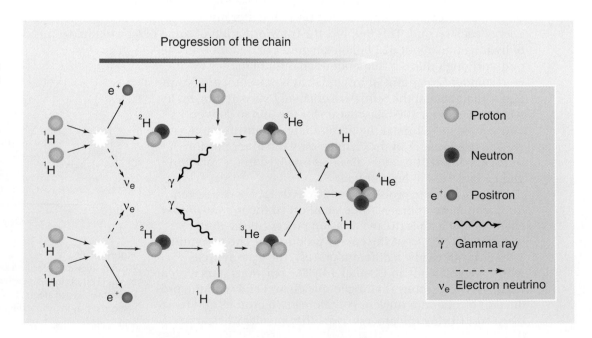

ADVANCING THE MODEL

Fission and Fusion Power on Earth

The dream of unlimited energy has been with us at least since the beginning of the industrial age, and humans wondered at the tremendous power of the Sun long before that. We now know that nuclear fusion reactions are the source of the Sun's energy, and just a few decades ago humans harnessed this energy (if *harnessed* is an appropriate word here) in the hydrogen bomb. The bomb's name comes from the fact that it uses a form of hydrogen (deuterium) as its fuel. The nuclear reaction in the H-bomb is similar to that in the Sun; it produces helium from hydrogen.

Peaceful uses of fusion power have not yet been developed, although much research has taken place over the last four decades and is still in progress. A hint at the problems of controlling fusion can be seen by considering the tremendous temperatures and pressures that are necessary to produce fusion in the centers of stars. On the other hand, such reactions have an essentially unlimited supply of fuel—deuterium—and therefore controlled energy production from fusion is a very attractive goal.

Although fusion is much more common than fission in the universe as a whole, humans developed fission first. Fission involves the release of energy when a large nucleus is broken into two medium nuclei, with mass being converted into energy in the process. The atomic bomb (poorly named, for it uses the energy of the nucleus rather than the energy of the outer atom) was the first application of fission power. Since the development of the A-bomb, we have learned to control this reaction, and today we use the energy of fission to produce electricity in nuclear power plants. Contrary to the ready availability of fuel for fusion, the uranium that must be used for fission power is definitely limited. Many people, scientists and nonscientists alike, question the wisdom of building and using fission power plants. Fission power must be viewed as only a temporary solution to the problem of finding a long-range source of energy on Earth.

Hydrogen exists throughout the world. Yet it does not, on its own, fuse into helium. The reason for this is that all nuclei have the same type of electric charge (positive) and therefore they repel one another. This electrical repulsion force acts over great distances, at least compared to the very short distances over which nuclear forces act. This means that the particles are unable to get close enough together for the attractive nuclear forces to take over unless they happen to be moving toward one another at a great speed (**Figure 11-8**). Since the temperature of a gas is determined by the speed of its particles, the particles of hot hydrogen are more likely to fuse than those of cool hydrogen. Significant fusion occurs only in high-temperature matter.

In addition, as we have seen, a great number of fusions of hydrogen nuclei must occur each second in order to produce the Sun's power. Thus, we know that the density of matter must be extremely high in the region of the Sun where fusion takes place. To see how and where this occurs, let us investigate the internal structure of the Sun.

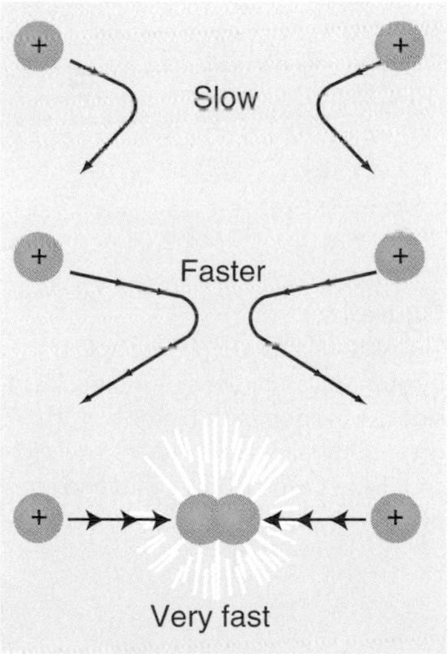

Figure 11-8
Nuclei moving toward one another at too slow a speed will be repelled because of their positive charges. However, if they are moving fast enough, electrical repulsion will not be strong enough to prevent them from colliding and fusing.

11-3 The Sun's Interior

Obviously, we cannot examine the interior of the Sun directly. However, astronomers have learned much about its interior by computer modeling, based on observations of the Sun's surface and our knowledge about the behavior of matter at the temperatures and pressures necessary to sustain fusion reactions. We know that at temperatures as high as the Sun's, matter must exist as a gas rather than as a solid or a liquid, and this fact makes the analysis easier, for gases are much simpler than liquids or solids. As will be discussed later, the temperature on the surface of the Sun is about 5800 K, and it increases greatly below the surface. At these temperatures, solids and liquids cannot exist and most electrons are stripped away from their nuclei. As a result, most of the material of the Sun's interior consists of free nuclei and free electrons. The behavior of this material, however, is similar to that of a simple gas, so we must first study properties of gases. The properties of importance to us are temperature, pressure, and *particle density.*

Pressure, Temperature, and Density

When you blow up a balloon, the gas inside exerts a pressure on the rubber of the balloon and supports it against its tendency to contract. The pressure exerted by the gas is the result of collisions of the individual gas molecules with the rubber surface. **Figure 11-9a** illustrates this. Each molecule, as it strikes the rubber wall and rebounds, exerts a tiny force on the wall. Although we cannot detect each individual bounce and the corresponding force, the overall force exerted by the gas is simply the total of all of these individual tiny forces. To see this, imagine tiny grains of sand being fired at a board by a great number of sand-throwers, as in Figure 11-9b. The force exerted on the board by a single grain of sand might seem negligible, but the overall result could be a force great enough to cause the board to move.

Notice that this discussion sometimes refers to force and other times to pressure. *Pressure* is defined as the amount of force exerted per unit area and might be expressed as newtons per square meter or pounds per square inch. So when we think of a single grain of sand or a single atom rebounding from something, it is more natural to speak of the force exerted by the particle. On the other hand, when we think of many sand grains or many atoms striking over a large area, we speak of the force exerted on each unit of area, or the pressure.

What determines the pressure of a gas, then? There are two factors: the speed of the molecules of the gas and their density. To see that each of these factors is important, think again of the gas molecules in the balloon. If we heat this gas while keeping the balloon's volume constant, then the molecules will start moving faster. Each collision with the inside surface of the balloon will be more violent, exerting more force on the wall of the balloon. In addition, a greater speed results in more collisions. For these two reasons, the total force exerted on one square centimeter of the wall will be greater; that is, the pressure exerted by the gas will be greater. On the other hand,

particle density The number of separate atomic and/or nuclear particles per unit of volume.

pressure The force per unit of area.

One newton is a unit of force (about 0.225 lb, the weight of an average apple).

Figure 11-9
The walls of a balloon (a) are held out by numerous collisions by molecules inside the balloon, in the same way that a great number of tiny sand-throwers (b) could exert a force on a board and topple it backward.

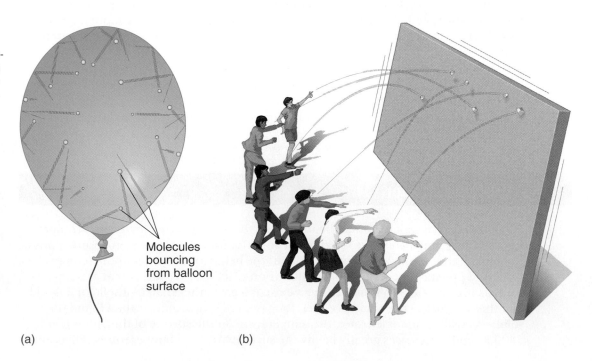

Molecules bouncing from balloon surface

(a) (b)

if the gas is cooled, it will exert less pressure. Thus we can conclude that the pressure and temperature of a gas are related. More rigorous analysis shows that in fact a direct mathematical relationship exists between them, so that one changes proportionately to the other.

To appreciate the effect of particle density on pressure, imagine that twice as much gas at a given temperature is somehow put into the balloon without allowing it to expand. If this is done, twice as many molecules will be striking the inside walls of the balloon, exerting twice as much pressure. This is what causes the balloon to expand when more gas is added without restraining the volume. Thus we can conclude that pressure and density of a gas are related. In fact, a direct mathematical relationship exists between them, so that one changes proportionately to the other.

Thus, pressure, density, and temperature are interrelated. If one changes, one or both of the others must change. Now let's turn back to the Sun and consider how these factors determine the character of the Sun's interior.

> Under ideal conditions, pressure is proportional to density times temperature.

Hydrostatic Equilibrium

The Sun is held together by gravitational force. The force of gravity holds the solar material to the Sun just as the force of gravity holds the atmosphere of the Earth near its surface. In this respect, the Sun is merely a big ball of gas. Our Earth's atmosphere is denser near the surface, not simply because the force of gravity is greater there than higher up, but also because the pressure exerted by the gas above compacts the lower layers. Gases lower in the atmosphere have to support the gases above. The same logic applies to the Sun. At any particular depth below the Sun's surface, the pressure of the gas at that point must be enough to support the gas above. Thus it is convenient to think of the Sun as having layers, like the various layers of an onion, as shown in **Figure 11-10a**. Keep in mind, however, that in the Sun there are no distinct boundaries between layers, but rather a continuous change as we move toward or away from the center.

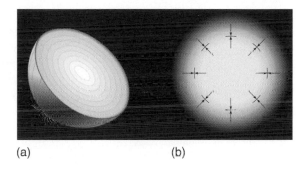

(a)   (b)

Figure 11-10
(a) You can think of the Sun as consisting of multiple layers, like those of an onion, except that there is no distinct boundary between the Sun's layers. (b) Within the Sun, the pressure upward on any layer must be the same as the pressure downward.

Since the Sun is in a state of equilibrium (that is, neither noticeably contracting nor expanding), the pressure downward on any thin layer must be equal to the pressure exerted upward on that layer (Figure 11-10b). This allows us to calculate the pressure at any depth below the surface, for knowing the total mass of the Sun, we can calculate the weight of gas above any particular layer. Knowing this, we can calculate the pressure that is needed within the layer to support the gas above.

The equilibrium conditions in the Sun are known as *hydrostatic equilibrium.* The name is almost self-explanatory: "hydro" refers to the fluid state. This is basically just a more complex case of the situation we discussed with the inflated balloon. In that case, the stretched rubber holds the air inside in a compressed state. As long as the outward pressure of the compressed air inside is enough to support the inward pressure of the rubber, equilibrium is maintained.

> **hydrostatic equilibrium** In a star or a planet, the balance between the downward pressure caused by the weight of material above a thin layer and the upward pressure exerted by material below.

Since the gas at the center of the Sun is supporting the weight of the gas all the way out to the surface, we should expect great pressures at the center. In fact, the pressure there is calculated to be about 2.5×10^{11} times that on the surface of the Earth. This tremendous pressure pushes protons close enough together that hydrogen fusion can take place. Only near the center of the Sun are the temperature and density of hydrogen great enough to support fusion. The solar core, where fusion is taking place, extends out to perhaps 25% of the radius of the Sun.

The fusion reactions in the core provide a heat source that obviously must be taken into account when calculating conditions within the Sun. Recall that when a gas is heated, it tends to expand. A balloon expands when its temperature rises, but then it stabilizes at a (different) equilibrium condition. Likewise, the Sun exists in a

> The density at the Sun's center is about 160 g/cm³, approximately 20 times the density of iron. The temperature there is about 15.7 million K.

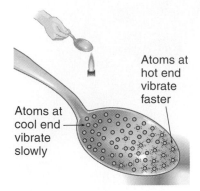

Figure 11-11
The fast-vibrating atoms at the end of the spoon in the fire cause atoms next to them to vibrate faster. This continues until atoms at the far end are also vibrating fast, meaning that this end also becomes hot.

Atoms at hot end vibrate faster

Atoms at cool end vibrate slowly

conduction The transfer of energy in a solid by collisions between atoms and/or molecules.

convection The transfer of energy in a gas or liquid by means of the motion of the material.

Figure 11-12
This figure illustrates the relative thicknesses of the three energy-transport zones of the Sun.

Can we tell what happens in the Sun's core by analyzing the light produced there?

state of equilibrium, with the force of gravity balanced by forces tending to expand the gas.

To see how hydrostatic equilibrium works, imagine that the Sun could somehow be compressed artificially. Under compression, the pressure within the Sun would increase. The fusion rate would then increase, raising the temperature and pushing the Sun back out to another equilibrium position. As our discussion of the life cycle of stars in a later chapter will explain, once the energy production of a star slows and the core cools, contraction of the core begins. However, as long as energy production is stable, the Sun remains in equilibrium.

To see what happens to the energy produced in the core of the Sun, we must look at energy transport within the Sun.

Energy Transport

We observe that energy is radiated from the surface of the Sun. However, the fusion reactions in the Sun occur at its core, where the temperatures and pressures are greatest. The energy produced at the core must then be transported out to the surface.

There are three possible methods by which energy can be transferred from the center of the Sun outward: conduction, convection, and radiation. These same three processes occur here on Earth and, indeed, everywhere in the universe.

If you put one end of a spoon on the burner (or in the flame) of a kitchen stove and hold the other end, you'll feel your end of the spoon gradually getting warmer. Energy is being transferred by vibration through the metallic crystal structure of the spoon. This method of transfer is called **conduction.** Imagine the atoms near the end of the spoon in the fire. As that end heats, the atoms vibrate at greater speeds (**Figure 11-11**). However, these atoms exert forces on adjacent atoms of the metal and cause those atoms to start vibrating faster. Gradually, the increased vibration spreads up the spoon until the atoms at the other end are also vibrating more rapidly than they were. Notice that in transferring energy by conduction, atoms do not move from one region to another, but vibrational energy—thermal energy—is transferred.

Conduction requires that the particles of the substance be in close contact, as are the atoms in a solid. In a star, this is not the case, except in some extremely dense stars, so conduction is not a significant factor in transporting energy from within the Sun.

Convection occurs when the atoms of a warm fluid (liquid or gas) move from one place to another. Put your hand about a foot above a hot stove burner. You will feel hot air rising from the burner. This takes place because heated air is less dense than cooler air and therefore the hot, less dense air rises. The result is that energy is transferred upward from the stove by the motion of the hot gas. In forced-air central heating/cooling systems, the motion of the hot (or cool) air is caused by fans. On Earth, convection currents (thermals) enable hang gliders and gliding birds to travel long distances, carried along with the rising air.

In a star, convection between adjacent layers is significant only when the temperature difference is great compared to the pressure difference. In the case of our Sun, this condition is met only in the region within about 200,000 kilometers of the surface (**Figure 11-12**). In this region, convection constantly mixes the solar material as hot gas rises and cooler gas descends. Deeper within the Sun, convection is almost inconsequential and mixing does not occur to a great degree.

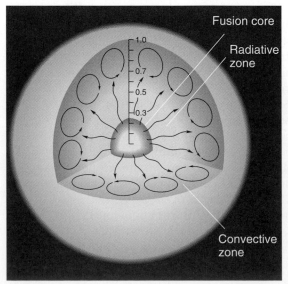

Fusion core

Radiative zone

Convective zone

1.0
0.7
0.5
0.3

The final method of energy transfer is by *radiation.* If you hold the palm of your hand exposed to the stove burner, you can tell that your hand is being heated by another method besides rising hot air. To emphasize this, hold your hand off to the side, where it is not in the stream of hot air, and you will still feel your hand being heated. Radiation of energy occurs in all portions of the electromagnetic spectrum. Its effect on another object depends only upon whether the receiving object absorbs the particular wavelengths of radiation emitted by the radiator. The air of your kitchen, for example, is transparent to most electromagnetic radiation produced by the stove burner, so it does not absorb the radiation and is not heated by it directly. Your hand, however, being opaque to the radiation, absorbs it and is heated by it.

Inside the Sun, and most stars, radiation is the principal means of energy transport. If the Sun were transparent, the electromagnetic radiation produced in the core would travel outward at the speed of light and reach the surface in about 2 seconds. In reality, the material of the Sun is nearly opaque, so the radiation that is emitted travels only a small distance before being absorbed. It is then reemitted, absorbed, reemitted, and so on, with a typical distance between successive absorptions of 1 centimeter (**Figure 11-13**). The reemissions occur in random directions, and as a result, energy that began perhaps as a gamma ray photon resulting from the proton-proton chain travels a very circuitous path and may take hundreds of thousands of years to reach the surface. This seems an impossibly long time for something traveling at the speed of light, but keep in mind both the multitude of absorptions and reemissions taking place and the extreme length of the roundabout path taken by the energy. As a result of the many absorptions and reemissions, any information about the Sun's core carried by the photons produced there is lost to us. This is similar to what happens when you forward a message by fax, using a fax you received from a friend, who did the same thing before you, and so on. The end result is that the final recipient will most likely not be able to read the message.

Figure 11-12 shows the various portions of the interior of the Sun: the core, the radiative zone, and the convective zone. Once the energy of the Sun reaches the surface, it is again radiated outward. The energy from the Sun is released primarily as ultraviolet, visible, and infrared radiation, but it also comes in two less familiar forms: charged particles and neutrinos. As we explain in a later section of this chapter, the charged particles flowing outward from the Sun have observable effects on Earth. Neutrinos, however, are very difficult to detect because the probability of them interacting with matter they pass through is very low. This presents astronomers with a major problem. However, since neutrinos have very little mass—the latest experiments suggest a mass of less than a billionth the mass of a proton—they move very close to the speed of light and thus can tell us about the current conditions in the Sun's core.

The Solar Neutrino Problem

There is almost no doubt about the fundamental ideas of the solar model just described, for the concepts of pressure and density are well understood, and we are confident that the Sun's energy is produced by nuclear fusion. However, we are less sure of the details of the workings of the Sun's interior. The generally accepted theory of the Sun is called the *standard solar model,* and it makes a prediction about neutrinos that can be checked to determine the validity of the model.

The standard solar model predicts that so many neutrinos flow from the Sun that about 65 billion of them pass through every square centimeter of your body each second. These neutrinos do not affect your body because neutrinos have a very low probability of interaction with whatever matter they pass through, but this same low interaction rate makes them difficult for astronomers to detect. Astronomers want to measure the number of solar neutrinos that reach the Earth in order to check the standard solar model.

In order to shield neutrino detectors from cosmic rays and natural radioactivity, the detectors must be located far underground. Otherwise, the other radiation would overwhelm the few reactions caused by neutrinos. In addition, since neutrinos react

radiation The transfer of energy by electromagnetic waves.

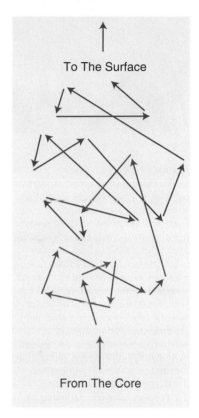

Figure 11-13
A photon is absorbed and reemitted numerous times as it travels from the Sun's core. Each reemission is in a random direction.

standard solar model Today's generally accepted theory of solar energy production.

Recall that the proton-proton chain produces two electron neutrinos for each helium atom produced.

(a)

Figure 11-14
(a) The Homestake neutrino experiment used 100,000 gallons of cleaning fluid in this tank, located nearly a mile under the hills of South Dakota. (b) Scientists use a boat to check the detectors in the Super-Kamiokande neutrino experiment. The tank holds 50,000 tons of ultrapure water when filled. (c) An artist's view of the SNO. The SNO detector consists of 1000 metric tons of ultrapure heavy water surrounded by 9600 light sensors, which detect tiny flashes of light emitted as neutrinos are stopped or scattered in the heavy water. The detection rate is about 1 neutrino/hour. It is the size of a 10-story building, 2 kilometers underground in Inco's Creighton Mine near Sudbury, Ontario.

(b)

(c)

> There is something wrong either with the Sun or with the neutrinos—or with what we think we know about them.
>
> *John Bahcall, co-leader of the Homestake mine neutrino experiment*

so seldom with matter, a neutrino detector must contain a large amount of material in order to get enough reactions to detect. The world's first solar-neutrino detector (**Figure 11-14a**) began operation in the late 1960s in the Homestake Gold Mine in Lead, South Dakota. This detector, operated by a group from Brookhaven National Laboratory led by Raymond Davis, Jr., and John Bahcall, holds 378,000 liters (100,000 gallons) of perchloroethylene (C_2Cl_4), a dry-cleaning fluid containing chlorine. When a solar neutrino strikes a chlorine atom with enough energy, a reaction occurs, transforming the chlorine into a radioactive isotope of argon that has a half-life of 35 days. The accumulated argon atoms are collected from the tank and counted, thereby revealing the number of neutrinos involved in the reactions.

During the 1970–1995 period, Davis's group conducted the experiment many times. Their results suggest that, on average, one radioactive argon atom was created in the tank every three days. This corresponds to only about one-third of the number of neutrinos predicted by the standard solar model. They have checked and rechecked the apparatus and procedure, but have found no fault with them.

In an effort to confirm or improve these results, other solar-neutrino experiments followed. The second such experiment, known as Kamiokande, began in 1986 in Kamioka, Japan, and operated until 1995. This detector had the ability to provide information on the direction of the neutrinos' travel. It confirmed that neutrinos do indeed come from the Sun, giving direct evidence that nuclear reactions do occur in the Sun. However, like the Homestake experiment, it found fewer neutrinos than theory predicts (about 50%).

The two detectors discussed thus far were sensitive only to high-energy neutrinos, which account for a very low percentage of the total neutrino production. Two other neutrino detectors, known as SAGE and GALLEX, employ the element gallium, which is able to detect the low-energy neutrinos produced by the dominant reactions in the Sun's core. In these experiments, the interaction between solar neutrinos and gallium atoms produces radioactive germanium atoms, which can be collected and counted. The first detector, SAGE, was a cooperative Russian-American experiment and operated between 1990–2001. It was housed in a tunnel under a mountain in the northern Caucasus, near a town called Neutrino City. The SAGE scientists did detect about 55% of the number of neutrinos predicted by the standard model.

Statistically more substantial results came from the GALLEX experiment, which was located in a tunnel in Italy and operated between 1991–1997. The GALLEX group reported detecting about 60% of the predicted neutrinos. This was closer to the expected value than SAGE's results, but not close enough to resolve the issue of the missing neutrinos once and for all.

The results of the experiments mentioned so far did not rule out the possibility that our theories concerning the detection of neutrinos were in error. But it is very unlikely that all experiments are wrong, since they all use very different detection techniques and have been thoroughly tested. Also, the neutrinos produced in the Sun's core have a wide range of energies, and the different detectors are sensitive to different energy ranges. The discrepancy between the observed and predicted numbers of neutrinos is commonly referred to as "the solar neutrino problem," and the neutrino deficit depends on the energy of the neutrino.

Until recently, neutrinos were thought to be massless and therefore moving at the speed of light. Since they rarely interact with matter, they are the best probe we have of "seeing" what is happening inside the Sun "now." The solar neutrino problem forced astronomers to examine several alternative explanations. An obvious possibility was that we did not correctly understand the structure and constitution of the Sun, and therefore the reaction mechanisms. This would have been a real blow for the standard solar model, which has otherwise been very successful. Many modifications to the model had been proposed, but none could satisfactorily solve the problem. One suggestion was that a previously unknown particle—called WIMP, for *Weakly Interacting Massive Particle*—was carrying off some of the energy thought to have been carried by neutrinos, thus cooling the core enough to lower the number of neutrinos produced.

The experiments mentioned so far were designed to detect electron neutrinos, the type created in the Sun's core. However, neutrinos come in three types (related to three different charged particles—the electron, and its lesser known relatives, the muon and the tau). Could it be possible that neutrinos change (oscillate) from one type to another during their 8-minute flight from the Sun's core? This is precisely what physicists Stanislaw Mikeyev, Alexei Smirnov, and Lincoln Wolfenstein proposed. However, for neutrinos to oscillate they must have mass. The so-called MSW theory was put to the test in experiments around the world.

The Super-Kamiokande experiment, a joint Japan–U.S. collaboration and a continuation of the Kamiokande experiment, started operating in 1996 (Figure 11-14b). It uses pure water, is mostly sensitive to electron neutrinos, and cannot distinguish between the different types. In 1998, it found evidence for neutrino oscillations and thus for the possibility that neutrinos may have mass. This experiment also found about 50% of the neutrinos predicted by the standard solar model. The Sudbury Neutrino Observatory (SNO), in Sudbury, Canada, is a Canada–Britain–U.S. collaboration that began operating in 1999 (Figure 11-14c). It uses the unique properties of heavy water (where the hydrogen has an extra neutron in its nucleus) to detect all types of neutrinos and is therefore able to detect neutrino oscillations and masses, if they exist. Indeed, by early 2002, SNO scientists were able to measure the total number of neutrinos of all three types reaching their detector. They found that the number of electron neutrinos observed is only about one-third of the total number reaching the Earth. This shows with great certainty that solar neutrinos change their type en route to Earth, and arrive as a mixture of electron neutrinos and the other two types. These are exciting results because they imply that neutrinos have mass, and that the mass differences between the three types can be calculated. These masses are minute, and for all practical purposes neutrinos can still be thought of as traveling at the speed of light. The total number of neutrinos observed by SNO is also in excellent agreement with calculations of the nuclear reactions powering the Sun,

SAGE is an acronym for *Soviet American Gallium Experiment* (the experiment and the acronym were devised before the political disappearance of the Soviet Union).

suggesting that the standard solar model is quite accurate in describing the inner workings of the Sun. This agreement is very impressive if we consider the fact that the predicted number of neutrinos depends on the 25th power of the temperature at the center of the Sun. We can now calculate this temperature to better than 1%.

Strong support for neutrino oscillations came in late 2002 from the KamLAND project. This detector is located at the Kamioka mine in Japan and measures antineutrinos (the antimatter equivalent of neutrinos) from all 17 nuclear power plants in the country. Since matter and antimatter are mirror images of each other, studying antineutrinos produced by the fission reactions in the power plants should be the same as studying neutrinos produced by the fusion reactions in the Sun's core. In this case, the amount of nuclear material generating these neutrinos is well known, and thus physicists can observe neutrino oscillations without making any assumptions about the properties of the source of the neutrinos. The results of this experiment clearly showed that neutrino oscillations are real.

The idea of neutrino oscillations seems to explain the solar neutrino problem. However, so far we have only measured about 0.005% of the total number of neutrinos emitted by the Sun. The remaining neutrinos are difficult to detect because they are at lower energies. Additional experiments have been proposed in a concerted effort to solve this very serious problem. For example, the Borexino experiment, in Gran Sasso, Italy, is scheduled to start operating in 2004. Its main goal is to observe neutrinos of specific energy created during the nuclear chain reactions involving beryllium. According to the standard solar model, this is the second most important neutrino production reaction after the basic proton–proton chain. The flux of these neutrinos is also predicted more accurately, and is about a thousand times greater than the flux measured by Super-Kamiokande and SNO.

There is at least one other possibility. The prediction of the total number of neutrinos produced by the Sun is based on the energy we receive from the Sun. However, as we have seen, the solar energy that comes to us as electromagnetic radiation requires millions of years to get from the core of the Sun to the photosphere where it is radiated away. Thus, the energy we receive had its beginning millions of years ago. But solar neutrinos from the core reach us in only about 8 minutes. Could it be that the core of the Sun has decreased its energy production since it released the electromagnetic energy we are now receiving? If so, the change in solar activity will not have an effect on Earth until thousands or millions of years from now, when it will cause another ice age.

Figure 11-15
This figure was made by combining as separate wedges six different photos of the Sun. Each photo was made using a different portion of the spectrum.

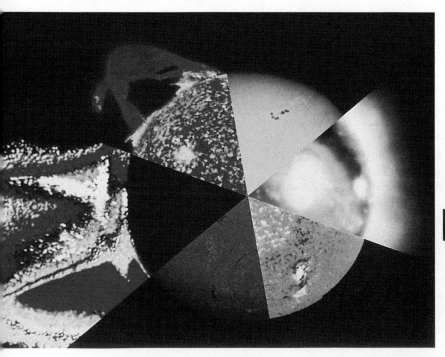

The solar neutrino problem is a very important problem in astronomy. At stake is not just the accuracy of a model for the Sun. If the standard solar model is accurate, which now seems to be the case, then we can be confident about our understanding of not only our Sun but of all stars. This obviously has ramifications for our understanding of galaxies and the overall evolution of the universe. The latest results suggest that neutrinos do have mass, and, thus, being the most numerous entities in the universe other than photons, they contribute to a small degree to its overall mass and influence its evolution.

11-4 The Solar Atmosphere

Figure 11-15 is a combination of six photos of the Sun, each taken by a different method and in a different portion of the spectrum. The yellow "pie wedge" at the top right position shows the Sun in visible light—the way it appears in a "regular" photograph. In this case we see what we call the Sun's surface, although the Sun does

TOOLS OF ASTRONOMY

Data Uncertainty in the Homestake Experiment

The discussion of the work of Tycho Brahe in Chapter 2 pointed out that he pioneered the scientific practice of reporting not only the direct result of a measurement, but also the amount of likely error in that measurement. The data reported by the researchers in the Homestake neutrino experiment serve as a modern example of this practice in science.

Refer to the graph in **Figure B11-2** and find the first dot on the left. Its location indicates that in this particular experiment (in late 1970), about 0.2 atoms/day were detected. In the next experiment, about 0.5 atoms/day were detected. However, the graph tells us more about each of the experiments. The lines that extend above and below each of the dots are called *error bars*, and they tell us the range within which the researchers are confident the "true" measurement lies. For example, the data taken during the first experiment revealed that about 0.2 atoms/day were produced, but this value may be in error such that the true value (if the experiment were perfect) may be as little as 0 or as great as 0.5 atoms/day.

The length of the error bars in each case is not simply a guess, but is based on statistical measurements of the way in which the data fluctuated day-to-day. Note that some error bars are much longer than others. Look at the report of the 1986 experiment. It shows a measurement of zero, but has an error bar that extends as high as 1.8 atoms/day.

Why are the error bars so long? This is the same as asking why the possible errors are so great. The answer lies in the difficulty in detecting neutrinos and the resulting low numbers that are detected. When few nuclear reactions are detected, researchers say that the "statistics are poor." A simple example: Suppose that we did not know that if a coin is flipped, the probability of it landing "heads" up is 50%, and that we are trying to determine

this percentage by experiment. If we flip a coin only 4 times, we will have poor statistics upon which to make a judgment. On the other hand, if we flip it 10,000 times, we will have good statistics (and our result will be very close to 50%).

The line labeled "Observations" is the weighted average of all of the experiments. The line labeled "Theory" is the best theoretical prediction by John Bahcall, a leading theorist working with the standard solar model. You can see that although the error bars of a few of the experiments indicate that those experiments do not necessarily disagree with the theory, the total experimental results disagree greatly with the theory. Hence, the solar neutrino "problem."

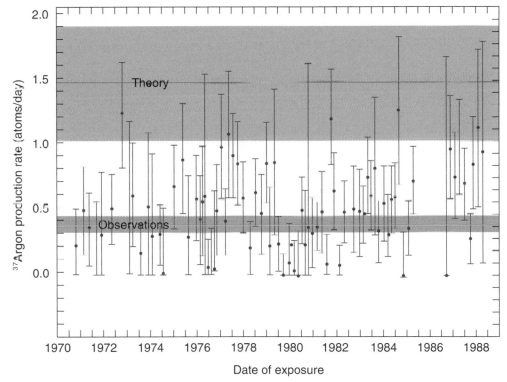

Figure B11-2
This graph compares the results of the Homestake experiment through the years with the value predicted by the standard solar model. The scale on the left indicates the average number of radioactive argon atoms detected in each experiment. (Each single experiment lasted several days.) The dates across the bottom indicate when the experiment was performed. The pink dots across the lower part of the graph show the measured value of the number of radioactive argon atoms produced per day in each experiment.

not have a surface in the sense that the Earth does. When we speak of the surface of the Sun, we are speaking of the part of the Sun from which we receive visible light, the *photosphere.*

Figure 11-15 shows that we receive different information about an object depending upon the wavelength used in taking a photograph. The photo (particularly the "pie wedge" to the left) shows that there is material beyond the Sun's surface that

photosphere The visible "surface" of the Sun. The part of the solar atmosphere from which light is emitted into space.

Figure 11-16
(a) The solar disk, the photosphere, in visible. Notice the limb darkening. (b) The light we receive from the center of the disk of the Sun originated at a greater depth than the rays we receive from near the edge. The three light rays travel an equal distance inside the photosphere. Therefore, the one from the limb must originate in the upper photosphere, which is relatively cooler and thus glows less brightly. **Warning: Never look directly at the Sun.**

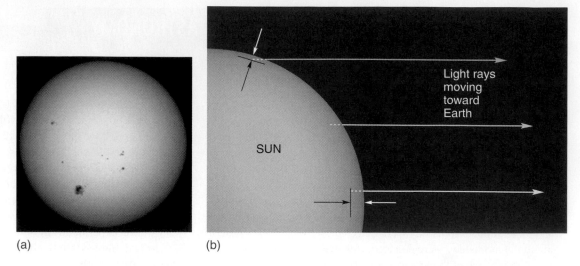

(a)   (b)

is not visible to our eye. This is the solar atmosphere. It is convenient to divide the atmosphere into three regions: the photosphere, the chromosphere, and the corona. We will discuss each in turn.

The Photosphere

The photosphere is a very thin layer (about 400 kilometers thick) from which the observed optical photons originate. This thickness implies that some of the light we receive from the Sun comes from one depth within the photosphere and other light comes from other depths. Saying that the photosphere is 400 kilometers thick means that we can see to that depth. When we look at an edge (the *limb*) of the Sun, we see that it appears to be "darker" than the center of the solar disk (**Figure 11-16a**). This limb darkening occurs because we see to a lesser depth as a result of observing the Sun at a grazing angle. Figure 11-16b shows this effect, which is important in that it allows us to analyze light from different depths within the photosphere and therefore to determine the temperature at different depths. We learn that the photosphere varies in temperature from about 6500 K at its deepest to about 4400 K near the outer edge. Overall, the light we receive from the photosphere is representative of an object whose temperature is about 5800 K. An intensity/wavelength graph of the radiation from the Sun (**Figure 11-17**) peaks near the center of the visible spectrum.

The pressure of the outer photosphere (calculated as for the inner layers, from knowing the gravitational force there and the amount of material above each layer) is only about 0.01 the pressure at the surface of the Earth. Knowing the temperatures

limb (of the Sun or Moon) The apparent edge of the object as seen in the sky.

Photosphere = "sphere of light."

The fact that the solar spectrum peaks near the center of the electromagnetic region visible to us is not just a coincidence, for our eyes evolved so as to use the available electromagnetic radiation efficiently.

Figure 11-17
The intensity/wavelength graph of light from the Sun reaches a peak about the center of the visible portion of the spectrum.

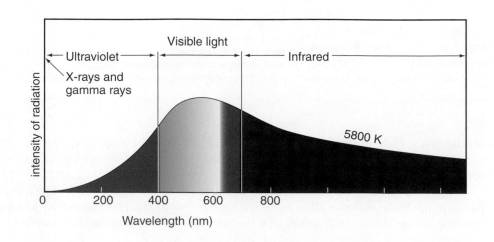

TOOLS OF ASTRONOMY

Helioseismology

In 1962, scientists discovered that the Sun is vibrating. Doppler shift measurements indicate that parts of the photosphere move up and down about 10 kilometers with a period of about 5 minutes. Since then, many other vibration frequencies have been discovered, all taking place at the same time. **Figure B11-3a** is a computer simulation illustrating a high-frequency vibration.

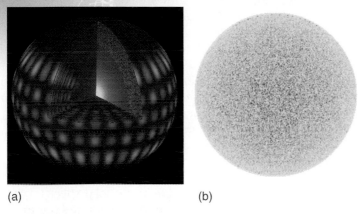

(a)   (b)

Figure B11-3

(a) A computer model of solar resonance that produces the observed vibrations of the photosphere. (b) An image of the actual solar surface showing oscillations; blue areas are moving toward us and red areas are moving away from us. The motions are mainly radial (inward and outward), as shown by the fact that the signal is strongest near the center of the imaged disk of the sun and weakest near the edge.

In this figure, the blue color represents regions where material is expanding at a given instant and orange represents areas where it is contracting. Thus, on the surface, the orange areas are moving inward and the blue areas outward. If this were a movie, in about 2 minutes the colors would reverse. The pulsations are caused by waves, similar to sound waves, produced by pressure fluctuations in the turbulent convective motions of the Sun's interior. When the waves reach the photosphere, they reflect back toward the interior. These inward moving waves refract because of the changing physical conditions and eventually return to the surface. These trapped sound waves set the Sun vibrating in millions of different patterns. The combination of waves coming out and going back in produces a resonating effect, like a "gong."

Just as geologists use earthquakes to study the interior of the Earth, helioseismology—the study of vibrations of the Sun—is beginning to give us insight into the Sun's interior. The Global Oscillation Network Group (GONG) has established a six-station network of telescopes around the world that obtain nearly continuous observations of the Sun's oscillations. The direct evidence we now get about the Sun's interior allows us to better test our theories of stellar structure and evolution and to find the role that magnetic fields play in the Sun's behavior. Also, a more accurate measurement of the helium abundance of the Sun put limits on cosmological models of the early universe and helped falsify a suggested explanation of the solar neutrino problem.

and gas pressures of the photosphere, we can calculate the density of particles there. We find that the density of the matter is only about 0.0005 of the density of air at sea level on Earth (even though the gravitational field there is 28 times what it is at the surface of the Earth).

When we observe the base of the photosphere (**Figure 11-18**), we see irregularly shaped bright areas surrounded by darker areas, a constantly changing patchwork with individual regions appearing and disappearing with a period of a few minutes. Recall from the discussion of the intensity/wavelength diagram in Section 4-4 that a hotter object emits more radiation than a cooler object of the same size. The

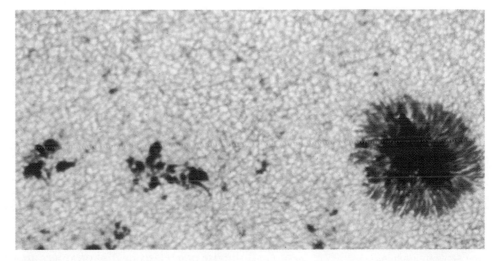

Figure 11-18

A photo of the granules of the photosphere. The bright center of each granule is material that is rising from the hotter inner portion of the Sun. A granule is about 1000 kilometers (600 miles) across, so each granule covers an area about the size of Texas. Supergranules can be 35,000 kilometers across. Supergranules appear to move across the Sun's surface faster than the Sun rotates; however, this is an illusion similar to "the wave" done by fans at a sporting event. The dark areas are sunspots of various sizes.

Figure 11-19
Granules are seen where hot material from below the photosphere rises. Where it descends after cooling slightly, we see the darker edges of the granules. The flow can reach speeds of up to 15,000 miles per hour, producing sonic "booms" and other noise that generates waves on the Sun's surface.

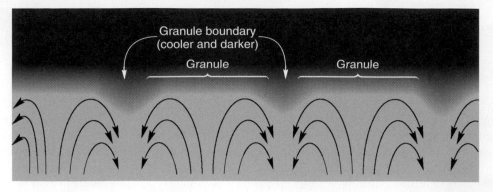

granulation Division of the Sun's surface into small convection cells.

These are percentages by mass, not by number of atoms.

Again, these percentages are by mass. About 92% of the atoms of the Sun are hydrogen and 8% are helium. The Sun's composition changes slowly over time as hydrogen is converted to helium in the core.

chromosphere The region of the solar atmosphere between the photosphere and the corona.

Chromosphere = "sphere of color."

spicule A narrow jet of gas that is part of the chromosphere of the Sun and extends upward into the corona.

corona The outermost portion of the Sun's atmosphere.

The word *corona* comes from the Latin word for crown.

brighter areas of the Sun are brighter simply because they are hotter. This *granulation* of the Sun's surface is the result of convection. Granules are areas where hot material (the light areas) is rising from below and then descending (the dark surroundings). **Figure 11-19** illustrates the effect and demonstrates that the photosphere is a boiling, churning region.

Using methods described in Chapter 4, we can determine the chemical composition of the photosphere. We find that it is about 78% hydrogen and 20% helium (by mass). The remaining 2% consists of some 60 elements. All of these elements are known on Earth and occur in about the same proportions on Earth as in the Sun's atmosphere, with a few exceptions. The exceptions are of two types: (1) elements such as helium that have masses so low that they would have escaped Earth if they were once here in abundance, and (2) elements found on Earth but whose characteristic spectra are such that they would not be detectable in the solar spectrum if the elements were as rare in the Sun as they are on Earth. As we pointed out in Chapter 7, this similarity of composition between objects as different as Earth and the Sun is not accidental, but results from the way the Sun and the solar system formed.

Notice that it is the composition of the Sun's atmosphere—not of the entire Sun—that we deduce from the solar spectrum. From our knowledge of nuclear fusion, we know that helium must be more abundant in the core of the Sun than in the atmosphere. Overall, the Sun is theorized to be about 73% hydrogen and 25% helium; this leaves only about 2% for the remainder of the elements.

The Chromosphere and Corona

The *chromosphere,* a region some 2000 kilometers thick lying beyond the photosphere, is not normally observable from Earth. It was first reported in the seventeenth century during a solar eclipse. It appears as a bright red flash, lasting only a few seconds, when the Moon has just covered the photosphere. During solar eclipses from 1842 to 1868, it was examined in more detail. Its spectrum was observed to be a bright line (or emission) spectrum because in viewing it we are seeing light from a hot gas with the dark sky behind it. Because the chromosphere is so much dimmer than the photosphere, it is only observed at the time of an eclipse, when the brighter portions of the Sun are blocked out (**Figure 11-20**).

Figure 11-21a is a photograph of the chromosphere that was taken at a wavelength that allows us to see its structure. The *spicules* that can be seen shooting upward into the corona typically reach a height of 6000 to 10,000 kilometers and last from 10 to 20 minutes.

Today the chromosphere and the region beyond it, the *corona,* can be observed by the use of a telescope that produces an artificial eclipse of sorts. With this instrument we can observe the Sun's atmosphere at various depths. We learn that as one moves outward from the photosphere, the temperature increases instead of diminishing, as we would expect. It is as high as 30,000 K in the outer portions of the chromosphere and continues to increase beyond the chromosphere into the corona, where it may reach

(a)   SUN   Moon   Earth

Figure 11-20

(a) The chromosphere is seen against the dark background of space during a solar eclipse. (b) At the higher temperatures in the chromosphere, hydrogen emits light that gives off a reddish color (H-alpha emission). This colorful emission can be seen in prominences that project above the limb of the Sun during total solar eclipses. (c) This is an image of the Sun taken with a filter that isolates H-alpha emission.

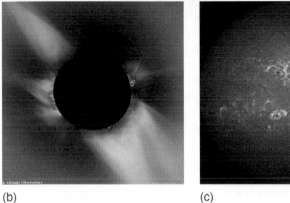

(b)   (c)

two million K. This change in temperature occurs rapidly in a transition region of about 300 kilometers between the chromosphere and the corona.

Most of the radiation emitted in these regions is in the X-ray portion of the spectrum rather than in the visible. (You can show this by using Wien's law.) As a result, you might think that these regions would be extremely bright because of their high temperature. However, the chromosphere and corona have a low density of matter, so hardly any matter is available to glow. The corona's density is less than one-trillionth that of the Earth's atmosphere. A simple example is to consider what would happen if you were to put your hand inside a hot oven. Even though the tem-

A telescope designed to photograph the atmosphere of the Sun (when there is no eclipse) is called a *coronagraph.*

(a)

Figure 11-21

(a) The chromosphere's spicules give it an irregular and ever-changing boundary as they shoot thousands of kilometers upward into the corona and then die down within a few minutes. (b) The "solar moss" (observed in the extreme UV) by the *Transition Region and Coronal Explorer* (*TRACE*) spacecraft.

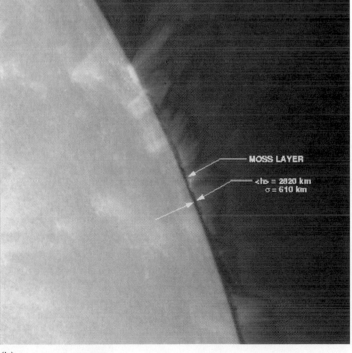

(b)

MOSS LAYER

$<b> = 2820$ km
$\sigma = 610$ km

perature is the same for the walls and air in the oven, you would get burned much easier by touching the walls—where the matter density is large and thus provides lots of thermal energy.

The reason for the high temperatures within the chromosphere and corona is not well understood. After all, the corona cools rapidly, losing its energy as radiation into space, and thus something must be pumping energy into it from below. The prevailing theory has been that the high temperature is the result of sound waves that are produced within the convective regions of the Sun and intensify as they pass outward until they are absorbed in the chromosphere and corona, heating these regions. Recent data, however, call this explanation into question, and many astronomers now believe that the heating is caused by an interaction between the Sun's magnetic field and its differential rotation. Figure 11-21b, taken by NASA's *TRACE* spacecraft in 1999, shows "solar moss," a sponge-like feature that is associated with regions where there is strong magnetic activity. The moss consists of very hot gas (about 1 million K), occurs in patches as large as 20,000 kilometers (12,000 miles) in extent, and sometimes reaches 5000 kilometers (3000 miles) high above the Sun's visible surface. This observation gives us a glimpse of how the magnetic field of the Sun becomes increasingly organized as we move from the photosphere into the corona. The transition region is very dynamic and changes as the magnetic field becomes complicated by the presence of the relatively cool jets from the chromosphere. The reason for the high chromosphere and corona temperature is just one of the unanswered questions astronomers have about the Sun.

The corona, a region extending for millions of kilometers from the Sun, has been observed during total solar eclipses for centuries, although many people used to claim that it was just an optical illusion caused by the sudden dimming of light as the Sun is eclipsed. **Figure 11-22** shows its extremely irregular appearance.

prominence The eruption of solar material beyond the disk of the Sun.

The photograph on the first page of this chapter and **Figure 11-23** show spectacular occurrences in the Sun's atmosphere. These are ***prominences,*** eruptions of solar material up into the chromosphere and corona. Some of these are relatively slow-moving and remain fairly stable for as long as a few days. They may reach as high as thousands of kilometers above the photosphere. Some move much more quickly, ejecting material from the Sun at speeds up to 1500 kilometers/second and reaching heights of nearly a million kilometers. Prominences are often associated with sunspots and the solar activity cycle to be discussed in the next section.

Figure 11-22
This is a composite photo of the Sun and its corona taken in March 1988. The surface of the Sun is a combination of X-ray and visible-light images. The photo of the corona was taken during the March 1988 eclipse. Note the irregularity of the corona and how it streams outward to form the solar wind.

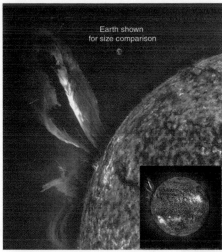

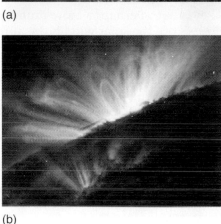

(a)

(b)

Figure 11-23
(a) A solar prominence. The Earth could easily fit under one of the loops shown in the picture. (b) Gas erupts in all directions from the Sun's surface in this photograph taken by the *TRACE* satellite. (c) This sequence of photos, taken from space, shows how this particular prominence progressed as charged particles were pushed from the Sun by its magnetic field.

(c)

When we divide the Sun into regions, we must remember that the boundaries between the regions are artificial, for we have named the various regions and distinguished between them on the basis of certain selected properties. For example, we consider that the process of energy transport is important, and we therefore talk of a radiative zone and a convective zone. If we emphasized another property, we might not make a division between these two parts of the Sun at all. In addition, remember that although the boundaries between various regions appear sharp in our drawings, in reality they are not as well defined. This is especially true of the outer limits of the corona, where the coronal material becomes the solar wind.

Yesterday I got a prominence photo good enough to prove the success of the method, and the result is that I am just now feeling pretty neat.
George Ellery Hale, upon taking a photo of the Sun in the light from calcium vapor, in 1891.

The Solar Wind

The *solar wind* is a continuous outflow of charged particles from the Sun, mostly in the form of protons and electrons. **Figure 11-24a** is an X-ray image of the Sun, in false color, showing the hottest, most dense regions as bright, and the cooler, less dense regions as dark. Such images indicate that X-ray emission is not uniform and that the active regions change appearances on a time scale of hours to days. The large dark area is called a *coronal hole,* because it has very little luminous gas. Coronal holes correspond to regions where the magnetic field lines are open, thus providing a corridor for charged particles to escape into space, generating the solar wind. The high coronal temperatures result in the wind particles escaping the Sun's strong gravity. These particles stream through space, taking mass away from

solar wind The flow of charged particles from the Sun.

coronal hole A region in the Sun's corona that has very little luminous gas.

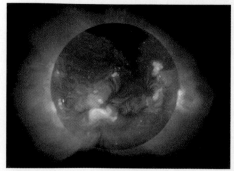

(a)

Figure 11-24

(a) An X-ray image of the Sun, in false color, taken by the *Yohkoh* satellite (a Japan/US/UK collaboration that has been providing important information about the Sun's corona since 1991). The large, dark area (corresponding to a cooler, less dense region) is a coronal hole. (b) A composite image of the solar wind and corona, in false color, with additional information on the solar magnetic field and speed of the solar wind. Data and images were taken by the *Ulysses* and *SOHO* spacecraft (ESA/NASA missions), and correspond to the 1994 period of sunspot minimum.

Recall the photo of an aurora on Saturn (Figure 9-24).

The darkest region of a sunspot may be as large as 30,000 km in diameter; this is about twice the Earth's diameter.

The splitting of spectral lines by strong magnetic fields is called the *Zeeman effect,* after the Dutch physicist who discovered it.

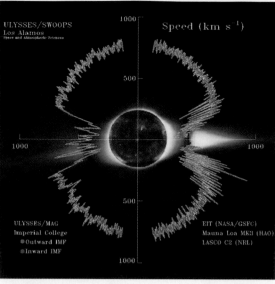

(b)

the Sun, about 6×10^{16} kilograms every year. This corresponds to only a tiny fraction of the Sun's total mass.

Figure 11-24b shows that the solar wind is not uniform, and its speed is higher over coronal holes. Near the Earth, the solar wind normally travels at about 400 kilometers/ second and has a density of about 2 to 10 particles per cubic centimeter. Recall that one effect of the solar wind is that it causes comet tails to point away from the Sun as its particles sweep comet material along with them.

Perhaps a more dramatic effect of the solar wind is the auroras seen near the poles of the Earth. Auroras (refer back to Figure 6-36) result when the solar wind creates tears in the Earth's magnetosphere, allowing energy and charged particles to enter. This is the driving force for space weather activity around Earth. Where the Earth's magnetic field lines converge toward the surface of the Earth, the electrically charged particles strike the molecules of the upper atmosphere and cause them to emit the beautiful, eerie glow we call an aurora.

11-5 Sunspots and the Solar Activity Cycle

Observations of dark spots on the Sun were reported by the Chinese as early as the fifth century B.C. It is sometimes possible to see very large sunspots with the naked eye if the Sun is viewed when it is very near the horizon. Europeans did not report sunspots until Galileo saw them with his telescope, perhaps because the Europeans did not have observers as astute as the Chinese, or perhaps—after Aristotelian thought was adopted—because Aristotle had proclaimed that the Sun was flawless. (We tend not to see what we disbelieve.)

In the late eighteenth century, Alexander Wilson hypothesized that sunspots were places where we were seeing through the outer surface of the Sun and into a cooler interior. William Herschel, the discoverer of Uranus (Advancing the Model, Chapter 9), even thought that the interior of the Sun might be cool enough to support life. Today's spectroscopic measurements of solar temperatures reveal that sunspots are indeed about 1500 K cooler than the surrounding photosphere. They are still very hot, however. At the central part of a sunspot the temperature may be as low as 3900 K. Using the Stefan-Boltzmann law (Section 4-4), we find that the radiation emitted by a sunspot is $(5800/4300)^4 \approx 3$ times less than that from the surrounding photosphere, and as a result the sunspot appears dark. Sunspots are temporary phenomena, lasting anywhere from a few hours to a few months.

The explanation for sunspots involves the magnetic field of the Sun, which can be measured using a technique discovered late in the 19th century. For an object in a magnetic field, the field can cause each emission line of the object's spectrum to split into two or more lines, and the strength of the magnetic field can be determined from the extent of the splitting. The splitting can be measured in the spectrum of light from individual parts of the Sun and is an important tool in studying the Sun.

Sunspots often appear in pairs, aligned in an east-west direction. Early in the 20th century it was found that the magnetic field in a sunspot is about 1000 times as strong as the magnetic field of the surrounding photosphere. In addition, we find that sunspot pairs have opposite magnetic polarities, one being north and the other south.

Sometimes the Sun contains a great number of sunspots, and sometimes few or no sunspots are seen. In 1851, Heinrich Schwabe, a German chemist and amateur astronomer, discovered that there is a fairly regular cycle of change in the number of sunspots and that the cycle lasts about 11 years. He found that although individual spots do not last long, about the same number are found on the Sun at any one part of its cycle. The cycle varies somewhat in period, but averages about 11 years between repetitions. **Figure 11-25a** is a graph showing how the number of sunspots has changed since 1880. The extended sunspot record shown in Figure 11-25b suggests that the Sun went through a period of inactivity during 1645–1715, when very few sunspots were seen on its surface (the Maunder minimum). Solar observations during this period were extensive enough that the lack of observed sunspots was well documented. This period corresponds to the "Little Ice Age" climatic period on Earth, and there is evidence that similar periods of inactivity existed in the more distant past. There is clearly a connection between the changes in solar activity and our climate on Earth.

When a large number of sunspots appears on the Sun's surface, its luminosity decreases by about 0.1%.

Modeling the Sunspot Cycle and the Sunspots

At a sunspot maximum, most spots occur about 35 degrees north or south of the equator. Then, as the cycle progresses, the spots are seen closer and closer to the equator. By the time they reach the equator, the cycle is at a minimum, and new spots are be-

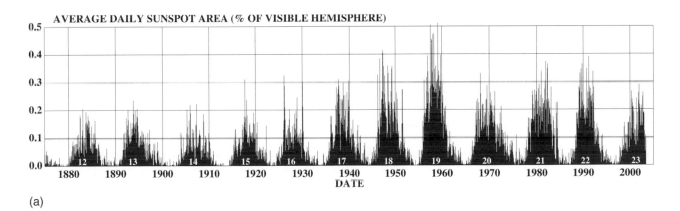

(a)

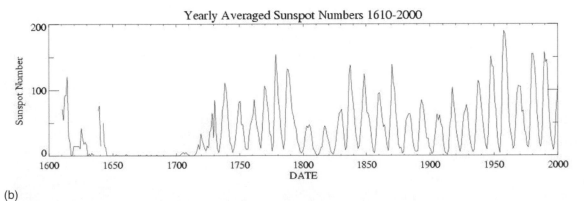

(b)

Figure 11-25

(a) The number of sunspots varies with a period of about 11 years, but there is a great difference in the maximum number during each cycle. It also appears that some cycles are double-peaked. (b) The Maunder minimum (1645–1715) corresponds to a period of inactivity for the Sun and the "Little Ice Age" for Earth.

Figure 11-26
This plot, called a butterfly diagram, shows the location and relative number of sunspots as the years pass. Notice that when each set of butterfly wings forms, the sunspots are at higher latitudes. Then, as new sunspots are formed, they move closer to the equator. Notice also that about every 11 years there is a period of very few sunspots.

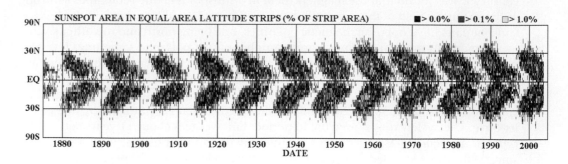

DAILY SUNSPOT AREA AVERAGED OVER INDIVIDUAL SOLAR ROTATIONS

> The reality of the [sunspot] minimum and its implication of basic solar change may be but one more defeat in our long and losing battle of wanting to keep the Sun perfect, and if not perfect, constant, and if not constant, regular. Why the Sun should be any of these when other stars are not is probably more a question for social than for physical science.
>
> *John Eddy, solar physicist, 1976.*

ginning to form again at greater latitudes. If we plot the location of the spots as time goes by, we get the pattern shown in **Figure 11-26**, called a *butterfly diagram* for obvious reasons. It is important to point out that a given sunspot does not move from higher to lower latitudes. (The lifetime of a sunspot can be up to a few months, much shorter than the 11-year solar cycle.) Instead, the diagram tells us that as time passes and old sunspots die out, new ones form closer to the Sun's equator.

The leading modern hypothesis explains the existence of sunspots and their 11-year cycle as being due to patterns of magnetic field lines, generated by the flow of the hot ionized gases, within the interior of the Sun, close to the boundary between the radiative and convective zones. The Sun's rotational velocity suddenly changes at this boundary, and this velocity shear drives the formation of the solar magnetic field. It is thought that groups of these lines form "tubes" threading through the Sun. When the tubes first form, they are relatively straight and buried deep within the Sun as shown in **Figure 11-27a**. The differential rotation of the Sun, however, causes the lines to wrap around the Sun, as shown progressively in parts (b), (c), and (d) of Figure 11-27. As the tubes become more and more twisted around the Sun, they are forced to the surface by convection. When they break through, we see a pair of sunspots, one with a north magnetic pole and one with a south magnetic pole.

An interesting feature of the 11-year cycle is that during one cycle, the leading sunspot of each pair in a given hemisphere of the Sun is a north magnetic pole and the trailing sunspot is a south pole. (In the other hemisphere, the opposite is true.) Then, during the next cycle, the pattern reverses, with the leading sunspot in that hemisphere being a south pole. The Sun's overall magnetic field also reverses from one cycle to another. During each cycle, the polarity of a leading sunspot in a given hemisphere is the same as the polarity of the Sun's magnetic pole for that hemisphere (as shown in Figure 11-27d). Thus, the entire magnetic cycle of the Sun has a 22-year period. The reversal of the Sun's magnetic poles happens at the middle of every 11-year sunspot cycle, during the period of sunspot maximum. Currently, the Sun's north magnetic pole points through the Sun's southern hemisphere, and it

Figure 11-27
It has been proposed that tubes of magnetic field lines form below the Sun's surface. The Sun's faster rotation near its equator then twists the tubes around the Sun. The insets show the concentrated magnetic field lines breaking through the solar surface resulting in sunspots.

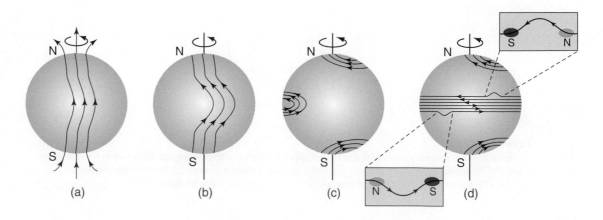

(a)   (b)   (c)   (d)

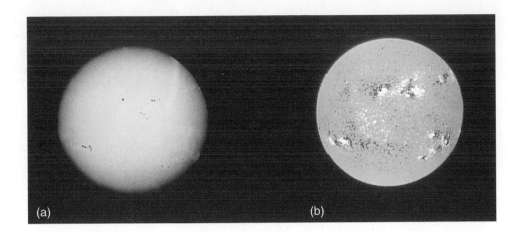

Figure 11-28
(a) A visible-light photo of the Sun, showing sunspots. (b) A magnetic map of the Sun on the same day shows where the magnetic field is strongest on the Sun.

will do so until the year 2012. **Figure 11-28** illustrates the relationship between sunspots and the Sun's magnetic field.

What causes the Sun's field to flip every 11 years? Observations made with the *Yohkoh* and *SOHO* satellites show that giant loops of hot plasma, extending into the Sun's corona, link each of the Sun's magnetic poles to sunspots of opposite polarity (trailing sunspots) near the Sun's equator. As flows from the equatorial regions of the Sun to the magnetic poles transport opposite magnetic flux, the Sun's magnetic field steadily weakens. At the same time, the leading sunspots at both hemispheres migrate toward the equator and, when they meet, their opposite magnetic polarities cancel each other out. (Helioseismology has allowed us to study the flows of differentially rotating magnetic bands inside the convection zone. Magnetic bands at high latitudes migrate toward the poles, and bands from low latitudes migrate toward the equator.) At the height of sunspot maximum, the magnetic poles change polarity and the Sun's magnetic field begins to grow in a new direction. During the flip, the field is very weak and uneven across the Sun's surface, which in turn becomes very active, shooting bubbles of hot gas and energy in every direction. At the time of sunspot minimum, the Sun's magnetic field resembles that of a bar magnet (just like Earth's), and is about 100 times stronger than Earth's (or about as strong as a refrigerator magnet).

Observations made by *SOHO* are helping us understand the stormy areas on the Sun's surface where sunspots appear (**Figure 11-29**). We mentioned before that the magnetic field is very concentrated in sunspot regions. But we know from experience, playing with magnets, that magnetic fields of similar polarity repel each other. The same should be happening in a sunspot region, and as a result the sunspot should dissipate quickly. Instead, we observe sunspots that last for weeks. How is this possible? We know that the strong magnetic field below a sunspot behaves like a "plug" that stops the normal upward convective flow from the hot solar interior. As a result, the material above the plug cools. This includes the observable sunspot region, and that is why it looks darker than its surroundings. But as this material cools, it becomes denser. According to the *SOHO* observations, this material then plunges downward at up to 5000 kilometers per hour, drawing the surrounding plasma and magnetic field inward toward the center of the sunspot in the process. This increases the strength of the magnetic field, which in turn prevents

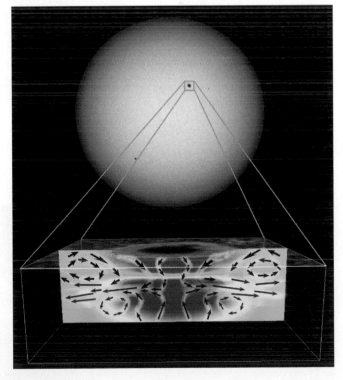

Figure 11-29
An artist's version of the region below a sunspot. Hot regions are shown in red, cool regions in dark blue.

more energy from reaching the solar surface. As the cooling material above the plug sinks, it draws more plasma and magnetic field inward, setting up a cycle that can last as long as the field is strong enough. As Figure 11-29 suggests, the region below the plug is hot, and this is just the opposite of the conditions at the surface. It also seems that the observed outflows at the surface are confined to a very narrow layer. The *SOHO* observations give us a better understanding of the overall structure of sunspots, but we still do not have a clear picture of the details, especially at the roots of the sunspots.

Sunspots can occur in large groups that cover an area tens of times the diameter of the Earth. As shown in Figure 11-18, the observed structure of a sunspot includes a dark area where the magnetic fields are locally vertical surrounded by a lighter area where the fields are locally horizontal. The release of magnetic energy can cause violent solar eruptions such as solar flares and *coronal mass ejections*, which we study next. The main sources for such activity are large clusters of sunspots. With the help of helioseismology, astronomers were able to observe strong circulation patterns near the Sun's surface that play a role in holding the clusters together. Measuring such winds may prove to be a powerful tool in predicting solar flare activity.

Solar Flares and Coronal Mass Ejections

The turbulent magnetic field of the Sun is responsible for the prominences discussed earlier, giving prominences their unique shapes, such as those in the chapter-opening photograph and Figure 11-23. It also causes the colossal flareups called *solar flares* that normally occur during sunspot maxima. Lasting from a few minutes to a few hours, and reaching up to 100,000 kilometers in length, a solar flare can release the equivalent energy of a few million of our largest nuclear weapons (**Figure 11-30**). According to the model presented in the last section, these flares occur when a great number of twisted tubes of magnetic field lines release their energy at once through the photosphere. They occur near sunspots, usually along the dividing line between areas of oppositely directed magnetic fields. In just a few seconds, flares can heat solar material to tens of millions of degrees and accelerate solar particles to very high speeds. In the case of the largest flares, these particles can reach the Earth in less than an hour. They are responsible not only for spectacular auroras but also for disruptions of earthly radio transmissions.

Radio disruption occurs when particularly energetic particles of the solar wind strike a layer of the Earth's atmosphere called the ionosphere. The ionosphere plays a part in radio transmission because it reflects radio waves back down to the surface of Earth. Normally, the Earth's magnetic field prevents particles of the solar wind from reaching the ionosphere by deflecting and trapping them, but the high-energy particles emitted by a solar flare are able to penetrate to the ionosphere. When the ionosphere is disrupted by these particles, we may experience static in radio reception or even complete loss of signal.

coronal mass ejection An event in which hot coronal gas is suddenly ejected into space at speeds of hundreds of km/s.

solar flare An explosion near or at the Sun's surface, seen as an increase in activity such as prominences.

Figure 11-30
(a) An X-ray image of a solar flare.
(b) A powerful flare observed by the *SOHO* spacecraft on July 2002. The flare was associated with an Earth-directed full-coronal mass ejection.

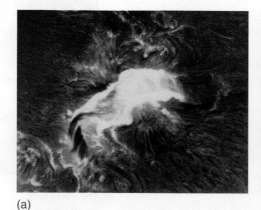
(a)

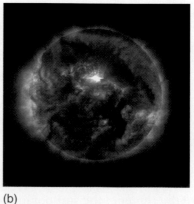
(b)

Coronal mass ejections (**Figure 11-31**) are often associated with solar flares and prominences, but can also occur in their absence. Such ejections and flares might be just different aspects of the same phenomenon. We know they are related, with some events showing more eruptive behavior, while others show more flare behavior. These mass ejections occur more frequently when the Sun is most active. They are huge bubbles of gas threaded with magnetic field lines that are ejected from the Sun over the course of several hours. The mass associated with these ejections can be in the billions of tons (more mass than Mt. Everest), and the speed of the charged particles can be up to 2000 kilometers per second.

Beautiful twisted structures can be seen in coronal mass ejections (**Figure 11-32**). The twists in the magnetic field probably originate below the solar surface, and, just like twisted coils of spring metal, these twisted structures contain energy that is used to blast the material into space.

We know that the active regions on the Sun's surface from which coronal mass ejections originate consist of a great number of loops filled with plasma. Such loops are shown in Figure 11-23b and in **Figure 11-33a**. The plasma that fills the coronal loops is not static. Instead, observations from the *TRACE* and *SOHO* satellites suggest that the plasma moves in the loops at great speeds, probably because of uneven heating at their bases.

Coronal mass ejections disrupt the flow of the solar wind and produce disturbances that strike the Earth with sometimes catastrophic results. When the disturbances reach Earth, they can create a major disruption in the Earth's magnetic field, which can knock out communication satellites and cause power surges. One such power surge overload resulted in a collapse of Quebec's hydroelectric system, cutting

Figure 11-31
A coronal mass ejection. This is one frame of an ejection in progress, observed in 1997 by *SOHO* using a coronagraph. This produces an artificial eclipse of the Sun by placing an "occulting disk" over the image of the Sun, allowing us to observe the Sun for long periods of time.

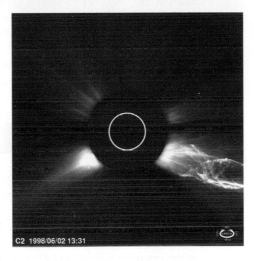

Figure 11-32
A coronal mass ejection showing twisted magnetic field lines. The field acquires this twist (or helicity) beneath the solar surface.

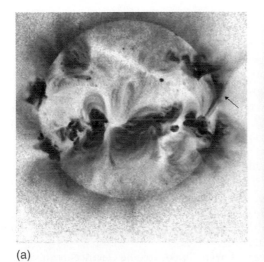

(a)

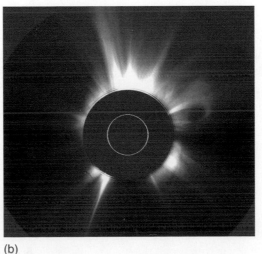

(b)

Figure 11-33
(a) A *Yohkoh* image of a loop in soft X-rays. The width of the arrow's tip represents the size of the Earth. Such loops are about 50,000 kilometers in size and 450,000 kilometers long (equivalent to 40 Earths side by side). (b) A *SOHO* image of the loop in part (a) as it erupts into space.

off six million people for 9 hours in March, 1989. In 1997, a TV communication satellite was lost due to a coronal mass ejection.

Do solar flares and coronal mass ejections have other effects on the Earth and on the lives of those of us who live on this planet? As we learn more about the Sun, we may find that what seem like small quirks in the Sun's behavior are actually of major importance to life on Earth.

We will mention just two possible connections. First, when high-speed protons from the Sun bombard the Earth's upper atmosphere they break up molecules of nitrogen gas and water vapor. The broken molecules form nitrogen and hydrogen oxides, which then react with ozone and reduce its amounts. Even though the overall effect of such a bombardment is minimal (on the order of 1% of the total ozone), it is important to understand and separate the natural effects on ozone from the human factor. Second, cycles of drought at the Yucatan peninsula and cycles of cooling and warming at other parts of the Earth, which can be tracked through radioactive carbon-14 dating, seem to be correlated with a known 206-year cycle in solar intensity. Small variations in the Sun's energy output, if sustained over a long period of time, could have catastrophic climate effects. The observations seem to suggest that some mechanism in our climate is amplifying these natural variations.

◆ As we explained in the discussion of the formation of the solar system, the solar wind had a major function in determining the nature of today's solar system.

Conclusion

Like our ancestors, we recognize the importance of the Sun to life on Earth. We also try to understand the changes we observe and how these changes influence our planet. The Sun's luminosity varies on time scales from milliseconds to billions of years. We think that some of these variations affect our climate, but we do not yet know how. Similarly, the ultraviolet light and X-rays emitted by the Sun heat up our atmosphere. The solar wind emanating from the Sun affects the Earth's magnetic field, pumps energy into the radiation belts, and can cause power surges. As we become more dependent on satellites, we will increasingly feel the effects of space weather and the need to predict it.

To astronomers, the Sun takes on even more importance because it is by far the closest star. Since astronomers must understand stars if they are to understand the workings of the universe, the Sun becomes critical in such a study. Many physical processes occurring elsewhere in the universe can be examined in detail on the Sun. Solar astronomy teaches us much about stars, planetary systems, galaxies, and the universe itself.

Study Guide

RECALL QUESTIONS

1. The Sun's energy is generated by
   A. gravitational contraction.
   B. nuclear fission.
   C. hydrogen fusion.
   D. helium fusion.
   E. chemical reactions.

2. The layer of the Sun that is normally visible to us is the
   A. corona.
   B. chromosphere.
   C. photosphere.
   D. core.
   E. solar wind.

3. Sunspots are areas on the Sun that are
   A. hotter than their surroundings.
   B. cooler than their surroundings.

   C. brighter than their surroundings.
   D. [Both A and B above.]
   E. [Both B and C above.]

4. The energy produced in nuclear reactions in the Sun results from
   A. friction as the nuclei crash together.
   B. heat produced from the electrical effects of the reactions.
   C. the increase in mass of the particles due to the reactions.
   D. the decrease in mass of the particles due to the reactions.

5. Why is a high temperature needed for energy production in the core of the Sun?
   A. Hydrogen will not combine with oxygen at a low temperature.
   B. Energy is needed to overcome electrical repulsion.
   C. Electrons will not recombine at low temperatures.

D. The force of gravity is greater at high temperatures.

E. Speeds are less at high temperature, so there is more time for reactions between nuclei.

6. We know that the Sun's energy does not result from a chemical burning process because
A. of the Doppler effect.
B. of the redshift.
C. the Sun would have burned up already.
D. [Both B and C above.]
E. [Both A and B above.]

7. The two forces producing hydrostatic equilibrium in the Sun to determine its size are
A. electrical forces and gravity.
B. nuclear forces and gravity.
C. electrical forces and gas pressure.
D. electrical forces and nuclear forces.
E. gravity and gas pressure.

8. As the Sun "burns,"
A. its total mass decreases very slightly.
B. its total mass increases very slightly.
C. its energy decreases, but the Sun's mass remains the same.
D. energy is produced, but the Sun's mass remains the same.
E. [None of the above.]

9. During a total solar eclipse, the Sun's atmosphere becomes visible. Why?
A. It is brighter during an eclipse because of light reflected from the Moon.
B. The light reemitted after absorption becomes visible because the brighter Sun is blocked out.
C. The atmosphere becomes hotter during an eclipse.
D. [The statement is not true.]

10. The total luminosity of the Sun can be calculated from its
A. rotation period and temperature.
B. rotation period and diameter.
C. diameter and distance from the Earth.
D. diameter and the solar energy at Earth's distance.
E. distance from Earth and the solar energy detected at Earth's distance.

11. Two factors that determine the pressure of a gas are
A. the speed of the molecules and the particle density.
B. the speed of the molecules and the gas's temperature.
C. nuclear reactions in the gas and its temperature.
D. chemical reactions in the gas and its temperature.
E. [Both C and D above.]

12. At any particular level within the Sun, the pressure outward is
A. less than the pressure inward.
B. equal to the pressure inward.
C. greater than the pressure inward.
D. [No general statement can be made.]

13. Granulation of the photosphere is a direct result of
A. heat conduction.
B. convection.
C. heat radiation.

14. A prominence is
A. a cool spot on the Sun.
B. the ejection of material from the photosphere.
C. a fairly permanent bulge on the photosphere's surface.
D. a reaction within the Sun's core.

15. The solar wind extends
A. about to Mercury's orbit.
B. about to Venus's orbit.
C. almost to Earth's orbit.
D. far beyond the Earth's orbit.

16. The 11-year cycle of sunspots corresponds to
A. the period of change in the magnetic field of the Sun.
B. the rotation period of the Sun near the equator.
C. the rotation period of the Sun near the poles.
D. the revolution period of Jupiter.
E. [None of the above.]

17. Which of the following is the thinnest layer of the Sun?
A. corona
B. chromosphere
C. photosphere
D. radiative layer
E. convection layer

18. Solar energy strikes the Earth at the rate of 1380 watts/m^2. It strikes a sphere that is 2 astronomical units from the Sun at a rate of
A. 345 watts/meter^2.
B. 690 watts/meter^2.
C. 1380 watts/meter^2.
D. 2760 watts/meter^2.
E. 5520 watts/meter^2.

19. The primary source of energy for the Sun is a series of nuclear reactions in which
A. four hydrogen nuclei fuse to form a helium nucleus
B. a helium nucleus fissions to form four hydrogen nuclei.
C. uranium nuclei fission to form several other elements.
D. two nuclei fuse to form uranium or plutonium.
E. oxygen nuclei combine to form more massive nuclei.

20. To begin nuclear fusion in a star, high temperatures are required in order to overcome the
A. nuclear force between the protons.
B. nuclear force between the electrons.
C. electrical force between the neutrons.
D. electrical force between the protons.

21. The photosphere is
A. the layer of the Sun where energy is created from mass.
B. the outermost layer of the Sun.
C. the layer of the Sun that we see when viewing the Sun.
D. the layer of the Sun in which we see granulation.
E. [Both C and D above.]

22. When four hydrogen nuclei fuse to form a helium nucleus, the total mass at the end is _____ the total mass at the beginning.
A. less than
B. the same as
C. more than

23. The sun emits its most intense radiation in which region of the electromagnetic spectrum?
A. Radio
B. Infrared
C. Visible
D. Ultraviolet
E. X-ray

24. Nuclear theory predicts that we should detect
A. fewer neutrinos than we do.
B. just the amount of neutrinos that we do, confirming the theory.
C. more neutrinos than we do.

25. What is meant when it is said the Sun has "differential" rotation?

26. Describe some evidence that shows that the source of solar energy cannot be chemical reactions.

27. Distinguish between chemical and nuclear reactions, giving an example of each.

28. What is it about the nucleus of an atom that distinguishes one atom from another?

29. Name the chemical element that is consumed and the element that is produced in the Sun. What produces the energy when the change occurs?

30. In what physical state is most of the material in the interior of the Sun?

31. Define and explain *hydrostatic equilibrium*.

32. How do we know how great the pressures are at certain depths below the surface of the Sun?

33. List the three methods of heat transfer, giving an example of each. What method(s) is important in which region(s) of the Sun?

34. If all electromagnetic radiation travels at the speed of light, why does radiated energy take so long to get from the center of the Sun to the surface?

35. Describe the thickness and temperature of the photosphere.

36. If the chromosphere and corona are so hot, why are they not brighter than the photosphere?

37. According to present theory, what causes sunspots?

QUESTIONS TO PONDER

1. Why do nuclear reactions occur only at the center of the Sun?

2. Explain, without using any math, how we measure the total energy output of the Sun.

3. What evidence do we have that the Sun's energy does not result from the burning of fossil fuels?

4. If the mass of the Sun decreases because of nuclear fusion, why don't we see the Sun decreasing in size?

5. Distinguish between force and pressure as defined in science, and give an example of units in which each can be expressed.

6. Describe the relationship among pressure, density of particles, and temperature in a gas.

7. What would happen to the Sun's core if the rate of fusion reactions decreased suddenly?

8. How do we know the pressure deep within the Sun?

9. What method of energy transport causes the handle of a poker to get warm when the business end of the poker rests in a fire?

10. We see the Sun not as it is now, but as it was eight minutes ago. The energy we detect, however, began millions of years ago. Discuss the implications this has on what we mean by the word "now." How does this provide a possible explanation for the neutrino problem?

11. What is the solar neutrino problem and why are astronomers so interested about it?

12. List and describe the layers of the Sun's atmosphere.

13. Historically, what has been the value of an eclipse in studying the Sun?

14. The neutrino problem is an example of an unsolved problem concerning the Sun. Another such problem exists in explaining temperatures within the solar atmosphere. Explain the problem.

15. If the magnetic field of the Sun reverses every 22 years, what is meant when we refer to the "northern" hemisphere of the Sun?

CALCULATIONS

1. Calculate the angular size of a 15,000-kilometer sunspot as seen from Earth.

2. Verify that the average density of the Sun is about 1.4 times larger than that of water.

3. If one kilogram of coal is burned in one second, it will produce 2 million watts of power. How many kilograms would have to be burned each second to produce the Sun's energy output? If the Sun were made of coal, how much time would pass before it would burn out at its present rate of energy production? (In fact, oxygen would be needed for the burning—nearly three times more mass of oxygen than of coal—so only one-fourth of the Sun's mass would be coal.)

4. Use Wien's law (Section 4-4) to calculate the wavelength that corresponds to the peak of the Sun's blackbody curve. Use 5800 K as the temperature of the Sun's photosphere. What color does this wavelength correspond to? Why does the Sun appear yellow to our eyes?

5. The average power requirement for the United States is about 6×10^{12} watts. With our current technology we can collect solar power at the surface of the Earth of about 200 watts/meter^2. If we were to build one huge circular solar collector, what would its radius be in miles? (Hint: The area of a circle of radius R is πR^2.) Do you think this idea could lead to a solution for our energy needs?

ACTIVITIES

1. Measuring the Diameter of the Sun

With simple equipment, you can measure the size of the Sun with a fair degree of accuracy. All you need is a sunny day, a piece of cardboard, a ruler, and the knowledge that the Sun is 150,000,000 kilometers from Earth. An Activity in Chapter 6 discussed observing an eclipse by pinhole projection; we will use pinhole projection here, also.

Punch a small hole (perhaps one-eighth inch) in a piece of cardboard and hold the cardboard so that the Sun shines through the hole onto a surface behind it. (Refer to Figure 6-47.) You may have to adjust the size of the hole to get an image bright enough to see clearly, and you might use additional cardboard to shield your screen from reflected sunlight.

Making sure that the screen is perpendicular to a line from the pinhole, measure the diameter of the image of the Sun and the distance from the pinhole to the screen. Now, as shown in **Figure 11-34**, the following ratio applies:

$$\frac{\text{diameter of Sun}}{\text{distance to Sun}} = \frac{\text{diameter of image}}{\text{distance from screen to image}}$$

Use this equation to calculate the diameter of the Sun.

To get a feel for the accuracy of your measurement, make several measurements with the screen at different distances. How closely do your various measurements agree? What is the largest source of error in this procedure? How does the value you obtained compare to that found in the Data Page?

2. Observing Sunspots

Sunspots were first observed by the naked eye, as described in the text, but such a method is not recommended. You would probably have to search the setting Sun for long periods of time over many

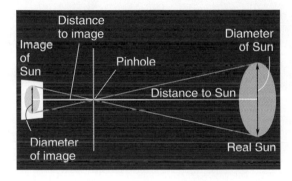

Figure 11-34
This drawing, obviously not to scale, illustrates the relationship between distances and sizes of the Sun and its image. Since the triangle at the left is similar to that at the right, the ratio of distance to size is the same for each.

years before you saw your first sunspot, and staring even at the setting Sun might damage your eyesight.

A more realistic way to view sunspots is with a telescope. ***Do not, however, look directly at the Sun with a telescope.*** You can obtain a solar filter to put over the front of the telescope but it is even better to use the telescope to project an image of the Sun on a screen, as described below.

First, a caution: The intensity of sunlight is so great that you run a risk of damaging your telescope. One way to decrease this risk is to cover the objective lens with a piece of cardboard with a hole cut in it smaller than the lens. Tape it down so it won't fall off, and it will block out some of the light. If your telescope has a finderscope, you should cover it by taping a piece of cardboard (without a hole) over its objective lens. This will not only prevent eye damage to someone who, out of habit, looks through it, but it will prevent the Sun from burning out the finderscope's crosshairs.

In using a telescope to view the Sun, set it up by first focusing on a distant object (**NOT THE SUN**). Then pull the eyepiece out just slightly. Now is the time to cover the finderscope and partially cover the objective. Point the telescope exactly at the Sun. This is not as easy as it sounds, and the best way is to move it until the shadow of the telescope tube is smallest. **DON'T LOOK THROUGH THE FINDERSCOPE.** When this is done, you should be able to see a spot on a screen held behind the eyepiece. Focus by moving the screen and eyepiece to various positions until you get the view you want. A cardboard shield around the telescope will shadow your image from direct rays and improve your view. If you have a star diagonal (which reflects the image off to the side), use it. Trace the image of the Sun and any spots you see. Then repeat your observation in a day or two and look for motion of the sunspots.

Figure 6-46 (in Activity #2 in Chapter 6) shows a small telescope being used to project the Sun's image during an eclipse.

Have fun, but be careful.

You can find information about current missions to study our Sun at NASA's home page (http://www.nasa.gov) and at ESA's homepage (http://www.esrin.esa.it).

1. "The Stellar Dynamo," by E. Nesme-Ribes, S. L. Baliunas, and D. Sokoloff, in *Scientific American* (August, 1996).

2. "How Stars Shine," by J. Trefil, in *Astronomy* (January, 1998).

3. "Where are the Solar Neutrinos?," by J. Bahcall, in *Astronomy* (March, 1990).

4. "SOHO Reveals the Secrets of the Sun," by K. Lang, in *Scientific American* (March, 1997).

5. "The Sun-Climate Connection," by S. Baliunas and W. Soon, in *Sky & Telescope* (December, 1996).

Quest Ahead to Starlinks:
www.jbpub.com/starlinks

Starlinks is this book's online learning center. It features **eLearning,** which contains chapter quizzes and other tools designed to help you study for your class. You can also find **on-line exercises,** view numerous relevant **animations,** follow a guide to **useful astronomy sites** on the web, or even check the latest **astronomy news** updates.

EXPANDING THE QUEST

STARLINKS

CHAPTER 12

Measuring the Properties of Stars

Mauna Kea Observatories, Mauna Kea, Hawaii. The summit (4200 meters high) houses the world's largest observatory for optical, infrared, and submillimeter astronomy.

ASTRONOMER SIDNEY WOLFF AND HER HUSBAND WERE INVOLVED in the building of the observatory on Mauna Kea, a 14,000-foot high mountain in the Hawaiian Islands, in the late 1960s. In an essay in *The Scientist*, she writes:

We had a wonderful time in those early days, developing a site without even such basic amenities as a source of water; a site where all power had to be generated locally because there was no power line, a site where blizzards raged in winter and where even in summer temperatures dipped to freezing every night. But we learned. We learned about altitude sickness and the best strategies for forcing our bodies to acclimate. We learned first aid so we could cope with accidents, since professional help was hours away. We learned how to handle heavy machinery and how to maintain generators. We learned more about telescope gears and worm drives and how to repair scored gears than we ever wanted to know. Nearly every one of us who was involved in those early days can tell—loves to tell—stories of being nearly trapped on the mountain during a blizzard, of hiking to the summit because the road was blocked by snow, of climbing to the top of the dome to remove snow so that not a moment of observing was lost. It was a great adventure, an adventure that surely I had not envisioned when I planned a life of research alone in my office.*

Only recently in human history have we even become aware of the astonishing fact that each of the thousands of stars we see is another sun similar to the one that rules our sky. This realization makes obvious the immense distances to those stars; just imagine how far away our Sun would have to be in order to appear as dim as one of these stars. Can we hope to learn much about such faraway objects? As we explained in Chapter 4, spectral analysis can be used to determine both the temperature of a star and the chemical composition of its atmosphere. Galileo and Newton would have been amazed that we can learn such things. However, temperatures and chemical compositions are only the beginning. In this chapter we discuss how parallax allows us to calculate the distances to many stars and how, once we know their distances, we can determine other properties, including their luminosities, motions, sizes, and masses. We describe relationships among the various properties that give us clues as to why one star differs from another, how stars are formed, how their lives progress, and how they die. We delay discussion of the life cycle of stars until the next chapter, turning our attention now to how we measure those properties of stars that divulge information about their life cycles.

■ *Sidney C. Wolff, National Optical Astronomy Observatories.

The stars, while differing the one from the other in the kinds of matter of which they consist, are all constructed upon the same plan as our sun, and are composed of matter identical at least in part with the materials of our system.
William Huggins, 1863

12-1 Stellar Luminosity

When we speak of the brightness of a star, we must be careful to distinguish between its apparent brightness (**Figure 12-1**) and its *luminosity*. In Section 11-2 we explained how we can calculate the luminosity of the Sun if we know the Sun's distance and the amount of solar radiation striking a given area of Earth in a certain amount of time. Suppose two stars differ in apparent brightness so that different amounts of light reach the Earth from the two stars. The cause for this might be any combination of three things: (1) one star may be inherently brighter than the other (its luminosity may be greater); (2) one star may be closer, making it appear brighter; or (3) there may be more interstellar material absorbing light from one star than from the other.

In our discussion of the two quantities, *brightness* refers to the apparent brightness seen from Earth and *luminosity* refers to the total amount of power emitted by a star. You may occasionally find the term *absolute luminosity*; this simply emphasizes that we are speaking of the radiation actually being emitted, not the radiation reaching Earth.

luminosity The rate at which electromagnetic energy is emitted.

Apparent Magnitude

One of the greatest astronomers of the pre-Christian era was Hipparchus, a Greek thinker who lived in the second century B.C. Hipparchus compiled a catalog of some 850 stars, listing each star's location in the sky, along with a number that designated its brightness. To indicate the brightnesses of stars, he divided all visible stars into six

Figure 12-1
Some of the stars of the constellation Orion appear bright because of their proximity, while others are inherently so luminous that they look bright from Earth even though they are very far away. Betelgeuse, the star that appears red at the upper left, is a variable star, about 430 light years away, with radius 660 times that of the Sun and luminosity 4500–15,000 times that of the Sun. Rigel, the star that appears white at the lower right, is about 730 light years away, with radius 60 times that of the Sun and luminosity about 39,000 times that of the Sun. Note that the stars appear to be different sizes in the photo. These differences are not due to the stars' actual sizes, but occur because brighter stars expose larger areas of the photographic plate.

◆ The quantity described here is apparent magnitude rather than absolute magnitude, which refers to the luminosity of a star and is discussed later.

apparent magnitude A measure of the amount of light received from a celestial object.

◆ Photometry and CCDs (charge-coupled devices) were discussed in Section 5-4.

groups, calling the brightest stars in the heavens *magnitude one* stars and the dimmest he could see *magnitude six* stars. Other stars fell in between, with differences between magnitudes representing equal differences in brightness. Thus the brightness difference between a third-magnitude star and a fourth-magnitude star on his scale was visually the same as that between a fifth- and a sixth-magnitude star. Although it may seem odd to assign the larger number to the dimmer star, we might appreciate his reasoning by thinking of the brighter star as a first-class star and the dimmest as a sixth-class star.

Today we use a slightly revised version of Hipparchus's ***apparent magnitude*** scale, for we measure the brightnesses of stars by photographic and electronic methods. Photographic techniques began to be used for such measurements in the mid-1800s. When these measurements were made, astronomers found that if two stars differ by one magnitude, we receive 2.5 times as much light from the brighter star as from the dimmer one. For example, a fifth-magnitude star is about 2.5 times brighter than a sixth-magnitude star. Moving up from fifth to fourth magnitude means that the brightness increases another 2.5 times, making a fourth-magnitude star 6.25 times brighter than a sixth-magnitude star ($2.5 \times 2.5 = 6.25$). A magnitude change of five would mean a change in brightness of 2.5 raised to the fifth power, 2.5^5, or about 100. Astronomers recreated the scale accordingly, *defining* a difference of five magnitudes as corresponding to a factor of exactly 100 in the light reaching us. This means that stars differing in apparent magnitude by one unit have a brightness ratio equal to the fifth root of 100, or 2.512.

Before electronic devices became common, astronomers measured light intensity from a star (photometry) by measuring the size and density of its image on a photograph (Figure 12-1). During the last few decades, the process used in photometry is similar to the way your automatic camera determines the brightness of the subject you are photographing. **Figure 12-2** illustrates the idea. Today, however, video imaging is becoming common, and CCDs are used to detect and measure light intensity.

Although the eye is able to distinguish only a few different classes of stars (Hipparchus distinguished six), photometric methods allow us to discern the difference in brightness between two stars that may appear identical to the eye. The ability to measure magnitudes accurately has changed Hipparchus's unit-step system into a continuous one so that the magnitudes of stars are now measured to fractional values, to an accuracy of 0.001 magnitude or better.

When the magnitude scale is defined as described, we find that some stars in Hipparchus's first-magnitude group are much brighter than others in that group. If a star were 2.5 times brighter than first magnitude, it would have to be assigned a magnitude of zero. A star 2.5 times brighter than this would have a magnitude of −1, a negative number. Sirius, the brightest star in the night sky, is about 10 times brighter

Figure 12-2
The basic idea of photometry is that the light from a star is focused onto a photocell, which measures the amount of light. Often a filter is used to allow only light of a certain wavelength range to enter. More modern methods use CCD technology rather than photocells.

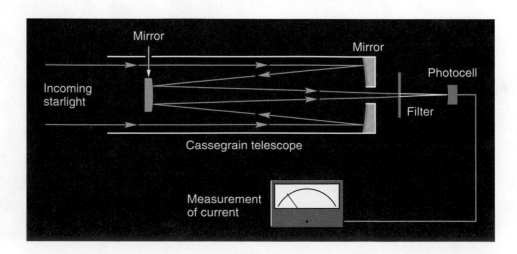

Table 12-1

Magnitude Difference versus Ratio of Brightness

Magnitude Difference*	Ratio of Light Received	
1.0	2.5	
2.0	6.3	(2.5×2.5)
3.0	16	(2.5^3)
4.0	40	(2.5^4)
5.0	100	(2.5^5)
10.0	10,000	(2.5^{10})

*If the difference in magnitude between two stars is 3, we receive 16 times more light from one star than from the other.

Figure 12-3
Apparent magnitudes of some familiar objects.

Apparent magnitude

BRIGHT

−25 —— Sun

−20

−15 —— 100-watt bulb (at 100 feet)
−10 —— Full Moon

−5 —— Venus (at brightest)

0 —— Sirius
—— Saturn (at brightest)

5 —— Naked-eye limit

DIM

10 —— Binocular limit

15 —— Pluto

20 —— Large telescope (visual limit)

25 —— Large telescope (photographic limit)

? Is our Sun the brightest star in the Milky Way galaxy?

than the average first-magnitude star and has an apparent magnitude of −1.43. **Figure 12-3** shows the approximate apparent magnitude of several objects, and Appendices E and F contain tables of the brightest and nearest stars, respectively, including magnitude values. As shown in Appendix E, in addition to our Sun and Sirius, two other stars in our night sky (Canopus and Arcturus) have negative apparent magnitudes.

Table 12-1 shows the ratio of light received for a given difference in apparent magnitude. The following example shows how to use the chart.

EXAMPLE

The star W Pegasi is within the square of the constellation Pegasus (**Figure 12-4**). This star varies in brightness, getting as bright as eighth magnitude. Fomalhaut (the brightest star in the constellation Pisces Austrinus, just south of Aquarius) is a first-magnitude star. Which star appears brighter, and how many times more light do we receive from that star than from the other?

SOLUTION To answer the first question, remember that the star with the lesser magnitude is the one that appears brighter. Thus Fomalhaut is the brighter star. The difference in magnitude provides the information needed to answer the second question. The difference is 7. Table 12-1 does not list a difference of 7, but the values in the table can be used to determine what the light ratio is. Note that a difference in magnitude of 5 corresponds to a light ratio of 100 and a difference of 2 corresponds to a ratio of 6.3. Thus a difference of 7 means that the ratio of light received is 100×6.3, or 630. We receive 630 times as much light from Fomalhaut as from W Pegasi (when the latter is at its brightest). [The magnitudes of Fomalhaut and W Pegasi are 1.17 and 7.9, respectively. For simplicity, we have rounded them off.]

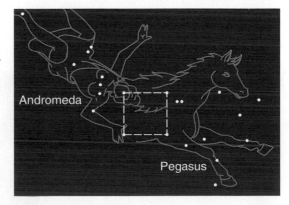

Figure 12-4
The great square of Pegasus is an obvious feature of the fall sky. The constellations Pegasus and Andromeda share one bright star, Alpheratz, shown as an eye of Andromeda.

TRY ONE YOURSELF

Barnard's star has an apparent magnitude of about 10. Suppose that on some particular night Mars is measured to have an apparent magnitude of 2. How many times more light do we receive from Mars than from Barnard's star?

As indicated in the example, telescopes allow us to see stars that are much dimmer than sixth magnitude. A telescope with a diameter of 12 centimeters (about 5 inches) might permit one to see, under perfect conditions, stars of about thirteenth magnitude. With a 5-meter telescope, one can photograph stars as dim as twenty-fifth magnitude.

It is valuable to know the brightnesses—the apparent magnitudes—of stars, but this is not a quantity that is inherent in the stars. Instead, it partially depends upon their distances from Earth and therefore upon the position of the Earth in the Galaxy. We would like to know the *actual luminosity* of each star, for this would tell us something about the star itself, independent of the Earth's location. To do this, however, we must know the distance to the star.

12-2 Measuring Distances to Stars

Of course, the observation of stellar parallax is now an argument in favor of the heliocentric system.

Recall from the earlier discussion of the heliocentric/geocentric debate (Section 2-3) that the absence of observable stellar parallax was an argument against the heliocentric system. Copernicus, however, held that it was not observed simply because the stars are too far away. Not until 1838 was stellar parallax first observed, for the maximum angular displacement of the nearest star is only 1.52 arcseconds. Although the angle between lines from opposite sides of the Earth's orbit toward the star is 1.52 arcseconds, astronomers define the **parallax angle** as half of that, or 0.76 arcsecond in this case. (Parallax angle is defined in this way so that there is a straightforward application of the properties of right triangles; see **Figure 12-5**.)

parallax angle Half the maximum angle that a star appears to be displaced due to the Earth's motion around the Sun.

The formula used to determine the distance to a star by parallax can be written as

$$\text{distance to star (light-years)} = \frac{3.26 \text{ light-years}}{\text{parallax angle in arcseconds}}$$

parsec The distance from the Sun to an object that has a parallax angle of one arcsecond.

Astronomers usually prefer to express parallax using a different distance unit, the **parsec**. With distance in parsecs, this equation simplifies to

$$\text{distance to star (parsecs)} = \frac{1}{\text{parallax angle in arcseconds}}$$

Figure 12-5
The nearest star has a displacement of about 1.5 arcseconds. The parallax angle is defined as half of this, using one astronomical unit as the baseline. The drawing is far out of scale.

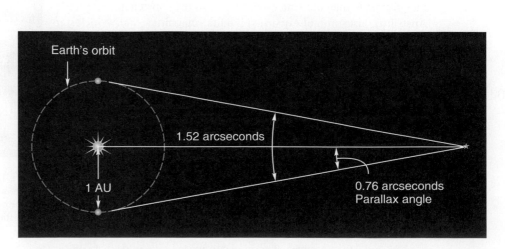

Earth's orbit

1.52 arcseconds

1 AU

0.76 arcseconds
Parallax angle

TOOLS OF ASTRONOMY

Naming Stars

Most of the names of the constellations are Latin translations of the original Greek names. Most of the names of stars, however, are of Arabic origin. Ptolemy's *Almagest* (his summary of Greek astronomy, including a catalog of over 1000 stars) was preserved and passed on by Arab astronomers, who assigned names to numerous stars. The Arabic translation of the English article "the" is "al," which explains why so many names of stars begin with those letters—Alcor (in the constellation Ursa Major), Aldebaran (in Taurus), and Alpheratz (in Andromeda; see Figure 12-4).

However, only the brightest stars have popular names. Stars are assigned "official" names by several methods. The brightest stars in each constellation are given a Greek letter according to their brightness. Thus Elnath, the second brightest star in Taurus,

is β *Tauri*. The brightest star in Taurus is Aldebaran, but since Aldebaran is part of a binary system of stars (to be discussed later in the chapter), it is named α *Tauri A*, and its dimmer companion is α *Tauri B*. Dimmer stars are given English letters followed by their constellation names (for example, *W Pegasi*, a star that was used in the example concerning stellar magnitudes).

We soon run out of letters in the Greek and English alphabets, of course, so most stars are known only by a catalog number. While the brightest star in Taurus is named *Aldebaran* and α *Tauri*, it is also known as *87 Tau*. Less distinguished stars don't have popular names and don't have letter designations. One near Aldebaran just goes by the tag *75 Tau*.

In fact, this form of the distance equation *defines* the parsec: One parsec (abbreviated pc) is the distance from the Sun to a star that has a parallax angle of one arcsecond. The parsec is the unit astronomers normally use to express stellar distances. One parsec corresponds to about 3.26 light-years, or 206,265 astronomical units.

Only stars within about 120 parsecs (400 light-years) have parallax angles large enough (0.008 arcsecond) to permit accurate parallax measurements from Earth. In 1989 the European Space Agency launched *Hipparcos* (*HI*gh *P*recision *PAR*allax *CO*llecting *S*atellite), which measured the positions and parallaxes of some 120,000 stars to an accuracy of 0.001 arcsecond, and more than 2 million stars to an accuracy of 0.02–0.03 arcsecond, before its power failed. *Gaia*, the successor mission to *Hipparcos*, is named after the Greek Earth goddess, and will be launched around 2010. Its mission is to give precise and detailed information about the billion brightest objects in the sky, thus giving us a large and precise three-dimensional map of the Milky Way. This map will allow astronomers to do an archaeological study of our Galaxy, uncovering its history and mysteries in the process. NASA is now planning the *Space Interferometry Mission* (*SIM*), scheduled for launch in 2009, which will be the first space mission with an optical interferometer as its primary instrument. *SIM* is designed to achieve an accuracy, over the whole sky, of 1 microarcsecond. This is 1000 times better than the accuracy achieved by the *Hipparcos* mission. Such accuracy would revolutionize the field of astrometry—the precise measurement of the positions of stars.

Having more accurate stellar distances will be very valuable to astronomers for two reasons. First, because knowledge of the distance to a celestial object is often the key to determining other properties of the object, such as its luminosity. Second, because it will help us determine the distance scale of the universe more accurately. In the next section we explain how knowing the distance to a star allows us to calculate the star's luminosity. In a later section we will explain why the accuracy of our measurements of distances to very remote objects often depends upon accurate knowledge of the distances to nearby stars.

0.001 arcsecond corresponds to the angular size of a golf ball viewed from across the Atlantic Ocean. *Hipparcos*'s mirror was so smooth that if it were scaled to the size of the Atlantic Ocean, its surface bumps would be less than 4 inches high.

TOOLS OF ASTRONOMY

Calculating Absolute Magnitude

We can find the apparent magnitude of an object by simply observing it, but its absolute magnitude is not as easy to measure. After all, we cannot just move an object to a distance of 10 parsecs from Earth in order to observe it. Instead we calculate the absolute magnitude. The relationship between an object's apparent magnitude (*m*), absolute magnitude (*M*), and distance (*d*, in parsecs) can be expressed as

$$m - M = 5 \times \log(d) - 5$$

The difference (*m* − *M*) is called the ***distance modulus.***

To illustrate the use of this equation, let's calculate the absolute magnitude of Sirius. Sirius has an apparent magnitude of −1.43 and is 2.6 parsecs from Earth. We substitute

these values into the equation and solve for the absolute magnitude.

$$-1.43 - M = 5 \times \log(2.6) - 5$$
$$-1.43 - M = 5 \times 0.42 - 5$$
$$M = 1.47$$

What is the physical meaning of this equation? Let's use Sirius as an example. According to the inverse square law, if we "move" Sirius from its actual distance of 2.6 to 10 parsecs from us, it would appear $2.6/10^2$ or 1/15 times as bright. From Table 12-1 we see that this corresponds to a change in magnitude of about 2.9. Therefore, Sirius's magnitude at 10 parsecs would be about (−1.43) + 2.9 = 1.47, exactly the same value obtained above.

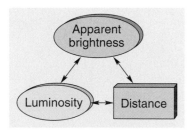

Figure 12-6
If any two of the three quantities are known, the other can be calculated. Interstellar absorption of light is generally so little that it can be ignored.

✦ Note the similarity of Figure 12-6 to the discussion in Section 6-1 of the parallax formula and the small-angle formula. Also note that distance is a key quantity in all of these relationships.

absolute magnitude The apparent magnitude a star would have if it were at a distance of 10 parsecs.

Absolute Magnitude

In Chapter 11 we explained how we can calculate the luminosity of the Sun, using the distance from Earth to the Sun and the value of the solar power striking a square meter of the Earth. In a similar manner, we can calculate the luminosity of any light source if we know the distance to the light source and the brightness of the source. (The power striking a given area of Earth determines brightness.) We all make subconscious judgments similar to this when we look at a distant light at night—a street light, for example—and decide that the lamp is dim or bright. We subconsciously take into account how bright the light appears, as well as its distance from us, in order to decide (qualitatively) the wattage of the bulb. If we had the tools to measure brightness and distance quantitatively, we could calculate the lamp's power in watts.

Figure 12-6 emphasizes that if we know any two of the following three factors we can calculate the third: apparent brightness (or, in astronomy, apparent magnitude), luminosity, and distance. The connection between these three quantities will be used several times in the remaining chapters. Brightness is most easily measured, and it is always one of the two known quantities. When the other known quantity is distance, we can calculate the luminosity; when the luminosity is known, we can calculate the distance.

The triple connection between brightness, distance, and luminosity allows astronomers to calculate the luminosity of some 1000 nearby stars; that many stars are close enough for us to determine their distance fairly accurately by parallax. If we want to know a star's luminosity in watts, we can use the procedure discussed for the Sun in Section 11-2. Usually, however, astronomers use another method to state the intrinsic luminosity of a star. The ***absolute magnitude*** of a star is defined as the apparent magnitude that the star would have if it were located 10 parsecs from the Earth. The absolute magnitude therefore becomes a way to compare stars' actual brightness; it is a very fundamental property of a star (**Figure 12-7**). A method of calculating the absolute magnitude of a star is illustrated in the Tools of Astronomy box above.

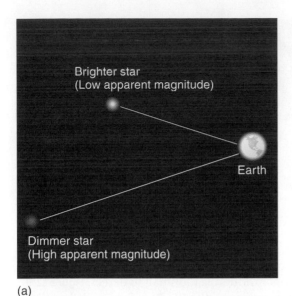

(a)

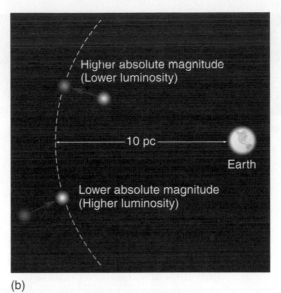

(b)

Figure 12-7
(a) As seen from Earth, one star may appear brighter than another star because it is closer, not because it is inherently brighter. (b) By calculating the magnitudes of stars at the same distance from Earth (10 parsecs), we can determine which stars have higher luminosity (lower absolute magnitude).

12-3 Motions of Stars

You sometimes hear references to the "fixed stars." In fact, the stars are not fixed, but are moving relative to the Sun. This motion is obviously not visible to the naked eye, for the constellations have retained their shapes fairly well over the centuries. However, in 1718, Edmund Halley discovered that stars do move with respect to one another and therefore constellations do gradually change their shapes.

Figure 12-8 shows two photographs, taken 22 years apart, of a magnified portion of the sky. The arrows point to a particular star, Barnard's star, that has moved noticeably during that time.

Barnard's star is the second closest star to the Sun and shows the greatest motion as observed from Earth. It moves at the rate of 10.3 arcseconds per year. Motion expressed as the angle through which a star moves each year (as seen from the Sun) is called the ***proper motion*** of the star. (You might speculate as to how a star could have improper motion, but the name comes from an old use of the word "proper" meaning "belonging to," for this is the motion that actually belongs to the star, as opposed

proper motion The angular velocity of a star as measured from the Sun.

(a)

(b)

Figure 12-8
(a) The arrow indicates a star named Barnard's star. (b) This photo, taken 22 years later, shows that Barnard's star has moved noticeably during that time. (The width of each photo is about 1 degree, so the full Moon would cover about half the width of the photo.)

| 50,000 years ago | Today | 50,000 years from now |

Figure 12-9
The center drawing shows the Big Dipper as it is today; the arrows indicate the proper motions of its stars. From these motions, we can conclude that it once had the shape shown in the left drawing and will some day have the shape shown in the right one.

UFO reports often state the speed of the unidentified object, although what is actually observed is the proper motion of the object. This is one example of how a poor understanding of astronomy can lead to inaccurate conclusions concerning celestial objects.

to observed motion due to Earth's movement.) Only relatively nearby stars show proper motion, so the background stars might still be called *fixed* stars, although they only appear to be fixed because their proper motion is too small to detect. **Figure 12-9** shows how proper motion has changed the shape of the Big Dipper over the past 50,000 years and how its shape will continue to change in the future. We normally identify constellations by the stars that appear brightest, but since these stars are often the closest ones, they have the greatest proper motion. Thus the constellations are gradually changing shape over the ages.

Proper motion does not tell us the actual velocity of a star in normal units of velocity, but we can calculate the velocity by using the small-angle formula. (We used this formula to determine the Moon's size in Section 6-1. Instead of the width of the Moon, we calculate the distance the star moved.) Of course, one must know the distance to the star to make such a calculation (**Figure 12-10**). In doing the calculation, only the speed of the star *across* our line of sight—its *tangential velocity*—is being computed.

The velocity of a star toward or away from the Earth—its *radial velocity*—is easier to detect and measure than its tangential velocity. We discussed in Section 4-7 how this measurement is made and emphasized that the Doppler effect measures only the star's radial velocity.

space velocity The velocity of a star relative to the Sun.

A star's actual motion relative to the Sun—its ***space velocity***—is a combination of its radial and tangential velocities. Since these two are at right angles to one another, we can use the Pythagorean theorem (**Figure 12-11**) to calculate the star's space velocity.

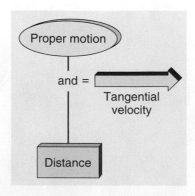

Figure 12-10
To calculate a star's tangential velocity, we must know both its proper motion and its distance.

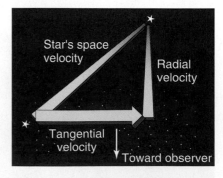

$$(\text{Tangential velocity})^2$$
$$(\text{Radial velocity})^2$$
$$(\text{Space velocity})^2$$

Figure 12-11
Since radial velocity and tangential velocity are at right angles, we can use the Pythagorean theorem to add them.

Naturally, the Earth's movement in its orbit affects the observed motions of stars and must be taken into account in calculating the velocity of a star. Once this is done, we have the star's velocity relative to the Sun. We will see later that there are methods for determining the Sun's movement relative to the distant, "fixed" stars. If this motion is taken into account, we can determine a star's velocity relative to the distant stars.

EXAMPLE

Barnard's star has a proper motion of 10.3 arcseconds per year and is about 1.82 parsecs away. Using the Doppler effect, we measure the star's radial velocity to be −111 kilometers/second. (The minus sign indicates that the star is moving closer to us.) Calculate the space velocity of Barnard's star.

SOLUTION   Using the small angle formula we find

$$\text{distance traveled across the sky per year} = \frac{10.3''/\text{yr}}{206,265} \times 1.82 \text{ pc}$$

But 1 parsec is 206,265 AU, 1 AU is about 1.5×10^8 kilometers and 1 year is about 3.15×10^7 seconds. Therefore, Barnard's tangential velocity is

$$\text{tangential velocity} = \frac{10.3''/\text{yr}}{206,265} \times 1.82 \text{ pc} \times 206,265 \frac{\text{AU}}{\text{pc}} \times \frac{1.5 \times 10^8 \text{ km/AU}}{3.15 \times 10^7 \text{ s/yr}}$$

$$= 89 \text{ km/s}$$

The star's space velocity is

$$\sqrt{89^2 + 111^2} = 142 \text{ km/s or about } 320,000 \text{ mi/h}$$

12-4 Spectral Types

Figure 12-12 is a photo of Orion taken in an unusual way. During the 30-minute exposure, the camera was held steady, but its focus was changed in steps so that the stars became more and more out-of-focus as they drifted by toward the west (the right). This causes each star to form a fan-shaped image, and it reveals the colors of the stars. Most of the brightest stars appear as blue. The star that appears as red at upper left is Betelgeuse, and the red "star" at lower center is the Orion nebula.

As we explained in Section 4-4, the color of a star is determined by its temperature. In fact, the wavelength at which a star emits most of its energy provides an accurate way to measure the star's temperature. Refer back to the blackbody curve (intensity/wavelength graph) of the Sun in Figure 11-17, which shows that the Sun's energy output peaks near the center of the visible region and indicates a temperature of about 5800 K for the surface of the Sun.

Another method of analyzing the spectrum provides an independent measure of temperature. This method depends upon the absorption of radiation at various wavelengths—the absorption spectrum. As we have seen, the absorption of certain wavelengths is what allows us to determine the chemical elements in a star's atmosphere. However, the absorption by an atom depends not only upon what element it is, but also upon the state of its electrons (that is, whether some electrons have been moved to higher energy levels or stripped from the atom). In a gas at higher temperature, more atoms are at higher energy levels, and the transitions that occur from

Figure 12-12
The colors of the stars are obvious in this photo of Orion. The photo was made by successively de-focusing the camera as the stars moved by toward the right. Rigel is at lower right.

Figure 12-13
Annie J. Cannon (1863–1941), a member of the Harvard College Observatory for almost 50 years, was the founder of the spectral classification scheme in use today.

Are all stars the same yellow-white color?

these atoms are different from the transitions that take place in atoms of a cool gas (whose atoms are at lower energy levels). This provides a method of determining temperature other than by the intensity/wavelength graph. The Tools of Astronomy box on page 404 explains this method in greater detail.

Annie Jump Cannon (**Figure 12-13**), an astronomer at Harvard College Observatory, devised a classification scheme for spectra and separated several hundred thousand stars according to the strength of the hydrogen lines in their spectra. The classes were first labeled alphabetically, with "A" having the strongest lines, then "B," and so forth. It was later understood, through the work of Cecelia Payne-Gaposchkin, that the classes represented stars of various temperatures. The classes were then rearranged in order of temperature, and some classes were dropped as redundant. The spectral classes used today are designated, from hottest to coolest, as O B A F G K M (**Figure 12-14**).

The hottest stars, the blue-violet O stars, range in temperature from about 30,000 to 60,000 K. The blue-white B stars have a temperature range of 10,000–30,000 K, while the white A stars range in temperature from 7500 to 10,000 K. The ranges for the yellow-white F stars, yellow G stars and orange K stars are 6000–7500, 5000–6000, and 3500–5000 K, respectively. The coolest stars are the red M stars, with temperatures less than 3500 K.

Within each spectral class, stars are subdivided into 10 categories, called spectral types, and are indicated by attaching an integer from 0 to 9 to the original spectral class. For example, our Sun is listed as a G2 star. You need not be concerned about these subdivisions, but you should remember the classification scheme from hottest to coolest as being O B A F G K M. The traditional mnemonic is "Oh, Be A Fine Guy/Girl, Kiss Me" although you may want to devise your own memory aid.

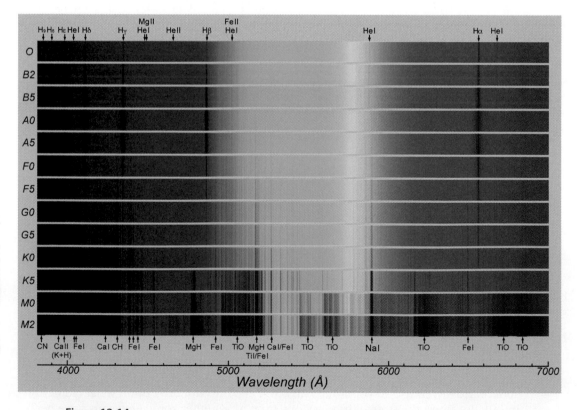

Figure 12-14
The spectra of stars of various spectral types, from O to M, with their classes and subclasses shown at left. A few of the major absorption lines are identified at the top and bottom. (Recall that 1 nm = 10 Å.)

The Hertzsprung-Russell Diagram

Suppose that an extraterrestrial being, unfamiliar with humans, somehow learned the age and height of each person in your neighborhood and then used these values to plot a graph, such as that shown in **Figure 12-15**. The first thing to notice in this graph is that a pattern exists: there is some relationship between height and age for humans. The graph also implies an important fact about the life cycle of humans, even though no one person had been analyzed through his or her entire life. The alien could logically hypothesize that an individual spends most of his or her life at about the same height, that height being the tallest achieved by the person. In our study of stars, we have a similar chart, called the ***Hertzsprung-Russell diagram.***

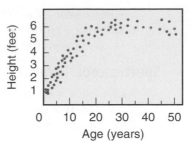

Figure 12-15
A plot of age versus height of people in a neighborhood might look like this. Each dot represents one person.

Early in this century, Ejnar Hertzsprung, a Danish astronomer, and Henry Norris Russell, an American astronomer at Princeton University, independently developed the diagram now named for them and often called simply the H-R diagram. The diagram shows that a pattern exists when stars are plotted by two properties: their temperature (or spectral type) and their absolute magnitude (or luminosity). **Figure 12-16** is similar to the diagram Russell plotted in 1913. The stars are not evenly distributed over the entire chart, but seem to group along a diagonal line. If we include many more stars than those on Russell's first diagram, we obtain a plot as shown in **Figure 12-17**. (Notice the Sun's position on the diagram.)

About 90% of all stars fall into a group whose properties are such that the plotted values form a band running diagonally across the H-R diagram. This band is called

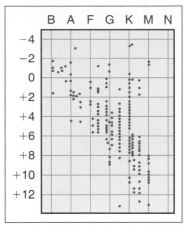

Figure 12-16
This graph is similar to the first H-R diagram, plotted by Henry Norris Russell in 1913. He plotted absolute magnitude versus spectral type and saw an obvious pattern in the distribution of stars. Notice that his category labeled N contained no stars.

Hertzsprung-Russell (H-R) diagram

A plot of absolute magnitude (or luminosity) versus temperature (or spectral type) for stars.

Figure 12-17
A modern H-R diagram shows that stars fall into various categories, including main sequence stars, white dwarfs, giants, and supergiants. Notice that temperature increases as you move to the left on the diagram.

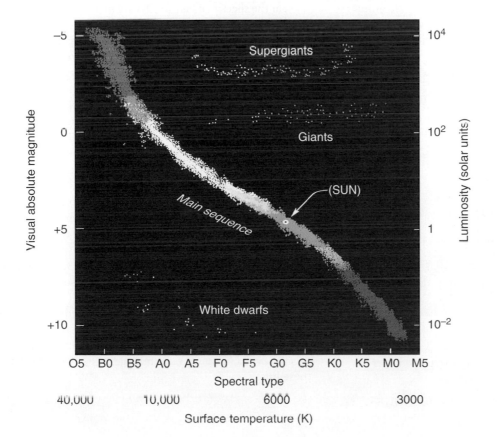

main sequence The part of the H-R diagram containing the great majority of stars; it forms a diagonal line across the diagram.

spectroscopic parallax The method of measuring the distance to a star by comparing its absolute magnitude to its apparent magnitude.

the *main sequence.* Other groups are named as shown in the diagram. The significance of the groups will be discussed later in this chapter and in the next few chapters, along with the life cycle of stars.

Spectroscopic Parallax

As we have seen, the distances to the nearest stars can be measured using stellar parallax. The H-R diagram provides another method of measuring such distances, one that is not confined to neighboring stars. Although this method does not involve parallax, it is called *spectroscopic parallax.*

As we will show later, it is possible to determine from the spectrum of a star whether that star is a main sequence star, a giant (or a supergiant; the term *giant* is often used for both), or a white dwarf. Suppose that we observe a particular star and

TOOLS OF ASTRONOMY

Determining the Spectral Type of a Star

Before studying this Tools of Astronomy box, you may find it helpful to reread a related box in Chapter 4, "The Balmer Series."

Hydrogen is the most common gas in the atmosphere of stars, so it will serve as our example here. Recall that each hydrogen atom has one electron in orbit around its nucleus. Suppose that the hydrogen gas near a star is cool enough that most of its electrons are in the ground state ($n = 1$ in **Figure B12-1**). When an atom with its electron in the ground state absorbs a photon, the photon's energy is such that the electron moves from the ground state to one of the other energy levels. These lines form the Lyman series, which is in the ultraviolet region of the spectrum;

therefore they do not appear in the visible part of the spectrum and cannot be seen in Figure 12-14.

Now consider a hotter star, one with a significant number of atmospheric hydrogen atoms with electrons in the second energy level. For one of these atoms to absorb a photon, that photon must be of the right energy to move the electron from level 2 to higher levels. Figure B12-1 shows that the wavelengths of these photons are part of the Balmer series and are in the visible range. Look at the spectra of Figure 12-14. The Hydrogen-α line is indicated at 656 nanometers, in the red part of the spectrum, far to the right of the other lines shown. The Hydrogen-β line is at 486 nanometers, in the blue part of the spectrum. The other two Balmer lines shown in Figure B12-1 (Hydrogen-γ, at 434 nanometers, and Hydrogen-δ, at 410 nanometers) are in the violet part of the spectrum. The spectrum of the cool K5 star does not show a pronounced absorption line for the energy associated with Hydrogen-β. The reason is that the star is not hot enough for many of its atmospheric hydrogen atoms to be in the level 2 state.

Look at the Hydrogen-β absorption line in the hotter stars. As one moves to hotter and hotter stars, the absorption becomes more intense. In fact, the line is most pronounced in A-type stars. In the spectra of the hottest O- and B-type stars, the Hydrogen-β absorption line is again less pronounced. The reason in this case is that in these very hot stars, most hydrogen atoms have so much energy that their electrons are at level 3. Thus there are few electrons at the second level to absorb photons that correspond to the Hydrogen-β absorption line.

Only one line of the Balmer series of hydrogen has been considered here. A complete analysis would require examining a number of lines of several different chemical elements. The spectral type of a star is determined by the relative intensities of these lines.

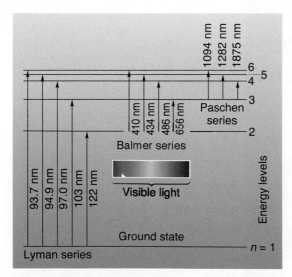

Figure B12-1
The energy levels of the hydrogen atom. If many atoms are in the $n = 2$ state, their absorption of photons will cause a dark line in the visible portion of the spectrum.

determine that it is a main sequence star. As we have seen in our discussion of the blackbody curve and Wien's law in Section 4-4, it is a fairly routine procedure to ascertain the temperature of a star. Let's suppose that our star has a temperature of 10,000 K. Refer to the H-R diagram of **Figure 12-18**. The region of the main sequence where stars have a temperature of 10,000 K is marked on the diagram. It is now a simple matter to determine that our chosen star has an absolute magnitude between about +0.5 and −0.5.

Recall the connection between absolute magnitude, apparent magnitude, and distance. If we know two of these, we can calculate the third. Use of the H-R diagram has allowed us to determine, within a small range, the absolute magnitude of the star. Since apparent magnitude is directly measurable, we can calculate the star's distance.

As indicated, we know the absolute magnitude only within a certain range, and therefore our precision in determining distance is limited. Keep in mind, however, that this is *always* the case with a measurement. In using spectroscopic parallax, the source of the error is obvious, but as we have seen, the determination of distance by trigonometric (stellar) parallax is also limited in accuracy, both because of the difficulty in measuring the small angles involved and because other motions must be taken into account, such as the proper motion of the star being measured. All measurements contain error. The important thing in science is to recognize how great the likely error is.

In practice, the temperature cannot be measured with absolute precision, so some error is introduced in this manner also.

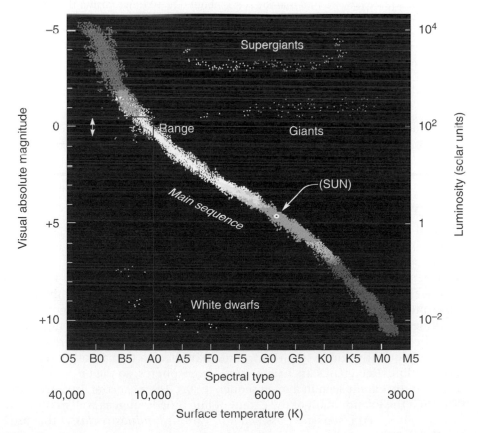

Figure 12-18
The absolute magnitude of a main sequence star with a temperature of 10,000 K can be determined from the H-R diagram.

Because of the limits in the accuracy of spectroscopic parallax, it is most useful when applied to groups of stars that are nearly the same distance from Earth. As we show in the next chapter, it is common for stars to be grouped with many other stars in large clusters. The stars of each cluster are very close to one another compared to their distance from Earth, and we use statistical methods to combine individual measurements of spectroscopic parallax to determine the distance to the cluster. The resulting value has much less error than the distance we calculate to any individual star.

Such averaging over a number of cases to improve accuracy is a valuable and common technique in science.

EXAMPLE

Spica, the brightest star in the constellation Virgo, is a B1-type main sequence star with a temperature of about 25,000 K. Its apparent magnitude is about 0.96. From this data, determine whether its distance from Earth is less than 10 parsecs, approximately 10 parsecs, greater than 10 parsecs, or much greater than 10 parsecs.

SOLUTION

Referring to Figure 12-18 and using the fact that Spica is a B1-type star, we determine that its absolute magnitude is between about −3 and −4.5 (although a safer range might be from −2.5 to −5.0).

Next, let's compare our absolute magnitude values with Spica's observed apparent magnitude. Spica's apparent magnitude is 0.96, and we can reason that since its apparent magnitude has a greater value than its absolute magnitude, Spica is farther than 10 parsecs away. (Remember, greater magnitude means a dimmer star, so its apparent brightness is less than its brightness would be at 10 parsecs.) In fact, from the difference between +0.96 and −3 or −4, we can conclude that it would have to be moved fairly far to bring it as close as 10 parsecs. How far? The difference corresponds to about 4 to 5 magnitudes. According to Table 12-1, this means that we receive 40 to 100 times less light than if the star were at a distance of 10 parsecs. Using the inverse square law, we find that the star is about 6 to 10 times farther than 10 parsecs, or between 60 and 100 parsecs from us.

Actually, Spica is about 80 parsecs, or 260 light-years, from us. Its absolute magnitude is about −3.6.

TRY ONE YOURSELF

The star 40 Eridani is a main sequence star of spectral type K1, which means that its temperature is 4500 K. Its apparent magnitude is +4.4. (Actually, this is the magnitude of the brightest star of a triple-star system. In the example above, Spica is a binary system. Multiple star systems are discussed later.) What can you conclude about the distance to 40 Eridani?

Luminosity Classes

Recall from Chapter 4 that a solid object emits a continuous spectrum because its atoms interact with one another and thereby distort their energy levels. This smears what would be separate, discrete wavelengths into a continuous range of wavelengths. Recall also that an absorption spectrum is produced when light passes through a star's atmosphere. In the 1880s, Antonia Maury discovered that absorption lines are also subject to a smearing effect and that, in general, they are not fine lines. This discovery has become very valuable in classifying stars.

The atmosphere of a main sequence star is fairly dense, and therefore its atoms collide frequently and stretch what would have been a thin absorption line into a broader line. Red giant stars have thinner atmospheres, so the broadening does not occur to the extent seen in a main sequence star. Their spectral lines are narrower. Through the examination of the extent of line broadening, as well as other subtle differences in spectra, stars are classified into various **luminosity classes** that are located on the H-R diagram as shown in **Figure 12-19**.

The example of spectroscopic parallax in the preceding section involved only main sequence stars. The classification of stars into luminosity classes allows spectro-

Maury worked at Harvard College Observatory with Annie Jump Cannon and Henrietta Leavitt, who will be discussed later in this chapter. Maury learned astronomy as a student of Maria Mitchell (see Chapter 10).

luminosity class One of several groups into which stars can be classified according to characteristics of their spectra.

scopic parallax to be used with any star. If we know a star's luminosity class and temperature, we can read its absolute magnitude (and thus luminosity) from the appropriate location on the H-R diagram.

Analyzing the Spectroscopic Parallax Procedure

Observe that in using the method of spectroscopic parallax to find a star's distance from us, we use two triple connections (**Figure 12-20**): By knowing the temperature of a star (that is, its spectral type) and its luminosity class, we can determine its absolute magnitude. Then, by knowing its absolute magnitude and its apparent magnitude, we can calculate its distance. This last triple connection is used in a different way than it was before. For nearby stars, distance and apparent magnitude were used to determine absolute magnitude.

Absolute magnitudes calculated for nearby stars enabled astronomers to draw the H-R diagram. Once the patterns of that diagram were known, they could be applied to stars that are too far away to permit distance measurement by parallax. We reasonably assume that these stars fit the same H-R diagram pattern as do nearby stars, so we can then use the diagram to determine the absolute magnitudes of these faraway stars.

The basic procedure outlined here is often used in astronomy (and in other sciences). By observing familiar objects (nearby stars, in this case), we see patterns and formulate laws (or statements of relationships). We then assume that these patterns and laws hold for more distant objects of the same type. This allows us to use the same patterns to learn more about those distant objects. At times it appears that the whole system is a house of cards ready to fall down, but usually cross-checks are available; measurements can be taken in several ways, thereby permitting us to verify theories by independent measurements.

Luminosity and the Sizes of Stars

Although the sizes of a few of the largest stars have now been directly detected by interferometry methods, in general a star is observed as only a point of light. It shows no size. How, then, can we determine the size of an object so distant that it appears to have no size?

Consider the group of stars in the lower left corner of the H-R diagram, the *white dwarfs*. These stars are hot, but their location in the lower part of the diagram indicates that they are intrinsically dim. One might think that a hot star would be bright, for this is not only the pattern seen in the main sequence, but it also follows from the intensity/wavelength graph: As an object becomes hotter, the wavelength at which the intensity of the emitted radiation is maximum becomes shorter, and the *total amount of radiation per unit area from the star increases*. How, then, can these stars be dim? The answer is that they are small. Being hot, each square meter of their surface emits more energy per second than a cooler star, but they simply have small total surface areas. (Recall our discussion of the blackbody curve and Stefan-Boltzmann law in Section 4-4.) The name "white dwarf" indicates their temperature (white-hot) and their size.

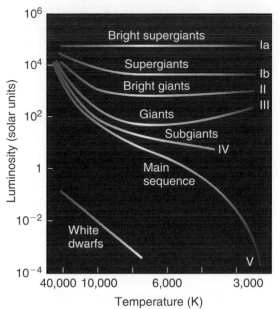

Figure 12-19
Stars can be classified according to luminosity class, which is determined by analysis of their spectra.

Do not confuse *luminosity class* with *spectral type* discussed earlier.

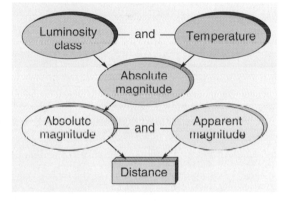

Figure 12-20
From the luminosity class and temperature of a star, its absolute magnitude can be determined. From this and its apparent magnitude, its distance can be determined.

white dwarf A very small, low-mass hot star. (Typical diameter is 0.01 that of the Sun—about the size of Earth.)

An everyday example might help here. You can adjust the burner of an electric stove to various temperatures. As the burner gets hotter, its color changes from dull red to orange-red (and perhaps to orange). At the same time, its overall brightness increases. Imagine, though, that you have a very small burner and you make it orange-hot. If it is small enough, it will emit less total radiation—it will be less bright—than a very large burner that is not as hot. The small, hot burner corresponds to a white dwarf in that its small size causes its overall luminosity to be less than average.

On the other hand, consider the *giants* and *supergiants.* How do we know that these are large stars? Simply because they are very bright in spite of their low temperatures. The giants at the upper right corner of the H-R diagram are often called *red giants,* paralleling the name given the white dwarfs. (A single stove burner set to "warm" does not yield much light, but if enough of these relatively cool burners are lit, they provide quite a lot of red light.)

In these examples of white dwarfs, giants, and supergiants, qualitative reasoning was used to learn something about their sizes. However, determination of stellar sizes is not limited to these classes of stars; nor is it limited to qualitative methods. Knowing the temperature of an object, we know the amount of energy emitted each second per square meter of its surface (that is, the energy flux). Then if we know the total energy emitted by the object each second (that is, the luminosity, by knowing the absolute magnitude), we can calculate the area of its surface and therefore its diameter. In other words,

$$\text{Luminosity} = \text{area} \times \text{energy flux} \propto \text{radius}^2 \times \text{temperature}^4$$

A star may be very luminous and large in size but have a relatively low temperature, or it may be very luminous with an average size if it has a high temperature.

giant star A star of great luminosity and large size (10 to 100 times the Sun's diameter).

supergiant A star of very great luminosity and size (100 to more than 1000 times the Sun's diameter).

The term "energy flux" is what we described in Section 4-4 as "energy emitted each second per square meter of area."

EXAMPLE

We observe the spectrum of a star and find its luminosity class and spectral type. In addition, using Wien's law and the star's spectrum, we find that it has a temperature twice that of the Sun. From the temperature and luminosity class of the star, we find, using the H-R diagram, its luminosity, which is about that of the Sun. What is the radius of this star?

SOLUTION

A star's luminosity (L) is related to its radius (R) and temperature (T):

$$L_{\text{star}} \propto R^2_{\text{star}} \times T^4_{\text{star}}$$

The above expression also holds true for the Sun:

$$L_{\text{Sun}} \propto R^2_{\text{Sun}} \times T^4_{\text{Sun}}$$

In this example we have $T_{\text{star}} = 2 \times T_{\text{Sun}}$. Also, $L_{\text{star}} = L_{\text{Sun}}$, which allows us to set the right-hand sides of the above two expressions equal to each other. We get

$$R^2_{\text{star}} \times (2 \times T_{\text{Sun}})^4 = R^2_{\text{Sun}} \times T^4_{\text{Sun}} \text{ or } R^2_{\text{star}} \times 16 = R^2_{\text{Sun}}$$

Therefore, the radius of this star is 1/4 that of the Sun.

TRY ONE YOURSELF

Find the radius of a star if its luminosity is 10,000 times that of the Sun and its temperature is about 3870 K. (Hint: The star's temperature is about 2/3 that of the Sun.)

Figure 12-21a illustrates the triple connection in the chain of reasoning. We find that stars on the main sequence range from about 0.1 times the Sun's diameter up to as much as 20 times the Sun's diameter. Non-main sequence stars, however, differ in size even more: white dwarfs may be as small as 0.01 of the Sun's size and supergiants may be 100 times larger in diameter than the Sun. The H-R diagram in Figure 12-21b is similar to that of Figure 12-17 but it also includes information about the radii of stars, calculated by using the Stefan-Boltzmann law.

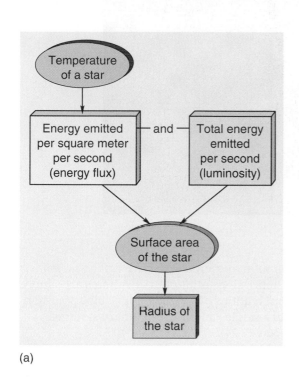

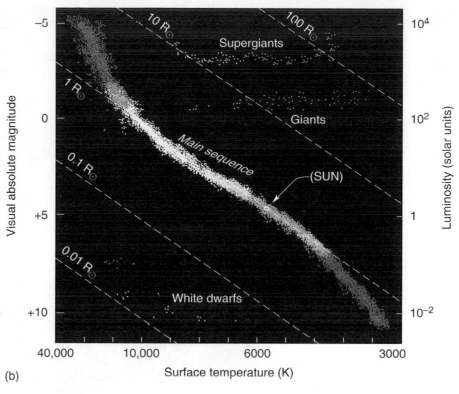

(a)

(b)

Figure 12-21

(a) If the temperature of a star and the total power emitted by it are known, the triple connection shown here can be used to calculate its radius. (b) This H-R diagram is the same as the one in Figure 12-17, but it now includes information about the radii of stars.

It is always desirable to have a second, separate method of measuring a quantity. This allows us not only to check our theories regarding the object being measured, but also to check our measurement techniques. Fortunately, there is a second method for measuring the sizes of stars, although only a few stars can be measured by this technique. To see how it works, some discussion of multiple star systems is necessary.

optical double Two stars that have small angular separation as seen from Earth but are not gravitationally bound.

12-5 Multiple Star Systems

A few decades ago, astronomy books reported that about one-fourth of the objects that appear to be single stars really contain two or more stars in a close grouping. Books published a decade ago reported this as about one-third. A few years ago, it was considered to be about half. Now we can safely say that *more than half* of what appear as single stars are, in fact, multiple star systems. These systems must be distinguished from pairs of stars that appear close together as a result of being nearly in the same line of sight from Earth (**Figure 12-22**). These *optical doubles,* as they are called, are merely chance alignments of stars. The stars in multiple star systems are gravitationally bound so that they revolve around one another. When two stars are gravitationally bound, they are said to be a *binary star system.* **Figure 12-23** illustrates how the two stars of a binary pair revolve about their common center of mass, in the same manner as discussed in Chapter 3 for a star and a planet. Although groups of more than two stars are

binary star system A system of two stars that are gravitationally bound so that they orbit one another.

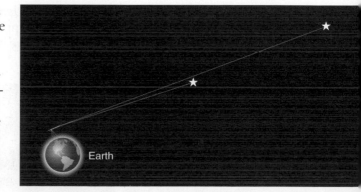

Figure 12-22

Two stars are said to form an optical double if they appear close together but have no actual relationship to one another.

Figure 12-23

Binary stars orbit their common center of mass. When one star is at *A*, the other is at *a*, then *B* and *b*, and so forth. Can you tell which is the more massive star?

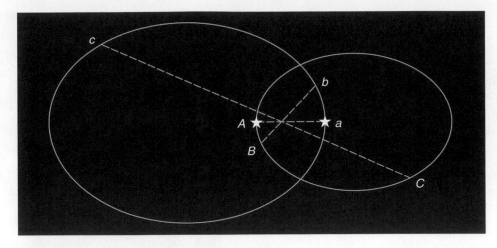

Figure 12-24

A triple-star system. The two stars at the left orbit their center of mass (point *m*). This pair and the star at right orbit the overall center of mass at *M*.

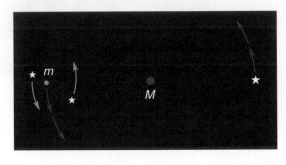

common, our emphasis here will be on binary systems, since multiple star systems are simply combinations of, for example, a binary system and a single star or of two binary systems (**Figure 12-24**).

In 1802, William Herschel obtained the first observational evidence that some double stars orbit one another. (This was also the first direct evidence of gravitational force at work outside our solar system.) Herschel had taken an interest in double stars in hopes of using them for parallax measurements, but his discovery of binary star systems turned out to be much more important to astronomy, for it is by such systems that we are able to determine the mass of stars.

Binary systems are classified into several categories according to how they are detected. We discuss each category in turn and then explore what we can learn from binary stars.

Visual Binaries

visual binary An orbiting pair of stars that can be resolved (normally with a telescope) as two separate stars.

A *visual binary* is a system in which the pair of stars can be resolved as two individual stars in a telescope. Using the largest telescopes, perhaps 10% of the stars in the sky are visual binaries. In a small telescope, only a small fraction of these can be resolved, but some are beautiful sights, particularly when the two stars are very different in color. **Figure 12-25a** is a photograph of Albireo, a visual binary in Cygnus, taken by an amateur astronomer. If you have access to a telescope, find Albireo during a clear summer night.

If two stars appear close together in the sky, how can astronomers tell if they are a binary system? The best way is simply to observe the pair over a period of time and look for signs of revolution. **Figure 12-26** shows a binary (at the right in the photograph) that reveals obvious orbital motion over a number of years. However, things are not usually this easy, because in order for us to be able to resolve a binary pair, the stars must be either very close to Earth or very far apart. The Albireo pair, for example, is separated by some 4100 AU. As indicated by Kepler's laws, a great distance of separation indicates a long

Figure 12-25

(a) Albireo is a binary pair that shows an obvious color difference between the two stars. (b) Albireo is the bottom star of the cross of Cygnus, about 386 light-years from Earth.

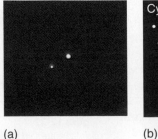

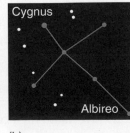

(a)　　　　　(b)

period of revolution, and the Albireo binary is thought to have a period of many thousands of years. Since no detectable orbital motion has occurred in the relatively short time over which we have photographic records, it is difficult to confirm that such a system is indeed a binary pair. We do know, however, that the two stars are about the same distance from us, and therefore we think that they are gravitationally bound.

A second method of determining whether a pair is indeed a binary system employs that powerful astronomical tool, the Doppler effect, which leads to the next type of binary system.

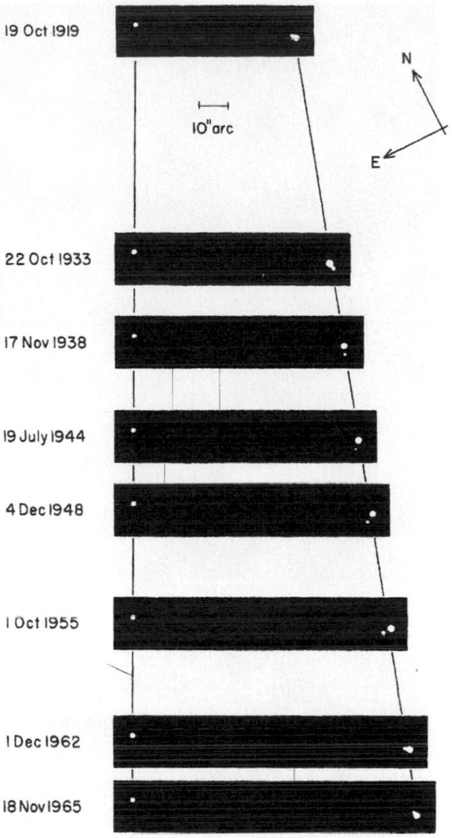

Spectroscopic Binaries

Figure 12-27 shows two spectra of the star κ (the Greek letter "kappa") Arietis, taken at different times. The upper spectrum contains more lines than the lower. Look closely and you will see that this is the result of each line in the lower spectrum having broken into two lines in the upper spectrum. Continuing observations of this star reveal that the spectrum repeatedly goes through a cycle in which each line gradually breaks into two, which spread until they reach a maximum separation and then come together again. The explanation for this is that we are seeing not one, but two stars. They are revolving around one another so that the Doppler effect causes us to see separate spectral lines when one star is moving toward us and the other away. Such a binary system is called a *spectroscopic binary.*

The Doppler effect shows only the radial components of the stars' motion; that is, the motion toward or away from us. If a binary pair is oriented so that its plane of revolution is perpendicular to the line of sight from Earth, no Doppler shift is observed in its spectral lines. On the other hand, if the plane of revolution is tilted directly toward the Earth, the Doppler effect allows measurement of the stars' actual velocities during the part of their orbits when one is moving directly toward us and the other directly away (**Figure 12-28**). At any other position or any other

Figure 12-26
The visual binary Kruger 60 can be seen during its 45-year period. In addition, its proper motion away from the star at the left can be seen.

spectroscopic binary An orbiting pair of stars that can be distinguished as two due to the changing Doppler shifts in their spectra.

19 Oct 1919

10" arc

N

E

22 Oct 1933

17 Nov 1938

19 July 1944

4 Dec 1948

1 Oct 1955

1 Dec 1962

18 Nov 1965

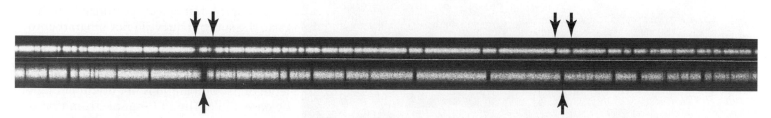

Figure 12-27
Two spectra of the spectroscopic binary κ Arietis are shown here. Notice that lines that are single in the bottom spectrum are split in the upper one. This is particularly evident for the lines to the right of center. When single lines appear, the two stars are moving at right angles to the line of sight. When the lines are double, one star is moving toward us and the other is moving away.

Figure 12-28
When the binary stars whose spectra are shown in Figure 12-27 are moving perpendicular to the line of sight, no Doppler shift is observed. The Doppler effect is seen only when the stars have a component of motion toward or away from us.

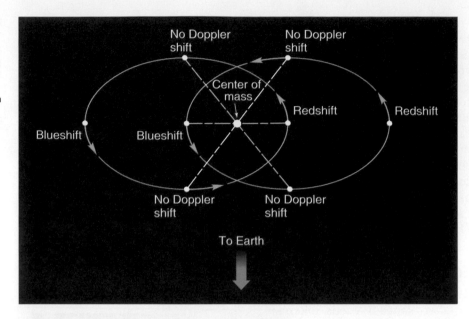

Figure 12-29
Even a small telescope reveals that Alcor and Mizar are three stars.

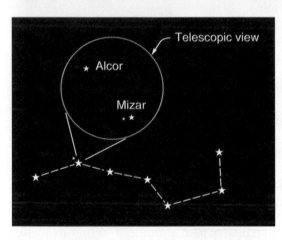

Mizar B is called a *single-line spectroscopic binary system* rather than a double-line system such as Alcor.

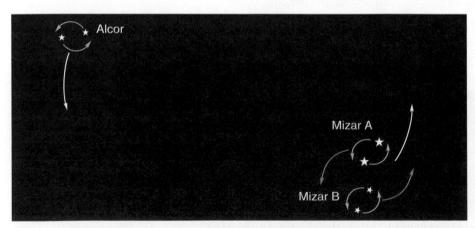

Figure 12-30
Spectroscopic evidence reveals that Alcor and Mizar actually form a six-star system.

orientation of the plane of revolution, we detect only the radial component of the stars' motions.

In the handle of the Big Dipper is a particularly interesting example of binary stars (**Figure 12-29**). As pointed out in Section 5-3, good eyesight reveals two stars at that location, the brighter one named Mizar and the dimmer one Alcor. Although it had been thought that Alcor and Mizar form an optical double rather than a gravitationally bound binary, more recent observations of their radial velocities indicate that they are orbiting one another. In any case, if you view Mizar and Alcor through a telescope, you will see that Mizar itself is two stars (Figure 5-10). Following the standard practice for naming double stars, we call the brighter Mizar A and the other Mizar B. Mizar A and B are a widely separated visual binary and have a period of at least 3000 years.

In 1889, a Harvard College astronomer, Edward Pickering, found that Mizar A is a spectroscopic binary with a period of only 104 days. When it was discovered that Mizar A was binary, Mizar B was scrutinized. In this case, the spectral lines did not separate into two parts, but instead moved back and forth; first they redshifted and then blueshifted. Mizar B is indeed a binary star, but the spectrum of its companion is not bright enough to be observable. The shifting spectrum that is observed is the spectrum of the brighter star of the pair. We deduce the existence of the companion from that motion.

Finally, it has been found that Alcor is a spectroscopic binary. Thus the dot in the handle of the Big Dipper is in fact six stars (**Figure 12-30**)!

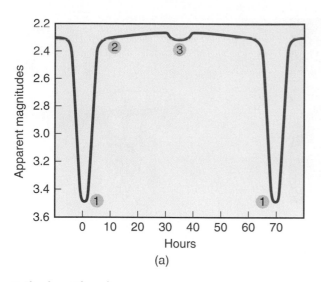

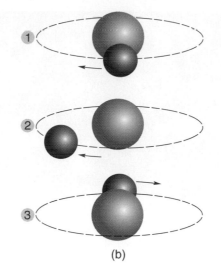

(a)

(b)

Figure 12-31
The light curve of Algol (a) is explained by the eclipsing of its components, one of which is much darker than the other. Positions 1, 2, and 3 in the light curve correspond to the numbers in part (b).

Eclipsing Binaries

A special case of binary stars occurs when their plane of revolution is along a line from Earth, so that one star moves in front of the other as they orbit. The star Algol, in the constellation Perseus, is a good example of an eclipsing binary. Its brightness changes periodically, and in 1783, John Goodricke, an amateur astronomer, explained the brightness change as being due to a dimmer companion passing in front of the brighter. **Figure 12-31a** shows a graph of the light received from Algol as time passes, called the system's *light curve*. Notice that every 69 hours the apparent magnitude of Algol changes from 2.3 to 3.5. This happens when Algol A is partially eclipsed, as shown in Figure 12-31b. Midway between these dips in the light curve we see smaller dips, caused by the dimmer companion being eclipsed.

Since Goodricke's discovery that Algol is a binary star, its nature has been confirmed from spectroscopic evidence. Like Mizar B, its companion is too dim to show its own spectrum, but the spectral lines of Algol A move back and forth in rhythm with its cycle of brightness changes. Like most eclipsing binaries, it can now be classified as both an eclipsing binary and a spectroscopic binary.

light curve A graph of the numerical measure of the light received from a star versus time.

Other Binary Classifications

There are at least two other ways to detect binary stars. Sometimes a star is seen to shift back and forth in its position among the other stars, indicating that it is revolving around an unseen companion. Several such systems, known as *astrometric binaries,* have been found.

If one star of a binary system is much hotter than the other, their spectra differ enough from each other that it is possible to ascertain that the spectrum we see is not from a single star, but is a composite of two spectra. Such a system is called a *composite spectrum binary.* This provides another method of detecting binary stars, but nothing can be learned of the motions of the stars in such a system.

astrometric binary An orbiting pair of stars in which the motion of one of the stars reveals the presence of the other.

composite spectrum binary A binary star system with stars having spectra different enough to distinguish them from one another.

12-6 Stellar Masses and Sizes from Binary Star Data

Binary stars are interesting in themselves. For example, astronomers speculate on the stability of a planetary system around a star that is part of a binary system and on how conditions would be different on such planets because of the extra sun. However, the major importance of binary stars is that they allow us to measure stellar masses. Recall from Section 7-2 that we can calculate the mass of a planet from the orbit of one of its moons and that likewise we can calculate the mass of the Sun from the orbits of the ob-

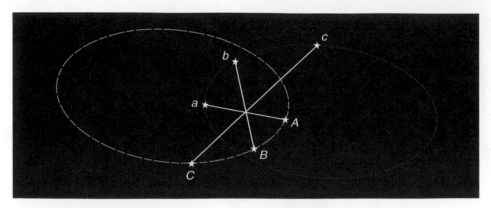

Figure 12-32
The observed orbits of a binary pair, along with the locations of the stars at three times. The center of mass must be located where the lines that connect the stars cross. Notice, however, that the center of mass is not at the focus of either ellipse. (Compare this to Figure 12-23.)

Figure 12-33
A circular shape, such as the basketball rim, appears elliptical when viewed at an angle.

jects circling it. The calculation involves Kepler's third law as revised by Newton. Stellar masses of binary systems are calculated in the same way.

In order to do such a calculation, we must know two things about the orbit of one (or both) of the stars: the size of the ellipse (its semimajor axis) and the period of revolution. The period is easy to ascertain in all cases except when it is extremely long. Determining the size of the ellipse is a little more complicated.

Figure 12-34
This figure shows the apparent path (above) and the true path (below) of a binary star. Although both shapes are ellipses, if you turn your book, you can see that point *M* is at a focus of the lower ellipse.

barycenter The center of mass of two astronomical objects revolving around one another.

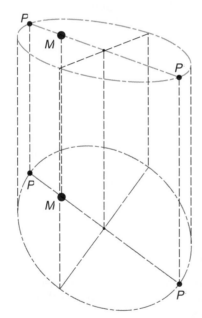

Figure 12-32 shows the orbit of a visual binary as it might be observed. The center of mass, or **barycenter,** of the two stars can be determined from the fact that it must always lie along a line connecting them (as was illustrated in Figure 12-23). Each of the orbits is an ellipse, but notice that the center of mass of the two stars does not lie symmetrically in either ellipse—it is not at any of the foci of the ellipses. The explanation for this is that we are not seeing the ellipses straight-on. A circle, when viewed at an angle, yields an elliptical shape (**Figure 12-33**), and an ellipse viewed at an angle also yields an ellipse, but one of a different shape than the original. The fact that the center of mass is not at one of the ellipses' foci can be used to determine how much the ellipses are tilted with respect to our line of sight (**Figure 12-34**).

Once the true shape of the ellipse is determined, its size can be calculated using the small-angle formula. (We have to know the distance to the pair in order to do this.) Knowledge of the size of one of the stars' ellipses, along with knowledge of the period of its motion, allows us to calculate the *total mass* of the two stars, using Kepler's third law. To determine how the mass is distributed between the two, we need only consider the ratio of the two stars' distances to the center of mass. This is analogous to the way that we can calculate the weight of each of the people on a seesaw if we know the total weight of the two people and know how far each is sitting from their center of mass if they are balanced (**Figure 12-35**). In Chapter 3 we worked out an example of a binary system in which we found the mass of each star by using this method.

Figure 12-35
Starting from the simple relationship $W_1 d_1 = W_2 d_2$, it can be shown that, if W is the total weight of the two people,

$$W_1 = \frac{W \times d_2}{d_1 + d_2}$$

This same method allows us to determine the mass of each star in a binary system if we know the total mass and the distance of each star from the center of mass.

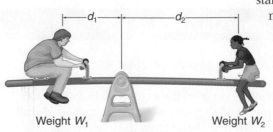

Weight W_1 Weight W_2

TOOLS OF ASTRONOMY

The Mathematics of the Mass-Luminosity Relationship

The graph of Figure 12-37 is empirical; that is, it is plotted from values that result directly from measurements, rather than from a theory of how mass and luminosity should be related. Since the points that represent main sequence stars lie along an approximately straight line, we can determine the mathematical equation for that line. The equation is found to be

$$L = M^{3.5}$$

where both L, the luminosity of a star on the main sequence, and M, the star's mass, are given in solar units. (Raising a number to a power of 0.5 means taking its square root. Raising it to a power of 3.5 means cubing it and then multiplying it by its square root.)

We can use this equation to calculate the luminosity of a star that has three times the mass of the Sun:

$$L = M^{3.5} = 3^{3.5} = 3 \times 3 \times 3 \times \sqrt{3} = 46.8$$

Thus this star's luminosity is about 47 times that of the Sun.

In the case of a spectroscopic binary, if we are confident that the plane of the stars' revolution lies very close to a line from Earth, we can do a similar calculation. In this case, however, instead of calculating the size of the orbit with the small-angle formula, we calculate it from a knowledge of the maximum speed of the star in orbit and the period of the orbit. Since it is difficult to know the inclination of the orbit, we cannot calculate the masses. However, we can obtain valuable information about *average* masses of stars in a great number of spectroscopic systems by assuming an average inclination of the orbits. (Since any orbital inclination from 0 degrees to 90 degrees is equally likely, if enough systems are included in the analysis, the average inclination will be 45 degrees.)

Eclipsing binaries that are also spectroscopic binaries provide us with a way of measuring not only the masses of the two stars, but also their sizes. Since the fact that they eclipse one another means that the inclination of their orbit with our line of view is zero (or very nearly so), their Doppler shift tells us their velocities. Knowing their velocities and the time it takes to complete an eclipse, we can calculate the size and luminosity of each star. **Figure 12-36** shows a simplified case that illustrates this calculation. Recall that we can also determine the size of a star if we know its luminosity, distance, and apparent magnitude. Eclipsing binary systems give us an independent method of measuring star sizes.

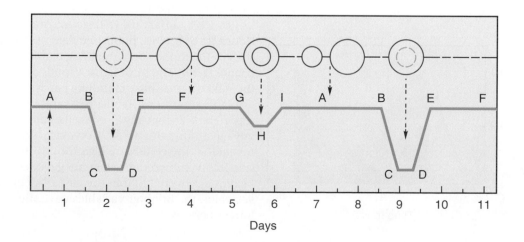

Figure 12-36

Suppose Doppler effect data tell us that the relative velocity of these two stars as they pass one another is 8.0×10^2 km/s (which is 6.9×10^7 km/day). Assume that in the leftmost drawing of the stars, the small one is moving to the right. At point B on the light curve, it began to be hidden by the large star. At point D it starts to emerge. From B to D is about 0.8 days, so it took the small star this long to cross the large one. The diameter of the larger star must then be (0.8 days) $\times$ (6.9 $\times$ 10^7 km/day), or 5.8×10^7 km. In a similar manner, using the time period between points B and C, we can calculate the diameter of the smaller star.

Figure 12-37
When the luminosities of stars are plotted against their masses, an obvious relationship is seen for most main sequence stars. The three at the bottom that do not fit here are white dwarfs. (Notice that the scale at the bottom is a logarithmic scale.)

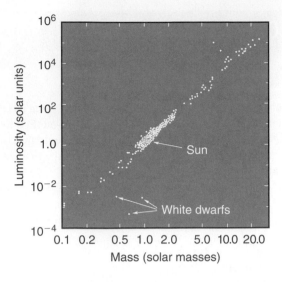

mass-luminosity diagram A plot of the mass versus the luminosity of a number of stars.

The mass-luminosity relationship holds only for main sequence stars. The mathematics of the relationship is in the Tools of Astronomy box on page 415.

12-7 The Mass-Luminosity Relationship

Suppose we plot a graph of the masses of several stars compared to their luminosities. **Figure 12-37** shows such a graph, plotting mass in solar masses along the horizontal axis (*x*-axis) and luminosity in solar units along the vertical axis (*y*-axis). There is an obvious correspondence for main sequence stars: More massive stars are more luminous.

This ***mass-luminosity diagram*** was produced from knowledge of binary stars that are close enough to Earth to yield the necessary data. However, it is reasonable to assume that more distant stars also follow this pattern. (Otherwise, we would be claiming that we live in some special place in the universe where stars behave differently.) The mass-luminosity relationship is valuable to astronomers in investigating less accessible stars and in constructing and evaluating hypotheses concerning the life cycle of stars. This will be discussed in the next chapter.

12-8 Cepheid Variables as Distance Indicators

δ is the Greek letter "delta."

Goodricke made his discovery when he was 19 years old, just two years before his untimely death.

Cepheid (pronounced SEF-e-id) variable One of a particular class of pulsating stars.

Eclipsing binary stars are sometimes called eclipsing *variables,* since their light intensity varies over time. Other types of variable stars are also found in the heavens. One particular type is of importance to us here because it provides a method of measuring distances.

In 1784, John Goodricke (the same astronomer who explained Algol's variations) discovered that the star δ Cephei varies in luminosity in a regular way, but that its variations cannot be explained by an eclipse of a binary companion. This star changes its apparent magnitude between 4.4 and 3.5 every 5.4 days, with a light curve as shown in **Figure 12-38**. Soon thereafter, other stars exhibiting this characteristic light curve—brightening quickly and then dimming more slowly—were seen. Doppler effect data show that such stars are actually pulsating—changing size—in rhythm with their changes in luminosity. It is fairly easy to identify this class of stars, now called ***Cepheid variables,*** or ***Cepheids.*** Each Cepheid has a very constant period of variation; the periods range from about one day to about three months for different Cepheids.

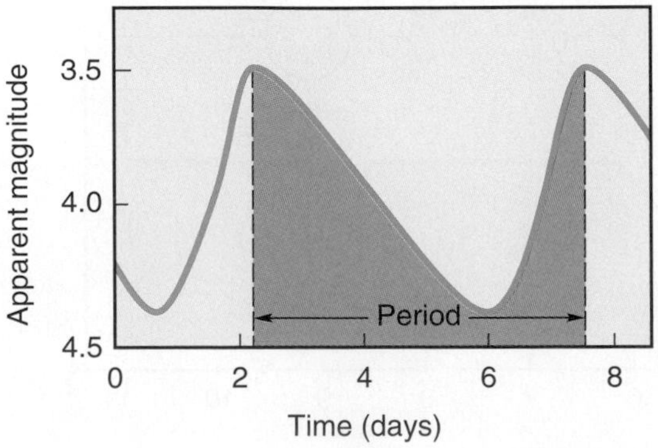

Figure 12-38
This is the light curve of δ Cephei, the prototype of a class of stars that have light curves with this shape and are therefore called Cepheid variables.

In 1908, a Harvard College astronomer, Henrietta S. Leavitt, published data on variable stars in the Small and Large Magellanic Clouds (**Figure 12-39**). The Magellanic Clouds appear to the naked eye as hazy cloudlike patches in the sky of the Southern Hemisphere and were first reported to Europeans by Magellan after his voyage around the world. However, telescopes reveal that the clouds are gigantic groups of stars, now known to be separate galaxies. Leavitt discovered that in the case of Cepheid variables, the brighter variables have the longer periods.

It may seem strange that period and *apparent* magnitude are related, for apparent magnitude is not a quantity intrinsic to the star; rather, it depends on our distance from it. The reason the relationship exists is that all of the stars of the Magellanic Clouds are about the same distance from us—at least, compared to the distance to the clouds. To see this, suppose that you are on a hill outside your town at night, observing the street lights of the town (each of which, we will assume, has the same absolute luminosity). The lights will *appear* to be many different brightnesses, depending upon each light's distance from you. Now sup-

Figure 12-39
The Large Magellanic Cloud is on the left of this photo and the Small Magellanic Cloud is on the lower right. (The bright star Achernar is at the top right.)

pose that the hill is high enough so that you can use a small telescope to observe the lights of a town 200 miles away. You will find that the lights of this town all appear about equally bright, since each of them is 200 miles away, give or take only a few miles.

The stars of the Magellanic Clouds do not all have the same intrinsic luminosity, but since they are all at about the same distance, their apparent magnitudes are related directly to their absolute magnitudes. For example, if the absolute magnitude of a certain star in the Large Magellanic Cloud is five magnitudes less than its apparent magnitude, the absolute magnitude of each star in the cloud will be five magnitudes less than its apparent magnitude.

Astronomers quickly realized the importance of the relationship between the magnitude and the period of Cepheids: the period (which is easy to determine) allows us to determine the absolute magnitude, and we can use this quantity along with apparent magnitude to determine the distance to the variable star (using the relationship between absolute magnitude, apparent magnitude, and distance that was explained in the Tools of Astronomy box on page 398). The problem was that only the *apparent* magnitudes of the Cepheids in the Magellanic Clouds were known. The absolute magnitudes of these stars could not be determined because the clouds are too far away for parallax to be observed. The apparent magnitudes of Cepheids in the clouds are around 15; they appear dim. However, it was obvious from Cepheid variables nearer the Earth that Cepheids, as a class, are very luminous stars. This provided qualitative evidence that the Magellanic Clouds are at a great distance from Earth.

Polaris is a Cepheid variable with a small magnitude variation of between 1.9 and 2.1 and a period of 3.97 days.

All that was needed was to find one Cepheid variable near enough to the Earth that its distance could be measured by parallax. But there is none. Beginning in 1917, Harlow Shapley (to be discussed more fully when we study galaxies) worked out a complex statistical method to determine distance to Cepheids within our Galaxy. Using such methods, by the early 1950s astronomers were confident that they had determined correct distances to several Cepheids. This allowed them to plot a graph of period against *absolute* magnitude, as shown in **Figure 12-40**.

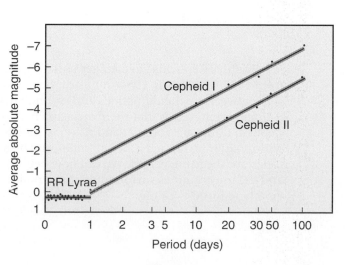

Figure 12-40
The period-luminosity diagram for Cepheid (and RR Lyrae) variables. Recall that luminosity and absolute magnitude are intricately related because they are two different ways of representing the same quantity: the true energy output of a star.

ADVANCING THE MODEL

Henrietta Leavitt

Henrietta Swan Leavitt (**Figure B12-2**), one of seven children, was born on July 4, 1868. She graduated from Radcliffe College (then known as the Society for the Collegiate Instruction of Women) in 1892. In her senior year she took a course in astronomy, and after teaching and traveling for a few years, she joined the Harvard College Observatory in Cambridge, Massachusetts, in 1895 as a student and volunteer research assistant. The quality of her work resulted in a permanent position on the staff in 1902. Soon she was named chief of the photographic photometry department.

Leavitt's primary work was with variable stars, and she discovered many of them. In 1908 she published a list of 1777 variable stars in the Magellanic Clouds; the list included a table of 16 Cepheid variables on which she had very precise data. Concerning these she inserted a comment: "It is worthy of notice that in Table VI the brighter variables have the longer periods."[1] This

brief comment went unnoticed by other astronomers. Later that year, because of ill health, she was forced to return to her family home in Wisconsin.

During her years of illness in Wisconsin, Leavitt continued her work (the observatory sent stellar photographs to her). When she returned to Cambridge in 1912, she published a report on 25 Cepheid variables. In it she stated, "A remarkable relation between the brightness of these variables and the length of their periods will be noticed."[2] She included a graph showing the relationship between period and brightness, but although she recognized the importance of the discovery, her duties at the observatory (and the nature of the work being done there) prevented her from following up on it.

Henrietta Leavitt was one of a group of some 40 women hired by Edward Pickering at the Harvard Observatory, starting in the 1880s, when it was unusual for women to work outside the home. Although women had made contributions to astronomy prior to this time, many still considered science an inappropriate field for them. Pickering was not really a progressive thinker in hiring the women; he did so because they would work for less money than men. Nonetheless, he was rewarded by the significant advances made by many members of the group, particularly Annie Jump Cannon, Antonia Maury, and Henrietta Leavitt.

Although her health was never good and her hearing was impaired from childhood, Leavitt worked at the Harvard Observatory until her death from cancer in 1921.

Figure B12-2
Henrietta Leavitt (1868–1921).

1. Henrietta Leavitt, "1777 Variables in the Magellanic Clouds," *Annals of the Harvard College Observatory* 60 (1908): 107.

2. Henrietta Leavitt, *Periods of 25 Variables in the Small Magellanic Cloud,* Harvard College Observatory Circular no. 173 (March 3, 1912).

Brighter Cepheids have longer pulsation periods. One of the reasons that it took from 1917 to the early 1950s for astronomers to determine the correct relationship between Cepheids' periods and absolute magnitudes (the *period-luminosity relationship*) accurately is that there are actually two different types of Cepheid variables. Cepheid I stars are younger and richer in metals than Cepheid II stars. The two groups differ from each other in luminosity by about a factor of four, or 1.5 magnitudes, for the same period. We can now detect the difference between the two types, and the uncertainties surrounding the period-luminosity relationship have been greatly reduced with time.

To show how the period-luminosity chart can be used to determine distances, suppose that a Cepheid I variable has a period of 10 days and an apparent magnitude of 14. Figure 12-40 shows that a Cepheid I with a period of 10 days has an absolute magnitude of about −4. Now that both apparent and absolute magnitudes are known, we can calculate its distance. (Since the star's apparent magnitude is greater than its absolute magnitude by 18, you can see that the star must be extremely far away. In fact, using the distance modulus relationship on page 398, we find a distance of about 40,000 parsecs.) Use of Cepheid variables tells us that the Large and Small Magellanic Clouds are, in fact, 48,000 and 56,000 parsecs away, respectively.

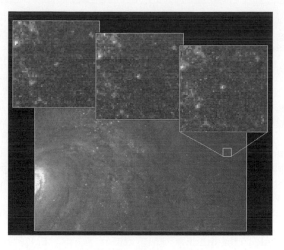

Figure 12-41
A Cepheid variable in the galaxy M100, seen by the *Hubble Space Telescope*. The measured period of 51.3 days indicates a distance from Earth of 56 million light-years.

Corresponding to normal metric usage, 1000 parsecs is one kiloparsec, abbreviated kpc, so the Large Magellanic Cloud is 48 kpc away.

Figure 12-40 includes another type of variable stars, the RR Lyrae stars. They are also used to measure distances, but since they are less luminous than the Cepheid stars, they can only be seen to smaller distances. They have periods ranging from a few hours to one day, and their luminosity is approximately constant. (In Chapter 14, we will examine the reason behind the variability of both Cepheid and RR Lyrae stars; see Figure 14-15.)

Although astronomers have detected only a few thousand Cepheid variables, their importance greatly outweighs their number. Because they are among the very brightest of stars, we can see them not only in distant parts of our own Galaxy, but in other galaxies, giving us a method of measuring distances to these faraway worlds (**Figure 12-41**).

Conclusion

As we have shown in this chapter, once we know the distance to a star, many other properties become available to us. Knowing distance and proper motion, we can calculate a star's tangential velocity. The Doppler effect allows us to measure radial velocity, and from these two velocities, we can determine a star's actual motion relative to the Earth and to other stars. From distance and apparent magnitude, we can calculate a star's absolute magnitude. This knowledge can be combined with knowledge of the star's temperature to calculate its size.

On the other hand, the patterns in the H-R diagram allow us to determine a star's absolute magnitude without first knowing its distance, so that we can use the distance versus absolute magnitude versus apparent magnitude relationship in reverse to determine stellar distances.

The discovery and analysis of binary stars have provided a second method of learning a star's size and also a method of measuring a very fundamental property of a star, its mass.

Finally, a particularly useful type of star, the Cepheid variable, provides a tool to measure even greater distances, opening up the entire field of galactic astronomy.

Study Guide

1. Suppose you observe a previously uninvestigated star and find its apparent magnitude. To determine its absolute magnitude, you need to know the star's
   A. distance.
   B. color.
   C. velocity.
   D. brightness as seen from Earth.
   E. Doppler shift.

2. The absolute magnitude of a star is
   A. the same as the apparent magnitude.
   B. equal to the greatest the apparent magnitude can be, in the case of a variable star.
   C. equal to the apparent magnitude if the star is 10 parsecs away.
   D. the size of a star from one side to the other.
   E. equal to the brightness of the star on the clearest night.

3. A magnitude change of 1 corresponds to a brightness change of about 2.5 times. A magnitude change of 2 corresponds to a brightness change of about
   A. 1.3 times.
   B. 2.5 times.
   C. 5 times.
   D. 6.3 times.
   E. 15.9 times.

For the next three questions, refer to the following table:

Star	Apparent Magnitude	Absolute Magnitude	Distance (parsecs)
Sirius	−1.5	—	2.7
Vega	0.0	0.5	—
Antares	0.9	−5.1	160.0
Fomalhaut	1.15	—	6.9

4. Which star appears dimmest to an observer on Earth?
   A. Sirius
   B. Vega
   C. Antares
   D. Fomalhaut
   E. [The answer cannot be determined from the information given.]

5. Which of the following is about the correct distance for Vega?
   A. 4.2 parsecs
   B. 8.1 parsecs
   C. 11.7 parsecs
   D. 26.7 parsecs
   E. [The answer cannot be determined from the information given.]

6. Fomalhaut's absolute magnitude is about
   A. 0.75.
   B. 2.
   C. 12.
   D. −2.
   E. −0.75.

7. Suppose that Star X is twice as far away as Star Y. The parallax angle of Star X is
   A. half that of Star Y.
   B. the same as that of Star Y.
   C. twice that of Star Y.
   D. four times that of Star Y.
   E. [The answer cannot be determined from the information given.]

8. The distances to nearby stars can be measured by
   A. comparing the apparent magnitudes of several stars.
   B. bouncing radar pulses from their surfaces.
   C. measuring the time it takes light to get here from them.
   D. measuring their shifting motion against background stars through the year.
   E. [Both B and C above.]

9. Star S and Star K have the same tangential velocity, but Star S is closer to Earth. Which has the larger proper motion?
   A. Star S
   B. Star K
   C. They both have the same proper motion.
   D. [The answer cannot be determined from the information given.]

10. Star S and Star K have the same tangential velocity, but Star S is closer to Earth. Which has the larger radial velocity?
   A. Star S
   B. Star K
   C. They both have the same radial velocity.
   D. [The answer cannot be determined from the information given.]

11. If a star is 100 light-years away, what is its approximate distance in parsecs?
   A. 3000 parsecs
   B. 900 parsecs
   C. 30 parsecs
   D. 9 parsecs
   E. 1/3 parsec

12. To use spectroscopic parallax, we must know
   A. the star's diameter.
   B. the star's temperature.
   C. the distance from the Earth to the Sun.
   D. the star's luminosity class.
   E. [Both B and D above.]

13. If the temperature of a star increases without a change in the star's size, its point on the H-R diagram moves
   A. up and to the left.
   B. up and to the right.
   C. down and to the left.
   D. down and to the right.

14. Stars on the main sequence that have a small mass are
   A. bright and hot.
   B. dim and hot.
   C. dim and cool.
   D. bright and cool.
   E. [Any of the above; there is no regular relationship.]

15. To observe spectroscopic binaries, we rely on
    A. knowing the composition of the individual stars.
    B. our knowledge of the distance separating the stars.
    C. our knowledge of the distances from Earth to the stars.
    D. the change in light intensity as the stars orbit.
    E. the Doppler effect.

16. The stars of binary star systems
    A. revolve around a point midway between their centers.
    B. revolve around a point somewhere between their centers (but not necessarily midway).
    C. revolve such that one star remains still, and the other revolves around it.
    D. do not revolve around each other.

17. Binary star systems are especially important to us because they allow us to calculate the _____ of stars.
    A. compositions
    B. proper motions
    C. radial velocities
    D. temperatures
    E. masses

18. Which type of binary system provides the most information about its component stars' masses and sizes?
    A. Eclipsing binaries.
    B. Spectroscopic binaries.
    C. Visual binaries.
    D. Composite spectrum binaries.
    E. [All of the above about equally.]

19. We can determine the size of stars by
    A. measuring the size of their image in a telescope.
    B. a measurement using a spectroscope.
    C. calculations based on temperature and absolute magnitude.
    D. Doppler effect measurements.
    E. mass measurements.

20. Cepheid variables can be used to find distances because their
    A. luminosity is related to their period.
    B. radial velocity is related to their mass.
    C. distance is related to their mass.
    D. magnitude is related to their color.
    E. period is related to their radial velocity.

21. If the orbital plane of a certain binary star system is not parallel to a line from Earth, the maximum radial velocity of one of its stars measured by the Doppler effect will be
    A. less than its true velocity.
    B. greater than its true velocity.
    C. either greater or less than its true velocity, depending upon other factors.

22. If a star has a parallax angle of 0.20 arcseconds, how far away is it?
    A. 0.20 parsec
    B. 1.0 parsec
    C. 2.0 parsecs
    D. 5.0 parsecs
    E. [The answer cannot be determined from the information given.]

23. When a star's spectrum is redshifted as a result of the Doppler effect, we know that the star is
    A. much cooler than average.
    B. slightly cooler than average.
    C. about average temperature.
    D. hotter than average.
    E. moving away from Earth.

24. Spectroscopic parallax allows us to measure
    A. the distances to stars using the Earth's orbital motion.
    B. the distances to stars using the Earth's rotational motion.
    C. the distances to stars using the H-R diagram.
    D. the temperatures of stars using their spectra.
    E. the radial speed of stars using their spectra.

25. Considering only stars on the main sequence, the most massive stars are the
    A. hottest and brightest.
    B. hottest and dimmest.
    C. coolest and brightest.
    D. coolest and dimmest.
    E. [No general statement can be made.]

26. In order to calculate (or judge) the velocity of an object moving in the sky, what quantities must we know?

27. Define and distinguish between radial velocity and tangential velocity.

28. Define and distinguish between apparent magnitude and absolute magnitude. To which is luminosity most closely related?

29. Describe the most direct method of measuring the distances to stars.

30. How is the radial velocity of a star measured?

31. What two quantities are plotted on an H-R diagram?

32. Sketch an H-R diagram, labeling the axes and showing the general location of the following: main sequence, red giants, supergiants, and white dwarfs.

33. Explain the method of spectroscopic parallax.

34. It would seem that a hot star should be a very luminous star. How can white dwarfs be dim stars?

35. Distinguish between optical doubles and binary stars. Why is one of these much more important to astronomers than the other?

36. Name and describe at least three methods of detecting binary stars.

37. Sometimes we can tell by the spectrum of a star that it is part of a binary system. Explain how this is done.

38. Describe the problem presented by the fact that the plane of revolution of a binary is unlikely to be parallel to a line from Earth.

39. Name two quantities that must be known in order to determine the absolute magnitude of a star (other than a Cepheid variable).

40. Explain how Cepheid variables are used to determine distances.

1. The apparent magnitude of the star Proxima Centauri (the closest nighttime star) is about 11, and the apparent magnitude of 61 Cygni B is about 6. Which star appears brighter? How many times more energy do we get from it than from the dimmer star?

2. Both "parsec" and "light-year" might sound—to non-astronomy students—to be units of time. Define each. Why do astronomers find "parsec" to be a useful unit?

3. The apparent magnitude of a star is fairly easy to determine. What else must be known in order to calculate absolute magnitude? Why is it important to know the absolute magnitude of a star?

4. Why do only nearby stars show measurable proper motion? In general, how does the tangential velocity of nearby stars compare to that of distant stars?

5. In Figure 12-26 (Kruger 60), the binary pair is getting farther from the single star. From this series of photographs it would be impossible to tell which has the proper motion. Still, the binary pair would be taken as the most likely candidate. Why?

6. Draw an H-R diagram, labeling the axes and including the names of at least three classes of stars.

7. It is possible for a binary star system to fall into two or more classifications (such as visual binary and eclipsing binary). Describe some situations in which this might be the case.

8. Explain why there is such a simple relationship between the apparent and absolute magnitudes of stars in the Magellanic Clouds.

9. Do you think that it is reasonable to use the H-R diagram, which is plotted from data on nearby stars, for stars much more distant from us? Why or why not?

10. In a group of stars that are gravitationally bound, we observe a red star and a blue star. Both seem to have the same radius. Which one will look brighter? Explain your answer.

11. A star's luminosity class plays an important role in determining its luminosity. What are we looking for in a star's spectrum in order to determine its luminosity class?

12. Why is spectroscopic parallax called "parallax" even though no angle measurement is involved?

13. Would a mass-luminosity diagram (see Figure 12-37) show the same relationship if the scales were independent of the Sun instead of being multiples of the Sun's values?

14. We observe a binary system composed of two stars of the same radius and temperature. We observe their elliptical orbits edge-on (so that we get full eclipses during each cycle). Sketch the light curve for this eclipsing binary. How would the light curve change if the two stars were of the same radius but one of them were blue and the other yellow?

15. Why was it important to find the distance to a Cepheid variable after Leavitt had plotted the data for Cepheids in the Magellanic Clouds?

1. What is the ratio of light received from stars that differ in magnitude by 15?

2. About how many times as much light reaches us from Antares, the brightest star in Scorpius (apparent magnitude +1.0), than from τ (the Greek letter "tau") Ceti (apparent magnitude +3.5)?

3. The star Ross 128 has a parallax angle of 0.30 arcsecond. How far away is it in parsecs? In light-years?

4. α (the Greek letter "alpha") Centauri is the nearest naked-eye star to the Sun. Its apparent magnitude is +0.01 and its distance is 1.35 parsecs. Is its absolute magnitude greater or less than 0?

5. Use data from the last question to calculate the parallax angle of α Centauri.

6. Altair, a star visible in the summer and fall, is on the main sequence, is spectral type A7 (8300 K), and has an apparent

magnitude of +0.77. Is Altair closer than 10 parsecs, about 10 parsecs away, somewhat farther than 10 parsecs, or much farther than 10 parsecs?

7. Rigel, a bright supergiant, is a B8 Ia star with a temperature of about 10,300 K. Its apparent magnitude is +0.14. What can you conclude about its distance from us?

8. A certain star has a mass four times greater than the Sun. How does its luminosity compare to the Sun's?

9. A binary system, observed face-on, has a period of 4 years and parallax 0.2 arcsecond. The length of the semi-major axis is 0.4 arcsecond. What is the total mass of the system? (Hint: You will need to first find the distance to the binary system. Then use the small-angle formula to find the length of the semimajor axis in AU. Finally, use Kepler's third law to find the total mass.)

1. "Origins of the Stellar Magnitude Scale," by J. B. Hearnshaw, in *Sky & Telescope* (November, 1992).

2. *Stars and Their Spectra,* by J. Kaler (Cambridge University Press, 1997).

3. "Reading the Colors of Stars," by C. Sneden, in *Astronomy* (April, 1989).

4. "Accretion Disks in Interacting Binary Stars," by J.K. Cannizzo and R.H. Kaitchuck, in *Scientific American* (January, 1992).

This is an *HST* image of the spiral galaxy NGC4603, about 108 million light-years away. It is the most distant galaxy in which Cepheid variables have been found.

(a)

(b)

(c)

(d)

Figure 13-1

Interstellar clouds of dust hide the stars behind them. (a) The dark C-shaped nebula is an interstellar dust cloud in the constellation Aquila. (b) The dark molecular cloud Barnard 68 is 300 light-years away in the constellation Ophiuchus. It has a total mass of about 1.5 times the mass of the Sun, at a temperature of only 10 K (one of the coldest objects in the universe). The cloud's average diameter is about 12,000 AU. Even though molecular clouds are stellar nurseries, this one seems to be stable (except for a long-period pulsation, probably due to the influence of a shockwave from an exploding star). (c) The Horsehead nebula results from clouds of dust blocking light from a bright nebula behind. Find the dark profile of a horse's head at the right center of the photo. This nebula is in the constellation Orion, about 1500 light-years from us. The bright star at the left center is Alnitak, the easternmost star in the belt of Orion, 251 parsecs (820 light-years) from us. (d) A composite color image (in the visual part of the spectrum) of the Horsehead Nebula and its immediate surroundings. The distance to this region is about 1400 light-years (430 parsecs). Red color is for the hydrogen (H-alpha) emission from the HII region, brown for the foreground obscuring dust, and blue-green for scattered starlight.

(a)

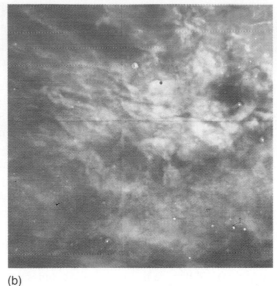

(b)

Figure 13-2
(a) A composite image, taken by *IRAS*, of the constellation Chameleon, at wavelengths of 12 micrometers (coded blue), 25 micrometers (coded green) and 60 micrometers (coded red). Stars are brighter at 12 micrometers and appear as blue dots in the picture. (b) The *Hubble Space Telescope* obtained this infrared image of the edge of the Orion Nebula. The image reveals the diffuse, patchy clouds of the nebula.

extinction of that light. Another effect results from the fact that while dust grains may absorb some light, they scatter much of it, and this scattering is more efficient for light of shorter wavelengths. As a result, blue light is scattered more than red light (**Figure 13-3**). Thus the light from distant stars is *reddened* by the dust through which it passes.

The reddening caused by scattering should not be confused with the redshift caused by the Doppler effect. In the case of reddening by scattering, the positions of the spectral lines are not changed, but in the case of Doppler redshift, the spectral lines actually shift. **Figure 13-4** illustrates the difference. Also note that some objects might be intrinsically red, and thus their color is not related to interstellar reddening.

How large are the interstellar dust grains? We know that they must be smaller than the wavelengths corresponding to visible light, because otherwise they would not scatter blue light more than red. This means that the dust grains are the size of particles of cigarette smoke and smaller. (You may have noticed that a cloud of cigarette smoke has a blue tint.)

What is the dust made of? Spectral analysis indicates that it contains silicate grains and carbon in the form of graphite, similar to the carbon that is found in comets. Beyond this we know little about its chemical makeup. In Chapter 15 we will

The effect by which starlight is blocked completely by interstellar material is called *interstellar extinction*. It is due to scattering and absorption. A similar phenomenon occurs when the headlights of oncoming cars appear dimmer while driving in fog.

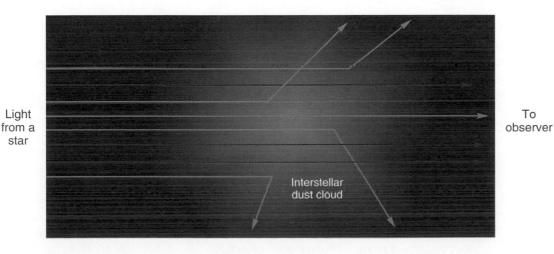

Light from a star

To observer

Interstellar dust cloud

Figure 13-3
Grains of dust scatter blue light more efficiently than red light, so a star seen through a cloud of dust appears redder than it would if the dust were not present. The amount of scattering and absorption depends on the number density of particles, the wavelength of light, and the cloud thickness.

ADVANCING THE MODEL

Holes in the Heavens?

Sir William Herschel, the discoverer of Uranus, proposed that the dark patches we see in the sky are simply large spaces between the stars that allow us to see into the dark void beyond. How do we know that this is not the case?

Today, of course, infrared imaging allows us to detect the dust that makes up the "holes in space." But before we ever made infrared images of the dust, an argument based on geometry gave astronomers reason to doubt Herschel's idea. Recall that stars are not all at the same distance from us; they are not on a "celestial sphere." Instead, they are spread out so that some are relatively nearby and others are very far away. Thus, for us to be able to see through gaps between the stars, the gaps would have to be simi-

lar to tunnels, and the tunnels would have to be perfectly aligned with Earth. Suppose you are deep in a forest. Even with open clearings among the trees, you might not be able to see beyond the forest. For you to see outside, there would have to be straight, open trails. In the case of stars, it is extremely unlikely that so many open trails ("tunnels") would be perfectly aligned with Earth. As astronomer Arthur Cowper Ranyard put it in 1894, "The probabilities against such a radial arrangement with respect to the Earth's place in space [seem] to my mind to conclusively prove that the narrow dark spaces are due to streams of absorbing matter, rather than to holes or thin regions in bright nebulosity."

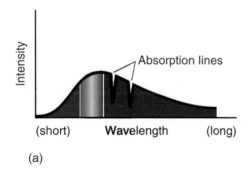

(a)

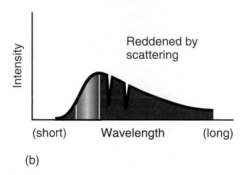

(b)

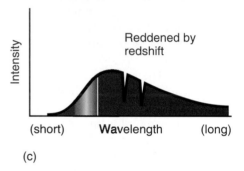

(c)

Figure 13-4
(a) A blackbody (intensity-wavelength) curve showing two (exaggerated) absorption lines. (b) This is the same source with its light reddened by scattering. Notice that the absorption lines remain in the same place. (c) Again, the same source but with its light reddened by the Doppler shift. The entire curve is shifted to the right in this case.

discuss the origin of the interstellar material and the related hypotheses about the composition of the dust.

Interstellar Gas

Dust in space accounts for only about 10% of the total mass of the interstellar material. The remainder is simply gas, which in most cases is extremely diffuse (so that for most purposes the volume it occupies may be considered a vacuum). However, in some cases the gas clusters together into giant clouds. The gas between the stars reveals its presence in several ways.

fluorescence The process of absorbing radiation of one frequency and re-emitting it at a lower frequency.

emission nebula Interstellar gas that fluoresces due to ultraviolet light from a star near or within the nebula.

• Clouds of gas can be seen in photographs of the sky. The red glow behind the Horsehead nebula in Figure 13-1c is a giant cloud of gas that is glowing due to ultraviolet light from hot stars within it. The glow is due to a *fluorescence* process; that is, the atoms of the gas absorb ultraviolet light from nearby hot stars, which causes the atoms to become energized. The atoms lose their extra energy by emitting light—*fluorescing*. Of particular importance is the Hydrogen-alpha line produced in the red part of the spectrum (see Figure B12-1). **Figure 13-5** shows other examples of such bright *emission nebulae.* Such nebulae are direct evi-

(a)

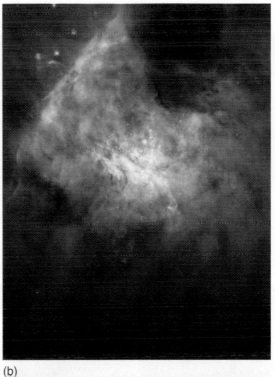

(b)

(c)

Figure 13-5
(a) The Lagoon nebula in Sagittarius is an emission nebula. (b) This is a mosaic of 45 images of the Orion nebula taken by the *HST*. Light emitted by oxygen is shown as blue, hydrogen emission is shown as green, and nitrogen as red. The overall color balance is close to what an observer near the nebula would see. (c) The Tarantula nebula is the largest emission nebula in the sky, a bit less than the size of the full Moon. It is located in the Large Magellanic Cloud, about 170,000 light-years from us, and measures more than 1000 light-years across. At the heart of the nebula is a hot, young star (2–3 million years old) whose UV light excites the emission nebula. The predominantly red light comes from hydrogen atoms (the H_α line), and the green-blue light comes from hydrogen atoms (H_β) and doubly ionized oxygen ions.

dence of gas atoms in the interstellar medium. A typical emission nebula temperature is about 10,000 K and a typical mass is between 100–10,000 solar masses. The term nebula (plural *nebulae*) has its origin in the adjective *nebulous*, which means "lacking definite form or limits." This is an apt description of an interstellar cloud, for such clouds have no definite boundaries. A cloud is called a nebula if it is dense or bright enough to show up in a photograph.

- Interstellar gas causes absorption lines in stellar spectra. Astronomers can distinguish these absorption lines from the absorption lines of a stellar atmosphere in three ways. First, absorption lines due to interstellar gas tend to be narrower than those produced by a star's atmosphere (**Figure 13-6**). This difference occurs because the atoms of a star's atmosphere are generally moving fast in random directions, causing a range of Doppler shifts. Second, since the star and the interstellar gas

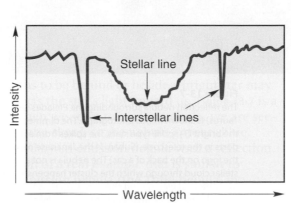

Figure 13-6
This graph is a highly magnified portion of the intensity-wavelength graph of light from a star. It shows three absorption lines. The outer two are much finer than the central one, indicating that they were caused by interstellar gas.

ADVANCING THE MODEL

Blue Skies and Red Sunsets

The same phenomenon that explains the reddening of starlight by interstellar dust also explains why the sky is blue and sunsets are red (or at least orange). As we pointed out in the discussion of interstellar reddening, dust scatters blue light more than it scatters red. A similar effect happens in the Earth's atmosphere, where the scattering of sunlight is what gives the daytime sky its color. The sky appears blue simply because particles in the air (and air molecules) scatter higher-frequency (blue) light more than lower-frequency light. It follows from this that light that reaches us directly from the Sun without scattering is somewhat deficient in these higher frequencies. Consequently, the Sun seen from Earth's surface appears redder than it appears when viewed from a vantage point in space. From Earth, the Sun appears somewhat yellow, but from space it is closer to white.

When we look at the Sun as it is setting, the sunlight that reaches our eyes has traveled even farther through the atmosphere than sunlight at midday. Thus even more of the high-frequency light is scattered away, and the Sun appears red. This effect is especially noticeable if the atmosphere contains significant amounts of dust. The dust that remained in the air from major volcanic eruptions in 1991 caused beautiful red sunsets for the next few years.

Figure 13-9
The reflection nebula NGC 1999 in Orion. It is illuminated by the bright, young star to the left of center. The dark nebula, located right of center and in front of NGC 1999, is a condensation of cold molecular gas and dust so thick that it blocks light.

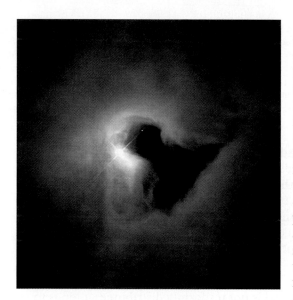

dark nebula A cloud of interstellar dust that blocks light from stars on the other side of it.

The Trifid nebula also contains dust lanes that absorb the light from the emission nebula, extinguishing this light and appearing dark. Dust clouds such as this may have densities ranging from as low as 1000 to perhaps 1,000,000 particles per cubic centimeter. (This is still not very dense compared to number densities on Earth, since air at sea level is still 20 trillion times more dense than the densest interstellar cloud.) A dust cloud is able to extinguish light from a star behind it only if the cloud is extremely large, perhaps hundreds of parsecs across. The photographs in Figure 13-1 and **Figure 13-9** contain other examples of *dark nebulae*. Although it is the dust of a dark nebula that is responsible for blocking starlight, most of the particles in such a nebula are atoms, about 75% hydrogen and 23% helium, with the remaining 2% being heavier elements. Temperatures of dark nebulae range from 10 to 100 K—low enough for hydrogen to form molecules.

If we take an inventory of the dust, gas, and stars within a radius of about a thousand parsecs from Earth, we find that interstellar matter contributes about 15–30% of the total mass. Even though astronomers are able to take an inventory only of our "neighborhood," they are unable to make this figure more precise primarily because determining the mass of the interstellar material is so difficult. Does interstellar matter make up this same percentage of total matter elsewhere in the Galaxy? We would like to know the answer to this question, but we are unable to make the necessary measurements.

As we will discuss later, the interstellar material is constantly in a state of flux, for it is being used up as new stars are formed and being replenished as old stars die.

13-2 A Brief Woodland Visit

Imagine that you are a visitor from another star system who has landed in a forest on Earth for a 2-day visit. What could you learn about the life cycles of the woodlands during your short visit? You would see small and large trees, but would you be able to determine that trees progress from small to large? In 2 days you would have no chance to see any growth. You would see decayed material on the forest floor, but it might not be obvious that this is what remains of once-standing trees. Perhaps you might be lucky enough to see a change take place. You might see a limb fall from a tree. Would you conclude that trees get smaller by dropping their branches? Or perhaps you might see an entire tree fall. What would that tell you?

Like an extraterrestrial forest visitor, we humans are brief tourists in the universe. During our short lifetimes, we see very little change in the heavens. As we measure time, the stars change slowly. We do have an advantage over the forest visitor in that previous generations have handed down their observations to us so that our learning cycle is longer than one lifetime. However, the earliest reports of astronomical observations are only a few thousand years old; this is a minute time period compared to the total life of the heavens.

It is possible, though, to learn about the life cycle of stars. We are fortunate that a few stellar events happen quickly enough for us to observe them over human lifetimes or (in some cases) over much shorter times. Although astronomers cannot experiment directly with their subject matter, they can observe tremendous numbers of stars in various stages of development. This, along with earthbound experiments on the material of which stars are made, allows scientists to develop theories of stellar life cycles in which we can have reasonable confidence.

13-3 Star Birth

Like other scientific theories, today's theories of star birth and star death developed slowly. Successful theories do not spring fully developed into a scientist's mind. Theories in astronomy result from long struggles with data from the heavens, often accompanied by experimentation here on Earth.

The story of today's understanding of the life cycles of stars starts early in this century, with astrophysicist Henry Norris Russell playing the major role. The H-R diagram is the key to understanding the lives of stars, but it is not easy to read. Russell's early theories held that stars begin their lives as red giants, become O- and B-type main sequence stars, and then move down the main sequence, gradually dimming as they live out their lives. When new evidence arrived, particularly from H-R diagrams of clusters and from new knowledge of nuclear energy processes, Russell realized that stars live most of their lives on the main sequence with very little change in their positions and that the red giant stage is not the beginning of a star's life but is near the end. We will start at the beginning.

The Collapse of Interstellar Clouds

Stars are born in the coldest places in the Galaxy, the giant molecular clouds (GMCs) of interstellar space. These clouds have a temperature of about 20 K (−420° F) and consist mainly of molecular hydrogen. The small temperature of a GMC implies that particles in it have low speeds and therefore the corresponding gas pressure is small. As a result, gravity can bring interstellar material together by overwhelming the pressure pushing this material apart. All that is needed is a relatively dense region so that atoms and dust particles are close enough for gravity to take over. The average number density of a GMC is about 200 molecules per cubic centimeter, and a typical cloud may be 50 parsecs across. In all, a GMC may contain as much as a million solar masses of material. The densest regions in a GMC form dark nebulae where new stars are born.

Figure 13-10b is a picture of a very small portion of M-16, the Eagle nebula (shown in **Figure 13-10a**). The dark pillar-like structures are part of a GMC, so they are made up of gas and dust. Intense ultraviolet radiation from hot, massive newborn

Astronomers estimate that there are 5000 giant molecular clouds in the Galaxy.

(a)

(b)

(c)

(d)

Figure 13-10

(a) This is the Eagle nebula, in the constellation Serpens, about 7000 light-years distant and 20 light-years across. (Can you see the nebula as a landing eagle?) (b) This 1995 *HST* image shows material from a very small part of the Eagle nebula being swept away from dense areas where new stars are forming. The pillar to the left is about 1 light-year long and each pillar is a bit wider than our solar system. (c) This infrared image of Eagle's fiery heart, taken by ESA's *Infrared Space Observatory*, is a combination of two exposures. The first at 7.7 microns is shown in blue, and the second at 14.5 microns is shown in red. The temperature of the dust in the Eagle, seen as a "bluish fog," is about 170 K. (d) This 1999 infrared *HST* image of the Eagle nebula shows that only a small fraction of the areas that are opaque in visible, as shown in (b), remain so in the infrared. Instead of the substantial columns of material implied by the optical images, the infrared images show only isolated clumps of dust and gas.

stars that lie at the upper right, beyond the photo, causes the columns. **Figure 13-11** illustrates the process. Prior to the sequence shown in the figure, a molecular cloud was fairly evenly spread throughout this region of space. In part (a) of the figure, ultraviolet radiation from the upper right is blowing the cloud back and evaporating gas outward from the cloud's surface. At the same time, the radiation illuminates the surface of the cloud, causing it to glow.

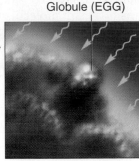

(a) The surface of a molecular cloud is illuminated by intense ultraviolet radiation from nearby hot stars. The radiation evaporates material from the cloud's surface.

(b) A denser-than-average globule of gas (an "EGG") begins to be uncovered. Because it is denser than its surroundings, it is not evaporated as quickly and is left behind. Young stellar objects begin to form within some EGGs.

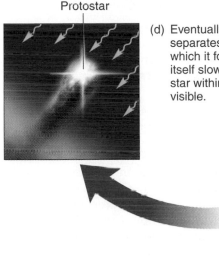

(d) Eventually, the EGG separates from the cloud in which it formed. As the EGG itself slowly evaporates, the star within it becomes visible.

(c) The EGG is now largely uncovered. The EGG protects a column of gas behind it, giving it a finger-like appearance.

Certain parts of the cloud are denser than average. In part (b) a dense globule of gas is about to be uncovered. Globules like this are called "EGGs," for "Evaporating Gaseous Globules." Radiation pushes the surrounding gas away from the EGG in much the same way that a strong wind on a beach blows sand along, uncovering shells and leaving trails of sand behind the shells. Part (c) of Figure 13-11 shows most of the nebula blown away, but a snake of gas and dust remains that has been protected from radiation by the EGG. Later, the nebula will blow still farther away, leaving behind a teardrop shape, shown in part (d). Even later, all of the nebula will disappear, leaving behind only the EGG.

What is an EGG? At least in some cases, it is the beginning of a star. Within the globule, gravitational forces between individual molecules and dust particles are sufficient to draw the particles together. The globule was in the process of this contraction when the ultraviolet light began blowing away the surrounding nebula. The contraction continues within the globule. When the center of the EGG gets hot enough to give off its own radiation, it will itself blow away any remaining parts of the nebula that have not fallen inward and a star will remain. Stars born in this manner probably do not have planets around them, for the radiation that causes the pillars probably disperses the material that would have ended up as planets.

The name EGG also fits the globules because they are embryonic stars.

Some stars form in more isolated conditions, without nearby hot stars blowing their raw material away. In these cases, parts of the nebula that remain near the embryonic star will continue to fall into it, perhaps resulting in a very large star, or perhaps revolving around the star to become a planetary system. Such isolated massive stars and compact ionized hydrogen clouds have been observed in a number of galaxies in the Virgo cluster, at the boundary between a galaxy's halo and the intercluster space, and high up in the halo of our own Milky Way.

One question remains before we consider how an EGG becomes a star. What triggers the collapse of parts of a giant molecular cloud to form globules? The answer may involve collisions between GMCs. Along the boundary of such a collision, molecules would be forced closer together, perhaps close enough that gravitation would take over and continue the compression. Another cause for collapse may be interstellar shock waves. As such a wave strikes the GMC, it forces parts of the cloud together enough for gravity to become dominant.

There are at least four possible sources of interstellar shock waves. We have seen one—radiation from hot, newly forming stars. Second, as we will explain in the next section, very massive stars undergo a period during which they release enormous quantities of material at great speeds. These bursts are a source of shock waves in the surrounding GMC that can trigger the birth of new stars.

A third source of interstellar shock waves is a supernova—the explosion of a star as it nears the end of its life. We will discuss supernovae in Chapters 14 and 15. Finally, as we will discuss in Chapter 16, we know that tremendous shock waves move around the entire Galaxy, forming its galactic arms. These waves may be the most common trigger of star formation.

Protostars

Whatever causes the material within the core of a giant molecular cloud to start to collapse, once the collapse begins, the increased gravitational force causes the molecules of the core to pick up speed as they fall inward. The condition of fast-moving molecules is synonymous with high temperature, so material near the center of the collapse becomes hot. Soon the higher pressure caused by the increase in temperature prevents further collapse of the central core. Material farther from the core continues falling inward, crashing down on the central core and causing a further increase in temperature.

The hot core is now called a ***protostar.*** It receives its energy from the infall of material, so the ultimate source of its energy is gravitational. This is similar to the way a hammer that is repeatedly lifted and dropped onto a piece of metal heats both itself and the piece of metal. Each molecule that falls onto the core converts gravitational energy into thermal energy.

As the cloud contracts and its temperature increases, about half of its gravitational energy is radiated away from the heating center. As the center gets hotter and hotter, it radiates more and more energy. In only a few thousand years, the center is as bright as the Sun. However, we cannot see the center directly because the outer portion, which continues to fall relatively slowly toward the center, blocks most of the radiation. This ***cocoon,*** or ***cocoon nebula,*** absorbs the radiation from the center, becoming warmer as it does so.

A warm object emits infrared radiation, and it is the infrared radiation emitted from the cocoon that we see from Earth and that gives evidence for the existence of protostars. **Figure 13-12** is an infrared photograph and is therefore shown in false color, with each color representing a different intensity of infrared radiation. Such false-color photographs of nonvisible sources are very useful tools in astronomy, but do not be fooled into thinking that the objects have colors like the photographs.

Evolution toward the Main Sequence

Figure 13-13 shows the progress of a protostar of one solar mass on an expanded H-R diagram. (Notice that the main sequence is now compressed into the left side of the figure.) The diagram had to be expanded beyond the coolest stars on the main se-

The material flowing from stars is commonly called the *stellar wind.* The *solar wind* that we detect (and that results in auroras on Earth) is far weaker than the stellar wind bursts described here.

protostar An object in the process of becoming a star, before it reaches the main sequence.

cocoon nebula The dust and gas that surround a protostar and block much of its radiation.

In speaking of evolutionary tracks and stellar evolution, astronomers are using the term *evolution* in a very different way than biologists do. One star is said to "evolve" as it lives its life. We are not speaking of an entire species evolving through generations.

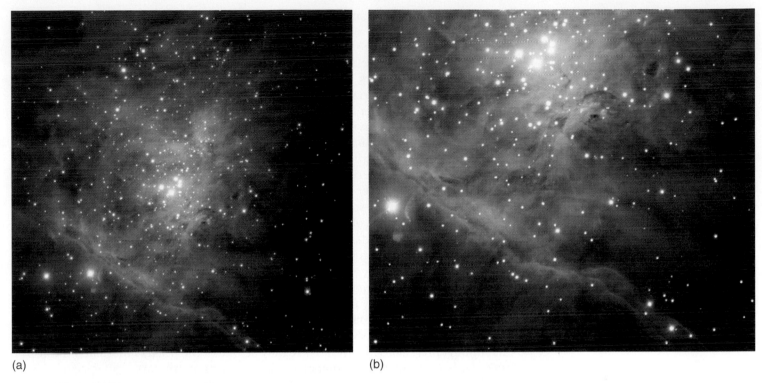

(a)   (b)

Figure 13-12

(a) A VLT color composite of 81 images of the central part of the Orion nebula. The image shows the Trapezium stars (near the center) and the associated cluster of about 1000 stars, about 1 million years old. This near-infrared image (1–2 microns) shows thermal radiation from the area's dust, which is heated by the intense UV and visible light from the stars. It also shows the warm, dense dust cocoons around the very young stars. (b) A detail of the region shown in (a), with the Trapezium stars at top center. The UV light and strong winds from these stars eat their way into the surrounding molecular cloud.

quence because a protostar begins its life much cooler than any main sequence star. The details of this part of the star's *evolutionary track* are not known, but the protostar begins as a cool, dim object, warms up by gravitational contraction, and moves toward the main sequence. Finally, as the cocoon continues to contract, its smaller size causes it to appear dimmer, so the star begins to move downward on the diagram as it nears the main sequence.

Motion on the H-R diagram does not mean the star is actually moving in space, of course. Recall from Chapter 12 the diagram of people's height versus age. As a person ages, his or her dot moves on this diagram. This is analogous to the motion of a star on the H-R diagram.

As time passes, the center of the protostar continues to shrink and become hotter, emitting more radiation all the time. This radiation gradually blows away the outer portions of the cocoon. Some of the cocoon may not be blown away but may condense to form planets, as discussed in Chapter 7. In some stars the cocoon is apparently blown away by sudden bursts of stellar winds. Certain stars, named *T Tauri stars* after the variable star T in the constellation Taurus, appear to be

evolutionary track The path on the H-R diagram taken by a star as its luminosity and color change.

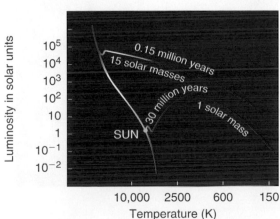

Figure 13-13

This H-R diagram shows temperatures much lower than on previous diagrams in order to include the early, cool stages of a star's formation. A more massive star ends up at a higher location on the main sequence because it is hotter and brighter. Since gravitational forces are greater on the more massive star, it spends much less time in its protostar stage.

T Tauri stars A class of stars that show rapid and erratic changes in brightness.

young stars undergoing some sort of instability that causes enormous flares. These flares are thought to play a part in blowing away the cocoon of newly forming stars, particularly K- and M-type stars, which, being at the bottom of the main sequence, are of low mass.

The most massive stars, the O- and B-type stars, follow a track on the H-R diagram toward the top of the main sequence (Figure 13-13). These stars are much fewer in number than stars of lesser mass. However, they are so energetic that they signal their presence by emitting ultraviolet light, which stimulates the hydrogen of their cocoon (and the nebula beyond the cocoon) to emit its own light, forming an emission nebula. **Figure 13-14a** is a photograph of the Orion nebula, an emission nebula lit by energy from new stars within it. The very hot, blue star in Figure 13-14b is experiencing a continuous mass loss from its outer layers. The two symmetric shells of material (emission nebulae) on either side of the star were formed during past vigorous outbursts.

The infalling particles of massive stars experience much greater gravitational force than do particles in low-mass stars, and massive stars reach the main sequence much faster. M-type stars remain protostars for hundreds of millions, perhaps billions, of years. The Sun, with more mass, spent about 30 million years in this phase. However, the most massive stars remain protostars for only tens of thousands of years. Recall that stars of low mass may undergo a period of instability just before joining the main sequence. Massive stars do likewise, but their instability is more violent; O and B stars blow off material at supersonic speeds during this time, creating a shock wave in the surrounding material. This shock wave may be one of the triggers that starts the collapse of other portions of the interstellar cloud to form more stars.

The great amount of radiation from massive stars is what limits how massive a star can be. Astronomers calculate that a star with a mass greater than about 100 solar masses emits radiation so intense that it prevents more material from falling into the star, thereby limiting the star's size.

On the other hand, protostars with masses less than about 0.08 solar masses do not develop the necessary internal pressure and temperature to start hydrogen fusion. They heat up due to gravitational contraction, but never become main sequence stars. Eventually, they contract as far as they can and then begin to cool, becoming planetlike objects—cold cinders in space—called brown dwarfs, which we will study in the next

The Horsehead nebula of Figure 13-1d is a smaller part of this same Orion nebula. New stars are thought to be forming in the dark horsehead.

Figure 13-14

(a) The Orion nebula is an emission nebula that contains a dark nebula where new stars are forming. The Orion nebula is easily visible in small telescopes or even binoculars. (b) The blue star HD 148937 (the brightest in a triple-star system in the constellation Norma) is experiencing a continuous mass loss from its outer layers. The emission nebulae (NGC6164, NGC6165 on either side of the star) were formed during past vigorous outbursts.

(a)

(b)

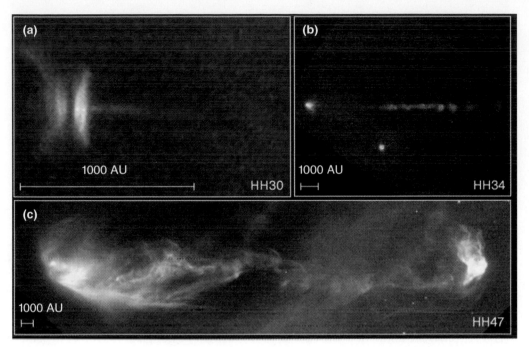

(a) 1000 AU HH30

(b) 1000 AU HH34

(c) 1000 AU HH47

Figure 13-15
Jets emanating from three different protostars. The designation HH stands for Herbig-Haro objects, the bright knots of hot, ionized material that appear to be moving away from their corresponding protostars. (a) An edge-on disk of dust around a protostar associated with object HH-30, about 450 light-years away in the constellation Taurus. The protostar is hidden behind the densest and opaque part of the disk but illuminates the disk's top and bottom surfaces. (b) The jet in object HH-34 shows a beaded structure, suggesting an episodic ejection of dense parcels of material. This system is about 1500 light-years away, close to the Orion nebula. (c) The protostar in object HH-47 is hidden inside a dust cloud (near the left edge of the image) while its light is reflected by the white filaments on the left. The jet structure suggests that the protostar might be wobbling. This system is about 1500 light-years away, at the edge of the Gum nebula. (All three composite *HST* images were taken in visible light.)

chapter. Recall that Jupiter emits about twice as much energy as it receives from the Sun and that this comes from gravitational energy that remains from its formation.

Recent infrared observations have revealed that it is very common for protostars to be surrounded by disks of gas and dust. These observations fit well with the theory of the formation of the solar system and lead us to believe that planetary systems are common around stars. In addition, streams of material (jets) have been observed flowing from the poles of many protostars. **Figure 13-15** clearly shows the presence of disks and outflows in protostars. Even though the *HST* images are of three different objects, when compared together they give us a general view of how jets from protostars behave as they propagate in space. A jet emanates from a disk surrounding a young star. The jet remains narrow and shows a beaded structure, and then creates a shock wave when it interacts with the surrounding interstellar medium.

accretion The process by which an object gradually accumulates matter, usually due to the action of gravity.

On one hand, a protostar's mass may increase by the process of *accretion*, as particles in the disk surrounding the protostar lose energy and spiral closer to it as a result of collisions within the disk. On the other hand, a protostar's mass may decrease as a result of the ejection of mass along the jets (**Figure 13-16**). (In some cases the outflow occurs along only one jet, while in other cases the outflow is along two oppositely directed jets.) Astronomers hypothesize that these outflows play an important part in reducing the angular momentum of stars. Recall from the earlier discussion of the formation of the solar system (Section 7-6) that theoretical models predict that the Sun should be spinning faster than it is. If, early in the Sun's formation, material were ejected from its poles, that ejection would have resulted in reducing the Sun's rotational speed to what is observed. Much remains to be learned about the part played by the ejection of matter from the poles of protostars.

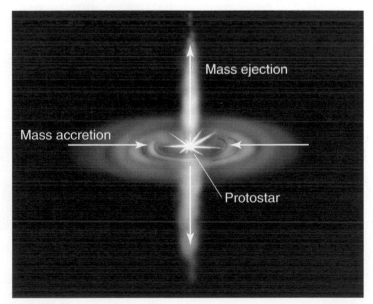

Mass ejection

Mass accretion

Protostar

Figure 13-16
A typical protostar strikes a balance between mass accreting to its surface from a surrounding disk of dust and gas, and mass ejected from the protostar in powerful jets of material.

Figure 13-17
(a) Close-up of protoplanetary disks in the Orion nebula. The field of view is 0.14 light-years across. (b) A very young star (between 300,000 and 1 million years old) surrounded by its protoplanetary disk. There is enough material in the disk to form at least seven planets like Earth. The disk is 7.5 times the diameter of our solar system.

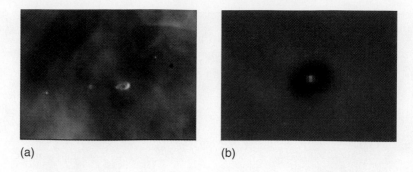

(a)

(b)

The problems of star formation are wonderfully complex. In a place like Orion you can see the whole region disintegrating. Supernovas, stellar winds, shocks are slamming into the molecular cloud, sweeping it back, igniting a conflagration. Things are happening wholesale. You have to be careful in applying simple models.

Patrick Thaddeus, radio astronomer

The presence of disks of gas and dust surrounding newly formed stars is also shown in **Figure 13-17a**. This *HST* image of a small portion of the Orion nebula shows five stars, four of which are surrounded by disks of gas and dust. These protoplanetary disks will probably evolve on to form planets. Figure 13-17b is an image of a very young star surrounded by a protoplanetary disk, seen in silhouette against the background of the Orion nebula. Not all stars form such disks; for example, it is likely that massive stars or some stars in binary systems do not form protoplanetary disks. However, as recent observations suggest, these disks are found around most young, low-mass stars. The duration of the different phases that a young star goes through as it evolves depends a lot on its environment. If a young star has nearby companions, their UV radiation and stellar winds may disrupt the young star's accretion disk. The longevity of the star's jets will be affected by whether or not planets start forming in the disk, clearing away gas that feeds these outflows. **Figure 13-18** shows two examples of young stars making their presence felt in their surrounding medium.

Many details of our theories of the life cycles of stars come from computer modeling. The image shown in **Figure 13-19** represents one stage in the formation of a star. As material collapses to form the star, it flattens into a pancake shape. Then the rapid rotation of the star causes it to shed its outer layers. Some of the material

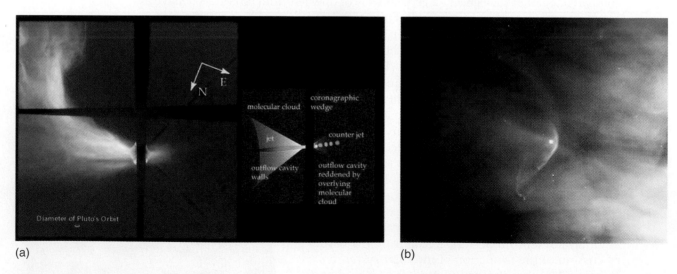

(a)

(b)

Figure 13-18
(a) A false color image of structures resembling gas jets from the young star SU Aurigae. White represents the brightest regions and dark red the dimmest. A jet is seen extending about 1200 AU approximately horizontally on the left side of the star. A counterjet is seen on the right, directed into a vast, dark molecular cloud that provided the matter for the star's formation. The bright white patch on the left side of the star is a void in the cloud, 1800 AU wide, filled with gas and dust that reflects the star's light. The star's age, about 4 million years, suggests that jets can persist for long periods of time. (b) A bow shock around the very young star LL Ori, in Orion's Great Nebula. Such shocks can be created in space when two streams of gas collide. The star emits a strong wind, which interacts with the surrounding medium in this intense star-forming region.

shown in blue will eventually collapse into the blue green area, forming a planetary system. The material shown in red will collapse further to become the star. Astronomers believe our solar system formed in a manner very similar to this.

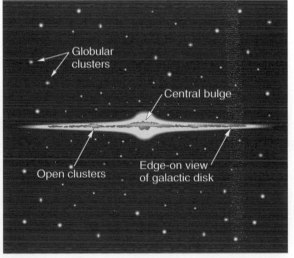

Figure 13-19
A frame from a videotape of a computer model showing a stage in the process of star formation.

Star Clusters

The Pleiades (Figure 13-7) is a small cluster of some 500 stars in the constellation Taurus. The brighter stars of the cluster are visible to the naked eye and are a beautiful sight even in a small telescope. **Figure 13-20** shows two other clusters of the same type, called *galactic clusters,* or *open clusters.* Galactic clusters are found primarily in the disk of the Galaxy. **Figure 13-21** shows another type of star cluster, called a *globular cluster.* Globular clusters contain hundreds of thousands of stars. They are not confined to the galactic disk, but orbit the center of the

galactic (or **open**) **cluster** A group of stars that share a common origin and are located relatively close to one another.

globular cluster A spherical group of up to hundreds of thousands of stars, found primarily in the halo of the Galaxy.

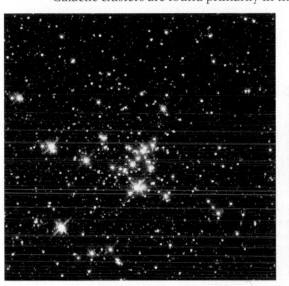

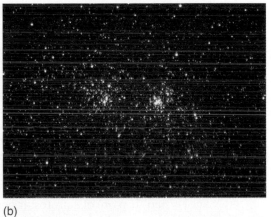

(a)

(b)

Figure 13-20
Many clusters are named by their number in the New General Catalog (NGC), which was compiled between 1864 and 1908. (a) The cluster NGC 457, in Cassiopeia, is called the Owl cluster and contains about 80 stars. The bright star is φ (Phi) Cassiopeiae. (b) A pair of open clusters (NGC869 and NGC884) in the constellation Perseus.

Globular is pronounced "glob" (as of mud) rather than "globe."

Figure 13-21
This is a globular cluster (M80). Globular clusters are gravitationally bound groups of hundreds of thousands of stars.

Figure 13-22
Galactic (open) clusters are found within the disk of the Galaxy, and globular clusters are found primarily in the halo surrounding the disk.

Figure 13-23
An H-R diagram of a very young cluster shows that low-mass protostars have not yet reached the main sequence.

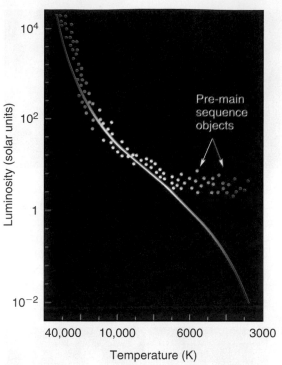

Galaxy in all directions so that they spend most of their time outside the disk. **Figure 13-22** shows the location of the two types of clusters relative to our Galaxy.

Clusters are important to astronomers for two reasons. First, all of the stars in a given cluster are at about the same distance from us. This means that their apparent magnitude is a direct indication of their absolute magnitude—those that look the brightest really are the brightest. Second, all the stars within a cluster formed at about the same time. (We will see that there is good evidence for this seemingly wild conjecture.) This means that they formed out of the same giant molecular cloud, and therefore it is probably safe to assume that all of them have about the same chemical composition.

Much of what we know about star formation has come from examination of clusters. The fact that all the stars in a cluster began forming at about the same time allows us to learn how stars of different mass have progressed at different rates. **Figure 13-23** is a representative H-R diagram of a very young cluster. The diagram indicates that stars of low mass have not yet reached the main sequence. All the stars of the cluster began their formation at essentially the same time, when their interstellar cloud began its collapse. H-R diagrams of clusters provide evidence that stars of low mass spend much more time in the protostar stage than do more massive stars. As we will explain in the next chapter, H-R diagrams of older galactic clusters can be used in a similar manner to reveal when stars end the main sequence part of their lives.

Thus far, we have tracked the evolutionary path of stars through the protostar stage. When the core of a protostar becomes hot and dense enough, nuclear fusion begins. The new star becomes stable and begins the main sequence portion of its life. We will discuss this stage of a star's life in Chapter 14.

As we observed earlier, there is evidence that all stars within a cluster began their lives at about the same time. The regularities we see in H-R diagrams such as Figure 13-23 prove the validity of this assumption, for what other explanation could there be for the regularities? In the next chapter we will continue to examine the H-R diagrams of star clusters, for they are very valuable tools in our quest to understand stellar life cycles.

Conclusion

Interstellar space is so barren of matter that for many purposes it can be considered a vacuum. However, vast quantities of dust and gas do exist between the stars, and this material is important to astronomers for two main reasons. First, radiation from stars passes through the interstellar medium on its journey to Earth, and if astronomers are to understand the information carried by starlight, they must understand how the interstellar medium affects radiation. Second, interstellar matter is the stuff of which stars are made. If astronomers want to understand the life cycles of stars, they must first understand the substance from which the stars form.

Clouds of dust and gas in interstellar space differ in density, composition, and size. Interstellar cirrus is extremely diffuse, but some dust clouds are dense and large enough to extinguish light from stars on the other side of them. In some clouds, the processes of fluorescence and reflection provide beautiful images that can be appreciated by astronomers and non-astronomers alike.

Technological developments during the past few decades—particularly in microwave and radio telescopes—have provided many answers to questions about the interstellar medium. But, as is usual in science, they have also created additional questions. The quest to understand how stars are formed from the interstellar medium is of special interest to astronomers, and many mysteries remain, particularly about the initial steps in the formation of a star.

Whatever serves as the trigger to begin the collapse of portions of a giant molecular cloud, mutual gravitational forces between the particles continue and accelerate the compression. As the core of the cloud collapses, gravitational energy is transformed into thermal energy, which raises the temperature of the protostar. Finally, the inner portion of the collapsing matter becomes so hot and dense that nuclear fusion starts. At that time, the cloud begins its life on the main sequence of the H-R diagram. Examination of star clusters has shown the importance of a star's mass in determining how the star develops, for more massive stars become hotter and more luminous than less massive stars. In the next chapter we will show that mass is not only important in determining the early development of a star, but that it is the single most important property in determining the fate of stars.

Study Guide

1. In the densest interstellar cloud, the particle density is _____ the particle density of air at sea level on Earth.
   A. much less than
   B. about the same as
   C. much greater than

2. The reddening of starlight as it passes through interstellar dust clouds is due to
   A. the Doppler effect.
   B. fluorescence.
   C. scattering.

3. A nebula that glows due to fluorescence is called
   A. a fluorescence nebula.
   B. a reflection nebula.
   C. an emission nebula.

4. Absorption lines due to interstellar gas can be distinguished from spectral absorption lines caused by a star's atmosphere because
   A. their Doppler shifts are different.
   B. different chemical elements are found in the two different places.
   C. [The question is misleading; interstellar gas does not cause absorption lines.]

5. A protostar is
   A. a newly forming star.
   B. a main sequence star.
   C. a star nearing the red giant stage.
   D. a black dwarf.

6. H-R diagrams of young star clusters show
   A. fewer than about 10 stars.
   B. very few of the most luminous stars.

C. that massive stars have not reached the main sequence.
   D. that low-mass stars have not yet reached the main sequence.

7. What is the primary difference between a star and a planet as each forms?
   A. age
   B. mass
   C. volume
   D. chemical composition
   E. velocity

8. Which of the following statements is true?
   A. Less massive protostars spend a longer time in the protostars stage than do more massive protostars.
   B. Protostars are surrounded by cocoons of gas and dust.
   C. Protostars radiate mainly in the infrared.
   D. [All of the above.]

9. Prior to reaching the main sequence, a star's energy comes from
   A. gravitation.
   B. nuclear fusion.
   C. nuclear fission.
   D. hydrogen conversion to helium.
   E. [Both B and D above.]

10. While a star is on the main sequence, its energy comes primarily from
    A. gravitational shrinking.
    B. nuclear fusion.
    C. nuclear fission.
    D. proton instability.
    E. chemical reactions.

11. As a star is forming by the condensation of gases, the gases
    A. cool as they fall.
    B. heat up as they fall.
    C. stay about the same temperature.
    D. [Any of the above, depending upon the mass involved.]

12. A star is considered to begin its main sequence life when
    A. it starts to collapse.
    B. its protostar life begins.
    C. nuclear reactions start.
    D. it begins to move off the main sequence.
    E. its planetary system has formed.

13. The contraction of an interstellar cloud to become a star is caused by
    A. magnetic forces.
    B. electric forces.
    C. nuclear forces.
    D. gravitational forces.
    E. [Both A and B above, for they are closely related.]

14. An evolutionary track on an H-R diagram reveals
    A. the changes that occur in a star's life.
    B. the changes that occur as one star evolves into another.
    C. the motion of a star relative to others in its cluster.
    D. the motion of a star relative to others in the entire galaxy.
    E. [Both C and D above.]

15. Which of the following choices is *not* considered a possible trigger to begin the collapse of an interstellar gas cloud?
    A. A shock wave from a supernova.
    B. A shock wave occurring during the formation of very massive stars.
    C. A shock wave resulting from radiation from nearby emission nebula.
    D. A shock wave passing around the galaxy.
    E. [All of the above are considered likely triggers.]

16. Clusters help us to learn about the evolution of stars because stars in a cluster
    A. are all about the same temperature.
    B. are all about the same mass.
    C. interact with one another and affect each other's evolution.
    D. are all about the same age.
    E. can be observed more easily than individual stars.

17. The H-R diagram for a single star cluster is different from the usual H-R diagram because all stars in the cluster have the same
    A. mass.
    B. temperature.
    C. diameter.
    D. rotational speed.
    E. age.

18. What is a protostar?

19. List three possible events that may be responsible for triggering the collapse of interstellar clouds.

20. Which stars take longest to complete their protostar stage? Why?

21. What is the source of the energy that powers protostars?

22. Why is there a lower limit to the possible mass of a star? An upper limit?

23. Explain how galactic clusters provide observational evidence for theories of stellar evolution.

QUESTIONS TO PONDER

1. Starlight that has passed through a cloud of dust is reddened. So is starlight from a star that is moving away from us. In each case, how can we tell that the light from a star has been reddened? Perhaps what we see is the actual color of the star.

2. If light is passing through space, why does the universe look black except where we see dust, gas, or an object?

3. Describe the processes that produce each of the following types of nebulae: dark nebulae (like the Horsehead nebula), pink nebulae (like the Lagoon nebula), blue nebulae (like the one that appears near the Trifid nebula).

4. Explain how an examination of the H-R diagram for a large group of stars can tell us the relative lengths of the various stages of a star's life.

5. Why do astronomers feel that there must be a triggering mechanism for the beginning of an interstellar cloud's collapse? Why couldn't it happen on its own?

6. Explain the formation of "pillars" like those in Figure 13-10b.

7. Why do we think that most stars form from a *rotating* disk of gas? Couldn't it just as well be non-rotating?

8. Explain why the mass of a star determines the amount of time that the star remains a protostar.

9. The primary difference between protostars and main sequence stars is their energy source. What is the source of energy of each type of star?

10. Explain the value of star clusters to astronomers in their attempt to understand stellar evolution.

CALCULATIONS

1. Along the line of sight to a particular star cluster, interstellar extinction allows only 20% of the light from the stars to pass through 1000 light-years of the interstellar medium. If this cluster is located 2000 light-years from us, what percentage of its light do we receive?

2. During our Sun's early stages of evolution, it had a surface temperature of 1000 K and luminosity 1000 times larger than it has now. What was our Sun's radius at this stage? Express this radius in astronomical units and as a multiple of the Sun's current radius. (Hint: 1000 K is about 1/6 of the Sun's current surface temperature. Consider the Stefan-Boltzmann law.)

3. A star in Orion is located 0.1 degree from the center of the nebula and is moving away from the center at 100 kilometers/second. How long ago did it get ejected from the center? (Hint: The distance to Orion is about 1500 light-years. Consider the small-angle formula.)

4. The temperature of dust grains in most interstellar clouds is about 300 K. Using Wien's law, calculate the wavelength at which most of the radiation is emitted by the grains. Is there a reason why astronomers are interested in using the infrared part of the electromagnetic spectrum to study these clouds?

Deep Sky Objects with a Small Telescope

Many beginning telescope users limit themselves to viewing the Moon and planets, but some interesting objects outside the solar system are accessible with a small telescope or even with binoculars. The star maps in this book can help you find the objects listed below. At the end of each listing is the time of year when the object is highest in the sky in the evening, but each may be seen in roughly the same place later at night, earlier in the year.

Orion nebula—A cloud of dust and gas 1500 light-years away. Its diameter is some 15 light-years. Find the Trapezium, four stars in a small dark area of the nebula (sometimes called the Fish Mouth). (January through March)

Alcor and Mizar—A naked-eye double star in Ursa Major. Your telescope will reveal that Mizar is itself a double star. (March through July)

Sagittarius—Such a rich area of the sky for interesting objects that if you slowly scan it with your telescope, you will find several clusters and nebulae. (July and August)

Albireo—A binary pair with obviously different colors (Figure 12-25). It is the "beak" star of the swan Cygnus, or the bottom star of the Northern Cross. (August through October)

Andromeda galaxy—A spiral galaxy similar to the Milky Way. Can you detect an oval shape? (October through December)

Pleiades—An open cluster visible to the naked eye. It contains a few hundred stars (although only six or seven are visible with the naked eye) and is about 400 light-years away. See if you can detect the nebula around the stars. (November through February)

h and χ (the Greek letter "chi") Persei—A pair of open clusters visible to the naked eye (though not as easily as the Pleiades). In the sword handle of Perseus, they are an excellent view in binoculars. (November through February)

To fully enjoy your telescope (or binoculars), a good star atlas is almost a necessity. The following are recommended:

The Monthly Sky Guide (5th ed.). Ian Ridpath and Wil Tirion. Ingram, 1999.

Whitney's Star Finder (5th ed.). C. Whitney. Ingram, 1989.

A Field Guide to the Stars and Planets (3rd ed.). D. Menzel and J. Pasachoff. Boston: Houghton Mifflin, 1992.

To keep up with developments in astronomy, you might write to the nonprofit Astronomical Society of the Pacific (390 Ashton Avenue, San Francisco, CA 94112), whose aim is to share the excitement of astronomy with a wider public. Its members include scientists, teachers, hobbyists, and thousands of people from around the world who enjoy reading about the universe.

1. "The Stuff Between the Stars," by G. Knapp, in *Sky & Telescope* (May, 1995).

2. "The Early Life of Stars," by S. W. Stahler, in *Scientific American* (July, 1991).

3. "Interstellar Molecules," by G. L. Verschuur, in *Sky & Telescope* (April, 1992).

4. "Spying on Stellar Nurseries," by R. Jayawardhana, in *Astronomy* (November, 1998).

5. "Deciphering the Mysteries of Stellar Origins," by C. Lada, in *Sky & Telescope* (May, 1993).

Quest Ahead to Starlinks: www.jbpub.com/starlinks

Starlinks is this book's online learning center. It features **eLearning,** which contains chapter quizzes and other tools designed to help you study for your class. You can also find **on-line exercises,** view numerous relevant **animations,** follow a guide to **useful astronomy sites** on the web, or even check the latest **astronomy news** updates.

ACTIVITIES

EXPANDING THE QUEST

STARLINKS

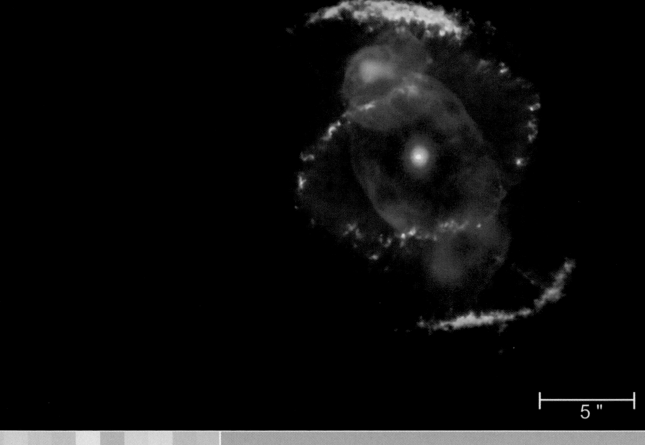

|— 5 " —|

CHAPTER 14

The Lives and Deaths of Low-Mass Stars

The "Cat's Eye nebula," officially named NGC 6543, as seen by the *Hubble Space Telescope* and the *Chandra X-ray Observatory*.

THIS BEAUTIFUL COMPOSITE IMAGE SHOWS THE "CAT'S EYE NEBULA." The image includes X-ray data (shown in purple) from the *Chandra Observatory* and optical data from the *Hubble Space Telescope*. The image reveals that the nebula contains concentric gas shells, jets of high-speed gas, shock-induced knots of gas, and a cloud of multimillion-degree gas. Because of its age (estimated to be 1000 years) the nebula is a visual "fossil record" of the late stages of evolution of a dying star, which is expected to collapse into a white dwarf in a few million years.

Astronomers hypothesize that the star might be a double-star system. The effects of two stars orbiting one another most easily explains the intricate structures, which are much more complicated than features seen in most nebulae that surround dying stars. The two stars are too close together to be individually resolved by *Hubble*, and so they appear as a single point of light at the center of the nebula. According to the most generally accepted model, a "fast stellar wind" of gas blew off the central star and created the elongated shell of dense, glowing gas. The areas that appear in green might indicate where high-speed gas collided with slower-moving gas from an earlier ejection.

The optical component of the color picture, taken with the Wide Field Planetary Camera-2, is a composite of three images taken at different wavelengths. Such photographs reveal the beauty of distant dying stars, but we hope that as we study the lives and deaths of stars, you will also admire the beauty of the theory that underlies these events.

Stars form from the material of the interstellar medium. As a portion of a giant molecular cloud collapses, the matter in its core becomes hotter and denser until nuclear fusion begins. When this happens, the object joins the family of stars, for it is then on the main sequence of the H-R diagram, where it will spend most of its life.

In this chapter we describe stars of relatively low mass, those toward the bottom of the main sequence (including our Sun). We will show that although all stars live their main sequence lives in much the same manner, they differ greatly in the way they die. Stars with very low masses seem to fade away, "giving up the ghost" gently. Others expand to become red giants and then puff away their outer layers, thereby revealing their previously hidden white-hot cores.

As we study the amazing events in the lives of stars, keep in mind that we are describing the results of various models of stellar interiors. We cannot examine the inside of a star directly. Instead, we observe the properties of the surfaces of stars and make mathematical models of what must be occurring inside them to produce our data. Such models have improved dramatically in the last 20 years as computer capabilities have increased; the calculations that produce today's models are numerous and complex.

Theoretical models cannot be divorced from reality, of course. They must correspond to observations. For example, the major supernova of 1987 served as a case history for testing models of supernovae. The supernova excited astronomers not only because it provided another opportunity to test their theories, but—as we will see—because it confirmed most of their predictions.

14-1 Brown Dwarfs

As we explained in the previous chapter, the mass of a collapsing object determines how long the object remains in the protostar stage. Once a star has completed its protostar stage and reached the main sequence, its mass continues to be the dominant property determining the steps it passes through during its life. Therefore, before discussing stellar life cycles, we should examine the question of what limits the mass of a star.

The maximum mass that a star can have is probably about 100 solar masses. If a protostar has more mass than this, it collapses quickly and develops tremendous amounts of energy and internal pressure. This pressure is so large that it overwhelms gravity and blows the star apart before it can establish equilibrium on the main sequence.

On the other hand, suppose that a very small portion of an interstellar cloud becomes compressed enough for the gravitational force to take over and continue the collapse. Calculations show that if this protostar has a mass greater than about 8% of the Sun's mass, the object's core becomes hot and dense enough for nuclear fusion to begin. In that case, the object joins the main sequence as a full-fledged star. But what if the protostar has less than enough mass for fusion to begin? In this case, the object still heats up due to gravitational contraction. **Figure 14-1a** shows hypothetical evolutionary trails of five protostars on an H-R diagram. The two with the most mass become stars on the main sequence. Two other protostars on the diagram— those with masses of 0.07 and 0.02 times the mass of the Sun—heat up so that they glow with a dull red color, but nuclear fusion never starts in their cores. Their dull color gives them their name: *brown dwarfs.* After a brief initial phase where some deuterium is burned, brown dwarfs continue to cool, releasing the heat left over from their birth, and become progressively fainter. The fifth object has the mass of

brown dwarf A star-like object that has insufficient mass to start nuclear reactions in its core and thus become self-luminous.

Figure 14-1

(a) Protostars with masses less than about 0.08 solar masses never reach the main sequence. Of the five stars whose evolutionary paths are shown here, two become main-sequence stars, two become brown dwarfs, and one becomes a planet. (b) The small object just right of center is GL229B, a brown dwarf, photographed by the *HST* in far red light. The dwarf is a companion to a much brighter star, Gliese 229, that is off the left edge of the image. The two objects are about 18 light-years away from us and are separated from one another by at least 4 billion miles.

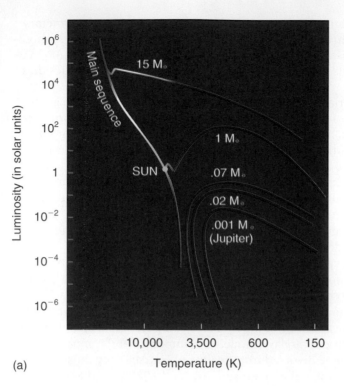

(a)

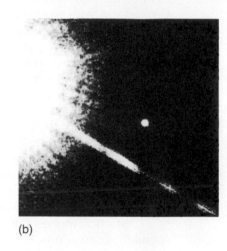

(b)

◆ Probable limits of mass for brown dwarfs are from 0.002 to 0.08 solar masses. We discussed these objects briefly in Section 7-7.

◆ A red dwarf is a small-mass star with low temperature and as a result, it appears red.

the planet Jupiter, which may once have emitted a small amount of visible light, but now is too cool to do so.

The question of how small an object can be and still be classified as a brown dwarf is open for discussion. Even an object as small as the planet Jupiter (0.001 solar masses) heats up as it forms. Such a small object probably does not deserve to be classified in a stellar category at all, so we will arbitrarily set the lower mass limit of brown dwarfs at twice the mass of Jupiter, or about 0.002 solar masses.

Brown dwarfs have been predicted since the mid-1900s, but until one was found, astronomers were not able to confirm that it is possible for such a small portion of an interstellar cloud to collapse. Until 1994, the only objects whose existence was confirmed that were near the required size were the planets of our solar system. The first brown dwarf was seen in October, 1994, using adaptive optics with the 60-inch telescope on Mount Palomar. It appears very close to another star, a red dwarf, and a year after the discovery, astronomers confirmed that the brown dwarf is actually gravitationally bound to the nearby star.

Figure 14-1b is an *HST* photograph of the brown dwarf, called GL229B. This is a false-color image, so the dwarf appears white rather than dull red. Its much brighter binary companion, GL229, is beyond the picture to the left, but GL229 is so bright that it floods the detector and causes the blur and the diffraction spike from the optical system. The brown dwarf is about 20 to 70 times the mass of Jupiter, but is about the same diameter as Jupiter. The two objects are about 18 light-years away in the constellation Lepus.

Brown dwarfs bridge the gap between stars and planets. Their properties reveal new insights into how stars and planets form. Even though brown dwarfs have been discovered orbiting stars, most of them seem to be isolated. How do brown dwarfs form? Two main scenarios have been proposed for isolated brown dwarfs. According to the first scenario, brown dwarfs form just like stars, by contraction in an interstellar cloud of dust and gas. Observations of excess near-infrared radiation from many brown dwarfs in the Orion nebula, which is thought to show the presence of dusty disks around them, support this scenario. If such disks exist around young brown dwarfs exactly as they do around real stars, then both types of objects form in the same way. After a period of time, the disk disperses and that is why we do not find them around older brown dwarfs. In the second scenario, when a very young star in a multiple star system gets ejected from the system, it loses access to the surrounding accretion disk; this effectively stops its further growth by accretion, and it ends up as a brown dwarf. Computer simulations that follow the collapse of an interstellar gas cloud support this scenario. Such simulations show that the process of star formation is chaotic; in new-born stellar groups the more massive objects gather more gas than lower mass objects, the latter being ejected from the groups so quickly that they do not manage to accrete enough gas to become stars.

After the discovery of brown dwarfs, two new spectral types were introduced below the long-known cool M dwarfs. The L and T types, loosely defined at present, cover the temperature range from 3500 to about 1000 K. Objects in this range are almost invisible at optical wavelengths, with most of their emission coming out in the infrared. As the surface of a brown dwarf cools below 1500 K, something dramatic happens that changes its appearance; large amounts of methane form, making the T-type brown dwarfs the coolest objects detected so far.

Notice that we have avoided referring to a brown dwarf as a star. The reason for this is that the word *star* is reserved for an object massive enough to sustain nuclear fusion in its core—that is, to have reached stellar maturity on the main sequence.

Current predictions suggest that there should be twice as many brown dwarfs as main sequence stars. As of early 2003, there are about 200 known brown dwarfs.

14-2 Stellar Maturity

When the center of a collapsing star becomes hot and dense enough, hydrogen fusion begins. Gravitational contraction serves as the energy source for protostars, heating them and causing the emission of radiation, but main sequence stars—including our Sun—have as their energy source the fusion of hydrogen into helium.

Do stars simply ignite and start fusion reactions immediately?

Stellar Nuclear Fusion

In Section 11-2 we described the fusion reactions that occur in the Sun. That series of reactions, known as the proton-proton chain, is the predominant reaction that provides the energy for stars of low mass, up to about 1.5 solar masses. For the sake of completeness, it is repeated in **Figure 14-2.**

Stars that have masses greater than about 1.5 solar masses have higher temperatures in their cores, and a different chain of nuclear reactions dominates. This series of reactions involves carbon, nitrogen, and oxygen and is called the **carbon cycle** or the **CNO cycle.** Its steps are shown in **Figure 14-3**. If you examine the steps of the CNO cycle, you will see that its overall effect is to transform four hydrogen nuclei into one helium nucleus, just like the proton-proton chain. Carbon-12 is one of the nuclei that participates in the reaction, but for each carbon-12

carbon (or CNO) cycle A series of nuclear reactions that results in the fusion of hydrogen into helium, using carbon-12 in the process.

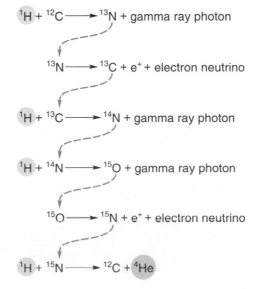

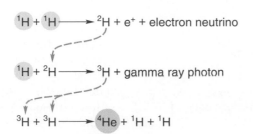

Figure 14-2 (left)
In the proton-proton chain, four hydrogen nuclei are changed to one helium nucleus, releasing energy in the process. The symbol e^+ denotes a positive electron, a positron. Dashed arrows indicate a nucleus taking part in the next reaction. In the third reaction, the second hydrogen-3 nucleus comes from a second occurrence of the two previous steps.

Figure 14-3 (right)
The CNO cycle changes four hydrogen nuclei into one helium nucleus, with an attendant release of energy.

nucleus that enters the reaction, one is produced at the end. Therefore there is no net change in the carbon; it participates only as a facilitator. (This is similiar to the role played by catalysts in some chemical reactions. The catalyst stimulates the reaction but is not used up in it.)

The Stellar Thermostat

Thermostats on the walls in our homes have two features: a thermometer to read the temperature and a switch to make the furnace or air conditioner turn on or off. If the thermometer indicates that the temperature is below some preset level, the switch turns on the furnace until the thermometer indicates that the temperature is high enough, whereupon the switch turns off the furnace. The core of a main sequence star has an analogous regulating mechanism, which controls the rate of consumption of the hydrogen fuel.

Suppose that somehow the rate of hydrogen fusion in a star begins to increase. The extra energy produced causes the temperature of the core to increase. However, when the temperature of a gas increases, the gas expands. This expansion serves as a switch to decrease the rate of fusion. This occurs because in the expanded gas, the average distance between hydrogen nuclei has increased, and the increase in distance means fewer collisions and therefore fewer fusions. The result is that the expansion causes the rate of energy production to decrease, which causes the expansion to stop. In this way equilibrium is reached, and fusion reactions occur at a uniform rate.

On the other hand, if the rate of fusion in a star's core somehow decreases, the core contracts, decreasing the distance between nuclei. This causes the fusion rate to increase. Another effect that occurs when the core contracts is the conversion of gravitational energy into heat, just as occurred when the original interstellar cloud collapsed. Therefore, energy comes both from an increase in fusion and from gravitational energy, and both of these energies contribute to the thermostat that brings the star once again to equilibrium. The overall effect of the stellar thermostat is that nuclear fusion proceeds at a rate that is just enough to balance the force of gravity that tends to compress the star.

Main Sequence Life of Stars

The stellar thermostat is important during the main sequence life of a star. In a main sequence star, hydrogen in the core of the star is continually being converted to helium as the nuclear reaction occurs. This causes the number of particles in the core to gradually decrease (since *four* hydrogen nuclei are converted to *one* helium nucleus), and this results in the core shrinking slightly. In turn, the contraction causes the temperature to rise within the core, and this increases the rate of fusion. Therefore, the core releases more energy, and as this energy flows outward, it causes the outer portion of the star to expand somewhat. These steps are outlined in **Figure 14-4**. We see the effects of these

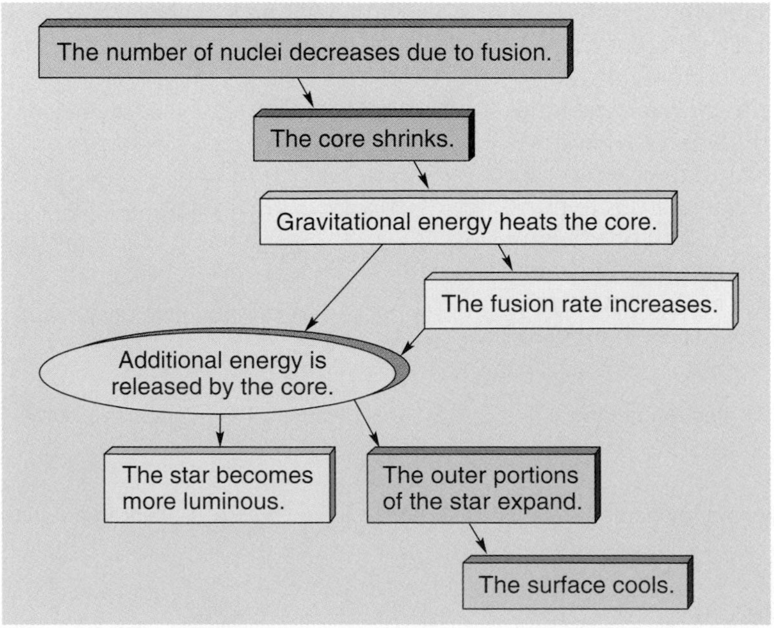

The analogy of a thermostat is an alternative description of hydrostatic equilibrium, which we discussed in Section 11-3.

When you pucker your lips and blow on your hand, your breath feels cold because it expands as it moves from inside your mouth to the outside air. Similarly, gases in a bottle of carbonated drink expand and cool when you open the bottle, creating a small cloud in the bottle's neck.

When you pump air into a bicycle tire with a hand pump, the compressed air gets warm, making the pump warm, too.

Figure 14-4
This diagram shows why a star becomes more luminous and cooler as it ages on the main sequence. Each arrow indicates that one event causes the next.

The number of nuclei decreases due to fusion.
↓
The core shrinks.
↓
Gravitational energy heats the core.
↓
The fusion rate increases.
↓
Additional energy is released by the core.
↓
The star becomes more luminous.   The outer portions of the star expand.
↓
The surface cools.

changes on the H-R diagram (**Figure 14-5**), for the increase in energy from the star means that it has a greater luminosity, so it moves upward on the diagram. In addition, the expansion of the star causes its outer layers to cool a little, and this causes the star to move to the right on the diagram, toward lower temperatures. These changes are the reason the main sequence has a perceptible width. Stars start their lives as *zero-age main sequence* stars on the left side of the strip, then move up and to the right as they age.

The changes that the Sun is undergoing will have drastic effects on the Earth. The Sun began its main sequence life about 5 billion years ago, at which time its chemical composition was about 74% hydrogen, 25% helium, and the remaining 1% other heavier elements. As a result of the nuclear reactions, the Sun's core has contracted a bit, the Sun's luminosity has increased by about 40%, and its core now contains more helium than hydrogen (about 64% helium and 35% hydrogen). There is still enough hydrogen in its core for our Sun to continue on the main sequence for another 5 billion years, at which time it will be twice as luminous as it is now. This will raise the average temperature of the Earth by about 20°C (36°F), enough to melt the polar caps and drastically change the Earth's climate.

Stars that are more massive than the Sun have a much greater fusion rate because their cores have greater pressures and higher temperatures—the thermostat is set higher in massive stars. This means they are more luminous. (Recall from our discussion in Section 12-7 the mass-luminosity relationship for main sequence stars: the more massive, the more luminous.) The greater fusion rate also causes these massive stars to use up their core hydrogen in a much shorter time. The most massive stars fuse hydrogen so quickly that their cores run out of hydrogen in only a few million years. Calculations for the least massive stars, on the other hand, show that they continue with hydrogen fusion on the main sequence for hundreds of billions of years.

Observational evidence for the shorter lifetimes of massive stars comes from galactic clusters. **Figure 14-6a** is the H-R diagram of the Pleiades. Although this is considered a young cluster, its most massive stars are already leaving the main sequence. Arrows on the figure indicate their evolutionary paths after they end their main sequence lives. Figure 14-6b is a similar diagram for the cluster M11, a cluster older than the Pleiades. Notice that the most massive stars have moved even farther to the right and that stars of less mass are now leaving the main sequence. Assuming that the stars in a cluster were all formed at approximately the same time, H-R diagrams such as those shown in Figure 14-6 allow us to find the age of a cluster. Stars at the *turnoff point,* which corresponds to the top of the group of stars still on the main sequence, are just now exhausting their core hydrogen and starting to move off the main sequence. Therefore, the main-sequence lifetime of the stars at the turnoff point is equal to the age of the cluster.

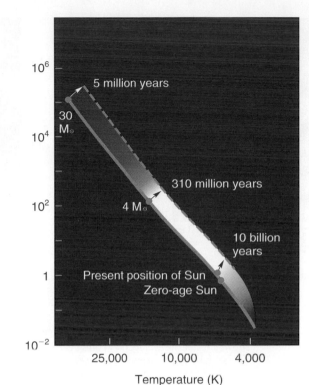

Figure 14-5

As stars age on the main sequence, they change slightly in temperature and luminosity. (The Tools of Astronomy box on page 453 shows how we calculate main sequence lifetimes.)

zero-age main sequence The main sequence of newly formed stars that just started hydrogen fusion in their cores.

The name M11 means that this cluster is number eleven in the Messier Catalog of Nebulae and Star Clusters, originally developed in 1781 by the French astronomer Charles Messier (1730–1817).

turnoff point The point on an H-R diagram of a cluster of stars where the stars are just leaving the main sequence.

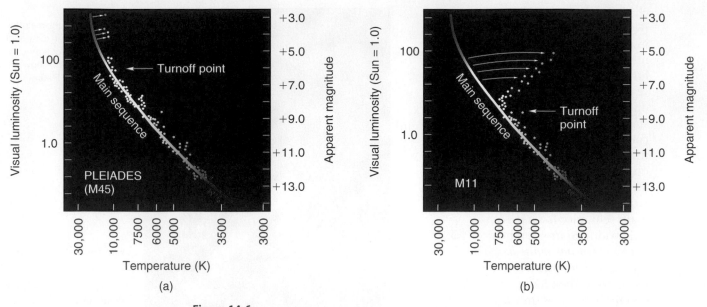

Figure 14-6
(a) An H-R diagram of the Pleiades reveals that its most massive stars have started leaving the main sequence. The arrows indicate the evolutionary paths they must have taken. (b) The cluster M11 is older than the Pleiades, and stars are turning off the main sequence at a lower point.

14-3 Star Death

Until the end of their lives on the main sequence, the primary difference between the evolution of stars of various masses is in the amount of time they spend as protostars and as main sequence stars. For example, a one-solar-mass protostar takes about 20 million years to become a main-sequence star, while a 15-solar-masses protostar takes only 20,000 years. The main-sequence lifetime of a one-solar-mass star is about 10 billion years, while that of a 15-solar-masses star is only about 10 million years. From this point on, however, the mass of a star determines which of several very different paths its life will take. We will discuss each of these paths in turn, beginning with stars of very low mass. Since mass classifications are fairly arbitrary and astronomers have not agreed on official names for the different classes, we will group them in the following categories: very low mass (less than 0.4 $M_\odot$), moderately low mass (0.4–4 $M_\odot$), moderately massive (4–8 $M_\odot$), and very massive stars (greater than 8 $M_\odot$), where $M_\odot$ is the mass of the Sun.

14-4 Very Low Mass Stars (<0.4 $M_\odot$)

Recall from the discussion in Section 11-3 of energy transport within the Sun that little convection takes place in the Sun except in the outer layers. That is why, once hydrogen is used up in the core of a star like the Sun, the core is not replenished with fresh hydrogen from outside. This causes the star to move from the main sequence.

However, in the least massive stars (those with a mass of less than about 0.4 solar masses), convection occurs throughout most or all of the star's volume (**Figure 14-7**). Therefore, hydrogen from throughout the star is cycled through the core, and the entire star runs low on hydrogen at the same time. When this

TOOLS OF ASTRONOMY

Lifetimes on the Main Sequence

The amount of time that a star spends on the main sequence depends on two things: the amount of hydrogen in its core and its rate of hydrogen consumption. The relationship can be expressed as a proportionality:

$$T \propto \frac{\text{amount of hydrogen}}{\text{rate of hydrogen consumption}}$$

where T represents the star's lifetime and the symbol "$\propto$" means "proportional to." Since all stars have roughly the same proportion of core hydrogen at equivalent points in their lives, the amount of hydrogen in a star is proportional to its mass. The rate at which the hydrogen is consumed depends upon the star's luminosity. Therefore

$$T \propto \frac{M}{L}$$

In a Tools of Astronomy box in Chapter 12 we explained that a star's luminosity (L) is proportional to its mass (M) raised to the power of 3.5. Thus

$$T \propto \frac{M}{M^{3.5}} \propto \frac{1}{M^{2.5}}$$

Using the lifetime of the Sun as a standard, so that T_{Sun} is the lifetime of the Sun on the main sequence and M is measured in units of the Sun's mass, the previous expression can be written as an equation:

$$T = \frac{T_{Sun}}{M^{2.5}}$$

Then the main sequence lifetime of a star with a mass of 3 solar masses can be calculated as follows:

$$T = \frac{T_{Sun}}{M^{2.5}} = \frac{T_{Sun}}{3 \times 3 \times \sqrt{3}} \approx 0.064 \times T_{Sun}$$

Thus, a star with three times the Sun's mass will live on the main sequence only about 6% as long as the Sun. Since the Sun's main sequence life (T_{Sun}) is about 10 billion years, this star's life will be about 640 million years. Similarly, a star with a third of the mass of the Sun will remain on the main sequence for 16 times longer than the Sun, about 160 billion years.

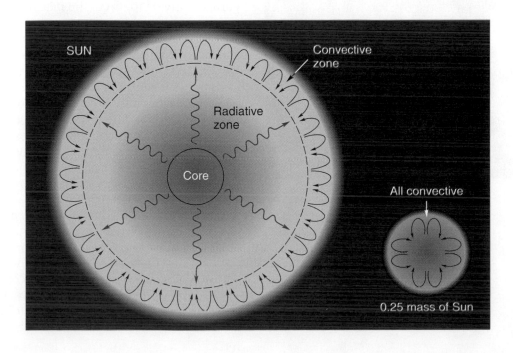

SUN

Convective zone

Radiative zone

Core

All convective

0.25 mass of Sun

Figure 14-7
The Sun (at left) contains a large radiative zone. A star of very low mass, on the other hand, consists of one convective zone, meaning that all of its material mixes.

Figure 14-8
Very low mass stars spend their main sequence lifetimes at the very bottom of the main sequence. Although the universe is not old enough for any to have completed that stage of their life, when they do, they will slowly become hotter and join the category of white dwarfs.

thermal energy Energy that is due to the random motions of molecules and atoms of a substance.

◆ In everyday language, the word "heat" is often used when "thermal energy" would be more correct. Strictly speaking, heat is the exchange of thermal energy.

white dwarf The burnt-out relic of a low mass star.

◆ The lowest mass stars may live as main sequence stars for as long as 100 billion years.

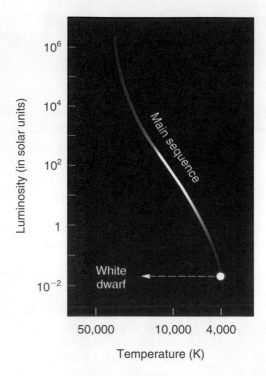

happens, the rate of fusion in the core decreases. As described earlier, this causes the core to contract and its temperature to increase as gravitational energy is converted into **thermal energy**. In a more massive star, this thermal energy is carried outward by radiation, but a very low mass star distributes its thermal energy by convection. The result is that the entire star contracts and heats up. On the H-R diagram, the star moves toward the lower left (**Figure 14-8**) and becomes a **white dwarf**.

Computer models indicate that a star with very low mass requires more than 20 billion years to complete the nuclear burning of its hydrogen fuel and end its main sequence life. Since this is greater than the age of the universe, the white dwarfs that astronomers see could not have originated in this manner. For this reason, we discuss white dwarfs again later in this chapter.

14-5 Beyond the Very Low Mass Stars: The Red Giant Stage

About 90% of the stars in the sky are on the main sequence. This tells us that a typical star spends 90% of its luminous lifetime there. In all cases except for stars of very low mass (those with mass less than about 0.4 solar masses), the next step for stars leaving the main sequence is essentially the same. They become red giants.

Figure 14-9
As the core of a star shrinks, it heats up, causing additional hydrogen fusion in a shell surrounding the core. (The core and hydrogen-fusing shell are actually a much smaller fraction of the star than is shown here.)

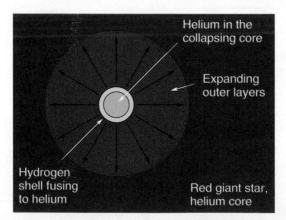

The process of becoming a red giant starts when the core begins to run low on hydrogen fuel. Until then, the nearly constant production of fusion energy kept the core from collapsing. Now, lacking a source of energy to fight gravitational collapse, the core starts to shrink dramatically. However, just as in the case of a protostar, the contraction converts gravitational energy into thermal energy and radiation. In fact, the energy produced in the core actually increases over what it was when the energy source was fusion. The resulting increase of radiation from the core causes the shell of material around

◆ The cooling of a gas as it expands is what causes the gas escaping from an aerosol spray can to feel so cold.

the core to heat up enough that hydrogen fusion begins there (**Figure 14-9**).

Now gravitational energy is being converted to thermal energy in the core and fusion is producing energy in a shell surrounding the core. These two sources of energy in the star's center result in a large increase in radiation from the center and cause the outer part of the star to expand. When a gas expands, it cools, and the outer portion of the star does just that. Thus the star's surface temperature decreases and the star moves to the right on the H-R diagram. At the same time, it moves upward, toward the red giant region, because its total luminosity is increasing due to the additional energy being produced in the core.

Table 14-1
A Typical Red Giant

Absolute magnitude	−1
Luminosity (solar units)	500
Mass (solar units)	1
Diameter (solar units)	200
Average density	10^{-7} g/cm^3 (10^{-7} Sun's average density)
Surface temperature	3600 K

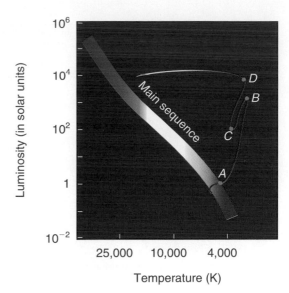

Figure 14-10
The Sun will enter the first stage of its death when it leaves the main sequence at point *A*. The helium flash (which we discuss in the next section) will occur when it reaches the luminosity and temperature of point *B*, and this will cause a quick decrease in its luminosity until it reaches point *C*. Helium fusion around the Sun's inner carbon core will then move it up to point *D*, where it will reach its greatest luminosity.

It may seem strange that a star can decrease its surface temperature and at the same time increase its luminosity, for when an object cools, it emits *less* radiation per square meter of its surface. The key to understanding this apparent contradiction is to realize just how large the red giant becomes. The Sun, at this stage in its life as a giant, will have expanded so that it encompasses the orbit of Mercury! It is true that each square meter of the Sun's cooler surface will emit less radiation, but the surface will have become so tremendously large that the total radiation emitted will be greater than before. **Figure 14-10** illustrates how the Sun will change its position on the H-R diagram. As the Sun expands, it will move from point *A*, its turn-off point from the main sequence, toward point *B*. At this point as a red giant star, our Sun will be about 1000 times brighter than it is today, even though its surface temperature will be only about 3500 K. The outer planets will lose their atmospheres because of the increase in brightness and some of the inner planets will be vaporized. **Table 14-1** lists some properties of a typical red giant.

Recall the Stefan-Boltzmann law (Sections 4-4 and 12-4), $L \propto R^2 T^4$ where L, R, and T are the luminosity, radius, and surface temperature of a star, respectively.

14-6 Moderately Low Mass Stars (0.4–4 M$_\odot$)

All main sequence stars become red giants (or supergiants, which we will study in the next chapter) when they leave the main sequence. How stars change during the red giant stage, however, depends upon how massive they are. We first consider the moderately low mass stars on the main sequence. This group includes stars from about 0.4 solar masses up to about 4 solar masses, and therefore it includes our Sun. Even among members of this group there are differences in the conditions that develop in their cores as they proceed to become red giants and in the way that the next cycle of nuclear fusion (helium fusion) begins in their cores. Because of these differences, we first consider stars with mass less than about 2 solar masses.

Electron Degeneracy and the Helium Flash

The core of a star at the end of its main sequence life consists of helium nuclei intermingled with electrons, a mixture that has properties similar to a regular gas. In this case, a simple relationship exists between pressure, temperature, and volume. For example, an increase in pressure exerted on a normal gas causes its volume to decrease and its temperature to increase. That is why gravitational pressure causes a star's core to contract and heat up when hydrogen fusion shuts down at the end of main sequence life.

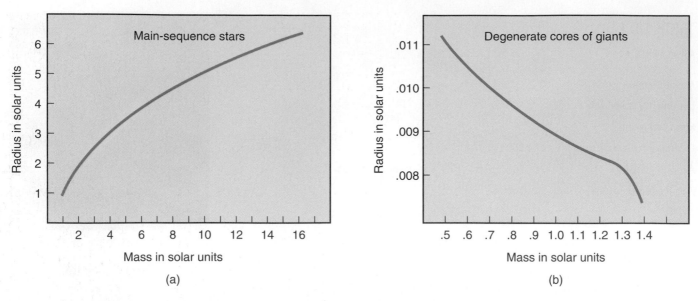

Figure 14-11
(a) In the case of normal, main sequence stars, the more massive the star is, the larger is its radius. (b) In the case of the degenerate core of a red giant, however, the more massive the core is, the *smaller* is its radius.

electron degeneracy The state of a gas in which its electrons are packed as densely as nature permits.

As a moderately low mass star becomes a red giant, some unusual conditions start to develop in its core. Under the increasing temperatures and pressures, the free electrons in the core are forced to get very close to each other. Once the density of the core reaches a certain value, the electrons are so closely packed together that any additional compression violates a very important natural law. According to this law, two electrons cannot have completely identical parameters describing their state. In a simple analogy, if we are playing a game of musical chairs, with each chair corresponding to a specific state for the electrons, only one electron is allowed to occupy each chair. As a result, the electrons produce a large pressure that resists any attempt to compress them further. At this point the core stops contracting. The matter of the core is now said to be ***degenerate,*** and the pressure that supports it against gravity is called *electron degeneracy pressure.*

When a normal gas is heated, its pressure increases and the gas expands. In degenerate matter, pressure does not depend on temperature; it depends only upon density. The properties of degenerate matter lead to a strange phenomenon. Refer to the graph in **Figure 14-11a**, which shows the relationship between the radius and the mass of a main sequence star. As you would expect, the more massive the star is, the larger is its radius. However, the degenerate core of a red giant does not act this way. Figure 14-11b shows what happens in this case. The more massive the core is, the smaller is its radius! This is very different from our experience with nondegenerate—regular—matter, and it has a profound influence on the next stage of the star's evolution.

As the hot, degenerate core of a red giant radiates its thermal energy outward, hydrogen continues to fuse to helium in a shell around the core. The hydrogen-fusing shell gradually works its way outward, all the time dumping its helium "ashes" onto the core. Because of the strange relationship between the mass and the radius of a degenerate core, the increased mass falling onto the core causes the core to shrink further and further. This, in turn, causes the core's temperature to continue to rise. Finally, the temperature of the red giant's core reaches a critical value, calculated to be about 100,000,000 K. At this point, shown as point *B* in Figure 14-10, helium nuclei begin to combine, forming carbon. The first steps in helium fusion reactions that

Table 14-2

Helium Fusion Reactions That Occur in Red Giants

Reaction	Explanation
$^4_2H + {}^4_2H \rightarrow {}^8_4Be + \gamma$	Two helium nuclei combine to produce a beryllium-8 nucleus and a gamma ray (γ).
$^4_2H + {}^8_4Be \rightarrow {}^{12}_6C + \gamma$	Another helium nucleus combines with the beryllium nucleus to form a carbon-12 nucleus and another gamma ray.
After sufficient carbon has been generated by the triple-alpha process, carbon nuclei can capture alpha particles to produce oxygen.	
$^4_2H + {}^{12}_6C \rightarrow {}^{16}_8O + \gamma$	Carbon in the core fuses with helium and produces oxygen and another gamma ray (and more energy).

produce carbon and oxygen from helium are shown in more detail in **Table 14-2** and **Figure 14-12.** Helium fusion does not occur at lower temperatures because helium nuclei repel each other with a much stronger force than hydrogen nuclei do, since each helium nucleus contains two protons, compared to only one proton for a hydrogen nucleus. At higher temperatures, helium nuclei move faster and therefore can get close enough for fusion to occur.

> A helium nucleus is also known as an *alpha particle*, and the conversion of three helium nuclei into a carbon nucleus is called the *triple-alpha process.*

Helium fusion does not proceed smoothly, however. Degenerate matter is a good conductor of thermal energy, so the additional energy produced by the new nuclear reactions spreads rapidly throughout the core. The helium in the core heats up and as a result helium fusion reactions occur faster. However, the pressure in the core does not change because the core is degenerate, and the electron degeneracy pressure does not depend on temperature. As a result, the core cannot cool by expanding and the continuously increasing core temperature causes the helium to fuse faster. The core "ignites" quickly and violently, a process known as the **helium flash.**

> **helium flash** The runaway helium fusion reactions that occur in the core during the evolution of a red giant.

You might think that the violence of the helium flash would have a drastic effect on the outer portion of the star. After all, the luminosity generated by the core during a helium flash can reach 100 billion times the luminosity of our Sun, comparable to the luminosity of our entire Galaxy. However, this huge energy release lasts for only a few seconds and most of the energy never makes it to the surface. Nevertheless, computer models show that the helium flash changes the core drastically. Most of the energy released during a helium flash goes into heating up the core and destroying the electron degeneracy, thus restoring the stellar thermostat. The energy that escapes the core is absorbed by the layers overlying the core, which are opaque, possibly causing some mass from the surface of the star to be lost.

After the helium flash occurs, signaling the start of helium fusion in the core and providing the star with a new energy source, the star contracts slightly and its surface temperature increases. You might find this to be a very counterintuitive result. Since the core is not degenerate anymore, it expands like a normal gas. As a result, the temperature around the core decreases, and this cooling slows down the hydrogen fusion reactions in the shell around the core. Since this shell provides most of the red giant's luminosity, the resulting decrease in energy output allows the star's outer layers to contract and heat up. This stage in the evolution of the star is shown in Figure 14-10 as the movement from point *B* to point *C*.

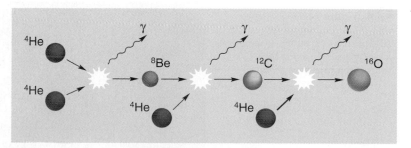

Figure 14-12
This diagram shows schematically the steps that occur as helium fuses into carbon in the core of a red giant.

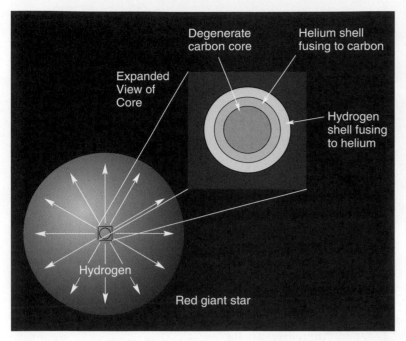

Figure 14-13
When a red giant reaches the stage where it has a carbon core, the heat from that shrinking core ignites helium fusion in a shell around it. At the same time, hydrogen is fusing in a second shell beyond the first. This activity occupies very little of the volume of the star.

The by-products of helium fusion reactions are carbon and oxygen nuclei. Our Sun will take about 100 million years to convert all the helium in its core to carbon and oxygen. At this point, since there is no energy source in the core, the core contracts until the electron degeneracy pressure stops the contraction again. As a result of the contraction, the temperature in a thin shell surrounding the core increases to the point that helium fusion reactions begin in this shell. While all this is happening close to the core, hydrogen fusion continues in a shell around the helium shell. The star now enters a new red giant phase. **Figure 14-13** shows the structure of a star like our Sun during this stage of its evolution. During this time, the star evolves from point *C* to point *D* on the H-R diagram of Figure 14-10. Its size is now even larger than before. When our Sun reaches this stage, about 5 billion years from now, it will have expanded to enfold Mercury, Venus, and the Earth (**Figure 14-14**). Its luminosity will reach 10,000 times its current luminosity.

What will happen to the Earth? As the outer layers of the red giant approach Earth, our planet will find itself speeding through the hot gas of the Sun's atmosphere. Earth's atmosphere will be ripped away. The planet will be slowed by friction and will begin falling toward the Sun's center. Meanwhile the Earth's crust will melt and then vaporize. Our planet will become one with the Sun. Mars and the Jovian planets will mostly evaporate away.

In this section we concentrated on stars with mass less than about 2 solar masses. Computer simulations indicate that stars more massive than about 2 solar masses do not experience a helium flash. The cores of these stars get hot enough toward the end of hydrogen burning that they make a smooth transition to helium burning without becoming degenerate. When their supply of helium is exhausted, however, their carbon core shrinks and becomes degenerate just as it does for lower mass, Sun-like stars. Therefore, in both cases the interior of the new red giant has the structure of Figure 14-13.

Stellar Pulsations

During its evolution, a star continuously tries to remain in equilibrium. As its outer layers contract and expand, responding to the changes occurring in the core, the star's luminosity changes drastically, and these changes are often periodic. Astronomers have been able to deduce this behavior by using spectroscopic observations and the Doppler effect, and they have catalogued more than 20,000 pulsating stars, with periods that range from a few seconds to hundreds of days. Two types of pulsating stars are especially important to astronomers because they can be used as distance indicators. In Chapter 12 we discussed one of these groups, the Cepheid variables, which are pulsating stars with periods between 1 and 100 days. In this section we explain the reason for the pulsations for both the Cepheids and another important group of distance indicators, the RR Lyrae variables.

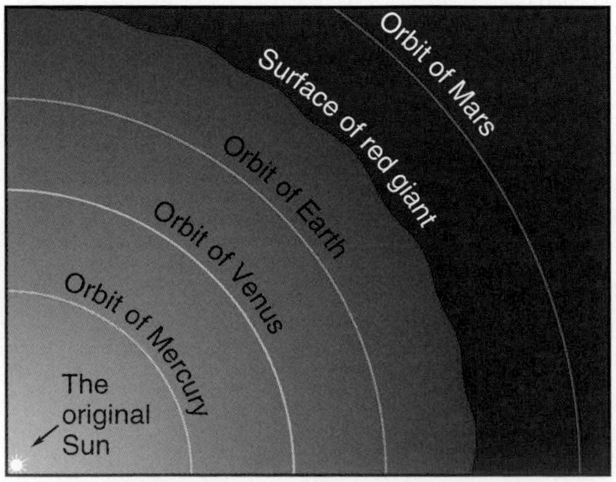

As for any periodic phenomenon, a particular mechanism is responsible for the oscillations of a star's outer layers. When helium

Figure 14-14
When the Sun becomes a red giant, it will expand until its surface is somewhere between Earth and Mars. Compare this to its present size.

gets compressed at the outer layers of a Cepheid, the compression ionizes the helium atoms instead of increasing their temperature. Since the resulting gas of ionized helium is opaque, the layers trap heat from inside the star. This increases the pressure under the layers, forcing them to move outward. As a result of the expansion, the outer layers cool, the helium atoms are not ionized any more, the helium gas in the layers becomes transparent, and the trapped energy is released. Finally, the outer layers fall inward, compressing the helium and starting the cycle again. The conditions necessary for such pulsations to occur exist in a narrow strip on the H-R diagram, called the ***instability strip*** (Figure 14-15).

After helium fusion reactions begin in the core, a *massive* star moves across the middle of the H-R diagram. As the star crisscrosses the red-giant region of the H-R diagram, it can become unstable and pulsate while inside the instability strip, becoming a Cepheid variable. After the helium flash stage, *low mass stars* also crisscross the instability strip, but through its lower end.

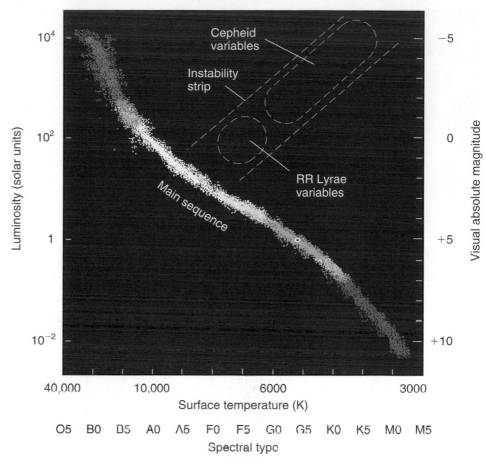

Figure 14-15
A section of the instability strip above the main sequence. Only two types of variable stars are shown. A star passing through this strip as it evolves becomes unstable and pulsates.

Some of these low mass stars become variable stars with periods shorter than a day and approximately constant luminosity (100 times that of our Sun). These low mass variables are called ***RR Lyrae variables.*** The relationship between the period and luminosity of RR Lyrae stars allows astronomers to find distances to globular clusters, in the same way that Cepheids are used to find distances to other galaxies.

In some cases the pulsations of a star can be so large that the speed with which the outer layers are moving outward is larger than the escape speed from the star. As a result, the star loses its outer layers to outer space, contributing to the recycling of interstellar material through the processes of star formation and stellar evolution.

instability strip A region of the H-R diagram where pulsating stars are found.

RR Lyrae variable Pulsating stars with periods shorter than a day.

Mass Loss from Red Giants

During its main sequence life, our Sun continually sheds material as the solar wind. Orbiting X-ray and ultraviolet telescopes such as the *High Energy Astrophysical Observatories* and the *International Ultraviolet Explorer* have detected emission spectra from around other main sequence stars, indicating that stars commonly have hot chromospheres and coronas just as our Sun does. It seems safe to assume that they also have stellar winds like the Sun. The solar wind carries away about 10^{-14} of the Sun's mass each year. This is an extremely small amount, for at this rate, the Sun will lose only 0.01% of its mass in 10 billion years. Mass loss is much more important for red giants.

We are able to detect a slight Doppler shift in emission lines from gas near red giants. Based on the amount of the Doppler shift, astronomers conclude that the gas is being blown off red giants at speeds between 10 and 20 kilometers/second, and that a

Figure 14-16
Star HD65750, a red supergiant, is losing material at a high rate. It is surrounded by a reflection nebula.

The fluorescence produces an emission spectrum, and the green color is due primarily to ionized oxygen. (Recall the fluorescence of a comet's coma and tail.)

planetary nebula The shell of gas that is expelled by a low mass red giant near the end of its life.

typical red giant loses about 10^{-7} solar masses per year. **Figure 14-16** shows mass loss from HD65750, a giant star surrounded by a reflection nebula.

One reason for the greater mass loss in red giants is that these stars are so much larger than Sun-like stars that they exert less gravitational force on material near their surfaces. In addition, the helium-burning process that provides much of the energy for red giants is unstable and varies greatly in response to small changes in temperature. As a result, a red giant undergoes instabilities and pulsations as helium fusion increases and decreases. In fact, during a portion of a red giant's life, it is common for the star to pulsate with periods of less than one day, as we discussed in the previous section.

Planetary Nebulae

The chapter opening figure and **Figure 14-17** show photographs of objects called *planetary nebulae.* Planetary nebulae were given that name decades ago because, when viewed in a small telescope, many of them have a blue-green color that resembles the color of Uranus and Neptune viewed through similar telescopes. In fact, planetary nebulae have nothing to do with planets. Spectroscopic analysis reveals—again by the Doppler effect—that these nebulae are made up of material moving outward from a very hot, low mass central star. Many planetary nebulae appear to be doughnut-shaped, but this is because we are looking through a shell of material when we view them, as explained in **Figure 14-18.** The material glows because ultraviolet radiation from the central hot star causes it to fluoresce.

At least two models attempt to explain planetary nebulae:

- One model holds that the pulsations in the core of a red giant continue to increase in intensity until the outer layers of the star also become unstable. With each pulse, part of the star's envelope is blown away. After perhaps 1000 years—a short time by stellar standards—the entire outer portion of the star will have been ejected, forming a shell around the original core.

- A second model, now gaining in favor, holds that the stellar winds emitted from a dying star occur in definite stages. At the earliest stage, the outer layers of the star are blown off as a slow, dense wind at speeds of perhaps 90,000 kilometers/hour. This finally leaves behind a hot, dense star that emits high-energy radiation. Now the character of the wind changes, so that it becomes a low-density but very fast wind, with speeds up to 18 million kilometers/hour. When this second wind

Figure 14-17
Planetary nebulae consist of gas ejected from a low mass red giant. (a) This *HST* image of the Ring nebula (M57), in the constellation Lyra, is in approximately true colors: blue for very hot helium; green for ionized oxygen; and red for ionized nitrogen. It seems as if we are looking down a barrel of gas ejected from a dying star (with surface temperature of about 120,000 K) thousands of years ago. The nebula is about 2000 light-years away and 1 light-year in diameter. (b) The Helix nebula (NGC7293), 450 light-years away, in the constellation Aquarius. It glows red in the light of hydrogen and nitrogen atoms energized by the ultraviolet light from the star. The rings are about 1.5 light-years across.

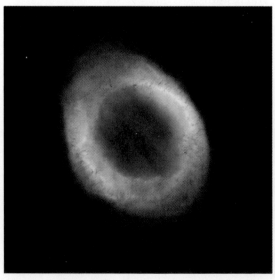

(a)

(b)

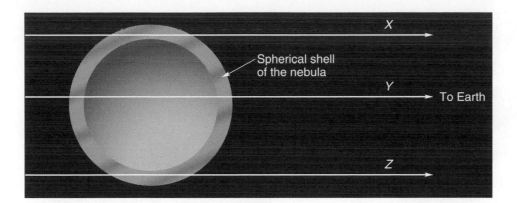

Figure 14-18
Three lines of sight through the spherical shell of a nebula. Lines *X* and *Z* pass through much more of the glowing nebula than does line *Y*; thus the nebula appears brighter here.

catches the first, it pushes the first forward, causing a region where there is a very dense wave. (To picture this, imagine that a group of walkers is being overtaken and run into by a group of runners. Where the two groups are colliding, the people are bunched very close together as the runners push the walkers into one another.) In addition, the intense radiation causes that dense area to glow, and this is the shell that we see as a planetary nebula. In the next chapter, when we discuss supernovae, we will see that some hourglass-shaped planetary nebulae have been discovered that can be explained using this model.

Whichever model is correct (or perhaps a combination of the two will be found to be the best explanation for the phenomenon), not all planetary nebulae appear as rings. The Dumbbell nebula of **Figure 14-19a** shows just a hint of a spherical structure. The *HST* image of a planetary nebula in part (b) reveals that the initial ejections from the star were periodic and fairly uniform (as evidenced by the concentric spheres). The bright inner regions of the nebula must have been produced by irregular ejections of material, and dense clouds of dust condensed from the ejected material. The dense clouds are shown here in yellow. The "twin jet" nebula of **Figure 14-20a** is a clear example of a bipolar planetary nebula. The measured speed of the ejected gas is more than 720,000 miles/hour. Observations show that the nebula's size increases with time, suggesting that it formed just 1200 years ago. The shape of the Hourglass nebula shown in Figure 14-20b is thought to result from the expansion of a stellar wind

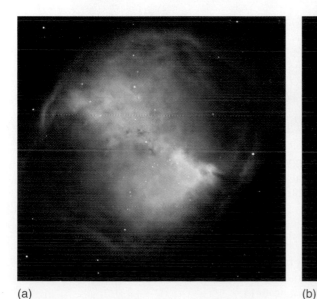

(a)

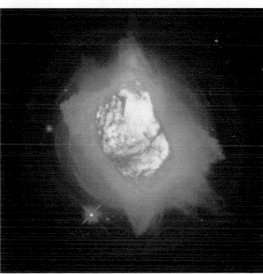

(b)

Figure 14-19
(a) A three-colored composite image of the Dumbbell Nebula (M27) taken by ESO's VLT. The nebula is about 1200 light-years away. (b) This *HST* image of NGC7027 is a composite of images in the visible and infrared. The central part is not solid, as it may appear in the false-color image.

Planetary Nebulae

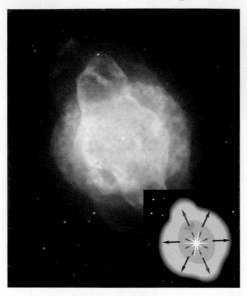

NGC3918 features an inner balloon of gas starting to break through the spherical outer envelope in two places.

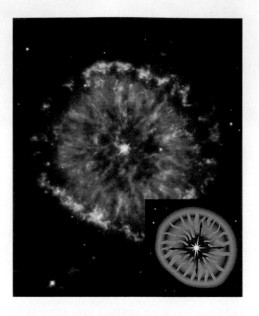

NGC6751 is 6500 light-years away. Hot blue gas forms a rough ring around the central star, pushing the cooler orange gas into an outer ring with streamers toward the central star.

Hubble 5 is a butterfly or bipolar nebula. The lobes are fast winds of gas extending beyond the disk formed early in the star's death.

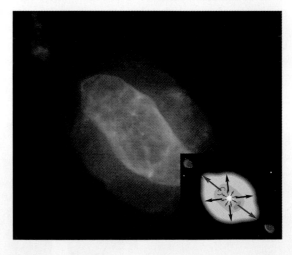

NGC7009 has a barrel-shaped envelope of material. Regions of low-density gas have shot out through each end of the barrel, glowing faintly.

QUICK FACTS

- Planetary nebulae have nothing to do with planets, but are clouds of glowing gas and dust thrown off from moderately massive stars changing from red giant stage to white dwarfs. The shapes we see depend on how the outflows interact with each other and with the star's magnetic field.

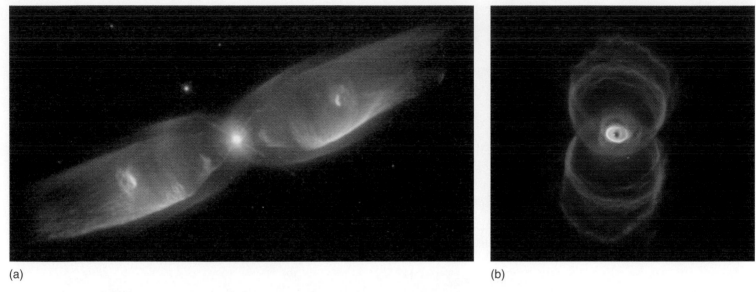

(a) (b)

Figure 14-20
(a) M2-9 is an example of a "butterfly" or bipolar planetary nebula. The central star is known to be in a very tight binary system. A disk that surrounds both stars is seen in some *HST* images and has a size 10 times the diameter of Pluto's orbit. The nebula is 2100 light-years away in the constellation Ophiuchus. In this *HST* image, red corresponds to neutral oxygen, green to singly ionized nitrogen, and blue to double-ionized oxygen. (b) This is a composite of three *HST* images of the Hourglass nebula (MyCn18). Red, green, and blue represent ionized nitrogen, hydrogen, and doubly ionized oxygen, respectively. The nebula is 8000 light-years away.

that moves fast inside a slowly expanding cloud (which is denser close to its equator than near its poles).

Only a few thousand planetary nebulae have been observed. Although this may seem a large number, it is very small compared to the number of stars, and this fact indicates that planetary nebulae do not last long. However, it is possible to make direct observations of cosmic evolution. **Figure 14-21** shows two examples of ***proto-planetary nebulae,*** which represent the transition phase between the last stages of a red giant star's life and the early stages of a planetary nebula after the star has ejected most of its mass.

proto-planetary nebula An object in transition between the last stages of a red giant star's life and its planetary nebula phase.

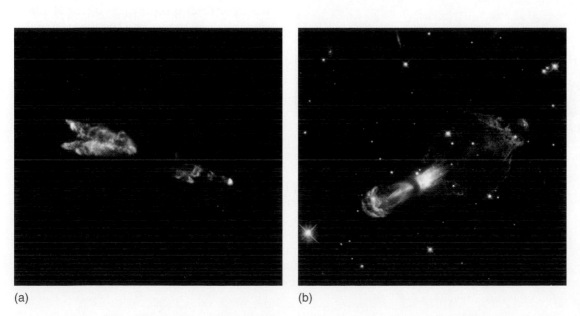

(a) (b)

Figure 14-21
(a) The proto-planetary nebula CRL 618. It evolves so rapidly that we can watch through the eyes of the *Hubble* telescope the hatching of a heavenly butterfly from its dusty cocoon. Red represents singly ionized sulphur, green represents neutral hydrogen, and blue-green comes from neutral oxygen. (b) The Calabash proto-planetary nebula. It is also known as the Rotten Egg nebula because of the large amounts of sulphur compounds present, which would produce an unpleasant smell if we could smell in space. The gas (shown in yellow) has a speed of up to 1.5 million kilometers/hour, and when it rams into the surrounding medium it forms shock-fronts on impact, which heat the surrounding gas. Light from hydrogen and ionized nitrogen arising from these shocks is shown in blue.

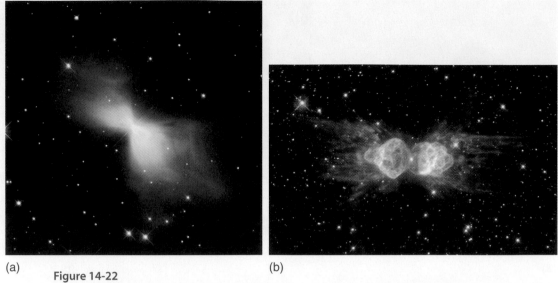

(a)   (b)

Figure 14-22

(a) The Boomerang nebula. It is a young nebula, in the constellation Centaurus, about 5000 light-years from us. The bow-tie shape of the nebula is created by a 500,000 kilometer/hour wind that results in a solar mass of material being lost every 1000 years (10 to 100 times more than other similar objects). The rapid expansion enables the nebula to be the coldest known region in the universe, just one degree warmer than absolute zero; as such, this nebula is the only object found so far that has a temperature lower than the background radiation. (b) The Ant nebula. This *Hubble* image shows the "ant's" body as two lobes protruding from the dying star.

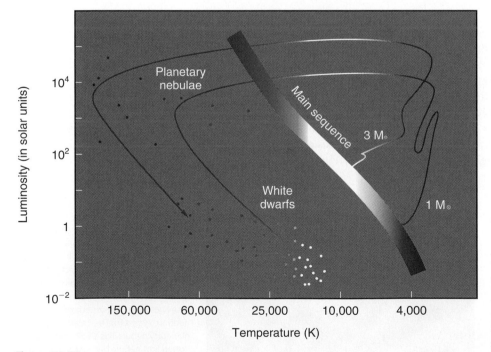

Figure 14-23

This H-R diagram has been expanded to the left—to higher temperatures—in order to include stars that are passing through the planetary nebula stage. The evolutionary tracks of a one-solar-mass star and a three-solar-mass star are shown.

The proto-planetary nebula phase is thought to last between a few hundred and 1000 years. Observations by the *Infrared Space Observatory* have found evidence that complex organic molecules form rapidly during this phase. It is likely that proto-planetary nebulae are huge factories of organic molecules. The images in Figure 14-21 show that the ejected material forms complex shapes and symmetries, implying that some powerful processes are at work. It is thought that magnetic fields probably play a crucial role in creating these shapes and symmetries; support for this was provided by observations in late 2002, which showed that the magnetic field close to a number of aging stars is 10 to 100 times stronger than our Sun's. **Figure 14-22** provides two more examples of the different shapes found among planetary nebulae. The material that we see around the central star quickly dissipates into space and becomes part of the interstellar medium.

As the nebula dissipates, the hot, bright core of the star begins to peek through. This core may have a temperature of 100,000 K, and its appearance causes the star to shift its position on the H-R diagram. To include this stage on an H-R diagram, the diagram must be extended toward higher temperatures than have been included in previous diagrams. **Figure 14-23** shows that the star moves far to the left. The core remains at its very luminous stage for a relatively short time; then it quickly moves down the H-R diagram and collapses to become a white dwarf.

14-7 White Dwarfs

White dwarfs were discovered long before astronomers predicted their existence. In 1844 the German astronomer Friedrich Bessel (1784–1846) noticed that the star Sirius (in the constellation Canis Major) wobbled back and forth slightly. He hypothesized that an invisible companion orbiting Sirius causes the wobble; that is, Sirius and another star form a binary pair. Bessel's calculations showed that the unseen companion (called Sirius B) has a mass about the same as Sirius (now called Sirius A). Eighteen years later—and sixteen years after Bessel had died—the American telescope maker Alvan G. Clark (1804–1877) discovered the companion of Sirius and, in doing so, became the first person to see a white dwarf. **Figure 14-24a** shows Sirius B as it appears in a much larger telescope. Part (b) shows seven white dwarfs in M4, a compact cluster of stars.

More than 300 stars have been identified as white dwarfs, but none is bright enough to be seen by the unaided eye. Astronomers estimate that about 10% of all stars are white dwarfs, but this is a rough estimate, since white dwarfs are difficult to see.

White dwarfs are the cores of red giants that remain after the outer layers of the original stars have blown away. White dwarfs that formed from low mass stars are made up of helium and carbon. White dwarfs formed from more massive stars contain oxygen, neon, and sodium and may contain atomic nuclei as massive as iron. Because white dwarfs consist of degenerate matter, the more massive a white dwarf is, the smaller it is. (Refer to the graph of Figure 14-11b.) As a white dwarf gets older, its temperature and luminosity decline without a significant change in its size, for electron degeneracy prevents it from contracting further.

The Chandrasekhar Limit

Even though electron degeneracy supports the white dwarf against collapsing completely, there is a limit to the amount of pressure that degenerate electrons can withstand. In 1930, a 19-year-old student on his way from his native India to graduate school in England calculated this limit, showing it to be 1.4 solar masses. Subrahmanyan Chandrasekhar (or, as he is known to astronomers, "Chandra") was awarded a share of the 1983 Nobel Prize in Physics for his discovery, and the limit is known as the *Chandrasekhar limit.*

Recall that a binary system in which the motion of one star reveals the presence of the other is called an astrometric binary. Bessel discovered, also in 1844, that Procyon, the brightest star in Canis Minor, is an astrometric binary.

Chandrasekhar limit The limit to the mass of a white dwarf star, above which it cannot be supported by electron degeneracy and cannot exist as a white dwarf.

Young people have made many major breakthroughs in science; Newton (age 23 in 1665) and Einstein (age 26 in 1905) are famous examples.

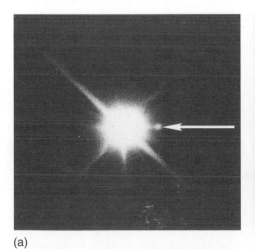

(a)

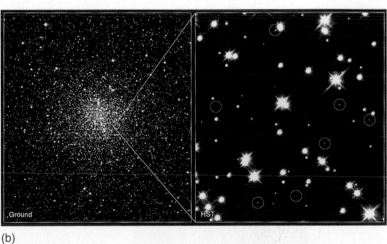

(b)

Figure 14-24
(a) Sirius is revealed to be two stars in this telescopic view. The spikes that radiate from Sirius A are caused by diffraction around mirror supports within the telescope. (b) The image at left is a cluster of stars 7000 light years away. The *HST* image at right is a small part of this cluster, only 0.63 light-years across. It contains seven white dwarfs (inside the blue circles) among the other much brighter stars.

Chandra's calculations showed that if a white dwarf becomes more massive than 1.4 solar masses, the pressure due to gravity at its surface is greater than the maximum pressure a degenerate gas can support. This means that white dwarfs must have masses less than 1.4 times the Sun's mass. Main sequence stars with masses of up to about 4 solar masses (which we have classified as moderately low mass stars) can end up as white dwarfs only because they lose mass during the red giant and planetary nebula stages. Stars more massive than these stars retain cores whose mass exceeds the Chandrasekhar limit and cannot form white dwarfs. Such moderately massive and massive stars end their lives in a dramatically different way than do less massive stars. We will discuss the more massive stars in the next chapter.

Characteristics of White Dwarfs

White dwarfs have been observed with surface temperatures from 4000 to 85,000 K, but computer models predict that it is possible for them to have even higher temperatures. The masses of white dwarfs range from perhaps 0.02 solar masses up to 1.4 solar masses. Since the typical white dwarf is comparable in size to the Earth, the material of these stars is tremendously dense. The density of a white dwarf is about 10^6 grams per cubic centimeter. A teaspoon of white dwarf material would weigh two tons! **Table 14-3** lists the properties of a "typical" white dwarf.

When the Sun becomes a white dwarf, its mass will be about 0.6 of its present mass. The remainder of its material will have been puffed away during its red giant stage and blown away during its planetary nebula stage. It will be nearly as small as the Earth (**Figure 14-25**), with luminosity about a tenth of its current luminosity. As time passes, our Sun's luminosity will keep decreasing and it will simply fade away.

To get to the white dwarf stage, Sun-like stars must go through the following stages: protostar, main sequence star, red giant, and planetary nebula. Protostars produce energy from gravitational contraction. Main sequence stars produce energy from nuclear fusion. Red giants produce energy from gravitational contraction of their cores (until the cores become degenerate) and from fusion within shells around the cores. White dwarfs, however, do not produce energy. They are hot because of leftover energy, and as they radiate this energy away, they get cooler. After billions of years, a white dwarf will have cooled enough that it no longer radiates in the visible region of the spectrum. It will appear on the H-R diagram at a position below and to the right of the bottom of the main sequence. As billions more years pass, it will cool further to become a ***black dwarf,*** the burned-out cinder of a once-proud star. It is unlikely that the universe is old enough for many, if any, black dwarfs to have formed.

black dwarf The theoretical final state of a star with a main sequence mass less than about 4 solar masses, in which all of its energy sources have been depleted so that it emits no radiation.

Figure 14-25
When the Sun becomes a white dwarf, it will be slightly larger than today's Earth and therefore much smaller than the present Sun. Its mass, however, will be 0.6 of what it is now.

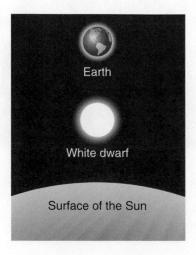

Earth

White dwarf

Surface of the Sun

Table 14-3
A Typical White Dwarf

Absolute magnitude	+11
Mass	0.8 solar mass
Diameter	10,000 km (3/4 Earth diameter)
Density	10^6 g/cm^3
Surface temperature	15,000 K
Surface gravity	400,000 times Earth's

Figure 14-26 reviews the evolutionary steps taken by very low mass and moderately low mass stars. All begin as protostars and end up as black dwarfs. The end state of more massive stars—the moderately massive and very massive categories—is even more exotic, and we will discuss it in the next chapter.

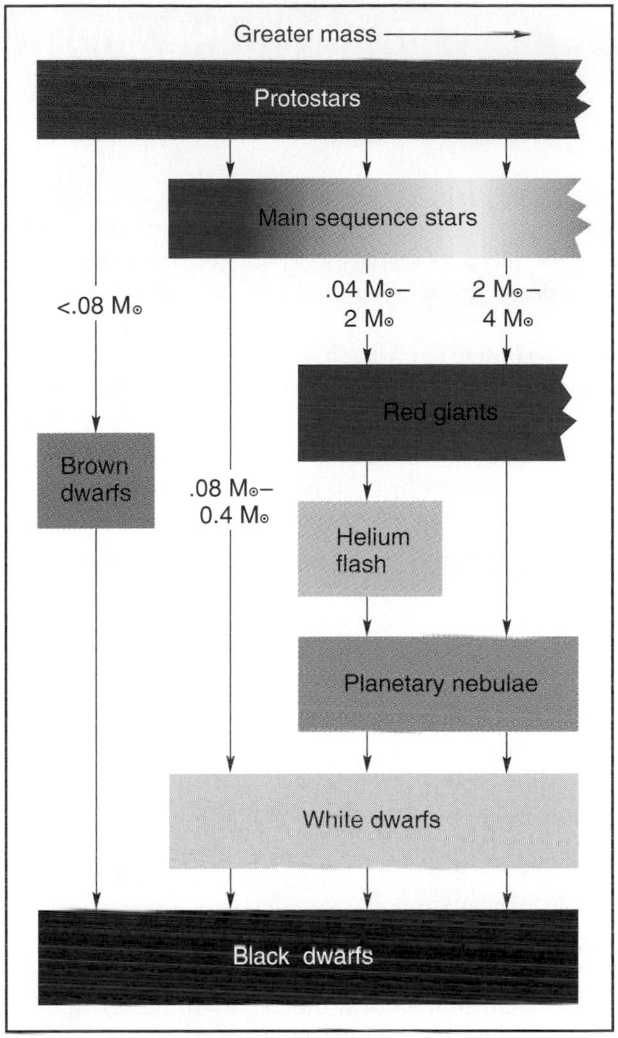

Novae

Ancient astronomers believed that the sky never changed. However, in Chapter 12 we showed that stars move and that the constellations gradually change shape. Other changes also occur in the heavens. In 1572, Tycho Brahe observed a "new star." His excitement at the discovery is revealed in his words in a nearby Advancing the Model box. We know now that the phenomenon Tycho observed, and similar phenomena observed by others since his time, are not actually new stars, but are a phenomenon associated with white dwarfs.

Consider a binary star system in which one star is more massive than the other. The more massive star will end its main sequence life before its smaller companion. The massive star will then become a red giant, and, if it is in our moderately low mass category, it will shed its outer portion and become a white dwarf. This will leave a white dwarf in orbit with a main sequence star, which then, in its turn, ends its fusion-burning life and begins to grow into a red giant.

At some point, the material of the outer portion of the new red giant is attracted to the white dwarf with more force than is exerted from its own core. This material is pulled away from the giant star. Although some might fall directly onto the white dwarf, analysis shows that most of the material pulled from the red giant will go into orbit around the white dwarf, forming an ***accretion disk*** (**Figure 14-27**). In the accretion disk, matter swirling around the white dwarf is heated as it falls inward, another example of gravitational energy being changed into thermal energy. Collisions within the disk cause its material—mostly hydrogen from the outer portions of the red giant—to fall inward onto the surface of the white dwarf. The hydrogen builds up there, becoming denser and hotter. When the temperature reaches about 10 million kelvin, hydrogen ignites in an explosive fusion reaction. The

Figure diagram

Greater mass →

Protostars

Main sequence stars

.04 $M_\odot$– 2 $M_\odot$ 2 $M_\odot$– 4 $M_\odot$

<.08 $M_\odot$

Red giants

Brown dwarfs

.08 $M_\odot$– 0.4 $M_\odot$

Helium flash

Planetary nebulae

White dwarfs

Black dwarfs

Figure 14-26
This diagram shows the evolutionary steps taken by the stars discussed in this chapter. Unless we include planets in the category, it is probable that no black dwarfs have ever formed, so their box could have been left off the chart. The boxes for the top three categories are open at the right to indicate that more massive stars also fall into these categories. Keep in mind that the limiting masses in each case are known only approximately.

In 1054 A.D., Chinese scholars recorded what they called a "guest star," a new star never before seen. In the next chapter we discuss this star, which was bright enough that it could be seen in the daytime.

accretion disk A rotating disk of gas orbiting a star, formed by material falling toward the star.

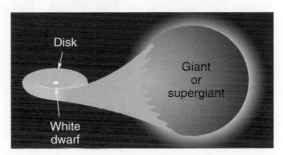

Figure 14-27
In a binary system composed of a white dwarf and a red giant, material from the growing giant will fall toward the dwarf and go into orbit around it.

ADVANCING THE MODEL

Tycho Brahe's Nova

Tycho Brahe had been taught that the heavens were perfect and unchanging. Therefore, when he saw a "new star"—a nova—he was very excited. (We know now that the "nova" observed by Tycho was actually a supernova.) His writings reveal his feelings:

In the evening, after sunset, when, according to my habit, I was contemplating the stars in a clear sky, I noticed that a new and unusual star, surpassing all the other stars in brilliancy, was shining almost directly above my head; and since I had, almost from boyhood, known all the stars of the heavens perfectly (there is no great difficulty in attaining that knowledge),

it was quite evident to me that there had never before been any star in that place in the sky, even the smallest, to say nothing of a star so conspicuously bright as this. I was so astonished at this sight that I was not ashamed to doubt the trustworthiness of my own eyes. But when I observed that others, too, on having the place pointed out to them, could see that there really was a star there, I had no further doubts. A miracle indeed, either the greatest of all that have occurred in the whole range of nature since the beginning of the world, or one certainly that is to be classed with those attested by the Holy Oracles.

Table 14-4	
A Typical Nova	
Luminosity increase	10,000 times
Absolute magnitude	-8, or 10^5 Suns
Time to brighten	A few days
Time to dim	6 months to one year

nova (plural **novae**) A star that suddenly and temporarily brightens, thought to be due to new material being deposited on the surface of a white dwarf.

There are various types of novae, including *classical novae* and *dwarf novae*, but these differences are not important here.

explosion, which occurs *only* on the surface of the white dwarf, blows off the outer layers of the white dwarf. Although the shell contains only a relatively tiny amount of mass (perhaps 0.0001 solar masses), it can cause the white dwarf to become 10,000 to 100 million times brighter (10 to 20 magnitudes) within a few days.

During Brahe's time, people thought such a newly brightened star was a new star and called it a ***nova,*** the Latin word for *new*. In actuality, the star is not new, but is simply the sudden brightening of an old star. **Figure 14-28** shows one of the brightest of recent novae, and **Table 14-4** contains data for a typical nova.

Because so little material is blown off during a nova, the explosion does not disrupt the binary system. The companion star soon resumes transferring matter to the white dwarf. Depending on how fast the hydrogen is transferred, fusion can be re-ignited as quickly as a few months later, or 10,000 to 100,000 years may be required for a recurrence of the nova. Because of their repeating nature, these novae are called recurrent novae.

Figure 14-28
These photos of Nova Cygni 1975 show its dimming from magnitude 2 at maximum light to magnitude 15.

14-8 Type I Supernovae

White dwarfs consist mostly of degenerate carbon (but they probably contain heavier elements as well). Imagine a white dwarf in a binary system accreting material from its companion. When the accretion brings the mass of the white dwarf above the Chandrasekhar limit, electron degeneracy can no longer support the star, and it collapses. The collapse raises the core temperature enough that carbon fusion suddenly begins. This is somewhat similar to the helium flash during the red giant stage, except that now carbon rather than helium is the fuel. Runaway fusion reactions result in carbon fusing to elements as massive as iron. The fusion reactions increase the temperature in the interior but the pressure remains the same, since the core matter is degenerate. As a result, the white dwarf cannot expand and cool. As the temperature increases, so does the rate of the fusion reactions, in a catastrophic fashion. At some point, the temperature becomes so high that the electrons are no longer degenerate; they start behaving like an ideal gas, resulting in an explosive expansion of the white dwarf. Whereas a red giant undergoing a helium flash has a surrounding envelope of gases to control its explosion, the white dwarf doesn't. The star explodes completely.

The destruction of a white dwarf in this manner is called a ***supernova.*** The name is unfortunate, for a supernova is not just a "super nova." A nova might reach an absolute magnitude of −8 (about 100,000 Suns), but a supernova attains a magnitude of −19 (10 *billion* Suns). When a supernova is observed in another galaxy, it may shine brighter than all the rest of the galaxy combined.

Astronomers classify supernovae into two categories, *Type I* and *Type II,* depending upon their spectra. The spectrum of a Type II supernova contains prominent hydrogen lines, but the spectrum of a Type I supernova does not contain any hydrogen lines. Furthermore, Type I supernovae are divided into three subclasses: Ia, Ib and Ic, again depending upon their spectra. Astronomers think that *massive* stars that have lost different proportions of their outer layers before exploding cause Types Ib and Ic. These supernovae are found only near regions of recent star formation. However, Type Ia supernovae are found even in regions where there is no star formation. Since white dwarfs contain no hydrogen, they are thought to be responsible for Type Ia supernovae. Type II supernovae result from the explosion of a single star; we will discuss them in the next chapter.

A Type Ia supernova reaches maximum brightness after a few days. It fades quickly for about a month and then declines in brightness more gradually until it dissipates in about a year. **Figure 14-29** shows the light curve of a Type Ia supernova. Theoretical models indicate that its energy (after the initial explosion) comes from ra-

supernova The catastrophic explosion of a star, during which the star becomes billions of times brighter

Photos of other supernova remnants appear in Chapter 15.

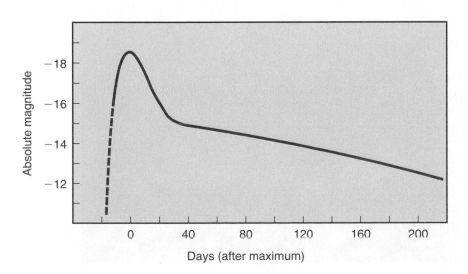

Figure 14-29
A Type Ia supernova, thought to be produced by the destructive explosion of a white dwarf, reaches peak brightness in a few days and then declines in brightness, first quickly and then more slowly. The gradual decrease is due to the decay of radioactive elements in the products of the explosion.

Figure 14-30
The left panel shows X-ray data of the supernova remnant DEM L71 taken by the *Chandra Observatory.* A hot, 10-million-K inner cloud of glowing silicon and iron is shown in aqua, surrounded by an outer blast wave. The right panel shows this blast wave at optical wavelengths. After a white dwarf explodes, the expanding material drives a shock wave in front of it and into the surrounding interstellar medium (the bright outer rim). The pressure behind this shock wave drives another shock wave inward that heats the expanding material (the cloud shown in aqua). The size and temperature of this remnant indicate that it is several thousand years old.

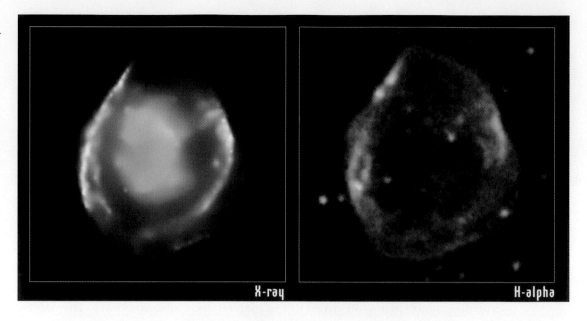

dioactive decay of nuclei produced in the explosion. The light curves and spectra of all Type Ia supernovae are very similar. This is surprising given the varied range of stars from which they start. It is likely that the slow process of accretion that brings these white dwarfs to the point of destruction erases most of the initial differences between the original stars. This similarity among the light curves of Type Ia supernovae is a powerful tool that astronomers use to trace the history of the expansion of the universe, as we will discuss in Chapter 18.

A Type Ia supernova remnant is shown in **Figure 14-30**. The X-ray spectrum of the remnant DEM L71 shows a high concentration of iron atoms relative to oxygen and silicon. The mass of the ejected material was found to be comparable to the mass of the Sun. These observations show that we are looking at the remnants of an exploded white dwarf.

What happens to the remains of a supernova? They disperse into space. **Figure 14-31** is a photograph of the Vela nebula, the remains of a Type II supernova that exploded

Figure 14-31
The Vela nebula is a supernova remnant.

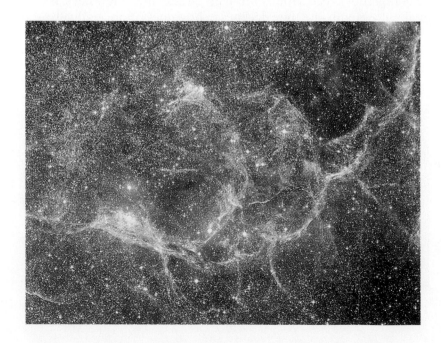

some 11,000 years ago. To calculate its age, astronomers determined the Doppler shift of the part of the remnants that is moving toward us and calculated when the explosion had to have occurred to produce the present speed. In Chapter 15 we will discuss Type II supernovae and describe examples of past and recent supernovae.

Conclusion

Mass is the single most important factor that determines a star's properties and how the star evolves. The least massive protostars never get hot and dense enough to join the main sequence and become hydrogen-fusing stars. These so-called brown dwarfs heat up from gravitational contraction during their protostar stage, but after that they have no heat source, so they cool to become dark, planet-size cinders. More massive protostars become stars on the main sequence, where they fuse hydrogen to helium (and, in more massive stars, to heavier elements). During their main sequence lives, stars become slightly more luminous and cooler, resulting in a small change in their position on the H-R diagram. The more massive a star is, the less time it takes to exhaust its fuel and end its main sequence life.

In stars massive enough to support fusion but with masses less than about 0.4 solar masses, their entire supply of hydrogen mixes as fusion takes place. Although the universe is not old enough for any of these stars to have had time to complete their main sequence lives, computer models show that when they run low on hydrogen, they will shrink and heat up until electron degeneracy sets a lower limit to their size. They will then live out their lives as white dwarfs until they, too, cool to burned-out cinders.

Moderately low mass stars fuse hydrogen in their cores without any intermixing with the outer portions. When these stars run low on fuel, they become brighter, larger, and cooler; that is, they become red giants. Their electron-degenerate cores heat up until they become hot enough to begin helium fusion. When the Sun and other stars of similar mass reach this point, a helium flash occurs; the stars then consume their helium quickly, getting hotter and dimmer for a time.

During the red giant stage, stars lose great amounts of matter from their outer layers because of intense stellar winds. At the end of the red giant stage, Sun-like stars blow away their outer shells and become beautiful planetary nebulae. Left behind when the nebula has dispersed into space is the star's white-hot degenerate core—a white dwarf. White dwarfs are dim and hard to see, but a white dwarf that is part of a binary system will advertise its presence as a nova if its companion, as it swells to become a red giant, leaks material onto the hot surface of the white dwarf.

When the mass of a white dwarf that is accreting matter from a companion becomes greater than the Chandrasekhar limit, the star explodes in a brilliant supernova. This completely destroys the star, dispersing its radioactive remnants through space.

Stars with masses greater than about 4 solar masses end the red giant portion of their lives in a different way than stars of lower mass. In the next chapter we describe the behavior of these stars, including the supernovae that they produce.

Study Guide

1. The length of a star's main sequence lifetime is determined by the star's
   A. carbon content.
   B. distance from the center of the galaxy.
   C. surface temperature.
   D. mass.
   E. spectral type.

2. Why do stars of larger mass live longer on the main sequence than stars of lesser mass?
   A. The massive stars have more hydrogen fuel.
   B. The massive stars use their fuel more slowly.
   C. The massive stars go through many stages of fusion.
   D. [More than one of the above.]
   E. [None of the above; the statement in the question is false.]

3. If the rate of hydrogen fusion within the Sun were somehow to increase, the core would
   A. collapse and the Sun would grow cool.
   B. collapse and heat up further.
   C. expand and therefore tend to slow the fusion.
   D. expand and therefore increase in temperature.
   E. stay the same size but become hotter.

4. Why does hydrogen fusion occur only in a star's center?
   A. Only near the center is there enough heat and pressure.
   B. Only near the center is there enough hydrogen that is not mixed with other elements.
   C. Only near the center is the speed of light favorable for the reaction to occur.
   D. Heat is transferred down to the center during the main sequence life of the star.
   E. [The statement is false; fusion occurs throughout the star's volume. This is what causes the surface to be bright.]

5. Why do massive stars run out of hydrogen in their cores faster than less massive stars?
   A. Their hydrogen fuses faster because of greater pressure.
   B. There is less hydrogen in their cores.
   C. The cores of less massive stars contain a greater percentage of helium, which slows hydrogen fusion.
   D. The cores of less massive stars contain a lesser percentage of helium, which slows hydrogen fusion.
   E. [The statement is false; more massive stars do not run out of hydrogen faster than stars of less mass.]

6. The Sun will at some time in the future become
   A. a red giant.
   B. a white dwarf.
   C. a black dwarf.
   D. [All of the above.]
   E. [None of the above.]

7. Which choice lists the stages in a star's life in correct order?
   A. Main sequence, protostar, white dwarf, red giant.
   B. Protostar, red giant, main sequence, white dwarf.
   C. Protostar, main sequence, white dwarf, red giant.
   D. White dwarf, protostar, main sequence, red giant.
   E. Protostar, main sequence, red giant, white dwarf.

8. Red giants are more luminous than white dwarfs because
   A. red giants are hotter.
   B. red giants are closer.
   C. red giants are larger.
   D. [All of the above.]
   E. [None of the above; the statement is not true.]

9. Why are all known white dwarfs relatively close to the Sun?
   A. White dwarfs are formed only in our neighborhood of the Galaxy.
   B. Light from distant white dwarfs has not yet reached the Earth.
   C. No white dwarfs are bright enough to be seen at great distances.
   D. Light from distant white dwarfs is too redshifted to be seen.
   E. [The statement is false; white dwarfs are seen at all distances from the Sun.]

10. What is the difference between the Sun and a one-solar-mass white dwarf?
    A. The Sun is larger.
    B. The Sun has more hydrogen.
    C. They have different energy sources.
    D. [All of the above.]
    E. [None of the above.]

11. One of the causes for the phenomenon called a nova is
    A. the fusion of iron in the core of a massive star.
    B. the infall of material onto a neutron star from a white dwarf.
    C. the transfer of material onto a white dwarf in a double star system.
    D. the collapse of a protostar.
    E. the death of a massive star and the formation of a black hole.

12. The nuclear reactions in a star's core are kept under control so long as
    A. the star's luminosity depends on its mass.
    B. the pressure of the gas in the core depends on its temperature.
    C. the star's density depends on its mass.
    D. the star's mass depends on its temperature.

13. A planetary nebula is
    A. the vastly expanded shell of a dying star.
    B. a cloud of gas out of which stars form.
    C. a cloud of cold dust in space.
    D. the same as a white dwarf.
    E. a circular ring around a black hole.

14. Compared to a young star of the same mass, an older star contains
    A. more hydrogen.
    B. more elements heavier than hydrogen.
    C. about an equal amount of each element.
    D. more of every element.
    E. [No general statement can be made.]

15. The mass of a white dwarf is
    A. always greater than 4 solar masses.
    B. always less than 0.5 solar masses.
    C. less than the mass of its original main sequence star.
    D. greater than the mass of its original main sequence star.
    E. [Two of the above.]

16. A brown dwarf is
    A. the final fate of all stars.
    B. the final fate of stars like the Sun, but not of less massive stars.
    C. a stage of a star's life prior to the white dwarf stage.
    D. a stage of a star's life after the white dwarf stage.
    E. a warm starlike object that has too little mass to support fusion in its core.

17. When more material falls into a white dwarf, its size
    A. increases.
    B. remains the same.
    C. decreases.
    D. [Any of the above, depending upon the nature of the white dwarf.]

18. When the core of a star shrinks after hydrogen fusion stops,
    A. the core cools and the star expands.
    B. the core cools and the star contracts.
    C. the core heats and the star expands.
    D. the core heats and the star contracts.

19. The Chandrasekhar limit is a limit to a star's
    A. size.
    B. volume.
    C. density.
    D. mass.
    E. [Both A and B above.]

20. Some supernovae occur when the Chandrasekhar limit is exceeded in a white dwarf. In this case, the energy producing the explosion is
    A. chemical in nature.
    B. nuclear fusion.
    C. nuclear fission.
    D. degenerate electrons.
    E. gravity.

21. The primary source of energy that powers a main sequence star is
    A. gravity.
    B. chemical energy.
    C. nuclear fission.
    D. nuclear fusion.
    E. fossil fuel.

22. A star is considered to begin its main sequence life when
    A. it starts to collapse.
    B. nuclear reactions start.
    C. its protostar life begins.
    D. it begins to move off the main sequence.
    E. its planetary system has formed.

23. When a star "moves off" the main sequence to become a red giant,
    A. it moves across the sky toward regions where stars are hotter.
    B. it moves across the sky toward regions where stars are cooler.
    C. it stays in the same place in the sky, but becomes hotter.
    D. it stays in the same place in the sky, but becomes cooler.
    E. [Either A or B above, depending on the mass of the star]

24. When a star ends its protostar stage, what determines where it ends up on the main sequence?
    A. Its location in the sky.
    B. Its radial velocity.
    C. Its transverse velocity.
    D. Its mass.
    E. [Both B and C above.]

25. Which stars have longer lifetimes on the main sequence?
    A. The most-massive stars.
    B. The least-massive stars.
    C. Stars in binary systems.
    D. Stars in star clusters.

26. A planetary nebula occurs
    A. in a newly forming planetary system.
    B. in an old, dying planetary system.
    C. in a white dwarf near the end of its life.
    D. at the end of the life of a red giant.
    E. in the core of a white dwarf.

27. A white dwarf (compared to main sequence stars) has
    A. a very high average density and a high surface temperature.
    B. a very low average density and a high surface temperature.
    C. a very high average density and a low surface temperature.
    D. a very low average density and a low surface temperature.
    E. [No general statement can be made.]

28. What property of a star is most important in determining the stages of its evolution?

29. Why do stars of the lowest mass not become red giants?

30. What causes the core of a star to heat up after its hydrogen fusion ceases?

31. List the stages in the life of a one-solar-mass star.

32. What will happen to the Earth when the Sun becomes a red giant? How large will the Sun become?

33. Starting at the center and going outward, list the various layers of a red giant that contains a carbon core.

34. Describe two hypotheses designed to explain what causes a planetary nebula. Describe the appearance of a planetary nebula. What causes it to glow?

35. Describe a white dwarf with regard to size, mass, luminosity, and temperature.

36. What supports a white dwarf from further collapse?

37. What happens to the size of a main sequence star when mass is added to it? What happens to the size of a white dwarf when mass is added to it?

38. Why does a recurrent nova continue to flare up time after time?

---

1. What is a brown dwarf? What determines if a protostar becomes a brown dwarf or a main sequence star?

2. Gravity causes protostars to collapse. What stops the collapse?

3. Massive stars have much more hydrogen in their cores than do less massive stars. Why, then, do they run out of hydrogen faster than stars of low mass?

4. We see relatively few white dwarfs compared to main sequence stars, but astronomers are confident that white dwarfs are very common. Explain this discrepancy.

5. What determines whether the carbon cycle will occur in a particular star?

6. Based on how stars evolve during their main sequence lifetime, explain why the main sequence on an H-R diagram should not be a thin line, but should instead have a thickness.

7. What causes a star to get larger as it becomes a red giant? What causes its surface to cool?

8. Explain why a Sun-like star, once it becomes a red giant, undergoes a helium flash, but a star a few times more massive than the Sun begins helium fusion gently.

9. Explain what causes the pulsations of RR Lyrae and Cepheid variables.

10. Explain how star clusters provide evidence that supports our theories of what happens to stars after their main sequence life ends.

11. What do planetary nebulae have to do with planets? What are they and how did they get the name "planetary nebulae"?

12. Compare parts (a) and (b) in Figure 14-6. Although these are generally true to the actual data, they were not drawn carefully from the data. Look at the plotted stars and explain what tips you off that the diagrams do not actually fit real data.

13. Explain the meaning of the two graphs of Figure 14-11, and explain what happens to make the two types of star act differently.

14. Why are novae and Type Ia supernovae not included in Figure 14-26, which shows the evolutionary steps of normal stars?

CALCULATIONS

1. The Sun's life on the main sequence is expected to be 10 billion years. What is the life expectancy of a star with a mass half that of the Sun? What is the life expectancy of a star with a mass 10 times that of the Sun? (Hint: See the Tools of Astronomy box "Lifetimes on the Main Sequence.")

2. Assume that you observe a circular nebula on the celestial sphere. Its angular size is 1.2 arcminutes and it is located about 3000 light years away. Material in the nebula is expanding at the rate of 20 kilometers/second. How long ago did the central star lose its outer layers?

(Hint: You will first need to use the small-angle formula to find the radius of the circular nebula.)

3. The star at the center of a planetary nebula has luminosity of 1000 times the Sun's and temperature of 100,000 K. What is the radius of the white dwarf in solar units? (Hint: Use the Stefan-Boltzmann law.)

4. We mentioned in the text that during a nova explosion, a white dwarf becomes 10,000 to 100 million times brighter. Show that this corresponds to a change in magnitude by 10 to 20.

5. The star Zeta Geminorum is a Cepheid variable with a period of about 10 days. Interferometric observations show that the star's angular diameter changes by about $4.7 \cdot 10^{-8}$ of a degree during its cycle. Doppler measurements show that the change in the star's diameter is about 8.55 million kilometers. Show that the star is about 338 parsecs away.

EXPANDING THE QUEST

1. "Planetary Nebulae," by N. Soker, in *Scientific American* (May, 1992).

2. "The Lives of Stars: from Birth to Death and Beyond," by I. Iben and A. V. Tutukov, in *Sky & Telescope* (December, 1997; January, 1998).

3. "Giants in the Sky: The Fate of the Sun," by J. Kaler, in *Mercury* (March/April, 1993).

4. "Stellar Metamorphosis," by S. Kwok, in *Sky & Telescope* (October, 1998).

5. "White Dwarfs: Fossil Stars," by S. Kawaler and D. Winget, in *Sky & Telescope* (August, 1987).

STARLINKS

Quest Ahead to Starlinks:
www.jbpub.com/starlinks

Starlinks is this book's online learning center. It features **eLearning,** which contains chapter quizzes and other tools designed to help you study for your class. You can also find **on-line exercises,** view numerous relevant **animations,** follow a guide to **useful astronomy sites** on the web, or even check the latest **astronomy news** updates.

A composite *HST* image of N 49, a suprenova remnant within the Large Magellanic cloud. This remnant harbors a spinning neutron star with an extremely strong magnetic field. The material in this remnant will be recycled into building new stars, in the same way that our solar system was formed by remnant material of supernovae explosions in our Galaxy billions of years ago. We study supernova explosions of massive stars (such as the one that resulted in this remnant) in the next chapter.

The Deaths
of Massive Stars

After and before (arrow) images of the region around the great supernova of 1987. SN1987A was discovered near the Tarantula nebula in the Large Magellanic Cloud, a companion galaxy to our Milky Way.

IT WAS WINDY AND THE ROOF WAS BROKEN THAT FEBRUARY night in 1987. Ian Shelton, resident astronomer at the Las Campanas Observatory in Chile, had to open the roof by hand in order to use the 10-inch telescope he had repaired. Ian was responsible for maintaining the larger 24-inch telescope, but seldom got to use it himself, since visiting astronomers kept it pretty busy. Shivering in the mountain air, he climbed a ladder and pushed the corrugated metal aside.

This night had started like most others. Winds up to 40 miles/hour pounded the cinderblock telescope shack, but once Ian had opened the roof, he was able to begin the work he liked: photographing the sky. This night he decided to photograph the Large Magellanic Cloud, which is not a cloud at all, but the nearest galaxy to the Milky Way. It and its companion, the Small Magellanic Cloud, are visible only from the Southern Hemisphere. To get a good bright photo, he decided to take a three-hour time exposure. (Long exposures are the norm for astronomers.) His telescope, however, was not like more modern telescopes that use motors and computers to guide them so that they automatically follow the stars. Instead, he had to look through the eyepiece constantly and guide the telescope manually.

At three o'clock the sky suddenly went dark; the wind had blown the roof shut and jarred the telescope. What a night! Thinking that his photograph had been ruined, he considered going to bed, but then decided to develop the photographic plate and see what he had.

In the dim light of the darkroom, he examined the result. The Magellanic Cloud was apparent, but when things start going wrong, nothing stops the slide—there was a flaw on the plate. A bright blotch appeared near the edge of the galaxy's image. Of course, a bright star looks just like a blotch, but Ian knew the sky well enough to know that there is no bright star where the spot appeared. He had been using photographic plates of the same type for some time, however, and he knew that a flaw of this size would be extremely unusual. What was the alternative? A new star—a supernova? None bright enough to be seen with the naked eye had appeared in the last 383 years. If this were a supernova, it certainly was a bright one. It couldn't be.

The graduate school dropout argued with himself for 20 minutes before going outside to check the sky with his own eyes. There it was, the first naked-eye supernova since 1604, an object that will be studied by astronomers for years and one that has already caused us to change our theories about the death of stars. The unpromising night had yielded the astronomical event of the century, and Ian Shelton was the first to observe it.

Within hours, nearly all the radio and optical telescopes in the Southern Hemisphere were focused on SN1987A, as it was named. The *International Ultraviolet Explorer* satellite was recording radiation from the supernova by the next day; the Japanese X-ray satellite *Ginga* was observing the area within weeks. The neutrino detection teams in Cleveland, Ohio, and Kamioka, Japan, ran to check their records of the past few days. High-altitude planes and balloons were launched to observe the region without interference from Earth's atmosphere. Said one astronomer, "It's so exciting, it's hard to sleep."

In Chapter 14 we described how stars of low mass those with masses less than about four solar masses—evolve from protostars to white dwarfs. In general, we might say that such stars end their lives in a fairly uneventful manner. Although all but the very low mass stars become red giants and some put on beautiful displays as planetary nebulae, all-in-all, low mass stars die not with a bang, but with a whimper. However, if a white dwarf happens to be part of a binary system, things may get explosive. The resulting novae and particularly supernovae give a preview of what happens when massive stars die.

You might say that the smallest stars don't die—they just fade away.

The death of a massive star is truly a major event. Instead of gently puffing away their outer layers, some stars blow away their outer portions in cataclysmic supernova explosions, such as the one described in the chapter opening. Many end up as superdense neutron stars, sending out powerful lighthouse beacons across space. The most massive stars, however, end their lives as something even more exotic—as black holes.

15-1 Moderately Massive and Very Massive Stars (> 4 M$_\odot$)

Although the final destinies of moderately massive stars (perhaps 4 to 8 solar masses) and very massive stars (greater than about 8 solar masses) are far different, they behave similarly during most of their lives, and so we consider them together for now.

Recall from Figure 14-7 that the internal structure of main sequence stars depends on their mass. In a star of mass greater than about 4 solar masses, the core temperature and pressure are larger than in stars of low mass, and so is the temperature difference between the core and outer layers. As a result, the energy created in

Figure 15-1
The internal structure of stars with mass larger than 4 solar masses. Energy transport from the core occurs by convection in the inner part of the star and by radiation in the outer part.

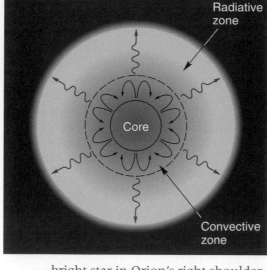

supergiant The evolutionary stage of a massive star after it leaves the main sequence.

the core by fusion reactions is transported outward by convection, while the low density in the outer layers allows energy transport by radiation (**Figure 15-1**).

The stars that have the most mass live very short main sequence lives. A star 15 times as massive as the Sun takes only about 10 million years to burn up its hydrogen. Like a less massive star, it then expands to become a red giant. During its red giant stage, the 15-solar-mass star changes gradually from a hydrogen-fusing central core to one that fuses helium. Because of its greater mass and corresponding greater core temperature, it becomes brighter than the standard red giant, and we call it a **supergiant**. The most prominent supergiant is Betelgeuse, the bright star in Orion's right shoulder (**Figure 15-2a**). A supergiant may be a million times brighter than the Sun and have an absolute magnitude of −10. Figure 15-2b shows the position of red supergiants on the H-R diagram, and **Table 15-1** lists some typical data for a supergiant.

As the core pressure and temperature increase inside a supergiant, the products of the previous cycle of fusion reactions become the fuel for the next cycle of nuclear reactions. Heavier elements are continuously produced, including neon, silicon, and even iron. As a result, successive cycles proceed faster, new shells of material are gen-

Figure 15-2
(a) This *HST* image of Betelgeuse is the first direct image of a star (other than the Sun). It reveals a huge atmosphere with a mysterious hot spot on the star's surface. In the constellation Orion (Figure 1-3), Betelgeuse is the bright star in the hunter's right shoulder. (b) Supergiants have about the same surface temperatures as red giants, but they are much brighter.

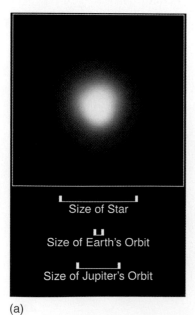

(a)

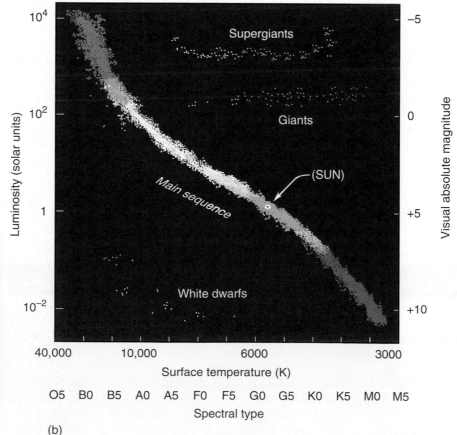

(b)

Table 15-1

A Typical Supergiant

Absolute magnitude	−5
Luminosity (solar units)	10,000
Mass (solar units)	12
Diameter (solar units)	1200
Average density	10^{-8} g/cm^3 (Sun's density = 1.4 g/cm^3)
Surface temperature	3600 K

Figure 15-3

(a) This H-R diagram shows the evolution of two massive stars. Massive stars do not experience a helium flash; instead, when core helium fusion reactions start, the evolutionary track turns sharply downward in the red-giant region. (b) Most of the volume of a 15-solar-mass red giant is hydrogen, but the central part is multilayered. The hypothesized mass of each layer is shown.

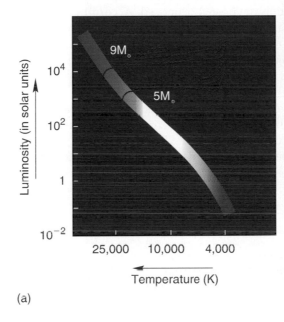

(a)

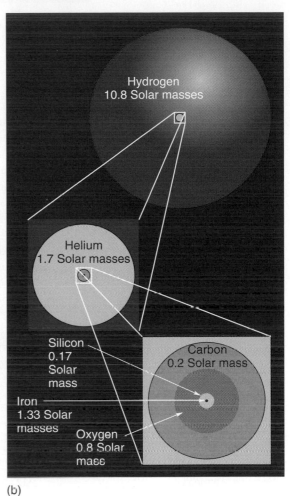

(b)

erated around the core, and new cycles of fusion reactions occur at the outer layers. The star continuously enters new red-giant phases and repeatedly moves back and forth on the H-R diagram (**Figure 15-3a**). The various reactions and their approximate longevity are shown in **Table 15-2** for a 15-solar-mass star. The core of the red supergiant now contains several layers, as illustrated in Figure 15-3b.

Table 15-2

The Evolution of a 15-Solar-Mass Star

Element Fused	Fusion Products	Time	Temperature
Hydrogen	Helium	10,000,000 years	4,000,000 K
Helium	Carbon	>1,000,000 years	100,000,000 K
Carbon	Oxygen, neon, magnesium	1000 years	600,000,000 K
Neon	Oxygen, magnesium	A few years	1,000,000,000 K
Oxygen	Silicon, sulfur	1 year	2,000,000,000 K
Silicon	Iron	A few days	3,000,000,000 K

15-2 Type II Supernovae

We classify supernovae into two types, based on their spectra. A Type I supernova has no hydrogen lines in its spectrum, and most supernovae of this type (Ia) are thought to be produced by white dwarfs in binary systems, as we described in Chapter 14. Type Ia supernovae are powered by nuclear energy released by runaway nuclear reactions as the white dwarf completely disintegrates. The spectra of Type II supernovae, on the other hand, reveal prominent hydrogen lines, indicating that these supernovae result from stars that still contain hydrogen in their outer layers. Type II supernovae (along with Ib and Ic) are powered by gravitational energy that is released as gravity continuously collapses the core. Differences between the two types of supernovae are outlined in **Table 15-3**; typical light curves are shown in **Figure 15-4**.

The process by which Type II supernovae occur is not well understood, but various models have been proposed. According to the leading model, once the core of a supergiant has reached the stage where silicon has formed and begins fusing to iron, things start to happen quickly. Computer modeling indicates that the silicon layer fuses to iron in only a few days. Considering that stellar lifetimes are measured in millions and billions of years, a few days is a remarkably short time. But after this, things change even faster!

Silicon fuses to iron in the center of the star. Previously, when a new central core (first of helium, then carbon, then oxygen, and finally silicon) was formed, the material of that core began to fuse to heavier elements when it got hot enough. Iron is different. All of the fusion reactions in Table 15-2 produce energy when they occur. When silicon nuclei fuse to form iron, for example, energy is produced. However, if iron were to fuse together to form even more massive nuclei, energy would be *absorbed* instead of released. In other words, if iron is to fuse, it must have a supply of energy. Therefore, as the iron core shrinks and heats up, it does not fuse to something more massive. Instead, once its mass reaches the Chandrasekhar limit, the core collapses violently. In much less than a second, the tremendous pressure generated by gravity in the core pushes pairs of electrons and protons together to form a neutron and a neutrino in each case. Most of these neutrinos escape from the core, carrying a large amount of energy, which causes the core to become cooler and therefore smaller and denser.

◆ We described neutrinos in Chapter 11, during the discussion of fusion reactions in the Sun.

Table 15-3

Typical Supernovae

	Type Ia	Type II
Spectrum	No hydrogen lines	Prominent hydrogen lines
Magnitude at peak	−19	−17
Light curve	Sharp peak	Broader peak
Expansion rate	10,000 km/s	5000 km/s
Mass ejected	0.5 solar masses	5 solar masses

Figure 15-4
Typical light curves from the two types of supernovae are shown. Type Ia supernovae get brighter and have a sharper peak. The curve is dashed at the beginning, during the time that the luminosity of the star is usually not observed.

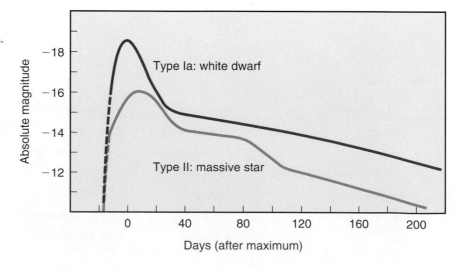

ADVANCING THE MODEL

Nucleosynthesis

Only hydrogen, helium, and lithium, the very lightest elements, were present in the very early universe. All the other heavier elements that we now see around us (up to iron) were produced inside stars through nuclear fusion reactions. However, the fusion process cannot produce elements heavier than iron. Instead, these elements can be produced in a process where neutrons are added to the nuclei of atoms. This addition creates different, heavier isotopes and can take place because neutrons are neutral particles; thus, they do not feel any electrical repulsion from the charged nuclei.

There are two different stellar environments where the process of neutron capture can happen. The first environment is inside very massive stars when they explode as supernovae. In such an event, the build-up proceeds via the "rapid-process." The explosion creates an enormous blast of neutrons that pulverizes atomic nuclei, which have no chance of decaying by emitting an electron and turning into a stable atom of the next atom in the Periodic Table. As a result, nuclei that are extremely neutron-rich are formed, which finally decay very fast.

A second environment is inside normal stars during the end of their lives when they burn their helium. Here the addition of neutrons is done rather gently via the "slow-process." About half of all the elements heavier than iron are believed to be formed by this process, which takes place during a specific stage of stellar evolution, just before an old star expels its outer layers and dies as a white dwarf. During the slow-process, neutrons enter the nucleus of an atom, turning it into an isotope. This neutron capture continues until there are too many neutrons inside the nucleus for the isotope to remain stable. Then the isotope emits an electron and becomes a stable atom of the next element on the Periodic Table.

Computer-based stellar models predict that the slow-process is very efficient in stars whose metal content is low (metal-poor stars). Such stars were born at an early stage in our galaxy and are therefore quite old. In these stars, the slow-process is expected to produce nuclei all the way up to the most heavy and stable ones, those with atomic numbers around 82 (the lead region). The discovery in late 2001 of metal-poor stars with high abundances of lead showed an excellent agreement between the predicted and observed abundances, reinforcing our current understanding of how the slow-process operates in the interiors of stars.

How does the slow-process operate? When a carbon-13 nucleus (which includes 6 protons and 7 neutrons) is hit by a helium-4 nucleus (which has 2 protons and 2 neutrons), they fuse to form oxygen-16 (which has 8 protons and 8 neutrons) while also releasing a neutron. A neutron is also released in a similar fusion reaction that involves a neon-22 nucleus and a magnesium-25 nucleus. It is these neutrons that become the building blocks for making the heavier elements. The carbon-13 isotope is in turn produced by fusion of a normal carbon nucleus (carbon-12) and a proton (a hydrogen nucleus). But in order for this process to work, there must be a region in the star where sufficient amounts of carbon and hydrogen coexist. This is problematic since most hydrogen nuclei have already fused to heavier nuclei, including carbon. Current models of stellar interiors and the discovery of "lead" stars suggest that there must be a moderate mixing between the outer hydrogen-rich stellar layers and the inner carbon-rich regions. It is not yet clear how this mixing process works.

As the core collapses, its density increases and very quickly reaches the density found inside nuclei, which is about 4×10^{14} grams/centimeter^3. It is very difficult to compress the core beyond this density. Computer simulations indicate that when the core reaches this stage, its innermost part rebounds—it stops contracting and expands outward—sending a strong pressure wave outward. Meanwhile, the outer layers of the core are falling inward to fill the gap left when the iron core shrank. The rebounding surface collides violently with the infalling material, and produces two effects: First, the collision is energetic enough to cause iron to fuse into heavier elements. (Recall that this requires an *input* of energy. The necessary energy comes from the collision.) Second, the collision sends shock waves outward that throw off the outer layers of the supergiant. Some astronomers question whether the shock wave is energetic enough to cause the star to explode. One theory holds that as the shock wave begins to lose its energy, it is heated by great numbers of neutrinos escaping the core. The heating action takes less than a second, and at the end of it, the newly energized shock wave has enough energy to blow away the outer portion of the star.

We know that a major fraction of the energy of a supernova appears in neutrinos. If this energy is included, Type II supernovae are by far more luminous than Type Ia.

Figure 15-5
Supernova 1993J was discovered in a spiral arm of the galaxy M81 on March 28, 1993. (a) It is identified by an arrow in the left photo, taken March 30, when it had an apparent magnitude between 10 and 11. (b) Notice that nothing appears at that spot in the right photo, which was taken earlier. The other dots in the images are stars in our Galaxy and are not associated with M81.

(a)

(b)

Supernova theory is still young, and advances will have to be made before it can be considered well founded.

Edwin Hubble (1889–1953), for whom the space telescope is named, was an important American astronomer whose name will appear often in the next two chapters.

Some Type II supernovae are more luminous than others, but a typical one reaches a peak absolute magnitude of −17. This is nearly a billion times brighter than the Sun! **Figure 15-5** shows a supernova that occurred in the galaxy M81 in March, 1993. The other dots in the photograph are not part of M81, but are stars in our galaxy. Except for the supernova, stars in M81 show up only as a haze.

Physicists have long recognized the special place that iron holds in the list of chemical elements. When lighter nuclei are fused to form heavier nuclei, energy is produced, but only so long as the newly formed nucleus is not more massive than iron. Elements heavier than iron cannot be produced without some source of energy, so they are not formed spontaneously in nature. What, then, is the source of the heavy elements that we find here on Earth? Supernovae are a part of the answer. Supernova explosions in the distant past produced some of the heavy elements and blasted them away into space to become part of the interstellar material from which the Earth was formed. In the Advancing the Model box on "Nucleosynthesis", we describe in more detail the processes involved in nucleosynthesis, the building up of heavier elements from lighter ones.

Detecting Supernovae

Three supernovae in our galaxy have been seen with the naked eye; they occurred in the years 1054, 1572, and 1604. The most spectacular on record occurred in the constellation Taurus on July 4, 1054, and was observed by Chinese astronomers, who reported that it was bright enough to be seen in daylight and to read by at night. It remained bright for a few weeks, then gradually faded until it disappeared from view after about two years. Invention of the telescope was centuries away, of course, so the Chinese of the eleventh century had no way to continue observing their "guest star" (as they called the object). In 1731, an amateur astronomer reported a small nebula in Taurus, and two hundred years later Edwin Hubble discovered that the nebula consists of material expanding at a rate such that it must have begun its expansion at the time the Chinese reported the guest star. **Figure 15-6** is a photograph of the supernova remnant, called the Crab nebula because of its shape. Telescopic observations taken over the last half-century show its growth; they reveal that its outer portions are still expanding outward at about 1400 kilometers/second and that it is now about 4.4 light-years in diameter.

Pulsar

Figure 15-6
The Crab nebula is the remnant of a supernova that was seen in the year 1054. The arrow indicates the position of the Crab pulsar.

(a)   (b)

Figure 15-7

(a) The Veil nebula, shown here, is part of the Cygnus loop, a supernova remnant 2500 light-years away. (b) This *Chandra* X-ray image of Cassiopeia A shows the regions of greatest intensity in low-, medium-, and high-energy X-rays in red, green, and blue, respectively. The red material on the left outer edge is rich in iron. The bright greenish white region on the lower left is rich in silicon and sulphur. In the blue region on the right edge, low- and medium-energy X-rays have been filtered out by a cloud of dust and gas in the remnant.

We have also been able to find remnants of the supernova seen by Tycho Brahe in 1572 (which provided evidence to him that the heavens were not unchanging) and the one seen by Johannes Kepler in 1604. Kepler's supernova occurred just before the invention of the telescope; since that time there have been no visible supernovae in our galaxy.

Table 15-4 lists some major supernova remnants. Images of two of these—the Cygnus loop and Cassiopeia A—appear in **Figure 15-7**.

Table 15-4

Some Supernova Remnants

Remnant	Distance (Light-Years)	Diameter (Light-Years)	Age (Years)
Cassiopeia A	10,000	15	320
Crab nebula	6000	4.4	950
Cygnus loop	2500	100	20,000
Vela nebula	300*	2300	11,000
Tycho's supernova	9800		420
Kepler's supernova	20,000		390

*This is the distance to the nearest part of the nebula.

15-3 SN1987A

When a new astronomical object is observed, a telephone call or telegram is sent to the International Astronomical Union's Central Bureau for Astronomical Telegrams, in order to establish priority of discovery. Telegram number 4316, received on February 24, 1987, read:

> W. Kunkel and B. Madore, Las Campanas Observatory, report the discovery by Ian Shelton, University of Toronto, of a mag 5 object, ostensibly a supernova, in the Large Magellanic Cloud. . . .

This supernova (**Figure 15-8**) is officially known as SN1987A ("A" because it was the first discovered that year).

One of the first things that astronomers did after the discovery of SN1987A was to examine recent photographs of the region where it occurred, to determine which star was its source. It is fortunate that the supernova occurred in a region of the Large Magellanic Cloud where new stars were known to be forming, for this meant that astronomers already had an interest in the region and many photographs of it could be found. Two very luminous blue stars were found very close together at the location where the supernova occurred. This caused confusion because it was thought that

Since the Crab nebula is 6000 light-years from us, the supernova actually occurred about 7000 years ago (6000 + the 950 years since 1054), but we speak of it as if it happened about 950 years ago.

The star that formed SN1987A (the "progenitor" star) was Sanduleak −69°202, so named because it is the 202nd star in the 69th degree south of the celestial equator (−69° declination) in a catalog compiled by Nicholas Sanduleak.

ADVANCING THE MODEL

Supernovae from Moderately Low Mass Stars?

A white dwarf that is accreting material from a binary companion continues to do so until its mass reaches the Chandrasekhar limit, whereupon it collapses and suddenly initiates carbon fusion. A Type Ia supernova is the result. Some stellar models indicate that the same process might be responsible for Type II supernovae in red giants of close to 4 solar masses. Consider the point in its evolution when such a star is a red giant with a carbon core. Surrounding the carbon core is a shell of helium that is fusing together into carbon and dropping the carbon onto the core. If this model is correct, a 4-solar-mass star would form enough carbon that the mass of its core could reach the Chandrasekhar limit. When this occurs, the carbon core collapses, heats up, and undergoes violent carbon fusion.

According to this model, the sudden detonation of carbon is powerful enough to blast the remainder of the star away in a supernova explosion. Since the outer portions of the red giant are made up primarily of hydrogen, the spectrum of the hot expanding shell of gas will contain hydrogen lines and will have the characteristics of a Type II supernova.

The mass of the star discussed here is only about 4 solar masses, which puts the star in the category of moderately low mass stars, a group we described in the last chapter as evolving into planetary nebulae and then ending up as white dwarfs. If the model presented here is correct, some may die in a much more explosive way.

red giants, not blue stars, explode as supernovae. The explanation appeared when a shell of gas was detected (by the *International Ultraviolet Explorer* satellite) about a light-year from the central star. It appears that the shell consists of material that was shed from what was originally a red supergiant with a mass about 20 times that of the Sun. The star ejected this material about 20,000 years ago, leaving behind its hotter, blue surface. Since 20,000 years is little more than an instant in a star's life, we can say that the red supergiant ejected its surface material shortly before it exploded as a supernova.

Supernova theory had predicted that a burst of neutrinos would be emitted by the explosion. As we discussed in Section 11-3, neutrinos are nuclear particles produced within stars and are the subject of research at various neutrino detectors around the world. Such devices make a record when they detect neutrinos. Our theory of supernovae was confirmed when neutrino researchers in the United States and Japan checked their records and reported that the number of neutrinos had increased a full three hours before the supernova was seen. The delay was due to the very low

Figure 15-8
The photo at the left was taken before SN1987A occurred. The supernova is obvious in the other photo.

tendency of neutrinos to interact with matter. After they were formed in the core, most of these neutrinos traveled the remaining volume of the star and escaped into space. However, the shock produced when the rebounding core of the star collided with its infalling outer layers took about 3 hours to reach the star's surface. Only when this shock reaches the layers closest to the surface can we see the increase in the star's luminosity. By this time the neutrinos were 3 hours ahead of the light and remained so for the rest of their travel between the star and Earth, a distance of about 169,000 light-years. Using the number of neutrinos detected and other factors (such as the sensitivity of the detector and the inverse square law of radiation discussed in Chapter 4, which applies to both light and neutrinos), astronomers were able to learn a great deal about this violent explosion. Over a period of about 10 seconds, SN1987A emitted a total of about 10^{58} neutrinos, which carried off an amount of energy equivalent to 100 times the energy that our Sun has emitted in its entire lifetime. As we mentioned earlier, if we take into account the energy carried off by neutrinos, Type II supernovae are by far more luminous than Type Ia.

> If neutrinos have no mass, they should have arrived at almost the same moment. The observed delay in the arrival times of the different neutrinos allowed astronomers to set an upper limit to the mass that they might have.

We continue to observe and learn from SN1987A. **Figure 15-9a** shows its light echo, the deflection of light by two dust-sheets near the supernova. In Figure 15-9b we see the three-ring structure of SN1987A and its evolution between 1994 and 1996. Figure 15-9c shows the glowing gas ring around SN1987A, heated by the collision of a shock wave from the explosion. Observations of SN1987A and other supernovae provide valuable information about these self-destructing stars and the influence they have on their surroundings, but there are still many mysteries about supernovae. Whatever their details, such events are singular in the lives of stars. In some cases, it appears that the entire star is blown apart, including the core; but in other cases, the core is left behind as a tiny remnant of the once-mighty star. The nature of this leftover core depends upon whether the star was originally in our moderately massive or our very massive class. In either case, a unique, peculiar object is formed.

> Besides its mass, other factors, such as the star's angular momentum, play a role in determining what the star leaves behind.

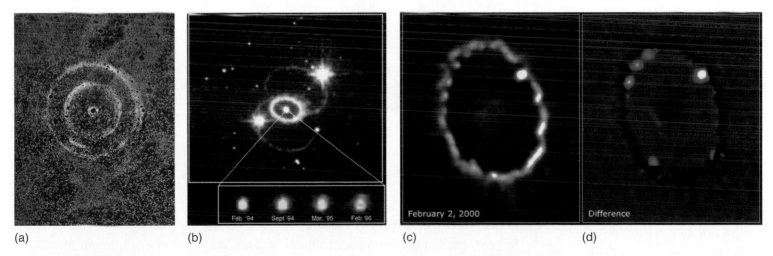

(a)   (b)   (c)   (d)

Figure 15-9
(a) The light echo of SN1987A. Some light from the explosion was deflected by two dust sheets (about 470 and 1300 light-years from the supernova). Since it covers a longer path to reach us, it is seen after the star has faded away. This image was made by subtracting images taken before and after the explosion. (b) This *HST* image was taken in light emitted by nitrogen gas, at 658 nanometers, for the central part and using a visible filter, at 550 nanometers, for the debris. It shows the structure of the explosion debris of SN1987A, which is expanding at about 6 million miles per hour. The debris is resolved into two blobs moving in opposite directions along the short axis of the inner ring surrounding the supernova. The hourglass-shaped shell made by the three rings is probably the result of the interaction between slow/fast moving winds produced by the star before/after it became a blue supergiant. (c) This *HST* image (in visible light) shows that the ring around SN1987A is glowing as it is heated by the collision of a 40 million mile per hour shock wave from the supernova explosion. (d) Image processing was used to create the image to the right in order to emphasize four new bright knots of superheated gas discovered in 2000. The brightest knot to the right was first seen in 1997.

15-4 Neutron Stars

The neutron was discovered in 1932 by James Chadwick, although its existence had been suspected for some years before that. Only two years later, in 1934, the first suggestion was made that some stars might collapse into a strange object consisting of nothing but neutrons.

Theory: Collapse of a Massive Star

In the previous few chapters we have described much of the theory concerning the lives of stars, but have only occasionally cited evidence for it. (As we explained in Chapter 14, much of the evidence of pre- and post-main sequence life comes from examining star clusters.) In addition, we have not detailed the many steps that accompanied the development of the theory. There is an interesting story, however, that illustrates a portion of the theory of stellar life cycles. It not only shows how the theory was confirmed, but also how the confirmation provided further knowledge—a common occurrence in science.

A hypothesis worked out in the 1930s predicted that after the mass of a star's core increases beyond the Chandrasekhar limit, the star will collapse further and its electrons and protons will combine to form neutrons, resulting in a ***neutron star***. Just as electron degeneracy prohibits a white dwarf from collapsing under gravity, neutron degeneracy does the same for a neutron star. The hypothesis predicted that the remains of a moderately massive star's collapsed core would become a neutron star—a tremendously compressed star with a mass between 1.4 and about 3 solar masses. **Table 15-5** shows the properties of such a star. To try to imagine it, picture the entire mass of a star larger than the Sun compressed into a ball the size of a small city, about 20 kilometers across. The particles in a neutron star are packed so tightly together that a cubic centimeter (about the size of a sugar cube) of the star would have a mass of about a trillion kilograms; also, the gravitational force at the star's surface is so powerful that a marshmallow hitting the surface would release the energy of a Hiroshima-type bomb.

Astronomers had little hope of finding such a star because, despite its high temperature, its extremely small size results in it being very dim. Thus the idea was put on astronomy's back burner as an interesting hypothesis but one that seemed beyond our ability to confirm or deny. In 1967, an accidental discovery changed this.

neutron star A star that has collapsed to the point at which it is supported by neutron degeneracy.

The diameter of a typical neutron star is only 0.2% of the diameter of a white dwarf, yet the neutron star is a billion times denser.

Six or eight weeks after starting the survey, I became aware that on occasions there was a bit of "scruff" on the records, which did not look exactly like a scintillating source, and yet did not look exactly like man-made interference either.

Jocelyn Bell, on first noticing a pulsar signal.

Table 15-5

A Typical Neutron Star

Mass	1.5 solar masses
Diameter	20 km (width of a small city)
Density	10^{15} g/cm^3
Temperature	10,000,000 K

Observation: The Discovery of Pulsars

In 1967, Jocelyn Bell (now Jocelyn Burnell), a graduate student in astronomy at Cambridge University in England, was working with Antony Hewish and a group of researchers who were searching for quasars, energetic stellar sources that we will discuss in Chapter 17. The radio telescope she was using for her research did not look at all like the giant radio telescope dishes normally associated with radio astronomy. Instead, it looked like a field of clotheslines covering a total of 4-1/2 acres (**Figure 15-10**). It was designed to detect faint radio sources and to see quick changes in their energy. In the course of research for her dissertation, Bell found a new, unexpected, and unexplained source of radio waves. The signal

Figure 15-10
Part of the radio telescope that first detected a pulsar.

ADVANCING THE MODEL

The Pulsar in SN1987A?

After the discovery of SN1987A in February, 1987, astronomers began to look for a pulsar at its center. They were not particularly surprised that the pulsar had not been visible immediately because debris from the explosion would block our view of it.

On January 18, 1989, nearly two years after the supernova was discovered, an international team of astronomers headed by John Middleditch (of Los Alamos National Laboratory) observed the supernova from Cerro Tololo Inter-American Observatory in Chile for nearly 7 hours, using a detector that is sensitive to visible and near-infrared light. The signals contained pulses! In order to check their procedure and their instruments, the experimenters turned the telescope to another object that they knew did not pulse. Indeed, no pulses were detected, and measurements of that object's brightness corresponded to previous brightness measurements. They announced that the pulsar in SN1987A had been discovered!

Astronomers were excited about the new pulsar for at least three reasons: First, the discovery confirmed the theory that links pulsars and supernovae. Second, the pulsar was pulsing 1968.629 times per second, indicating a tremendously high rotation rate. The star had to be spinning every 0.0005 seconds, much faster than any other known pulsar. Although astronomers expected that a new pulsar would spin faster than an old one, this great rotation rate was difficult to explain, and it indicated that more theoretical work was necessary. Third, the pulses varied slightly in frequency over the seven-hour observation. This also could not be explained using present supernova theories.

Unfortunately, when the astronomers repeated their observations a week later in an attempt to confirm the discovery, they could find no pulses in the signal. Continued observation of the supernova through the remainder of 1989 failed to reveal pulsations. The only explanation seemed to be that, during the January 18 observation, light from the pulsar had reached Earth through a temporary gap in the debris cloud around the pulsar.

As 1989 passed, theoreticians worked to adapt pulsar models to the new findings, and experimental astronomers continued to look for the pulsar's reappearance. As time passed, concern rose that perhaps the initial observation was flawed. The data and methods of the observation were examined over and over, but no flaw could be found. Then, at the February 1990 meeting of the American Association for the Advancement of Science, John Middleditch reported that he and Tim Sassen, a graduate student at the University of California at Berkeley, had solved the problem. They found an old video camera in the observatory and discovered that if the camera was left on during an observation, it sometimes caused an electronic pulse with a frequency of 1968.629 cycles per second—exactly that found for the pulsar. Apparently the camera had been on the night of the pulsar's "discovery," and conditions had been just right for its signals to join the signals received from the sky.

There are lessons to be learned from the would-be discovery of the SN1987A pulsar. Although skeptics might point to this as a failure of science, it is actually an example of a successful use of the scientific method, for it illustrates how science corrects itself. Although Middleditch showed courage in reporting the mistake, his embarrassment would have been greater if someone else had found it. The supposed discovery prompted re-analysis of our theories of pulsars, and we should be better prepared for whatever we find when—and if—the pulsar of SN1987A is finally discovered.

---

from it pulsed rapidly, about once every 1.3 seconds. This was a much more rapid pulsation than had ever been observed from a stellar energy source.

The researchers' first thought was that the waves Bell had received were of terrestrial rather than celestial origin. A check of local radio transmitters, however, failed to indicate a source. In addition, the signal was detected about four minutes earlier each night than the previous night. Recall that a given star sets four minutes earlier each night as a result of the Earth's moving around the Sun. The researchers concluded therefore that the source was in the sky and was not of human origin.

Their next thought was that a signal from an extraterrestrial race had been detected. In fact, the source was referred to for a short while as an LGM (the initials for "Little Green Men," a reference to science-fiction-type extraterrestrials). This speculation was abandoned for a couple of reasons. First, the pulsations continued in a very regular fashion, instead of changing as they would if they contained a message. More convincing, however, was the discovery by Jocelyn Bell of three more such sources in

Figure 15-11
The chart of pulses from the first pulsar indicates their regularity. The difference between pulse duration and pulse period is illustrated. The pulse duration shown is exaggerated.

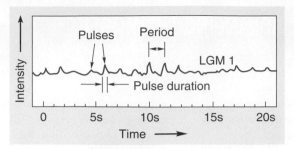

pulsar A pulsating radio source with a regular period, between a millisecond and a few seconds, believed to be associated with a rapidly rotating neutron star.

◆ The observed stretching out of any sudden change in the Sun would not be due to our distance from it, but rather to its size. Be sure to understand this idea; it is a useful tool in astronomy that tells us the maximum size of some objects.

Figure 15-12
The arrows indicate light that left the Sun at the same time. Since point X is about two light-seconds closer to Earth than points Y and Z, its light reaches us two seconds sooner. Thus if the Sun were to increase in brightness instantaneously, we would see the light increase gradually over two seconds.

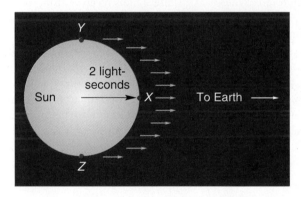

other directions in the sky, each with its own characteristic rate of pulsation. It was highly unlikely that several different civilizations were sending such signals toward us at the same time, so the sources had to be natural ones. They were renamed *pulsars.*

The first pulsar detected had a period of 1.3373011 seconds. Such great precision is possible in measuring the rate of a regularly repeating phenomenon because we can measure over a great number of cycles and then divide by that number, obtaining the time for a single cycle. **Figure 15-11** shows a record of pulses from this pulsar and indicates that although they are extremely regular, they do vary in intensity.

All of the pulsars that were found had a pulse with a duration of about 0.001 second. (Figure 15-11 also illustrates the difference between pulse duration and period.) This immediately revealed to the astronomers an upper limit to the size of the object emitting the signals. The objects could be no greater than about 0.001 light-seconds in diameter, which corresponds to 0.001 second × 300,000 kilometers/second = 300 kilometers. To see how astronomers could reach such a conclusion, even before the nature of the pulsation was known, consider what we would observe if the Sun were to brighten instantaneously. We would not see this instantaneous brightening as being instantaneous at all, for light from the part of the Sun closest to us would reach us about two seconds before light from its limb (**Figure 15-12**). Since the Sun is about two light-seconds in radius, we would see the intensity of light build up over two seconds. Likewise, if the Sun suddenly shut off, the dimming would appear to take two seconds, rather than appearing to happen all at once.

The smallest stars known in 1967 were the white dwarfs, but these are Earth-size objects, not small enough to emit pulsations that last only 0.001 second; at least, not if the pulsation is caused by a change in the light emitted from the entire object. Thus the pulsar must be even smaller than a white dwarf. How could a star be so small? Enter the hypothetical neutron star.

15-5 The Lighthouse Model of Neutron Stars/Pulsars

Today, more than 1000 pulsars have been detected, most with pulses between 0.1 and 4 seconds. As with any scientific theory, a model that explains pulsars must work for all these various cases.

Theory: The Emission of Radiation Pulses

Let us consider how an object might emit pulses of radiation. One way would be for its surface to vibrate up and down. (Some kinds of variable stellar objects are known to do this, including Cepheid variables, as we discussed in Chapter 14.) Not only did the short duration of flashes from pulsars seem to indicate that they were not white dwarfs, but when astronomers considered the nature of the material of a white dwarf and the gravitational force on its surface, calculations showed that the surface of a white dwarf could not vibrate as quickly as once per second. Neutron stars are much denser than white dwarfs and have a much greater gravitational force on their surface, so their surface could beat up and down more quickly. In fact, calculations showed that they should be unable to oscillate as *slowly* as once a second. That is, vibrations of white dwarfs were eliminated as a possibility because they have periods that are too long to explain the observed faster pulses, while vibrations of neutron stars were eliminated because they have periods that fall below the range of the slower observed pulses.

A second possible way for an object to emit radiation in pulses is by an eclipsing binary process, with a bright object orbiting a dark one. However, the radiation curve of pulsars is different in nature from one that can be explained by eclipses. In addition, Kepler's third law (as modified by Newton) for an eclipsing binary with a period of 1 second or less suggests that the average distance between the two stars must be less than a few thousand kilometers. This distance is even smaller than the diameter of a white dwarf. Now, neutron stars are small enough that two of them could orbit each other with a period in agreement with those observed for pulsars. However, this scenario actually leads to the neutron stars spiraling closer together (a result of Einstein's general theory of relativity, which we discuss later in this chapter). This would decrease the orbital period of the neutron stars, which contradicts the observed increase in the periods of the pulsars. Therefore, eclipsing binaries must be ruled out.

A final mechanism for producing pulses is by radiation that comes from a small part of the surface of a rotating object. This is the mechanism that causes sailors at sea to observe pulses of light from a lighthouse. On a foggy night, they might see the lighthouse beam sweeping through the fog, but on a clear night the sailors would see the light only when it shines directly at them. This makes it appear to blink on and off. Could a star rotate with a period as short as that observed for pulsars? A star the size of the Sun would be torn apart by such fast rotation, but white dwarfs or neutron stars would have two advantages in this regard: Their smaller size would mean that less force would be needed to retain their surfaces under fast rotation, and their small size and great mass would result in a much greater gravitational force on their surfaces than is experienced on the surface of the Sun. Calculations showed that a white dwarf might be able to withstand the forces involved in rotating with a period of one second, and perhaps with a period of one-fourth second, but certainly no faster than that. A neutron star, on the other hand, would have no difficulty rotating with a period of a fraction of a second.

It is easy to see what would cause either a white dwarf or a neutron star to rotate so fast. Recall that as a spinning object decreases in size, its rotation rate increases. White dwarfs, and especially neutron stars, are so small that they would be expected to be spinning very fast.

Logic seemed to be pointing more and more to the neutron star as the explanation for pulsars. In analogy with a sailor's lighthouse, the model developed to explain how neutron stars create pulses is called the ***lighthouse model.***

Recall that the Sun has a magnetic field. When the theory of the existence of neutron stars was developed back in the 1930s, it was suggested that such a star might have an extremely strong magnetic field, since the star is the compacted core of a main sequence star that presumably would have had a magnetic field. During its main sequence period, the star's magnetic field was spread out over its entire surface and was "frozen-into" the star's ionized gases. As the star evolved and collapsed into a neutron star, its surface area was reduced by many orders of magnitude, and thus its magnetic field was concentrated on a much smaller area. As a result, the magnetic field of a neutron star is very strong. This strong magnetic field is a necessary part of the lighthouse model.

As we discussed when describing the solar system, it is common for the magnetic poles of a planet to be out of alignment with the axis of the planet's rotation. Refer back to Figure 6-34 to see the location of the Earth's magnetic pole in the Northern Hemisphere, some 1400 kilometers from the north geographic pole. Recall also that the magnetic field of the Earth traps charged particles and that these particles result in the

Compare Figure 15-11 to a light curve for an eclipsing binary— Figure 12-36, for example.

As we discussed in Section 7-6, the lack of a shorter rotation period for the Sun posed problems for astronomers in understanding the formation of the solar system from a nebula.

lighthouse model The theory that explains pulsar behavior as being due to a spinning neutron star whose beam of radiation we see as it sweeps by.

Even though a neutron star is mostly made up of neutrons, the presence of a magnetic field requires the existence inside the star of protons and electrons. It is thought that inside a neutron star, neutrons flow without any friction and protons move around without any resistance.

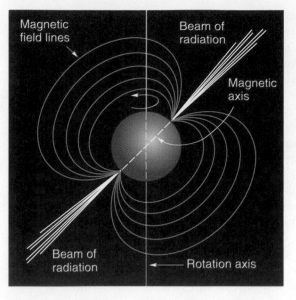

Figure 15-13
A beam of radiation is emitted near each magnetic pole of the pulsar. As the star rotates, the beam sweeps through space.

auroras seen near the Earth's magnetic poles. In the case of a fast rotating neutron star, the rapidly changing strong magnetic field gives rise to an electric field that continuously rips charged particles from the star's surface. The charged particles are quickly accelerated near the neutron star's magnetic poles. These charged particles spiraling around magnetic fields produce a narrow cone of radiation in the forward direction. The overall picture is that of two oppositely directed beams of radiation emanating from the star. The energy associated with these beams is extremely intense, and each beam (of radio waves and other radiation) is emitted from each magnetic pole. If the star's magnetic poles were located off the rotation axis, these beams would sweep through space as the star spins (**Figure 15-13**). Then, if the Earth were located in the path of a beam, we would see a pulse of radio waves each time the beam sweeps by us. In other words, according to this model, the observed pulses do not correspond to a "pulsing" source. Instead, energy is continuously being emitted by the pulsar, and we simply observe a pulse every time the star's beam sweeps by Earth.

Notice that the lighthouse model also predicts the existence of many pulsars we could not observe from Earth, for we would see only those whose lighthouse beam happens to sweep by us. Since the duration of every flash from a pulsar is very short compared to its period, we can conclude that the angular size of the beam is small (**Figure 15-14**). This means that from Earth we would see only a small percentage of

Figure 15-14
The angular width (A) of the beam from a pulsar determines how long we see its pulse. No pulsar has a long-duration signal, which indicates that pulsars' angular beam widths are narrow and therefore that we are seeing only a small fraction of the pulsars that exist.

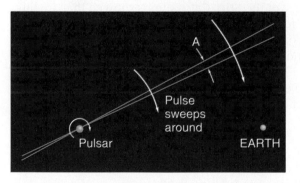

the pulsars that exist. It also means that we may never see a pulsar in SN1987A, depending on how the leftover rotating neutron star aligns with respect to Earth.

In 1967, the lighthouse model seemed to be a logical explanation for the pulsar observations, but in order to test it, more pulsars would have to be found. Good evidence for the model would be finding a pulsar related to the remnants of a supernova, where neutron stars are theorized to be.

Observation: The Crab Pulsar and Others

It was only a matter of months after the discovery of the first pulsar that one was found in the Crab nebula (shown in Figure 15-6). It might have been found sooner, but its rate of pulsation is much faster than was expected; it has a period of 0.033 second, so that it blinks 30 times per second. This short period finally ruled out completely the possibility of a spinning white dwarf as the source of pulses; a white dwarf would tear apart if it rotated 30 times per second.

Not only was the Crab pulsar flashing more frequently than others yet discovered, but it was emitting great amounts of energy in the radio region of the spectrum—the radio luminosity of the Crab's pulse is about 0.0025 of the total energy emitted by the Sun. Astronomers at the University of Arizona then observed that it also pulses in visible light, with the same pulse rate as it does in radio waves. Since that time, astronomers have found that the Crab pulsar emits radiation in all regions of the spectrum, from radio waves to X-rays. Adding up the energy emitted by the Crab in all the various regions of the spectrum, it was found that the total energy from the Crab nebula is

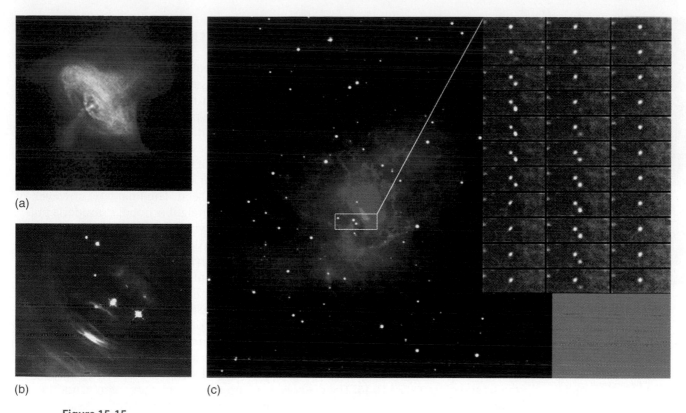

(a)

(b)

(c)

Figure 15-15

(a) This X ray image clearly shows the Crab pulsar, the rings that surround it and the jet-like structures. The X-rays from the Crab nebula are produced by fast-moving particles spiraling around magnetic fields in the nebula. The diameter of the inner ring is larger than 200 times the diameter of our solar system. (b) The Crab pulsar and other features seen in this *HST* image. The knot may be a "shock" in the jet, while the ring may be the boundary between the polar wind and jet and an equatorial wind that powers a larger torus of emission that surrounds the pulsar. (c) A time sequence for the Crab pulsar is shown in context against an image taken using an optical filter The mosaic of 33 time slices is ordered from top to bottom and from left to right, with each slice representing about one millisecond in the period of the pulsar. The brighter, primary pulse is visible in the first column; the weaker, broader inter-pulse can be seen in the second column.

more than 100,000 times the energy from the Sun. **Figure 15-15a** is an X-ray image (obtained by the *Chandra X-ray Observatory*) of the Crab pulsar and its neighborhood. We clearly see tilted rings of high-energy particles to a distance of a light year from the pulsar, and jet-like features emanating from the pulsar. The *HST* image shown in Figure 15-15b indicates the presence of a knot on one side of the pulsar and a ring-like "halo" on the opposite side. The knot, pulsar and center of the halo line up with the direction of a jet of X-ray emission.

Astronomers wondered what the source of the Crab pulsar's energy was. This question had been asked even before its pulsar had been discovered. Astronomers had long been puzzled by the luminosity of the nebula itself; it emits more energy than was thought possible. Both of these energy problems were solved when it was discovered that the Crab pulsar is slowing down; its pulses were found to be growing slightly less frequent, corresponding to an increase in its period by about 15 microseconds a year! It was hypothesized that the source of the energy that powers the nebula is the rotational energy of the pulsar. As the pulsar spins, its magnetic field propels electrons out into the nebula. These electrons are the cause of the nebula's great luminosity, but in being swept from the pulsar, the electrons, in turn, slow down the pulsar. If this hypothesis is correct, the amount of rotational energy lost as the object slows its spinning should correspond to the amount of energy emitted by the nebula.

The pulse rate for a pulsar is sometimes observed to increase suddenly (this is called a *glitch*). This occurs because as the neutron star slows its spinning, forces on its surface change. It is possible that these changes cause "starquakes" and sudden changes in the pulsar's shape. It has also been suggested that perhaps the glitches result from the interaction between the pulsar's crust and the layer below it.

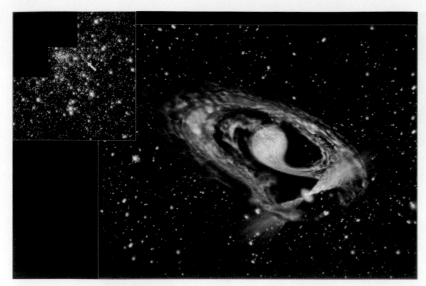

Figure 15-16

The millisecond pulsar system in the globular cluster NGC6397 in Ara (the Altar). In the *Hubble* image insert, the arrow marks the companion star. The artist's impression shows the pulsar (in blue with two radiation beams) and its bloated red companion star. Once the pulsar has been spun up, it can no longer accrete gas from its companion.

Calculations showed that the two amounts of energy did indeed correspond. The theory was confirmed quantitatively.

The pulsar in the Crab nebula is spinning faster than most others because the supernova in which it had its start was so recent; it occurred only 950 years ago. As time passes, this pulsar will gradually slow down. As it loses its rotational energy, the intensity of its pulses will also decrease, and it will no longer emit X-rays. Finally, in tens of thousands of years, it will be just another radio pulsar with its nebula spread so far that it can no longer be seen. The final fate, however, of a pulsar depends on its neighborhood. Consider a pulsar in a binary system, with its companion being a low-mass star. As the pulsar slows down with time, while the companion star evolves and expands toward becoming a red giant, the pulsar might accrete material from the outer layers of the companion. As this material strikes the pulsar's surface at an angle and at high speed, it will cause the pulsar to spin faster. Astronomers believe that a small group of about 100 pulsars (with periods of 1 to 10 milliseconds) have been "spun up" by such transfer of matter. The companion in such a binary system might not survive the strong radiation of the pulsar and might eventually be destroyed. An example of such a millisecond pulsar system is shown in **Figure 15-16**. The observations show that the companion is not the expected white dwarf (the end stage of a low mass star), but a bloated red star. There is also a large amount of gas in the system, suggesting that we are seeing a millisecond pulsar just after it has been "spun up" by its companion star.

Neutron stars in binary systems give rise to two other exotic phenomena. As the neutron star captures material from the companion star, its strong magnetic field funnels this material onto the magnetic poles of the neutron star. The resulting high-speed impact creates extremely hot regions that emit large amounts of X-rays. As the neutron star rotates, the rotating beams of X-rays sweep the sky and may be observed as pulses of X-rays. On the other hand, if the magnetic field of the neutron star is not strong enough, the infalling material may be distributed more evenly over the surface of the neutron star. Since most of this material is hydrogen, the large pressures and temperatures in this layer result in fusion reactions that create a layer of helium under a layer of infalling hydrogen. When the conditions are such that helium fusion reactions begin, the surface of the neutron star gets very hot and a sudden burst of X-rays is emitted that lasts for a few seconds. As a result of this emission, the star cools, but as new hydrogen flows onto the neutron star this process repeats itself every few hours or days. This is an explosive thermonuclear process similar to a nova (where the degenerate star was a white dwarf, as we discussed in Chapter 14).

More than 1000 pulsars have been discovered; a few have periods less than 0.1 second, but most have periods between 0.1 and 4 seconds. Normally, no nebula is found surrounding pulsars, for the nebula has long since dispersed. To further confirm that pulsars are indeed the neutron stars predicted to be left behind in supernovae, astronomers looked for more instances of pulsars associated with expanding nebulae. Since the Crab's pulsar was found, another has been found in the Vela nebula (Figure 14-31). The bizarre image of the Vela pulsar, shown in **Figure 15-17a**, suggests that after the star exploded, jets with unequal thrust along the neutron star's poles accelerated it like a rocket. Figure 15-17b is another spectacular image of a young, oxygen-rich supernova remnant with a pulsar at its center.

Astronomers are now confident that they have found the neutron stars that theory predicted 70 years ago. However, locating the pulsar associated with a given super-

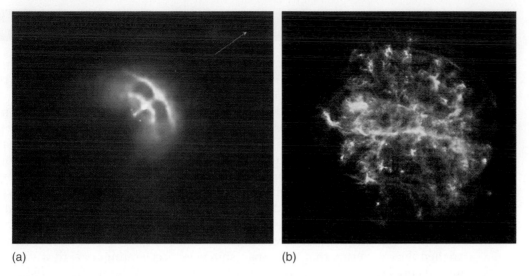

(a)   (b)

Figure 15-17

(a) A *Chandra* image of the compact nebula around the Vela pulsar. The two bows are thought to be the near edges of rings of X-ray emission from energetic particles produced by the pulsar at the center. The rings represent shock waves due to matter moving away from the pulsar. The jets emanating from the pulsar are perpendicular to the bows and are along the direction of the pulsar's motion, as indicated by the arrow. (b) A *Chandra* image of G292.0+1.8, a young pulsar surrounded by a rapidly expanding shell of gas, 36 light-years across, that contains large amounts of oxygen, neon, magnesium, silicon, and sulfur. The estimated distance is about 20,000 light-years, and the age about 1600 years.

nova remnant is not easy. When the star explodes, any asymmetry in the explosion might kick the resulting pulsar away from the site of the explosion. In other cases, the pulsar's beams might not sweep past the direction of the Earth, so it is never seen. It is also possible, as is suspected for SN1987A, that the stellar core might collapse into a black hole. The Advancing the Model box on "The Pulsar in SN1987A?" in this chapter describes what was thought to be its discovery in 1989.

Figure 15-18

These are the steps a moderately massive star takes from protostar to its final stages.

15-6 Moderately Massive Stars—Conclusion

Figure 15-18 reviews the steps a moderately massive star takes from protostar to pulsar/neutron star. However, only a small fraction of all stars are in this category, for two reasons. First, the lifetimes of such stars are short compared to the lifetimes of stars with lower mass. Thus, even if they are being formed at the same rate, not as many will be in existence at any given time. Second, only stars in a small range of mass end up with a core too massive to be supported by electron degeneracy and of low enough mass to be supported by neutron degeneracy. We know that neutron stars, which are the final ends of moderately massive stars, are more massive than the Chandrasekhar limit of 1.4 solar masses, but we are not sure of the upper limit to the mass of a neutron star. (We certainly do not have any neutron-degenerate matter on Earth with which to experiment.) Theory indicates, however, that the limit falls somewhere around 3.2 solar masses. Thus, only stars that end up with masses greater than—but not much greater than—1.4 solar masses can ever become neutron stars and send out their characteristic lighthouse beam of radiation.

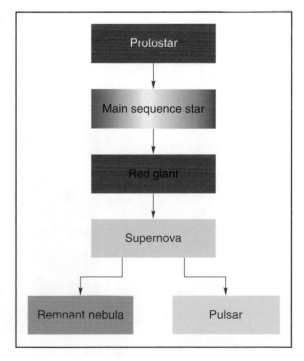

15-7 General Relativity

special theory of relativity A theory developed by Einstein that predicts the observed behavior of matter due to its speed relative to the person who makes the observation.

general theory of relativity A theory developed by Einstein that expands special relativity to accelerated systems and presents an alternative way of explaining the phenomenon of gravitation.

principle of equivalence The statement that effects of acceleration are indistinguishable from gravitational effects.

If gravity bends light, shouldn't we see it curve here on Earth?

Albert Einstein stated that in his youth he wondered whether, if he were moving at the speed of light, he could see himself in a mirror. Such questions led him in 1905 to develop the *special theory of relativity,* which we discussed in a Tools of Astronomy box in Chapter 3. This theory makes the perception of electromagnetic waves independent of the motion of the observer. It allows us to answer Einstein's mirror question with a "yes," but it reveals a link between the nature of space and time that does not appear in our everyday perception of the universe. Special relativity has some interesting effects, but it is Einstein's expansion of the theory to the *general theory of relativity,* or *general relativity,* that must be used to describe the fate of massive stars.

In Section 3-10 we described how the general theory of relativity provided a different way of explaining what we normally call gravitation. We began the discussion with a "thought experiment" involving a woman in a spaceship far from Earth. We reached the conclusion that if the spaceship were accelerating upward with an acceleration equal to the acceleration due to gravity on Earth, the woman could not distinguish her situation in the accelerating spaceship from one in a stationary spaceship on Earth. That is, an object would seem to fall in the accelerating spaceship just as if the ship were in a gravitational field. General relativity is based on the *principle of equivalence,* which states that there is no way to distinguish between an accelerating reference frame in a location where there is no gravitational field and a nonaccelerating reference frame that is in a gravitational field.

Now we extend the thought experiment to include electromagnetic radiation. Suppose that the woman is in a spacecraft on the surface of the Earth, not accelerating. The spacecraft has a window on one side, and through the window a beam of light enters parallel to the floor. The principle of equivalence leads to a prediction that the beam will bend downward as it crosses the spacecraft. To see why this is so, imagine that the spacecraft is not on Earth but is in deep space, with an acceleration upward equal to the acceleration due to gravity. A quick burst of light enters the craft, as shown in **Figure 15-19**. As the craft accelerates upward, gaining speed all the time, the light takes the path shown in parts (b) and (c) of the figure. Notice that from our point of view, in the unaccelerated frame of the page in the book, the light continues in a straight line. From the point of view of the woman in the accelerating craft, however, the light will have bent downward, as shown in **Figure 15-20**.

The principle of equivalence says that the same bending should happen when the spacecraft is stationary on the surface of the Earth. In this case, however, the bending will appear to be due to gravity.

Figure 15-19
A burst of light entering the accelerating spaceship continues in a straight line (a), but since the spaceship is accelerating upward (b), the light hits the floor (c).

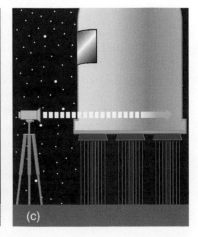

Burst of light

Click! Click!

(a)   (b)   (c)

Figure 15-20
As seen by a person in the spacecraft (a), the light beam bends downward (b) as if falling to the floor (c) The acceleration of the beam would be measured to be exactly equal to the acceleration of the ship.

The drawings that show the bending of light exaggerate the amount of bending, of course, for in the time required for a beam of light to cross a room-sized spacecraft, its acceleration would not result in enough curvature of the light's path to make it measurable. But if the woman's craft were in an extremely strong gravitational field, where the acceleration due to gravity is 10 billion times greater than on Earth, the bending would be appreciable.

If the principle of equivalence is valid, light should bend in the presence of a massive object. This prediction was made by Einstein's theory in 1915, but the predicted amount of bending near the Earth was very small. No experimental check of the prediction was done until a solar eclipse in 1919, after the end of the First World War. When the Sun is totally eclipsed by the Moon, stars can be seen in the sky, and this provides an opportunity to observe the bending of light that originated at a distant star as the light passes near the Sun.

Suppose a total eclipse occurs while the Sun—as seen from Earth—is between two bright stars. **Figure 15-21a** shows the stars as they normally appear. During the eclipse, shown in part (b) of the figure, light from the stars must pass near the Sun before reaching Earth. The theory predicts that the light will be bent and will therefore make the stars appear slightly farther apart, as shown in part (c). In practice, the bending was predicted

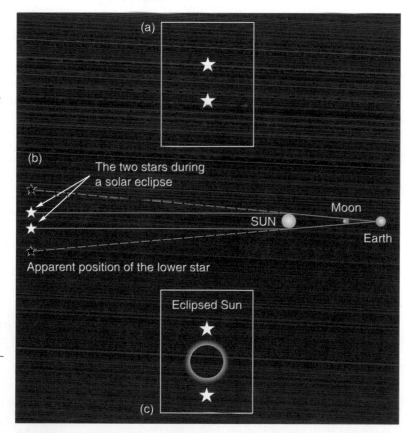

Figure 15-21
Light from the two stars is bent as it passes near the Sun, causing the stars to appear farther apart.

to be very little, only 1.75 arcseconds. This would produce very little change in the apparent position of a star close to the Sun, much less than indicated in the figure. However, the eclipse of 1919 did produce results in agreement with the general theory of relativity, and it provided the first experimental confirmation of the theory. Since then, similar measurements have been taken during other eclipses and have always confirmed the predictions of Einstein's theory.

Newton's theory predicts no gravitational effect on light, since light has no mass. And, indeed, light is not observed to respond to gravity in our everyday world. Only when very strong gravitational fields are involved is the bending of light observed. Since Newton's theory had never been checked in such strong fields, no one realized that it makes incorrect predictions in those cases.

Thus the general theory of relativity presents a different way of looking at the phenomenon we call gravitation. In Chapter 3 we described how the theory explains gravitation not as a force but as the result of the curvature of space. The particular result of interest to us here is that the theory predicts that electromagnetic radiation responds to this curvature in a way that makes it seem to be responding to the force of gravity. The present discussion will continue to speak of the force of gravity in Newtonian terms, but we will accept the Einsteinian prediction that electromagnetic radiation, including visible light, appears to respond to this force.

General relativity has survived every test to which it has been put (some of which were described in Chapter 3), and it is a well-accepted theory—not a hypothesis, as is sometimes suggested by the media.

A Binary Pulsar

In 1974, Joe Taylor, a professor at the University of Massachusetts, and Russell Hulse, a graduate student working with him, were using the giant radio telescope at Arecibo to study pulsars. They discovered a very unusual pulsar, whose pulse period of about 59 milliseconds seemed to be changing in a regular way. Sometimes the pulses were received a bit later than expected, sometimes a bit earlier, and these variations repeated with a period of about 7.75 hours. After nearly a month studying the object, they deduced that the changes were due to the Doppler effect as this pulsar revolved around a companion in a binary system. The companion, however, could not be detected. As we explained in previous chapters, binary star systems are useful in determining the masses of stars, but when the spectrum of only one star is observed, the object's mass cannot be calculated. However, astronomers were very interested in determining the masses of the objects in this case, for it would provide the first measurement of the mass of a pulsar and thereby serve as a check on the neutron star/pulsar theory.

Fortunately, the short period of revolution of this binary system (about 7.75 hours) allowed Hulse and Taylor to observe many orbits over a few weeks. This permitted them to determine that the pulsar's orbit was precessing at the rate of about 4.2 degrees/year, a precession much greater than Mercury's because of the great mass of the pulsar and its nearby companion.

The amount of precession predicted by general relativity depends only on the masses of the two objects in orbit and their distance of separation. If the masses of the objects had been known, the binary pulsar would have served as one more experimental check on general relativity. On the other hand, if the applicability of the theory is assumed, general relativity can be used to calculate the masses of the two objects. By 1978, observational data suggested that both stars have masses very close to 1.4 solar masses (the Chandrasekhar limit), strengthening the conclusion that the companion star was also a neutron star. After nearly two decades of observing the system, the orbits of the two objects are known very precisely, allowing a precise calculation of their masses. The pulsar has a mass of 1.441 solar masses and its companion's mass is 1.3874 solar masses. The system has been perfect for studying the

If we were to use Newtonian mechanics to describe the motion of a photon as a particle (of mass m and speed c) in a gravitational field, we would find that the photon's path is deflected by an angle that is half the size predicted by the general theory of relativity.

Joe Taylor and Russell Hulse won the 1993 Nobel Prize in Physics for this work.

We discussed the precession of Mercury's orbit in Section 3-11.

TOOLS OF ASTRONOMY

The Distance/Dispersion Relationship

In Chapter 12 we described various methods to measure distances to stellar objects. Pulsars provide another method, which relies on two phenomena: each burst of radiation from a pulsar contains an entire spectrum of wavelengths, and different wavelengths of electromagnetic radiation travel through space at slightly different group speeds. For many purposes, it can be assumed that all wavelengths travel at exactly the same speed in space—the so-called *speed of light*—but in practice, since space is not a perfect vacuum, longer wavelengths travel at slightly lower group speeds than shorter wavelengths. Recall from Section 5-2 that this difference in speed—called *dispersion*—is the same property that causes chromatic aberration in lenses. Its effect here is that as each pulse of radiation from a pulsar travels through space, the longer wavelengths get slightly behind.

When we stated in the text that the pulses from a pulsar are typically only 0.001 second in length, we were referring to pulses of a single wavelength. If we consider, for example, a pulsar that emits visible light as well as radio waves, we find that the visible-light portion of each pulse reaches us before the radio portion, although for any one wavelength of visible light, or any one wavelength of radio energy, the pulse length is the same—perhaps 0.001 second. The relative amount of dispersion might be stated as the time that elapses between the detection of a given wavelength of visible light and the detection of a given wavelength of radio energy in the same pulse.

Two factors determine the amount of dispersion: the distance the pulse travels through the interstellar medium and the dispersion properties of that medium. We have here another triple connection: distance to the pulsar, dispersion of the pulse, and the dispersive nature of the interstellar material (**Figure B15-1**). The amount of dispersion can be measured directly. Thus, to the degree that one of the other quantities is known, the final one can be calculated. If the dispersion properties of the interstellar matter between us and a pulsar are known, the distance to the pulsar can be determined. On the other hand, if distances can be determined by another method, this provides a means of learning more about the interstellar material.

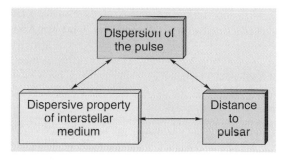

Figure B15-1
If any two of the three quantities are known, the third can be calculated.

predictions of general relativity: it was the first instance of the use of general relativity to calculate a stellar property and the first precise determination of the mass of a neutron star.

What makes this binary system special, however, is that it indirectly confirms another of general relativity's predictions. As we discussed before, according to the general theory of relativity, matter tells space how to curve and space tells matter how to move. If the distribution of matter in a volume of space changes (for example, because of the motion of two stars in a binary system), the resulting changes ("ripples") in the curvature of the surrounding space may propagate outward as a ***gravitational wave,*** carrying away energy and angular momentum. As a result, general relativity predicts that the two stars will spiral closer together. Since the masses of the stars were reasonably well measured by 1978, general relativity predicted that the orbital period should get about 75 microseconds shorter every year. By 1983, the observed decay rate gave a value of 76 ± 2 microseconds/year, in excellent agreement with theory. The separation of the two stars becomes smaller by about 3 millimeters/orbit and the two stars will come together in about 300 million years. Even though we still have not directly observed gravitational waves, this binary system has made it very hard to doubt their existence.

The *Laser Interferometer Gravitational-wave Observatory* (LIGO) began operating in late 2002. It should be able to detect gravitational waves from collapsing cores of supernovae or colliding neutron stars up to 70 million light years away.

gravitational wave Ripples in the curvature of space produced by changes in the distribution of matter.

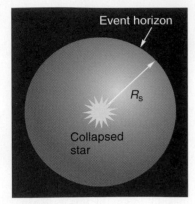

Figure 15-22
Once a star has shrunk to inside the Schwarzschild radius, light can no longer escape its surface. The Schwarzschild radius forms a sphere whose surface is called the event horizon.

Schwarzschild radius The radius of the sphere around a black hole from within which no light can escape.

black hole An object whose escape velocity exceeds the speed of light.

15-8 The Fate of Very Massive Stars

Now that we have reviewed general relativity, let's return to the examination of the evolution and death of very massive stars. A very massive star proceeds through its life in basically the same manner as a moderately massive star, although each stage occurs more quickly. Very massive stars differ from moderately massive ones primarily in what happens to them when their core is compressed to a greater density than electron degeneracy can support. When this occurs in a moderately massive star, the resulting supernova leaves a neutron star at its center. In a very massive star, an even more spectacular event happens: the core becomes a black hole. The general theory of relativity plays a very important part in understanding black holes.

Black Holes

Neutron degeneracy cannot support a neutron star whose mass is greater than about three solar masses. Such a star will collapse. How far it will collapse is still an open question, but there is no known force capable of keeping the force of gravity from collapsing a star to zero size. This is an unimaginable situation, for we cannot envision matter having no size whatsoever, especially an object several times more massive than the Sun. Fortunately, we do not have to answer the question of how far the star collapses in order to predict how it will appear from outside.

Shortly after Einstein introduced general relativity, Karl Schwarzschild calculated that when a star collapses to a dimension equal to or less than what is now called the *Schwarzschild radius,* light is unable to escape the object. Recall the escape velocities of the various solar system objects that were noted earlier (in Section 7-5). The escape velocity of an object is the minimum velocity a projectile must have in order to escape the gravitational field of the object. (The escape velocity from the Earth's surface is about 11 kilometers/second.) As a star decreases in size, the escape velocity from its surface becomes greater. When its radius becomes so small that its escape velocity is greater than the velocity of light, the star has reached the Schwarzschild radius (**Figure 15-22**).

The size of the Schwarzschild radius depends upon the mass of the star, for it is this mass that determines the force of gravity. The approximate radius (R_S) is given in kilometers by the formula

$$R_S = 3M \ (R_S \text{ in km}; M \text{ in solar masses})$$

where M is the mass of the star expressed in multiples (or fractions) of the Sun's mass. Thus a star with a mass five times greater than the Sun's will have a Schwarzschild radius equal to about 15 kilometers. If we had the power to squeeze the mass of our Sun ($M = 1$) into a sphere of radius $R = 3$ kilometers, then it would become a black hole. This in no way would affect the Earth's orbit, since at 1 AU our planet is safely far away and Newtonian physics would still be all we need to describe its motion. The only problem, of course, would be the lack of sunlight and its negative effects on life on Earth.

A star whose radius is less than the Schwarzschild radius is called a *black hole;* "black" because no light escapes from it, and "hole" because matter—or light—that falls into it can never be retrieved. A black hole exists if the star's radius is equal to or less than the Schwarzschild radius, but the radius of the black hole is considered to be the Schwarzschild radius.

The previous equation indicates that if matter falls into a black hole, the Schwarzschild radius becomes larger. This is reasonable, for as matter falls into the black hole, more mass is inside exerting more gravitational force and increasing the escape velocity at any given distance.

A name given to the surface of the sphere formed at the Schwarzschild radius illustrates an important and interesting feature of black holes. The surface is called the *event horizon.* Just as we cannot see beyond the Earth's horizon, we cannot see inside the event horizon. More than that, there is no way we can know about an event inside that sphere. Nothing that happens there is accessible to us and may just as well be in another universe, since we can have no experience of it. As far as we know, once a star collapses inside its event horizon, no force in the universe can stop the star from collapsing to a single point of infinite density. This point, at the center of the black hole, is called the **singularity.** Here the strength of gravity—and thus the curvature of space and time—is infinite. Time and space do not exist as separate entities at this point. Random things can happen here but do not affect the world outside the event horizon. It is the unpredictability of events and our lack of understanding as to what happens at a singularity that has forced astronomers to accept the proposition that singularities not surrounded (or shielded) by an event horizon cannot exist in nature. Thus, the structure of a black hole that is not rotating is easily described by the singularity and the event horizon.

Properties of Black Holes

Although the theory of black holes involves general relativity and therefore might be considered complicated, black holes themselves are very simple. A black hole can be described completely by only three numbers: one for its mass, one for its electric charge, and one for its angular momentum. A full description of any other object in the universe would involve a long list of properties, including size, color, chemical composition, texture, and temperature (in addition to the three properties named above). For a black hole, these other properties have no meaning. For example, chemical composition is meaningless because it does not matter what original material condensed to form the black hole. The matter no longer exists as a chemical element in the black hole. Whatever the properties of the material that formed the black hole, that information is forever removed from the universe.

The mass of a black hole is, in principle, easy to measure. It would be measured the same way as the mass of any other celestial object. That is, we would observe an object that is in orbit around the black hole and apply Kepler's third law to calculate the total mass of the black hole and the object.

A black hole might have an electric charge, either positive or negative. In principle, this charge could be measured by detecting its effect on charged objects near the black hole. In reality, a black hole is not expected to have an electric charge, because if it had a positive charge, for example, the black hole would draw in negative charges until it became neutral. For this reason, electric charge is not normally considered in discussing black holes.

The final property that a black hole may have is angular momentum. Recall that as an object shrinks in size, it increases its rotation rate, and therefore a black hole would be expected to spin rapidly. Results of the theory of general relativity show that this rapid rotation would change space-time around the black hole. In effect, the spinning black hole would drag nearby space-time around with it. Because of this, light passing near a rotating black hole on one side behaves differently than light passing by the other side. On one side the light is moving along with space-time, and on the other side it is moving against the motion of space-time. This, of course, is not easy to imagine, for motions of space-time are not part of our everyday experience. If instead of light we used two spaceships orbiting the black hole in opposite directions, they would have two different orbital periods. Thus we could measure the black hole's angular momentum by comparing these orbital periods.

Because a black hole can be described completely by just three properties, astronomers say that "a black hole has no hair." Any properties of objects that fall into the black hole are forever gone from the universe. They leave behind only the properties of mass, electric charge, and angular momentum. A black hole is a simple, "hairless" thing.

event horizon The surface of the sphere around a black hole from within which nothing can escape. Its radius is the Schwarzschild radius.

singularity The center of a black hole, where density, gravity, and the curvature of space and time are infinite.

Describing the structure of a *rotating* black hole requires the idea of an additional region surrounding the event horizon. In this region, the dragging of space and time is so severe that nothing can remain stationary. However, objects could travel through it without being sucked into the black hole.

In my talk, I argued that we should consider the possibility that at the center of a [supernova remnant] is a gravitationally completely collapsed object. I remarked that one couldn't keep saying "gravitationally completely collapsed object" over and over. One needed a shorter descriptive phrase. 'How about black hole?' asked someone in the audience. . . . Suddenly this name seemed exactly right.
John Archibald Wheeler, usually credited with coining the term "black hole"

Detecting Black Holes

Astronomers predicted the existence of black holes in the 1930s when they realized that a star's mass may cause it to collapse beyond neutron degeneracy. A prediction of something as unusual as a black hole certainly calls for observational verification. But how?

There is, of course, no hope of seeing a black hole directly, for we cannot see something from which no light escapes. If, however, matter were falling into a black hole, we should expect some of that matter to orbit the black hole in a manner similar to the way matter falling into a white dwarf orbits it (causing a nova as it falls onto the white dwarf). Since the gravitational field near a black hole is so strong, the orbital speed of nearby matter would be extremely great. As collisions among particles turned the regular orbital motion to random thermal motion, the matter would reach temperatures of hundreds of millions of degrees. **Figure 15-23** illustrates material being pulled into orbit around a black hole from a companion binary star. Such hot material would radiate great amounts of energy and since it is not yet inside the event horizon, we should be able to detect it. We can predict the characteristics of the radiation, and we know that the object should appear as an X-ray source.

Numerous X-ray sources have been found in the heavens, particularly by NASA's *HEAO*s and by the Japanese and the European orbiting X-ray observatories. Are all of these black holes? Probably not. Only if one of these sources is found to be associated with a particularly massive star can we hope that it is a black hole. When we wish to know the mass of a star, we search for a binary system. Then if we find a binary system in which one of the stars is invisible with a mass greater than four or five solar masses, we can conclude that the star must be collapsed (or otherwise it would be visible). Finally, if the star emits X-rays characteristic of those predicted for a black hole, we would have good evidence for claiming to have found a black hole.

In the 1960s, astronomers discovered an X-ray source in the constellation Cygnus. Because it was the first X-ray source found in that constellation, it was named Cygnus X-1. Emission from this source is highly variable and the time scale for the flickering can be as small as 0.01 second. Recall that such a time scale sets an upper limit to the size of the object emitting the signals. In this case, the size of the emitting region cannot be larger than 0.01 light-second or about 3000 kilometers, which is smaller than the size of the Earth. Then, in 1971, they discovered that the location of Cygnus X-1 corresponds with a ninth-magnitude star named

◆ Actually, Pierre Simon Laplace proposed in 1798 that a very massive star might have such a strong gravitational force that light could not escape it. This was 100 years before Einstein's theory linked gravitation with the travel of light!

Figure 15-23
If a black hole and red giant or supergiant form a binary system, material will be pulled from the giant (left) and will swirl around the black hole, causing X-rays to be released from the heated material in the disk.

HDE226868, a blue supergiant of spectral type B0. A periodic Doppler shift of the spectrum of the supergiant indicates that it is part of a binary system with a period of 5.6 days, but its companion is invisible. A B0-type supergiant is expected to be a very massive star. Calculations reveal that if HDE226868 has the expected mass of about 30 solar masses, its companion must have a mass greater than 3 solar masses. If so, the companion is probably a black hole.

Cygnus X-1 was the first candidate for a black hole. For years astronomers sought other explanations for the behavior of the object. For example, if the mass of HDE226868 is less than normal for a B0-type supergiant, its companion could have a mass less than the Sun. If this is so, the companion may be a neutron star instead of a black hole. Such explanations cannot be made to fit all the data, however, and it appears that Cygnus X-1 is indeed a black hole.

Since the discovery of Cygnus X-1, other black hole candidates have been found. A few are even part of eclipsing binary systems, thus providing us with further information about them.

In 1989, the star V404 Cygni, a G- or K-type star in our own galaxy, called attention to itself by erupting with a powerful X-ray flare. In 1992, three European astronomers reported that Doppler shift analysis of V404 Cygni shows that it orbits an unseen companion with a period of 6.47 days (**Figure 15-24**). In this case, the mass of the visible star does not present a problem, for even if the calculation is based on its least possible mass, the dark companion must have a mass of at least 6.3 solar masses, and probably 8 to 12. V404 Cygni is now considered almost surely to be a black hole. It further convinces astronomers that black holes are a reality.

Another black hole candidate is the flickering X-ray source A0620-00, which is also a member of a spectroscopic binary system. In this case, the visible companion (a K5 main sequence star) is relatively faint and orbits the X-ray source every 7.75 hours. Since we can observe the shifting spectral lines of both stars, the minimum mass of the X-ray source is about 3.6 solar masses and thus almost certainly a black hole. In Chapter 17 we discuss black hole candidates that astronomers are finding in the centers of some galaxies. The discovery of black holes not only confirms the theory of the death of the most massive stars but also serves as another confirmation of the theory of general relativity. Although the general public may still think that black holes are on the fringes of science, these objects are in fact firmly entrenched in astronomical theory.

Black-hole candidates have been observed to have masses that span the entire spectrum, from just barely above the minimum 3.2 solar masses or so that separates the neutron stars from black holes to billions of solar masses at the cores of galaxies. Since the early 1990s, orbiting X-ray observatories have been providing important data that allow astronomers to distinguish between black hole candidates and neutron stars. When a neutron star accretes matter, energy is released when the infalling matter hits the solid surface of the star. However, when a black hole accretes matter, there is no surface for the matter to hit. Instead, both the matter and the energy is lost once they cross the event horizon. A small amount of energy can escape just before the matter crosses the event horizon, but not as much as in the case of a neutron star. Observations made by X-ray satellites (*ROSAT, RXTE* and *Chandra*) support the idea of the event horizon around a black hole by showing that the X-ray emissions from neu-

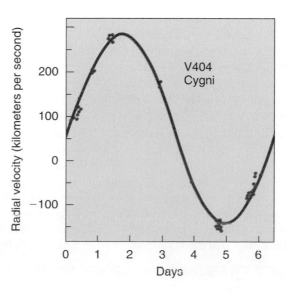

Figure 15-24

Applying Doppler shift analysis to the spectrum of V404 Cygni shows that its radial velocity changes with a period of 6.47 days. The dots indicate measured velocities (by the Doppler shift) at different times, and the curve is drawn to fit the data.

ADVANCING THE MODEL

Black Holes in Science, Science Fiction, and Nonsense

Black holes are fantastic objects and are fruitful subjects for science fiction as well as for a lot of nonsense. We will try to separate the science from the nonsense and the science fiction.

Nonsense

There is a common misconception that a black hole is a giant vacuum cleaner, sweeping up matter across wide portions of space. In fact, the gravitational force of a black hole is unusually great only near the black hole. To make this clear, suppose that the Sun could magically become a black hole without losing any mass. If it did so, the gravitational force it exerts on the Earth would not change. The force exerted on the Earth would remain the same and the Earth would continue in its same orbit. The only difference to us on Earth would be that no radiation would arrive from the Sun. Newton's law of gravity states that the force of gravitational attraction between two objects depends only upon the masses of the two objects and the distance between them. Although predictions from Einstein's theory differ from those of Newton's, these differences show up only where the forces are extremely great, so Newton's theory can still be used to discuss Earth's orbit. The strength of Newton's gravitational force does not depend upon the Sun's size, only its mass.

Where is the increased gravitational field, then? To understand this, note that with the Sun in its present state, the closest you can get to it (and still be outside) is its surface—some 700,000 kilometers from the center. If you could go down inside the Sun, the force of gravity on you would become *smaller*, for there would then be a gravitational force back toward the matter near the surface (**Figure B15-2a**). In fact, at the center of the Sun, you would be weightless, for you would be attracted equally in all directions.

The difference in the case of the solar-mass black hole is that now you can get closer to the center of the star while remaining outside its surface (Figure B15-2b). For such a black hole (only 6 kilometers across), you could get within a few kilometers of the center, and—as predicted by Newton's law of gravity—the force of gravity would be very great at these small distances. (Actually, Newton's law would not make accurate predictions this close to the black hole, but it correctly predicts that the force would be extremely large.)

Predictions from Science

Stephen Hawking made some interesting predictions concerning black holes. Based on calculations of the density of the universe at its beginning (Chapter 18 again), he predicted that it may be possible that *mini* black holes formed at the time. These might have been the size of a pinhead and have had the mass of an as-

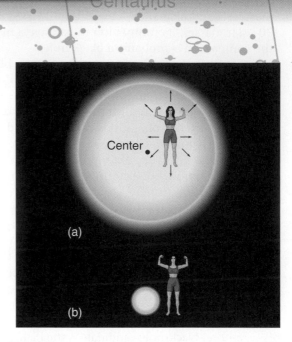

Figure B15-2
(a) If a person could exist inside the Sun, she would weigh less than she did on the surface, for gravitational forces would be exerted on her in all directions by parts of the Sun. (b) If the entire Sun could be shrunk to a small enough ball, a person could be the same distance from its center as in part (a) and still be outside its surface. She would then weigh much more than in part (a).

teroid. Alas, later theorizing by Hawking showed that if such mini black holes ever existed, they would have long since evaporated by radiation.

Unusual phenomena—by earthly standards—occur as an object falls into a black hole. For one thing, as an object nears the black hole, it is pulled apart by tidal forces. Recall that tidal forces result because of the difference between gravitational forces on one side of an object and the other. An object approaching a black hole would feel a much stronger gravitational force pulling on its side nearer the hole than on the other side. The force difference would pull the object apart. It would be impossible for a person to fall into a black hole and remain intact.

Even stranger would be the observation of something falling into a black hole. Forget for now the destruction caused by tidal forces. Einstein's theories of relativity tell us that if we could watch the object fall, we would never see it reach the event horizon. We would see it getting closer and closer to the event horizon and getting redder and redder (Doppler shift-like), but because of the distortion of time that is predicted by the theory of relativity, it would take forever—according to our observation—for the object to reach the event horizon. As time is reckoned on the object, however, it would fall into the hole very quickly.

ADVANCING THE MODEL

Black Holes in Science, Science Fiction, and Nonsense *(Continued)*

Science Fiction

Where does an object go when it falls into a black hole? Writers have speculated that it may appear elsewhere or "elsewhen"—at another place or another time. Such travel through "hyperspace" or through time has lent itself to numerous science fiction plots. If it indeed occurs, we should see "white holes" where matter and energy are appearing out of nowhere. (In the language used, the matter and energy enter a black hole, pass through a "worm hole," and emerge from a white hole.) No such phenomenon has been observed.

Even more speculative is the idea that the matter may come out in another universe—not in another galaxy, but in a parallel universe. Since by definition we have no contact with such a universe, we have no way of verifying such speculation. Thus it is not in the realm of science at all. (Recall that a hypothesis must be verifiable to be classified as scientific.) While the hypothesis of white holes in our universe might perhaps qualify as a scientific hypothesis, the speculation of a parallel universe must remain pure science fiction.

---

tron stars and black holes are different. **Figure 15-25** illustrates these differences based on observations of X-ray novae, so named because they occasionally erupt as brilliant X-ray sources and then settle into decades of dormancy.

Figure 15-26 reviews the steps taken by very massive stars as they progress from protostars to black holes.

Black holes can also be formed by accreting white dwarfs and/or neutron stars if they accrete enough material. Two dead stars in a binary system coming together to form a single object may also give rise to a black hole.

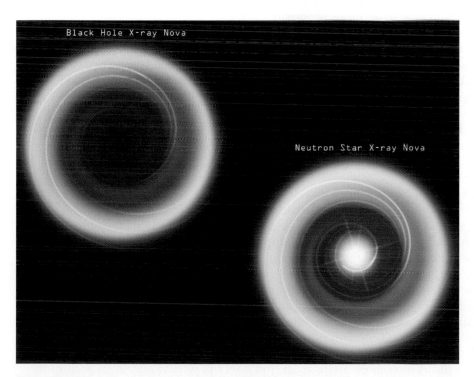

Figure 15-25
Gas from the companion star is drawn by gravity onto the compact object (black hole or neutron star) following a swirling pattern. As it nears the event horizon of the black hole (or the surface of the neutron star), a strong gravitational redshift makes the gas appear redder and dimmer. In the case of the black hole, when the gas finally crosses the event horizon, it disappears from view and the central region is black. In the case of the neutron star, when the gas strikes the star's solid surface, it glows brightly.

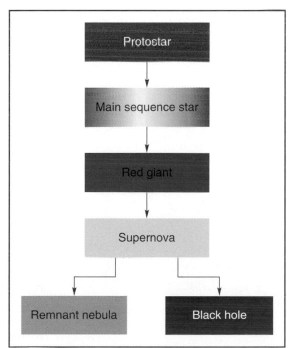

Figure 15-26
The steps in the life of a very massive star.

15-9 Our Relatives—The Stars

The idea that the stars are our relatives might be surprising, but in fact astronomers can show that humans and stars are related—albeit distantly.

Astronomers divide stars into three classes, called *Population I* stars, *Population II* stars, and *Population III* stars. The three populations are distinguished by the amount of heavy elements they contain. The Population III class has been theorized to include the very first generation of stars formed in the cosmos. These stars formed using only hydrogen and helium, the only elements available in the early history of the universe. Population II stars contain very little material in their atmospheres other than hydrogen and helium. Heavier elements are produced in the cores of stars as fusion takes place, so heavy elements do exist in their cores, but little is seen in their atmospheres. The spectra of Population I stars, on the other hand, reveal that their atmospheres contain heavier elements. The separation of stars into these three groups is somewhat arbitrary, since a continuum actually exists from stars that have no elements beyond hydrogen and helium in their atmospheres to those that contain the greatest amounts of heavy elements; nonetheless, the distinction is convenient.

The existence of different amounts of the heavy elements in different stars can be easily explained. Recall that stars are formed from interstellar clouds of gas and dust and that most stars end their lives by blowing (or exploding) much of their mass back out into space. Therefore, the material they expel into space contains some of the heavier elements produced within the star. Thus Population II stars are those old stars that were formed from interstellar material long ago in the history of the universe, before the interstellar material became enriched in heavy elements. Population I stars, on the other hand, are young stars formed from material that contained the remains of previous generations of stars.

Theoretical calculations for Population III stars suggest that these stars lived for only about a million years before extinguishing themselves and showering the metals they had created in their cores into space. How then could we confirm their existence? We could be able to glimpse some signs of their existence by looking in the most distant realms of space. The proposed *James Webb Space Telescope*, the successor to the *HST* (and a collaboration between ESA and NASA), could see these stars exploding as supernovae. ESA's *Integral* satellite, a gamma-ray observatory currently in orbit around Earth, could also provide clues about these stars by observing some of the most violent events known in the universe: the gamma-ray bursts, described in a nearby Advancing the Model box. Some astronomers suspect that some of these bursts are created by the death of these first-generation stars.

In which population does our Sun belong? We can answer this question without knowing anything about the spectrum of the Sun. Recall that the solar system was formed from material that did not fall all the way into the Sun as the interstellar cloud collapsed; the planets were formed from the same material that made the Sun (Section 7.6). The fact that Earth (and the other planets) contains heavy elements means that the cloud from which the planets formed contained those materials. The Sun is a Population I star (although it contains less heavy material than some other stars).

Recall that fusion in the cores of stars continues to release energy as heavier and heavier elements are formed until iron is produced. In order for iron nuclei to fuse with other nuclei to form heavier elements, there must be an *input* of energy. The reaction that forms heavier elements from iron does not release energy; it absorbs it. Although most of the matter of which the Earth is made is less massive than iron, many elements are more massive than iron. Such matter could not have been formed by fusion within the core of a star. It was instead formed during a supernova explosion, when tremendous amounts of energy were available. Most of the energy of a supernova is used up in releasing radiation (especially neutrinos) and in blasting

ADVANCING THE MODEL

Gamma-Ray Bursts

Gamma-ray bursts (GRBs) were discovered by accident in the late 1960s by satellites designed to observe clandestine nuclear detonations in space. By the mid-1980s, most astronomers thought that the bursts originated on nearby neutron stars in our Galaxy. This was based on spectroscopic observations that suggested the presence of intense magnetic fields. In 1991, observations by the *Compton Gamma Ray Observatory* showed that the GRBs were distributed isotropically, with roughly the same number of bursts in any direction in the sky.

A burst may last from a hundredth of a second to tens of minutes, and a few bursts are observed each day from somewhere in the universe. For a short time period, some bursts become the brightest objects in the gamma-ray sky (**Figure B15-3**). The distances to these energetic events were not known for over 20 years, until early 1997 when the Italian-Dutch satellite *Beppo-SAX* provided X-ray measurements that allowed astronomers to pinpoint the position of a burst to within a few arc minutes in the Orion constellation. This enabled them to take additional data using optical and other telescopes. These observations confirmed that gamma-ray bursts are very distant and, therefore, that extremely energetic explosions must cause them. Two years later, astronomers tracked the visible glow of another GRB while it was still emitting gamma rays. The intensity of the eruption was measured to be equal to that of millions of galaxies.

What is the source of this incredible amount of energy? In the first of the two leading scenarios, GRBs are produced during the merger of a neutron star and a black hole, or a pair of neutron stars or black holes. In the second scenario, the core of a very massive star (more than 30 solar masses) collapses to form a rapidly rotat-

ing black hole surrounded by a disk, while the star's outer layers are ejected to form a supernova. A short time later, the black hole-disk system produces energetic jets of high-speed particles that shoot out (**Figure B15-4**). The jets create a traffic jam close to the explosion, which quickly heats up and explodes. A shock wave is generated from this fireball and travels at a speed very close to the speed of light. If the jet and shock wave are aimed toward Earth, we observe the emitted gamma and X-rays as a burst. When the jet interacts with the expanding supernova shell, it produces an afterglow in X-rays and other frequencies, which can last for days or even months. Support for this scenario was provided by a 21-hour *Chandra* observation in August 2002 of the afterglow of GRB020813, which revealed the presence of elements characteristically ejected by the supernova explosion of a massive star. Additional support was provided in April 2003 by observations of the afterglow of GRB030329 by the Multiple Mirror Telescope.

In either scenario, a burst of gamma rays seems to signal the birth of a new black hole. While the origin of GRBs is still a mystery, an analysis of 2000 bursts supports the idea that long bursts (those lasting more than 2 seconds) originate from explosions of very massive stars (more than 30 solar masses), while short bursts (those lasting less than 2 seconds) result from mergers.

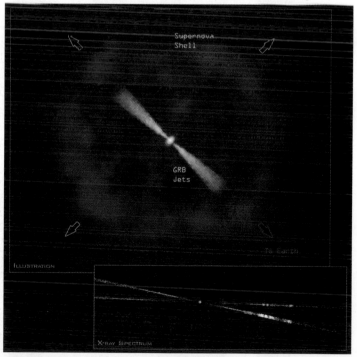

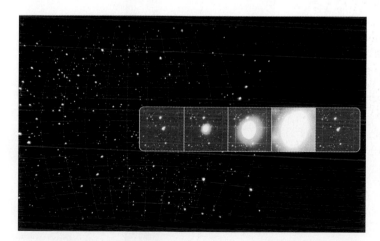

Figure B15-3
An artist's impression of a GRB illustrating how it can flare dramatically over a short time period. There is no way to predict when or where a GRB will occur.

Figure B15-4
An illustration that shows the connection between a GRB and a supernova explosion.

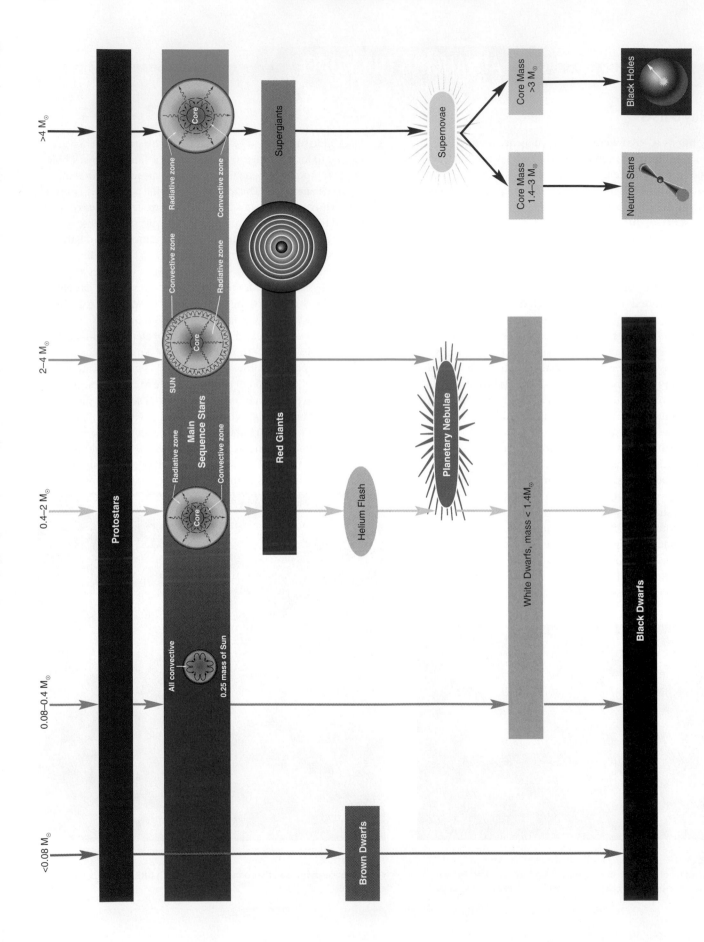

Figure 15-27

A summary of the steps in the life cycles of stars of various masses. The limits in mass are not well known.

away the outer parts of the star, but a small fraction is absorbed by the material of the star, where light elements are fused into heavy elements.

This discussion has implications for us humans. The material that makes up our bodies is from the Earth. Where was it before it was part of the Earth? In an interstellar cloud. And before that? In a star!

Harlow Shapley, former director of the Harvard College Observatory, whose work we will discuss later, lists cosmic evolution as one of the ten revelations that have most affected modern humans' life and thought. He says, "Nothing seems to be more important philosophically than the revelation that the evolutionary drive, which has in recent years swept over the whole field of biology, also includes in its sweep the evolution of galaxies, and stars, and comets, and atoms, and indeed all things material."*

We are star stuff.

Astronomer Carl Sagan

Conclusion

All stars get their starts when shock waves compress parts of cold interstellar dust clouds. Most differences between stars—from the various stages through which each star will progress as it lives and dies, to how long each of those stages lasts—result entirely from differences in mass. The universe itself has not existed long enough for the least massive stars to have ended their lives, but the most massive stars have short lives, and at the end they become the most astonishing things in nature: black holes. **Figure 15-27** summarizes the life cycles of all of the stars: very low mass, moderately low mass, moderately massive, and very massive stars.

With this chapter we complete our examination of the stars. Just as our horizons expanded many times over when our focus changed from the solar system to the stars, they will expand again in the next chapter, where we begin to study the largest class of objects in the universe: the giant galaxies of stars.

*Harlow Shapley, *Beyond the Observatory* (New York: Scribner, 1967), pp. 15–16.

Study Guide

1. The core of a mature supergiant
   A. is layered, with heavier elements in the center.
   B. is layered, with lighter elements in the center.
   C. is uniformly composed of hydrogen and helium.
   D. is uniformly composed of helium.
   E. is composed primarily of iron and carbon.

2. Which type of supernovae have hydrogen lines in their spectra?
   A. Type I (from binary systems).
   B. Type II (from single stars).
   C. [Neither type.]

3. The most massive element that can be formed by nuclear fusion with the liberation of energy is
   A. helium.
   B. carbon.
   C. oxygen.
   D. iron.
   E. lead.

4. The most recent naked-eye supernova was observed
   A. in 1054 by the Chinese.
   B. in the 1500s by Tycho Brahe.
   C. around 1600 by Kepler.
   D. in 1929 by Hertzsprung and Russell.
   E. in 1987 by Ian Shelton.

5. SN1987A was the explosion of a
   A. red giant.
   B. blue supergiant.
   C. white dwarf.
   D. pair of stars.

6. The pulse rate of pulsars is slowed because
   A. they convert rotational energy into radiation.
   B. they drag companion stars around.
   C. of friction with the interstellar medium.
   D. of the conservation of angular momentum.

7. The first candidate for a black hole was
   A. Cygnus X-1.
   B. Algol (the "Demon Star").
   C. Sigma Xi.
   D. Centaurus.
   E. Xi Ursa Majoris.

8. Theory predicts that a neutron star should spin fast because
   A. it was given increased speed by the supernova explosion.
   B. it was given increased speed by a companion star.
   C. it conserved mass as it collapsed.
   D. it conserved angular momentum as it collapsed.
   E. [The statement is false; neutron stars are not predicted to spin fast.]

9. Which object is the same as a neutron star?
   A. The Crab nebula.
   B. A Cepheid variable.
   C. A dark nebula.
   D. A pulsar.
   E. [None of the above.]

10. Pulsars are
    A. larger and more massive than neutron stars.
    B. larger but less massive than neutron stars.
    C. smaller but more massive than neutron stars.
    D. smaller and less massive than neutron stars.
    E. the same as neutron stars.

11. Pulsars emit sharp bursts of pulses, each lasting for less than one second. From this we can conclude that pulsars
    A. have a great amount of energy.
    B. have little energy.
    C. are small in size.
    D. are moving rapidly away from us.
    E. are far away from Earth.

12. Which is the correct sequential order for the life cycle of a star?
    A. red giant, white dwarf, main sequence, protostar.
    B. protostar, main sequence, red giant, white dwarf.
    C. white dwarf, protostar, main sequence, red giant.
    D. protostar, main sequence, white dwarf, red giant.
    E. white dwarf, red giant, main sequence, protostar.

13. The first pulsar was discovered by
    A. a professional American astronomer using a radio telescope.
    B. a professional American astronomer using a visible-light telescope.
    C. a professional British astronomer using a radio telescope.
    D. a professional British astronomer using a visible-light telescope.
    E. a British graduate student using a radio telescope.

14. According to present theory, the pulses of radiation from a pulsar are due to
    A. pulsations of the surface of the star.
    B. pulsations from within the core of the star.
    C. eclipses of the star by a binary companion.
    D. rotation of the star.
    E. [Any of the above, depending upon the particular pulsar.]

15. What remains after a supernova?
    A. A main sequence star.
    B. A white dwarf.
    C. A neutron star.
    D. A black hole.
    E. [Either C or D above, depending upon the mass of the star.]

16. Which of the following choices lists the stages in the life of very massive stars after they leave the main sequence?
    A. Supergiant, supernova, black hole.
    B. Supergiant, supernova, neutron star.
    C. Supergiant, white dwarf, black dwarf.
    D. Supergiant, planetary nebula, white dwarf.
    E. Planetary nebula, supergiant, black hole.

17. When more material falls into a black hole, the diameter of its event horizon
    A. increases.
    B. remains the same.
    C. decreases.

D. [Any of the above, depending upon the nature of the black hole.]

18. In which of the following categories are the objects the least dense?
    A. Main sequence stars.
    B. Nebulae.
    C. Pulsars.
    D. Red giants.
    E. White dwarfs.

19. Which objects have the greatest absolute magnitude when they are at their brightest?
    A. Black holes.
    B. Protostars.
    C. Pulsars.
    D. Supernovae.
    E. White dwarfs.

20. Which objects were first detected by Jocelyn Bell, a graduate student at Cambridge University in England?
    A. Black holes.
    B. Protostars.
    C. Pulsars.
    D. Supernovae.
    E. White dwarfs.

21. What do Cygnus X-1 and V404 Cygni have in common (other than that they are in the same constellation)?
    A. Both are neutron stars.
    B. Both are planetary nebulae.
    C. Both are white dwarfs.
    D. Both are probably black holes.
    E. Both are supergiants.

22. Which of the following statements is true?
    A. Black holes have been detected based on their gravitational effect on another star.
    B. Black holes have been detected based on radiation emitted from material falling into them.
    C. Black holes cannot be detected.
    D. [Both A and B above.]

23. The principal factor that determines the evolution of a star is
    A. its location in the sky.
    B. its radial velocity.
    C. its transverse velocity.
    D. its mass.
    E. [Both B and C above.]

(The following question is from an Advancing the Model box):

24. If the Sun could magically and suddenly become a black hole (of the same mass), the Earth would
    A. continue in its same orbit.
    B. be pulled closer, but not necessarily into the black hole.
    C. be pulled into the black hole.
    D. fly off into space.

25. Compare the luminosity of a supernova to that of a massive main sequence star.

26. What happens to the size of a main sequence star when mass is added to it? A white dwarf? A neutron star? A black hole?

27. Outline the stages in the lives of moderately massive and very massive stars. What causes them to end their lives differently?

28. What leads us to believe that the object seen by the Chinese in 1054 was a supernova?

29. Describe how pulsars were discovered. What led astronomers to conclude that they were neutron stars?

30. When astronomers were searching for the nature of pulsars, why, if one assumes that the pulses from a pulsar are caused by vibrations in their surfaces, did the observed rates of pulsation seem to rule out both white dwarfs and neutron stars?

31. Describe the lighthouse model of pulsars. What observation(s) led astronomers to select this model from among other suggested models?

32. What is a black hole? How can we expect to observe one?

33. Define *Schwarzschild radius* and *event horizon*.

34. What general statement can be made about the escape velocity from a black hole?

35. If the Sun (magically) became a black hole, what would be the effect on the Earth? Explain.

36. What is the observational evidence that black holes exist?

37. List in order the steps in the evolution of the most massive stars.

---

1. The fusion of lightweight nuclei is said to produce energy, yet fusion cannot occur until a high temperature is reached. Doesn't the high temperature indicate that energy is being supplied to the reaction rather than being produced by it? Explain.

2. Explain why, if a star could instantaneously change its luminosity, we would not see it as an instantaneous change. Explain how the time observed for the change allows us to calculate something about the star's size. Does this method yield the minimum or maximum size of the star?

3. When astronomers were looking for pulsars in supernova remnants, they automatically searched the Crab nebula, for the result of a recent supernova. If the lighthouse model is correct, why was it unlikely that we would find a pulsar there (and therefore lucky that one was found)?

4. The Schwarzschild radius is not the radius of the matter that makes up the black hole. Of what is it the radius?

5. Explain why, if HDE 226868 has less mass than expected, Cygnus X-1 would have too little mass to be a black hole. Why is this not a problem with V404 Cygni?

6. Explain why main sequence stars, white dwarfs, neutron stars, and black holes respond differently with regard to changes in their sizes when matter is added to them.

7. The two parts of Figure 15-5 were not taken with exactly the same orientation and are not printed with the same magnification. Match the stars in the two figures.

8. Figure 15-11 is not an actual chart of the first pulsar, but is an artist's conception. The actual pulses had a duration of about 0.001 second. How would you change the figure to make it accurate in this regard?

9. The Veil nebula is very faint in Figure 15-7a. Find it and describe where it is located in the figure.

10. Explain why the observation that radiation from the first pulsar was appearing four minutes earlier each night led astronomers to conclude that the pulses were not of earthly origin (or, if they were, that astronomers had produced them).

11. The rotation of the Crab pulsar slowed due to the electrons sweeping away from it. Compare this to the problem of the rotational speed of the Sun that was discussed in Section 7-6.

12. In order to calculate how fast a white dwarf (or neutron star) would pulse by vibrating its surface, we must know the strength of the gravitational field at its surface. What other property of the object must be known?

13. It is predicted that supernovae explode at the rate of one per second over the entire universe. Why don't we see more of them?

---

1. What is the value of the Schwarzschild radius of a star of 7 solar masses?

2. If the mass of a black hole is doubled, how does the size of its event horizon change?

3. Betelgeuse is a red giant 428 light-years away. Tycho's supernova was 9800 light-years away. If Betelgeuse were to become a supernova, how many times brighter than Tycho's supernova would it appear?

4. In section 15.5 we stated that Kepler's third law (as modified by Newton) for an eclipsing binary

with a period of 1 second or less suggests that the average distance between the two stars must be less than a few thousand kilometers. Show that this is a correct statement. (Hint: What is the maximum mass that each star could possibly have?)

5. In section 15.5 we stated that the strength of the magnetic field of a neutron star is orders of magnitude larger than that of its progenitor. Show that this statement is correct. (Hint: Would the radius of the progenitor star be larger than the radius of our Sun? How big is a neutron star? Compare the corresponding surface areas.)

---

1. "The Great Supernova of 1987," by S. Woosley and T. Weaver, in *Scientific American* (August, 1989).

2. "Pulsars Today," by F. Graham-Smith, in *Sky & Telescope* (September, 1990).

3. "Binary Neutron Stars," by T. Piran, in *Scientific American* (May, 1995).

4. "Do Black Holes Exist?" by J. McClintock, in *Sky & Telescope* (January, 1988).

5. "Wormholes and Time Machines," by P. Davies, in *Sky & Telescope* (January, 1992).

16 The Milky Way Galaxy

The *Cosmic Background Explorer (COBE)* satellite provided this false-color image (left) of the inner part of the Galaxy. The wide-angle view shows the nuclear bulge and the dense central plane near the bulge. The Milky Way seen in radio wavelengths (right). Images of the Milky Way at different wavelengths were shown in Figure 5-36.

SERENDIPITY CAN BE DEFINED AS "MAKING A FORTUNATE and unexpected discovery by accident." An example occurred in 1933, when Karl Jansky of Bell Telephone Laboratories was using an antenna he had built in order to study radio static, hoping to improve radio telephone service to Europe. Jansky was studying the direction of arrival of static from thunderstorms. He detected radio waves he could not account for, and after much work he determined that they were coming from the center of the Galaxy. Years later, in a letter to a colleague, Jansky wrote:

> As is quite obvious, the actual discovery, that is, the first recording made of galactic radio noise, was purely accidental and no doubt would have been made sooner or later by others. If there is any credit due me, it is probably for a stubborn curiosity that demanded an explanation for the unknown interference and led me to the long series of recordings necessary for the determination of the actual direction of arrival.*

*Quoted by W. T. Sullivan III in *Serendipitous Discoveries in Radio Astronomy,* ed. K. Kellermann and B. Sheets (Green Bank, WV: National Radio Astronomy Observatory, 1983).

Serendipitous discoveries are fairly common in science, and analysis has shown that almost all of them have certain features in common. First, as Jansky wrote in his letter, the discoverer has a "stubborn curiosity that demand(s) an explanation for the unknown. . . ." The data that led to Jansky's discovery were only a series of tiny blips on a chart. Jansky could easily have put the chart aside and ascribed the unexpected blips to equipment problems or to any number of possible "unexplainable" sources. Instead, his stubborn curiosity led him to investigate.

A second feature common to serendipitous discoveries is that the discoverer is extremely knowledgeable about the subject he or she is investigating. The fact that the discoveries are unexpected does not mean that just anyone can make them, for important discoveries are not made by the unprepared. Instead, they are made by competent observers engaged in serious scientific investigation.

Throughout this text we have seen how our perception of the universe has grown from a small universe centered on our immediate environment to a larger universe whose center was thought to be located farther and farther away. As we show in the next chapter, it now appears that "the center" simply does not exist: the universe has no center. Hand-in-hand with this receding center has come a tremendous expansion of our realization of the size of the universe. Galileo caused controversy not only because he placed Earth off-center, but also because the Earth played such a small part in his universe.

Yet Galileo did not realize that the Sun that rules our sky is just one of countless suns—that each of the stars he saw in the sky is another sun. When astronomers shift their focus from the solar system to the stars, their entire scale of thinking must expand. For example, as we have seen, the unit of measure used in the solar system, the astronomical unit, is very inconvenient for stellar distances; therefore astronomers use the light-year and parsec to describe such distances.

Galileo's contemporary, Giordano Bruno, did propose that the stars were suns, and he paid for his belief with his life; in 1601 he was burned at the stake for heresy.

Another mental leap in our thinking about distances and sizes is required when we consider stars not as individual objects, but in galaxies. The leap is so great that it may be impossible to truly understand the distances involved. It is fun to try, however. We'll see that our Galaxy contains about a trillion solar masses. Give some thought to what such a number means.

When we study objects at greater and greater distances from us, our knowledge of their nature and properties is more recent and less certain. In this chapter we discuss our galaxy of stars, of which the Sun is such an insignificant part, and describe how astronomers measure galactic properties. We then compare two different theories that attempt to explain the structure of galaxies like ours. Finally, we describe how our Galaxy formed and evolved to its present state. We'll see that in the case of galaxies, measurements are much less precise than we might be comfortable with.

16-1 Our Galaxy

Figure 16-1 shows part of the faint band of light that stretches around the sky, encircling the Earth at an

Figure 16-1
This wide-angle photo shows the Milky Way in Sagittarius (left) and Scorpius (right). Comet Halley is visible at left center in this photo taken March 13, 1986.

Figure 16-2
The Sun is located about one-third of the way out along the disk of the Galaxy, shown here both edge-on (top) and face-on (bottom). Recent evidence indicates that our Galaxy may not have a symmetric nucleus as shown here. (See the Advancing the Model box "The Milky Way: A Barred Spiral Galaxy?")

?

Do some of the stars we see at night belong to other galaxies far, far away?

Milky Way Galaxy The galaxy of which the Sun is a part. From Earth, it appears as a band of light around the sky.

Supernova SN1987A (Chapter 15) occurred in the Large Magellanic Cloud.

Figure 16-3
This image shows the region around the Galactic Center near the border between Sagittarius and Scorpius. You can see the absorbing dust lanes that block the view of the galactic center and some old bulge stars (yellow). The labels mark 17 Messier objects—including the nebulae Eagle (M16), Omega (M17), Lagoon (M8), and Trifid (M20)—globular clusters (M25, M22, M28, M54, M69 and NGC6723), open clusters (M6, M7, M23, M21, and M18), and the star cloud M24. Also marked are the dark Pipe nebula and Baade's Window (the region with the clearest view into the central bulge of our Galaxy).

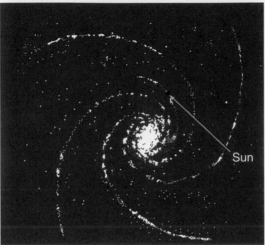

angle of about 63° with respect to the celestial equator. The ancient Greeks named this hazy band *galaxies kuklos,* meaning "the milky circle." The Romans called it *via lactae,* "milky way." You are encouraged to find a good clear night sky away from city lights and look at the beautiful Milky Way, a sight many of us in the modern world never get a chance to experience. The Milky Way completely encircles the Earth, passing through the constellations Sagittarius, Aquila, Cygnus, Cassiopeia, and Auriga and between Gemini and Orion on the northern hemisphere of the celestial sphere. Then it passes through Monoceros, near Canis Major, and through Vela and Crux on the southern hemisphere.

We know today that the haze of the Milky Way results from the many stars in the disk of the galaxy in which we live. The Sun is one of about 200 billion stars that make up what we call the **Milky Way Galaxy,** the Milky Way, or simply the Galaxy. Most stars in the Galaxy are arranged in a wheel-shaped disk that circles around a bulging center (**Figure 16-2**). The diameter of the Galaxy is about 50,000 parsecs (160,000 light-years), and the Sun is about a third of the way out from center (8000 parsecs or 26,000 light-years). The Galactic center is in the direction of Sagittarius (**Figure 16-3**) in our sky.

With only a few exceptions, every naked-eye object in our sky is part of the Galaxy. The Magellanic Clouds (see Figure 12-39) are exceptions that are visible from the Southern Hemisphere. These small galaxies are close to the Milky Way Galaxy, but are not part of it. An exception that those of us in the Northern Hemisphere can view in the fall sky is the Andromeda galaxy (**Figure 16-4**) in the constellation Andromeda. The Andromeda galaxy is the oval patch on the September sky chart at the end of this book. Our Galaxy is probably similar in appearance to the Andromeda galaxy.

The discovery that we live in a galaxy of hundreds of billions of stars was made fairly recently. It began, though, four hundred years ago when Galileo turned his telescope to the Milky Way and found that it is not made up of haze, but of stars far too numerous to count. The Milky Way appears misty because so many of the stars are so far away that the naked eye cannot distinguish individual stars and sees only the overall illumination from them.

When Galileo looked through his telescope, he saw more haze behind the many stars his telescope revealed, and he concluded that it was caused by even more stars too faint to see individually.

William Herschel, discoverer of
Uranus, wrote that through a telescope

> We find [the stars] crowded be-
> yond imagination along the ex-
> tent of [the Milky Way]; . . . so
> that, in fact, its whole light is
> composed of nothing but stars of
> every magnitude from such as
> are visible to the naked eye
> down to the smallest points of
> light perceptible with the best
> telescope.

Figure 16-4
The Andromeda galaxy (M31) is a spiral
galaxy about 2.9 million light-years
from our Galaxy. This galaxy can be seen
with the naked eye as a fuzzy spot in
the constellation Andromeda.

The telescopic view of the Milky Way led astronomers to conclude that we live in
a disk of stars. When we view the Milky Way in the sky, we are looking along that
disk (**Figure 16-5**), and when we are looking in other directions, we are looking out of
the disk.

As the Milky Way is viewed from Earth, it appears at first glance that we are at its
center, for to the naked eye, despite some local variations, the Milky Way seems to be
about as bright in one direction as in another. In the 1780s, in order to determine
where the Sun lies relative to the disk, William Herschel and his sister Caroline made
star counts in nearly 700 selected regions distributed around the sky. They reasoned
that if more stars were found in one direction than in another, that direction could be
assumed to be toward the center of the disk. Their conclusions not only confirmed
the disk-like shape of the Galaxy but also indicated that the Sun is indeed at the center, for they saw about the same number density of stars in all directions in the Milky
Way. **Figure 16-6** shows the shape they arrived at for the Galaxy. Notice that the Sun is
nearly at its center.

In the early part of the 20th century, Jacobus C. Kapteyn sought to find the Sun's
location by analyzing the density of stars in various directions from the Sun. He did
this by measuring not only the number of stars in each direction, but also their distances from us. He found that the density of stars decreases in every direction from
the Sun, and it was logical for him to conclude, like William and Caroline Herschel,
that the Sun is at the center of the disk.

The conclusion that the Galaxy centers on the Sun was viewed with skepticism,
as you might expect. After all, we once thought that the Earth was the center of the
universe, only to find that it circles the Sun. Were we now discovering that the Sun is
the center? The evidence from the Herschels and from Kapteyn pointed to an affir-
mative answer, but the finding seemed to be contrary to the trend that began before
written history and continued through Copernicus and Galileo.

Today we know that the Sun is not at the center. To understand why the
Herschels and Kapteyn obtained their erroneous results, imagine yourself standing
in a large forest. Suppose you try to decide whether or not you are at the center of

Jacobus Kapteyn (1851–1922) was
one of 15 children. From the age of
27 until he retired at 70, he held a
position at the University of
Groningen in the Netherlands.

Figure 16-5
When we see the Milky Way in the sky, we are looking along
the disk of the Galaxy. Otherwise we are looking out of the
disk.

Figure 16-6
The Herschels' counting of stars led them to conclude that
the Galaxy is shaped like this. The Sun is located at the bright
spot within the Galaxy.

the forest by counting trees in all directions. Unless you are very near an edge, you will see the same number of trees in all directions even though you are nowhere near the center. The reason is that the trees themselves prevent you from seeing beyond a certain distance. If you cannot see to the edge of the forest, this method of determining your location is not valid.

The situation in the case of the stars is somewhat different, for the stars do not fill our view as do the trees in a forest. When we look out among the stars, though, interstellar dust and gas place a limit on how far we can see. The Herschels were unaware of the existence of this material. They assumed that they could see all the way to the edge of the group of stars within which our Sun lies. Since they could only see a limited distance and since the Sun is not near an edge, they concluded that it was at the center.

Likewise, interstellar dust kept Kapteyn from counting the stars correctly. The density of stars at greater and greater distances from the Sun appeared to decrease because when Kapteyn counted stars at great distances, the interstellar dust prohibited him from seeing them all. In Figure 16-1, the dark areas stretching along the Milky Way are dust and gas clouds.

These investigators reached erroneous conclusions simply because one of their assumptions—that there is nothing in space to block the view of distant stars—was wrong. Incorrect assumptions cause trouble not only in science but also in everyday life. Often we are not even aware of what our assumptions are, and this prevents us from even accepting the possibility that our conclusion may be in error.

> Another analogy: If you are surrounded by dense fog, the fact that you can see the same distance in every direction is not evidence that you are in the center of the fog.

Globular Clusters

As we discussed in Chapter 13, some stars begin their lives in galactic clusters. These clusters are called "galactic" because they are found within the disk of the Galaxy. The Pleiades (Figure 13-7) is the most prominent example of this type of cluster, which may typically contain hundreds of stars. **Figure 16-7** shows two other galactic or open clusters. **Figure 16-8** shows two examples of a much larger type of cluster, a **globular cluster.** These beautiful, symmetrical clusters may look as though they have solid centers, but they are actually groups of hundreds of thousands of stars. The stars are so densely packed in the center of the cluster that we simply see a white area, not the individual stars. The average separation of stars near the center of a globular cluster is about 0.5 light-year. (In the Sun's region of space, stars are sepa-

globular cluster A spherical group of up to hundreds of thousands of stars, found primarily in the halo of the Galaxy.

Figure 16-7
The photo shows two open clusters in Perseus.

Figure 16-8
(a) The globular cluster 47 Tucanae, near the Small Magellanic Cloud, likely contains a million stars. At a distance of 16,000 light-years it is about the same size as the full moon in the sky. Stars near the center of the cluster are so closely spaced that they look as though they are touching. In 2001, astronomers announced the discovery of ionized gas and more than 20 pulsars in the central region of the cluster.
(b) The globular cluster M15, 37,000 light-years away, is the most tightly packed cluster of stars in our Galaxy. The *HST* distinguished hundreds of stars in a small area at the core of M15, whereas ground telescopes see a single blur of light. The observations hint at the presence of either a massive black hole or a "core collapse" that is driven by the intense gravitational pull of so many stars in such a small volume of space.

(a)

Globular Cluster M15 - HST WFPC2 - ST ScI OPO

(b)

rated by an average distance of about 4 to 5 light-years.) Globular clusters are not confined to the disk of the Galaxy but are seen outside the disk.

While Kapteyn was seeking to determine the Sun's location in the Milky Way by studying star locations, Harlow Shapley was trying to do the same using globular clusters. However, he had a problem determining the distances to globular clusters, for they are much farther away than the stars Kapteyn was analyzing. Just a few years earlier, Henrietta Leavitt had discovered the relationship between the periods and the apparent magnitudes of Cepheid variables in the Magellanic Clouds. Shapley observed Population II Cepheids and RR Lyrae variable stars, which he considered to be like Cepheids but with shorter periods. Population II Cepheids are about 1.5 magnitudes dimmer than the classical (Population I) Cepheids of the same period observed by Henrietta Leavitt (Section 12-8) and thus are less luminous by a factor of 4. RR Lyrae variables are commonly found in globular clusters; their average luminosity is about 100 times that of the Sun, and their periods are less than one day. Since these stars are easily identified because of their periodic changes in luminosity, it is relatively easy to find their distances using their period-luminosity relation and the distance modulus (**Figure 16-9**). The distances to these variable stars correspond to the globular cluster distances.

Figure 16-9
Once Shapley had determined the relationship between the periods and absolute luminosities of RR Lyrae variable stars, he could use a sequence like the one shown here to determine the distance to any RR Lyrae variable.

I did not doubt that the center of the entire galactic system must coincide with that of the globular clusters, and that it must have a mass very much greater than that of the Kapteyn system in order to prevent the high-velocity stars and the clusters from escaping altogether.

Jan Oort

◆ Bertil Lindblad (1895–1965) was a Swedish astronomer who was president of the Royal Swedish Academy of Sciences for 22 years and was chair of the Nobel Foundation when he died in 1965.

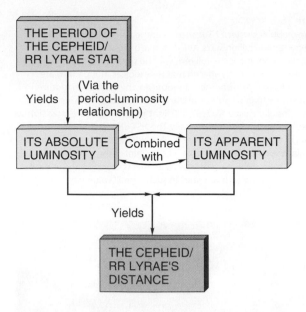

In 1917, Shapley published results of his survey of distances and directions to the then-known 93 globular clusters. He showed that they are not distributed evenly around the sky, but tend to be located more on one side, centered about the constellation Sagittarius. In fact, they seemed to be distributed in a sphere centered on a point about 50,000 light-years away from the Sun. **Figure 16-10** shows the approximate distribution of globular clusters compared to the Herschels' model of the Galaxy. Shapley assumed that the clusters revolve around the center of the Galaxy and therefore concluded that the Galaxy's center lies at the middle of the group of globular clusters. This meant that the Galaxy is much larger than indicated by the Herschels' model. It also meant that the Galaxy is not centered about the solar system; that is, our Sun is not at the center of the Galaxy.

In the 1920s, further evidence indicated that the Sun is not at a unique position in the Galaxy. Jan Oort (who proposed the comet cloud that bears his name) and Bertil Lindblad studied the motions of great numbers of stars near the Sun. They found that there is a pattern in the velocities of stars, depending upon their directions from the Sun. Kepler's third law, when applied to stars revolving around the center of the Galaxy, predicts that stars closer to the center should move faster and those farther from the center should move slower. This is the reason for the pattern of velocities shown in the Oort-Lindblad analysis. They concluded, as had Shapley, that the center of the Galaxy was thousands of light-years away in the direction of Sagittarius.

It should be pointed out that Oort and Lindblad saw a pattern only after analyzing very great numbers of stars, for stars have random motions along with their pattern of motion around the Galactic center. This is similar to the way that cars on a multilane freeway have a pattern of motion in one direction, though at any given time certain cars may be changing lanes or otherwise deviating from the pattern. Oort and Lindblad ignored individual stellar motions and concentrated on patterns of average motions.

Finally, in 1930, the interstellar dust was discovered. This resolved the conflict between the conclusions of Herschel and Kapteyn on the one hand and of Shapley, Oort, and Lindblad on the other, for the interstellar dust had prevented Herschel and Kapteyn from seeing to the edge of the Galactic disk. It is interesting that interstellar extinction was the reason why both Kapteyn and Shapley were wrong. Since Kapteyn

Figure 16-10
(a) The distribution of globular clusters as determined by Shapley is shown relative to the Sun and to the Herschels' model of the Galaxy. (b) The distribution of globular clusters and the nuclear bulge of the Galaxy shown through the interstellar dust. The lines mark the area blocked from view by the dust.

(a)

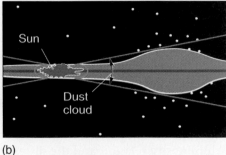

(b)

ADVANCING THE MODEL

The Shapley-Curtis Debate

In the late 1910s, there was considerable controversy about the size of the universe and the nature of the spiral nebulae. Several scientific papers were published on these topics, and the controversy attracted national attention. Harlow Shapley of Mount Wilson Observatories and Heber Curtis of Lick Observatory became the primary spokesmen for two opposing views. In April, 1920, the National Academy of Sciences invited these two astronomers to present papers before a meeting of that group in Washington, D.C., and to engage in a public debate concerning their research and conclusions.

At that time, Shapley had calculated the diameter of the Galaxy to be about 300,000 light-years. The reason that his result was too large is that he did not know about the interstellar medium. Another error led him to calculate the distance to the Magellanic Clouds to be only 75,000 light-years—less than his calculated diameter of the Galaxy. He was using Cepheid variables to determine the distance to the clouds, but he did not know that there are two types of Cepheids and that he was seeing the unknown type in the Magellanic Clouds.

Adriaan van Maanen, a Dutch astronomer who was a friend of Shapley, had published results showing that he could observe rotation of the Andromeda nebula. In seeing the rotation, he was observing proper motion of parts of the nebula. Since we cannot observe proper motion within an object unless it is relatively close to us, this indicated that the nebula is nearby. (Van Maanen's observation was simply in error.) Because of his own observations and those of van Maanen, Shapley concluded that not only are the Magellanic Clouds and the Andromeda nebula part of our own system of stars, but that other spiral nebulae are also part of that system.

Curtis held the opposite view. He believed that the group of stars that includes the Sun is much smaller than Shapley claimed and that the Andromeda nebula and other spiral nebulae are outside that group and are other "island universes." One bit of evidence he used to back his view was that Vesto Slipher of Lowell Observatory had investigated 15 spiral nebulae and found that 11 of them have significant redshifts. The redshifts indicated that they are moving away from us at great speeds, and this would make it unlikely that they are nearby. (These redshifts will be of major importance in Chapter 17.)

It is interesting that although Shapley thought that spiral nebulae are part of our Galaxy, he held that the universe is larger than Curtis envisioned. In his recollections of the debate, Shapley says that the subject assigned to the debaters was the size of the universe, but that Curtis turned the topic to the nature of the spiral nebulae.

Most of the two or three hundred people present at the debate were members of the National Academy of Sciences, but the debate made headlines in the *New York Times* and excited public interest. When, just a few years later, Hubble discovered Cepheids in the Andromeda nebula and showed conclusively that it was indeed another "island universe" (as galaxies were called), it looked as though Shapley had been entirely wrong in the debate. In his own mind, however, Shapley felt that the score was even, for he had been more correct in predicting the overall size of the universe.

observed regions mostly in the Galactic disk (where extinction is strongest), he could not observe the far regions in our Galaxy, thus underestimating its size (to about 10,000 parsecs in diameter). A simple analogy is the case of a person standing in a dense fog, with limited visibility, trying to see the surrounding area. In Shapley's case, he observed globular clusters that are mostly above or below the plane of our Galaxy, in directions where extinction is not very strong. However, Shapley calibrated the period-luminosity relation by using variable stars (RR Lyrae and Population II Cepheids) near the galactic plane, where extinction due to dust in the disk dimmed the starlight, thus increasing the apparent magnitude of the stars. Shapley incorrectly concluded that the dimness of these stars implied large distances and thus calculated a diameter of about 100,000 parsecs. Although Shapley's original values for the size of the Galaxy had to be revised downward by a factor of about 2 when it was discovered that there are two types of Cepheid variables, his basic deductions were correct. He had shown that the Galaxy was much larger than previously thought. The boundaries of the universe were again pushed back.

Harlow Shapley (1885–1972) quit school after the fifth grade, returned to school at age 16, and earned a Ph.D. in astronomy from Princeton at age 27.

16-2 Components of the Galaxy

We can describe four components of the Galaxy: the disk (which contains the Sun), the nuclear bulge, the halo, and the galactic corona (**Figure 16-11**). We describe each of these components here, and then discuss some of them in more detail in later sections.

The **disk** is the large, flat part of the Galaxy that rotates in a plane around its center. The disk contains individual stars, clusters of stars—particularly open clusters—and almost all of the gas and dust found in the Galaxy. Most of the stars in the disk are young, metal-rich, Population I stars. The disk appears bluish because of the presence of the hot O and B main sequence stars, which suggests that there is active star formation in the disk. The edges of the disk are not well-defined, and therefore its width and thickness cannot be stated exactly. As we noted earlier, its diameter is about 50,000 parsecs. Stars are most crowded near the plane of the Galaxy and become less crowded as we move from that plane. The disk is generally considered to be about 1000 parsecs thick, which makes its thickness about 2% of its diameter. Thus, it has the shape of about two compact disks stacked on one another.

Since star formation is more likely to occur where the interstellar material is most dense, stars are born with greatest frequency near the galactic plane. Gradually, their motions result in them wandering away from the plane. Very massive stars have short lifetimes, so we would expect to find few of them far from where they were formed. This is indeed the case, for almost all of the O-type stars lie within about 100 parsecs of the galactic plane. Only less massive stars live long enough to move very far from the plane.

It is difficult to determine the structure of the galactic disk from our position inside it because interstellar dust obstructs our view of distant locations. This dust is what limited early investigators' attempts to determine the Sun's position in the Galaxy. In Section 16-3

disk (of a galaxy) The flat, dense portion of a spiral galaxy that rotates in a plane around the nucleus.

O-type stars lie within about 300 light-years of the center of a disk that is about 3000 light-years thick.

Figure 16-11
The Galaxy consists of a nuclear bulge, a rotating disk, and a halo. The halo probably extends farther than indicated here.

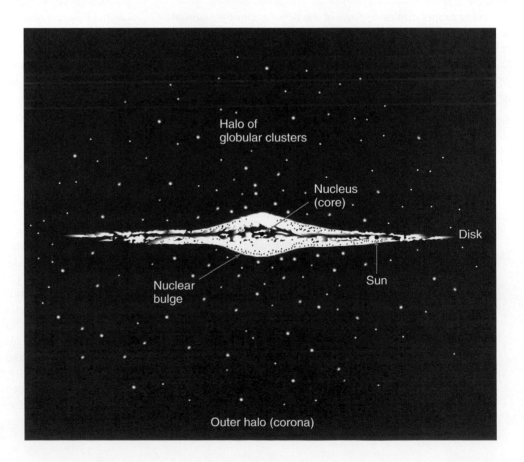

we will describe how radio astronomy has provided evidence for the spiral arms, and in an Advancing the Model box we outline new evidence that the Galaxy may not be a standard spiral galaxy after all.

The *nuclear bulge* of the Galaxy is about 2000 parsecs (6500 light-years) in diameter. Thus, if you imagine the disk to be a stack of two compact disks, you must add a peanut at their center to represent the bulge. This is necessary in order to represent the Galaxy correctly, because the bulge is wider along the disk than it is across its other dimension. Stars, dust, and gas are much more densely packed inside the nuclear bulge than they are anywhere else in the Galaxy. The bulge contains both young and old stars, and appears reddish because of the presence of many red giants and supergiants. We will discuss the bulge and its mysterious core in Section 16-5.

Figure 16-11 shows the Galactic *halo,* which contains the globular clusters that caused Harlow Shapley to conclude that the Sun is not at the center of the Galaxy. Observe that a person viewing the globular clusters from the position of the Sun would see many more in some directions than in others, and that their distribution is centered at the nuclear bulge. Besides the globular clusters, which mostly consist of old, metal-poor Population II stars, the halo contains small amounts of gas and dust.

Globular clusters feel a gravitational force toward the center of the Galaxy and thus cannot simply hover at their positions in the halo. Instead, they orbit the Galactic nucleus, passing through the disk twice during each orbit. **Figure 16-12** indicates their motion. About 150 globular clusters are known to be associated with the Milky Way, and the orbits of 75 of them have been calculated. Of these, 63 are in the halo and 12 are confined to the disk.

In the next chapter we will discuss an unresolved problem in galactic astronomy: the motions of galaxies indicate that much more gravitational force is exerted on galaxies than can be explained by the amount of mass that we can see. Astronomers conclude that there is more invisible mass in the universe than visible mass—by a factor of about 6! It is now thought that much of this mass exists in *extended halos* that surround galaxies. Therefore an outer halo was included in Figure 16-11. We do not know what the corona consists of—small black holes, cool dwarf stars, and/or great numbers of neutrinos have been hypothesized. The first direct detection of such cool dwarf stars was announced in 2001. Thirty-eight previously unseen cool white dwarfs were observed within about 450 light-years of Earth. Observations suggest that, at most, 35% of the unseen matter in our Galaxy is made of normal matter, composed of neutrons, protons, and electrons. It is thought that the corona extends to perhaps two or three times the radius of the disk and halo.

nuclear bulge The central region of a spiral galaxy.

halo (around a galaxy) The outermost part of a spiral galaxy; fairly spherical in shape, it lies beyond the spiral component.

The outer halo is sometimes called the galactic corona.

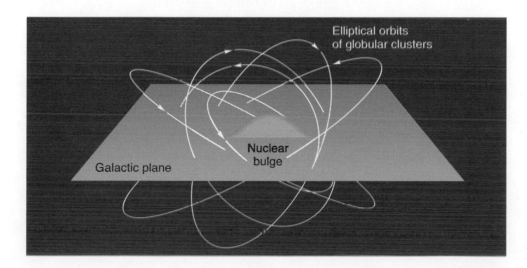

Figure 16-12
Globular clusters orbit the Galactic nucleus, passing through the disk twice during each orbit.

Elliptical orbits
of globular clusters

Nuclear
bulge

Galactic plane

Table 16-1 shows today's values for various Milky Way properties. Keep in mind, though, that the numbers are approximate, not only because every measurement is uncertain to some extent, but also because there are no specific boundaries for the various parts of the Galaxy. The total diameter of the halo is little more than a guess.

Table 16-1

Galactic Data

Radius of disk	80,000 light-years
Radius of nuclear bulge	3000 light-years
Total radius of halo	200,000 light-years
Sun's distance from center	26,000 light-years
Sun's orbital period	250,000,000 years
Thickness of disk	3000 light-years
Number of stars	200 to 400 billion

Galactic Motions

One method for determining the motion of the Sun around the Galactic center relies on the observation that the orbits of globular clusters seem to be randomly distributed around the center of the Galaxy. If we assume that the average velocity of all of the clusters, relative to the nucleus, is zero, we can measure their velocities relative to the Sun and attribute the average motion that is observed to the motion of the Sun. An analogy would be a person in a boat drifting on the ocean far from land and trying to measure the boat's speed. Suppose that the person is able to measure the speed of planes flying overhead, and that equal numbers of planes are seen flying eastward and westward. If the person can assume that planes fly eastward at the same speed as planes fly westward, she would know that the average speed of all the planes is zero. But if her measurements show that relative to her the planes'

The Doppler effect is used to measure the radial speeds of the globular clusters.

Figure 16-13

(a) If all the mass of the Galaxy were concentrated at its center, the galactic rotation curve would follow Kepler's laws, so that objects farther and farther from the center would move at lower and lower speeds, as shown by the dashed line in this graph. If the mass were distributed so that most of it is somewhat near—but not at the center—a rotation curve such as the one shown by the solid line would result. (b) This is the observed rotation curve of the Galaxy. It indicates that large amounts of mass orbit the center far beyond the Sun's orbit. The almost vertical segment of the rotation curve close to the center implies that matter in that region revolves as if it were a rigid body.

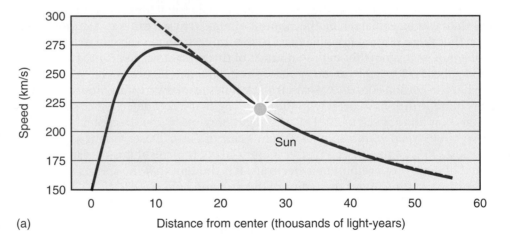

(a)

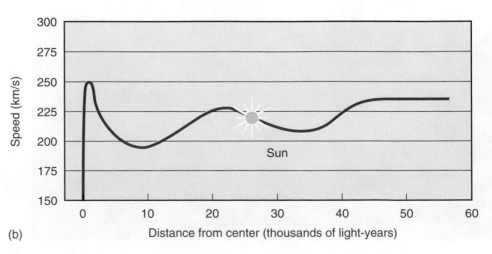

(b)

average speed is 10 miles per hour toward the east, she could conclude that her boat is moving 10 miles per hour toward the west.

Scientists always look for different methods to measure a quantity, and in this case another method exists. Recall that Oort and Lindblad studied the motions of great numbers of stars relative to the Sun and applied statistical methods to determine the motion of stars in the Sun's neighborhood. Similar studies yield a value for the speed of the Sun. The best measurements show that the Sun is traveling in a nearly circular path around the Galactic center at a speed of about 220 kilometers/second (135 miles/second). It is now moving toward the constellation Cygnus. Knowing that the radius of the Sun's orbit is 8000 parsecs (26,000 light-years), we can calculate the circumference of the Sun's path and find that it takes about 230 million years to complete one revolution. Although this seems a tremendously long time, the Sun has completed some 20 orbits during its 5-billion-year lifetime.

If the Galaxy had almost all of its mass concentrated at its center, its parts would rotate according to Kepler's third law, just as the planets do in the solar system. The dashed line in **Figure 16-13a** shows the *galactic rotation curve* that would be expected in this case. If the mass were distributed so that most of it is fairly near the center, but not in the center, the curve would look more like the solid line in Figure 16-13a. In fact, the Galactic rotation curve is somewhat like Figure 16-13b. This indicates that the Galactic mass is unevenly distributed, and the fact that the speeds of objects located farther from the center than the Sun are not lower than the Sun's speed indicates that large amounts of mass lie beyond the Sun's distance. The amount of matter we "see" in our Galaxy because of the light it emits amounts to about 10% of the total mass necessary to give the observed rotation curve. The remaining 90% of the mass of the Galaxy is not visible. In the next chapter we will discuss the possible nature of this unseen matter.

As we might expect, however, the orbits of stars are not perfectly circular. Along with the general orbiting motion of each portion of the Galaxy, each star has its own peculiar motion—perhaps having a component of motion from one side of the disk to the other or perhaps moving closer to the galactic center at one time and moving farther away at another time as it follows an elliptical path. In general, however, the paths of stars seem to be nearly circular.

The Mass of the Galaxy

In Section 3-5 we showed that Isaac Newton revised Kepler's third law to read

$$\frac{a^3}{P^2} = (m_1 + m_2)/M_{Sun},$$

where a = average radius of the orbit (in AU)

P = period of orbit (in years)

m_1 and m_2 = the masses of the two objects

Later (Section 7-2) we showed that the relationship holds not only for objects in orbit about the Sun, but also for satellites in orbit around the planets. Finally, in Section 12-6 we showed the law's application to binary star systems.

The discovery (in 1927) by Oort and Lindblad that the Galaxy in the Sun's neighborhood undergoes differential rotation meant that Kepler's third law might be applied to calculate the masses involved. In the case of a star revolving around the Galaxy's center, one of the masses in the equation is the mass of the star. The other mass is the mass of the *entire inner Galaxy,* including all objects in the Galaxy that are closer to the center than that star is. This may seem strange, since the inner portion of the Galaxy is made up of many objects rather than one, as in the case of the Earth's orbit around the Sun. It can be shown, however, that this is the correct application of the equation.

The first step in using Kepler's third law to calculate the mass of the part of the Galaxy inside the Sun's orbit is to express the Sun's distance from the center

In practice, wind speed would affect the speed of the planes, and her assumption would not be valid.

Saying that the Sun is moving toward Cygnus does not mean that it is getting closer to Cygnus; it means that the Sun is moving in that direction through space. Besides, each star of the constellation Cygnus has its own motion—some toward us and some away from us.

galactic rotation curve A graph of the orbital speed of objects in the galactic disk as a function of their distance from the center.

(26,000 light-years) in astronomical units. Since one light-year is about 63,200 AU (from Appendix A), the Sun is

$$26,000 \text{ LY} \times \frac{63,200 \text{ AU}}{1 \text{ LY}} \approx 1.6 \times 10^9 \text{ AU}$$

away from the Galactic center. Using this value, along with the Sun's period of 230,000,000 years, in Kepler's third law, we find

$$m_1 + M_{Sun} = \frac{a^3}{P^2} \times M_{Sun} \approx \frac{(1.6 \times 10^9 \text{ AU})^3}{(2.3 \times 10^8 \text{ yr})^2} \times M_{Sun} \approx 10^{11} M_{Sun}$$

The value obtained is the total of the mass of the Sun and the mass of the inner Galaxy. The mass of the Sun, of course, is negligible compared to the value obtained, so this answer is simply the mass of the part of the Galaxy that lies within the Sun's orbit. This value, 100 billion solar masses, must be taken as approximate. Recent analysis of the pattern of rotation in the outer parts of the Galaxy indicates that the total mass of the Galaxy is about 10^{12} (one trillion) solar masses, or about 10 times more mass than we calculated for the inner Galaxy, using Kepler's third law. At present, the nature of this additional mass is unknown, for there are not enough stars to account for that much mass. We will see later that a similar problem exists in the case of other galaxies.

> Refer to the discussion of dark matter in Chapter 17 and again in Chapter 18.

16-3 The Spiral Arms

> **spiral galaxy** A disk-shaped galaxy with arms in a spiral pattern.

The *spiral* nature of the Galaxy is not obvious from observations in visible light because we can see only limited distances along the plane of the Galaxy. In 1951, however, astronomers at Yerkes Observatory detected that as one looks either toward or away from the galactic center, the distribution of O- and B-type stars is not uniform. They seem to be clustered at certain distances. **Figure 16-14** illustrates the concentrations of this type. This was the first hint of the spiral nature of the Galaxy. Unfortunately, since such stars are best observed in visible light, interstellar extinction allows us to observe them only in the vicinity of the Sun. The same is true for the red emission nebulae found near many of these hot, luminous, blue main sequence stars. However, more evidence for the spiral nature of our Galaxy came with the discovery during that same year of a specific emission line of neutral hydrogen.

You might guess, correctly, that if we could map the distribution of hydrogen clouds in our Galaxy, we should be able to infer its structure. However, most of the emission lines of hydrogen are at visible and ultraviolet wavelengths. Such radiation gets scattered or is absorbed easily by interstellar dust and this limits the range at which we could observe these clouds. Since the longer the wavelength, the easier it is for radiation to travel through interstellar dust without being scattered or absorbed, we need to look for radiation at infrared wavelengths. It turns out that cool hydrogen gas does emit radiation of such wavelength, 21.1 centimeters, in the radio portion of the spectrum.

The 21-centimeter emission photon results from a transition that a hydrogen atom makes from a higher energy level to a lower one (**Figure 16-15a**). To understand this transition, consider an approximate model in which the electron and proton of the hydrogen atom are small, charged spheres that spin around their axes. Since charges in motion create magnetic fields, we could visual-

Figure 16-14
O- and B-type stars are found at greatest concentrations in the shaded areas.

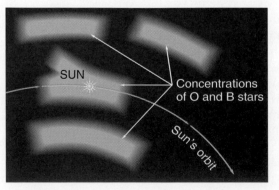

SUN

Concentrations of O and B stars

Sun's orbit

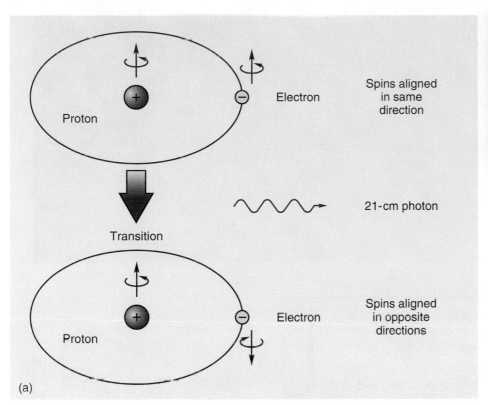

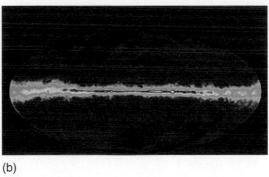

(b)

Figure 16-15
(a) A 21-cm wavelength photon is emitted when the spins of the electron and proton in a hydrogen atom make the transition from being aligned in the same direction to being aligned in opposite directions. (b) Our Galaxy in 21-cm emission of hydrogen. This is a composite of several surveys with ground-based telescopes in the northern and southern hemispheres. Hydrogen gas is concentrated along the plane of our Galaxy (which extends horizontally across the image). The range of hydrogen density is shown from black/blue (lowest) to red/white (highest).

ize both the electron and proton as tiny magnets. When both magnets have their poles aligned in the same direction (as in the case of the electron and proton spinning in the same direction), they repel each other. When the two magnets are aligned in opposite directions (as in the case of the electron and proton spinning in opposite directions), they attract each other. The first configuration has a bit more energy than the second. Thus, when the transition occurs, a 21-centimeter photon is emitted. Of course, absorption can also occur that excites the hydrogen atom into aligning the spins of its electron and proton in the same direction.

Radio telescopes can use **21-centimeter radiation** to detect high concentrations of cool hydrogen, such as exist in interstellar clouds. Since hydrogen is the main component of interstellar material—and the entire universe—this is an ideal method of detecting cool hydrogen clouds at great distances. Figure 16-15b is a false-color image of our Galaxy in 21-centimeter emission from interstellar hydrogen. The 21-centimeter emission traces the "warm" interstellar medium, which on a large scale is organized into diffuse clouds of gas and dust that have sizes of up to hundreds of light-years.

As we have seen, new stars arise from gas and dust to which some of the material of a massive star returns at the end of the star's lifetime. Therefore hydrogen gas clouds detected by 21-centimeter radiation are located at the same places as newly forming stars. Recall that O and B stars are very massive and are therefore short-lived. This means that they are found only where stars have recently formed and that we might expect to find hydrogen clouds at locations identified by the astronomers at Yerkes Observatory as having high concentrations of O- and B-type stars.

It may seem that it would be impossible to determine the distance of a source of 21-centimeter radiation, for a radio telescope would seem to reveal only the direction of the source of the waves. In a sense this is true. However, the Doppler effect allows us to determine the *radial* motion of the hydrogen with respect to us. For example, suppose that a radio telescope is pointed in a direction across the Galaxy, as shown in

twenty-one-centimeter radiation
Radiation from atomic hydrogen, with a wavelength of 21.1 centimeters.

Figure 16-16
If hydrogen gas were distributed uniformly in the galactic disk, 21-centimeter radiation from the Sun's forward direction would show a fairly regular blueshift (c).

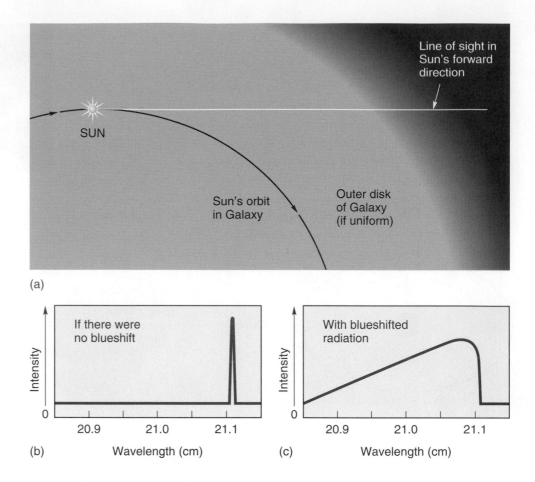

(a)

(b)   Wavelength (cm)

(c)   Wavelength (cm)

Figure 16-16a. Because of differential rotation, we would be moving toward the hydrogen along this direction, so we would see a blueshift of the 21-centimeter radiation. If the hydrogen were distributed uniformly, the radiation would be Doppler shifted so that its wavelength would range from just less than 21.1 centimeters to quite a bit less. A graph of radio intensity versus wavelength might then appear as shown in part (c) of this figure.

On the other hand, consider how the radiation would appear if the hydrogen is located in spiral arms, as shown in **Figure 16-17a**. In this case, each spiral arm would

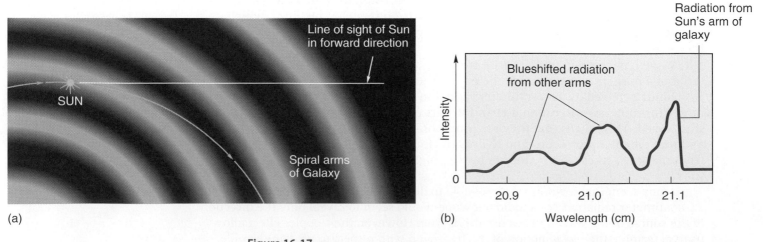

(a)

(b)   Wavelength (cm)

Figure 16-17
Twenty-one-centimeter radiation observed in the Sun's forward direction actually shows peaks of blueshifted radiation (b). The peaks occur because the hydrogen that emits the radiation is concentrated in spiral arms (a).

(a)   (b)

Figure 16-18
If we could get outside our Galaxy and view it face-on, it would probably look like one of these galaxies. The bright stars in each image are part of our Galaxy. Because these stars are much closer to us than the other galaxy, we see them as we see flies on a car windshield when we look through it. Our Galaxy probably has four primary arms, with irregular branches extending from each. (a) NGC2997 is a spiral galaxy in the southern constellation Antlia. (b) The galaxy M83 in Hydra (also a southern constellation) has a slightly elongated nucleus.

have a fixed value for its Doppler shift, and the graph would appear as shown in Figure 16-17b. This is how it appears in an actual case, so 21-centimeter radiation provides us further evidence for the spiral nature of the Galaxy.

Mapping the Galaxy in 21-centimeter radiation contains plenty of room for error, but there is no doubt that ours is a spiral galaxy. As we will show in the next chapter, a spiral appearance is common for galaxies. **Figure 16-18a** shows a spiral galaxy that may be similar to the Milky Way, although recent work indicates that our Galaxy may have an elongated nucleus, more like the galaxy of Figure 16-18b. We discuss this idea in the Advancing the Model box on page 530.

16-4 Spiral Arm Theories

It may seem at first glance that spiral arms could be explained by the fact that stars near the center of the Galaxy complete a circle in less time than those farther out. This would occur even if the stars do not orbit according to Kepler's third law, simply because stars near the center have less distance to cover to complete their orbits. Consider a group of stars that at a given time are in a straight line across the center of the Galaxy, as shown in **Figure 16-19a**. As the stars move around the center (even if they all move at the same speed), the line of stars will wind up into a spiral. The problem is that they would wind up too much. Recall that the Sun has made some 20 revolutions around the Galaxy. As Figure 16-19b, c, and d show, the line of stars would wind up so much that it would not be distinguishable as a line. The

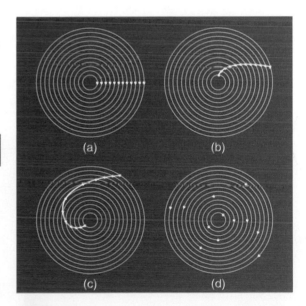

(a)   (b)

(c)   (d)

Figure 16-19
Stars that start out in a straight line from the Galaxy's center (a) would begin to show a spiral pattern (b) and (c), but after a few revolutions, the pattern would be lost (d).

Galaxy would appear as a fairly uniform disk. Yet we do perceive a spiral pattern in our Galaxy and in many other galaxies. Therefore this simple differential rotation hypothesis cannot be correct.

Currently, there are two competing theories to explain the spiral nature of galaxies: the density wave theory and the self-propagating star formation theory.

The Density Wave Theory

density wave theory A model for spiral galaxies that proposes that the arms are the result of density waves sweeping around the galaxy.

density wave A wave in which areas of high and low pressure move through the medium.

The *density wave theory* was first proposed in 1940 by Bertil Lindblad and holds that what we see in a spiral arm of a distant galaxy is not a simple fixed line of stars but a line formed by the brightest stars and the glowing nebulae surrounding them. The theory holds that stars revolve around the Galaxy independent of the spiral arms and that the arms are simply areas where the density of gas is greater than at other places. According to this theory, there are almost as many stars per unit volume between the arms as in the arms, but the arms contain more of the brightest stars and a higher density of gas and dust. The areas of denser gas move around the Galaxy in *density waves,* causing the formation of new stars and glowing emission nebulae. This is best explained by an analogy, illustrated in **Figure 16-20**.

In a spiral galaxy, are all the stars located in the spiral arms?

Suppose cars are traveling on a long highway at a speed exceeding the speed limit—perhaps leaving campuses for spring break. Also traveling along the road, at less than the speed limit, is a traffic patrol car with its radar on. Observing from a helicopter high in the sky, we see the cars fairly evenly distributed along the highway except around the police car. For a short distance behind and in front of the police vehicle, the cars are bunched up. As a given car approaches the police cruiser from behind, the car slows down, slowly passing the feared patrol car. Then when the driver feels that his or her car is safely in front of the police cruiser, it again speeds up.

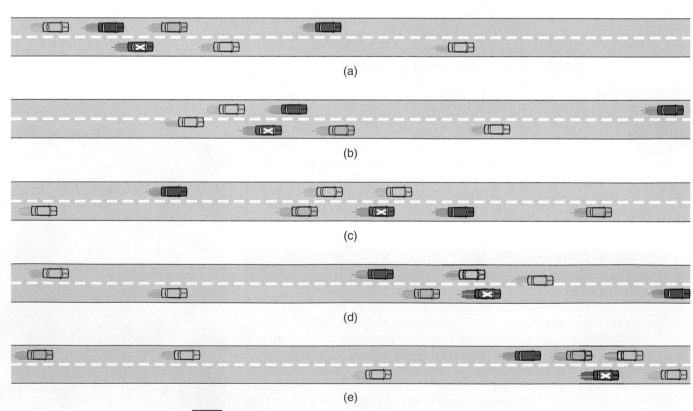

(a)

(b)

(c)

(d)

(e)

Figure 16-20

An overhead view of a highway, seen as time passes. Car X is a police car, moving just slower than the speed limit. Bunched up around that car are other slow-moving cars that only gradually pass the police car. Thus, as time goes by, the cars near X are different, but a high density of cars remains there.

As a result, there is constantly a high density of cars around the police car, even though the cars making up that group change all the time. The police car moves along, seeming to carry its high-density group along with it, in what we see from the helicopter to be a density wave.

Density waves are common here on Earth, for every sound wave is a density wave, with regions of high and low density making up the wave. In a galaxy, the density wave consists of a region of gas and dust that is denser than normal, perhaps by 10% to 20%. The major difference between sound waves and the density waves of a spiral galaxy is that in the atmosphere of Earth, a sound wave travels faster than the particles of the gas itself. However, in the near-vacuum of a galaxy, the wave travels more slowly than the particles. The gas and dust particles—as well as stars— catch the wave from behind and pass through it in much the same way that cars in the analogy passed through the density pulse around the police car. **Figure 16-21** illustrates the idea.

Now recall that new stars are formed when an interstellar cloud of gas collapses. The density wave theory holds that the trigger for this collapse is the wave. As inter-

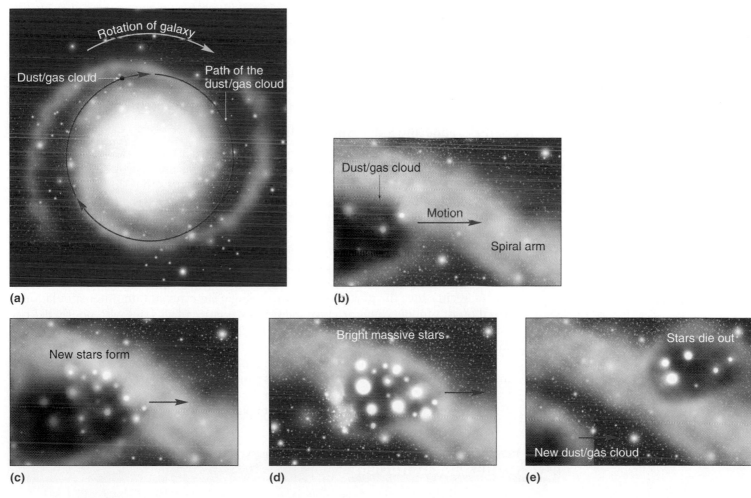

Figure 16-21
(a) The orbits of objects in the Galaxy are very nearly circular. Since their orbital speed is greater than the speed of the density wave that forms the spiral pattern, they pass in and out of the spiral arms. (b) A cloud of dust and gas is about to enter a spiral arm. (c) The higher density of gas in the arm causes the cloud to compress, resulting in star formation. (d) As the cloud passes through the spiral arm, bright, massive stars make the arm itself appear brighter than the rest of the galaxy. (e) The bright stars burn out before they exit the spiral arm, as another cloud of dust and gas begins to enter the arm.

(a)

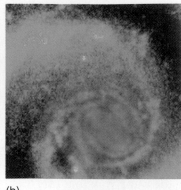

(b)

Figure 16-22
(a) The Whirlpool galaxy (M51), seen here in visible light, is an excellent example of a spiral galaxy. (b) This infrared image of the central portion of the Whirlpool galaxy indicates that the spiral arms penetrate farther into the nucleus than was previously thought possible. The finding conflicts with predictions made by the density wave theory.

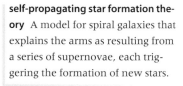

self-propagating star formation theory A model for spiral galaxies that explains the arms as resulting from a series of supernovae, each triggering the formation of new stars.

stellar clouds approach the density wave from behind, they are compressed and stars are formed. Although stars of all masses are formed along the edge of the density wave, the brightest, most massive stars have ended their lives before they pass far from this edge. This means that when we look at a spiral galaxy, the spiral arms are obvious to us because they are the areas containing the bright stars.

One problem with the density wave theory is the question of how the density wave is sustained through the life of the galaxy. Computer simulations indicate that it would die out. In addition, a recent infrared image of the center of the Whirlpool galaxy (**Figure 16-22b**) indicates that spiral arms penetrate much farther into the nucleus of a galaxy than had previously been thought. The density wave theory had predicted that a galactic nucleus absorbs density waves, preventing the waves from extending into it. The new information from the Whirlpool galaxy is causing a reexamination of the density wave theory.

It is possible that gravity is the driving mechanism behind the density wave. If the nucleus of a galaxy is not symmetrical (for example, if it is bar-shaped, as it might be the case for our Galaxy), then the asymmetric gravitational field might generate density waves by its interaction on stars and interstellar matter. Another possibility is that gravitational interactions between galaxies can generate and sustain density waves. Even if we resolve the issue of the driving mechanism, there is another problem. The spiral arms produced by density waves are well-defined (for example, Figure 16-18). However, some galaxies have poorly defined arms. For such galaxies, another theory for the formation of spiral arms has been proposed: the self-propagating star formation theory.

The Self-Propagating Star Formation Theory

According to the ***self-propagating star formation theory*** of galactic spiral arms, the triggers that start the collapse of most interstellar clouds are nearby supernova explosions. Then, as the more massive stars finish their lives and become supernovae, they trigger more star formation, and so on. A simple analogy is a forest fire, with the flames jumping from one tree to the next. Thus, the formation of new stars is confined to areas where this process is taking place. Now differential rotation enters the picture. Computer analysis shows that at the rate at which massive stars would be formed, differential rotation would cause spiral arms to be formed and sustained.

The self-propagating star formation model is able to explain how a spiral arm would begin, since the process starts with a single supernova. Computer simulations based on this model can reproduce the spiral structure of galaxies with poorly defined arms, but the model is not successful in reproducing the structures observed in spiral galaxies with well-defined arms.

The two theories we presented for the formation of spiral arms are very different. In the density wave theory the spiral arms cause star formation, while the opposite is true in the self-propagating star formation theory. It is probable that we need a combination of these two theories in order to explain the observations.

16-5 The Galactic Nucleus

Observations by Shapley, Oort, and Lindblad in the early part of the 20th century revealed that the center of the Milky Way Galaxy lies in the direction of Sagittarius. Since the presence of dust and gas in the Galactic plane dims visible light from the nucleus by about 28 magnitudes, astronomers had to await the development of nonoptical telescopes to learn more about that nucleus. In order to observe the galactic nucleus, we use wavelengths in the infrared/radio part of the spectrum or in X-rays/gamma rays. Our observations helped us construct a picture of the nucleus of our Galaxy that includes a massive black hole and a history of violent events.

The *Hubble* image in **Figure 16-23**, taken in the infrared, shows a pair of the largest young star clusters, located less than 30 parsecs from the Galactic center. The observed number density of stars is indeed increasing as we get closer to the

Figure 16-23
These clusters are ten times larger than typical young star clusters in our Galaxy; they will be destroyed in a few million years by the gravitational tidal forces in our Galaxy's core. The cluster on the left is so dense that 100,000 of its stars could fill the spherical volume between our Sun and its nearest neighbor star (about 4 light-years away). The cluster on the right is less dense; it has stars at the stage of becoming supernovae, and is the home of the brightest star in our Galaxy (the Pistol star). The false colors in this *HST* image correspond to infrared wavelengths. Galactic center stars are white, red stars are located behind (or are surrounded by) dust, and blue stars are foreground stars (located between us and the Galactic center).

Galactic center, down to about 2 parsecs from the center. For distances closer than 2 parsecs, observations of the velocities of stars suggest that there is a great deal of mass in a very small volume in the galactic nucleus (less than 0.5 parsec in radius). It seems that the Galactic nucleus harbors a massive black hole of mass about 3.4 million solar masses.

Radio observations show that a nuclear disk of neutral gas occupies a region between a few hundred parsecs to a thousand parsecs from the center. **Figure 16-24a**

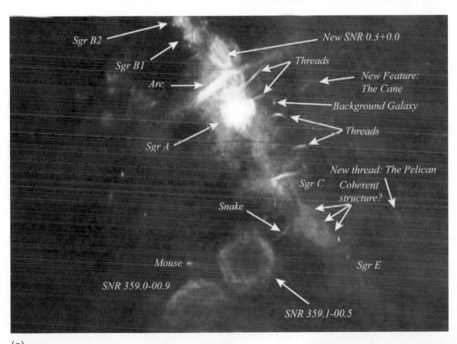

(a)

Figure 16-24
(a) This image was constructed from 1-meter radio data obtained by telescopes of the Very Large Array. It shows signs of a violent environment, including supernova remnants and filaments that may correspond to mass outflow from the Galactic center (located at the edge of the object Sgr A). (b) This image (in 20-cm) shows the inner region, about 60 parsecs across, of the Galactic nucleus. The nucleus is at the center of the strongest emission. The filament, running perpendicularly to the galactic plane, may be associated with magnetic fields. (c) Ammonia emission (at 1.3 cm) in yellow contours in the region around the Galactic center. The ring as seen at 3 mm is shown in the background. Streamers that possibly interact with the ring are numbered. The star labels the position of the massive black hole (Sgr A*).

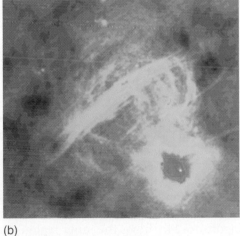

(b)

(c)

ADVANCING THE MODEL

The Milky Way: A Barred Spiral Galaxy?

In Chapter 17 we will show that there are two types of spiral galaxies. About half are standard spiral galaxies, like the one shown in Figure 16-18a. The other half have an elongated central bulge that appears as a bar across the center, like the one in **Figure B16-1**. Until recently, astronomers thought that the Milky Way is a standard spiral galaxy, but evidence is accumulating that our Galaxy is a barred spiral galaxy with an elongated nucleus that is probably shorter in proportion to the rest of the galaxy than the one in Figure B16-1.

The new evidence comes primarily from two sources. First, we are learning more and more by analyzing 21-centimeter radiation, which penetrates the clouds of interstellar space and allows us to detect the motion of clouds of cool hydrogen gas near the nuclear bulge. Second, in 1989 NASA launched the *Cosmic Background Explorer (COBE)*. It was designed to measure short-wavelength radio waves from space (for reasons we will discuss in Chapter 18).

Suppose the central bulge is elongated with its long axis oriented at some oblique angle to the line of sight, as in **Figure B16-2**. The Sun (and Earth) is located at the point indicated in the figure. Line C extends to the center of the Galaxy, and lines L and R are drawn at equal angles on each side of line C. Notice that the line of sight on the near side of the nucleus (the left side here) passes through part of the nucleus, but the equivalent line of sight on the other side does not. If the nucleus is shaped like this and has this orientation, astronomers should detect an asymmetry in the radiation from near the nucleus. Data from *COBE*, as well as data from Japanese astronomers, correspond to what would be expected from such an elongated nucleus.

Further information comes from analyzing motions of the gas near the nucleus. The gas moves around the nucleus in a noncircular pattern, corresponding to what would be expected from a barred nucleus. The data indicate that the bar is a fairly short one and that it is oriented almost in line with the Sun, so that measurements like those illustrated in Figure B16-2 are very difficult to make.

Astronomers are not in complete agreement on the interpretation of the new data, but because of the importance of knowing the basic shape of our Galaxy, the subject has become a hot topic of research.

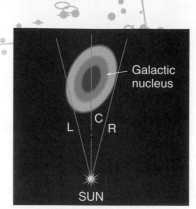

Figure B16-2
Opposite sides of an elongated nucleus—a bar—do not appear the same when viewed from the Sun, because lines that make equal angles with the central line pass through different volumes of the bar. Data indicate that the nucleus of the Milky Way is slightly elongated with its long axis almost aligned with the Sun.

Figure B16-1
The Milky Way may be a barred spiral galaxy similar to this one.

shows the region near the Galactic center, located at the edge of the bright object labeled Sagittarius A (Sgr A). The Galactic plane runs diagonally across the image, which includes supernova remnants (SNRs), along with many elongated threads of ionized gas, almost perpendicular to the disk. One of the most intriguing features of this region is shown in Figure 16-24b. It is likely that this feature is the result of magnetic fields, with the radio emission being due to charged particles spiraling around the magnetic field lines. In the Sgr A region we find a large molecular ring, with an inner radius of 2 parsecs and outer radius of 8 parsecs, whose mass is at least 20,000 solar masses. The material in this ring is much warmer and denser than material in molecular clouds elsewhere in the Galaxy. The velocities of the ionized gas in this region increase from about 100 kilometers/second at the inner edge of the ring to 700 kilometers/second at 0.1 parsec from the Galactic center. These measurements, along with the presence of Sgr A*—a very small, unresolved (less than 20 AU in size) radio source inside the Sgr A region—suggest that at the center of our Galaxy we have a massive black hole of 3.4 million solar masses. This is consistent with the results obtained using stellar velocities in this region. Figure 16-24c shows the inner region of the

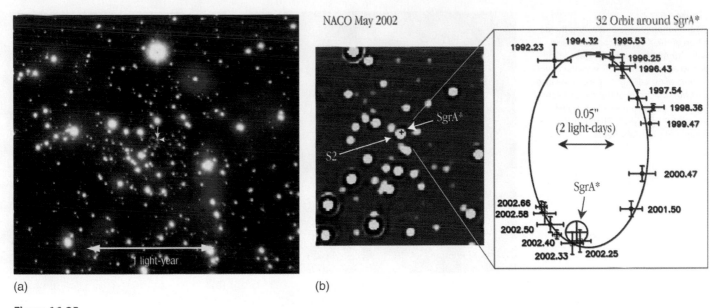

(a) (b)

Figure 16-25
(a) The innermost area of our Galaxy, shown as a combination of three infrared frames (between 1.6 and 3.5 μm). The compact objects are stars; blue denotes hot and red denotes cool stars. There is also infrared emission from dust between the stars. The arrows mark the position of Sgr A*. (b) An infrared image of an area 60 × 60 light-days, centered on the position of Sgr A*. The inset shows the positions of a star, designated S2, during the period 1992–2002. The solid curve is the best-fitting elliptical orbit, with Sgr A* at one of the foci.

Galactic nucleus (with resolution of details as small as 0.13 light-year at the center). Narrow "streamers" can be seen connecting giant molecular clouds of gas and dust to the ring. The observed increase in velocity along a streamer suggests that gas/dust is being accreted from the molecular clouds and "fed" to the ring and then to the black hole as a result of its strong gravitational pull. Infrared observations of the innermost area of the Galactic nucleus, obtained in 2002 by ESO's VLT and shown in **Figure 16-25a**, show emission from interstellar dust between a great number of stars around the position of Sgr A*. In addition, 10 years of observations allowed astronomers to trace the orbit of a star around Sgr A* (Figure 16-25b). The characteristics of this orbit leave little doubt that a supermassive black hole is present at the center of our Galaxy.

The case of a large black hole at the center of our Galaxy is also supported by observations of gamma rays emanating from a very small region in the Galactic nucleus. In addition, in late 2000, X-ray variability was observed from Sgr A* that lasted less than 10 minutes. As we discussed in section 15-4, variability is a great indicator of the size of the radiating region since nothing can travel faster than light. This observation suggests that the hot, radiating gas could not occupy a region bigger than the distance between the Earth and the Sun. The most probable scenario is that the force that powers Sgr A* is a black hole, orbited nearby by a great number of stars. As matter falls into the black hole, collisions cause energy to be released. The release of energy here is similar to the energy released as matter falls into stellar black holes, which we discussed in Chapter 15. In the case of the Galactic nucleus, however, a *massive* black hole is at the center, and it is surrounded by a great amount of matter, causing enormous quantities of energy to be released. **Figure 16-26** shows the intensity of high-energy gamma-ray emission in our Galaxy.

Black holes are common in the Galaxy. A few isolated stellar-mass black holes have been detected indirectly by the way their gravity bends the light of a more distant star behind them. We discuss this phenomenon in Chapter 17. Stellar-mass black hole candidates have been observed both in the Galactic disk and halo by the effect they have on the motion of their companion normal star.

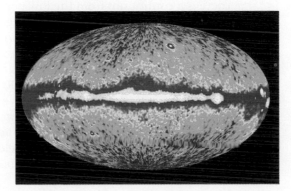

Figure 16-26
This image was taken by the *Energetic Gamma-Ray Experiment Telescope* (*EGRET*) on board the *Compton Gamma-Ray Observatory*. Our Galaxy is a diffuse source of gamma-ray light. Superimposed on this image are several gamma-ray pulsars and some active nuclei of other galaxies.

Figure 16-27
(a) A *Chandra* image that marks the position of the X-ray source associated with Sgr A*. Large lobes of gas (20 million K) are also shown. The image covers an area of 60 x 60 light-years. (b) A close-up *Chandra* image of the region around Sgr A*, showing a possible X-ray jet of length 1.5 light-years. The image covers an area of 9 x 9 light-years.

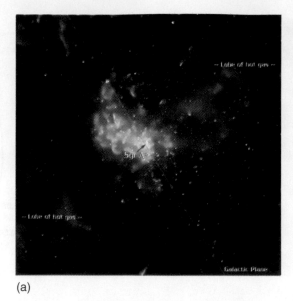

(a)

(b)

An X-ray look at our Galaxy's central black hole has revealed numerous outbursts and some large explosions. **Figure 16-27a** shows large lobes of hot gas (20 million K) in the innermost region around Sgr A*. These lobes show that many enormous explosions have occurred during the past 10,000 years. A possible jet of particles, shown in Figure 16-27b, suggests that the observed X-ray flaring activity from the vicinity of the black hole has been occurring for many years.

Some other spiral galaxies also have tremendously powerful energy sources at their center, and we see signs of violent activity near those centers. The massive black hole hypothesis is rapidly gaining support among astronomers and seems to be the best explanation for the energy source in our Galaxy and in similar galaxies. As we discuss in the next chapter, some galaxies produce even more energy in their nuclei. We are far from understanding the nucleus of our Galaxy, and further knowledge must come from observations of other galaxies as well as our own.

16-6 The Evolution of the Galaxy

The first step in developing a theory of the formation of the Galaxy is to consider the chemical content and the age of the Galaxy's components. Once that is done, the pieces can be put together to give us a picture of how the Galaxy developed.

Age and Composition of the Galaxy

In Section 15-9 we discussed the division of stars into Population I, Population II, and Population III according to their percentage of heavy elements. Although in reality there is a continuum from the low percentage of heavy elements in extreme Population II to the greatest percentage in extreme Population I, the grouping is still convenient. On the average, Population II stars contain only about 1% of the heavy elements of Population I stars. An extreme case is the star HE0107-5240, 36,000 light years in the southern constellation Phoenix, which has the lowest abundance of heavier elements ever observed, 1/200,000 that of the Sun.

Population I stars are stars that have formed from the remains of previous generations of stars, and most of them are found in the disk of the Galaxy. In fact, stars with the greatest percentage of heavy elements lie close to the nuclear bulge. The farther from the bulge, the lower the fraction of heavy elements. Most globular clusters, on the other hand, are made up of Population II stars. Of the known globular clusters in the Milky

Astronomers say that a star with a large abundance of heavy elements is *rich in metals,* or *metal rich.*

The abundance of heavy elements decreases by a factor of 0.8 for each thousand parsecs from the center of the disk.

Way, about 70% have an average heavy-element content of about one-twentieth of that of the Sun. The remaining 30% contain about one-third the heavy-element content of the Sun.

Star clusters provide a convenient method of measuring the age of stars. Recall that all stars in a cluster were formed at about the same time. Knowing this, astronomers can determine a cluster's age by analyzing its H-R diagram. **Figure 16-28** is the H-R diagram of a cluster old enough that its heaviest stars have left the main sequence.

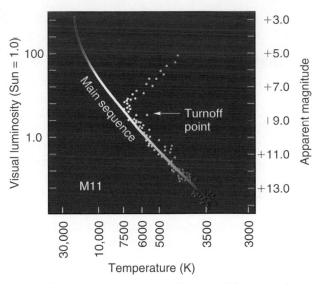

Figure 16-28
The H-R diagram of a galactic cluster shows a definite main sequence turn-off point. From the location of the turn-off point on the main sequence, astronomers can determine the age of the cluster.

The position of the main sequence turnoff point permits a calculation of the age of the cluster.

The classification of stars in a specific portion of the Galaxy can provide a clue to the age of that portion. For example, there are no white dwarfs in the part of the Galactic disk near the Sun. Since white dwarfs do not form until after stars have gone through the main sequence and red giant stages, this sets an upper limit of about 10 billion years for the age of the Sun's portion of the disk.

Within the disk, O- and B-type stars are found primarily near the Galactic plane. Since these stars have short lifetimes, the fact that they are not found at great distances from the Galaxy's plane tells us that star formation does not occur to any great extent except along that plane.

Stars in the halo of the Galaxy are older than those in the disk. While the oldest stars in the portion of the disk where the Sun is located are about 10 billion years old, the *youngest* stars in globular clusters are about 11 billion years old. Thus, globular clusters formed before the disk portion of the Galaxy. The oldest globular clusters are about 13 billion years old, and since these are the oldest stars we find, we take this to be the minimum age of the Galaxy. Globular clusters were probably much more numerous in the early days of our Galaxy, but tidal interactions between a cluster and the Galaxy could tear the cluster apart. In 2002, astronomers announced the detection of a stream of stellar debris emanating from a globular cluster that is being torn apart by our Galaxy. Such streams provide a new way to determine how the mass of the dark matter halo of our Galaxy is distributed.

The age of stars in the nuclear bulge is difficult to determine because the bulge is obscured by dust. It can be studied only in the X-ray, infrared, and radio portions of the spectrum. However, observations through a few gaps in the dust have allowed astronomers to sample the outermost bulge population, and they have learned that the stars of the nuclear bulge are very rich in heavy elements. The stars of the bulge all appear to be red giants of types K and M. This tells us that the nuclear bulge must be very old, according to the following logic: First, the stars themselves are old, for K- and M-type stars live long lives on the main sequence before becoming red giants. Second, they were formed from the remains of previous generations of stars.

The existence of a Galactic corona of hot gas has been confirmed based on data by NASA's *Far Ultraviolet Spectroscopic Explorer* (*FUSE*) satellite. Observations from *FUSE* show the presence in the corona of a specific species of highly ionized oxygen atom, which indicates the existence of coronal gas with temperatures close to a quarter million kelvin. The atoms observed by *FUSE* are believed to have been heated by supernova explosions to very high temperatures and pushed outward from the galactic plane into the halo. The clouds of ionized oxygen appear in almost

In Section 14-2 we discussed how we determine the age of a star cluster.

FUSE was launched into Earth orbit in June 1999 and is designed to sample light (look at spectra) in the UV part of the spectrum. This light can only be observed above the Earth's atmosphere.

all directions and extend as much as 5 to 10 thousand light-years away from the Galactic plane.

The Galaxy's History

Although astronomers disagree about the specifics, a general picture of the evolution of our Galaxy is emerging. The process is somewhat parallel to the way stars form. The Galaxy began as a tremendous cloud of gas and dust, bigger than the present Galactic halo (**Figure 16-29**). Mutual gravitation between the cloud's parts gradually pulled it together. The center portion was the first to become dense enough for stars to form, and it soon became a hotbed of star formation. Soon thereafter, stars began forming in dense pockets, hundreds of light-years across, which were in orbit around the center. They formed what are now the globular clusters.

The initial giant cloud had some rotation, and as it contracted, it spun faster. Just as occurs during a star's protostar stage, the rotating matter formed into a disk. In the case of a star, almost all the orbiting material becomes caught in the disk, but in the Galaxy, the density of matter in the disk is low enough that globular clusters pass through it without being captured. Finally, density waves formed in the Galaxy's disk, creating the spiral arms where star formation continues today.

Even though this scenario is very sketchy, not all astronomers agree with all of it. For example, in some models, several separate clouds of gas merge to form the Galaxy rather than one. High-velocity atomic-hydrogen clouds have been observed since 1963. They have the mass of a small galaxy and are about 15,000 parsecs across. Several hundred of these clouds have been mapped, and their orbits do not follow the approximately circular orbits of most objects in the Galaxy. If such clouds are indeed the building blocks of our Galaxy, then they represent the earliest structures that formed in the universe. They are also one of the few places where the material has not been heavily contaminated by later star formation. The theory that such clouds are the seeds of the Milky Way makes several predictions. First, since these clouds formed long ago, the abundance of heavy elements should be much lower than in our Galaxy. Second, the clouds should be large and of low pressure. Third, if these clouds are a part of the local group of about 30 galaxies (of which Andromeda and our Galaxy are members), they should show little of the fluorescence seen in the relatively nearby atomic hydrogen clouds. Fourth, if these clouds collide with the

◆ As discussed earlier, the density wave theory is not the only possible explanation for the spiral arms.

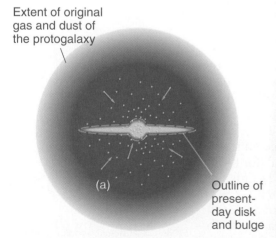

Extent of original gas and dust of the protogalaxy

(a)

Outline of present-day disk and bulge

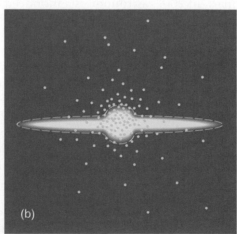

(b)

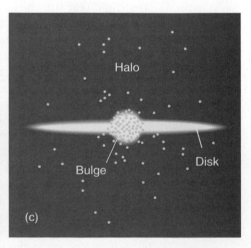

Halo

Bulge

Disk

(c)

Figure 16-29
(a) According to a leading theory of galactic evolution, the Galaxy began as a sphere of gas and dust about 13 billion years ago. (b) As the sphere collapsed toward the center, stars and clusters formed, so that the Galaxy appeared like this some 12 to 13 billion years ago. (c) The Galaxy reached its present state after the remaining gas and dust formed into a disk and spiral arms developed.

galaxies in the local group, there should be evidence of a large cloud of hot, ionized gas around our Galaxy and Andromeda.

Recent observations seem to support this theory of gradual formation of a galaxy through accretion of large hydrogen clouds. **Figure 16-30** is a composite image showing a high-velocity cloud "raining" material into the

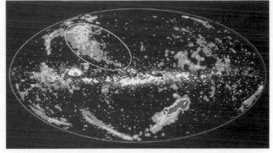

Figure 16-30
An all-sky composite image. The false-color orange-yellow clouds contain neutral hydrogen and are observed in 21-centimeter radiation, while the rendition of our Galaxy is in visible light. The edge-on Galactic plane appears as a white horizontal strip. *HST*'s spectrograph measured the composition and velocity of the encircled cloud.

Galaxy, supplying it with material to make new stars. Such evidence suggests that these clouds play a crucial role in the chemical evolution of our Galaxy; they supply it with metal-poor gas that counteracts the buildup of heavy elements in stars and in the interstellar medium of our Galaxy. This explains why most stars in the Galactic disk have similar concentrations of heavy elements, independent of their age. It also explains how our Galaxy can form about a new star every year without running out of its supply of gas after a tenth of its lifetime.

It is also possible that our Galaxy, like others, grew from smaller galaxies by collisions. This "hierarchical" picture seems to fit better with current theories that try to explain the present structure of the universe as it evolved through time. Data from the *Hipparcos* satellite support the idea of a dwarf galaxy colliding with the Milky Way some 10 billion years ago. As a result of the collision, and over a long period, the dwarf galaxy broke up into streams of stars that keep a common memory of their origin. Evidence for two such streams of stars was found in the *Hipparcos* data; they are among the oldest stars in our vicinity and spend most of their time in the Galactic halo. In early 2003, astronomers announced the discovery of a ring of stars around our Galaxy, approximately 120,000 light-years in diameter (**Figure 16-31**). The presence of the ring is an indication that at least part of our Galaxy was formed by collisions of many smaller galaxies billions of years ago. If indeed our Galaxy has accreted small nearby satellite galaxies, then many of its stars have originated in different places. More studies are needed of the distribution of the halo stars (some of the oldest in the Galaxy), their chemical composition, and their motion, so that we can better understand when and how the halo formed.

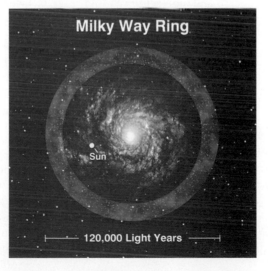

Figure 16-31
A ring of stars around our Galaxy was discovered by the Sloan Digital Sky Survey. The ring contains about half a billion stars, has a thickness of about 10,000 light-years, and encircles us like a giant doughnut.

Our knowledge of the Galaxy has only recently progressed to the point where we can formulate meaningful models of its formation. Today's models will have to be developed, and perhaps combined, before we can be confident of their validity. In the next chapter we expand on our discussion of the formation of galaxies.

Conclusion

Our Sun is one of some 200,000,000,000 stars in an enormous spiral galaxy that is itself just one of countless galaxies in the observable universe. The nature of our Galaxy was discovered only rather recently because our vision is obscured by gas and dust along the Galaxy's disk. Although measurements of the Galaxy's characteristics are necessarily imprecise, some of its properties can be measured using techniques dating from the time of Isaac Newton, such as the calculation of the Galaxy's mass using Kepler's third law. Detection and measurement of many other features of the Galaxy must rely on more modern methods and equipment, such as radio astronomy, which has opened many doors to understanding the Milky Way.

Analysis of 21-centimeter radiation leads us to conclude that our Galaxy is a spiral one, like many others we observe. Two principal theories have been proposed to explain the spiral arms of a galaxy: the density wave theory and the self-propagating star formation theory.

The nuclear bulge of our Galaxy contains a high density of stars. The tremendous energy that pours from the nucleus, as well as the rapid motions of objects near it, are consistent with the hypothesis that a massive black hole is at the center of the Galaxy.

Data that are accumulating about the Galaxy permit us to develop models of Galactic formation, but these models, like many of astronomy's theories about the Milky Way, are very tentative and await further research by today's and tomorrow's scientists.

Study Guide

RECALL QUESTIONS

1. Our Sun is located
   A. at the very outer edge of the Galaxy.
   B. far from the center of the Galaxy, but not at the outer edge.
   C. near the center of the Galaxy, but not right at the center.
   D. at the center of the Galaxy.

2. Astronomers of the eighteenth and nineteenth centuries thought that the Sun was near the center of the Milky Way Galaxy because they counted the same number of stars in the disk of the Galaxy in every direction. The reason they were not correct is that the Galaxy
   A. is an irregular galaxy with a chaotic shape.
   B. contains dust that obscures its distant regions.
   C. has the shape of a tube with the Sun near one end.
   D. has two kinds of Cepheid variables, so that all distance measurements until recently were incorrect.
   E. has a giant black hole at its center.

3. The center of the Galaxy was first discovered by observing
   A. X-rays that it emits.
   B. radio waves that it emits.
   C. the intense light that comes from it.
   D. the bulge around it.
   E. its yellow glow.

4. When we look at the sky on a dark, clear night, sometimes we can see a faint band of light stretching across it. What we are observing is
   A. the plane of the Milky Way Galaxy.
   B. other galaxies.
   C. the stars of some other galaxy.
   D. the center of the Milky Way Galaxy.
   E. clusters of other galaxies that are drawn to the Milky Way.

5. The Sun is one of about how many stars in the Milky Way Galaxy?
   A. 100 to 200 million
   B. One billion
   C. 100 to 400 billion
   D. One trillion
   E. [We don't have any idea how many stars are in the Galaxy.]

6. Evidence for the spiral nature of the Galaxy comes primarily from
   A. Cepheid variable data.
   B. Doppler shift data of stars.
   C. 21-centimeter data.
   D. observations of globular clusters.
   E. [Both A and D above.]

7. We have learned that the Galaxy is a spiral galaxy by
   A. the reflection of light from nebulae.
   B. the Doppler shift of radio waves.
   C. observations of globular clusters.
   D. comparing ages of various clusters.
   E. plotting the stars of various clusters on an H-R diagram.

8. Why is observation of distant parts of the Galaxy limited in the visible region?
   A. Interstellar dust blocks visible light.
   B. The redshift of distant stars makes them difficult to detect.
   C. Opacity of the atmosphere blocks visible light.
   D. Relative motion of the spiral arms makes them difficult to detect.

9. Radio observations can provide information about the Galaxy that we cannot learn from observations in visible light because
   A. radio waves are able to pass through clouds of dust.
   B. stars emit a greater amount of energy in radio wavelengths than in visible wavelengths.
   C. the velocity of radio waves is greater than that of light.
   D. our Galaxy is spiral in nature.

10. William and Caroline Herschel counted stars in various directions in an attempt to determine the Sun's position in the Galaxy. Later, Jacobus Kapteyn tried to improve on their determination by taking into account
    A. the interstellar dust.
    B. radio waves from stars.
    C. bright nebulae between the stars.
    D. redshift data on stars in various directions.
    E. the density of stars at various distances.

11. Oort and Lindblad determined the motion of the Sun relative to other stars by analyzing
    A. the interstellar dust.
    B. radio waves from stars.
    C. bright nebulae between the stars.
    D. redshift data on stars in various directions.
    E. the density of stars at various distances.

12. A calculation of the mass of the inner Galaxy can be made based on
    A. Kepler's laws.
    B. observations of nebulae in spiral arms.
    C. star counts.
    D. distances to Cepheid variables.

13. According to the density wave theory of the spiral arms,
    A. the number of stars per volume is about the same between the spiral arms as in them.
    B. the number of stars per volume is much greater between the spiral arms than in them.
    C. the number of stars per volume is much less between the spiral arms than in them.
    D. [The theory predicts nothing about the number of stars per volume of space.]

14. How does the density wave theory of spiral arms explain why they are brighter than the regions between the arms?
    A. There are more stars per volume in the arms than between them.
    B. Brighter stars are in the spiral arms because they follow along with the arms as they move.
    C. Brighter stars are in the spiral arms because they are formed there and are short-lived.

15. According to the self-propagating star formation theory of spiral galaxies' arms,
    A. stars revolve around the galactic center in perfect circles.
    B. stars obey Kepler's third law as they move around the galactic center.
    C. stars are about evenly spread around the galaxy.
    D. [Two of the above.]
    E. [None of the above.]

16. A black hole is thought to be at the center of the Galaxy because
    A. no light comes from that area of the Galaxy.
    B. large amounts of energy come from a small source there.
    C. motions of objects near the center indicate a large mass there.
    D. [Both A and C above.]
    E. [Both B and C above.]

17. Serendipity is the ability to
    A. detect small changes in data.
    B. make unexpected discoveries by accident.
    C. do mental calculations quickly and accurately.
    D. visualize situations in three dimensions.
    E. analyze data quickly and accurately.

18. The self-propagating star formation theory of spiral arms holds that
    A. nuclear reactions occur in cores of stars in chain reactions.
    B. the wave moves around the galaxy in chain reactions.
    C. new stars that form the arms are the result of shock waves from supernovae.

19. Use an analogy to explain the nature of a density wave.

20. Sketch the Galaxy both edge-on and face-on, showing its prominent features and the approximate location of the Sun.

21. Describe the Milky Way as seen from the Earth by naked eye.

22. Distinguish between galactic clusters, globular clusters, and clusters of galaxies.

23. Name four principal components of the Galaxy and describe the location of each.

24. What two observations give evidence of spiral arms in the Galaxy?

25. What leads us to conclude that globular clusters are very old?

1. If stars revolve around the Galactic center independently of the spiral arms, it would seem that there should be as many stars per unit volume between the arms as inside them. Yet the text states that there is *almost* the same density of stars between the arms as inside them. Why would there be a difference at all?

2. Explain how Kepler's third law is used to calculate the mass of the Galaxy. Discuss the limitations of this method.

3. Explain why the hypothesized Galactic rotation curve of Figure 16-13a (solid line) extends down to very low speeds near the zero point on the bottom axis, while the dashed-line curve of Figure 16-13a does not.

4. If a radio telescope cannot reveal directly the distance to a source of 21-centimeter radiation, how can this radiation be used to reveal spiral arms in the Galaxy?

5. Write a report on various models of the formation of the Galaxy based on the article "How the Milky Way Formed," in *Scientific American*, January, 1993, pp. 72–78.

6. What is the evidence that stars of the nuclear bulge formed before stars in other parts of the Galaxy?

7. The Milky Way looks more prominent in July than in December. Why?

8. How did William and Caroline Herschel seek to determine the Sun's position in the Galaxy? What measurements did Kapteyn make to answer the same question?

9. Explain why 21-centimeter radiation reveals the parts of the Galaxy in which O- and B-type stars are located.

10. We have never observed a globular cluster pass through the disk of the Galaxy. How do we know that they do?

CALCULATIONS

1. Show how you can calculate the period of the Sun's motion around the Galaxy, given that the Sun's distance to the center is 26,000 light-years and its speed is 220 kilometers/second.

2. Suppose that a portion of another galaxy is observed to move with a speed of 250 kilometers/second and to be at a distance of 40,000 light-years from the center of that galaxy. What is the period of revolution of that portion of the galaxy?

3. Use the data from Question 2 to calculate the mass of that galaxy that is inside the radius of the portion described.

4. Suppose we wish to construct a physical model of the Galaxy. If the Sun is represented by a small grain of sand 0.1 millimeter across, what will be the diameter of the model? (Hint: See the Activity.)

ACTIVITIES

The Scale of the Galaxy

Suppose that you wish to construct a scale drawing of our region of the universe. The Sun has an actual diameter of about 1,500,000 kilometers, and you represent it by a dot the size of a period on this page, about 0.5 millimeters across. The average distance between stars in our region of the Galaxy is about 5 light-years. Therefore, you must calculate what the corresponding distance would be on your drawing, given that one light-year is equal to 9.5×10^{12} kilometers.

First, set up the ratio of the distance on your scale to the actual distance:

$$\frac{\text{distance on scale}}{\text{actual distance}} = \frac{0.5 \text{ mm}}{1.5 \times 10^6 \text{ km}}$$

Next, calculate the distance between stars in kilometers:

$$5 \text{ LY} \times \frac{9.5 \times 10^{12} \text{ km}}{1 \text{ LY}} = 4.8 \times 10^{13} \text{ km}$$

Now, use that value as the actual distance and solve for the distance in the drawing that represents the separation of stars:

$$\frac{\text{distance on scale}}{4.8 \times 10^{13} \text{ km}} = \frac{0.5 \text{ mm}}{1.5 \times 10^6 \text{ km}}$$

$$\text{distance on scale} = 1.6 \times 10^7 \text{ mm}$$

Finally, change this to more appropriate units:

$$1.6 \times 10^7 \text{ mm} \times \frac{1 \text{ m}}{10^3 \text{ mm}} = 1.6 \times 10^4 \text{ m}.$$

This is 16 kilometers! It means that on your drawing, stars (end-of-sentence periods) must be 16 kilometers, or 10 miles, apart.

The Galaxy is about 160,000 light-years in diameter, or 32,000 times the average distance between stars. This means that on your scale, the Galaxy would be 512,000 kilometers across, which is larger than the Earth-Moon distance! This is beyond imagining, and distances to other galaxies have not yet been calculated. Therefore you will have to try another scale.

Suppose you take the average distance between stars on your revised scale to be 5 centimeters. Although the stars would be too small to see on an actual drawing to this scale, you might cheat and make each star a very tiny speck.

On your drawing, 5 centimeters would correspond to 5 light-years. Now calculate the size of the Galaxy, the distance to the Andromeda galaxy (which is actually about 2 million light-years away), and the distance to the most distant objects we can see (which, as discussed in the next chapter, are about 13 billion light-years away).

EXPANDING THE QUEST

1. "Meet the Milky Way," by V. Trimble and S. Parker, in *Sky & Telescope* (January, 1995).

2. "Searching for Dark Matter," by M. Mateo, in *Sky & Telescope* (January, 1994).

3. "Destination: Galactic Center," by R. Jayawardhana, in *Sky & Telescope* (June, 1995).

4. "How the Milky Way Formed," by S. van den Bergh and J. E. Hesser, in *Scientific American* (January, 1993).

STARLINKS

Quest Ahead to Starlinks: www.jbpub.com/starlinks

Starlinks is this book's online learning center. It features **eLearning,** which contains chapter quizzes and other tools designed to help you study for your class. You can also find **on-line exercises,** view numerous relevant **animations,** follow a guide to **useful astronomy sites** on the web, or even check the latest **astronomy news** updates.

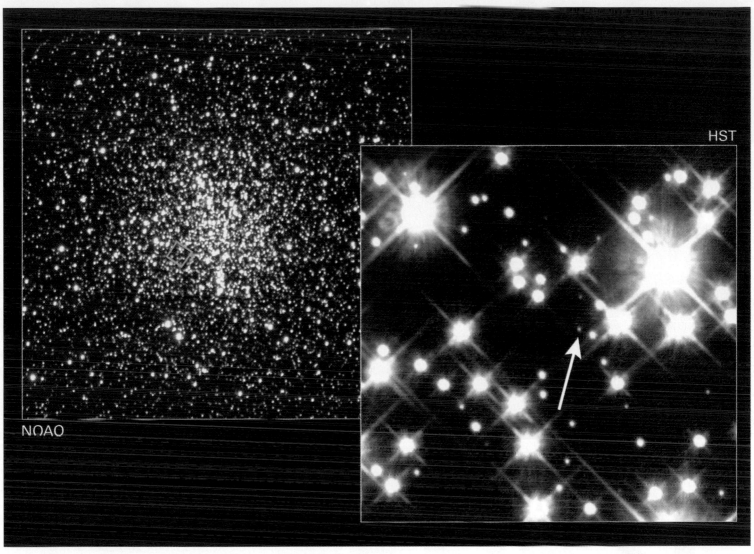

The image at left is a NOAO image of the 13-billion-year-old globular cluster M4, 5,600 light-years away in the constellation Scorpius. The *HST* image at right is a close-up view of the location marked by the green box. The arrow marks the position of a white dwarf, which has two companions. One of the companions is a pulsar. The identity of the third companion was a mystery for many years until the *HST* observations helped astronomers to measure its mass, about 2.5 time the mass of Jupiter. This object, therefore, is the oldest and farthest known planet.

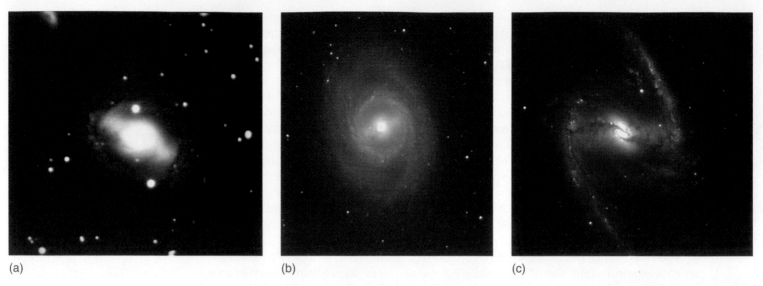

(a)   (b)   (c)

Figure 17-2
(a) The Sba barred spiral galaxy NGC4650 is in the constellation Centaurus. (b) M95 is an SBb barred spiral galaxy, possibly ringed, part of a cluster of galaxies at a distance of about 38 million light-years. (c) NGC1365 is a SBc barred spiral galaxy, part of a cluster of galaxies about 60 million light-years from our Galaxy.

Galaxies (and other objects) are named by their listing in the Messier catalog (prefix *M*) or in the New General Catalog (prefix *NGC*).

arms (type a) also have the most prominent nuclear bulges. **Figure 17-3** shows an Sa galaxy and two Sb galaxies. The galaxy in Figure 17-1 is an Sc galaxy. When a spiral galaxy is seen edge-on, it often displays the lane of dust and gas clouds that we see in our own Galaxy.

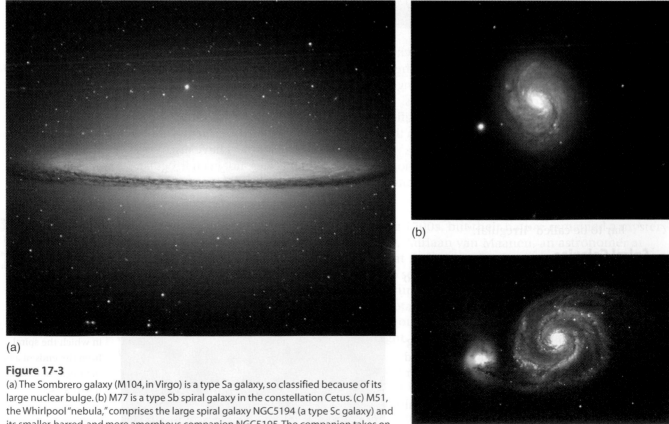

(a)

(b)

(c)

Figure 17-3
(a) The Sombrero galaxy (M104, in Virgo) is a type Sa galaxy, so classified because of its large nuclear bulge. (b) M77 is a type Sb spiral galaxy in the constellation Cetus. (c) M51, the Whirlpool "nebula," comprises the large spiral galaxy NGC5194 (a type Sc galaxy) and its smaller, barred, and more amorphous companion NGC5195. The companion takes on a reddish tinge because it is behind the dust-filled arm connecting it to NGC5194.

Barred spiral galaxies are also classified according to how tightly their arms are wound (Figure 17-2). Up to two-thirds of all spiral galaxies contain bars. The bar systems provide an efficient mechanism for fueling star births at the centers of SB galaxies. A few galaxies seem to have the nuclear bulge and disk of a spiral galaxy, but no arms. Hubble called these S0 galaxies.

Differences between the three categories (a, b, and c) might be related to how much gas and dust they contain. Sc and SBc galaxies contain much more gas and dust than Sa and SBa galaxies, resulting in a larger proportion of their mass being involved in star formation. This may explain why Sc and SBc galaxies have larger disks and smaller bulges than Sa and SBa galaxies.

Most spiral galaxies are from 50,000 to 2,000,000 light-years across and contain from 10^9 to 10^{12} stars. (We will see later how such measurements are made.) For comparison, recall that the Milky Way is about 160,000 light-years across and contains at least 2×10^{11} stars.

Elliptical Galaxies

Figure 17-4 shows some elliptical galaxies and makes clear why they were given that name. Various elliptical galaxies show different eccentricities in their elliptical shape, depending in part on their orientation to Earth. (For example, a football appears round if viewed end-on.) The actual eccentricity of an elliptical galaxy is difficult to

An elliptical galaxy is, of course, a three-dimensional object, not an ellipse. The correct name for its three-dimensional shape is an ellipsoid.

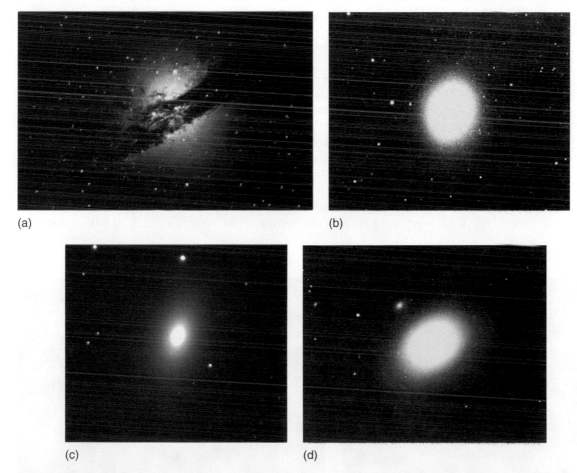

(a)

(b)

(c)

(d)

Figure 17-4

(a) NGC5128 (or Centaurus A) is a type E0 peculiar elliptical galaxy. It is one of the most luminous and massive galaxies known. The dark band across the galactic disk is debris from a smaller dusty galaxy that is being absorbed by Centaurus A. (b) M32 is a type E2 dwarf elliptical galaxy and is one of the companions to the giant Andromeda Galaxy M31. (c) M59 is a type E5 elliptical galaxy in the constellation Virgo, one of the many members of the Virgo Cluster. (d) M86 is a type S0 galaxy, a lenticular galaxy, although some prefer to call it an elliptical galaxy and a member of the large Virgo Cluster of galaxies.

determine because its orientation is unknown. Therefore, elliptical galaxies are classi-fied based on how they appear, from round (E0) to very elongated (E7).

Most of the galaxies in existence are ellipticals, but most galaxies listed in catalogs are spirals. The reason for this is that although a few giant elliptical galaxies are larger than any spiral galaxy (having 100 times more stars than the Milky Way), most are small and dim (with one-millionth as many stars as the Milky Way). Dwarf elliptical galaxies are visible from Earth only if they are relatively nearby.

Irregular Galaxies

Hubble found that some galaxies did not fit into either of the above categories, nor did they exhibit other common characteristics. **Figure 17-5** indicates why these were called "irregular" galaxies. Fewer than 20% of all galaxies fall in the category of irreg-ulars, and they are all small, normally having fewer than 25% of the number of stars in the Milky Way.

The Magellanic Clouds are usually classified as irregular galaxies, although some astronomers think that the Large Magellanic Cloud is a barred spiral that has been dis-rupted by its proximity to the Milky Way and perhaps by a past collision with the Small Magellanic Cloud. Collisions between galaxies should not be unusual, because on the average they are separated by distances only about 20 times their diameter. Stars within a galaxy, on the other hand, are separated by millions of times their diam-eter and therefore collide infrequently. However, as a result of their great distances from Earth, galaxies exhibit no proper motion (motion across our line of sight), so we have to deduce past collisions from their present appearance. **Figure 17-6a** shows two galaxies, called "the Antennae," that are in the process of colliding. Computer simula-tions show that gravitational forces between the galaxies produce odd formations, such as the two "antennae."

Computer simulations also show that colliding galaxies actually pass through one another with few collisions between individual stars, since distances between stars are very large. However, the large sizes of the interstellar dust and gas clouds in the colliding galaxies make them more likely targets. The collisions between clouds of interstellar gas result in greater gas density and therefore increased star forma-tion. Such "bursts" of star formation may even occur as a result of gravitational in-teractions among neighboring galaxies, as is the case for M82 (Figure 17-6b). In addition, gravitational forces drastically alter the shapes of the two colliding galax-ies. Figure 17-6c shows a grazing encounter of two spiral galaxies, in which the

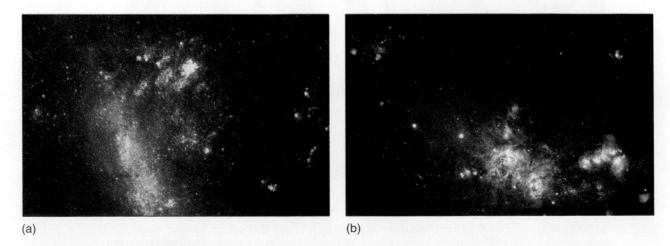

(a)   (b)

Figure 17-5

(a) The Large Magellanic Cloud, the nearest galaxy to the Milky Way at only 160,000 light-years away, is an irregular galaxy. (b) This is another irregular galaxy, NGC1313, that is visible from the Southern Hemisphere.

Edwin Hubble*

Like many famous people, Edwin Hubble has been the subject of several myths. He is said to have been a professional-quality boxer, a wounded World War I hero, and a lawyer, but no records exist to support any of these claims. Embellishments aside, one thing that we know for certain is that Hubble was a very capable and energetic scientist (**Figure B17-1**).

Hubble was born on November 20, 1889, in Marshfield, Missouri, where he was one of seven children. In high school he showed promise in both academics and athletics, becoming a track star and placing in the top quarter of his class. When he was sixteen, he entered the University of Chicago, where he continued to excel at science and math while lettering in track and basketball. After graduation he received a Rhodes scholarship and enrolled at the University of Oxford, where he studied law

Figure B17-1
Edwin Hubble showed conclusively that the Andromeda nebula is in fact made up of stars and is therefore another galaxy. His discovery expanded our concept of the universe tremendously.

(his family hoped he would become a lawyer). By 1914, Hubble was bored with the study of law and decided instead to do graduate work in astronomy at Yerkes Observatory in Wisconsin.

Early in his graduate career, Hubble attended a presentation by Vesto M. Slipher. A hot topic in astronomy at the time was whether spiral nebulae are part of our Galaxy or are galaxies in their own right. Slipher's presentation included observational data that supported the latter hypothesis, but not everyone was convinced. Possibly inspired by that talk, Hubble began photographing nebulae using the Yerkes's 24-inch reflecting telescope. His Ph.D. thesis, entitled "Photographic Investigations of Faint Nebulae," grew out of this earlier work. Although this work was not scientifically rigorous, it led Hubble closer to the conclusion that the nebulae were extragalactic objects separate from the Milky Way.

Before Hubble could reach any final conclusions, the United States entered World War I. Three days after receiving his Ph.D., Hubble enlisted and eventually rose to the rank of major in the 86th "Black Hawk" Division. Despite his success in the military, Hubble never saw any fighting because the Armistice had been declared by the time his division reached Europe. Hubble was disappointed, but

went instead to Mount Wilson Observatory, where he returned to his study of the nature of "nongalactic nebulae."

Using the 100-inch reflector at Mount Wilson, Hubble began observing NGC6822. By 1923, he had found 12 variable stars within it, as well as several smaller nebulae. In 1924 he married Grace Burk Leib, who later became responsible for some of the myths that now surround Hubble. Upon his return to work, Hubble began observing M31 (the Andromeda nebula). Hubble distinctly resolved six variable stars in the Andromeda "nebula." This discovery convinced him that M31 is a galaxy outside our own, and he concluded that other spiral nebulae must also be separate, distinct galaxies. In the course of this work, Hubble observed several Cepheid variables in the galaxies M31 and M33. By comparing each star's luminosity with his observations of the star's apparent brightness, Hubble was able to deduce that M31 and M33 are about 930,000 light-years distant. (Hubble's measurement was off by more than a factor of two because he did not know that there are two types of Cepheids.) This distance placed them far outside the known boundaries of the Milky Way, so Hubble had even more proof to support his hypothesis that there are other galaxies outside our own. Hubble also attempted to classify the galaxies that he observed, which led to his so-called tuning fork diagram, which is still widely used today.

In the mid-1920s, Hubble started investigating the expanding universe hypothesis. As he observed galaxies such as M31 and M33, his colleague Milton Humason measured their radial velocities. By combining these radial velocities with the measured distances to the observed galaxies, the two men deduced what is called the Hubble law of redshifts: $v = H_0 d$. The law was published in a 1929 paper on the expansion of the universe. It sent shock waves through the astronomical community. The Hubble law indicates that the universe is expanding because the law predicts that velocities of galaxies increase at increasing distances from any chosen point. From 1931 to 1936, Hubble concentrated on extending this law to increasingly greater distances.

As if all this weren't enough, Hubble next began researching how the density of galaxies changes with distance. This work wasn't as groundbreaking as his earlier work, but he still strongly influenced the direction of astronomical research and wrote books on astronomical subjects for the general public. In the last few years of his life, he pushed for the construction of a 200-inch reflecting telescope at Mount Palomar in California. During World War II, he joined the staff of the U.S. Army's Ballistics Research Laboratory, where he used his early astronomical training to lead a group calculating artillery-shell trajectories. After the war, he continued his work at the now-completed Palomar Observatory until he died of a stroke in 1953. His memory lives on with the many researchers who have built upon his earlier work, as well as the work of the space telescope named for him.

*Much of the material for this box comes from Osterbrock, Gwinn, and Brashear, "Edwin Hubble and the Expanding Universe," *Scientific American* (July, 1993). That article is highly recommended.

Figure 17-8
The photo at the left is the Andromeda galaxy (M31), and two of its close companion galaxies. The rectangle in the photo is enlarged at the right, and the lines indicate two Cepheid variables near the edge of the galaxy.

and nearby galaxies, astronomers have learned that all have approximately the same range of luminosity. In addition, large globular clusters and supernovae are of consistent brightness from one galaxy to another. This allows us to use these bright stars and clusters as distance indicators in more distant galaxies.

The logic here is interesting. By determining the luminosities of nearby Cepheid variables, astronomers determined the period-luminosity relationship for Cepheids. Then they assumed that Cepheids in other galaxies are basically the same as the ones in our Galaxy, so they used that same period-luminosity relationship in reverse to calculate the absolute luminosity and thereby the distances to Cepheids in the nearer galaxies. Knowing these distances, they learned that the brightest stars and globular clusters in galaxies of each type have approximately the same luminosity. That is, the brightest 50 stars in one spiral galaxy have approximately the same average luminosity as the brightest 50 in every other spiral galaxy. Likewise for globular clusters. Now astronomers can use these stars and clusters as distance indicators to learn the distances to galaxies in which they can see the individual objects. **Figure 17-9** summarizes this chain, which allows us to determine distances as far as about 1000 million light-years.

At distances at which we can no longer see individual objects within a galaxy, the distance indicator becomes the galaxy itself. This method, as you might suspect, is extremely imprecise. For example, suppose that we see a very distant spiral galaxy. Since we know the range of luminosities spiral galaxies have, we can make some

Here again we assume—with successful results—that the same laws of science that apply here on Earth also apply in distant galaxies.

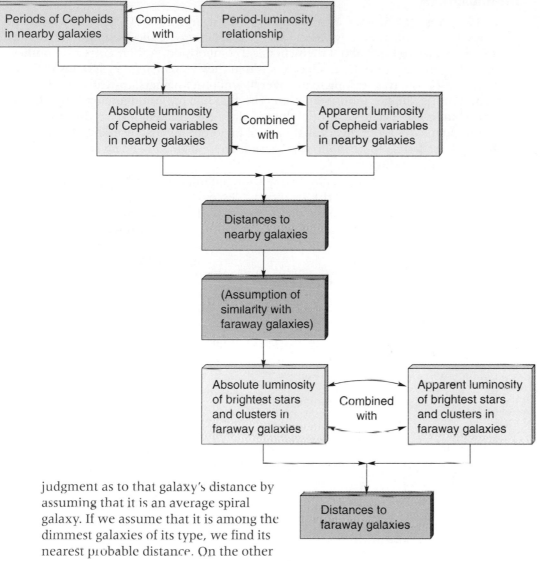

Figure 17-9
Starting with the period-luminosity relationship of Cepheids, astronomers are able to follow a chain of reasoning and observation that allows them to determine the distances to galaxies too far away for their Cepheids to be visible.

judgment as to that galaxy's distance by assuming that it is an average spiral galaxy. If we assume that it is among the dimmest galaxies of its type, we find its nearest probable distance. On the other hand, if we assume that it is among the brightest, we find its farthest probable distance. In this way, we find the range of distances within which we can be fairly certain the galaxy falls.

When applied to an individual galaxy, the whole-galaxy method of assessing distance is very imprecise. Fortunately, the method need not be restricted to individual galaxies. As we will discuss later, galaxies exist in clusters, which contain from a few galaxies to thousands of galaxies per cluster. If we consider a cluster that contains a great number of spiral galaxies, we can logically assume that the brightest spiral galaxy in that cluster has about the same luminosity as the brightest spiral in another cluster. Thus, we use the brightest galaxies as a distance indicator to the cluster.

Notice that one measurement builds on another in a series of steps. As a result, if there is an error in a beginning step, the error will propagate up through the chain of steps. For example, if somehow there is an error in our understanding of the period-luminosity relationship for Cepheids, the stated distances to the farthest galaxies will have to be adjusted. This is why constant checks are made as new data arrive.

The above analysis makes an important assumption: Galaxies in our neighborhood of the universe are basically the same as those farther away. This may seem reasonable, but remember that we are seeing distant galaxies not as they are today but as they were in the past. Light coming from a galaxy 100 million light-years away has been traveling 100 million years, and we cannot rule out the possibility that galaxies have changed in that time.

The Hubble Law

In 1912, Vesto M. Slipher, an astronomer working in Percival Lowell's observatory in Flagstaff, Arizona, was assigned the task of examining the spectrum of some of the spiral nebulae to learn about their chemical composition. A then-current hypothesis regarding the nature of these objects was that they were planetary systems in formation, and Lowell was seeking to discover life on other planets. Instead of finding evidence for life, Slipher found something else that had profound implications. He found a redshift in the spectra of most of the nebulae he examined. If the redshift were due to the Doppler effect, it meant that most of the other nebulae were moving away from us—as fast as 1800 kilometers/second. There seemed to be no reasonable explanation for this strange finding.

In 1924, Edwin Hubble showed that the nebulae are, in fact, galaxies. Then he and Milton Humason used the 100-inch telescope on Mount Wilson to photograph the spectra of these objects. Their work not only confirmed the findings of Slipher, but showed that there is a pattern in the speeds with which galaxies are receding from us.

Lowell's telescope on which Slipher worked had only a 24-inch aperture. The 100-inch telescope used by Hubble and Humason therefore had 16 times more light-gathering power.

Figure 17-10
This graph, based on the data collected by Hubble and Humason in the 1920s, shows a relationship between the recessional velocities of galaxies and their distances.

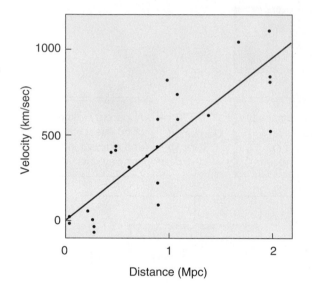

cosmology The study of the nature and evolution of the universe as a whole.

Figure 17-11
The data on Type Ia supernovae by Riess, Press, and Kirshner (1996) provide a dramatic confirmation of the relationship found by Hubble and Humason. Reevaluations of distances have caused the slope of the line to change.

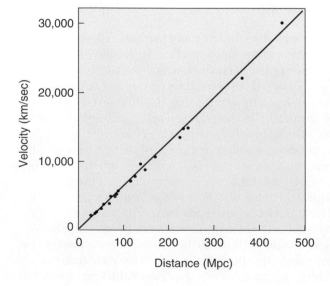

Figure 17-10 is a diagram by Hubble and Humason on which they plotted the distances to a number of galaxies against the galaxies' recessional velocities. It indicates that the more distant a galaxy is, the faster it is moving away from us. These data, taken during the 1920s, used distance indicators that relied on incorrect Cepheid variable values, so the data had to be adjusted when more was learned about Cepheids. However, as **Figure 17-11** indicates, a better understanding of standard candles dramatically confirmed the relationship.

The implications of Hubble's findings for our understanding of the past and future of the universe are very important, for they imply that the universe is expanding. In fact, Hubble's work has become the foundation for today's theories of *cosmology,* which we will study in Chapter 18. In that chapter we will also discuss the evidence for—and the implications of—the idea that the universe is expanding. We will show that the redshift that Hubble observed is not really due to the Doppler effect. The difference is not important here, however, because we are interested in using Hubble's findings only as a tool to measure distances to galaxies. What is important is that astronomers observe a regular relationship between redshift and distance, a pattern that is valuable to them in determining distances to faraway galaxies, as we will show.

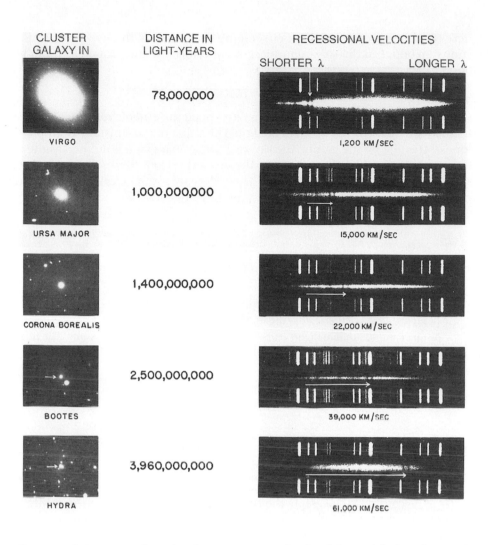

CLUSTER GALAXY IN	DISTANCE IN LIGHT-YEARS	RECESSIONAL VELOCITIES
VIRGO	78,000,000	1,200 KM/SEC
URSA MAJOR	1,000,000,000	15,000 KM/SEC
CORONA BOREALIS	1,400,000,000	22,000 KM/SEC
BOOTES	2,500,000,000	39,000 KM/SEC
HYDRA	3,960,000,000	61,000 KM/SEC

Figure 17-12
The photos at the left in the figure are of galaxies in clusters at various distances from us. At the right, the spectrum of each galaxy is shown. (The spectrum is the white, horizontal streak between the reference lines.) Arrows indicate how far the two darkest absorption lines have shifted in each case. (λ = wavelength)

Because it is convenient (and common) to think of the redshift as being due to the Doppler effect, that practice will be followed here.

Figure 17-12 shows photographs, all taken with the Palomar telescope, of five different galaxies and their spectra. Compare the location of the two prominent absorption lines in each of the spectra, noting that the dimmer the galaxy, the farther to the right (toward longer wavelengths) is the pair of lines. The distances to the galaxies, determined by methods described in the previous section, show that—as we would expect—the dimmer galaxies are farther away.

Using the distance values given and interpreting the redshift as due to the Doppler effect, **Figure 17-13** shows a "Hubble plot" of the galaxies' distances versus their recessional velocities. A dot on the diagram corresponds to the recessional velocity and distance of each of the five galaxies. The dots appear to be in a straight line (allowing for errors in measurement), so a straight line has been drawn to represent the trend they show. This graph confirms the relationship shown in the graphs in Figures 17-10 and 17-11.

Whenever data can be represented by a straight line on a graph,

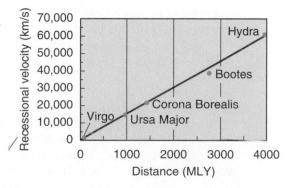

Figure 17-13
A graph of recessional velocity versus distance for the five galaxies of Figure 17-12 shows that their recessional velocities are proportional to their distances. Each galaxy is identified by the constellation it is in.

Hubble constant The proportionality constant in the Hubble law; the ratio of recessional velocities of galaxies to their distances.

Hubble law The relationship that states that a galaxy's recessional velocity is directly proportional to its distance.

there is a direct proportionality between the quantities on the two axes. Thus, the relationship between the two quantities, radial velocity, v, and distance, d, can be written as

$$v = H_0 d$$

where H_0 is the constant of proportionality, called the **Hubble constant** in honor of the man who discovered this **Hubble law.** The value of the constant is simply the slope of the line on the graph. The slope of a line on a graph is defined as the ratio of the change in the quantity plotted on the vertical (y) axis to the change in the quantity plotted on the horizontal (x) axis. The following example calculates the Hubble constant as an example of determining the slope of a line.

EXAMPLE

Calculate the slope of the line in the graph of Figure 17-13, thereby calculating the Hubble constant.

SOLUTION

Calculating the slope requires a comparison of the changes in the two quantities represented on the graph. Thus, we must choose two points on the data line of Figure 17-13. Observe that the line happens to cross the point where the velocity is 30,000 kilometers/second and the distance is 2000 million light-years. This will be one of the chosen points. The line also passes through the point where both quantities are zero; this will be the second point. Now we calculate the slope as

$$\text{slope} = \frac{30{,}000 \text{ km/s} - 0}{2000 \text{ MLY} - 0} = 15 \text{ km/(s} \times \text{MLY)}$$

TRY ONE YOURSELF

Calculate the slope of the line in the 1931 Hubble-Humason graph (Figure 17-10) to find the value of the Hubble constant as determined with that data.

Astronomers usually use Mpc, megaparsecs (million parsecs), as their distance unit. Fifteen km/(s × MLY) is 50 km/(s × Mpc).

Thus, according to the graph, the Hubble constant is about 15 km/(s × MLY) (read as "kilometers per second per million light-years"). It is important to see what Hubble's constant means and what its units represent. A value of 15 km/(s × MLY) means that for each million light-years a galaxy is farther from us, its speed is greater by 15 kilometers/second. Refer again to the graph to check that a galaxy at a distance of 1000 million light-years has a speed of 15,000 kilometers/second. In like manner, a galaxy 3000 million light-years away has a speed of 45,000 kilometers/second.

The slope of the graph of Figure 17-11, which represents modern data, yields a Hubble constant of about 20 km/(s × MLY). The slope of the 1931 Hubble-Humason graph is much greater than this. Their calculation for H_0 was about 140 km/(s × MLY), about 7 times the more modern value. This was due to their misunderstanding of the luminosity of Cepheids, as we discussed earlier.

Data compiled by Allan Sandage of the Mount Wilson Observatory and Gustav Tammann of the University of Basel, who have worked for years on the question, indicate a value of about 17 km/(s × MLY) for the Hubble constant. Another group of researchers uses slightly different methods for determining distances and obtains values close to 25 km/(s × MLY). In the late 1990s, astronomers announced that *HST* observations of Cepheids in 18 galaxies out to 65 million light-years from us suggest a Hubble constant of about 22 km/(s × MLY), while *HST* observations of Type Ia supernovae suggest a value of about 19.6 km/(s × MLY). The difference between the two values is about 10%, similar to the errors in the measurements. It is obvious that the values of H_0 obtained by different methods through the years are converging. What is important here is not to memorize all the different values but to consider all the efforts in finding the value of H_0 as an example of how science is done. Today, astronomers are very confident that the true value of H_0 lies between 15 and 25 km/(s × MLY).

As we will see in Chapter 18, the latest observations of the radiation left over from the hot big bang indicate a value of 71 ± 4 km/(s × Mpc) for the Hubble constant, or about 21.8 ± 1.2 km/(s × MLY).

The Hubble constant does not remain constant as time goes on. As we will see in Chapter 18, H_0 is the slope of the line in the graph of recessional velocity versus dis-

ADVANCING THE MODEL

Milton Humason, Mule Driver/Astronomer

Milton Humason (1891–1972) was born in Minnesota. He grew up in southern California, however, where he liked the mountain on which he lived, Mount Wilson, more than he liked school. When he finished the eighth grade, he dropped out of high school to wander the mountain. Later, when George Hale chose that mountain as the site of a large observatory with a 60-inch telescope, Humason was hired as a mule driver to transport equipment up the steep slopes.

Humason learned about the observatory by talking to the construction workers, and when it was completed, he was hired as the first janitor for Mount Wilson Observatory. A young man with great curiosity, he constantly talked to the astronomers about their work, and eventually he was allowed to help them develop photographic plates in the darkroom. He then progressed to helping them take photographs with the large telescope.

Humason was aware of the inadequacy of his education and persuaded Harlow Shapley and Seth Nicholson, another astronomer at the observatory, to teach him math, including calculus. In a few years he was hired as a full-time night assistant and gave up his janitorial duties. Finally, in 1930, when he was 29 years old, he was appointed to the professional staff of the observatory, which by then had built a 100-inch telescope. Humason worked closely with Edwin Hubble, using the 100-inch telescope on Mount Wilson, but since Hubble was the leader in their research, some astronomy books fail to even mention the former mule driver who worked with him.

When the 200-inch Hale telescope was completed in 1948 on nearby Mount Palomar, Humason used that instrument in his research. His contributions were so significant that he was awarded an honorary doctor's degree from the University of Lund, Sweden.

It is interesting that Milton Humason almost discovered Pluto. At the same time that Percival Lowell was attempting to find a planet beyond Neptune, W.H. Pickering was doing similar calculations in an attempt to predict its position. He asked Humason to take photographs of certain areas of the sky where he predicted the planet to be, but Humason was unable to locate the planet. Later examination of his plates showed that he had photographed Pluto twice, but in one case it was too close to a star to be visible, and in the other it happened to coincide with a flaw on the photographic plate.

---

tance (for example, Figure 17-11) as measured during *this* period of time in the life of the universe. The Hubble constant is very difficult to determine with accuracy. We can calculate the distance to galaxies in nearby clusters fairly accurately using established distance indicators, but the motion of a galaxy within a nearby cluster is significant compared to the motion of the cluster (so that the galaxy may even be approaching us). Thus, the velocity of a nearby galaxy is due in large part to random motions within its cluster. It therefore does not fit the Hubble law. A distant galaxy, on the other hand, is in a cluster whose motion away from us is much greater than the galaxy's individual random motion, so its redshift follows the Hubble law. However, for these galaxies, determination of the distance to the cluster is much less precise. So in cases where we can measure the distance accurately, the value obtained for velocity is less meaningful, and where we can have faith in our velocity measurement, the distance measurement is imprecise.

It is important to know the value of the Hubble constant because knowledge of the expansion rate of the universe is fundamental to our understanding of the universe as a whole. We will discuss this idea again in the next chapter.

The Hubble Law Used to Measure Distance

As we've discussed, distances to nearby galaxies can be measured using Cepheid variables as distance indicators. However, as we progress to more distant galaxies, Cepheids can no longer be seen, and we must use globular clusters and the brightest stars as distance indicators. At even greater distances, these cannot be seen, and the brightest galaxies serve as distance indicators to clusters of galaxies. Finally, at the farthest distances, even this method cannot be used, and we turn to the Hubble law to indicate a galaxy's distance. We illustrate this idea with an example.

Suppose a very faint source is observed to have a redshifted spectrum that indicates a recessional speed of 120,000 kilometers/second. Assuming that the Hubble law applies to this object, calculate its distance.

SOLUTION   First, we must decide which value of the Hubble constant to use. Here the distance is calculated using both 15 km/(s × MLY) and 25 km/(s × MLY), to give a range of distances within which we can be fairly confident the object lies. Using 15 km/(s × MLY):

$$v = H_0 d$$

$$120,000 \text{ km/s} = 15 \text{ km/(s} \times \text{MLY)} \times d$$

$$d = 8000 \text{ MLY}$$

Now, using 25 km/(s × MLY),

$$120,000 \text{ km/s} = 25 \text{ km/(s} \times \text{MLY)} \times d$$

$$d = 5000 \text{ MLY}$$

Thus the object is between 5000 and 8000 million light-years away. This is 5 to 8 billion light-years!

You may be uncomfortable with such great uncertainty in an answer, but remember that without the Hubble law and this calculation we would have no idea whatsoever of the distance to the object. The Hubble law does provide us with an idea of its distance.

Suppose an object is observed with a redshift that indicates a speed of 90,000 kilometers/second. What is the range of distances within which we can feel reasonably sure that the object lies?

ADVANCING THE MODEL

Observations, Assumptions, and Conclusions

It is always important in science to separate observations from the conclusions that are based on the observations. Figure 17-12 serves as a good example of this rule. The figure shows simply that the dimmer a galaxy is, the greater the redshift of its spectrum. Although the figure only indicates this relationship for five galaxies, the relationship has been found to be true in innumerable cases and has been confirmed by a number of researchers. It is based directly on many *observations* and is considered indisputable.

The middle column of the figure shows the distance of each galaxy from Earth and allows us to conclude that the farther away a galaxy is, the greater its redshift. This conclusion, however, is *not* based directly on observations, but upon a conclusion about the distances to galaxies. Its validity depends upon how accurate the distance determinations are. One would suppose that the decreasing brightnesses of the galaxies in the figure are due to greater distance, but that is not *proven* by the figure. Actually, each of the galaxies in the figure is part of a cluster of galaxies, and the distance to the cluster has been measured using a variety of distance indicators. We are confident that the dis-

tances given are fairly reliable, although no one would pretend that they are exact.

Even though the distances to the galaxies are known with some accuracy, it is important to remember that the relationship between redshift and distance is a *conclusion*, not a direct observation. Now we take the next step: we assume that the redshift is caused by the Doppler effect, and therefore that it is caused by the galaxies' motions relative to the Earth. The Hubble relationship between recessional velocity and distance depends upon this *assumption*. Is there any other possible explanation for the redshift? In Chapter 18 we will show that there is, although the explanation is similar to that of the Doppler effect and does not negate the Hubble law.

Scientists must be on guard to remember the difference between what is observed—the evidence—and the conclusions that they reach based on those observations. Between an observation and a conclusion lies one or more assumptions, and the validity of the conclusion is based not only on the accuracy of the observations, but upon the validity of the assumptions.

ADVANCING THE MODEL

The Precision of Science

It is sometimes said that science is an exact study. If those who make such statements mean that the measurements of science are exact, they are wrong. It is obvious that measurements of distances to galaxies are not exact. What is not so obvious is that *every* measurement in science is to some degree an approximation. No measurement is absolutely exact. What scientists attempt to do is to be aware of how *inexact* their measurements are. For example, in calculating the distance to a galaxy, parallel calculations are made to determine the probable error in the measurement. Measurements may show that a galaxy is 200 million light-years away, with a likely error of 50 million light-years. This means that the galaxy is measured to be between 150 and 250 million light-years away. Calculating the likely error is a

common practice in all natural sciences. Thus, scientists attempt to be specific about their inexactness.

In addition, the previous Advancing the Model box stated that scientists must take into account assumptions that are consciously or unconsciously included in the measurements. For example, in the case of determining distances to faraway galaxies, astronomers assume that there is negligible intergalactic material that might diminish the light from those galaxies. Early measurements of the size of the Galaxy were in error because they did not take into account the interstellar dust and gas. A scientist tries to be aware of the assumptions involved in each measurement. There is no claim that a measurement is exact. In fact, there is little room in science for the word "exact."

The Tully-Fisher Relation

In the 1970s, astronomers Brent Tully of the University of Hawaii and J. Richard Fisher of the National Radio Astronomy Observatory discovered that spiral galaxies with wider 21-centimeter lines have greater absolute luminosities. The reason for the relationship is that a galaxy's spectral line is wide if the galaxy is rotating fast, because the Doppler effect causes a redshift from one side of the galaxy and a blueshift from the other side. Thus, the faster the rotation, the wider the line. In addition, more massive galaxies would be expected (1) to rotate faster, and (2) to be brighter. **Figure 17-14** shows the logic that indicates that a wide spectral line would be associated with a bright galaxy.

Tully-Fisher relation A relation that holds that the wider the 21-centimeter spectral line, the greater the absolute luminosity of a spiral galaxy.

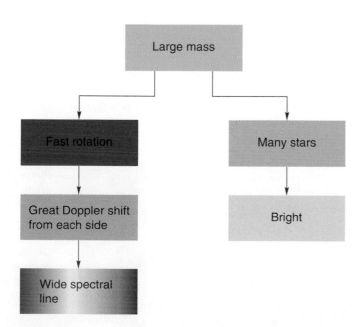

Figure 17-14
A more massive galaxy would be expected to be bright and to have a wide 21-centimeter spectral line. Similarly, a galaxy of low mass is dim and has a narrow line. Tully and Fisher discovered this relationship experimentally.

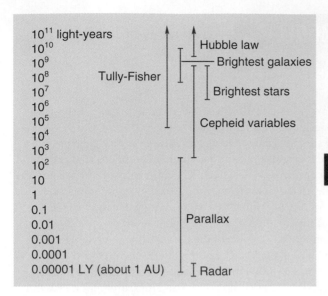

Figure 17-15
A review of the various methods of measuring distance in astronomy. Approximate limits are indicated for each method.

Using the **_Tully-Fisher relation,_** astronomers can determine the absolute magnitude of a galaxy and use it as a distance indicator. The new relationship is being used to check other methods of distance measurement and to improve our knowledge of distances to galaxies.

Figure 17-15 reviews the various methods used to measure distances in astronomy, from radar within the solar system to the Hubble law for the most distant objects.

17-3 The Masses of Galaxies

A galaxy's mass can be determined in several ways, all of them limited in precision, primarily because of the tremendous distances involved. One method of measuring the mass of a galaxy is by observing the rotation periods of some parts of it. Naturally, we cannot wait for part of a galaxy to complete a revolution, for this takes millions of years. Instead, we use Doppler shift data to measure the velocity of part of a galaxy. Then, knowing the distance of that part from the center, we can determine the period of revolution of the stars located there. As in the case of our own Galaxy, Kepler's third law then allows us to calculate the mass of the material within the orbit of the part of the galaxy being studied.

Another method of measuring galactic masses is similar to that used to measure stellar masses. There are many cases of a pair of galaxies revolving around one another. Again, we cannot wait the long times necessary to measure the period of revolution, but we can use the Doppler shift to gauge the speeds of the galaxies. The problem with this method is that it is difficult to determine the angle of the plane of revolution to our line of sight (**Figure 17-16**), and knowledge of this angle is necessary for an accurate measurement. However, making measurements for great numbers of such binary galaxies allows us to determine an average mass for a given type of galaxy.

Clusters of Galaxies; Missing Mass

Most galaxies are part of clusters rather than lonely nomads drifting through space. **Figure 17-17a** shows a typical cluster of galaxies. Thousands of such clusters are known, and the largest clusters may contain as many as 10,000 galaxies. Figure 17-17b shows one of the most distant clusters known, about 8 billion light-years away. Thirteen of the galaxies in this cluster are remnants of recent collisions or pairs of colliding galaxies, the largest number of colliding galaxies found in a cluster so far.

Figure 17-18 is a sky chart that includes the constellations Virgo and Leo. Instead of showing stars, the chart shows galaxies to the sixteenth magnitude. The large group

◆ Be careful with the terminology here. A *galactic cluster* is an open cluster of stars. It is not a *cluster of galaxies.*

Figure 17-16
In order to use the Doppler shift to measure the speeds of two galaxies that orbit one another, we would have to know the tilt of their plane of revolution to our line of sight.

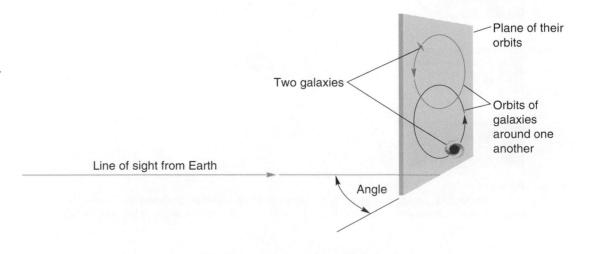

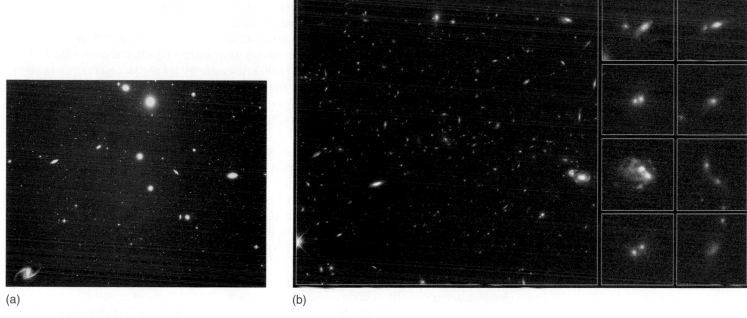

(a)   (b)

Figure 17-17
(a) Several galaxies are visible in this cluster called the Fornax cluster. It is in the constellation Fornax in the Southern Hemisphere. (b) This *HST* image shows MS 1054-03, a cluster of at least 81 galaxies, 13 of which are either remnants of recent collisions or pairs of colliding galaxies. As a result of tidal interactions, streams of stars can be seen being pulled out of the galaxies (insets).

of galaxies just above the center of the chart includes the Virgo cluster. It contains some 2500 galaxies! However, not all of the galaxies in the clump on the diagram are part of the Virgo cluster, for some are nearer to us and some farther from us than the cluster is. To determine whether a particular galaxy is part of a given cluster, as

Notice in Figure 17-18 that galaxies are not evenly distributed at all. If they were, we wouldn't see clusters of galaxies.

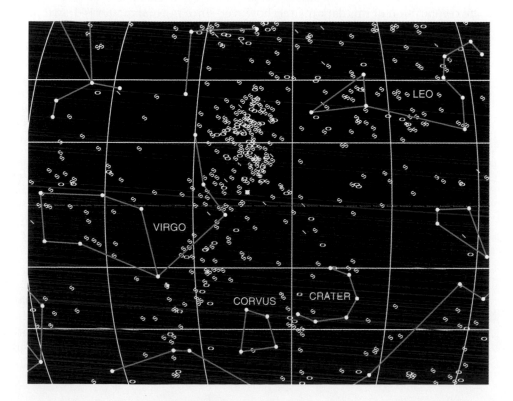

Figure 17-18
This figure shows galaxies to the sixteenth magnitude. The dashed line passing through the center of the map is the celestial equator. Spiral galaxies are designated by a slanted *S* and ellipticals by an oval. The Virgo cluster of galaxies is in the center just north of the celestial equator. (This map was produced by the Voyager software for the Macintosh.)

> Recall that all galaxies of a cluster have nearly the same redshift.

> **local group (of galaxies)** The cluster of 30 or so galaxies that includes the Milky Way Galaxy.

> In April, 1997, the discovery of a new member of the local group was announced. Named Antlia, it has only about one million stars.

> **missing mass** The difference between the mass of clusters of galaxies as calculated from Keplerian motions and the amount of visible mass.

> **dark matter** Matter that can be detected only by its gravitational interactions; it appears to be quite abundant throughout the universe.

> ESA's *Infrared Space Observatory* operated from 1995 until 1998.

tronomers must determine the galaxy's distance, because a cluster of galaxies is not just a group of galaxies that are near each other on the celestial sphere (like a constellation), but a gravitationally linked assemblage.

Our Milky Way Galaxy is part of a group of about 30 galaxies, including the Magellanic Clouds and the Andromeda galaxy, that form a cluster called the *local group.* The Andromeda galaxy and the Milky Way are by far the largest members of this cluster, which appears to contain only one other spiral galaxy. Our Galaxy is currently on a collision course with the Andromeda galaxy, closing the gap between the two at about 500,000 kilometers an hour. Within 5 billion years, the two galaxies will merge.

A third method of measuring the masses of galaxies takes advantage of their clustering. There is only one force that could hold a galaxy within its cluster: the gravitational force. If we observe a galaxy near the outside of a large cluster and assume that it is in a relatively circular orbit, we can use the Doppler effect to measure its speed and therefore its period. This allows us to calculate the total mass of the galaxies within its orbit. Calculating the mass of the cluster based on measurements for only one galaxy would not be meaningful because there would be too many uncertainties; but the same measurement, repeated for several galaxies on the outer portion of the cluster, gives a useful value for the mass of the cluster and therefore an average value for the masses of the galaxies within the cluster.

When measurements are made by the various methods, we find that clusters have much more mass than is accounted for by the visible stars within the galaxy. Within our own Galaxy, the interstellar gas and dust account for at least 10 to 20% of the total mass, and the rotation curve of the Milky Way indicates that even considering the interstellar matter, we can account for as little as one-tenth of the total mass of the Galaxy. (Recall that the rotation curve leads us to hypothesize a massive galactic corona.) Assuming that other galaxies have similar amounts of interstellar matter, there is still a large amount of *missing mass.* Observations suggest that only 20% of the "normal" matter (such as electrons, protons, and neutrons) in the early universe converged to form the stars and galaxies we see today. There has been no accounting for the remaining 80% of this normal matter. A related mystery concerns *dark matter*— matter that can be detected only by its gravitational interactions. We do not yet know the composition of this dark matter; however, since normal matter is drawn to the gravity of the dark matter, we can trace the location of dark matter through space. Different lines of evidence show that the universe is only about 4% normal matter and 23% dark matter (the remaining 73% being dark energy, which we will discuss in Chapter 18). The quest for an understanding of the nature of dark matter is a very active field in astronomy today because—as we will discuss in Chapter 18—it is fundamentally important to cosmology. Even though we still do not know the nature of the dark matter, we know that it is distributed in galaxies and clusters of galaxies in a way similar to visible matter. The evidence for this comes from the rotation curves of galaxies and from the way their gravity bends the light from an object farther away.

We know that intergalactic dust and gas exist between galaxies in a cluster, but this is also not nearly enough to account for the calculated mass. There are indications that the halo of our own Galaxy contains more material than previously thought, and perhaps galaxies' halos account for much or all of the matter that makes up the missing mass. For example, observations of 63 spiral and dwarf galaxies in the nearby Virgo cluster (chosen to be representative of the galactic population in the local universe) unveiled the presence of large amounts of very cold dust particles surrounding the galaxies. The data, taken by ESA's *Infrared Space Observatory* and announced in 2002, also show that up to half the total energy output in normal galaxies has been converted from visible into infrared light. This implies that the total amount of emitted starlight is much greater than previously suspected. In addition, astronomers announced in 2003 that they discovered a warm fog in the local group of galaxies. The temperature of this diffuse fog is between 100,000 and 10 million K,

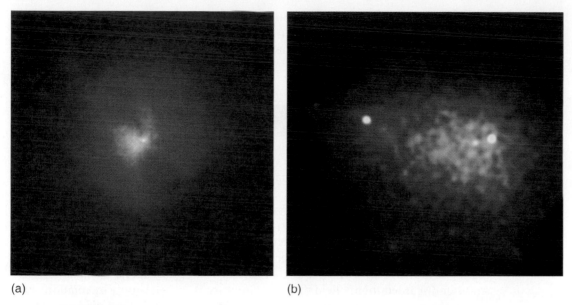

(a)   (b)

Figure 17-19

(a) This *Chandra* X-ray image reveals the presence of a large cloud of hot gas throughout the Hydra A galaxy cluster. The temperature of the gas decreases from about 40 million K in the outer parts to 35 million K in the inner region of the cluster. The gas cloud is several million light-years across. The cluster contains a few hundred galaxies and is 840 million light years from Earth. (b) This *Chandra* image shows the inner region of the Coma cluster. A 100 million K gas cloud surrounds the cluster, which contains thousands of galaxies. The image, about 1.5 million light-years across, shows the concentrations of cooler gas (10 to 20 million K) around the large galaxies NGC4874 (left) and NGC4889 (right).

causing it to shine faintly in X-rays. The data, taken by the *FUSE* satellite and the *Chandra X ray Observatory*, suggest that this warm fog contains as much as a trillion solar masses. This amount of matter is enough to gravitationally bind together the galaxies in the local group. Though vast, this material is only a part of larger hot gas filaments that exist between all the galaxies in the universe. The presence of hot gas in intergalactic regions is shown in **Figure 17-19**.

If stars are grouped into galaxies and galaxies are grouped into clusters, do clusters group into something bigger? Yes, we call them ***superclusters***. Our local supercluster includes the local group and the Virgo cluster (Figure 17-18) as well as other clusters. It has a diameter of perhaps 100 million light-years and contains some 10^{15} (a thousand trillion) solar masses. Between superclusters are great voids with no galaxies. Patterns of clusters, superclusters, and voids are clear in **Figure 17-20**,

supercluster A group of clusters of galaxies.

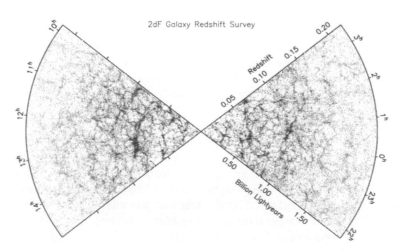

Figure 17-20

The cosmic web revealed by the Two Degree Field (2dF) Galaxy Redshift Survey. The universe contains long filamentary chains of galaxies, superclusters, and empty voids up to 200 million light-years across. The map contains about 230,000 galaxies, has a depth of 4 billion light-years, and covers 1/20th of the whole sky. The Milky Way is at the middle of this sky distribution.

which shows the sky distribution of about 30,000 galaxies, with the Milky Way at the center. It seems that matter in the universe forms a cosmic web in which galaxies are formed along filaments of normal and dark matter, and clusters are formed at the intersections of these filaments.

17-4 The Origin of Galactic Types

Astronomers once thought that Hubble's tuning fork diagram might represent the evolutionary sequence of galaxies. Perhaps spirals evolve into ellipticals or vice versa. This hypothesis can be dismissed quickly. First, we know that ellipticals do not evolve into spirals because elliptical galaxies contain very little gas and dust, while spirals contain significant amounts. Ellipticals have already used up their gas and dust forming stars. What about evolution the other way? This proposal presents mechanical problems. Once a disk has formed in a galaxy, there is no way that the galaxy, on its own, could disperse its disk and acquire a symmetric elliptical shape.

Another discarded theory held that the difference is caused simply by rotation: spiral galaxies evolved from dust clouds that were spinning fast and ellipticals came from slowly spinning clouds. However, this theory cannot account for the fact that elliptical galaxies have completed their star formation and have little interstellar material.

Today, there are two leading theories that explain why galaxies exist in various types. We discuss each in turn.

The Cloud Density Theory

Early in the history of the universe, before galaxies, some gas/dust clouds must have been denser than others. One theory of galactic formation holds that elliptical galaxies are those that formed from the densest clouds. Because of the great density of these clouds, star formation would have proceeded quickly and would have used up all dust and gas before a disk had a chance to form. Once stars have formed, they act as individual particles that orbit the center of the galaxy and have little interaction with one another. This is why galactic clusters are able to pass through the disk of our Galaxy without being captured. So, according to this theory, early star formation explains the lack of a disk in elliptical galaxies.

Clouds that had a lower density of gas and dust would have formed stars less frequently when the cloud contracted, and the dust and gas would have collapsed into a disk before star formation used it all up. The reason that a rotating cloud of gas and dust forms into a disk is that gas acts like a fluid, where the particles interact a great deal. When two fluids collide, energy is dissipated and the fluids do not pass through one another. (Recall that in the case of star birth, a disk is a common occurrence.) Within a galactic disk, star formation proceeds slowly. Instabilities cause density waves that assist star formation, so that new stars are still being born in spiral galaxies today.

Various hypotheses exist to explain why some spiral galaxies have bars and others do not. One idea is that faster-spinning clouds form elongated nuclei, or bars. Another hypothesis notes that barred spiral galaxies generally do not have a halo of globular clusters. It proposes that the lack of a halo results in instabilities that cause a bar to form.

The Merger Theory

A more recent theory proposes that spiral galaxies formed before elliptical galaxies, and that ellipticals result from mergers of spirals. The direction of spin of galaxies is random, so merged galaxies would often have little or no overall rotation (which is what is observed for ellipticals). The small quantities of interstellar matter in ellipticals is explained by the fact that the merger would compress interstellar clouds and would cause strong density waves. During the merger, then, star formation would be frequent and interstellar matter would be used up.

Evidence for the lack of interstellar material in ellipticals comes from the Infrared Astronomy Observatory, which detects very little infrared radiation from ellipticals. Interstellar material radiates in the infrared region of the spectrum.

These theories are not listed in the marginal glossary because they have no generally accepted names. The names here are unofficial.

A puff of wind cannot pass through still air without carrying the still air along (and reducing the wind speed in the process).

This theory is supported by the observation that ellipticals make up a greater percentage of the galaxies in large clusters, where galaxies are packed close together. In these clusters, mergers would have been frequent and would have produced numerous elliptical galaxies. In loosely packed clusters of galaxies, on the other hand, ellipticals are fairly rare.

You may have noticed that nothing has been said about irregular galaxies. Neither theory explains these oddballs well. Astronomers know that some galaxies that were once classified as irregulars are actually pairs of galaxies that are in the process of collision. Others are spiral galaxies with large, dense dust clouds that hide their spiral patterns. Perhaps all irregular galaxies can be explained like this. In any case, irregulars make up a small fraction of all galaxies.

It should be obvious that the question of why some galaxies are of one type and others are of another type is far from settled. This is just one of the unanswered questions that make extragalactic astronomy a lively source of research today.

Look-Back Time

As new observational techniques are developed, we are seeing objects farther and farther away. We have now detected objects that may be as far away as 13 billion light-years. If these objects are truly this far, the light we see left them 13 billion years ago. We are seeing far into the past. Astronomers speak of **look-back time**, a term that emphasizes this idea. An object 13 billion light-years away has a look-back time of 13 billion years.

One of the problems with using large galaxies as distance indicators relates to the idea of look-back time. When we see these galaxies, we are not seeing them as they are today but as they were in the past. There is some evidence that within large clusters of galaxies, some galaxies combine to form supergalaxies. That is, some galaxies are "cannibalized." If this occurs, the largest galaxies in nearby clusters are not the same size as the largest galaxies in distant clusters because we are seeing nearby clusters at a later stage of their life than we see distant ones. In the distant clusters, not enough time has passed for galaxies to have gobbled up one another. This would invalidate the assumption that distant clusters are similar to nearby clusters and would call into question the practice of using galaxies as distance indicators. Although this is a possible problem, the Hubble law seems to apply equally as well to faraway clusters, where galaxies are used as distance indicators, as it does to nearby galaxies, where other methods are used for determining distance. This indicates that the dilemma posed by look-back time is not a major obstacle in this case.

Among the objects with the greatest look-back times, we find a class of objects not yet discussed: active galaxies. These objects play a major part in helping us to determine the scale of the universe.

> Notice that the word *theory* is being used here for what we should probably call a *hypothesis*. There is no sharp dividing line between the terms.

> We are referring now to distant "objects" because, as we will explain later, we cannot be sure that they are galaxies.

> **look-back time** The time light from a distant object has traveled to reach us.

17-5 Active Galaxies

Until now we centered our discussion in this chapter on what are called *normal galaxies*, those that fit the Hubble classification. Now we turn to objects that do not fit into those categories—objects that were discovered since the late 1940s by radio astronomy. Before then, our understanding of the distant universe was based on observations in the visible part of the spectrum. Radio observations opened new windows on our understanding of the cosmos by introducing us to several exotic objects and phenomena.

All galaxies emit some radio waves; the Milky Way Galaxy radiates them from its nucleus (see Figure 16-24). However, the radio waves from a normal galaxy constitute only about 1% of that galaxy's total luminosity. In the late 1940s, astronomers began observing strong extragalactic radio sources, and in 1951 it was discovered that

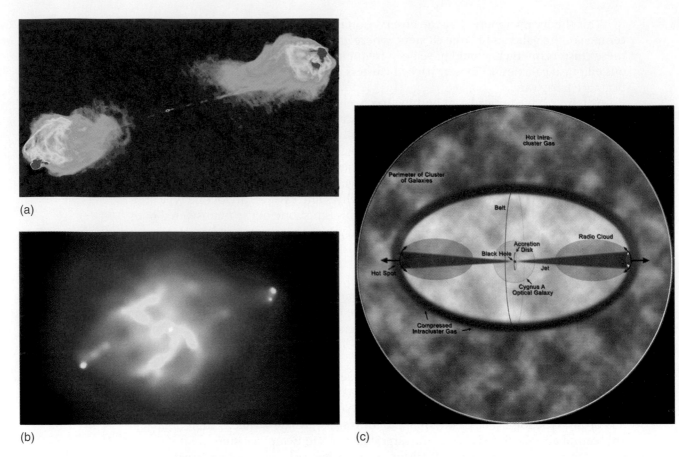

(a)

(b)

(c)

Figure 17-21

(a) This false-color radio image of Cygnus A (about 700 million light-years away) shows a jet extending from the source (an AGN) toward one of the radio lobes. Regions of brightest (fainter) radio emission are shown in red (blue). (b) This *Chandra* image shows X-ray emitting hot gas (bright orange region) that surrounds Cygnus A, and within it a giant cavity (the yellow and light orange inner region) created by the two jets emitted from the galaxy's central black hole region. The low-density hot gas (tens of millions K) fills the space between the galaxies in the cluster containing Cygnus A and provides enough resistance to slow down the jets. The jets terminate in radio and X-ray emitting "hot spots" some 300,000 light-years from the center of the galaxy. The pressure of the particles and magnetic fields in the jets inflates the cavity. (c) An illustration of the Cygnus A system.

a radio source in the constellation Cygnus is actually a double source associated with a galaxy (**Figure 17-21**). Cygnus A, as the source is called, emits about a million times more energy in radio waves than does the Milky Way.

Since the discovery of Cygnus A, many more such **radio galaxies** have been discovered, typically emitting millions of times more energy in radio waves than does a normal galaxy. **Figure 17-22** shows Centaurus A. Observe that the visible galaxy in this case is an elliptical galaxy with a prominent lane of interstellar gas and dust, which is bisected at an angle by opposing jets of high-energy particles blasting away from the nucleus. The double-lobed radio image seen in Centaurus A is common in radio galaxies, and most of the galaxies associated with double-lobed radio sources are either giant ellipticals or spirals. The radio lobes are enormous, sometimes extending 15 million light-years from the visible galaxy. In general, radio lobes mark the positions where the outflows start interacting with the intergalactic medium.

Radio galaxies often appear unusual when viewed in visible light. **Figure 17-23** shows a jet of hot gas being emitted from Virgo A (M87). This feature is seen in several cases. More recently, galaxies have been observed that have properties like those of radio galaxies, but have their primary emission at wavelengths other than the radio region of the spectrum. Astronomers therefore classify radio galaxies as one type of a group of high-energy galaxies called **active galaxies.**

radio galaxy A galaxy having its greatest luminosity at radio wavelengths.

Recall that the Milky Way is about 160,000 LY, or 0.16 MLY, in diameter. From end to end, large double lobes are 150 times this size!

active galaxy A galaxy with an unusually luminous nucleus.

CHANDRA X-RAY   DSS OPTICAL   NRAO RADIO
                                            CONTINUUM   NRAO RADIO
                                                               (21-CM)

Figure 17-22
A composite image of Centaurus A (about 10 million light-years away) made with X-ray (blue), radio (pink and green), and optical (orange and yellow) data. The compact nucleus weighs more than 200 million solar masses, suggesting the presence of a supermassive black hole. Two large arcs (part of a projected ring 25,000 light-years in diameter) of X-ray emitting, multimillion-degree hot gas exist in the outskirts of the galaxy on a plane perpendicular to the jets. The observations suggest that about 100 million years ago Centaurus A merged with a small spiral galaxy, which led to a burst of star formation in the galaxy's nucleus. About 10 million years ago a huge explosion occurred that produced the jets and a galaxy-sized shock wave that moved outward at about 450 kilometers/second. Radio observations show that parts of the jet have speeds of about half the speed of light.

Jets of material, such as those shown in Figures 17-21a and 17-22 and at the end of this chapter, seem to be a universal phenomenon. They are the natural byproducts of accretion onto a compact object, emanating at right angles to the disk that surrounds this compact object. They occur not only in radio galaxies, but also in the entire range of objects from young stellar objects to extremely powerful galaxies. These "outflows" are mostly very well collimated (that is, they form straight beams, with very little spreading over their length) and can extend from a few to hundreds of thousands of light-years. They transfer energy, matter, momentum, and magnetic fields from the central region to the surrounding environment. The speeds of the observed outflows seem to remain constant or, in some cases, even increase along most of the length of the jets. Some outflows have speeds that are clearly close to the speed of light. Such speeds signify the presence of an acceleration mechanism that works along the entire length of the outflow. Magnetic fields in the disk and along the length of the jets, coupled with the rotation of the central object, offer a very likely mechanism for the formation, acceleration, and collimation of these outflows.

During the first decade of observing active galaxies, many types were discovered, and various hypotheses were put forward to explain them. Then, less than 10 years

Because the energy of an active galaxy comes from its nucleus, astronomers often refer to active galactic nuclei (or *AGNs*) rather than *active galaxies*.

Figure 17-23

(a) This visible-light photo shows Virgo A (M87), a giant E0 elliptical radio galaxy, 50 MLY away in the constellation Virgo. (b) A short-exposure infrared image by the *Hubble Space Telescope* shows a jet of material coming from the nucleus of the galaxy. The nucleus is the bright white spot toward the left. *HST* observations show a spiral-shaped disk of hot gas in the nucleus. The rapid rotation of the disk suggests the presence of a black hole of about 3 billion solar masses, concentrated in a space no longer than our solar system. (c) VLBA measurements show that the jet is formed within a few tenths of a light-year from the core of the galaxy. The jet's opening angle is initially large (about 60°), but a few light-years away it is reduced to only 6°. Material in the jet is seen to be moving at apparent speeds greater than that of light. However, this superluminal motion is a geometric illusion as we will describe later in this chapter.

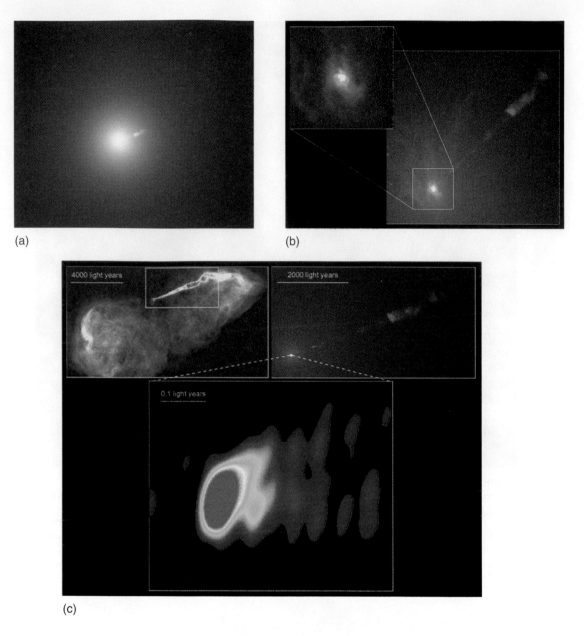

(a)

(b)

(c)

after the discovery of active galaxies, the mystery deepened with the discovery of what appeared to be an entirely new type of object.

Quasars

Prior to 1960, astronomers thought that intense radio waves like those from active galaxies necessarily come from large galaxy-sized objects. Individual stars are such weak sources of radio waves that before 1960 the Sun was the only star from which radio waves had been detected. In that year, however, two radio sources were found that were so small that they appeared to be stars. Recall that planets' sizes can be measured when they occult a star. The same phenomenon can be used in reverse to measure the size of a distant object. In the case of one of these radio sources, 3C 273 (so named because it is the 273rd object listed in the third Cambridge catalog), its visual and radio images were observed as it was occulted by the Moon. By observing 3C 273 as its visible light and radio waves were blocked out by the Moon and again as they reappeared on the other

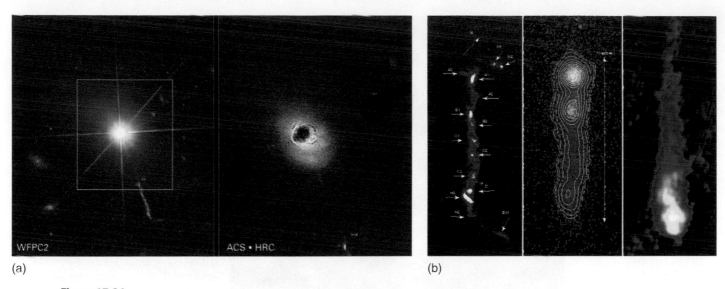

(a)   (b)

Figure 17-24
(a) The *HST* image at left (taken by WF/PC2) shows the brilliant quasar 3C 273 (about 3 billion light-years in the direction of the constellation Virgo). The image at right provided by *HST*'s new Advanced Camera for Surveys, is the clearest view yet in visible light of the quasar. After blocking the light from the central quasar, astronomers discovered that its host galaxy is more complex than previously thought. A spiral plume wound around the quasar, a red dust lane, and a blue arc and clump in the jet's path are clearly seen. (b) The jet emanating from the core of quasar 3C 273, as seen at different wavelengths. From left to right, the images are optical (*HST*), X-ray (*Chandra*), and radio (MERLIN). The scale on all three images is the same (about 10 arcseconds from one end of the jet to the other).

side, details of the source could be determined (**Figure 17-24**). The object appeared to be very small—more like a star than a galaxy. In addition, it had a small jet protruding from it, like the jets that have been observed from some radio galaxies.

The radio waves from 3C 273 seemed to have two sources: the jet and the main body of the object. We have seen that double-lobed radio sources are common for active galaxies.

The spectra of both 3C 273 and 3C 48 (the other source mentioned above) were found to be extremely unusual. Although *emission* spectral lines were prominent, the lines could not be identified with any known chemical element. Were we seeing objects that had entirely different chemical elements than are known to us? Because of their unusual nature (their star-like appearance and strong radio emission), the objects were called *quasi-stellar radio sources* or *quasars.*

In 1963, Maarten Schmidt of the California Institute of Technology found the solution to the puzzle of the unusual spectrum of 3C 273. He found that the prominent spectral lines that had been seen are simply hydrogen spectral lines that are very greatly redshifted. Each wavelength is redshifted to a value about 16% greater. (That is, the ratio of the change in wavelength to the nonshifted wavelength from a stationary source is 16%; $\Delta\lambda/\lambda_0 = 0.16$.) **Figure 17-25** illustrates the situation. Schmidt's colleague Jesse Greenstein examined the spectrum of 3C 48 and found that its spectral lines are shifted even farther—by about 37%.

If the redshifts of the two quasars are caused by the Doppler effect, one quasar is moving at 15% of the speed of light and the other is moving at 30% of the speed of light (or 90,000 kilometers per second). These were speeds far greater than

quasar (quasi-stellar radio source)
A small, intense celestial source of radiation with a very large redshift.

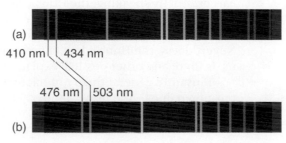

(a)

410 nm   434 nm

476 nm   503 nm

(b)

Figure 17-25
(a) An "at rest" spectrum. The wavelengths of two lines are indicated.
(b) The same spectrum shifted 16%. Observe that the pattern of lines is the same. The *average* redshift of a quasar is determined from measurements of a number of spectral lines.

Figure 17-26
An artist's view of a quasar's core. The material in brown and yellow corresponds to the disk of dust and gas that hides the central black hole. The black hole's gravity pulls in this material, which swirls as it moves closer. The resulting friction heats the gas and makes it shine brightly.

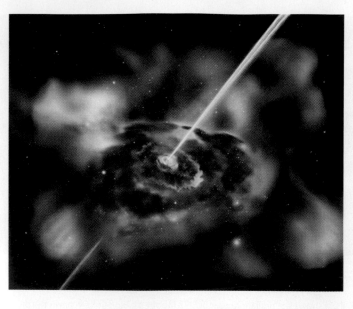

The familiar expression $z = \Delta\lambda/\lambda_0 = v/c$ relating redshift and speed should only be used for small speeds. After all, it implies that an object with redshift larger than 1 moves faster than the speed of light, which is not possible. The correct expression is

$$z = \frac{\Delta\lambda}{\lambda_0} = \sqrt{\frac{c+v}{c-v}} - 1 \,,$$

which for speeds less than about 0.1c can be approximated by $z \approx v/c$.

The largest redshift observed today for a galaxy is $z = 6.58$. It implies a distance of 13 billion light-years, based on recent research indicating that the universe's age is 13.7 billion years.

Since $v = H_0 d$ we get $d = \dfrac{c}{H_0} \cdot \dfrac{v}{c}$, where the ratio v/c can be calculated from the redshift.

any yet encountered for celestial objects (or for terrestrial objects, other than nuclear particles).

If these large redshifts follow the Hubble law, the closest of the two quasars must be as far away as the then known farthest galaxies. Yet it is brighter than those galaxies, even though in size it is more like a star than a galaxy.

Since the early 1960s, more than about 23,000 quasars have been discovered. Unlike the first two, most (about 90%) are not sources of radio waves. Most quasars are bluish-white objects and are X-ray emitters. In addition, many vary in intensity in an irregular way, typically changing intensity in weeks or months. This latter observation confirms their small size; they cannot be larger than a few light-weeks or light-months in diameter. A few have been found with intensity variations of less than a day.

The first two quasars, seen with redshifts of 16% and 37%, have smaller redshifts than most. The smallest observed redshift is about 6%. The quasar SDSS J1148+5251, in the direction of the constellation Ursa Major, has the largest redshift observed to date ($z = 6.41$). It indicates that the quasar's speed is about 96% of the speed of light and that its distance is about 13 billion light-years away. We are seeing this quasar when the universe was very young, only about 800 million years after the big bang. The amount of dust observed in the quasar's region suggests that galaxies and stars formed earlier in the history of the universe and at a faster rate than previously thought. The black hole at this quasar's core has a mass of 3 billion solar masses (**Figure 17-26**).

If quasars' redshifts fit the Hubble law, the objects are extremely far away. The Hubble constant is known with limited precision, but if we calculate distances to quasars using the smallest and largest current values for the constant, we obtain distances of about 12.6 and 14.1 billion light-years, respectively, for the most distant quasar.

Now consider what the luminosity of such an object must be for us to be able to detect it at that distance. We can calculate the luminosity by using the apparent brightness of the quasar and the inverse-square law of radiation. We find that a bright quasar is 1000 times more luminous than a galaxy like ours. Finally, remember that we are speaking about a small object—far smaller than a galaxy.

Competing Theories for the Quasar Redshift

When astronomers were faced with the prospect of an object being so small and yet so luminous, they were forced to reexamine their assumptions. As was pointed out earlier, strictly speaking, the Hubble law is not due to the Doppler effect, though the basis for the law is mathematically equivalent to that effect. In Chapter 18 we will discuss the theorized cosmological reason for the validity of the Hubble relationship. Nevertheless, the Hubble law indicates that quasars are at tremendous distances from us. Is there any other possible explanation for the redshift?

Einstein's general theory of relativity predicts that light leaving a massive object is redshifted due to the mass of the object. Calculations show that in order to produce redshifts such as those seen in quasars, the gravitational field near the object must be far greater than that near the most massive neutron star, and well-grounded nuclear theory tells us that there is a limit to the mass of a neutron star. After a certain mass is reached, a neutron star cannot exist, and a black hole is formed instead. The theory of relativity simply cannot be used to explain the redshift.

Perhaps quasars are nearby objects with enough local motion that they do not fit the Hubble law. (Recall that galaxies within the local group do not fit the Hubble law either.) **Figure 17-27** shows fast-moving objects ejected from our Galaxy; some are moving at nearly the speed of light. If this is what quasars are, then they do not have such tremendous luminosities after all.

The problem with this *local hypothesis* is not only that we can imagine no source for such objects, but that we must also ask why we do not see similar objects from other galaxies. We cannot expect our Galaxy to be unique in this regard. Yet if other galaxies emit such fast-moving objects, we would see some of them with tremendous *blue*shifts as they move toward us. In fact, however, such blueshifted objects are not seen, and the local hypothesis is not considered a likely explanation.

Perhaps there is some other explanation. However, such an explanation might require drastically different laws of physics than those of today. We never can disregard such a possibility, but we must progress with what we know until it is definitely ruled out. The redshift *can* be explained by the Hubble law, even though that law leads us to conclude that quasars are at distances greater than we are accustomed to and are therefore much more energetic than we are accustomed to. Almost all astronomers now agree that quasars' redshifts do indeed fit the Hubble law.

Seyfert Galaxies

Recall that one of the first two quasars seen (3C 48) has a double-lobed radio source and a jet from its center. In this regard, it resembles an active galaxy. As astronomers found more and more quasars, they discovered that most of the nearer ones are associated with clusters of galaxies and have essentially the same redshifts (and thus distances) as these galaxies. The quasars are more luminous than the galaxies, but the association causes us to look for more similarities.

One particular type of spiral galaxy, a *Seyfert galaxy* (**Figure 17-28**), has a very luminous nucleus that—although it is not as luminous as a quasar—in some cases varies in intensity in time periods even shorter than those of quasars. Similarities in the spectra of Seyfert galaxies and quasars point to a further link between galaxies and quasars, indicating that perhaps a quasar is a galactic nucleus.

Recent high-resolution photos of nearer quasars show a fuzz on the image near the quasar. Again, it appears that quasars may be at the nuclei of some type(s) of galaxies and that the fuzz is caused by the stars of the galaxies.

Quasars and Gravitational Lenses

Twin quasars were discovered in 1979 in Ursa Major. The two quasars are very close together, separated by only 6 arcseconds. They have the same luminosity, the same redshift ($z = \Delta\lambda/\lambda_0 = 1.4$), and identical spectra. The explanation for these identical twins lies with the theory of general relativity. That theory predicts the possibility of a large mass bending light from a more distant object so that two images of the distant object appear. The predicted phenomenon is called a *gravitational lens*, but until the discovery of the twin quasars, not much attention was paid to the prediction. Since

Figure 17-27
The local hypothesis holds that quasars are relatively nearby objects that have been ejected from the Galaxy at great speeds. If quasars are emitted by galaxies, however, some should be moving toward us from nearby galaxies and these would show blueshifts. No observed quasar has a blueshift.

local hypothesis A proposal stating that quasars are much nearer than a cosmological interpretation of their redshifts would indicate.

Seyfert galaxy One of a class of spiral galaxies having active nuclei and spectra containing emission lines.

Figure 17-28
The core of the nearby active galaxy Circinus, which belongs in a class of mostly spiral galaxies called Seyferts. The galaxy is only 13 million light-years away and most of the gas in its disk is concentrated in two rings (with diameters of 1300 and 260 light-years). At the center of the inner ring (located on the inside of the green disk) is a supermassive black hole that is accreting surrounding gas and dust.

gravitational lens The phenomenon in which the gravity due to a massive body between a distant object and the viewer bends light from the distant object and causes it to be seen as two or more objects.

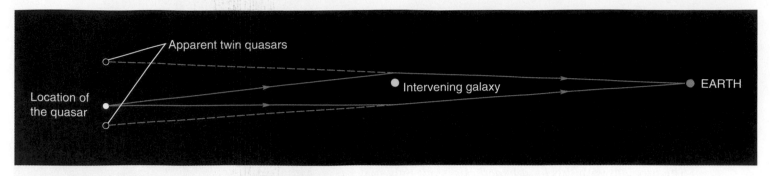

Figure 17-29
Light from the quasar at the left is bent as it passes by the intervening galaxy at the center. This causes the light to appear to come from twin quasars, one on either side of the galaxy. Such gravitational lenses are predicted by the theory of general relativity.

Seyferts and quasars are the same phenomenon, just with the rheostat turned up.

Astronomer Allan Sandage

the discovery of the twin quasars, the intervening galaxy has been found, and we are now confident that the twins are actually one quasar that has been made to appear double by the galaxy that lies between it and us (**Figure 17-29**). Many examples of gravitational lensing have been found since 1979. One interesting case is shown in **Figure 17-30a**. The two yellow-orange spots in the image are twin images of a single quasar behind a massive object that acts like a gravitational lens. The ring is caused

Figure 17-30
(a) This false-color image of object MG 1131 + 0456 was made by radio interferometry with the Very Large Array. The ring is called an "Einstein Ring." This 1987 image is the first such ring ever observed. (b) In 1995 the *Hubble Space Telescope* was used to discover this case of quadruple gravitational lensing. (c) Abell 2218 is a rich galaxy cluster. The *HST* captured this arc-like pattern spreading across the image like a spider web, a great example of gravitational lensing. Abell 2218 has a total of seven multiple systems. (d) This *HST* image shows five images of the same galaxy, distorted from a normal spiral to an arc-shaped object. The lensing cluster is 5 billion light-years away, and the blue-shaped galaxy is 2 times farther away.

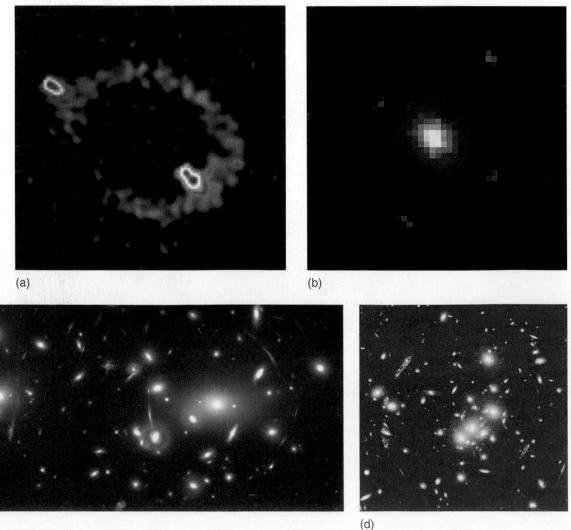

(a)

(b)

(c)

(d)

by another object behind whatever object causes the lensing effect. This object forms a ring because it is so well aligned that light passing on every side of the intervening massive object is bent toward Earth, making the distant quasar appear to be on all sides of it at once.

Arcs, such as in Figure 17-30a, are the most common type of lensing, although most cases are not as complete as that arc. Pairs of lensed objects are the next most common. Figure 17-30b is a 1995 *HST* image showing a much rarer case, one with a quadruple pattern. Figure 17-30c reveals numerous arcs, which are the distorted images of a galaxy population located 5-10 times farther than the cluster of galaxies doing the lensing. Figure 17-30d shows several blue, loop-shaped objects, which are multiple images of the same galaxy seen through the "lens" of the cluster of yellow,

(a)

elliptical, and spiral galaxies near the center of the image.

The importance of the discovery of gravitational lenses is not only that they provide another confirmation of the general theory of relativity, but that they indicate that quasars are indeed very distant—that their redshift fits the Hubble law. They really are the most distant objects yet observed.

Suppose that we plot the number of known quasars versus their distances from our Galaxy. The graph would not be very instructive because as the distance from our Galaxy (or from any point) increases, the volume of space in a shell increases. This happens because the area of a large sphere is greater than the area of a small sphere (**Figure 17-31a**). Therefore, even if

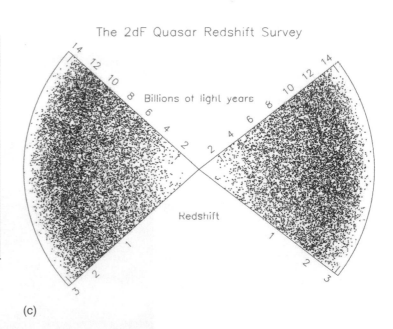

(b)

(c)

Figure 17-31
(a) If two equally thick shells are drawn at different distances from us, the more distant shell will contain more space. Even if quasars (dots in the figure) were equally spaced throughout the universe, more would be seen in the faraway shell. (b) The graph shows the relative abundance of quasars at various distances from the Milky Way. The distance scale is based on a Hubble constant of 23 km/(s × MLY). (c) As seen in this two-degree field (2dF) survey of about 23,000 quasars, the quasars are clustered to the same extent as local, optically selected galaxies.

Figure 17-32
This diagram shows how an object moving at a speed slower than the speed of light can appear to be moving at a speed faster than the speed of light due to projection effects.

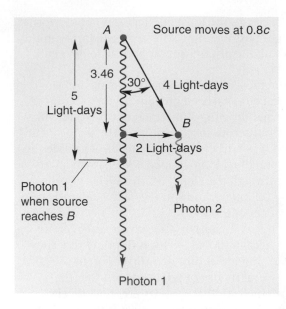

blazars (or **BL Lac objects**) Especially luminous active galactic nuclei that vary in luminosity by a factor of up to 100 in just a few months.

superluminal motion Motion that appears to occur at speeds faster that the speed of light.

quasars are evenly distributed throughout the universe, we would expect to see more of them at greater distances.

To truly illustrate the distribution of quasars, the graph should show the *density* of quasars—the number of quasars per unit of volume of space—versus their distance from us. Figure 17-31b is such a graph. The distance shown on the horizontal axis depends on the value chosen for the Hubble constant, and other choices are possible. Nevertheless, even if a different value were chosen for the Hubble constant, the shape of the graph would remain the same. The graph leaves no doubt that quasars appear at a fairly specific distance from the Milky Way. Since distance is proportional to time, this indicates that quasars existed during a relatively short period in the distant past.

Quasars, Blazars, and Superluminal Motion

In the early 1970s, astronomers discovered an important clue in their efforts to understand quasars. They found a new class of objects, called ***blazars or BL Lac objects.*** A blazar seems to be star-like and very luminous, just like a quasar. Its few faint emission lines show high redshifts, suggesting a very large distance, just like for a quasar. Blazars are unusual because they are highly variable—their luminosity can change by up to 30% in a day and by a factor of 100 in a few months. Radio observations showed that there is faint radio emission around a bright core, suggesting that a blazar is a double radio source oriented in such a way that one jet is coming straight (or nearly so) at us. This was soon supported by observations of ***superluminal motion,*** which is also observed in some quasars.

Figure 17-33
The top panel shows the jet emanating from the galaxy's nucleus (at far left; see Figure 17-23). The box indicates the location of the observed superluminal motion. The bottom panel includes the sequence of *HST* images taken between 1994 and 1998. The slanted lines track the observed moving features in the jet, and the apparent speeds of these features are given in units of the speed of light.

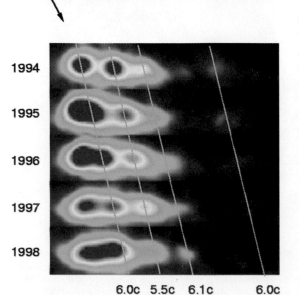

Superluminal Motion in the M87 Jet

Figure 17-32 shows a simple explanation for superluminal motion. Suppose that a source moves at 80% of the speed of light, along a line that makes an angle of 30° with the line of sight from Earth. At point *A*, it emits a photon toward us. After traveling for 5 days, it reaches point *B*, where it emits a second photon toward us. The real distance traveled by the source between points *A* and *B* is $5 \times 0.80 = 4$ light-days, and thus the source has moved toward us along the line of sight by about 3.46 light-days. In the meantime, the first photon has traveled a distance of 5 light-days toward us. Therefore, the first photon is ahead of the second by $5 - 3.46 = 1.54$ light-days. From our perspective, we see the source moving a horizontal distance on the sky (as a result of projection effects) equal to 2 light-days. Since this is the horizontal distance traveled by the source in the time period during which we receive the two photons, we conclude that the source is moving at a speed of $2/1.54 = 1.3$ times the speed of light! **Figure 17-33** shows the observed superluminal motion in the M87 jet.

17-6 The Nature of Active Galactic Nuclei

The first question that has to be answered in trying to solve the puzzle of the nature of active galactic nuclei is, "Why are these strange objects not found among nearby galaxies?" Here we use the term "active galactic nuclei" (AGNs) to mean the bright energy source at the centers of radio galaxies, Seyfert galaxies, quasars, and blazars. To answer this question, we invoke the idea of look-back time: AGNs are far away because they existed only in the distant past and are not part of today's universe. Therefore, any theory about the nature of these mysterious objects must explain why they existed long ago but not today.

According to today's prevalent theories, the tremendous amount of energy that comes from an AGN is caused by an immense black hole at the nucleus of the galaxy. The black hole is surrounded by an accretion disk that is extremely hot because material falls into it with a very great speed. The temperature of the accretion disk explains the enormous energies emitted by the objects. What remains to be explained is why some AGNs are so much more luminous than others, why some emit intense radio waves and others do not, why the luminosity of some of them changes so rapidly, and, in general, why there appear to be so many different types of these peculiar objects.

Various theories have attempted to explain the many observations about AGNs, but none is completely satisfactory. The leading theory provides a link between the different types of objects. This "unification theory" holds that they are all basically the same, and that they appear different depending upon their orientation with respect to us. **Figure 17-34b** illustrates the hypothesized object, which is an active galactic nucleus. At its center is a supermassive black hole of many million (or even a few billion) solar masses. It is surrounded by an accretion disk, perhaps a few light-months across.

A much larger ring, shaped like a fat doughnut (a *torus*), surrounds the accretion disk in the same plane. The torus consists of relatively cool gas and dust. It may be several light-years across and is dense enough that light from the accretion disk cannot penetrate it. Light from the disk does reach the outside world, however, for it shines out of the "doughnut hole" in two directions, forming two cones of light. Radiation in all other portions of the electromagnetic spectrum, including high-energy wavelengths, pours out along with the light. A jet of material is ejected by some unexplained process along the axis of each of the cones. Finally, irregular clouds of gas and dust move in orbit around the center.

Now suppose that the active nucleus in the figure is oriented so that we see it edge-on, so that the torus hides the accretion disk from our view. Telescopes would reveal infrared energy coming from the clouds, but would not show the high-energy radiation from the accretion disk. In some cases the jets cause emission of visible light, but in other cases they do not. If the active nucleus has energetic jets, astronomers would observe large radio-emitting areas at a great distance on either side of the galaxy. **Figure 17-35** (and 17-21a) shows the radio lobes and jets of some active galaxies.

When the torus is seen obliquely, radiation from the accretion disk reaches us, and the object appears as a very energetic source that emits radiation across the spectrum, including X-rays. These are the objects we call quasars. On the other hand, suppose the jet is aimed directly toward us. Again it would appear as a small, energetic source, but in this case it might vary quickly in intensity as clouds of dust move across our line of sight. These are the objects we call blazars.

How could we test this "unification theory"? The best test is by detecting radiation from AGNs that is not blocked by the dust in the torus and is not affected by how the torus is seen by the observer. These requirements are fulfilled by far-infrared radiation. Astronomers announced in early 2001 that observations made by ESA's *Infrared Space Observatory* showed that very hot and luminous quasar cores are found even in weak radio galaxies at large distances. These and additional

Be aware of the tremendous scale of extragalactic astronomy; a few million light-years is a very small distance.

That period [1960-1966] was incredibly electric. Every time you went to the telescope and came down you had a major new discovery.
Astronomer Allan Sandage

Recall from Chapter 13 (and from Figure 13-15) that jets of material have been observed coming from the poles of protostars. The galactic case is likely a similar phenomenon on a much larger scale.

There are at least two theories for the visibility (or the lack thereof) of the jets. Both theories depend upon the amount of material that lies along the trail of the jet.

Figure 17-34
(a) The leading model for AGNs holds that they are basically the same type of object, often involving a supermassive black hole. This photo of quasar PG 1012+008 could be explained by a supermassive black hole in one galaxy attracting matter from a nearby companion galaxy. (b) Active galactic nuclei appear to be different because they are oriented at different angles to our line of sight.

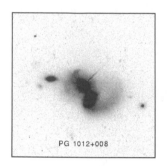

PG 1012+008

(a)

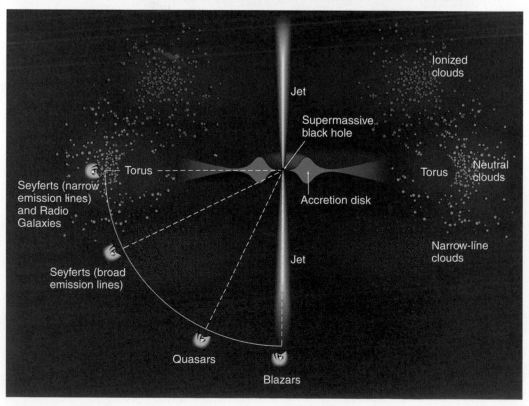

(b)

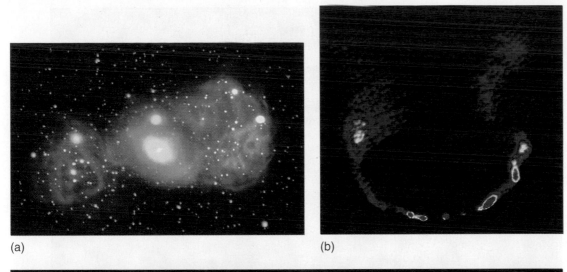

(a)

(b)

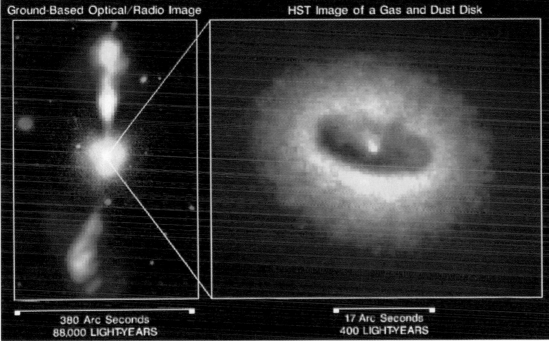

(c)

Figure 17-35
(a) A false-color image of the radio emission (red) is superimposed on a visible-light image (blue-white) of the radio galaxy NGC1316 (Fornax Λ). Each radio lobe is about 600,000 light-years across. The two faint pink extensions just visible in the center show the paths of the outflows (jets). (b) A false-color image of the jets in the radio galaxy NGC1265 (3C 83.1). The red circle denotes the galaxy's center. Red (blue) shows regions of intense (fainter) radio emission. The galaxy moves through the intergalactic medium at ∼2000 km/s, resulting in the "U" shape of the jets. (c) (Left panel) A composite image of the giant elliptical galaxy NGC4261 in the Virgo Cluster. In visible light (white) the galaxy appears as a fuzzy disk of hundreds of billions of stars; the radio emission (orange) shows two jets emanating from the nucleus. (Right panel) A giant disk of gas and dust fuels a possible black hole at the core.

observations support the conclusion that all AGNs are similar objects that appear different depending on how they are oriented with respect to us.

How does this model explain the observation that active galaxies are not found in our neighborhood? There is a limited amount of material near the black hole in the galactic center, and after that material spirals into the black hole, the galaxy calms down to become a more standard galaxy. In today's universe, no galaxy remains with a nucleus that is still in the quasar stage. This means that previous quasars and active galaxies are the ancestors of today's galaxies.

Figure 17-36a provides further evidence that galaxies have progressed from having quasars or blazars at their centers, to Seyferts or radio galaxies, to normal spiral or elliptical galaxies. The three galaxies in the figure are classified as normal galaxies, but they all show characteristics of having black holes at their centers. A census of 27 nearby galaxies carried out by the *Hubble Space Telescope* and ground-based telescopes in Hawaii suggests that nearly all galaxies may harbor supermassive black holes that

The most powerful object in the local universe is quasar PDS 456. With a redshift $z = 0.184$, it is only about 2.2 billion light-years away. It is powered by a black hole of about a billion solar masses.

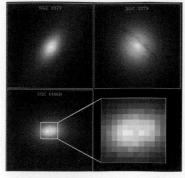

(a)

Figure 17-36
(a) These three active galaxies are believed to contain supermassive black holes in their nuclei. One has a double nucleus. (b) Examples of different home sites of quasars. The images in the left column represent normal galaxies, in the center, colliding galaxies, and in the right, peculiar galaxies.

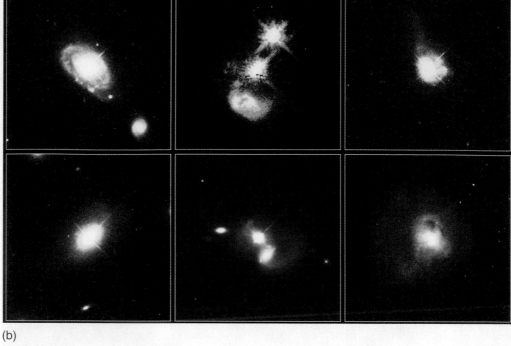

(b)

> With increasing distance our knowledge fades, and fades rapidly, until at the last dim horizon we search among ghostly errors of observations for landmarks that are scarcely more substantial. The search will continue. The urge is older than history. It is not satisfied and it will not be suppressed.
>
> *Edwin Hubble*

once powered quasars. It seems that supermassive black holes are so common that nearly every large galaxy has one. The *HST* images shown in Figure 17-36b give examples of quasars found in normal, colliding, and peculiar galaxies.

In addition, the black hole's mass is proportional to the mass of the host galaxy. This suggests that the growth of the black hole is linked to the formation of the galaxy in which it is located. Finally, the number and masses of the black holes found are consistent with what would have been required to power the quasars. Observations in X-rays (using *Chandra*) show that black holes at the centers of AGNs were much more active in the past than at present, while infrared observations (using *ISO*) show hundreds of young active galaxies with high rates of star formation. The ancestors of today's galaxies were much more active than previously thought. By combining observations in ultraviolet and X-rays, astronomers can study the link between violent stellar activity and accretion by a supermassive black hole at the nucleus of an active galaxy (**Figure 17-37**). Perhaps a giant step has been taken in our understanding of the evolution of galaxies. Time will tell.

Conclusion

In this chapter we presented knowledge about which astronomers are fairly confident, as well as knowledge that is very tentative. Galaxies can be classified with confidence, and Hubble's classification is based simply on appearance. Although the tuning fork diagram may appear to indicate an evolutionary sequence, it does not. At least two competing theories are available to explain why galaxies exhibit such different appearances.

Not much can be known about a galaxy unless we first know its distance, so the study of galactic distances is important and various distance indicators are in use today. To measure distances to the farthest galaxies, a new tool, the Tully-Fisher relation, has been added to the Hubble law.

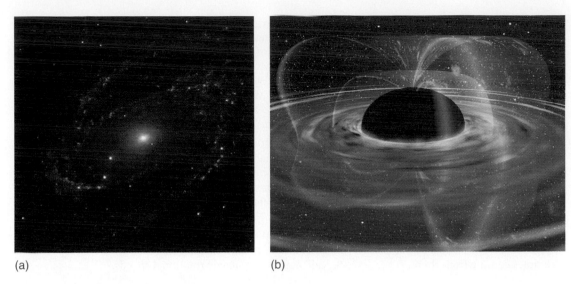

(a)   (b)

Figure 17-37
(a) The spiral galaxy M81 in UV (obtained by ESA's *XMM-Newton* in 2001). Strong UV emission is a feature of star formation, supernova explosions, and accretion by black holes. The emission from the galactic nucleus (the bright, point-like, white region) could support the idea of a mini-quasar or intense and violent stellar activity at the center. Blue denotes regions where young, hot stars are formed. Red denotes cool regions (corresponding to older, less massive stars). The bright, red, point-like objects are foreground stars in our Galaxy. (b) An artist's impression of the central region of the spiral galaxy MCG-6-30-15. Its spectrum (obtained by ESA's *XMM Newton* in 2000) shows that the number of measured photons and their energies far exceed the predictions of standard models for accretion disks around supermassive black holes. It is possible that strong magnetic fields exert a breaking effect on the black hole, extracting some of its rotational energy and heating the surrounding medium in the process.

Astronomers can calculate the mass of a galaxy by applying gravitational theory to the measured motions of parts of the galaxy. They measure the mass of a cluster of galaxies in a similar manner. Both calculations show that we can account for only a small fraction of the mass that must exist, and the mystery of the missing mass has not been solved.

A busy area of research today is the study of active galactic nuclei. These distant, energetic objects that have been discovered by the methods of radio astronomy have brought about major changes in extragalactic astronomy. They seem to produce new questions almost as prodigiously as they produce energy.

Study Guide

1. Cepheid variables are important in calculating
   A. the distances to galaxies.
   B. the composition of stars.
   C. the composition of the interstellar medium.
   D. the ages of galaxies.
   E. the temperatures of stars.

2. About how many galaxies are within range of the largest telescopes?
   A. 20
   B. 103
   C. 1000 to 2000
   D. 1 to 4 million
   E. More than 100 billion

3. The Magellanic Clouds are
   A. galaxies.
   B. extremely high atmospheric clouds.
   C. globular clusters.
   D. supernovae.
   E. of unknown nature.

4. The local group consists of
   A. about 100 nearby stars.
   B. about 15 to 30 nearby stars.
   C. about 100 nearby galaxies.
   D. about 15 to 30 nearby galaxies.
   E. the closest planetary nebulae.

RECALL QUESTIONS

5. Which of the following objects are used as a distance indicator to galaxies?
   A. Cepheid variables.
   B. Globular clusters.
   C. The brightest supernovae.
   D. [All of the above.]
   E. [None of the above.]

6. The masses of galaxies are determined by using the Doppler effect to measure the speeds
   A. of parts of individual galaxies.
   B. of individual galaxies in a cluster.
   C. of galaxies that are part of a binary pair.
   D. [All of the above.]
   E. [None of the above.]

7. Hubble found that the objects that were first called spiral *nebulae* are instead spiral *galaxies* by observing
   A. supernovae in them.
   B. Cepheid variables in them.
   C. radiation characteristic of black holes coming from them.
   D. black holes in their centers.
   E. that their spectra are characteristic of stellar spectra.

8. If galaxy X is four times more distant than galaxy Y, then according to the Hubble law, galaxy X is receding
   A. 16 times faster.
   B. 4 times faster.
   C. 2 times faster.
   D. 1.6 times faster.
   E. [No general statement can be made.]

9. Measurements of the total mass in galactic clusters indicate that they contain far more mass than we see. The "missing mass" is now known to be
   A. tiny black holes.
   B. cool gas and dust between the galaxies.
   C. black dwarfs within the galaxies.
   D. [None of the above; we don't know what it is.]

10. Which of the following objects is the largest?
    A. A galactic cluster.
    B. A cluster of galaxies.
    C. The Milky Way.
    D. The Large Magellanic Cloud.
    E. [Both A and B above; they are the same.]

11. The Tully-Fisher method of measuring distances to galaxies depends on the relationship between a galaxy's absolute magnitude and
    A. the galaxy's size.
    B. the galaxy's redshift.
    C. the galaxy's apparent color.
    D. the width of a line in the galaxy's spectrum.
    E. [Both A and C above.]

12. The Hubble law relates
    A. absolute magnitude and temperature.
    B. apparent magnitude and temperature.
    C. proper motion and radial velocity.
    D. proper motion and distance.
    E. distance and radial velocity.

13. The Hubble constant refers to
    A. the fact that the speed of light never changes.
    B. a number in a formula that will never change its value.
    C. the rate at which the universe is expanding now.
    D. the amount of mass in the universe.
    E. the amount of mass in galactic cores.

14. Suppose that we know that a galaxy 5 million light-years away is receding at 20 miles per second. What would be the Hubble constant based on this single galaxy?
    A. 4 mi/(s × MLY)
    B. 15 mi/(s × MLY)
    C. 25 mi/(s × MLY)
    D. 100 mi/(s × MLY)
    E. [None of the above.]

15. The look-back time of an object is
    A. how long light from the object takes to reach Earth.
    B. numerically equal to the object's distance in light-years.
    C. larger for more distant objects.
    D. [All of the above.]
    E. [None of the above.]

16. The Hubble law is based primarily upon
    A. calculations of galactic masses.
    B. calculations of galactic rotational speeds.
    C. Doppler shift data.
    D. knowledge of galactic evolution.
    E. data concerning the number of stars in galaxies.

17. Which of the following statements is true about quasars?
    A. They seem to be extremely large compared to most galaxies.
    B. They seem to be very small for the energy released.
    C. If they are as far away as they seem, they are very energetic.
    D. [Both A and C above.]
    E. [Both B and C above.]

18. The evidence for small size of quasars comes from
    A. the amount of energy they release.
    B. their distance from us.
    C. the rapidity of their luminosity changes.
    D. comparison with Cepheid variables.
    E. the magnitude of their blueshift.

19. We cannot judge quasars' distances by using their absolute luminosity as we do in the case of galaxies because
    A. they have no absolute luminosity.
    B. there are no nearby quasars with which to compare distant ones.
    C. their luminosity is too great.
    D. their recessional speeds are unknown.

20. The radio waves from a radio galaxy can come from an area many times bigger than the visible object.
    A. True
    B. False
    C. No general statement can be made concerning this.

21. Which of the following statements is an observation rather than the result of theory?
    A. Light from most galaxies is redshifted.
    B. Most galaxies are receding from us.
    C. The rate of expansion of the universe is slowing.
    D. Galaxies farther away are moving away faster.
    E. Quasars emit more energy than most galaxies.

22. If the local hypothesis for quasars is true, we would expect
    A. that quasars would be less luminous.
    B. that quasars would be more luminous.
    C. to observe quasars with blueshifted spectra.
    D. quasars to be emitted from galaxies other than ours.
    E. [Both C and D above.]

23. What is meant by a distance indicator and what serves as one for nearby galaxies? For more distant galaxies?

24. Describe three methods for measuring the mass of a galaxy other than the Milky Way.

25. What is the local group?

26. What is meant by *dark matter* in galactic astronomy?

27. The chapter stated that we know that elliptical galaxies do not evolve into spiral galaxies. The reason given was that we observe that spirals contain gas and dust while ellipticals do not. Explain why the observation proves the statement.

28. State the Hubble law. What is the currently accepted range of the Hubble constant? Include units in your answer.

29. Describe some difficulties encountered in determining the Hubble constant with accuracy.

30. Explain how the Hubble law can be used to determine the distance to a galaxy. Discuss the limitations of this method of determining distance.

31. What is meant by *look-back time?*

32. In what way are active galaxies "active"?

33. After the spectra of quasars were understood, why were the objects still so difficult to explain?

34. What do we mean when we say that a quasar has a redshift of 25%? Which quasar is moving away faster: one with a redshift of 25% or one with a redshift of 30%?

35. What is the present theory of the origin of the energy of quasars?

36. Why are there no nearby quasars?

1. Explain how we determine the distances to galaxies that are too far away to allow us to see Cepheids in them.

2. The difficulty in determining the orientation of the plane of revolution of binary galaxies limits the usefulness of Doppler shift measurements of velocity. Why does the same problem not exist in using Doppler shift data to determine the speed of part of an individual spiral galaxy?

3. Discuss the limitations on the accuracy of measurements of distances to galaxies. Doesn't the lack of precision in these measurements make galactic astronomy less a science than other sciences?

4. Compare the use of bright galaxies as distance indicators to grading a class "on the curve."

5. If the slope of the Hubble graph were greater (with distance plotted on the *x* axis), would the Hubble constant be larger or smaller? Explain.

6. What is a gravitational lens and what does the fact that they have been observed tell us about quasars?

7. Why is a spectral line of a galaxy broadened by the galaxy's rotation?

8. In describing what was called the *merger theory* of galaxy formation, no explanation was given for why some spiral galaxies have bars and others do not. Propose an explanation.

9. List similarities and differences between the collapse of an interstellar cloud to form a star and the formation of a galaxy.

1. Determine the slope of the graph shown in **Figure 17-38**. (The graph might represent the changing speed of a race car.) Be sure to include units in your answer.

2. If the data for the Hubble expansion were as shown in **Figure 17-39**, what would be the value of the Hubble constant?

3. Suppose that an object is seen with a redshift that indicates a speed of 100,000 kilometers/second away from us. If this redshift fits the Hubble law, how far away is the object? (Assume some reasonable value for the Hubble constant.)

4. Suppose that Alex is moving away from you at a constant speed of 15 miles/hour and is 60 miles away at this moment. If he has maintained this speed since leaving you, how long ago were you together?

5. For the motion in question 4, calculate the "Alex constant" (analogous to the Hubble constant).

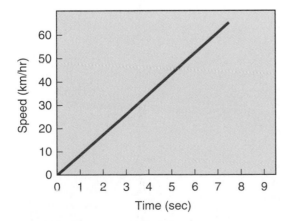

Figure 17-38
Graph for Calculation #1.

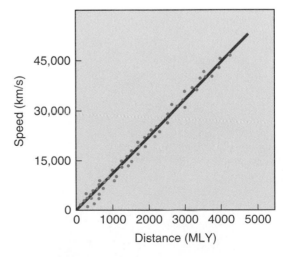

Figure 17-39
Graph for Calculation #2.

6. Figure 17-5a shows the Large Magellanic Cloud, a satellite galaxy of the Milky Way. It is 50 kiloparsecs from us, and has an angular size of 650 arcminutes by 550 arcminutes. Use the small-angle formula (in Chapter 6) to calculate the dimensions of the Large Magellanic Cloud in kiloparsecs.

7. **Figure 17-40** shows NGC4535. Its angular size is 6.8′ by 5′. Assuming that its longest dimension is about 100,000 light-years, use the small-angle formula to calculate its distance from us.

8. The existence of supermassive black holes is actually less exotic than you might think. The average density of an object is given by the ratio of its mass to its volume. For a spherical object of radius R, density = $M / (4\pi R^3/3)$. Writing the radius in km and the mass in solar masses show that the expression for density is given by $4.8 \times 10^{20} \times (M/R^3)$ kg/m^3. Recall that the radius of a black hole (in km) is related to its mass (in solar masses) by the Schwarzschild expression: $R = 3M$. Now show that the density necessary to make a black hole of mass M is $1.8 \times 10^{19} / M^2$ kg/m^3. What density is needed to make a black hole of a billion solar masses? A hundred million solar masses?

Figure 17-40
Galaxy NGC4535, Calculation #7.

1. "Understanding the Hubble Sequence," by G. Lake, in *Sky & Telescope* (May, 1992).

2. "Galaxies in the Young Universe," by F. D. Macchetto and M. Dickinson, in *Scientific American* (May, 1977).

3. "When Galaxies Collide," by J. Roth, in *Sky & Telescope* (March, 1998).

4. "Active Galactic Nuclei: Sorting Out the Mess," by A. Finkbeiner, in *Sky & Telescope* (August, 1992).

5. "Massive Black Holes in the Hearts of Galaxies," by H. Ford and Z. I. Tsvetanov, in *Sky & Telescope* (June, 1996).

6. "The Most Distant Radio Galaxies," by G. K. Miley and K. C. Chambers, in *Scientific American* (June, 1993).

7. "A New Look at Quasars," by M. Disney, in *Scientific American* (June, 1998).

8. "The Evolution of Galaxy Clusters," by J.P. Henry, U.G. Briel, and H. Böhringer, in *Scientific American* (December, 1998).

9. "Unmasking Black Holes," by J.-P. Lasota, in *Scientific American* (May, 1999).

10. Gale E. Christianson, *Edwin Hubble, Mariner of the Nebulae* (Farrar, Straus and Giroux, 1995).

Quest Ahead to Starlinks:
www.jbpub.com/starlinks

Starlinks is this book's online learning center. It features **eLearning,** which contains chapter quizzes and other tools designed to help you study for your class. You can also find **on-line exercises,** view numerous relevant **animations,** follow a guide to **useful astronomy sites** on the web, or even check the latest **astronomy news** updates.

This three-color composite image shows an interacting system of galaxies in the constellation Pavo (the Peacock). The system includes the barred spiral galaxy NGC6872 (of type SBb) and the smaller galaxy IC4970 (of type S0, just above the center). The distance to this system is about 300 million light-years and the size of NGC6872 is about 750,000 light-years (from tip to tip). The interaction between the two galaxies is cleary shown by the disturbed upper left spiral arm of NGC6872. The bright object in the image (to the lower right of NGC6872) is a star in our Galaxy.

CHAPTER 18

Cosmology: The Nature of the Universe

The cosmic microwave background (CMB) sky over Mt. Erebus in Antarctica. In this fanciful picture, the CMB sky, as seen by the *BOOMERANG* project, is shown behind the prelaunch preparations for the balloon that carried the equipment. The images of the early universe have been overlaid onto the sky to indicate what size the fluctuations would appear if a standard 35mm camera were sensitive to microwave light. The color map has been changed to match the rest of the picture.

THROUGH THE CENTURIES, ASTRONOMERS HAVE STUDIED THE EARTH–MOON SYSTEM, the solar system, the stars, and the galaxies, recognizing the need for a new scale of size with each step. What is left to study? The universe itself is the subject of cosmology, and now we must come to terms with its nature. Yet even here, our understanding is based on observation and deduction, in the spirit of science and not of guesswork.

The scientific study of the universe is a fairly recent field; only since the 1940s have astronomers realized that we can make quantitative predictions about the nature and behavior of the universe and verify these predictions through observations. One of the first physicists to point out the connections between particle physics (whose subject is subatomic particles) and cosmology was Stephen Weinberg, who shared a Nobel Prize in 1979 for his work on the forces between elementary particles. He applied the theories of high-energy physics to the conditions of the entire universe at the time of its beginnings, describing the sequence of events in the "First Three Minutes" of the universe. He concludes his book as follows:

> The universe will certainly go on expanding for a while. As to its fate after that, the standard model gives an equivocal prophecy: It all depends on whether the cosmic density is less or greater than a certain critical value. . . . [W]hichever cosmological model proves correct, there is not much of comfort in any of this. It is almost irresistible for humans to believe that we have some special relation to the universe, that human life is not just a more-or-less farcical outcome of a chain of accidents reaching back to the first three minutes, but that we were somehow built in from the beginning. . . . It is very hard to realize that [our world] is just a tiny part of an overwhelmingly hostile universe. It is even harder to realize that this present universe has evolved from an unspeakably unfamiliar early condition, and faces a future extinction of endless cold or intolerable heat. The more the universe seems comprehensible, the more it also seems pointless.
>
> But if there is no solace in the fruits of our research, there is at least some consolation in the research itself. Men and women are not content to comfort themselves with tales of gods and giants, or to confine their thoughts to the daily affairs of life; they also build telescopes and satellites and accelerators, and sit at their desks for endless hours working out the meaning of the data they gather. The effort to understand the universe is one of the very few things that lifts human life a little above the level of farce, and gives it some of the grace of tragedy.*

Weinberg has since expressed some regret over the somewhat pessimistic tone of these words, but he is certainly not alone in feeling the enormity of the universe around us. It is easy to feel lost in the vastness of space. However, an increasing understanding of how the universe works can give us a more uplifting sense of how we fit into existence. We can also turn to religion as a way of dealing with our place in the universe; we discuss this briefly in an Advancing the Model box on Science, Cosmology, and Faith. However you feel about the size of the universe, the very contemplation of it seems to be a uniquely human endeavor.

In this chapter we ask three simple questions: What is the nature of the universe? What was its past? What will be its future? These questions are the subject of cosmology, which studies the universe as a whole rather than its individual parts. Many of the ideas from Chapter 17 will be useful for answering cosmology's questions, but the questions are not easy to answer and today's answers must be regarded as tentative.

Nature has provided us with some clues in our quest to understand the universe. Analyzing these clues, astronomers have been able to reach very meaningful answers to our three questions—questions that must have been among the first asked by intelligent humans.

18-1 The Search for Centers and Edges

Throughout the first 17 chapters of this book, a story has unfolded about changes in our basic ideas of the nature and extent of the universe. These changes have produced different answers to the question of whether or not humans are located at the center of the universe. Ptolemy's model (Chapter 1) placed the Earth at the center. **Figure 18-1** represents the idea of Ptolemy's universe (although the planets and their complicated epicycles are not included). The stars in Ptolemy's geocentric model were believed to be on a sphere surrounding the Earth.

Copernicus (Chapter 2) moved the Earth from the center of the universe and replaced it with the Sun, but the stars remained on a sphere surrounding the solar system. The sphere of stars was generally considered to lie just beyond the most distant

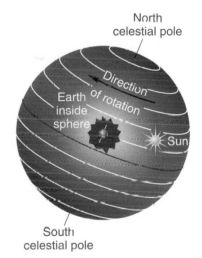

Figure 18-1
Ptolemy's universe placed the Earth at the center of a sphere of stars. This drawing does not include the planets, which would move between the Earth and the celestial sphere.

Dates are given in marginal notes for review and comparison.

Ptolemy: c. 150 A.D.

Copernicus: 1473–1543.

---

* Steven Weinberg, *The First Three Minutes* (Basic Books, 1988).

planet, and this was the entire physical universe. However, some important observations did not fit Copernicus's heliocentric model. The model could not explain why the spinning Earth did not leave behind objects thrown upward, nor could it explain the lack of stellar parallax as the Earth moved around the Sun. A half century later, Galileo (Chapter 3) introduced the concept of inertia to explain motion, and he argued that stellar parallax was not observed because the stars are simply too far away. For this to be true, the universe had to be much larger than people had suspected.

Isaac Newton (Chapter 3) proposed that the universe is infinite in extent. He argued that if it were not, the force of gravity would cause the universe to collapse to its center. An infinite universe would not be in danger of collapse because each object in it would be pulled equally hard in all directions. Newton's idea did not catch on, however. Others claimed that the gravitational force has limited range and therefore cannot cause the universe to collapse. An infinite universe simply conflicted too strongly with "common sense."

During the eighteenth century, William Herschel's discoveries (Chapter 16) led to a better understanding of the Milky Way, again enlarging the boundaries of the known universe. Herschel's data, however, indicated that our solar system is at the center of the Galaxy, thereby returning the human race to center. Herschel proposed that the many fuzzy nebulae visible in telescopes are not part of the Milky Way, but are other galaxies like ours. Like Newton's idea of an infinite universe, Herschel's proposal was not accepted, and by the end of the 19th century, most astronomers were convinced that what we call the Milky Way Galaxy comprised the entire universe.

In 1917, Harlow Shapley (Chapter 16) used globular clusters to conclude that the Sun is not at the center of the Galaxy, and in the next decade other astronomers calculated the Sun's motion. This showed finally that the Sun is not "special," either in nature or in position.

In 1923, Edwin Hubble (Chapter 17) discovered Cepheid variables in the Andromeda "nebula." This pushed back the limits of the universe tremendously, for astronomers recognized that our Galaxy of hundreds of billions of stars is just one of numerous galaxies. People perceived the size of the universe to be on a vaster scale than ever before imagined, and humans were far removed from any central position.

Einstein's Universe

While Hubble and others were using observations to advance our knowledge of the universe, several theoreticians were approaching the problem from a different perspective. Einstein's general theory of relativity (1915) substituted curved space for gravitational force. The theory holds that space itself is curved near a massive object and that the curvature causes the acceleration that previously had been attributed to the force of gravity. In Chapters 3 and 15 we described several observations that confirm Einstein's theory and show that it explains many phenomena better than does Newton's law of universal gravitation.

Newton knew that gravitational forces would cause a finite universe to collapse. The same situation appears in general relativity, but it is couched in different language. In the absence of any exotic forms of energy, if there is enough matter in the universe to eventually cause its collapse (because of the attractive nature of the gravitational force), space is curved enough that the universe is said to be **closed**. A closed universe has a positive curvature. Curvature of space is difficult to imagine because it involves a fourth dimension. The world of our direct experience is a three-dimensional one, and the best we can do to imagine a closed universe is to think of a two-dimensional plane being curved so that it forms a sphere (**Figure 18-2a**).

In Chapter 3 we compared our existence in a three-dimensional world that is curved into a fourth dimension to the situation of imaginary "flatfleas" on a two-dimensional surface that is curved (Figure 18-2b). If the creatures in the figure travel far enough in what they perceive as a straight line, they will return to their original

Galileo: 1564–1642.

Newton: 1642–1727.

Herschel: 1738–1822.

At the time, galaxies were called *island universes*, a name suggested by German philosopher Immanuel Kant.

Shapley: 1885–1972.

Hubble: 1889–1953.

Einstein: 1879–1955.

Does a closed universe mean that nothing can get in or out?

closed universe The state of the universe if its total mass and energy density is larger than a specific value, called the critical density.

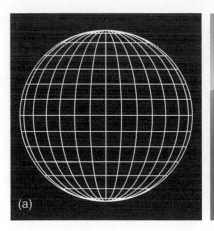
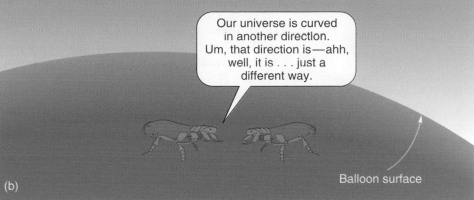

Balloon surface

Figure 18-2
(a) A closed universe can be thought of as a two-dimensional plane curved into the third dimension so that it becomes a sphere. (b) Imagine that fleas on the surface of the sphere think that they live in only two dimensions. That is, the surface appears flat to them. Yet, if they travel far enough in one direction they will end up where they began.

positions. Likewise, in a closed universe, a beam of light sent in one direction will travel around the universe and return to its starting point (if the universe lasts long enough). Like the finite two-dimensional surface of a sphere, such a universe would have no boundaries, even though it would have a limited volume and a limited number of stars.

The alternative to a closed universe is an ***open*** universe that does not curve back onto itself. In the absence of any exotic forms of energy, the amount of matter in an open universe is not enough to slow down its expansion, and such a universe will continue expanding forever. An open universe has negative curvature; this can be visualized in a two-dimensional analogy as a saddle shape (**Figure 18-3a**). An open universe is infinite and continues on forever.

At the boundary between an open and closed universe is the special case of a ***flat universe***. In this case the geometry that describes the universe is the same Euclidean geometry that you learned in high school (Figure 18-3b). The universe is infinite and has just the "right" amount of matter and energy (the ***critical density***) so that it shows no overall curvature; that is, it is flat.

Einstein believed that the universe is closed and finite. He would not accept the possibility that the universe was anything but static and unchanging (on a large scale). When he applied his equations of general relativity to the universe, he obtained a solution from which he concluded that such a universe would collapse on itself, just as Newton had concluded. (At the time, Einstein was unaware that his equations give rise to other solutions. One such solution, discovered by other scientists, corresponds to an expanding universe.) This seemed an unacceptable situation to Einstein, and he decided to adjust his theory to eliminate that conclusion. To do so, he inserted into the equations a ***cosmological constant*** (Λ), a term that adds a cosmological repulsive force to support the universe against collapse. This force would tend to make the universe as a whole expand. However, this force would be very unusual, for instead of weakening with distance, it would become greater

open universe The state of the universe if its total mass and energy density is less than a specific value, called the critical density.

flat universe The state of the universe if its total mass and energy density is exactly equal to a specific value, called the critical density.

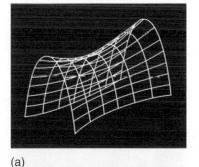

(a)

(b)

Figure 18-3
(a) An open universe can be represented as a plane curved into the shape of a saddle. The figure shows only a portion of the shape, for it extends forever in every direction (b) A flat universe can be represented as a plane.

critical density The average mass and energy density of the universe at which space would be flat. It is equal to $3 \cdot H_0^2 / (8 \cdot \pi \cdot G)$ and is equivalent to about 5.5 hydrogen atoms per cubic meter of space.

cosmological constant A term (denoted by the Greek capital letter "lambda," Λ) in the equations of general relativity that corresponds to a force throughout all space that helps the universe expand.

Something is said to be *ad hoc* if it is designed for one specific purpose. An ad hoc committee, for example, is one that is created for a specific job.

with distance. This explains why we do not experience the force at the small distances we normally deal with. The hypothetical cosmological repulsive force is significant only at large distances.

Einstein regretted having to adjust his theory in such an arbitrary, ad hoc manner. Adding the term violated the principle that a theory should be aesthetically simple. Nevertheless, he felt that the adjustment was necessary if his equations were to apply to a static universe. If he had rejected the cosmological constant, he probably would have concluded from his theoretical work that the universe is expanding. Instead, the expansion was discovered observationally.

Einstein abandoned the idea of a cosmological constant when evidence for the expansion of the universe and the big bang became almost completely accepted by the astronomical community. However, recent observations suggest the presence of an exotic form of energy, which results in a repulsive force helping the universe expand at an accelerating rate as time goes on. That is, the idea of a (positive) cosmological constant is supported by these observations. We do not really know yet the characteristics of this exotic form of energy.

18-2 The Expanding Universe

Figure 18-4 shows the life spans of the scientists who have been mentioned in this chapter.

In 1929, Edwin Hubble announced that the pattern of redshifts of distant galaxies indicates that they are moving away from us and that the more distant the galaxy, the faster it is moving away. In Chapter 17 we explained how the resulting *Hubble law* has become a valuable tool to measure the distances to faraway galaxies. In addition, Hubble's law has tremendous implications concerning the nature of the universe.

First, if other galaxies are moving away from ours, does this not mean that our Galaxy must be at the center? Have we finally discovered that we are at a special location after all? Hubble published his findings in 1929, 12 years after Shapley's work with globular clusters had taken our Sun out of the central position within the Galaxy. By this time it had become a working premise that our location within the cosmos is not central. To see why Hubble's discovery does not, in fact, conflict with this basic premise, consider the following analogy.

Imagine that you are a trainer in a flea circus and that you put a number of educated fleas on a balloon that your assistant is blowing up. You have instructed these fleas to hold their positions on the balloon. **Figure 18-5** shows the balloon being blown up with the fleas in place.

Now imagine what a particularly intelligent flea sees when it looks out toward neighboring fleas. (Assume either that the flea can see around the curvature of the

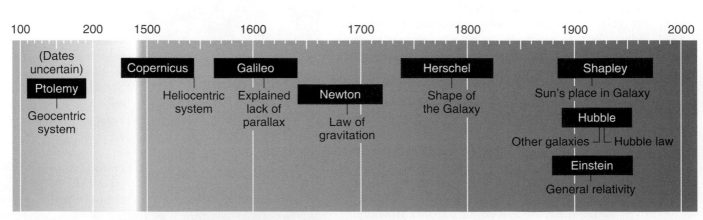

Figure 18-4
The life spans of the scientists whose cosmological contributions were described in the last few pages.

balloon, or that the fleas are close enough together that the balloon seems flat to them.) The intelligent flea sees every other flea getting farther and farther away as the balloon is blown up. In addition, the more distant fleas are moving away at a greater speed than those nearby. The more distant a given flea is from the observer, the faster it is moving.

The important point is that the same result would be obtained no matter which flea is the observer. While the balloon is being blown up, every flea sees every other flea moving away with a velocity that depends upon the flea's distance from the observer.

The analogy shows that if galaxies are part of an expanding universe, we would see exactly what Hubble and Humason observed: Every galaxy is seen to be moving away from us at a speed that depends upon how far away it is. Many astronomers since Hubble have surveyed galaxies at greater and greater distances and have come to the same conclusion: The universe is expanding.

In Chapter 17 we included an Advancing the Model box concerning the differences between observations and conclusions. The distinction is important in this case. The pattern of redshifts is an *observation*. All spectral lines of distant (dim) galaxies show the redshift, not just the lines of the visible spectrum. The observations are 100% in agreement with what would be predicted for light from a receding object. Thus, we *conclude* that the galaxies are getting farther away from us. We do not, however, *observe* this.

What Is Expanding and What Is Not?

Saying that the universe is expanding certainly does not mean that the solar system is expanding. Nor does it mean that stars within our Galaxy are getting farther apart. In fact, it does not even mean that all galaxies are getting farther apart.

The observed redshift leads us to conclude that other *clusters* of galaxies are moving away from ours. Some individual galaxies are actually moving toward us. This occurs in two ways. First, galaxies in our local group are moving randomly within the group as each responds to the overall gravitational force of the others. Thus, the Andromeda galaxy and a half-dozen others in the local group are at this time moving toward the Milky Way Galaxy.

Second, in nearby clusters, the same random motion of individual galaxies results in some of the galaxies moving toward us at the present time, even though the cluster in which each galaxy exists is moving away from the local group. A simple example can show how it is possible for an individual galaxy to show blueshifted lines even though it belongs to a cluster moving away from us. For a Hubble constant of, say, 65 km/(s × Mpc), a cluster at a distance of 2 megaparsecs from us will be moving away from us at 65 × 2 = 130 kilometers/second, due to the expansion of the universe. However, the random speeds of individual galaxies can easily exceed 130 kilometers/second. Therefore, if we were looking at an individual galaxy in this cluster with a random speed of 200 kilometers/second, we could be seeing this galaxy moving at speeds between 130 + 200 = 330 kilometers/second and 130 −200 = −70 kilometers/second. In the latter case, this galaxy will show blueshifted lines, even though its overall motion is away from us as a member of a cluster. No individual cluster—as far as we can tell—is expanding.

It is the clusters of galaxies that are moving farther apart. Thus, although we often say that other galaxies are moving away from us, we should really say that other clusters are moving away from our cluster. Anything that is gravitationally bound is not expanding as a result of the expansion of the universe. This includes objects (like us) that are held together by forces stronger than the gravitational attraction between their individual components.

The Cosmological Redshift

After the redshift of light from distant galaxies was discovered, astronomers assumed that the Doppler effect was the cause. Using the Doppler effect, they calculated the velocities of the galaxies. The Hubble law is a statement of the relationship between a

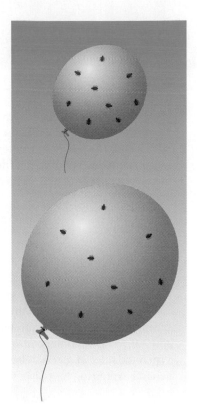

Figure 18-5

Fleas on a balloon get farther apart when the balloon is blown up.

The argument presented here also explains why we do not use the Hubble law to find distances to nearby galaxies.

ADVANCING THE MODEL

Cosmological Redshift

Objects have been discovered that have a redshift (z) greater than 5. A redshift of 5 means that the ratio of the shift in a particular wavelength ($\Delta\lambda$) to the unshifted wavelength (λ_0) is equal to 5. In the standard application of the Doppler effect, this means that the velocity of the object (v) is equal to five times the velocity of light (c), because

$$z = \frac{\Delta\lambda}{\lambda_0} = \frac{v}{c}$$

This cannot happen! An object cannot travel at a speed greater than the speed of light. The resolution to this problem lies in the fact that the standard expression for redshift ($z = v/c$) is valid only in the approximation of speeds much less than the speed of light. For objects moving at large speeds (v larger than about $0.1c$), astronomers must take into account special relativity. When the rules of special relativity are used, we get the following general expression for redshift:

$$z = \frac{\Delta\lambda}{\lambda_0} = \sqrt{\frac{c + v}{c - v}} - 1$$

This expression is valid for all speeds. For small speeds, it gives us our familiar expression $z = v/c$, while for large speeds (and thus large redshifts), the calculated speed of a receding cluster of galaxies is always less than the speed of light.

The fallacy here is that special relativity is not applicable in a universe where general relativity is the controlling rule. Special relativity can be applied only where space can be considered to be exactly flat. In the case of distant clusters, the curvature of space must be taken into account, so special relativity cannot be used.

To review this relationship between laws, recall that Newton's law of gravitation should not be considered *wrong*. Rather, it is a law with limited applicability that must be replaced by general relativity in certain cases. However, it is accurate enough for most everyday uses. Likewise, our standard concept of speed (where we do not worry about objects moving at or near the speed of light) works in everyday situations. Special relativity needs to be used only when speeds approach the speed of light. However, special relativity is also a law with limited applicability. It can only be used for nonaccelerated objects, and it cannot be used when space curvature is a factor.

When we try to combine physical laws that are not consistent with one another, we violate the principle that a theory should be aesthetically pleasing. Consistency is important in applying scientific principles.

Thus the Doppler effect cannot be used as the explanation for the cosmological redshift, since Doppler shifts are caused by an object's motion *through* space. The correct explanation lies in the idea of the expansion *of* space, as we discuss in the text.

The Doppler shift is still the correct explanation of other observations in astronomy.

galaxy's distance and its (Doppler effect) velocity, as we showed in Chapter 17. However, modern cosmology explains the redshift in an entirely different manner.

An expanding universe does not mean that clusters of galaxies are rushing *through* space. Instead, *space itself is expanding*. **Figure 18-6** illustrates the difference, using the balloon analogy again. This time, tiny seeds have been glued to the balloon, each seed representing a cluster of galaxies. Part (a) shows the Doppler interpretation of the expansion of the universe. The balloon is expanding against a background grid that represents space, indicating that the galaxies move through space. Part (b) shows the modern cosmological interpretation of the expanding universe. The grid that represents space is part of the surface of the balloon here. As the balloon expands, space itself expands.

Why don't individual galaxies expand along with space? General relativity shows that objects that are held together by their own gravity, such as the Earth, the solar system, the Galaxy, and the local group, do not expand as space expands. The force of gravity (or, in terms of general relativity, the local curvature of space) holds them together just as each individual seed on the balloon holds itself together and does not expand with the balloon.

This interpretation of the expansion of the universe means that clusters of galaxies do not actually have a velocity through space and therefore that the Doppler effect does not explain the redshift in their spectra. A Doppler redshift is caused by the relative motion of an object that is emitting a wave. The ***cosmological redshift*** has a different cause. Long ago, when a distant galaxy emitted some electromagnetic

cosmological redshift The shift toward longer wavelengths that is due to the expansion of the universe.

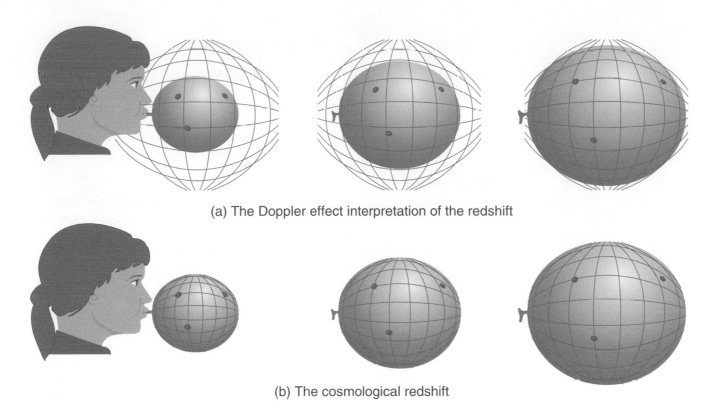

(a) The Doppler effect interpretation of the redshift

(b) The cosmological redshift

Figure 18-6

Each tiny seed glued to the balloon represents a cluster of galaxies. (a) This is the old, incorrect interpretation of an expanding universe in which the universe expands through space. (b) This is the correct cosmological interpretation. Space expands, dragging the clusters of galaxies along with it.

waves, those waves were not redshifted. But as space expanded, the waves expanded—that is, they lengthened—along with space. The more time the radiation has traveled, the more its waves have stretched in wavelength. **Figure 18-7** illustrates the difference between the two explanations.

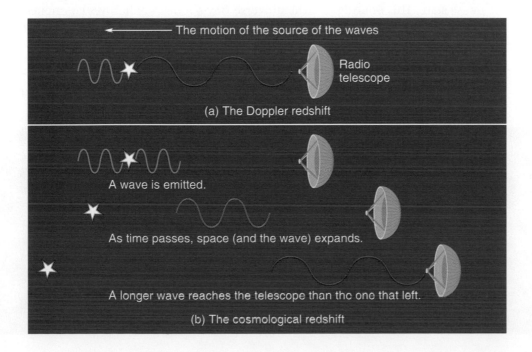

The motion of the source of the waves

Radio telescope

(a) The Doppler redshift

A wave is emitted.

As time passes, space (and the wave) expands.

A longer wave reaches the telescope than the one that left.

(b) The cosmological redshift

Figure 18-7

(a) A Doppler redshift results from the motion of the source. The wave traveling backward toward the radio telescope is lengthened by the motion of the source. (b) When a wave is emitted by a nonmoving object, it has the same wavelength in all directions. As space expands, that wave lengthens. If the wave takes long enough to get from the source to the telescope, it will have lengthened significantly. This is the case for galaxies that exhibit the cosmological redshift.

When the Hubble law was written, astronomers assumed that the observed redshift was caused by the now-rejected Doppler explanation. The law, which states the connection between a galaxy's distance and its velocity, is still valuable, however, if we think of the velocity in the Hubble equation as being caused by the expansion of space, rather than by velocity *through* space. Although astronomers sometimes use more modern ways of stating the redshift/expansion-rate relationship without using the concept of velocity, in this text we will continue to use the traditional Hubble relationship.

In equation form, the Hubble law is $v = H_0 d$ where $v =$ the velocity of the object, $d =$ the distance to the object, and H_0 is the Hubble constant.

Olbers's Paradox

Why is the sky dark at night? This appears to be a very simple question, but it actually has important cosmological implications. If stars (or galaxies or clusters of galaxies) are randomly distributed throughout space, and if the universe were infinitely old, the sky should not be dark, because no matter in which direction you look, the line of sight should end on a star (**Figure 18-8**). Light from any single star in the universe would have had infinite time to reach us. It is true that distant stars appear dimmer than nearby ones, but if we assume that stars are evenly distributed in our universe, then it is clear that there are more stars at great distances, and this eliminates the dimming effect. If you are deep within a forest, you see a tree no matter in which direction you look. Distant trees look smaller, but there are more of them.

Perhaps the absorption of light by dust clouds prevents the sky from being bright. But dust clouds heat up as they absorb light, and since they absorb light from an infinite number of stars, they should get hot enough to glow. Dust clouds cannot be used to explain the dark sky.

The paradox between what is observed and the argument that the sky should not be dark was known long before it was popularized by Wilhelm Olbers in 1823. It is called ***Olbers's paradox***. The problem exists whether the universe is open or closed. An open universe extends without limit, so the line of sight should extend through space until it hits a star. In a closed universe, the line of sight could possibly extend all the way around until the surface of a star is encountered.

The resolution of Olbers's paradox—the reason that the sky is dark—is based on two related ideas: one, that the universe has a finite age, and two, that it is expanding. First, a universe that has been expanding for billions of years, as suggested by observations, most likely had a beginning at one time in the past, emerging from a very dense state. This finite age for the universe (say about 14 billion years) implies that we cannot see any objects that are beyond a distance of 14 billion light-years from us. If we think of ourselves as being inside a sphere of radius 14 billion light-years, then our entire *observable* universe is located inside this sphere. Light from an object outside this sphere would have taken more than the age of the universe to reach us and thus we do not see it now. As a result, independently of whether the universe is finite

Olbers's paradox An argument showing that the sky in a static universe could not be dark.

The resolution of Olbers's paradox due to the finite age of the universe was first pointed out in 1848 by the American writer Edgar Allan Poe.

Figure 18-8
In an infinite, static universe, your line of sight would end at a star no matter in which direction you looked.

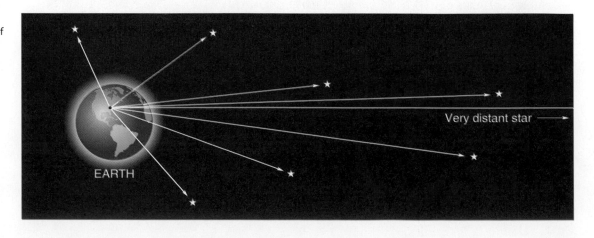

Very distant star ⟶

EARTH

or infinite, our line of sight does not always hit a star, which partially explains why the night sky is dark. Second, as a result of the expansion of the universe, light from distant galaxies is redshifted, and thus it also loses energy. This occurs because the energy of a photon of light depends directly on the light's frequency, so a photon of lower frequency and thus larger wavelength (due to redshift) has less energy. This decrease in energy of the light we receive also helps explain (to a lesser degree) why the night sky is dark.

The resolution to the paradox depends upon the fact that the universe is expanding and that it had a beginning. This latter point is at the heart of the major theory of modern cosmology, the *big bang theory*. Before discussing this theory, however, we need to describe the fundamental assumptions of modern cosmology.

> We discussed photons in Section 4-6. Recall that the frequency of a photon is inversely proportional to its wavelength.

18-3 Cosmological Assumptions

The assumptions of cosmology cannot be proved, but evidence indicates that they are reasonable. The first assumption is that the universe is **homogeneous**. This means that no matter where an observer is positioned in the universe, the universe looks essentially the same. This does not mean, of course, that there is another Earth from which an observer can look to see another Mars in the sky. It does not even mean that there is another local group of galaxies similar to our local group. It means, rather, that on the largest scale, the universe has about the same density and composition of matter at one location as it has at another. As we have described, humans have come to learn that they are not in a favored place in the universe. In cosmology, this becomes an underlying assumption.

> **homogeneous** Having uniform properties throughout.

The second assumption is somewhat related to the first. The assumption of **isotropy** (I-SOT-rah-pee) states that the universe looks the same in all directions. Although this sounds like homogeneity, it is actually a different quality. The ocean in **Figure 18-9** is homogeneous, for no matter where you are *located* on the surface of the ocean, the view is the same. However, you see a different view when you look across the waves than when you look along the waves. This means that the ocean is not isotropic.

> **isotropy** The property of being the same in all directions.

Like homogeneity, isotropy applies only on the largest scale. Recall that when we look out from Earth, we can see farther when we look out of the galactic disk than if we look along the disk. The universe is not isotropic on this small scale. To apply the idea of isotropy, we must imagine being outside the Galaxy and even outside the local group of galaxies. In this case, the assumption is reasonable.

These two assumptions of cosmology are so much at the heart of the subject that they form what is called the **cosmological principle**. This principle formalizes the idea that we are not located at a special place in the universe. It brings to final completion the Copernican revolution that proposed that the Earth is not central. The cosmological principle not only states that the Earth is at no special place in the universe, but that there *is* no special place in the universe.

> **cosmological principle** The basic assumption of cosmology that holds that on a large scale the universe is the same everywhere.

An additional cosmological assumption is **universality**. This means that the same physical laws apply everywhere in the universe. Although this may seem so obvious that it need not be stated, recall that it was only fairly recently in human history that Isaac Newton found the first law of nature that could be shown to apply beyond the Earth. Now we have vastly more evidence for this assumption as we analyze radiation from the most distant objects, but we must remember that it is still an *assumption* that this principle holds throughout the universe.

Figure 18-9
Although the ocean surface is homogeneous, you see something different when you look in different directions because the surface is not isotropic.

> **universality** The property of obeying the same physical laws throughout the universe.

These cosmological assumptions show our biases about the nature of the universe, but they are also necessary. Without them we would not be able to make any progress in understanding the universe. Unlike other disciplines, in cosmology we cannot control our subject, the universe, or make comparisons with another similar subject. There is only one observable universe. Based on these assumptions, our models have proven to be quite successful in describing the structure and evolution of the universe.

18-4 The Big Bang

The present motion of galaxies leads us to believe that at one time in the past, the matter in the universe must have been much closer together. If we run the "movie" of the expansion of the universe backward, there must have been a time when the density of matter was extremely high and some sort of "explosion" initiated the expansion of the universe. This explosion that started the expansion of the universe is called the *big bang*, an irreverently trivial name for the event that began the universe. In its most basic sense, the standard big bang model is simply the idea that every bit of the matter and energy in the universe was once compressed to an unimaginable density. In the big bang, the material exploded apart.

We must be careful not to think of the material of the big bang as existing at a certain place, at one point in the universe. It *was* the universe—the entire universe. It did not exist as part of the universe.

Understanding the details of what happened in the big bang requires an understanding of particle physics, sometimes called "high-energy physics" because elementary particles are used in high-speed collisions to help us better understand what they consist of. Certainly the big bang represents the highest concentration of energy ever reached in the history of the universe. We describe some of the events that immediately followed the big bang in the Advancing the Model box "The Early Universe" on page 593.

After the initial explosion, the steady decrease in the temperature of the universe allowed the nuclear particles to form into atoms of low mass (hydrogen and helium). These atoms then clustered under the force of gravity into stars, clusters of stars, galaxies, and clusters of galaxies. The clusters of galaxies are still moving apart. Again, we must be careful not to think of the universe as expanding *into* empty space around it, for space does not exist apart from the universe. The universe is all there is, and its parts are moving away from one another.

A common *misconception* is to consider the big bang as being similar to the situation shown in **Figure 18-10**. Suppose a large firecracker is placed at the center of a small stack of sand. Soon after the explosion of the firecracker, grains of sand are at various distances from their original position, and—here is the important point—grains that are farther from their respective starting points are moving faster. That is how they got farther from one another. No matter which grain one considers, all others are moving farther from it. Although this example is useful to illustrate that the fastest-moving galaxies would naturally be the farthest away, the example is misleading if you do not remember that in the case of cosmological expansion, galaxies do not move *through* space, as the grains of sand do. Because we have no experience with the expansion of space in our everyday lives, Earthly examples must involve motions through space.

big bang The theoretical initial explosion that began the expansion of the universe.

Calvin, of the cartoon Calvin and Hobbes, prefers the name the Horrendous Space Kablooie instead of the big bang.

It's interesting to note that progress in understanding how the universe itself began depends on understanding the forces that act between the tiniest particles we know of, and how these forces affect the behavior of these particles.

Figure 18-10
(a) A firecracker about to explode inside a small pile of sand. (b,c) The sand shown later. The sand grains that are farthest from the initial explosion are moving fastest.

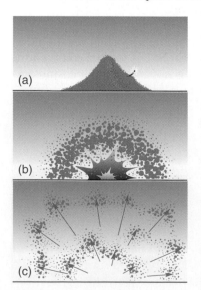
(a)
(b)
(c)

ADVANCING THE MODEL

The Steady State Theory

The cosmological principle holds that we are at no special place in space. It does, however, allow for us being at a special place in time; for at the present moment, the universe is less dense than it was in the past and more dense than it will be in the future because expansion reduces its density. In 1948, *a perfect cosmological principle* was proposed by astronomers Fred Hoyle, Hermann Bondi, and Thomas Gold. It states that our location in time is no more special than our location in space, and that the universe of the past was the same as it is now and the same as it will be in the future. Since the overall universe does not change in this cosmology, it is called the *steady state theory*.

Before looking at the implications of the perfect cosmological principle, consider why such an idea might be proposed. Remember that a theory should be aesthetically pleasing. The perfect cosmological principle is neater because it expands the nonspecialness to include time as well as space. The perfect cosmological principle carries the nonspecialness idea through both space and time.

It seems to be almost undeniable that the universe is expanding. How can this be squared with the steady state theory? It seems that if clusters of galaxies are getting farther apart, the universe will necessarily be less dense in the future than it is now. Hoyle, Bondi, and Gold proposed that matter is created in the space between galaxies and that this new matter is sufficient for the universe to maintain its density (**Figure B18-1**). Such spontaneous creation of matter would violate a long-established principle of physics: the conservation of mass/energy, which holds that the total amount of mass and energy in the universe cannot change.

The principle of the conservation of mass/energy is dear to scientists, but it has changed in the past. Before this century, there were two such conservation principles: the conservation of mass and the conservation of energy. It was thought that neither matter nor energy could be created or destroyed. Then, in 1905, Einstein's theory proposed that mass could be changed into energy—a prediction that was dramatically demonstrated in the atomic bomb. Einstein combined the two principles into one, the conservation of mass/energy, which sees mass and energy as two forms of the same thing and holds that the total mass and energy in the universe remains the same.

The amount of mass that would have to be created in order to fill the gaps between galaxies as they move apart would be very small—only one new hydrogen atom in each cubic meter of space every 10 billion years or so. This means that if one new atom were created every 10,000 years in the volume of the Houston Astrodome, the density of the universe would remain the same. The theory holds that as millions of years pass, newly formed atoms collect to form new stars and galaxies, and that the expanded space is thereby refilled, so that the universe is no different now than it was in the past and than it will be in the future.

The fact that the steady state theory violates the principle of conservation of mass/energy should not be considered a strike against the theory. First, according to the predictions, the necessary creation events are so rare that they would be far below any possible detection limit and therefore far below the accuracy to which the principle has been verified. Second, the big bang theory may also involve a violation of the principle. In this case it is a gigantic one-time violation, as all matter and energy are created in one big burst at the beginning.

One of the first predictions made by the steady state theory involved the number of faint (very distant) radio sources. In this theory, the number density of such sources (that is, the number of radio sources per unit volume) remains constant, and so does the average energy radiated away by a radio source per unit time. This leads to a very specific relationship between the number density and the energy received per unit time per unit area. However, observations made in the 1950s implied a different relationship, suggesting that either the number density or the average energy increases with distance. This, of course, implies that the universe is evolving, in contradiction to the steady state theory.

Another argument against this theory involves the amount of helium observed in the universe. This amount is much greater

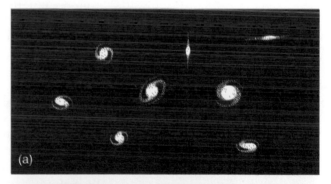

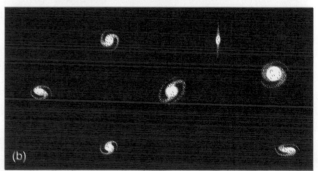

Figure B18-1
If the universe does not decrease in density as galaxies move apart, matter must be created in the space between galaxies.   *(Continued)*

ADVANCING THE MODEL

The Steady State Theory *(Continued)*

than what can be expected from stellar evolution. Therefore, we will have to accept that either helium is also spontaneously created or that there was a special period in the universe's life during which a great amount of helium was created, along with other elements.

Finally, the discovery of the cosmic background radiation made it impossible to reconcile its existence with the steady state theory. Even if we assume an unknown mechanism that can produce such radiation in the steady state theory, its temperature would have to be constant, independent of redshift. However, the measured temperature of this radiation depends linearly on red-shift. That is, at redshift z, the measured temperature is greater by a factor $(1 + z)$. This implies that the universe is evolving.

Hoyle, Burbidge, and Narlikar proposed a quasi-steady state theory in an effort to bring the theory closer to an agreement with the observations. However, none of the many versions of this theory have been successful so far. The historical development of both the big bang and steady state theories provides another example in our discussion of how science is done. At the end, the theory that is accepted the most is the one that is simple and elegant, explains the current observations, makes predictions that can be tested, and allows itself to be shown wrong.

Evidence: Background Radiation

The idea of the big bang arose from the evidence for the expansion of the universe. We must ask, however, if there is other evidence to support the theory. The first evidence was found quite accidentally when a prediction made by the big bang theory was found to be correct. As scientists learned more about nuclear processes in the 1940s, they were able to make predictions about the character of the early universe shortly after the big bang. The material of the big bang was originally extremely hot, but it cooled as it expanded outward, in much the same way that a gas cools when it expands. The hot matter of the very early universe was opaque to radiation, but when it cooled to about 3000 K, it became transparent. This happened when the universe was about 380,000 years old, at a redshift of about $z = 1000$. At this point, the gas that made up the universe was emitting radiation that had the characteristics of radiation from an object at 3000 K, and this radiation existed over the entire universe.

The universe has remained transparent to radiation since that time, so it was predicted that this radiation should still be in existence. Remember that as we look at objects at greater and greater distance, we are looking back in time. If we were to detect the radiation left over from the big bang, it would be coming from far back in time and therefore from a great distance. According to the Hubble law, then, it would be greatly redshifted. The original 3000-K radiation peaked in the infrared region of the spectrum, but calculations showed that it would by now have been redshifted to the microwave region, with wavelengths of a millimeter or so. Such radiation would not be characteristic of radiation from a 3000-K object, but it would appear the same as radiation from a very cold object—one at about 3 K ($-270°C$). It was predicted that this ***cosmic microwave background radiation*** (CMB) should be striking Earth from all directions, but that it should be very faint.

cosmic microwave background radiation (or **CMB radiation**). Long-wavelength radiation observed from all directions; thought to be the remnant of radiation from the big bang.

In 1948, when the prediction of the CMB radiation was made, there was no way to detect such weak waves. The prediction was laid aside and forgotten. Then in the mid-1960s, a group of physicists at Princeton University were studying the big bang and again predicted the existence of the CMB radiation, not realizing that the prediction had been made some 15 years before. They were confident that they could build a radio receiver that would be able to detect the radiation, and they set about building one on the roof of the Princeton biology building. They were too late.

The Early Universe

Stephen Weinberg's book *The First Three Minutes* was the first book written for the general public that describes the model cosmologists have developed to explain what happened at the beginning of the universe. Some of the events in this model have been refined over the past few decades as more data have accumulated from observations by satellites and balloon-borne instruments. But the sequence of events has become accepted enough to be called the "standard model of cosmology."

The model starts at about 10^{-43} seconds after the big bang itself. Why 10^{-43} seconds? Well, according to our present understanding of space and time, there is a limit to how closely you can look at tiny distances and still know what you're looking at. You know that if you look at a piece of paper, like a page of this book, through a microscope, it will no longer appear smooth, but will show a mixed-up jumble of wood fibers, along with filler that holds the fibers together. If you keep going in your imagination, you might think of the jumbled-up atoms that make up the page and the particles that make up the atoms. But if you keep going, eventually space itself becomes jumbled instead of smooth. Scientists have calculated the smallest distance that we can imagine before this jumbling occurs, coming up with a distance of about 4×10^{-35} meters. Light would take about 10^{-43} seconds to cross this distance, which makes this amount of time the smallest interval for which we can talk about the universe.

What was the universe like at this time? There were no atoms and everything was scrunched up into a far smaller size than any particles we're familiar with. All we can say with any confidence is that the universe was hot, about 10^{32} K, and it was expanding. As it expanded, it cooled, very quickly. At about 10^{-35} seconds, a period of inflation may have occurred, lasting for only 10^{-32} seconds, but having tremendous implications for the history of the universe, as we discuss in Section 18-6.

By the time of one millionth of a second (10^{-6} s) after the big bang, the temperature had dropped to about 10^{13} K, which was cool enough (relatively speaking, of course!) for protons and neutrons to form. How do we know this? Because according to present theories, at temperatures higher than 10^{13} K, these particles break up into even smaller particles (called quarks; you may have heard about these littlest of all particles). Once you get below this temperature, the quarks don't have enough energy to move away from each other and stay bound together in protons and neutrons.

In order to form nuclei more massive than hydrogen, the temperature of the background radiation must be low enough so that it does not break the nuclei apart. The first element more massive than hydrogen (whose nucleus is just a proton) is helium. Nuclei of helium consist of two protons and two neutrons

or two protons and a single neutron. The easiest way of forming helium nuclei is by first forming deuterium nuclei, which consist of a single proton and a single neutron. However, for the first 10 seconds after the big bang, the temperature of the background radiation was too high (more than 5 billion K) for deuterium to form. As the universe cooled and deuterium nuclei formed, they led to the formation of helium nuclei. This continued for about 15 minutes, by which time the temperature had dropped to below about one billion K and no further nuclei were formed. Again, how do we know this? The evidence is in the proportions of hydrogen, helium, and deuterium we have observed in the universe. Other sequences of events in the early universe have been proposed, but they all predict formation of the light elements in different proportions than what we observe.

The formation of light elements (hydrogen, deuterium, helium, and lithium) in the early universe is called "Big Bang nucleosynthesis." Elements heavier than lithium were later formed in stars, as we described in an Advancing the Model box on "Nucleosynthesis" in Chapter 15.

Figure B18-2 shows how the predicted abundances of deuterium, helium, and lithium depend on the density of ordinary matter in the early universe. The circles correspond to the observed abundances. The predicted abundance for helium-4 is

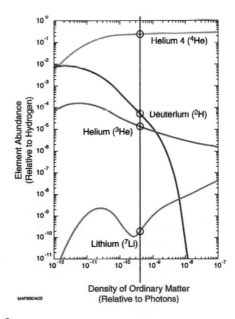

Figure B18-2
The curves show the dependence of the predicted abundances of helium, deuterium, and lithium on the density of ordinary matter in the early universe.

(Continued)

ADVANCING THE MODEL

The Early Universe (Continued)

about 24% of the ordinary matter in the universe, which agrees very well with the observations. In addition, the predicted abundances of the other light elements also agree with the observations if the overall density of ordinary matter is about 4% of the critical density. As we discuss in Section 18-7, astronomers announced in 2002 that data collected by the *Wilkinson Microwave Anisotropy Probe* (*WMAP*) show that the universe is indeed made of about 4% ordinary matter, 23% dark matter, and 73% dark energy.

What happened next? Nothing much. The universe, dominated by radiation, continued to expand and to cool. However, as the universe expanded, matter became increasingly dominant. About 2500 years after the big bang, when the temperature dropped to about 75,000 K (at redshift of about $z = 25,000$), the universe became matter-dominated and became increasingly so ever since. The expansion and cooling of the universe continued for the next few hundred thousand years. Only then, when the temperature dropped below a few thousand degrees, did electrons slow down enough to be captured by nuclei and form atoms. At this point, about 380,000 years after the big bang, at redshift of

about 1000, an amazing thing happened: the universe became visible. Before this time, the background photons were energetic enough that they would not allow electrons and protons to get together and form atoms. Instead, the photons were continuously and vigorously colliding with the charged particles, resulting in a very opaque universe. As the temperature of the background radiation dropped below about 3000 K (380,000 years after the big bang), the photons did not have enough energy to keep electrons and protons from forming hydrogen atoms. Once hydrogen atoms formed, the photons could now move freely through space, resulting in a transparent universe. It is these photons that make up the cosmic microwave background we observe today.

How reliable is this picture of the early universe? Some of it is conjecture, some of it is debatable; certainly none of it is proven. There is no photograph or eyewitness account of the big bang. But the description given here, while somewhat simplified, fits all the data we have so far acquired about the early universe. As more data become available, we will continue to revisit and refine our ideas about what happened in the beginning.

At the same time the predictions were being made at Princeton, only a few miles away two employees of Bell Telephone Laboratories, Arno Penzias and Robert Wilson (**Figure 18-11**), were doing applied research on microwave transmission, in hopes of improving the transmission of messages. They were frustrated, however, by some low-intensity radio waves they observed to be coming from all directions toward their receiver.

As you might suspect by now, the radiation they had found turned out to be the CMB radiation predicted by the physicists at Princeton and by astronomers years before—the prediction that was based on the big bang theory. Penzias and Wilson were awarded the Nobel Prize in 1978 for their discovery of the radiation—radiation for which they had not been searching and the cosmological implications of which they were unaware.

Recall that at the time of the big bang, the condensed material made up the entire universe. This is why the CMB radiation now comes to us from all directions.

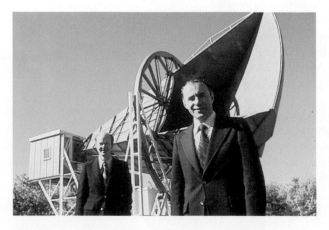

Figure 18-11
Arno Penzias (right) and Robert Wilson discovered the cosmic background radiation using Bell Laboratories' horn-shaped radio antenna, seen behind them.

This radiation fills the universe just as it did at the beginning. Matter and energy comprise the entire universe, and no space is left over.

Until the discovery of the CMB radiation, there was another popular cosmological theory (see the Advancing the Model box "The Steady State Theory"), but this discovery established the big bang theory as the accepted explanation for the beginning of the universe.

In 1989, NASA launched the *Cosmic Background Explorer* (*COBE*, pro-

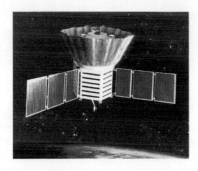

Figure 18-12
The *Cosmic Background Explorer* (*COBE*) was launched in 1989 and for 4 years provided important data regarding the CMB radiation.

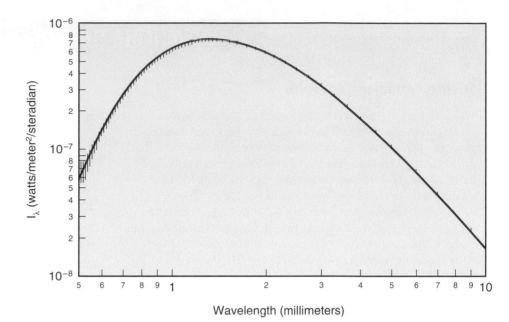

Figure 18-13
The theoretical intensity-wavelength graph for the cosmic microwave background radiation is shown in red. The tiny black lines represent data from *COBE*. The fit is remarkable!

nounced "KOH-bee," **Figure 18-12**). Its purpose was to measure the CMB radiation at various wavelengths and to map the sky in those wavelengths. *COBE's* results provide an astounding corroboration of the accuracy of predictions concerning cosmic background radiation. **Figure 18-13** is an intensity-wavelength graph like those used to describe the temperature of stars in Chapters 12 through 15. The red line on the figure is the predicted intensity of radiation at each wavelength. The tiny black lines represent data points measured by *COBE*. Each data point is depicted as a line, which takes into account possible errors in the measurement. The longer lines at the left indicate that the measurements there—those at shorter wavelengths—are less accurate. The data fit is extraordinary.

The *COBE* data show that the temperature of the background radiation is 2.728 K (with an uncertainty of 0.004 K). This matches theory to within 0.03%, which is 1000 times better than the best data before *COBE*.

Additional Evidence for the Big Bang

The big bang theory is the best current model for the universe. Observational data support the theory but do not prove it, since scientific theories are not proven. We have already mentioned some of these observations: the darkness of the night sky, the Hubble law, homogeneity, and isotropy at large scales.

Another argument in favor of the big bang involves the time dilation of distant supernova light curves. This dilation is the result of the standard interpretation of redshift. A supernova that takes 20 days to decay will appear to take $20 \times (1 + z)$ days to decay when observed at redshift z. Such dilation has been observed, providing a very strong confirmation of the cosmological nature of redshift.

Some additional observations show the evolution of the universe and are consistent with the big bang theory. For example, we mentioned in the Advancing the Model box on "The Steady State Theory" that the observed relationship between the number density of distant radio sources and quasars and the energy received from them per unit time per unit area shows that the universe evolved. The existence of the CMB and the dependence of its temperature on redshift also show that the universe evolved from a hot and dense state.

Additional evidence that supports the big bang comes from the observed proportions of light chemical elements (such as deuterium, helium, and lithium). These proportions could not have formed in stars. However, they are consistent with a process of nuclear fusion in the first few minutes of a hot, young universe. (See the Advancing the Model box on "The Early Universe.")

ADVANCING THE MODEL

Science, Cosmology, and Faith

Where did the matter and energy of the big bang originate? Science cannot answer that question at present. In fact, it seems possible—perhaps likely—that it will never be able to answer that question. For if matter existed in atomic form before the big bang, it did not carry information with it through the big bang. Thus matter itself can tell us nothing about its pre-big bang history. Actually, it is meaningless for us to talk about things "before" or "at the moment of" the big bang. Time did not exist until after the big bang. Because of the extreme conditions involved during the birth of our universe, the known laws of physics do not allow us to start describing its evolution until 10^{-43} seconds after the big bang. After this moment, we can consider the nature of space and time in the same way as we do today.

Does this then mean that science can say that there must be a creator, and that science has, after all, found a need for God? An individual scientist may believe this, but science itself cannot use God as the explanation for the big bang. Science cannot use God for any explanation. It has been said that science avoids God. It does, indeed. The reason that science does not use God for explanations is basically that science has been successful in explaining the material world without reference to a God. Science, by intention, uses natural causes to explain natural effects.

We say that science is successful in its method because scientific explanations of the workings of the material world have led us to further understanding of that world. The fact that the success has come without reference to God indicates that the material universe seems to be describable by completely natural principles.

What about cases where science is unable to find an answer? If science, when it comes to something that it cannot explain at the time, were to explain it by reference to God, the search for an explanation would end. If Newton had used God as an explanation for why things fall to Earth, he would never have developed the theory of gravitation. If we use God as an explanation for the big bang, there would be no reason to look further for a natural explanation. Use of supernatural explanations would shut down science. History, however, tells us that it is profitable to look for natural explanations. Supernatural explanations cannot be used in natural science.

Testability

There is another reason that science cannot use God for an explanation, and this relates to the reason that traditional science does not accept creationism as a science: A theory of science must be able to be shown to be wrong. A theory must be testable. Every theory must be regarded as tentative, as being only the best theory we have at present. It must contain within itself its own possibility of destruction. The 1948 prediction concerning the cosmic microwave background radiation was such a case. If the background radiation had not been found, the big bang theory would have had to have been adjusted, or—if enough such contradictions appeared—the theory would have had to be dropped and replaced with another. This has happened many times in science. It happened to the steady state theory and to other theories we presented in this text.

On the other hand, if science relied on a creator to explain the inexplicable, there would be nowhere to go, no way to prove that explanation wrong. The question would have already been settled.

This is not to say that God might not be the explanation. Many people believe that a creator is the ultimate explanation of everything. Perhaps some questions, like the origin of the material for the big bang, simply cannot be answered without reference to a creator. But in that case, science simply cannot answer the question. The question is beyond the realm of science. Science does not deny the existence of God. God is simply outside its realm.

Finally, as we discuss later in this chapter, the observed anisotropies in the CMB radiation support a big bang model that includes an inflationary period and is dominated by dark matter and dark energy.

The Age of the Universe

If the big bang theory is correct, we can use the Hubble constant to determine how long ago the big bang occurred, for the constant tells us how fast galaxies are spreading apart. Here we make the simplifying assumption that the rate of expansion, given by the Hubble constant, has remained constant through time. Even though the assumption is not exactly correct, it allows us to find an approximate age for the universe. Instead of writing v for speed in the Hubble law, substitute it with the definition of average speed: the ratio of distance to time.

$$v = H_0 d$$

$$\frac{d}{t} = H_0 d$$

Now choose two widely separated galaxies and let d be the distance between them. Thus, t is the time taken to go that distance, or the age of the universe. Solving the equation for t gives

$$t = 1/H_0$$

This indicates that the age of the universe is simply the reciprocal of the Hubble constant. This constant is normally expressed as km/(s × MLY) or km/(s × Mpc). Before we can substitute numbers to calculate the age of the universe, both distances must be expressed in the same units, rather than one in kilometers and the other in millions of light-years. One million light-years is about 9.5×10^{18} kilometers, so a Hubble constant of 15 km/(s × MLY) corresponds to 1.6×10^{-18} km/(s × km). The age of the universe is the reciprocal of this, or

$$t = \frac{1}{\dfrac{1.6 \times 10^{-18}}{s}} = 6.3 \times 10^{17} \text{ s} = 20 \times 10^9 \text{ yr.}$$

This is 20 billion years. On the other hand, using 25 km/(s × MLY) as the value of the Hubble constant gives an age for the universe of 12 billion years. The Hubble constant does not remain constant as time goes on. It is the slope of the line in the graph of recessional velocity versus distance (for example, Figure 17-11) as measured during the current period of time in the life of the universe.

This calculation assumes that galaxies have continued to move apart at the same rate back through their history; that is, that the Hubble constant does not change over time. In fact, we would expect the speeds of galaxies to change. Because of the gravitational force they exert on each other, the recessional speeds of galaxies should be decreasing. This would mean that in the past, they were moving apart faster than they are now and that the age of the universe is therefore less than calculated above. For this reason, and because the Hubble constant is not known precisely, 20 billion years is taken as the *upper limit* to the age of the universe. The universe is this age or younger.

As we noted in Chapter 17, the most distant quasar yet found has a redshift of 6.41. This indicates a look-back time of about 96% of the age of the universe. Although the age of the universe is not known exactly, Doppler shift data are accurate enough that we can be confident that the farthest quasar thus found has a look-back time of about 96% of whatever is the correct value for the age.

The most recent *WMAP* data yield a value of about 71 ± 4 km/(s × Mpc) or about 21.8 ± 1.2 km/(s × MLY) for the Hubble constant. This puts the age of the universe at about 13.7 ± 0.2 billion years.

18-5 The Future: Will Expansion Stop?

What came before the big bang? As we discuss in the preceding Advancing the Model box, this question may not be a scientific one. But it might. To see how scientists might get a hint at what came before the big bang, we look into the *future*.

If galaxies are moving apart, they will obviously be farther apart in the future than they are now. Where does this stop? There *is* an agent to stop it: gravity. The

Figure 18-14
(a) An open universe would continue to grow in size indefinitely. (b) A closed universe would someday stop growing and begin to contract, assuming that there is no cosmological constant. (c) Could it be possible that a closed universe could oscillate, going through consecutive big bangs and big crunches?

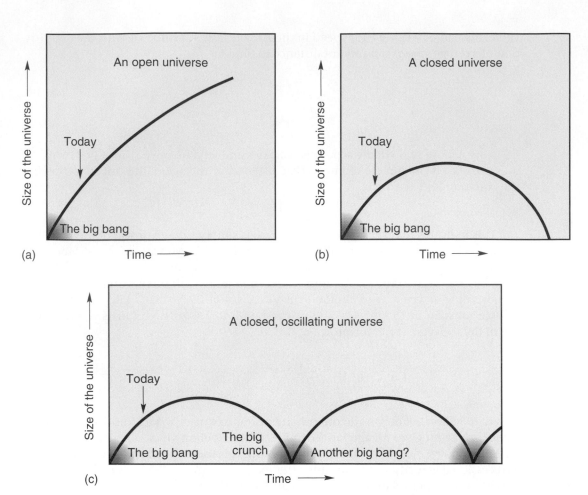

(a)

(b)

(c)

A closed universe will stop its expansion only in the absence of an exotic form of energy. An open universe won't stop at all.

oscillating universe model A big bang theory that holds that the universe goes through repeating cycles of explosion, expansion, and contraction.

force of gravity must be slowing the expansion of the universe. The question is, will gravity slow expansion enough to stop it and bring the galaxies back together?

There are two possible answers: Either the expansion stops or it doesn't. First, suppose the expansion does not stop. This means, simply, that the clusters continue to get farther apart. **Figure 18-14a** illustrates a universe that continues to grow. In this universe, the stars use up the hydrogen fuel that powers them, the glowing stars become fewer, the glow fades, and the universe fizzles out. Not a very attractive idea, but a possibility.

On the other hand, suppose the expansion stops, as shown in Figure 18-14b. Gravity begins to pull the galaxies back toward one another. Intelligent beings (our descendants?) living on some other planet will be able to use the Doppler effect to observe that the galaxies are then getting closer together. The infall will continue until all matter in the universe is condensed into a tremendously dense ball—the "big crunch." What might happen then? Perhaps another big bang, another universe.

If this is the future, perhaps it was the past. Perhaps this is what preceded the big bang: a previous universe containing the same matter and energy as ours. If so, then today's universe may be just one phase of a number of oscillations, each having its own big bang. Such a scenario is called an *oscillating universe* (Figure 18-14c). It is a much more exciting prospect than a fizzling-out universe.

Evidence: Distant Galaxies and High-Redshift Type Ia Supernovae

There are two ways to search for an answer to the question of whether the universe will stop expanding. One way is to see if we can measure how much the expansion is slowing with time as a result of the gravity of the matter in the universe. Of course, we cannot use the few years of our observations and expect to see a change. However, we are seeing distant galaxies as they were in the past, not the present. When we look at the most distant galaxies—at active galaxies and quasars—we are detecting light that left them billions of years ago, and the Doppler shift of that light tells us their speed then, not now. Therefore we should be able to tell how much they are slowing down by comparing their speeds to the speeds of galaxies nearer to us. If we can determine this, we can calculate whether their rate of slowing is enough to bring them to a stop.

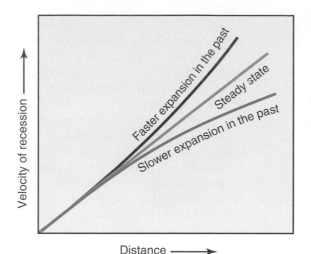

Figure 18-15
If the Hubble data fall along the straight line, a steady state universe would be indicated. If the data fall along the red curve, it would indicate that the universe was expanding faster in the past and has slowed its expansion. If the data fall along the blue curve, it would indicate that the universe was expanding slower in the past and has speeded up its expansion.

In practice, the observation is tough to make. The problem is that in order to look far enough back in time to get a significant change in speed, we are looking at objects so distant that the light is extremely dim. The Doppler effect is easily observable, but we cannot get accurate measurements of the distances to these galaxies. The primary reason for this is that distance is determined from magnitude, and we cannot be certain how stellar populations within galaxies have changed over time. If there were significantly more brighter (or dimmer) stars in galaxies when they were younger, our calculations of distances to the faraway galaxies would be incorrect.

Figure 18-15 is a graph of velocity versus distance for galaxies. One line is drawn to show the relationship as predicted by the steady state theory. This is the straight line, for the theory holds that the expansion is exactly the same now as it was in the past (and will be in the future). On the other hand, if gravity is slowing the expansion, the actual case will be a line that curves, perhaps like the red line. Along this line, the velocities of the most distant galaxies are faster than the steady state theory would predict, for the velocities were greater in the distant past.

The curved line in **Figure 18-16** represents the borderline case (flat universe) between a universe that will expand forever (an open universe) and one that will stop its expansion and start falling inward on itself (a closed universe). Suppose we plot measured values on the graph. If they fall on the straight line, they will indicate that we live in a steady state universe. If they fall in the green area of the graph, it will indicate an open universe, one that will expand forever. On the other hand, if the data indicate a line with a curvature greater than that of the curved line (that is, a line in the red area), the universe is closed.

As we described before, the borderline case between an open and a closed universe is called a flat universe. Inflationary universe theories, which we discuss in the next section, predict that the universe is exactly flat and that the data should fall on the curved line. If the universe is flat, and assuming no cosmo-

The value of the Hubble constant is uncertain (by about 5%) and it changes with time. As a result, astronomers refer to times in the past in terms of redshift, not years. Since the time that light left an object at redshift z, the universe has expanded by a factor $1 + z$. At that time, distances between clusters of galaxies were $1/(1 + z)$ as great as they are now.

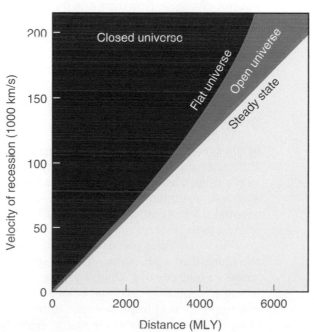

Figure 18-16
If the data fall in the green area, the universe is expanding too fast to stop, and we live in an open universe. If the true case is in the red area, this indicates that the expansion will stop.

logical constant, it will gradually slow its expansion, but will stop only after an infinite amount of time; therefore it will never fall back inward. While a closed universe is represented in three dimensions by a sphere and an open universe by the shape of a saddle, the corresponding representation of a flat universe is simply a plane, like the top of a desk. If the universe is flat, it is one that Newton would understand. In order for the universe to be flat, however, it must meet the exact conditions to be on the borderline of being open and closed.

This borderline is described by the value of the critical density. If the total mass and energy density of the universe is greater than the critical density, then the universe is closed; if it is less than the critical density, then the universe is open. For simplicity, astronomers prefer to work with the ratio of the total mass and energy density of the universe to the critical density. This ratio is called the **density parameter** (Ω_0). If Ω_0 is greater than 1, the universe is closed. If it is less than 1, the universe is open, and if it is equal to 1, the universe is flat.

Contributions to the density parameter come from three sources: matter (Ω_m), radiation (Ω_{rad}), and any possible form of exotic energy (Ω_Λ). However, as we discussed in the Advancing the Model box on "The Early Universe," the contribution of radiation is negligible to that of matter. Also, the observational data we describe below and in Section 18-7 do strongly support the existence of some form of exotic energy. This energy, called **dark energy**, cannot be detected by its gravitational effects and emits no radiation. It provides a strong negative pressure that, just like the cosmological constant, is currently accelerating the expansion of the universe. Therefore, we can write $\Omega_0 = \Omega_m + \Omega_\Lambda$.

Is gravity or dark energy the main factor in determining the future of the universe? In order to answer this question astronomers study the relationship between distances and recessional velocities for distant galaxies and Type Ia supernovae. If the rate of expansion has increased (due to the effects of dark energy) or decreased (because of gravity), we will find deviations from the Hubble law at great distances, as shown in Figure 18-15.

Figure 18-17a shows the results of the HST Key Project to measure the Hubble constant. The overall goal of the project was to measure H_0 based on a Cepheid calibration of a number of independent standard candles (such as Type Ia and Type II supernovae and the Tully-Fisher relation). The data show a linear relationship between recessional velocities and distances. This implies that the rate of expansion has not changed. However, the data go only as far as 400 Mpc, which corresponds to about 1.3 billion light-years into the past (only about 10% of the age of the universe). Thus, the observations suggest that the rate of expansion has remained approximately constant over the past 1.3 billion light-years. The value of the Hubble constant obtained from the data is about 72 ± 8 km/(s × Mpc). If we also take into account the current best estimates of the ages of globular clusters (about 12.5 billion years), these results favor a flat universe ($\Omega_0 = 1$) dominated by dark energy, with $\Omega_m = 0.3$ and $\Omega_\Lambda = 0.7$.

The data shown in Figure 18-17b go as far as 10 billion light-years (about 75% of the age of the universe). This graph shows the observed magnitude versus redshift of a number of nearby and distant Type Ia supernovae. These objects have very similar absolute magnitudes and thus are great standard candles. Their slow march to a sudden explosive end seems to erase most of the differences among the original stars. Recall from Chapter 12 that the apparent magnitude of an object is related to its distance through the distance modulus. Also recall that redshift is related to recessional speed. These relationships are not linear. However, the greater the redshift of a supernova, the greater is its recessional velocity; the greater the apparent magnitude of a supernova, the greater is its distance. Therefore, Figure 18-17b is also a graph of recessional velocity versus distance. The low-redshift points in the diagram ($z < 0.1$) imply a linear Hubble law. This is consistent with Figure 18-17a since the data go only as far as 2.5 billion light-years in the past. However, the

density parameter The ratio (denoted by the Greek letter "omega," Ω_0) of the total mass and energy density of the universe to the critical density.

dark energy An exotic form of energy whose negative pressure currently accelerates the expansion of the universe.

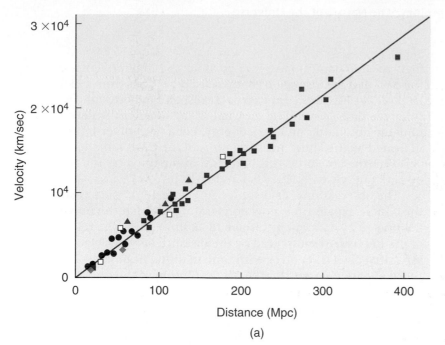

Figure 18-17

(a) Recessional velocities versus distance for different standard candles calibrated by Cepheids. (The candles are shown by different symbols; filled squares correspond to Type Ia supernovae.) These are the final results of the HST Key Project (published in 2001). They imply a Hubble constant of 72 ± 8 km/(s × Mpc). (b) The Hubble diagram for Type Ia supernovae. Supernovae at redshifts > 0.5 appear dimmer (and thus farther away) than their redshifts would suggest if the universe were coasting or slowing down due to gravity. The data suggest that the cosmic expansion has been accelerating since the universe was more than half its current age. The low-redshift data ($z < 0.1$) were published in 1993 (Hamuy et al.) and the high-redshift data in 1998 (by the Supernova Cosmology Project and the High-Z Supernova Search). The red curves correspond to models with $\Omega_\Lambda = 0$ (no dark energy) and range from an empty universe ($\Omega_m = 0$) to a universe of critical density ($\Omega_m = 1$). For a flat universe, the best fit is the blue line. It corresponds to a universe dominated by dark energy, with $\Omega_m = 0.3$ and $\Omega_\Lambda = 0.7$.

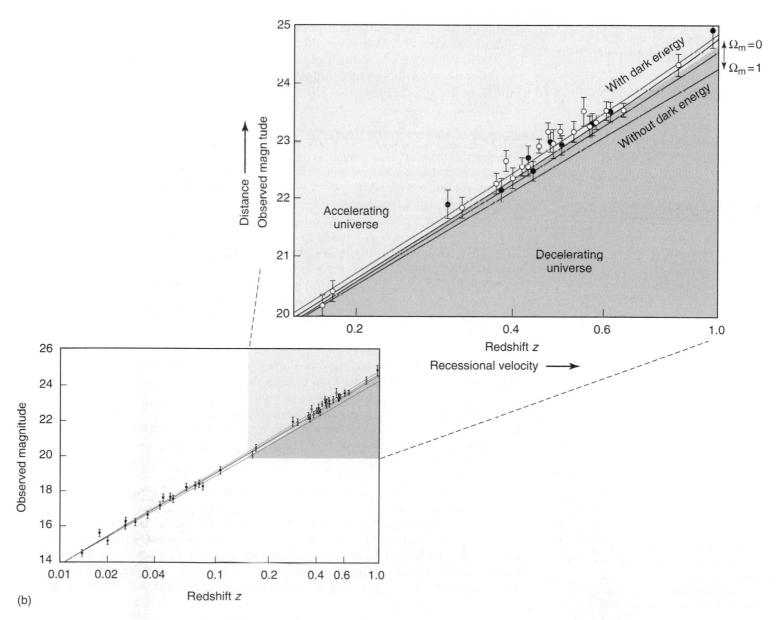

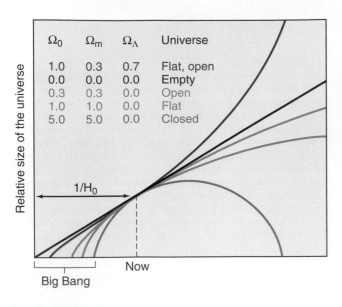

Ω_0	Ω_m	Ω_Λ	Universe
1.0	0.3	0.7	Flat, open
0.0	0.0	0.0	Empty
0.3	0.3	0.0	Open
1.0	1.0	0.0	Flat
5.0	5.0	0.0	Closed

Figure 18-18
The evolution of the universe in different models. There is growing evidence that our universe is following the red curve.

The theory of general relativity (Section 3-10) predicts that both matter and energy cause what we call gravitational force.

Nature has played a trick on astronomers. We thought we were studying the universe; now we know we are studying only the small fraction that is luminous.
Astronomer Vera Rubin

Since particle physicists often use the term "baryons" to refer to massive particles such as protons and neutrons, ordinary "dark" matter is sometimes called "baryonic dark matter."

high-redshift data (up to $z = 1$) go as far as 10 billion light-years in the past. It is in this region that we see deviations from the linear Hubble law. The different curves correspond to different cosmological predictions, depending on what we assume for the density parameters for matter and dark energy. For a flat universe (as indicated by the latest measurements of the CMB radiation), the best fit is the blue line. This implies that the universe is dominated by dark energy, with $\Omega_m = 0.3$ and $\Omega_\Lambda = 0.7$.

The observational data shown in Figure 18-17 clearly show that the cosmic expansion is now accelerating and that the universe is dominated by dark energy. **Figure 18-18** shows how the relative size of the universe (represented by the distance between two widely separated objects) evolves with time in different models. In the absence of dark energy ($\Omega_\Lambda = 0$), the evolution of the universe is determined by its geometry. For $\Omega_0 = 1$, the universe is flat and will expand forever, but only barely so. For $\Omega_0 > 1$, the universe is closed and it will eventually collapse. For $\Omega_0 < 1$, the universe is open and will expand forever. In the presence of (positive) dark energy, the situation is a bit more complicated. Using the most likely values for the density parameters, observations suggest that our universe is flat and will expand forever with an ever-increasing rate.

We have no clues as to the identity of the dark energy. Distinguishing among the different possibilities will require more accurate data, going further back in time than the data in Figure 18-17b. The future is bright for this age of empirical cosmology.

The Density of Matter in the Universe

The second method of seeking an answer to the question of whether the universe will stop expanding is to determine the overall density of matter and energy in the universe. If there is enough mass/energy in each volume of space, there will be enough gravitational force to stop the expansion.

Several observations have led to the conclusion that a great portion of the matter in the universe is nonluminous. First, as we discussed in Chapter 17, galaxies in large clusters are observed (from the Doppler shifts in their spectra) to be moving at high speeds. However, the clusters seem to be gravitationally bound. From the observed speeds and cluster sizes we can estimate the required amount of mass for a cluster to remain bound. This required mass is several times greater than the mass measured by other independent methods. Second, as we discussed in Chapter 16, the rotation curves of spiral galaxies suggest that there is a great amount of nonluminous mass at the outer reaches of the galaxies. Third, observations of elliptical galaxies show dark halos around their luminous regions. These halos are composed of faint shells that extend out a few times farther than most of the starlight. Finally, when astronomers compare the total mass of a system to its total luminosity, the ratio is greater than 1. For stars in the solar neighborhood, the ratio is about 3. For clusters of galaxies, the ratio is about 300. For the universe, assuming it is at critical density, the ratio is about 1000 solar masses for every solar luminosity.

What does this nonluminous matter consist of? The search for the answer to this question is one of the most active areas of research in astronomy today. Several possibilities have been proposed, and the actual situation may be a combination of several kinds of matter.

1. Ordinary "dark" matter In previous chapters we discussed objects such as white dwarfs, which no longer generate their own energy and are no longer easily observable, and brown dwarfs, which have never generated enough energy to become easily observable. These are common examples of ordinary "dark" matter, as are planets, meteors and comets, and the vast dust clouds of interstellar space. All this matter consists of particles with which we are familiar, primarily atoms containing protons, neutrons, and electrons.

2. Hot dark matter Recent experiments seem to indicate that neutrinos might have a small amount of mass. (We discussed neutrinos in Chapter 11.) There are so many neutrinos zipping around the universe that even a tiny mass could contribute to the mass of the universe. However, observations suggest that this contribution is not significant (otherwise the CMB radiation could not look as it does). Astronomers refer to this mass as "hot" dark matter because the particles are moving at high speeds, comparable to the speed of light. Other particles have also been proposed for the category of hot dark matter, including particles that have been hypothesized in several theories of particle physics, but which have not been detected.

3. Cold dark matter This is an exotic form of matter that is not made up of the protons, neutrons, and electrons of ordinary matter. Particles that might be moving through space at relatively slow speeds make up "cold" dark matter. Such particles have been given various names, such as photinos, axions, or neutralinos, but have not been detected. We surmise their existence because estimates of ordinary "dark" matter and hot dark matter do not make up the amounts of mass we can detect by gravitational effects but cannot see. In addition, various theories of particle physics and of galaxy formation in the early universe indicate that this kind of matter exists.

4. Black holes Astronomers speculate that a great number of black holes of all sizes could contribute to the dark matter.

Several research teams are presently investigating the nature and amount of dark matter. One promising approach is to search the galactic halo for stars that grow suddenly bright due to gravitational lensing as dark objects pass between them and the Earth. Such "microlensing" events have been observed, but the nature of the dark objects is not known for all events. Other experiments assume that the particles of dark matter have a small probability of interacting with the ordinary matter of Earth, much like neutrinos do, and can be detected. However, the sensitivity required of the detectors is presently still beyond our capability. Advances over the next few years might enable us to begin setting limits as to the nature and amount of dark matter that interacts with the ordinary matter of Earth.

If astronomers are incorrect in their prediction that great amounts of dark matter exist throughout the universe, then the universe is open and will continue to expand forever. If dark matter does exist in the amount predicted by gravitational studies, however, then the universe is just barely dense enough to lie on the boundary between open and closed. The density that the universe would have to have to be perfectly flat, the *critical density*, is equivalent to about six hydrogen atoms per cubic meter of space. Today's best calculations indicate that the universe's density is somewhere between 0.1 of the critical density and 2 times that density. Therefore the universe appears to be flat or almost flat.

Some astronomers refer to "dark" matter in the galactic halo (which might be brown dwarfs or stellar-mass black holes) as "massive compact halo objects," or MACHOs. Their name for the more exotic particles being looked for is "weakly interacting massive particles," or WIMPs. When you're looking for objects whose very existence has never been proven, it helps to have a sense of humor.

18-6 The Inflationary Universe

Scientists do not like coincidences, and the standard big bang model is unable to explain two cosmological observations that appear to be coincidences. Attempts to solve these two problems, as well as others, led astronomers to propose a modification of the standard model.

The Flatness Problem

The universe seems to be either flat or very nearly flat. This would be an extreme coincidence because a flat universe is a special case—a universe that has a density that puts it exactly on the border between open and closed. Of all the possible densities that the universe might have, why would it have this particular one? The universe could be a little more dense or billions of times more dense, and in either case it

Figure 18-19
The fleas on the small balloon (a) can tell that their universe is curved, but if the balloon is large enough, it will appear flat to the fleas (b).

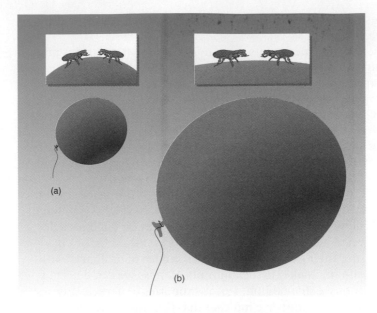

flatness problem The inability of the standard big bang model to account for the apparent flatness of the universe.

inflationary universe model A modification of the big bang model that holds that the early universe experienced a brief period of extremely fast expansion.

 If the diameter of a helium nucleus expanded by a factor of 10^{50}, the nucleus would be 40 billion billion light-years across!

would be closed. (If it were much more dense than it is, it would be short-lived, so that planets—and life—would not have had time to form.) On the other hand, it could be slightly less dense or billions of times less dense, in which case it would be open. (If it were much less dense than it is, gravity would not have gathered the diffuse matter into stars and planets, so again life would not exist.) The **flatness problem** asks, "Of all the possible conditions of the universe, why is it so nearly flat?"

The flatness problem is even worse than just described, however, because if the universe's density had not been exactly equal to the critical density in the beginning, the curvature of the universe would have increased as time passed, and the curvature would be even greater now. Consider the ratio of the actual density of the universe to the critical density. If, immediately after the big bang, that ratio was equal to 1 (that is, the universe's density was equal to the critical density), then the ratio would still be equal to 1 today. If, however, the ratio at the big bang was less than 1, calculations show that it would be *much* less than 1 today. And if the ratio just after the big bang had a value greater than 1, the ratio would have become even greater today. For today's ratio to be somewhere between 0.1 and 2, it must have been extremely close to 1 at the beginning. Could this be coincidence?

The **inflationary universe model** has an explanation for the flatness of the universe. The model provides details of the first fractions of a second after the big bang, and according to its predictions, during the short time interval between 10^{-35} second and 10^{-32} second after the big bang, the universe suddenly expanded at an extremely fast rate. In fact, it grew by a factor of 10^{50} during that time. To understand how this solves the flatness problem, we must return to the analogy of fleas on a balloon. If the fleas are on a small balloon, the curvature will be obvious to them. If the balloon is very large, however, the fleas will not be able to detect its curvature (**Figure 18-19**). The inflationary universe model holds that an "inflation" of the universe took place so quickly that any portion of it that had a great curvature before the inflation became flat as a result of the inflation.

Like any analogy, the balloon story does not work perfectly, for even a large balloon has some curvature. The inflationary model predicts that curvature was completely removed during the quick expansion. Therefore the universe's flatness is not a coincidence, but a necessary consequence of the brief period of inflation very early in history.

The Horizon Problem

The cosmic microwave background radiation is remarkably uniform, no matter in what direction one looks. *COBE* data reveal that, after correcting for the motion of the Earth, the radiation varies in temperature by less than one part in 100,000. This would indicate that the temperature of the universe was extremely uniform at the time that the CMB radiation was emitted. This does not seem possible in the standard model of the big bang because in that model the various portions of the universe could not have been in "communication" with one another when the radiation was emitted. To see why this is so, return to the balloon analogy one more time. First, imagine a balloon that is not expanding, and suppose that thermal energy is added to one portion of the balloon. That energy will spread around the balloon, but this will not happen instantaneously. The amount of time required for the energy to distribute all over the balloon depends on how fast the energy travels. Nevertheless, the energy will eventually distribute uniformly around a nonexpanding balloon.

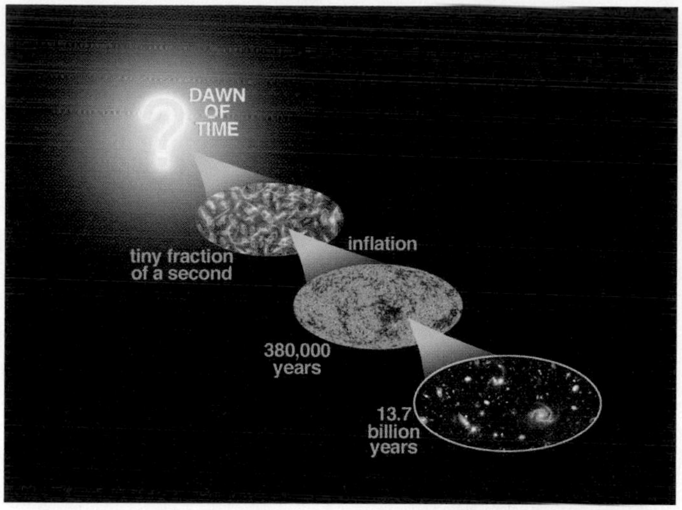

Now suppose that the balloon is expanding when energy is added to one portion of it. If the balloon is expanding rapidly enough, the energy will not be able to reach distant portions of the balloon because those portions are moving away too fast. There is a "horizon" beyond which the change in energy cannot be transmitted. In this rapid-expansion case, any portion of the balloon that experiences a change will not communicate this change to the rest of the balloon, and therefore the properties of the balloon will not be uniform.

In the case of the real universe, a change in one portion can be communicated to another portion with a maximum speed equal to the speed of light. The standard big bang model predicts that at the time that the CMB radiation was emitted, the universe was expanding at such a rate that the "horizon" distance was much less than the distance across the universe. Therefore any little change in one part of the universe could not be communicated to faraway parts. This means that there would be no reason for the entire universe to be at the same temperature and astronomers should not observe the same temperature in the background radiation in various directions. Hence, the *horizon problem*.

According to the inflationary universe model, before inflation occurred, the universe was smaller than the horizon distance. Therefore, any change in one portion of the universe was transmitted everywhere, and it was possible for the entire universe to be at a uniform temperature. Then inflation occurred. Two parts of the universe that were originally within each other's horizons were outside those horizons after inflation. They were still at the same temperature, however, as a result of their condition before inflation. This is why all parts of today's observable universe show a uniform background radiation.

Figure 18-20 is a NASA drawing that summarizes the inflationary big bang theory. At upper left is the universe at the big bang and then just after the big bang. At the latter time, the universe contained extremely small irregularities. The universe then expanded rapidly during the inflationary period. The central image is from the *WMAP* and its production is explained in the next section. It pictures the universe 380,000 years after the big bang. By that time, the universe had cooled down enough to become transparent to radiation. Galaxies and stars began to form. The bottom right part of the image shows today's universe, with each dot representing a cluster of galaxies.

Figure 18-20
A summary of the big bang theory. The central image—the *WMAP* sky map—is a map of the CMB, the first light to break free in the infant universe. This light emerged 380,000 years after the big bang. Its appearance is caused by minute temperature variations in the universe at the time the CMB was released. The observed patterns are the seeds for the development of the structures of galaxies we now see billions of years after the big bang.

horizon problem The inability of the standard big bang model to account for the directional uniformity of the background radiation.

The inflationary universe model was proposed in the early 1980s by Alan H. Guth of MIT, and has been further developed by Guth and other cosmologists, including Stephen Hawking. Along with dark matter, it is now the best explanation we have for the development of the universe.

18-7 The Grand Scale Structure of the Universe

The success of the inflationary theory at first caused a problem in explaining the formation of galaxies, for in a completely uniform universe, there would be no irregularities around which matter could group into galaxies and clusters of galaxies. Not

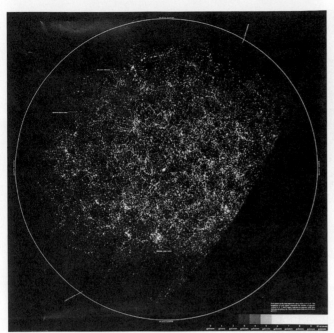

Figure 18-21
Each dot on the diagram is a cluster of galaxies. The figure shows that clusters are clumped together in an irregular way.

The transition from an opaque universe to a transparent one was not instantaneous. It lasted for a time interval that corresponds to a change in redshift of about 100.

only did galaxies form, however, but the discovery of extremely old quasars indicates that matter clumped together relatively soon after the big bang.

Discoveries made during the 1980s added to the problem. As we accumulated data on the distribution of clusters of galaxies (**Figure 18-21**), we were able to construct three-dimensional drawings to show their locations. The drawings showed that clusters of galaxies are not distributed uniformly at all, but are arranged as if to form bubbles, with a void between the walls of the bubbles. To visualize this, think of soap bubbles in a sink. Galaxies make up the surfaces of the bubbles, and where the bubbles join, we find the largest clusters of galaxies.

How is the problem of the grand scale structure of the universe resolved? The inflationary model includes a random distribution of density irregularities (perturbations) during the earliest moments of the universe. The amplitudes of these perturbations were smoothed out by the sudden and fast expansion of the universe during inflation. This model explains why the original perturbations would have left such small-scale anisotropies in the CMB radiation.

To understand the observed features in the CMB radiation, consider the universe during its infancy. For the first 380,000 years, the density and temperature of the universe were so great that it was opaque to electromagnetic radiation. The hydrogen and helium nuclei that were formed in the first few minutes of the universe's life, electrons, and radiation were all tightly coupled together through scattering and electromagnetic interactions. They behaved as a single "baryon–photon fluid" with a single temperature. In this fluid, the baryons provided the inertia, the photons provided the pressure, and the density perturbations that remained after the big bang provided a mechanism for oscillations. The oscillations were caused by a competition between two factors: gravity and radiation pressure. As a region of excess mass would begin to collapse under its own gravity, becoming even denser, radiation pressure would become sufficient to stop the collapse and cause a rebound. As the universe aged, the horizon grew and oscillations of greater wavelength were included inside the horizon. These oscillations in the baryon–photon fluid sent out pressure waves that propagated at the local speed of sound.

About 380,000 years after the big bang, the universe became cool enough (about 3000 K) that protons were able to capture electrons and form neutral hydrogen. It is at this stage, at redshift of about $z = 1000$, that matter and radiation decoupled from each other. It is also at this stage that the radiation last interacted strongly with matter and the universe became transparent. This radiation is what we see today as the CMB. After decoupling, the interaction between the radiation and the now neutral matter was not significant. As a result, the temperature of the radiation evolved differently from that of matter. The photons were free to propagate in the universe carrying on them the signature of the density perturbations just before the decoupling. Meanwhile, with the release of their pressure support, the baryons started collecting under the influence of the gravity of the dark matter. Recall that dark matter does not interact with radiation and therefore any dark matter formed early on would have been free to gravitationally collapse much earlier. It is these regions, where baryons collect under the influence of dark matter, that will grow into the structures we see today.

If we imagine a surface around us at about $z = 1000$, this surface represents the region where radiation last interacted strongly with matter. Since then, the wavelengths of these photons have stretched so much due to the expansion of the universe that we see them as the CMB photons. These photons reach us from all directions and can tell us a great deal about the early universe. Photons that started their journey from very dense regions on this surface lost more energy escaping gravity than photons from less dense regions. Since energy is related to temperature, this implies that a certain density perturbation on this surface gives rise to a certain temperature perturbation. As a result, the CMB radiation carries with it information about the early universe in the form of temperature anisotropies.

In 1992, the *COBE* team reported finding tiny irregularities (variations) in the CMB radiation. The irregularities amounted to a deviation of only 0.00003 K! In order to detect these tiny irregularities, thought to be due to tiny differences in the temperature of the early universe, the *COBE* team first had to subtract irregularities caused by known sources of radiation. These include the effect of the Sun's radiation, the effect caused by known radiators within the Milky Way, and a slight Doppler shift in the radiation that is caused by our motion through space. **Figure 18-22** represents *COBE* data showing the near uniformity of the

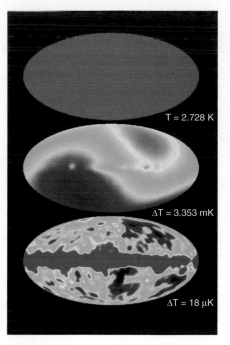

Figure 18-22
COBE data showing the near uniformity of the CMB brightness (top image), then on a scale intended to enhance the anisotropy due to the motion of our solar system relative to distant matter in the universe (middle image), and finally following subtraction of this anisotropy (bottom image). ΔT corresponds to temperature variations measured in mK (10^{-3} K) or μK (10^{-6} K). As we move from the upper right to the lower left of the middle image, the variation is between relatively hot and relatively cold areas. The signal attributed to this variation corresponds to only one thousandth of the brightness of the sky.

CMB brightness (top image); then the data on a scale intended to enhance the anisotropy due to the motion of our solar system relative to distant matter in the universe (middle image); and finally following subtraction of this anisotropy (bottom image). Subtracting emission from our Galaxy, we get **Figure 18-23**, which represents the *COBE* map of the early universe. The blue and red spots represent temperature variations and thus correspond to regions of greater or lesser density in the early universe; they give us a sense as to how matter and energy were distributed before matter gave rise to the structures we see today. These tiny irregularities are the roots of the structures we see in today's universe. Without the anisotropy, there would be no galaxies, no stars, no Earth, no us.

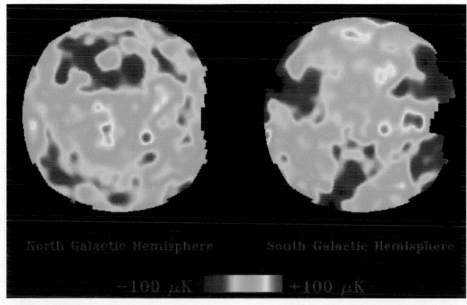

Figure 18-23
The *COBE* map of the early universe. This false-color image shows tiny variations in the intensity of the CMB radiation. The features traced in this map stretch across the visible universe. The smallest feature in this map is larger than the largest features seen by optical telescopes.

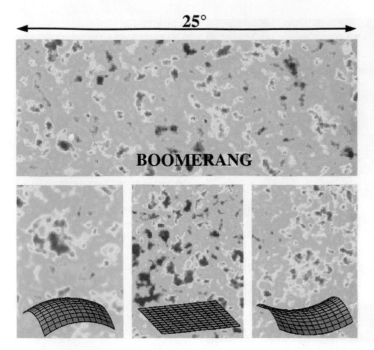

Figure 18-24
The *BOOMERANG* images allow us to determine the geometry of space. The image (top) is compared to images generated by computer simulations for a closed (bottom left), flat (bottom center) and open (bottom right) universe. The observations suggest that our universe is very nearly flat.

BOOMERANG stands for *Balloon Observations of Millimetric Extragalactic Radiation and Geophysics*. The project involved cosmologists from Canada, Italy, the UK and the US.

The *BOOMERANG* telescope flew for 10.5 days (Dec. 1998 to Jan. 1999) around the Antarctic continent on a 800,000 m³ balloon at an altitude of 36 km, above 99% of our atmosphere. It was launched again in early January 2003.

The CMB radiation gives us a view of the early universe, and we can use it as a tool to probe the conditions prevalent at the time. When we look at the CMB radiation, we are looking back to a time before the first stars and galaxies formed, when the universe was only about 380,000 years old. The spectrum of the CMB radiation and its almost perfect uniformity in all directions can only have one source: a hot, uniform, dense early universe. Five years after *COBE*, which had an angular resolution of about 7°, the *BOOMERANG* project gave us a map of the early universe that was about 40 times more detailed.

The *BOOMERANG* images, like the one shown in the chapter-opening figure, bring the CMB into sharp focus. The observed variations in the temperature of the CMB radiation are typically only 100 millionths of a degree Celsius (0.0001°C). This sharp snapshot of the early universe offers us a powerful tool in distinguishing among competing theoretical models on the evolution of the universe. Recall that for the first 380,000 years after the big bang, the universe was too hot for atoms to form and photons were trapped in a dense soup of charged particles. The density was so high that pressure waves reverberated through the mass like sound waves in water. The energetic photons were freed when the universe cooled enough for protons and electrons to form neutral hydrogen atoms and the universe became transparent to radiation. This happened when the universe was about 380,000 years old, when it had cooled down to about 3000 K. As the universe expanded, these photons cooled down to become the CMB radiation, while carrying with them the pattern of the "sound waves" at the moment the universe became transparent. We now use these waves as a cosmic ultrasound to image the conditions in the early universe. The tiny temperature variations we observe correspond to tiny density fluctuations in the primordial soup of particles. These density fluctuations are thought to grow by gravitational attraction into the structures we see today. The visible patterns in the *BOOMERANG* images confirm predictions of the patterns that would result from sound waves racing through the early universe.

The sizes of the hot and cold spots in the *BOOMERANG* data reflect directly on the geometry of space (its curvature) and thus its fate. **Figure 18-24** includes a *BOOMERANG* image (top) and three computer-simulated images, differing only in whether the universe is closed, flat, or open (bottom). Computer simulations suggest that in a universe whose geometry is flat, the images of the CMB radiation should be dominated by hot and cold spots of about 1° in size (bottom center). In a universe whose geometry is not flat, the curvature of space will distort the images. Specifically, in a universe that is closed, the images will be magnified by the curvature and structures will appear larger than 1° on the sky (bottom left). In a universe that is open, the images will appear smaller (bottom right). **Figure 18-25** shows how the geometry of space affects the size of the spots in the CMB radiation. In a flat universe, lines that are initially parallel remain parallel when extended. In a closed universe, parallel lines converge, while in an open universe, parallel lines diverge. The *BOOMERANG* images indicate that our universe is very nearly flat.

The best explanation for a flat universe is provided by the inflationary theory. Recall that according to this theory, moments after the big bang our universe went through a period of extreme inflation. As a result, whatever strong curvature might have existed in the early universe was removed by this immense stretching of space. For a simple analogy, consider a large balloon. If you balance yourself on the balloon, you will feel the curvature beneath your feet. However, if we expand the balloon to the size of the Earth, what you feel is nothing but flat space. If we now blow up the

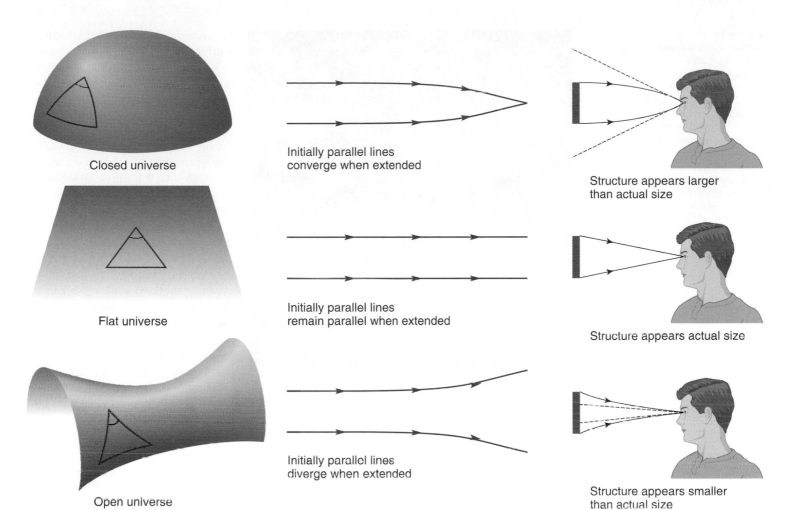

Figure 18-25
The geometry of space affects the angular size of an object because the curvature of space determines the path followed by a light beam. If the universe is closed, a structure in the CMB will appear larger than actual size. If the universe is flat, the structure will appear actual size. If the universe is open, the structure will appear smaller than actual size.

balloon to a cosmic scale, you can imagine how inflation could vastly flatten the observable universe.

The *BOOMERANG* results were not only confirmed, but also extended to finer resolution by *MAXIMA*. This was another balloon-borne project, which was more sensitive than *BOOMERANG* but looked at a smaller area of the sky. The *Wilkinson Microwave Anisotropy Probe* (*WMAP*), launched in June 2001, has made an accurate map of the relative CMB temperature over the full sky with angular resolution of about 0.3° and sensitivity of about 20 μK. The probe measures temperature differences between points about 140° apart on the sky using five different frequency bands. The *WMAP* image of the *CMB* is much sharper than the image obtained by *COBE*, as shown in **Figure 18-26**.

What do the tiny temperature fluctuations tell us about the universe? Recall that the CMB radiation is the first light that broke free in the infant universe. The observed temperature fluctuations imprinted on this light reflect the conditions set in motion just after the big bang. That is, these fluctuations are like a fingerprint that we could use to compare with fingerprints predicted by various cosmic theories. The results from the projects observing the CMB all point to the same match. Our universe has about 4% normal matter (which makes us and the stars), about 23% dark matter

MAXIMA stands for *M*illimeter *A*nisotropy *E*xperiment *Im*aging *A*rray. The project includes scientists from five countries. The balloon last flew in 1999. The project's sensitivity was from 4° to 10′.

WMAP observes the sky from an orbit about a semi-stable point on the line connecting the Earth and Sun (1.5 million km from Earth in the direction opposite the Sun). As the probe follows the Earth around the Sun, it observes the full sky every 6 months. *WMAP* is a partnership between Princeton and NASA and will continue observations until at least 2006.

Figure 18-26
WMAP's image of the infant universe brings the *COBE* image into sharp focus. It shows temperature fluctuations, with red spots being the hottest and blue spots being coldest. Foreground radiation from our Galaxy, which *WMAP* distinguishes by observing at five microwave frequency bands, has been subtracted from the equatorial band. These fluctuations correspond to the seeds that grew to become galaxies. The new image provides strict limits to a number of cosmological parameters.

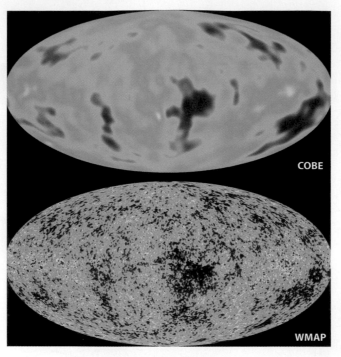

COBE

WMAP

An imprint of dark energy on the CMB radiation was found in 2003 by correlating millions of galaxies in the Sloan Digital Sky Survey and CMB temperature maps from *WMAP*. This provides physical evidence for the existence of dark energy.

(a mysterious, nonluminous form of matter that is indirectly detected by its gravity), and the remaining 73% is dark energy (an exotic form of energy that is responsible for accelerating the expansion of the universe). Also, the *WMAP* results suggest that the big bang occurred 13.7 ± 0.2 billion years ago, and that the CMB is a snapshot of the universe when it was only about 379 ± 8 thousand years old. The Hubble constant is found to be about 71 ± 4 km/(s $\times$ Mpc), and a number of other cosmological parameters are given with high precision.

How do we compare the various CMB observations with theory? **Figure 18-27** shows such a comparison. It is a graph of the strength of the CMB temperature fluctuations (temperature-fluctuation power) versus their angular size. The data points (including error bars) correspond to observations from three different projects. The curve shows the best theoretical fit to these data plus a number of other observations. The angular size of a temperature fluctuation on the sky is related to the quantity "ℓ" on the horizontal axis approximately as $180°/\ell$. For example, since *WMAP*'s angular resolution is about $0.3°$, the probe cannot follow the power spectrum beyond about $\ell = 180°/0.3° = 600$. In Figure 18-27, the temperature-fluctuation power for a given value of ℓ measures the average temperature difference (squared) between points on the sky separated by an angle of about $180°/\ell$.

The locations, heights, and shapes of the observed peaks and troughs in the angular power spectrum serve to constrain the cosmological parameters. Recall that be-

Figure 18-27
The angular power spectrum of the temperature fluctuations in the *WMAP* full-sky map. The curve shows the best theoretical fit to the observational data. The shaded area corresponds to irreducible cosmic variance.

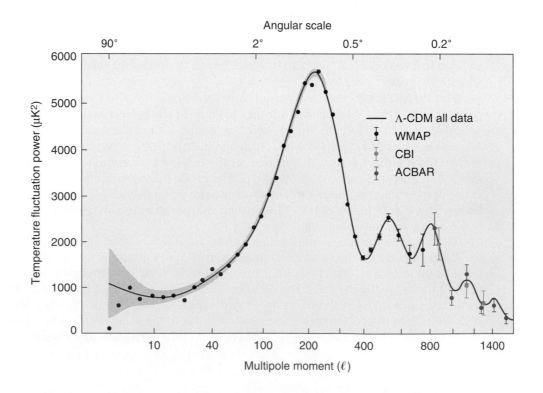

fore decoupling, normal (baryonic) matter and photons behaved as a single fluid. In this fluid, density perturbations provided a mechanism for oscillations. After decoupling, normal matter and photons went their own way. But the CMB radiation that we observe today carries with it information about this period of the early universe when the radiation last interacted strongly with matter and the universe became transparent. The temperature fluctuations we observe today correspond to the density and velocity components of the oscillations during this period (**Figure 18-28**). An oscillation at its maximum density during the decoupling will appear brighter and hotter than average, while an oscillation at its minimum density will appear colder.

With this in mind, it is now possible to understand the peaks in Figure 18-27. The first peak corresponds to a perturbation that crossed the horizon and reached its maximum density just when the universe became transparent (stage "b" in Figure 18-28). The remaining odd peaks correspond to perturbations that underwent an additional integral number of oscillations by this time. The even peaks correspond to perturbations at their minimum density (for example, stage "d" in Figure 18-28); these perturbations give rise to power peaks of smaller heights because the rebound must fight against gravity. The troughs (for example, stage "c" in Figure 18-28) correspond to positions of maximum velocity. The amplitude of the fluctuations is reduced as ℓ increases (on the smallest scales) because the transition from an opaque universe to a transparent one was not instantaneous. The shape of the spectrum in Figure 18-27 depends then on the details of the oscillations, and it is a direct result of the conditions that existed at the time of decoupling. Therefore, this spectrum serves to constrain the cosmological parameters. The curve in Figure 18-27 depends on the values of seven different cosmological parameters that were adjusted to fit the data.

The location of the first peak in Figure 18-27 is related to the geometry of the universe. That is so because the physical scale associated with the first peak is the horizon size at the time of decoupling. At this time, the universe became cool enough (about 3000 K) that protons were able to capture electrons and form neutral hydrogen. This stage clearly occurs at about $z = 1000$, and the angle subtended by the physical scale at this redshift depends on the geometry of the universe. The first peak occurs at $\ell \approx 200$, which corresponds to an angular size for the temperature fluctuations of about $180°/\ell \approx 1°$. This is consistent with a flat universe, as predicted by the inflationary universe theory.

Astronomers have hailed the *WMAP* results as the beginning of precision cosmology. Observing temperature fluctuations at the 10^{-6} K scale is not easy. Imagine that a band is playing random tunes at a location far away from us. Between the band and us there are many sources that generate noise. Our job is to understand the nature of these sources, remove their contributions to the overall sound we receive, and inter-

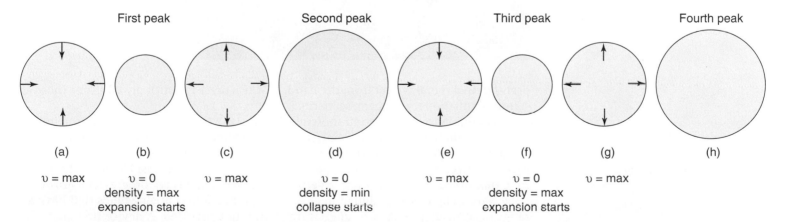

Figure 18-28
This illustration shows the relationship between the peaks and troughs in the angular power spectrum and the density and velocity components of the oscillations during the decoupling era.

Figure 18-29
This illustration demonstrates how *WMAP* sees polarized light. The polarization of the CMB radiation was produced by the scattering of the cosmic light when it last strongly interacted with matter.

Polarization: How It Works

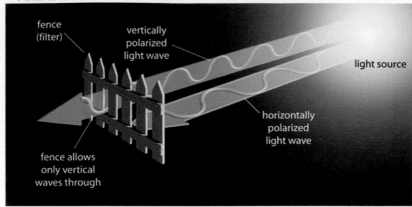

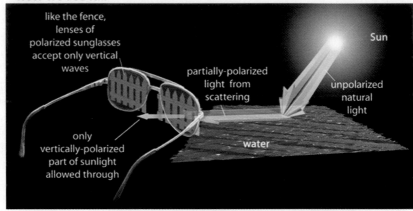

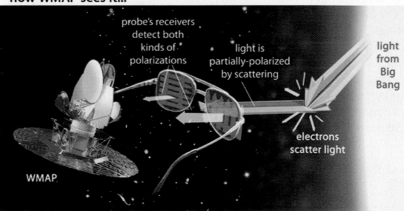

linear polarization The situation when the electric field in an electromagnetic wave oscillates in a fixed plane.

pret the final signal we get from the band. *WMAP*'s precision strongly confirms the standard inflationary big bang scenario.

In addition to the spectrum shown in Figure 18-27, *WMAP* also mapped the intensity and direction of ***linear polarization*** everywhere on the CMB sky. Recall our discussion of light as an electromagnetic wave in Chapter 4. We say that light is linearly polarized when its electric field oscillates in a fixed plane, just like a length of rope that moves up and down in a vertical plane while the wave progresses horizontally. Natural light is a combination of random and rapidly changing fields that have a range of wavelengths. As a result, natural light is not polarized. A device that takes natural (unpolarized) light and transforms it into polarized light is a polarizer, as do certain sunglasses, for example. However, there are a number of natural processes that produce polarized light, one of which is scattering. The illustration in **Figure 18-29**

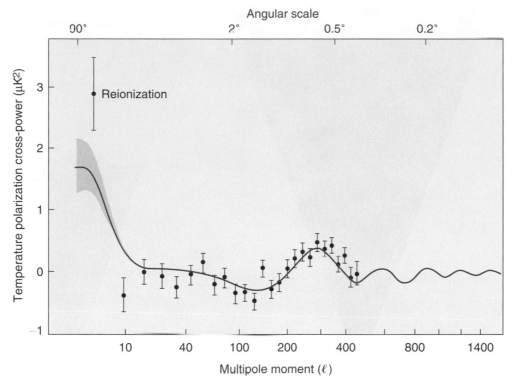

Angular scale

Figure 18-30

The cross-power power spectrum that correlates the temperature fluctuations and the polarization in the CMB radiation between points in the sky separated by an angle of about $180°/\ell$. The curve is a *prediction* based on the observed temperature anisotropies in the CMB radiation. The rise in the curve at low ℓ values indicates that the first stars in the universe formed very quickly.

Figure 18-31

A sequence of events in the life of the universe. The first frame shows temperature fluctuations (as color differences) in the CMB radiation observed by *WMAP*. These temperature fluctuations correspond to slight density perturbations. The second frame shows matter condensing to regions of higher and higher density as a result of gravitational attraction. The third frame shows the period of formation of the first stars, about 200 million years after the big bang. The fourth frame shows a greater number of stars, and galaxies forming along the filaments first seen in frame two. The last frame shows the modern era, billions upon billions of stars and galaxies that grew from the seeds planted in the early universe.

makes the use of polarizing sunglasses to show how *WMAP* "sees" polarized light. The results of *WMAP*'s polarization measurements are shown in **Figure 18-30**. It is a cross-power spectrum, analogous to that seen in Figure 18-27, which correlates the temperature fluctuations and the polarization in the CMB radiation between points in the sky separated by an angle of about $180°/\ell$. Note that the curve in Figure 18-30 is *not* a fit to the data; it is actually a *prediction* based on the observed temperature anisotropies in the CMB radiation. The overall shape of the cross-power spectrum contains a wealth of information about the conditions in the early universe. The rise in this curve at low ℓ values (large angles, about 90°) indicates that the first stars in the universe formed very quickly, about 200 million years after the era of decoupling.

Even though *WMAP* does not see the light of the first stars directly, the observed polarized light serves as the signature of the ultraviolet light released by the first stars. This UV light began the reionization of the neutral hydrogen that formed after decoupling. **Figure 18-31** shows a sequence of events in the history of the universe, from the initial density perturbations to the formations of billions upon billions of stars and galaxies. **Figure 18-32** is a more detailed illustration of important events in the early stages of the universe.

The *WMAP* results, along with results from other CMB projects and a range of different and independent observations (such as the galaxy and supernovae surveys), support the inflationary scenario. **Figure 18-33** shows how the combination of data from different projects allows us to limit the range of values for the average density of matter in space Ω_m (on the horizontal axis) and the density of dark energy in space Ω_Λ (on the vertical axis). The results from CMB projects support cosmological models whose parameters lie in the blue region. This curve is concentrated near the diagonal red line, which corresponds to a flat universe. The results from the supernovae studies support cosmological models whose parameters lie in the yellow region. The

Figure 18-32

Important events in the early stages of the universe, including the formation of deuterium and helium 100 seconds after the big bang. During the era of decoupling (about 380,000 years after the big bang), radiation last interacted strongly with matter and the universe became transparent. This "surface of last scatter" for the CMB radiation is analogous to the light coming through the clouds to our eyes on a cloudy day.

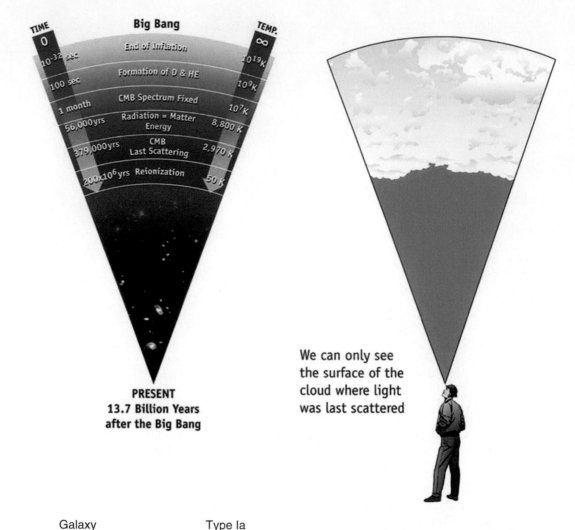

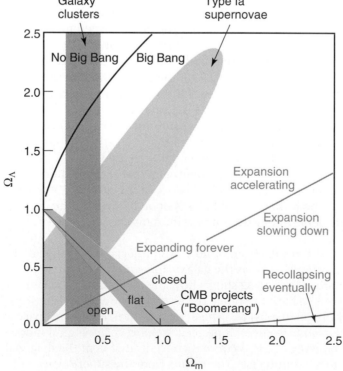

We can only see the surface of the cloud where light was last scattered

Figure 18-33

A comparison between results from CMB projects, surveys of galaxy clusters, and studies of Type Ia supernovae suggests that our universe is flat, started with a big bang, and will continue expanding. The horizontal axis shows Ω_m, the average density of matter in space compared to the critical density. Such matter (normal and dark) slows down the expansion of the universe. The vertical axis shows Ω_Λ, the density of the dark energy in the universe compared to the critical density. Such dark energy causes the expansion of the universe to accelerate. The latest results suggest that $\Omega_m = 0.27 \pm 0.04$, $\Omega_\Lambda = 0.73 \pm 0.04$, and $\Omega_0 = 1$ with an uncertainty of 2%.

results from surveys of galaxy clusters support cosmological models whose parameters lie in the orange region. If all measurements are correct, then only models whose parameters lie in the green overlap area are allowed. This region indicates that our universe has a flat geometry, started with a big bang, and will continue expanding.

Our efforts to look back in time and understand the conditions of the early universe are continuing through a number of space missions and other projects, three of which are shown in **Figure 18-34**. We are slowly putting together a photo album of the universe's life. The CMB projects are giving us a snapshot of the universe during its infancy. Observations with space and ground telescopes are providing us with pictures of its early adolescent years. These observations strongly support the idea that star formation started early on in the universe's life at an incredibly fast rate. Even though stars continue to be formed today, the rate of star formation has been declining since the early years. Recent (2001) infrared observations using ESO's VLT reveal that many rather large galaxies (some with spiral structure similar to that seen in nearby galaxies) existed at a time when the universe was extremely young, less than 1 billion years old. These galaxies appear to have already formed most of their stars and they probably account for at least half the total luminous mass of the universe at that time. Also, these galaxies seem to cluster close to each other. This supports the idea that the universe has a spongy structure (**Figure 18-35**), with galaxies forming along filaments.

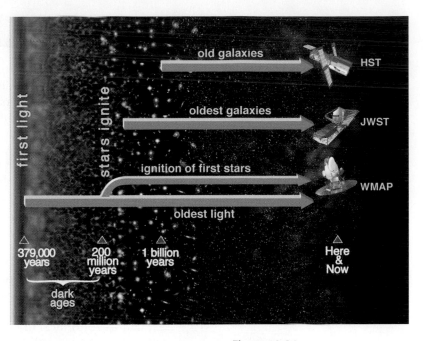

Figure 18-34
Looking back in time with the *HST*, the *WMAP*, and the *JWST* (to be launched in 2010, the year NASA will end *HST*'s mission).

All the recent observations confirm the generally accepted cosmological model that astronomers constructed over the past decade. We live in a universe that is dominated by dark matter and dark energy, about which we know very little; what we see of the universe corresponds to a small fraction of it. However bizarre such a model might seem, it is spectacularly confirmed by the observations, even though it is problematic. For example, since it is probable that ordinary matter and dark matter originate through different mechanisms, why is the abundance of ordinary matter similar (within an order of magnitude) to that of dark matter? Also, since the total abundance of matter is rapidly changing compared to that of dark energy as the universe expands, why are the two so similar (within an order of magnitude)? Finally, theoretical estimates for the most likely candidate for dark energy are orders of magnitude different than what the observations suggest; it is also possible that dark energy, which is currently positive and thus accelerates the expansion of the universe, will gradually become negative, causing the universe to become unstable and then collapse. Even though there are still uncertainties, do not think that the model is pure speculation, for each of the various modifications of the model makes predictions about today's universe, and the predictions can be tested against observations.

For example, a simple testable prediction of inflationary theories is the relative fractions of the chemical elements that would be produced and that should be found today. Observations support these predictions. Cosmologists are in fairly general agreement with the basic idea of the inflationary theory. The details, however, are tentative. As new observations bring us a flood of new data, we will undoubtedly have to revise the models. Some aspects of them will be discarded, but some will live on to be tested again and again until they join the body of firmly established science.

Science progresses best when observations force us to alter our preconceptions.
Astronomer Vera Rubin

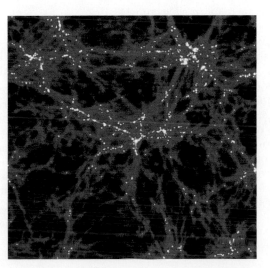

Figure 18-35
A computer model of the universe at an age of about 2 billion years. Gravity causes matter to arrange itself in thin filaments. Yellow, red, and blue correspond to high, medium, and low gas density, respectively. Stars will form in the yellow regions and slowly stream along the filaments. When the stars meet at the intersections of the filaments, they gradually form the galaxies we know today.

Conclusion

Future data from *WMAP* should help astronomers choose among competing inflationary models. The European-based *Planck Surveyor* is scheduled for launch in 2007 and will map the CMB in unprecedented detail.

The big bang model states that in the past, the universe was once much smaller, hotter and denser. The model is supported mainly by three observations. First, the expansion of the universe. Second, the CMB radiation. Third, by the observed proportions of light elements (such as deuterium, helium and lithium). The latest results are consistent with a universe that started with a big bang, is about 13.7 billion years old, has flat geometry, and will continue expanding.

In *The Nature of Reality*,* Richard Morris writes, "The scientific conception of the universe has changed so much that about the only thing that has remained the same is the word itself." In this chapter—and this book—we have shown how much our perception of the universe has changed. Nevertheless, we might pose the following question to a cosmologist: "What practical use can be made of the study of cosmology?" The answer is simply that not all pursuits need to have practical or useful consequences. We engage in some, like music, art, philosophy (and cosmology), for the sheer intellectual and artistic pleasure they provide. Curiosity is part of human nature, and questions like those raised in this chapter are among the most fundamental that can be asked—and perhaps answered—by science.

Our astronomical quest began 18 chapters ago in our home—the solar system. As you look back on the geocentric and heliocentric theories presented in the first few chapters, we hope that you see them as small, beginning steps in the development of today's view of the universe. After studying the solar system, we traveled outward to the stars and then much farther to galaxies. Finally, our study of cosmology considered the universe as a whole. We hope that this has given you a broader perspective on the natural world and a better appreciation of how the process of science helps us to understand the magnificent universe of which we are a part. We also hope that your quest for the universe will continue throughout your life.

* New York: The Noonday Press, 1987.

Study Guide

RECALL QUESTIONS

1. An open universe is one that
   A. will eventually collapse to form "a big crunch."
   B. will never stop expanding.
   C. [Either A or B above, depending upon its density.]
2. Einstein's conviction that the universe is closed and finite
   A. resulted directly from his general theory of relativity.
   B. resulted directly from his special theory of relativity.
   C. was a belief that caused him to make a change in his general theory of relativity.
   D. was a belief that caused him to make a change in his special theory of relativity.
3. What *evidence* leads us to believe that the universe is expanding?
   A. The big bang theory.
   B. The inflationary universe theory.
   C. The expansion of the solar system.
   D. The redshift of most galaxies.
   E. [All of the above.]

4. The assumption of isotropy states that the universe looks the same
   A. at all times.
   B. in all locations.
   C. in all directions.
   D. [All of the above.]
5. The cosmological principle refers to
   A. a comparison of our location in the universe to other locations in the universe.
   B. the redshift used to determine distances.
   C. Cepheid variables used to determine magnitude and therefore distance.
   D. the Doppler effect used to determine velocities.
   E. [Both B and C above.]
6. The existence of 3-degree background radiation supports
   A. the big bang theory.
   B. the steady state theory.

C. the big bang theory and the steady state theory about equally.

D. neither the big bang theory nor the steady state theory, for neither can explain it.

7. Big bang theories predict _____ speed for the most distant galaxies than do steady state theories.
   A. a lesser
   B. a greater
   C. the same

8. The Hubble law is a relationship between galaxies'
   A. redshifts and colors.
   B. distances and redshifts.
   C. redshifts and spectral types.
   D. colors and spectral types.

9. If present data are correct, the universe is
   A. closed, but not oscillating.
   B. closed and oscillating.
   C. flat.
   D. open.

10. A better knowledge of which quantity would most aid us in determining whether the universe is open or closed?
   A. The age of the Earth.
   B. The age of the solar system.
   C. The size of the Galaxy.
   D. The average density of the universe.
   E. [All of the above would help about equally.]

11. What *evidence* indicates that an explosion took place at the beginning of the universe?
   A. The density of matter in the universe.
   B. Low-intensity radiation from all directions.
   C. The big bang theory.
   D. The inflationary universe theory.
   E. The existence of supernovae.

12. If the average density of the universe is less than the critical density, the universe
   A. will expand forever.
   B. will eventually collapse.
   C. was produced in a big bang.
   D. is an inflationary universe.
   E. violates the laws of energy conservation.

13. Suppose that you read that astronomers calculate the age of the universe to be 15 billion years. Their calculation is based on
   A. calculations of the density of matter in the universe.
   B. the Hubble law.
   C. the age of rocks on Earth and Moon.
   D. the inflationary universe theory.
   E. [All of the above.]

14. The Hubble law is based on
   A. the big bang theory.
   B. the oscillating theory.
   C. redshift data.
   D. inflationary theories.
   E. [None of the above.]

15. Which of the following forces may halt the expansion of the universe?
   A. nuclear forces.
   B. electrical forces.
   C. gravitational forces.

16. The cosmological redshift indicates that
   A. most stars in the Galaxy are moving away from the Sun.
   B. most planets are moving away from ours.
   C. stars at greater distances are hotter.
   D. space is expanding.
   E. [All of the above.]

17. The horizon problem and the flatness problem
   A. are solved by the standard big bang theory.
   B. are solved by the inflationary universe theory.
   C. have not yet been solved by modern cosmology.

18. Which of the following is/are *assumptions* of cosmology?
   A. The cosmological redshift.
   B. The homogeneity of the universe.
   C. Universality of physical laws.
   D. [Both B and C above.]
   E. [All of the above.]

19. The discovery of the cosmic background radiation was made by
   A. Jocelyn Bell, a British graduate student.
   B. cosmologists early in the twentieth century.
   C. two Bell Telephone employees.
   D. the *COBE* satellite.
   E. Edwin Hubble.

20. When we talk about the cosmological expansion of the universe, what is expanding?
   A. The planets are getting farther apart.
   B. The stars of the Galaxy are getting farther apart.
   C. All galaxies are getting farther apart from one another.
   D. Clusters of galaxies are getting farther apart from one another.
   E. [All of the above, for the entire universe is expanding.]

21. The *BOOMERANG* project
   A. had as its purpose a search for gravitational lensing.
   B. had as its purpose a search for the edge of the universe.
   C. examined the CBM before the *COBE* satellite was launched.
   D. examined the CBM after *COBE* data was all received.

22. It is said that the universe is expanding. Explain just what this means.

23. Explain how, if almost all galaxies are moving away from ours, we are not necessarily at or near the center.

24. Name and explain three basic assumptions of cosmology.

25. What is background radiation and how does it enter the discussion of cosmological theories?

26. How is the age of the universe calculated?

27. Describe two methods of determining how quickly the expansion of the universe is slowing.

28. What is the oscillating universe?

29. Explain how the *COBE* results of Figure 18-22 relate to the universe of today.

QUESTIONS TO PONDER

1. Describe at least six steps in the historical progression of our ideas of the universe from Ptolemy's geocentric universe to Einstein's relativistic space curvature.

2. Quasars do not exist in our part of the universe. Why is this not a violation of the homogeneity principle?

3. Explain why a flat universe would be such a special case that it would seem to be an extremely unlikely state of the universe.

4. Why did Einstein add a term in his equation that predicted a repulsion force?

5. Popular science videos that illustrate the Big Bang commonly show an explosion as seen from outside. What would it mean to be "outside" the Big Bang? Comment on the noise that often accompanies such illustrations.

6. Discuss the observation of the CMB radiation as an example of the confirmation of a theory, comparing it to the prediction of stellar parallax in the 1600s.

7. The age of the universe calculated from the Hubble constant assumes a constant velocity of recession for galaxies. If, in fact, galaxies were moving faster at an earlier time, is the universe older or younger than calculated? Show your reasoning.

8. Where in the universe did the big bang occur? Explain.

9. It has been said—perhaps only half in jest—that when astronomers come upon a problem they cannot solve, they use a black hole. Give two examples that may have led someone to say this.

10. Explain why the horizon problem exists in the standard big bang model.

11. The horizon problem results from difficulties explaining how the background radiation can be so uniform in all directions. However, Figure 18-22 shows that it is not perfectly uniform. Explain.

12. Explain how the *COBE* results of Figure 18-22 relate to the universe of today.

CALCULATIONS

1. If the Hubble constant is 20 km/(s × MLY), what is the maximum age of the universe? Repeat for a Hubble constant of 17 km/(s × MLY).

2. Using Wien's law (Section 4-4), show that the blackbody curve for an object at temperature 3 K peaks in the microwave region of the electromagnetic spectrum. At what part of the spectrum does the blackbody curve of a 3000-K object peak?

3. The observed redshift of a quasar is $z = 5$. Show that its speed due to the expansion of the universe is about 94.6% of the speed of light.

EXPANDING THE QUEST

1. "Surveying Space-Time with Supernovae," by C.J. Hogan, R.P. Kirshner, and N.B. Suntzeff; "Cosmological Antigravity," by L.M. Krauss; and "Inflation in a Low-Density Universe," by M.A. Buchner and D.N. Spergel, all in *Scientific American* (January, 1999).

2. "Mapping the Universe," by S.D. Landy, in *Scientific American* (June, 1999).
   Many books for non-scientists have been written about cosmology. Some of the classics and some of the more recent ones follow.

3. S. Weinberg, *The First Three Minutes* (Basic Books, 1988).

4. S. Hawking, *A Brief History of Time* (Bantam Books, 1988).

5. M. Rees, *Before the Beginning* (Addison Wesley Longman, 1997).

6. M. Rowan-Robinson, *The Nine Numbers of the Cosmos* (Oxford University Press, 1999).

7. D. Goldsmith, *The Runaway Universe* (Perseus Books, 2000).

STARLINKS

Quest Ahead to Starlinks: www.jbpub.com/starlinks

Starlinks is this book's online learning center. It features **eLearning,** which contains chapter quizzes and other tools designed to help you study for your class. You can also find **on-line exercises,** view numerous relevant **animations,** follow a guide to **useful astronomy sites** on the web, or even check the latest **astronomy news** updates.

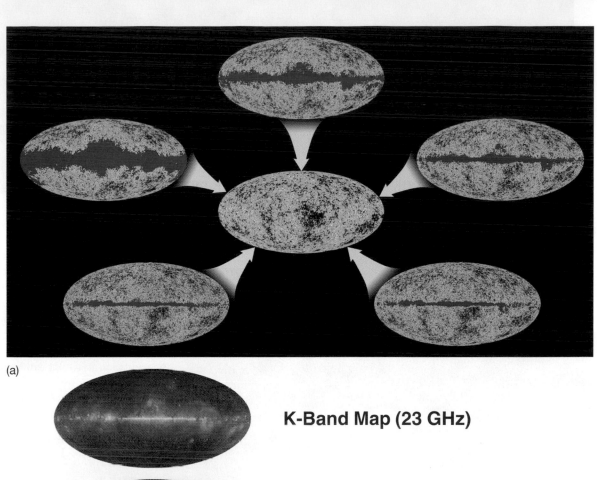

(a)

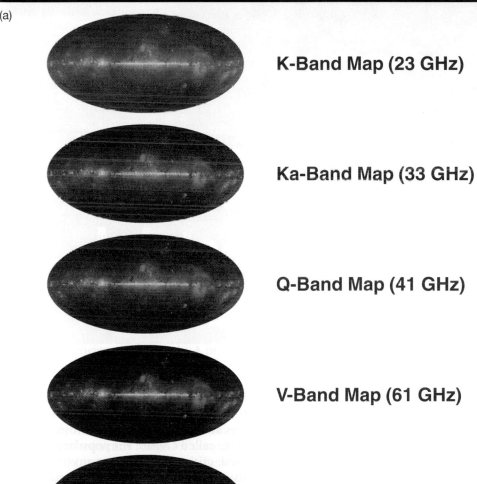

K-Band Map (23 GHz)

Ka-Band Map (33 GHz)

Q-Band Map (41 GHz)

V-Band Map (61 GHz)

W-Band Map (94 GHz)

(b)

(a) *WMAP* observes in five frequency bands to eliminate from the cosmic signal the microwaves from our Galaxy. The frequencies are (clockwise from top left) 23 GHz, 33 GHz, 41 GHz, 61 GHz, and 94 GHz. (b) These three-color foreground signal maps show synchrotron radiation in red, free-free in green, and thermal dust in blue. The horizontal streak across the middle of each map is the signal from our Galaxy. The corresponding frequencies for the images are the same as in part (a), from top to bottom.

and we could easily distinguish an ordered message from the random natural noise usually found at that wavelength.

Radio telescopes are designed and built to detect and measure radio waves coming from objects in space. If there are intelligent beings out there, these telescopes will also be useful both in detecting their presence and in communicating with them. Radio telescopes have been used on occasion to search the heavens for evidence of radio signals from intelligent beings. In 1960, American astronomer Frank Drake used a telescope of the National Radio Astronomy Observatory to search for signals from two nearby stars. The search, which Drake called Project Ozma, involved looking for unusual patterns in radio signals—patterns that were different from the signals emitted by inanimate objects such as stars and galaxies.

Since that time, several astronomers have conducted searches. For example, in an experiment conducted in the mid-1970s, more than 600 nearby stars were watched for about 30 minutes each. In 1971, a group of astronomers and engineers made plans for an elaborate array of radio telescopes to be devoted to a search. Their proposal, called Project Cyclops, would have cost billions of dollars to put into operation. Due to budget pressures, Congress canceled a search begun by NASA after $58 million had already been spent. This project was taken over by the SETI Institute, a private research group, and is now called Project Phoenix.

The Arecibo Radio Telescope in Puerto Rico receives more radio signals than we can search through for signals from extraterrestrial intelligence, even with the help of a large supercomputer. In May, 1999, a project called *SETI@home* began using individual computers in people's homes and offices to examine the data. As of April, 2003, 4.5 million people in 226 countries had downloaded the necessary software and had contributed computer time equal to 1.5 million years. The project is planned to continue until late 2003. Future plans include a *SETI@home* southern hemisphere search using the Parkes radio telescope in Australia.

Plans have been made in case we detect a signal that we verify as coming from an extraterrestrial being. There is even a protocol for involving international bodies in decisions about possible replies.

You can find the latest on SETI at http://www.seti-inst.edu/ and the latest on *SETI@home* at http://setiathome.ssl.berkeley.edu/

19-2 Communication with Extraterrestrial Intelligence

If our search for radio signals from a race of intelligent extraterrestrials is successful, it might succeed by detecting their stray, wasted radio signals. After all, we have been transmitting radio signals into space for decades now. If an extraterrestrial civilization exists nearby, say 20 light-years away, they have already received all the episodes of the "I Love Lucy" series, along with other radio and TV programming of the 50s and 60s. (That qualifies as a very good reason why they might not want to communicate with us in the first place!) On the other hand, those beings may already be transmitting messages into space with the purpose of announcing their presence and telling others something about themselves. The same radio telescopes that are used to receive signals from space can be used to transmit radio signals. Considering the probable differences between beings in different parts of the Galaxy, what could one race of these beings communicate to another? The study of this question is called CETI, for *C*ommunication with *E*xtra*t*errestrial *I*ntelligence.

First, we must point out that if another intelligent race were found, the tremendous distances between stars would prohibit a dialogue. The nearest star is nearly five light-years away from us. It would require five years for our radio signals to reach a planet circling that star and another five years for the signal to return from beings on that planet. And the likelihood that life is so common that it exists around the nearest star is extremely remote. If extraterrestrial life exists, the closest life sites are likely to be much farther away, making it impossible to get a reply to our signal during the span of one generation on Earth.

In Carl Sagan's science fiction novel *Contact* (later made into a movie), extraterrestrials first find out about civilization on Earth by detecting the TV broadcast of the 1936 Olympic games in Berlin, which was the first television transmission on Earth using significant power.

At first glance, the language problems might seem insurmountable. However, we do have something in common with every other race of beings that might exist: the physical universe and its mathematical laws, which we believe to be the same everywhere. Every intelligent race knows that hydrogen is the most common element and that an atom of hydrogen is made up of one proton and one electron. The prime numbers—those numbers that cannot be divided evenly by any other numbers but one and themselves—are the same in every language. Those scientists studying the problem have concluded that an understandable message could indeed be sent if enough time were devoted to its transmission.

Figure 19-1
The Arecibo Radio Telescope has been used to send a message to anyone out there listening. It has also been used to search for messages from any other intelligent sources.

We Earthlings have already sent a very short message. In 1974, the reconditioned reflecting surface of the Arecibo telescope (**Figure 19-1**) was rededicated, and at the ceremony, the telescope was used to transmit a message toward a cluster of 300,000 stars in the constellation Hercules. This transmission lasted only about 10 minutes, so the information that could be sent was very limited. The signal consisted of a series of Morse Code-type pulses containing 1679 data points. The number 1679 was chosen because, except for 1 and 1679, only one pair of numbers can be multiplied to obtain 1679: 23 and 73. If the data points are arranged into a rectangular array, there are only two ways to do it: either 23 across and 73 down, or vice versa. One array produces no pattern, but the other array makes a pattern that includes crude pictures and numbers to tell the beings that intercept the signal a little about those who sent the message.

If extraterrestrial beings happen to detect our 10-minute message, will they be able to decipher it? Who knows? Although we cannot know how much of it they will be able to understand, we can be confident that they will know that it comes from an intelligent source rather than an inanimate object. If more time were available to transmit messages, a slower development of language could be used so that understanding would be much more likely; but with the time limitation that existed, scientists believed this message was the best that could be done.

When might we receive a reply? The cluster toward which the message was sent is 26,000 light-years away, so we need only wait 52,000 years for an answer!

19-3 Letters to Extraterrestrials

In addition to our radio signals, we have sent "letters" into space in the form of a plaque and records.

The *Pioneer* Plaques

In late 1971, Carl Sagan learned that the trajectories of the *Pioneer 10* and *Pioneer 11* spacecraft would take them out of the solar system and that it might be possible to include on

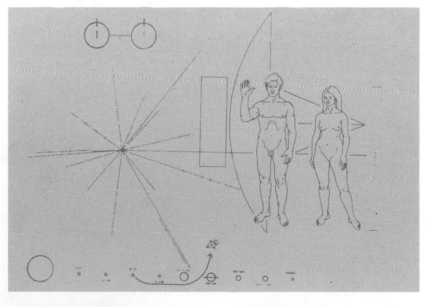

Figure 19-2
This plaque aboard *Pioneers 10* and *11* carried a message to extraterrestrials—and to Earthlings.

them a message to extraterrestrials. Sagan called NASA authorities and within three weeks got approval to put a plaque on both *Pioneer 10,* which was scheduled to be launched the following March, and *Pioneer 11,* which was to be launched a year later. Carl and his wife Linda, along with Frank Drake, designed the message, which was etched on a 6 × 9-inch gold-anodized aluminum plaque (**Figure 19-2**).

Some of the message on the plaque is easily recognizable (to humans, at least). The umbrella-shaped object behind the man and woman is an outline of the *Pioneer* spacecraft, drawn to scale with the people. Thus the finder can determine the size of a human. (The man on this scale is 5 feet 9-1/2 inches tall.)

The two circles at the upper left represent hydrogen atoms emitting radiation of a particular wavelength. Binary numbers on the plaque (not visible in this small rendition) use this wavelength to show the size of the woman and of the solar system at the bottom, where the *Pioneer* spacecraft is shown leaving the third planet.

The spidery-looking feature at the left shows the directions from the solar system to pulsars, compact celestial objects that emit beams of radio waves that when sweeping past the Earth are observed as radio pulses with regular periods (Section 15-5). The binary numbers along the lines represent the periods of the pulses, expressed in terms of the period of the wave from hydrogen. By analyzing the directions and periods of the pulsars, the finders could tell where our solar system is located in the Milky Way. In addition, since pulsars slow down with time, they could tell when the craft was launched.

Could extraterrestrials interpret the message? Curiously enough, since observations of hydrogen atoms and pulsars are common to all creatures of the universe, it is thought that they would be more likely to interpret the pulsar sketch than the human figures, which we recognize immediately.

The *Voyager* Records

In 1977, *Voyager 1* and *Voyager 2* were launched into space to rendezvous with Jupiter and Saturn. Plans were later adjusted to allow *Voyager 2* to continue to Uranus and Neptune. Both *Voyagers* are now headed for the outer boundary of the solar system, somewhere between 8 and 23 billion kilometers (5 and 14 billion miles) beyond the Sun. It is possible that the *Voyagers* will still be transmitting data when that boundary is encountered sometime in the first quarter of the 21st century.

With more time available to plan a message, a group of people, including the three who prepared the *Pioneer* plaque, designed a much more complete message to go aboard *Voyagers 1* and *2*. The messages on *Voyager* are contained on two copper phonograph records (**Figure 19-3**), designed to be played at 16-2/3 revolutions/minute. Instructions for playing the records are included in pictures on the cover. On the records are 90 minutes of music from around the world, 118 pictures, and greetings in nearly 60 languages (including one nonhuman message from a humpback whale).

The music on the *Voyager* includes parts of compositions from Bach (*Brandenburg Concerto No. 2* and *The Well-Tempered Clavier*), Beethoven (*Symphony No. 5, String Quartet No. 13 in B-flat*), and Mozart (*The Magic Flute*). In addition, it includes

Figure 19-3
The *Voyager* spacecraft carried these phonograph records beyond the solar system and into outer space.

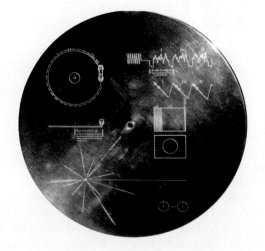

Chuck Berry's "Johnny B. Goode," Australian Aborigine songs, a Peruvian wedding song, and numerous other selections from the cultures of our world.

The records contain pictures as well as sound. The pictures include many photographs of scenes of nature here on Earth, of the Earth from space, of the Earth as changed by humans, of people in various activities, and of the biology of humans.

Will the Message Be Found?

The messages were not put aboard the *Pioneer* and *Voyager* craft with high hopes of them ever being found. They were sent in much the same manner as someone might put a message in a bottle and throw it into the sea. In fact, the space messages have even less chance than the bottle of being found. Even if the Galaxy abounds in intelligent life, the emptiness of space and the slow speed of the spacecraft make the chances slim indeed. The soonest any of the craft will pass within two light-years of a star is 40,000 years!

The messages are really a symbol of hope, a shout that "we are here." As stated in the book *Murmurs of Earth*—which tells the story of the *Voyager* records—after the Earth has been reduced to a charred cinder by an expanded and brighter Sun, the messages will continue to travel through space, "preserving a murmur of an ancient civilization that once flourished—perhaps before moving on to greater deeds and other worlds—on the distant planet Earth."

The quote is from the book *Murmurs of Earth,* by Carl Sagan, F. Drake, A. Druyan, T. Ferris, J. Lomberg, and L. Sagan (New York: Ballantine Books, 1978).

19-4 The Origin of Life

The search for extraterrestrial intelligence is interdisciplinary by nature. It involves physics, astronomy, geology, chemistry, and, of course, biology. Biology is an integral component of our search, since it concerns itself with the question of how did life begin. The answer to this question is important not just for biology, but for astronomy as well, since it helps us determine the probability of the existence of extraterrestrial life.

See the Tools of Astronomy box in Chapter 8 for a discussion of the search for life on Mars conducted by the *Viking* landers.

The theory of evolution, under development since the last half of the nineteenth century, explains how higher forms of life evolve from more primitive forms; but this still did not answer the question of the beginnings of life. Then, in the 1920s and early 1930s, J. B. S. Haldane, a Scottish biochemist, and A. P. Oparin, a Russian biochemist, proposed that soon after the Earth's formation, the necessary chemical elements were present for complex molecules to form—molecules that are needed for life. At the time, it was thought that the seas of the early Earth were composed primarily of water, methane, ammonia, and hydrogen. The two scientists hypothesized that these molecules would spontaneously collect into more complex, organic (carbon-based) molecules.

In the beginning, this world was nothing at all, Heaven was not, nor earth, nor space. Because it was not, it bethought itself: I will be. It emitted heat."
Ancient Egyptian text

The Haldane/Oparin hypothesis remained an interesting conjecture until 1953, when Harold Urey (the Nobel Prize winner who discovered deuterium) suggested that his graduate student, Stanley Miller, test the hypothesis by simulating the conditions of the early Earth. Miller put a mixture of water, hydrogen, methane, and ammonia into a sealed container (**Figure 19-4**). He heated the liquid—the young Earth was hot—and used an electrical source to produce sparks in the gas above the liquid. The sparks simulated lightning, which must have been common in the Earth's early atmosphere.

After Miller's apparatus had heated and sparked for a week, the mixture had turned dark brown. Analysis showed that it now contained large amounts of four different amino acids (organic molecules that form the basis for proteins). He also found fatty acids and urea (a molecule that is necessary in many life processes).

The Miller-Urey experiment showed that given the right chemicals and a source of energy, chemical reactions could occur that produce the building blocks of life.

For those who are studying aspects of the origin of life, the question no longer seems to be whether life could have originated by chemical processes involving nonbiological components but, rather, what pathway might have been followed.
National Academy of Sciences

Figure 19-4
The Miller-Urey experiment showed that organic molecules necessary for life could form easily from the compounds and conditions of the young Earth.

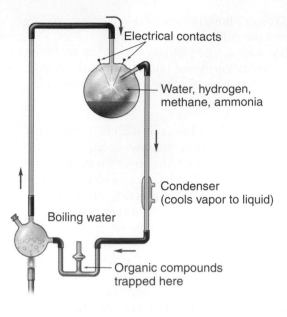

Electrical contacts

Water, hydrogen, methane, ammonia

Condenser (cools vapor to liquid)

Boiling water

Organic compounds trapped here

Since then, researchers have learned that the early Earth's atmosphere was made up primarily of carbon dioxide, nitrogen, and water vapor, rather than the four compounds used in Miller's experiment. When these compounds are used in the sealed apparatus, even more complex organic molecules are produced. Furthermore, if instead of using electrical sparks for energy we use ultraviolet light (which strikes Earth from the Sun), we achieve the same results.

It must be emphasized that the Miller experiment did not produce life, only organic molecules. As yet, we have not been able to cross the gap between chemical evolution and biological evolution. However, the experiment confirmed the Haldane/Oparin hypothesis, and showed that such molecules can form easily in a short time. The results of the experiment would indicate that the early Earth must have contained seas made up of organic goo.

More recently, traces of amino acids have been found in some meteorites. In addition, there is spectroscopic evidence that organic molecules exist in interstellar dust and gas clouds. Thus the molecular building blocks of life seem to be common in the universe. How does life form from these blocks? We don't know, but it seems that nature's deck may be stacked in the direction of life.

19-5 The Drake Equation

So far we have not had any success in receiving a transmission from an extraterrestrial civilization. What are the chances that one day we may detect such transmissions? Astronomer Frank Drake proposed an equation for *estimating* the number (N) of technologically advanced civilizations in our Galaxy whose signal we might be able to detect. The equation has been written in a variety of forms, including this one:

$$N = R_* \times f_p \times n_e \times f_l \times f_i \times f_c \times L$$

where

R_* = the rate at which solar-type stars form in our Galaxy
f_p = the fraction of these stars that have planetary systems
n_e = the average number of planets per such system that are Earth-like enough to support life
f_l = the fraction of those Earth-like planets on which life actually develops
f_i = the fraction of those life-forms that evolve to intelligence
f_c = the fraction of those intelligent species who are interested in interstellar communication and develop and use the necessary technology for it
L = the average lifetime of such a technologically advanced civilization.

The factors in the equation, as you might surmise, are far from well known. Let's take a quick look at each factor.

Based on the average lifetime of stars and on the number of stars in the Galaxy, we have a pretty good idea of the rate of solar-type star formation; the first term on the right is known to be in the range of 1 to 10 solar-type stars per year. The emphasis on "solar-type" stars is necessary. We should probably exclude stars with masses

The hypothesis that life originated on Earth through moss-grown fragments from the ruins of another world may seem wild and visionary. All I maintain is that it is not unscientific.

—*Lord Kelvin*

larger than about 1.5 M$_\odot$ because their main-sequence lifetimes are shorter than the time it took for intelligent life to evolve on our planet. Stars less massive than the Sun have longer main-sequence lifetimes than our Sun, but they are also dimmer. This implies that only planets close enough to such stars would be suited for life. But short distances imply strong tidal interactions. It seems that the best candidates are stars with spectral types between F5 and M0. Other factors tend to limit the search for suitable stars to a range of distances from the center of a galaxy, sometimes called "Habitable Zones" or "Life Zones" (see the chapter-opening figure). Closer than the habitable zone, stars are too close together for stable planetary orbits and are bathed in X-rays and gamma rays from the galactic center. Farther out than the habitable zone, stars don't have enough heavy elements to form solid Earth-like planets.

The more we learn about how planetary systems are formed, the better we can estimate what fraction of stars have planets orbiting around them. Better knowledge of the origin of our own system will help in this regard, and as we explained in Section 7-6, we are just beginning to understand the formation of our solar system. In addition, evidence is accumulating that planetary systems are common (Section 7-7). It is then reasonable to assume that for solar-type stars, $f_p \approx 1$. We might suspect that if this is so, many systems probably have life-supporting planets. (After all, our solar system *almost* has three planets capable of supporting Earth-type life.) Thus, a reasonable range for the value of n_e is between 0.1 and 1. We see that we can make reasonably confident estimates of the values for the first three terms in the equation. Beyond this, things get fuzzier.

As we described earlier, the Miller-Urey experiment indicates that the most basic molecules from which life is made are formed easily and naturally under conditions expected to be prevalent on some young planets. Many biologists believe that, given the right conditions and enough time, life will develop from these components, but they have no way of testing this hypothesis. Thus, a reasonable estimate for f_l is about 1.

The last three terms are even less well known. Does life evolve easily to intelligence, and if so, is it common for intelligent species to be interested in communication with aliens? (Humans seem to enjoy fiction on this topic, but we hesitate to spend money for a search.) Finally, how long does the typical civilization last? How long will ours last? Even if we assume that evolution naturally leads to intelligence, which is far from certain, and that an intelligent species would naturally try to communicate with other intelligent species in our Galaxy, the last factor introduces the largest uncertainty in the estimated value of N. Since the invention of nuclear weapons, we have come very close to rendering ourselves extinct, so a reasonable value for L might be only 100 years. On the other hand, it is certainly possible that L might have a value in the thousands or even millions of years.

Various astronomers have inserted their best guesses into the equation, and—depending upon whether they are optimists or pessimists—have come up with answers anywhere from "only one" or "only a few" to "millions." However, the value of the equation lies less in its mathematical answer than in showing us what we must learn in order to calculate an answer. It shows that we have a lot to learn.

> Our obligation to survive is owed not just to ourselves but also to that Cosmos, ancient and vast, from which we spring.
> *Carl Sagan*

19-6 Where Is Everybody?

The great size of our Galaxy and the tremendous number of stars in it suggest that the number of stars that could support a planet with intelligent life is very large. So it is possible that the value of N in the Drake equation is quite large.

However, there is an opposing point of view. The argument goes as follows: If the formation of life is common in the Milky Way, there should be a large number of civilizations like ours. We would expect many of the civilizations to be millions of years older than ours, because the Sun was formed fairly recently in the Galaxy's history.

The older civilizations would be scientifically advanced enough to be able to travel to other stars. In fact, some would have been forced to leave their home planet when their star reached the end of its main sequence life and moved toward the red giant stage. These civilizations would have moved to hospitable planets near other stars and would have colonized space. (Aren't we likely to do so within the next million years, assuming our species lives that long?)

Therefore, if life begins and evolves as readily as some thinkers would have us believe, the Galaxy should be teeming with life. We should have encountered not only radio signals from extraterrestrial beings, but Earth should have been visited. The question is, "Where is everybody?" Although some people say that UFOs of extraterrestrial origin have visited Earth on many occasions, no report of UFOs controlled by extraterrestrial beings has been substantiated to the degree demanded for acceptance in science. Few astronomers consider the reports credible.

The argument presented here leads to either of two conclusions: (1) Earth has been (and continues to be) visited by extraterrestrials who for some reason do not want to reveal themselves, or (2) intelligent life is rare in the Galaxy. If we refuse to accept the first option, we are left with the second, and the money we spend searching is wasted. Or so the argument goes.

It is possible that conservative estimates for the parameters in the Drake equation are closer to reality. However, this would imply that we are somehow special, a view that would contradict the Copernican attitude that characterizes today's science. It is also possible that advanced civilizations outgrow technology and decide not to engage in interstellar communication. Maybe there is such a thing as "the prime directive," and other advanced civilizations are not interfering with our technological and biological evolution. The most likely answer to the question "Where is everybody?" is that the lifetime of a technological civilization is short. An order of magnitude calculation suggests that unless a technological civilization survives through some 10,000 to 10 million years of technology, we would not expect to find another such civilization still surviving in our Galaxy at the same time as we are.

There is another completely different argument against spending money on the search for extraterrestrial intelligence (SETI). Recall that a good theory in science must have within itself the seeds of its own destruction. The theory must be capable of being proved wrong. Some scientists point out that no matter how much we spend over the next decade in searching for extraterrestrial radio signals, it is very possible that we will find nothing. This would not prove that extraterrestrials are not broadcasting radio signals, however, but only that we have not found them. Will those in favor of the search continue to ask for more money? When do we stop? This line of thinking leads to the conclusion that SETI is bad science and is not worth the considerable amounts of money being proposed for it. What do you think?

This question was first asked by the great physicist Enrico Fermi in 1950.

Absence of evidence is not evidence of absence.
Frank Drake

Conclusion

Efforts to discover extrasolar planetary systems are intensifying. In the next few years we might be able to answer the question "Are there Earth-like planets orbiting other stars?" If the results of these efforts turn out to be promising, searches for extraterrestrial intelligence will also intensify.

The potential benefits of knowledge and experience that would result from interstellar contact are immense. Beyond such benefits, SETI-type efforts will continue simply because we want to know what is out there, whether or not we are somehow unique. We want—and maybe need—to know where we came from and what it means to be human. Even if we never find the answers to these questions, it might still be worthwhile pursuing this quest. As in most cases, what is important at the end is the journey, not the destination.

Study Guide

1. Which of the following facts is the major factor that causes many astronomers to believe that extraterrestrial life is probable?
   A. UFO's that have been sighted on Earth.
   B. Laser signals that have been received from space.
   C. Radio signals that have been received from space.
   D. Signs of life that we have found on neighboring planets.
   E. The tremendous number of stars in the universe.

2. The most likely method of contact with extraterrestrial intelligence is likely to be
   A. radio waves.
   B. sound waves.
   C. light waves.
   D. actual visits.

3. The most likely wavelength at which to search for extraterrestrial intelligence is thought to be about
   A. 21 centimeters, the wavelength used to detect cool hydrogen clouds.
   B. 140 centimeters, the wavelength used to detect water.
   C. 1400 centimeters, the wavelength used to detect carbon.
   D. [Any of the above, for no wavelength is more likely than another.]

4. *SETI@home* is a project in which non-astronomers use
   A. dish antennae to search for extraterrestrial life.
   B. regular TV antennae to search for extraterrestrial life.
   C. radios to search for extraterrestrial life.
   D. antennae to send out signals to extraterrestrial life.
   E. computers to analyze data in a search for extraterrestrial life.

5. All of the following are projects that involve extraterrestrial life EXCEPT
   A. Project Phoenix.
   B. *COBE*.
   C. *SETI@home*.
   D. Project Ozma.
   E. *Voyager* records.

6. If intelligent life exists around a nearby star, it may be able to detect which of the following "signals" that we have sent?
   A. Sound waves emitted by loudspeakers at outdoor concerts.
   B. Radio and TV waves intended for communication on Earth.
   C. Radio waves sent from Arecibo intended to make contact.
   D. [All of the above.]
   E. [B and C above, but not A.]

7. Assuming that radio contact is established some day, which of the following do professionals in the field believe is true?
   A. Extraterrestrials will be able to figure out our message, but we will not be able to decipher theirs.
   B. We will be able to decipher their message, but extraterrestrials will be not able to understand ours.
   C. Neither we nor the extraterrestrials we contact will be able to understand the others' message.
   D. Communication will be monologues rather than dialogs because of the tremendous time lag.
   E. [Both C and D above.]

8. We have intentionally included messages to extraterrestrials on which of the following categories of spacecraft?
   A. The Space Shuttle.
   B. *Pioneer* and *Voyager*.
   C. *Magellan*.
   D. *Viking*.
   E. [All of the above.]

9. Assuming that extraterrestrial life is common, which of the following is true concerning the likelihood that one or more of our spacecraft will be picked up by extraterrestrials?
   A. It is so likely that it probably has already happened.
   B. It is likely to happen in the next few hundred years.
   C. It is likely to happen in the next few thousand years.
   D. It is unlikely to happen for many tens of thousands of years, if ever.

10. The *Voyager* record contains which of the following?
    A. Greetings in many languages.
    B. Music.
    C. Pictures of scenes on Earth.
    D. Pictures illustrating the biology of humans.
    E. [All of the above.]

11. The Miller-Urey experiment showed that if the chemical elements of the early Earth are put in a sealed container and energy is added,
    A. living matter in the form of viruses is formed.
    B. living matter in the form of bacteria is formed.
    C. some organic molecules are formed.
    D. [Two of the above.]
    E. [None of the above.]

12. The Drake equation shows us the factors that we need to know in order to calculate
    A. the rate at which stars form in the galaxy.
    B. the number of planets in an average planetary system.
    C. the number of Earth-like planets in the Galaxy.
    D. the number of places in the Galaxy that contain the most basic life.
    E. the number of places in the Galaxy that contain technologically advanced civilizations sending out signals.

13. Which of the following factors in Drake's equation is least known?
    A. The rate at which solar-type stars form in our Galaxy.
    B. The fraction of solar-type stars that have planetary systems.
    C. The fraction of Earth-like planets on which life actually develops.
    D. The average lifetime of technologically advanced civilizations.
    E. [All of the above are about equally known.]

14. Most astronomers who hold the "Where is Everybody?" argument conclude that
    A. extraterrestrials must be hiding from us.
    B. extraterrestrial intelligence is rare or non-existent.
    C. extraterrestrial communication is impossible because of technological constraints.
    D. there are not many Earth-like planets in the Galaxy.

15. What does "SETI" stand for? CETI?

16. Explain why the average lifetime of a technologically advanced civilization is a necessary factor in Drake's equation.

QUESTIONS TO PONDER

1. After years of listening at many frequencies, we have not picked up any artificial signals. Is this a proof that life has not evolved elsewhere? Is this a proof that technologically advanced life has not evolved elsewhere? Explain your answer.

2. What factors might limit the lifetime of a technologically advanced civilization? Explain your answer.

3. What would the consequences be if tomorrow we receive a radio signal from a civilization on a planet 10 light-years away, asking for a two-way communication? Explain your answer and consider both short- and long-term consequences.

4. Throughout our history, there have been many instances of devastation of certain cultures after contact with more "advanced" cultures. How does this historical fact affect your opinion on the searches for extraterrestrial intelligence?

CALCULATIONS

1. Assume that there are 10,000 technologically advanced civilizations in a galaxy of 200 billion stars. How many stars would astronomers in this galaxy have to observe in order to have a reasonable chance of finding one such civilization?

2. Assume that in a galaxy that is about 10 billion years old, a thousand to a million technologically advanced civilizations have shown up at various times. On the average, for how long must a civilization last in order to communicate with another civilization?

3. Choose a reasonable value for each of the parameters in the Drake equation and estimate the value of N. Explain your reasoning behind your choices.

4. In the text we mentioned that stars of mass larger than about 1.5 $M_\odot$ have main-sequence lifetimes that are shorter than the time it took for intelligent life to evolve on our planet. Show that this is true.

EXPANDING THE QUEST

The websites for the SETI Institute (http://www.seti-inst. edu) and SETI at the University of California, Berkeley (http://seti.ssl.berkeley.edu) offer the latest information on the search for extraterrestrial intelligence.

1. "The Search for Extraterrestrial Intelligence," by C. Sagan and F. Drake, in *Scientific American* (May, 1975).

2. *Murmurs of Earth*, by Carl Sagan, F. Drake, A. Druyan, T. Ferris, J. Lomberg, and L. Sagan (Ballantine Books, 1978).

3. "No Greater Discovery," by F. Drake and D. Sobel, and "Surfing The Cosmos," by A. K. Dewdney, in *Taking Sides*, 4th ed. (McGraw-Hill, 2000).

4. "The Evolution of Life on the Earth," by S. J. Gould, and "The Origin of Life on the Earth," by L. E. Orgel, in *Scientific American* (October, 1994).

5. *Extraterrestrials—Where Are They?*, by B. Zuckerman and M. Hart, eds., 2nd ed. (Cambridge University Press, 1995).

6. *Rare Earth: Why Complex Life is Uncommon in the Universe*, by Peter Douglas Ward and Donald Brownlee (Copernicus Books, 1999).

7. *Here Be Dragons: The Scientific Quest for Extraterrestrial Life*, by Simon Levay and David W. Koerner (Oxford University Press, 2000).

8. Searching for Extraterrestrials: "Where Are They?" by I. Crawford; "Where They Could Hide," by A. J. LePage; and "Intergalactically Speaking," by G. W. Swenson, Jr., in *Scientific American* (July, 2000).

STARLINKS

Quest Ahead to Starlinks: www.jbpub.com/starlinks

Starlinks is this book's online learning center. It features **eLearning,** which contains chapter quizzes and other tools designed to help you study for your class. You can also find **on-line exercises,** view numerous relevant **animations,** follow a guide to **useful astronomy sites** on the web, or even check the latest **astronomy news** updates.

Appendices

Appendix A

Units and Constants

Metric Prefixes	
nano (n)	$= 10^{-9}$
micro (μ)	$= 10^{-6}$
milli (m)	$= 10^{-3}$
centi (c)	$= 10^{-2}$
kilo (k)	$= 10^{3}$
mega (M)	$= 10^{6}$

Metric–English Conversion	
1 kilometer (km)	= 0.6214 miles
1 meter (m)	= 39.37 inches
2.54 centimeters (cm)	= 1 inch
1 kilogram (kg)	= 1000 g (weighs 2.2 pounds)
1 gram (g)	= 0.001 kg (weighs 0.035 oz)

Units of Length	
1 astronomical unit (AU)	$= 1.495979 \times 10^{11}$ m
	$= 92.96 \times 10^{6}$ miles
1 light-year (LY)	$= 6.324 \times 10^{4}$ AU
	$= 9.461 \times 10^{15}$ m
	$= 5.879 \times 10^{12}$ miles
1 parsec (pc)	= 206,264.81 AU
	$= 3.086 \times 10^{16}$ m
	= 3.262 LY
1 nanometer (nm)	= 10 Angstroms (Å)
	$= 10^{-9}$ m

Constants	
Speed of light	$= 2.99792458 \times 10^{8}$ m/s
Electron mass	$= 9.1094 \times 10^{-31}$ kg
Proton mass	$= 1.6726 \times 10^{-27}$ kg
Planck's constant	$= 6.6261 \times 10^{-34}$ J·s
Stefan-Boltzmann constant	$= 5.6705 \times 10^{-8}$ W·m^{-2}·K^{-4}

Temperature	Kelvin (K)	Celsius (°C)	Fahrenheit (°F)
Absolute zero	0	-273.15	-459.7
Freezing point of water	273.15	0	32
Boiling point of water	373.15	100	212
		$T_K = T_C + 273.15$	$T_F = 1.8 \times T_C + 32$

Appendix B

Solar Data

	Value	Ratio to Earth
Diameter	1,392,000 km	109.2
Mass	1.9891×10^{30} kg	332,980
Average density	1.41 g/cm^3	0.255
Surface gravity	274 m/s^2	28.0
Escape speed	618 km/s	55.2
Effective temperature	5778 K	
Luminosity	3.846×10^{26} watts	
Absolute magnitude	4.83	
Tilt of equator to ecliptic	7.25°	
Sidereal rotation period		
Equator	25.05 days	
40° latitude	27.40 days	
80° latitude	33.83 days	

Appendix C

Planetary Data

Physical Data

Planet	Equatorial Diameter* (km)	Equatorial Diameter* (Earth = 1)	Oblateness	Mass** (Earth = 1)	Density (g/cm^3)	Surface Gravity*** (Earth = 1)	Escape Speed (km/s)
Mercury	4,879	0.383	0	0.055	5.43	0.38	4.30
Venus	12,104	0.949	0	0.815	5.24	0.91	10.36
Earth	12,756	1	0.00335	1	5.52	1	11.19
Mars	6,794	0.533	0.0065	0.107	3.93	0.38	5.03
Jupiter	142,984[†]	11.2	0.065	317.83	1.33	2.36[†]	59.5
Saturn	120,536[†]	9.4	0.098	95.2	0.69	0.92[†]	35.5
Uranus	51,118[†]	4.0	0.023	14.5	1.27	0.89[†]	21.3
Neptune	49,528[†]	3.9	0.017	17.1	1.64	1.12[†]	23.5
Pluto	2,390	0.2	0	0.002	1.75	0.06	1.1

*Equatorial diameter of Earth = 12,756 km.
**Mass of Earth = 5.9736×10^{24} kg.
***Surface gravity of Earth = 9.78 m/s^2.
[†]At 1 bar.

Orbital Data, Axis Tilt and Rotation Period

Planet	Semimajor Axis (10^6 km)	Semimajor Axis (AU)	Sidereal Orbital Period*	Mean Orbital Speed (km/s)	Orbital Eccentricity	Inclination to Ecliptic (degrees)	Equatorial Tilt to Orbital Plane (degrees)	Sidereal Rotation Period**
Mercury	57.9	0.387	87.97 d	47.9	0.2056	7.0	0.01	58.65 d
Venus	108.2	0.723	224.7 d	35.0	0.0067	3.39	177.36	−243.02 d
Earth	149.6	1	365.256 d	29.8	0.0167	0	23.45	23.9345 h
Mars	227.9	1.524	1.881 y	24.1	0.0935	1.85	25.19	24.623 h
Jupiter	778.6	5.204	11.86 y	13.1	0.0489	1.3	3.13	9.925 h
Saturn	1433.5	9.582	29.46 y	9.7	0.0565	2.49	26.73	10.66 h
Uranus	2872.5	19.20	84.01 y	6.8	0.0457	0.77	97.77	−17.24 h
Neptune	4495.1	30.05	164.79 y	5.4	0.0113	1.77	28.3	16.11 h
Pluto	5869.7	39.24	247.7 y	4.7	0.2444	17.2	122.5	−6.39 d

*The symbols "h,""d," and "y" imply "hours,""days," and "years," respectively. One day is 24 hours; one year is 365.256 days.
**The "−" signs indicate retrograde rotation.

Appendix D

Planetary Satellites

The Moon	Value	Compared to Earth
Equatorial diameter	3476 km	0.27
Mass	7.35×10^{22} kg	0.0123
Density	3.35 g/cm^3	0.61
Surface gravity	1.62 m/s^2	0.166
Escape speed	2.4 km/s	
Revolution period	27.322 days	0.213
Sidereal rotation period	27.322 days	
Synodic period (phases)	29.531 days	
Surface temperature	−170°C to 130°C	
Albedo	0.11	0.36
Tilt of equator to orbital plane	6.68°	
Orbit		
Average distance from Earth	384,400 km	
Closest distance	363,000 km	
Farthest distance	405,500 km	
Eccentricity	0.055	

Chapter 7

Try One Yourself (Measuring Distances in the Solar System)

1/2 of 4.7 minutes is 141 seconds

$d = v\,t$

$= (3 \times 10^5 \text{ km/s}) \times 141 \text{ s} = 4.23 \times 10^7 \text{ km}$

$(4.23 \times 10^7 \text{ km}) \times (1 \text{ AU}/1.5 \times 10^8 \text{ km}) = 0.282 \text{ AU}$

Venus is 0.72 AU from the Sun, and $1 - 0.72 = 0.28$, in agreement with our answer.

Try One Yourself (Measuring Mass and Average Density)

$(a^3/P^2) = K\,M_{\text{Earth}}$
$(3.844 \times 10^8 \text{ m})^3/(2.36 \times 10^6 \text{ sec})^2 = 1.69 \times 10^{-12}\,M_{\text{Earth}}$
$M_{\text{Earth}} = 6.03 \times 10^{24}$ kilograms

This is close to the value in the appendix, 5.97×10^{24} kilograms. The difference is caused by our assumption that the Moon's mass is negligible. In fact, the value that we calculated is the total mass of the Earth and Moon.

Try One Yourself (Calculating Average Density)

Refer to the solution to the "Try One Yourself" in Chapter 6, where we calculate the diameter of the Moon to be 3485 kilometers.

Radius $= 1.74 \times 10^6$ meters

Volume of Moon $= (4\pi/3)\,R^3 = (4\pi/3)\,(1.74 \times 10^6 \text{ m})^3 =$
$\qquad\qquad 2.207 \times 10^{19} \text{ m}^3.$

Density = mass/volume = $(7.35 \times 10^{22} \text{ kg})/(2.207 \times 10^{19} \text{ m}^3)$

$\qquad = 3330 \text{ kg/m}^3$, in good agreement with the value 3350 kg/m^3 in the appendix.

Recall Questions

2. C	4. A	6. A	8. A	10. A	12. B
14. D	16. A	18. C	20. B	22. E	24. C

26. The total mass of the two objects.

28. Jupiter is the largest planet, with a diameter 11 times Earth's and a mass 318 times Earth's.

30. The masses of the planets that have moons were calculated using Kepler's third law. Mercury's and Venus's masses were calculated based on their gravitational effect on passing asteroids and comets, and later by their gravitational effects on spacecraft that passed near them.

32. All of the planets revolve around the Sun in the same direction. All except Venus, Uranus, and Pluto rotate on their axes in the same direction as they revolve.

34. Compared to the Jovians, the terrestrials are nearer the Sun, smaller, less massive, more solid, slower rotating, more dense, and have thinner atmospheres and fewer moons. No terrestrial planet has a ring, while all Jovian planets have rings.

36. Earth is just slightly more dense than Mercury and Venus, somewhat more dense than Mars, and much more dense than the Jovian planets and Pluto.

38. The temperature of a gas is related to the speed of its molecules, and at the same temperature, the molecules of a more massive gas move more slowly.

40. Nonvolatile elements condensed in the inner solar system, but volatile elements were swept outward by the solar wind.

42. The asteroids are planetesimals that were prohibited from forming a planet by the effects of Jupiter's gravitational force. The Oort cloud is hypothesized to have resulted when small objects in the outer solar system were thrown outward by gravitational forces when Jupiter or Saturn passed near them.

Calculations

2. 77.2 AU

4. 9.0×10^8 km, which is 6.0 AU

6. Mass of Mars: 6.4×10^{23} kilograms. We assumed that the mass of Phobos is negligible compared to that of Mars. Density of Mars: 3.9×10^3 kilograms/meter^3, which is 3.9 grams/centimeter^3.

Chapter 8

Try One Yourself (Mercury via *Mariner*—Comparison with the Moon)

Time to cool is proportional to radius (and to diameter), so since the diameter of Jupiter is 11.2 times that of Earth (from Appendix C), it takes Jupiter 11.2 times as long to cool as Earth does.

Recall Questions

2. B	4. B	6. D	8. C	10. E	12. D
14. C	16. B	18. D	20. B	22. D	24. A

26. Venus is most like Earth in size; Mars is most like Earth in rotation period.

28. Mercury's rotation period is exactly 2/3 of its period of revolution. This has occurred because the mass of the planet is not evenly distributed through its volume. The planet has an eccentric orbit, and the gravitational forces toward the Sun have changed its rotation rate so that resonance occurs between its rotation and revolution.

30. There are two reasons for the extremes in temperature. First, day and night periods are very long (88 Earth days each), providing long periods of time for the surface to heat and cool. Second, there is no atmosphere to block sunlight during the day and to retain heat at night.

32. The greenhouse effect is the phenomenon of infrared radiation being prohibited from escaping through a planet's atmosphere, thereby causing a high temperature on the planet. In a florist's greenhouse an additional effect occurs: the hot air is trapped inside the greenhouse by the walls and roof.

34. The tilt of Mars's axis is very similar to that of Earth's. Mars's equator is tilted 25.2 degrees from its orbital plane, while Earth's is tilted 23.5 degrees. (Tables normally show equatorial tilt rather than axis tilt; they are numerically the same.)

36. The polar caps of Mars are made of frozen water and frozen carbon dioxide.

38. The temperature near the equator of Mars varies from −135 degrees Celsius to +30 degrees Celsius—a much greater difference than ever experienced on Earth. The reason for the difference is that Mars's atmosphere is very thin and therefore does not shield sunlight during the day or provide a greenhouse effect to blanket the planet at night.

Calculations

2. Jupiter's volume is 11^3, or 1331, times Earth's volume.

4. The smallest feature that can be observed is about 750 km across. Caloris Basin is 1400 km in diameter, about twice the smallest feature that can be seen.

6. $a^3/P^2 = 1$ when units of AU and years are used.

$1.524^3/a^2 = 1$

$a = 1.881$ years

Chapter 9

Recall Questions

2. B	**4.** C	**6.** A	**8.** C	**10.** D	**12.** D
14. B	**16.** B	**18.** C			

20. The band at Jupiter's equator has that rotation rate.

22. The outer atmosphere is primarily hydrogen and helium at a relatively low pressure. As we go deeper, pressure increases until the material would be judged a liquid. About 20,000 kilometers below the cloudtops, liquid metallic hydrogen is found, and it extends down to whatever heavy-element core exists.

24. Saturn's rings are aligned with its equator, which is tilted at about 27 degrees to the planet's orbital plane. Thus as Saturn orbits the Sun, we on Earth (which is relatively near the Sun) sometimes see the southern side of Saturn and its rings and sometimes the northern side. Midway between these views, the rings are edge-on to our view and are nearly invisible.

26. 1) Knowing the distance to a planet and the planet's angular size, we can use the small-angle formula to calculate the diameter. 2) When a planet passes in front of a star, observation of the amount of time that the planet blocks the star's light, along with knowledge of the planet's speed, gives us another method to calculate the diameter.

28. Uranus's axis tilt (98 degrees) is such that each of its poles nearly points toward the Sun at one time during its revolution period.

30. When a moon changes its distance from its planet, the amount of tidal force exerted on the moon changes. This causes the moon to flex, and this produces heat. The more massive the planet, the greater the effect on its moon, and the closer the moon is to its planet, the greater the effect. Thus the inner moons of Jupiter that have a non-circular orbit experience tidal heating the most. Io is the moon that shows the most prominent effect of tidal heating.

Calculations

2. Its density would be 3/5, or 0.6, of Earth's.

4. Period = 6 days and the circumference of the spot is about 60,000 kilometers. Thus the speed is 10,000 kilometers/day, or 400 kilometers/hour, or 0.1 kilometers/second.

6. One percent of its mass is 8.9×10^{20} kilograms. At 1000 kilograms/second, it would take 8.9×10^{17} seconds, or about 10^{13} years. This is 1000 billion years, very much greater than the age of the solar system.

Chapter 10

Recall Questions

2. E	**4.** C	**6.** D	**8.** D	**10.** D	**12.** C
14. B	**16.** B	**18.** C	**20.** D	**22.** B	**24.** C

26. The most accurate values of Pluto's size come from observations of eclipses of its moon, Charon.

28. Before an asteroid is named, its orbit must have been determined accurately. The orbits of most observed asteroids have not been calculated.

30. Gravitational pull from Jupiter disrupts the orbits of asteroids so that the weaker gravitational forces between them cannot pull them together.

32. The nucleus of a comet is its solid core. Comet nuclei are irregular in shape and are typically a few kilometers across. The coma is made up of gas and dust surrounding the nucleus. It has a very low density and may be as large as 100,000 kilometers in diameter. A comet has two tails, one made up of ions and one of dust. Tails are typically from 10^7 to 10^8 kilometers long.

34. One tail is made up of ions swept away by the solar wind. It is straight because the ions move away from the coma at great speeds. The other tail consists of dust pushed away slowly by radiation pressure. The curvature results from the fact that the material moves away from the coma very slowly compared to the coma's speed.

36. First, the vast majority of long-period comets are in elongated elliptical orbits around the Sun. They must therefore spend most of their time far from the Sun. Second, comets are regularly being captured in the inner solar system. The existence of the cloud explains their origin.

38. A meteor shower results when the Earth encounters a swarm of meteoroids. As the Earth moves into the swarm, the resulting meteors seem to originate in the direction in which the Earth is moving.

Calculations

2. At 2×10^{-9} kilograms/meter^2 per day, it would take about 700,000 years.

4. The answer is simply the ration of areas, 2.6×10^{-4}, which is about one chance in 4000.

6. Warning time = 10^6 seconds, which is less than 12 days. It would pass through the atmosphere in 10 seconds.

Chapter 11

Try One Yourself (Solar Energy)

$$\text{Area of sphere} = 4\pi R^2 = 4\pi (2.3 \times 10^{11}\text{ m})^2$$
$$= 6.64 \times 10^{23}\text{ m}^2.$$

$$\text{Energy/area} = 3.9 \times 10^{26}\text{ watts}/6.64 \times 10^{23}\text{ m}^2$$
$$= 587\text{ watts/m}^2$$

Recall Questions

2. C	4. D	6. C	8. A	10. E	12. B
14. B	16. A	18. A	20. D	22. A	24. C

26. The most convincing evidence is that if a chemical process were the source of the Sun's power, the Sun would burn out over a few centuries, and we know that it has released energy fairly uniformly for (at least) many millions of years.

28. The number of protons in nuclei distinguish one element from another.

30. Most of the Sun is made up of gas.

32. The pressure at any depth within the Sun is caused by the weight (and therefore by the mass) of gas above that layer. To calculate the mass of gas above a given point within the Sun, one needs to know the density at levels above that point. This density is calculated from knowledge of pressures and temperatures. (Since the pressures and densities at each layer depend upon pressures and densities of layers above, the calculations are best done by a computer.)

34. Protons from the Sun's core are continuously absorbed and reemitted by material within the Sun. Since a photon is absorbed after traveling only about one centimeter and since photons are reemitted in random directions, the radiation travels a very great distance before arriving at the photosphere.

36. First, most of their radiation is in the X-ray portion of the spectrum. Second, they have a very low density and therefore there is not much material in them to emit light.

Calculations

2. Volume of Sun $= (4/3)\pi R^3 = (4/3)\pi (6.95 \times 10^8\text{m})^3 = 1.4 \times 10^{27}\text{ m}^3$.

   Mass/Volume $= 1.99 \times 10^{30}\text{ kg}/1.4 \times 10^{27}\text{ m}^3 = 1.4 \times 10^3$ kg/m^3. The density of water is 1.0×10^3 kg/m^3.

4. 500 nm. This is green (or yellowish green). It appears yellow because it has passed through the Earth's atmosphere, which scatters more light from the violet end of the spectrum than from the red end.

Chapter 12

Try One Yourself (Apparent Magnitude)

The difference in apparent magnitude between the two objects is 8. Referring to Table 12-1, we see that a difference of 5 corresponds to a ratio of 100 and a difference of 3 corresponds to a ratio of 16. Therefore we receive 1600 times more light from Mars than from Barnard's star.

Try One Yourself (Spectroscopic Parallax)

Using the H-R diagram in Figure 12-17 and the spectral type K1, we find that the absolute magnitude is between about +6.2 and +7.5. Since these numbers are greater than the star's apparent magnitude, we know that the star appears brighter than it actually is (because the greater the number, the dimmer the star). Thus the star must be closer than 10 parsecs from us.

We can use the equation in the "Calculating Absolute Magnitude" box to get a better answer, as follows:

$m - M = 5 \times \log(d) - 5$ where m = apparent mag., M = abs. mag., and d = distance.

Use the 6.2 absolute magnitude:

$4.4 - 6.2 = 5\log(d) - 5$
$\log(d) = 0.64$
$d = 4.4$ pc

When we use the same method with the other limit of absolute magnitude, 7.5, we get

$d = 2.4$ pc

So we conclude that the star is between 2.4 parsecs and 4.4 parsecs from us.

(Actually it is 3.2 parsecs away, close to the middle of our range.)

Try One Yourself (Luminosity and the Sizes of Stars)

Using solar units, Luminosity = radius^2 × temperature^4
$10,000 = r^2 \times (2/3)^4$
$r^2 = 50,625$
$r = 225$

Therefore, the radius of this star is 225 times that of the Sun. (If a star's radius is 225 times the Sun's, its diameter is also 225 times the Sun's.)

Recall Questions

2. C	4. D	6. B	8. D	10. D	12. E
14. C	16. B	18. A	20. A	22. D	24. C

26. We must know the object's distance as well as its angular velocity.

28. Apparent magnitude is a scale of the amount of light received from an object, while absolute magnitude is a scale of the amount of light emitted by an object. Luminosity is closely related to absolute magnitude, whereas brightness relates more closely to apparent magnitude.

30. The Doppler effect is used to measure a star's radial velocity.

32. Figure 12-17 is such a sketch.

34. If white dwarfs were as large as most stars, they would be very luminous. Their small size results in their being dim.

36. (1) A visual binary is visible as two stars (normally only with the use of a telescope).

    (2) A spectroscopic binary is a binary that can be detected by periodic Doppler shifts of the two stars. If the spectra of both are visible, each spectral line periodically splits into two lines.

(3) An eclipsing binary system changes brightness as one star eclipses the other.

(4) An astrometric binary system is one in which only one star can be seen, and that star is observed to move in a periodic motion that indicates the presence of an orbiting companion.

(5) A composite spectrum binary can be detected because of great differences in the spectra of the two stars.

38. If the plane of revolution of a binary is along a line from Earth, the maximum speed of each component that is observed by the Doppler effect is actually the speed of the star. If the plane of revolution is not along a line from Earth, however, that observed speed is less than the star's actual speed, and unless the angle between the plane of revolution and the line from Earth is known, the actual speed of the star cannot be calculated.

40. From the period of a Cepheid's variation, its absolute magnitude can be determined. Then, knowing its absolute magnitude and its apparent magnitude, its distance can be calculated.

Calculations

2. They differ in magnitude by 2.5, so we receive between 6.3 (which corresponds to a difference of 2) and 16 (which corresponds to 3) times as much light from Antares than from τ Ceti. (The actual ratio is about 9.9, which we calculate by raising the number 2.5 to the power of the difference in magnitudes—which happens in this case to also be 2.5.)

4. Moving α Centauri to a distance of 10 parsecs would entail moving it more than 7 times farther away than it is now, thereby making it much dimmer. Therefore its absolute magnitude must be greater than zero. (Its actual absolute magnitude is 4.4.)

6. It is closer than 10 parsecs. (Actual distance: 5.1 parsecs.)

8. Using Figure 12-37, we can see that it is more than 100 times as luminous as the Sun. Using the equation in the Tools of Astronomy Box, "The Mathematics of the Mass-Luminosity Relationship," we find that $4^{3.5} = 128$, so it is actually 128 times as luminous as the Sun.

Chapter 13

Recall Questions

2. C   **4.** A   **6.** D   **8.** D   **10.** B   **12.** C
14. A   **16.** D

18. A protostar is a star in the process of formation; it is heated by gravitational energy as its material falls inward.

20. Less massive stars take longer, because the lesser mass means that there is less gravitational force pulling each particle toward the center.

22. There is a lower limit because if a star does not have some critical amount of mass there will not be enough pressure at its center for nuclear fusion to be sustained.

More massive stars release more radiation as they join the main sequence, and stars above about 100 solar masses probably emit radiation so intense that no more material can fall onto them.

Calculations

2. Luminosity = radius² × Temperature⁴ when solar values are used for each quantity. (We could also use the Sun's diameter, since diameter is twice the radius.)

$$1000 = r^2 \times (1/6)^4$$
$$r = 1140$$

Thus the Sun at that time had a radius 1140 times greater than today's Sun.

This is 5.3 AU, which means the Sun would have extended to Jupiter's present position.

4. $\lambda = 2{,}900{,}000/T = 2{,}900{,}000/300 = 9700$ nm.

Visible light extends from about 400 to 700 nanometers, and 9700 nanometers is in the infrared region of the spectrum.

Chapter 14

Recall Questions

2. F   **4.** A   **6.** D   **8.** C   **10.** D   **12.** B
14. B   **16.** E   **18.** C   **20.** B   **22.** B   **24.** D
26. D

28. The star's mass.

30. After hydrogen fusion ceases, the core collapses and gravitational energy causes further heating.

32. The Earth will be absorbed in the outer part of the Sun. The Sun will expand so that its photosphere is somewhere between the present orbits of Earth and Mars.

34. 1) Pulsations in the core of a red giant increase in intensity until the outer layers of the star become unstable and are blown away.

2) The stellar winds blow away the outer portions of the red giant, and two different stages of the wind overlap to cause the glowing shell.

36. Electron degeneracy.

38. Hydrogen is being deposited on a white dwarf from a binary companion. When enough pressure builds up, this hydrogen erupts in a fusion reaction. After it is fused to helium (and some of it is blown away) more hydrogen continues to build up again until fusion starts again.

Calculations

2. The small angle formula yields 1.08 light-years. At 20 kilometers/second, this required 5.11×10^{11} seconds, or 16,200 years.

4. A magnitude change of 5 corresponds to a factor of 100 in brightness. Thus a change of 10 corresponds to a factor of 100×100, or 10,000. A change of 20 corresponds to a factor of 100^4, or 100 million.

Chapter 15

Recall Questions

2.	B	4.	E	6.	A	8.	D	10.	E	12.	B
14.	D	16.	A	18.	B	20.	C	22.	D	24.	A

26. When mass is added to a main sequence star, the star becomes larger. White dwarfs and neutron stars become smaller when mass is added, however. If the size of a black hole is considered to extend to its event horizon, added mass makes a black hole larger.

28. The fact that the Chinese "guest star" was very bright and that today we see an expanding cloud of debris at its location leads us to conclude that it was a supernova.

30. Pulsars vibrate faster than the surface of a white dwarf could vibrate. On the other hand, they vibrate more slowly than neutron stars could vibrate.

32. A black hole is an object that is so dense that its gravitational force is great enough that light cannot escape. We might observe a black hole when material falls toward it, for that material would heat up as it falls and emit radiation. Another way to observe one is by observing the motion of another star around which the black hole is revolving.

34. The escape velocity from inside the event horizon of a black hole is greater than the velocity of light.

36. Radiation has been observed that has the characteristics predicted to occur when material falls toward the event horizon of a black hole. Other black holes have been detected by observing a star that is in orbit around an unseen companion, when calculations show that the companion is massive enough that if it were not a black hole it would be emitting light.

Calculations

2. The Schwarzschild radius is proportional to the mass. Thus if the mass is doubled, the radius is doubled.

4. Section 15.6 indicates that the maximum mass of a neutron star can be 2 to 4 solar masses. Using 4 solar masses in Kepler's third law:

$$a^3/P^2 = (M_1 + M_2)/M_{sun} = 4\,M_{sun}\,/\,M_{sun} = 4$$

Using $P = 1$ second $= 3.2 \times 10^{-8}$ years, $a = 1.6 \times 10^{-5}$ AU = 2400 km

Chapter 16

Recall Questions

2.	B	4.	A	6.	C	8.	A	10.	E	12.	A
14.	C	16.	E	18.	C						

20. Figure 16-2 is what your drawing should look like. The exact location of the spiral arms is not important, however.

22. Galactic clusters are groups of up to a few dozen stars that share a common origin and are located relatively near one another. Most are found within the galactic disk. Globular clusters are groups of up to hundreds of thousands of stars. They are found primarily in the halo of the Galaxy. Clusters of galaxies are groups of individual galaxies that are held near one another by gravitational force.

24. The brightest stars, O- and B-type stars, are clustered at certain distances from the Sun. Better evidence is provided by 21-centimeter radiation, which indicates that cool hydrogen gas is located in spiral arms.

Calculations

2. 40,000 light-years = 3.8×10^{17} kilometers. The circumference of the path is then 2.4×10^{18} kilometers. Dividing this by 250 kilometers/second, we obtain 9.6×10^{15} seconds, or 3×10^8 years.

4. Diameter of Galaxy = 100,000 light-years = 9.5×10^{17} kilometers and the Sun's diameter is 1.5×10^6 kilometers. Setting up a ratio, we find that the scale model would be 63,000 kilometers across!

Chapter 17

Try One Yourself (The Hubble Law)

Figure 17-10 indicates that a distance of 90 million light-years corresponds to a speed of 17,000 kilometers/second.

slope = (17000 km/s – 0)/90 MLY = 190 km/(s $\times$ MLY)

Try One Yourself (The Hubble Law Used to Measure Distance)

Using a Hubble constant of 15 km/(s $\times$ MLY),
$v = Hd$
90,000 km/s = 15 km/(s $\times$ MLY) $\times$ d
$d = 6000$ MLY

Recall Questions

2.	E	4.	D	6.	D	8.	B	10.	B	12.	E
14.	A	16.	C	18.	C	20.	A	22.	C		

24. (a) Using Kepler's third law with measured values of rotation rates of various parts of a galaxy. (b) Application of Kepler's third law to binary pairs of galaxies. (c) Calculations of the mass necessary to prevent a galaxy from leaving the cluster of galaxies in which it is observed.

26. Not enough mass is directly observed in clusters of galaxies to hold the clusters together. Since they do hold together, more mass must exist than we can observe. Since it is not observed, it must be dark matter.

28. The Hubble law states that the velocity of recession of a distant galaxy is proportional to the galaxy's distance. The Hubble constant seems to be between 15 and 25 km/(s $\times$ MLY), which is between 50 and 80 km/(s $\times$ MLY).

30. To use the Hubble law to determine the distance to a galaxy, we measure the recessional velocity of the galaxy and then calculate its distance based on a value of the Hubble constant. Although recessional velocity can be accurately measured, some of that velocity may not be due to Hubble expansion. The greatest problem is that the Hubble constant is not known accurately.

32. An active galaxy emits more than normal amounts of radiation from its nucleus.

34. For a quasar with a redshift of 25%, the ratio of the amount of shift in wavelengths of its spectral lines to the unshifted wavelenghts is 0.25. The quasar with a redshift of 30% is moving away faster.

36. If they were nearby and moving at the speeds indicated by their redshifts, they would have had to have been produced recently. It is thought, instead, that quasars do not exist in the universe of today, and we see them only because they are very distant and therefore have a great look-back time.

Calculations

2. The slope, 11 km/(s × MLY), is the value of the Hubble constant based on Figure 17-33.

4. 4 hours

6. 0.80 kiloparsecs × 0.95 kiloparsecs

8. Density = d. Also, m is mass in solar masses and r is the radius in kilometers.

   $d = M/(4\pi R^3/3) = (m \times M_{sun})/(4\pi r^3\ 1000^3/3)$
   $= 4.8 \times 10^{20}\ (m/r^3)$

   For $r = 3$ m, we get $d = 1.8 \times 10^{19}/m^2$
   For $m = 10^9$, $d = 18$ kg/m^3
   For $m = 10^8$, $d = 1800$ kg/m^3 (The density of water is 1000 kg/m^3.)

Chapter 18

Recall Questions

2. C	4. C	6. A	8. B	10. D	12. A
14. C	16. D	18. E	20. D		

22. Clusters of galaxies are getting farther apart from one another. Theory tells us that space is expanding.

24. Homogeniety: The properties of the universe are the same throughout. On a sufficiently large scale, the universe looks the same everywhere. Isotropy: The universe is the same no matter in what direction we look. Universality: The same physical laws apply throughout the universe.

26. The Hubble constant must first be determined. The age of the universe is equal to the inverse of the Hubble constant, once million light-years (or million parsecs) have been converted to kilometers.

28. If the universe is an oscillating one, the present expansion will finally stop and compression begin. After compression leads to maximum density, the big bang will repeat, etc.

Calculations

2. $\lambda_{max} = 2,900,000/T$ where λ is in nanometers and T is in kelvin.

   Substituting 3 for the temperature, $\lambda_{max} = 9.7 \times 10^5$ nanometers. Figure 4-5 indicates that this is in the infrared. 3000 K yields 970 nanometers, in the near infrared.

Chapter 19

Recall Questions

2. A	4. E	6. E	8. B	10. E	12. E	14. B

16. The longer the average civilization lasts, the more likely a civilization is to be in existence when we are observing. Analogy: Suppose that on the night of the Fourth of July, 1000 sparklers are lit on my street. If a sparkler stays lit only 5 seconds, a person in a low-flying plane may not have a chance to see one as he/she flies over, but if they stay lit 30 minutes, several will be lit at any particular time.

Calculations

2. If 1000 civilizations, $10^{10}/1000 = 10$ million years on the average. If there are a million advanced civilizations, this is 10,000 years.

4. From the Chapter 14 Tools of Astronomy Box "Lifetimes on the Main Sequence,"

   $T = T_{sun}/M^{2.5} = 10^{10}/1.5^{2.5} = 3.6$ billion years

Glossary

absolute magnitude The apparent magnitude a star would have if it were at a distance of 10 parsecs.

absorption spectrum A spectrum that is continuous except for certain discrete frequencies (or wavelengths).

accelerate To change the speed and/or direction of motion of an object.

acceleration A measure of how rapidly the speed and/or direction of motion of an object is changing.

accretion The process by which an object gradually accumulates matter, usually due to the action of gravity.

accretion disk A rotating disk of gas orbiting a star, formed by material falling toward the star.

achromatic lens (or **achromat**) An optical element that has been corrected so that it is free of chromatic aberration.

active galaxy A galaxy with an unusually luminous nucleus.

adaptive optics A technique that relies on an active optics system that monitors and changes the shape of a telescope's objective to produce the best image.

albedo The fraction of incident sunlight that an object reflects.

altitude The height of a celestial object measured as an angle above the horizon.

angular momentum An intrinsic property of matter. A measure of the tendency of a rotating or revolving object to continue its motion.

angular separation The angle between lines originating from the eye of the observer toward two objects.

angular size (of an object) The angle between two lines drawn from the viewer to opposite sides of the object.

annular eclipse An eclipse in which the Moon is too far from Earth for its disk to cover that of the Sun completely, so the outer edge of the Sun is seen as a ring.

aphelion/perihelion The point in its orbit around the Sun where a planet, or other object, is farthest from/closest to the Sun.

Apollo asteroids Asteroids that cross the Earth's orbit and have semi-major axes larger than Earth's.

apogee/perigee The point in the orbit of an Earth satellite where it is farthest from/closest to Earth.

apparent magnitude A measure of the amount of light received from a celestial object.

asteroid Any of the thousands of minor planets (small, mostly rocky objects) that orbit the Sun.

asteroid belt The region between Mars and Jupiter where most asteroids orbit.

astrometric binary An orbiting pair of stars in which the motion of one of the stars reveals the presence of the other.

astrometry The branch of astronomy that deals with the measurement of the position and motion of celestial objects.

astronomical unit (AU) A unit of distance equal to the average distance between the Earth and the Sun (about 150 million kilometers or 93 million miles).

astrophysics Physics applied to extraterrestrial objects.

aurora Light radiated in the upper atmosphere due to impacts from charged particles.

autumnal and vernal equinoxes The points on the celestial sphere where the Sun crosses the celestial equator while moving south and north, respectively.

barred spiral galaxy A spiral galaxy in which the spiral arms come from the ends of a bar through the nucleus, rather than from the nucleus itself.

barycenter The center of mass of two astronomical objects revolving around one another.

big bang The theoretical initial explosion that began the expansion of the universe.

binary star (system) A pair of stars that are gravitationally bound so that they orbit one another.

binary system A system of two objects orbiting each other due to their mutual gravitational attraction.

blackbody A theoretical object that absorbs and emits all wavelengths of radiation, so that it is a perfect absorber and emitter of radiation. The radiation it emits is called **blackbody radiation.**

black dwarf The theoretical final state of a star with a main sequence mass less than about 4 solar masses, in which all of its energy sources have been depleted so that it emits no radiation.

black hole An object whose escape velocity exceeds the speed of light.

blazars (or **BL Lac objects**) Especially luminous active galactic nuclei that vary in luminosity by a factor of up to 100 in just a few months.

blueshift A change in wavelength toward shorter wavelengths.

Bohr atom The model of the atom proposed by Niels Bohr; it describes electrons in orbit around a central nucleus and explains the absorption and emission of light.

brown dwarf A star-like object that has insufficient mass to start nuclear reactions in its core and thus become self-luminous.

capture theory A theory that holds that the Moon was originally solar system debris that was captured by Earth.

carbon (or **CNO**) **cycle** A series of nuclear reactions that results in the fusion of hydrogen into helium, using carbon-12 in the process.

Cassegrain focus   The optical arrangement of a reflecting telescope in which a convex secondary mirror is mounted so as to intercept the light reflected from the objective mirror and reflect the light back through a hole in the center of the primary.

catastrophe theory   A theory of the formation of the solar system that involves an unusual incident, such as the collision of the Sun with another star.

celestial equator   A line on the celestial sphere directly above the Earth's equator.

celestial pole   The point on the celestial sphere directly above a pole of the Earth.

celestial sphere   The imaginary sphere of heavenly objects that seems to center on the observer.

center of mass   The average location of the various masses in a system, weighted according to how far each is from that point.

centripetal force   The force directed toward the center of the curve along which the object is moving.

Cepheid variable   One of a particular class of pulsating stars.

Chandrasekhar limit   The limit to the mass of a white dwarf star, above which it cannot be supported by electron degeneracy and cannot exist as a white dwarf.

charge-coupled device (CCD)   A small semiconductor chip that serves as a light detector by emitting electrons when it is struck by light. A computer uses the pattern of electron emission to form images.

chemical differentiation   The sinking of denser materials toward the center of planets or other objects.

chromatic aberration   The defect of optical systems that results in light of different colors being focused at different places.

chromosphere   The region of the solar atmosphere between the photosphere and the corona.

closed universe   The state of the universe if its total mass and energy density is larger than a specific value, called the critical density.

cocoon nebula   The dust and gas that surround a protostar and block much of its radiation.

coma   The part of a comet's head made up of diffuse gas and dust.

comet   A small object, mostly ice and dust, in orbit around the Sun.

composite spectrum binary   A binary star system with stars having spectra different enough to distinguish them from one another.

conduction   The transfer of energy in a solid by collisions between atoms and/or molecules.

conservation of angular momentum   A law that states that the angular momentum of a system does not change unless there is a net external influence acting on the system, producing a twist around some axis.

constellation   An area of the sky containing a pattern of stars named for a particular object, animal, or person.

continental drift   The gradual motion of the continents relative to one another.

continuous spectrum   A spectrum containing an entire range of wavelengths, rather than separate, discrete wavelengths.

convection   The transfer of energy in a gas or liquid by means of the motion of the material.

core (of the Earth)   The central part of the Earth, consisting of a solid inner core surrounded by a liquid outer core.

corona   The outermost portion of the Sun's atmosphere.

coronal hole   A region in the Sun's corona that has very little luminous gas.

coronal mass ejection   An event in which hot coronal gas is suddenly ejected into space at speeds of hundreds of km/s.

correspondence principle   The idea that predictions of a new theory must agree with the theory it replaces in cases where the previous theory has been found to be correct.

cosmic microwave background radiation (or CMB radiation)   Long-wavelength radiation observed from all directions; thought to be the remnant of radiation from the big bang.

cosmological constant   A term (denoted by the Greek capital letter "lambda," Λ) in the equations of general relativity that corresponds to a force throughout all space that helps the universe expand.

cosmological principle   The basic assumption of cosmology that holds that on a large scale the universe is the same everywhere.

cosmological redshift   The shift toward longer wavelengths that is due to the expansion of the universe.

cosmology   The study of the nature and evolution of the universe as a whole.

critical density   The average mass and energy density of the universe at which space would be flat. It is equal to $3H_o^2/(8\pi G)$ and is equivalent to about 5.5 hydrogen atoms per cubic meter of space.

crescent (phase)   The phase of a celestial object when less than half of its sunlit hemisphere is visible.

crust (of the Earth)   The thin, outermost layer of the Earth.

dark energy   An exotic form of energy whose negative pressure currently accelerates the expansion of the universe.

dark matter   Matter that can be detected only by its gravitational interactions; it appears to be quite abundant throughout the universe.

dark nebula   A cloud of interstellar dust that blocks light from stars on the other side of it.

density   The ratio of an object's mass to its volume.

density parameter   The ratio (denoted by the Greek letter "omega," Ω_0) of the total mass and energy density of the universe to the critical density.

density wave   A wave in which areas of high and low pressure move through the medium.

density wave theory   A model for spiral galaxies that proposes that the arms are the result of density waves sweeping around the galaxy.

deuterium   A hydrogen nucleus that contains one neutron and one proton.

differential rotation   Rotation of an object in which different parts have different periods of rotation.

spectrometer   An instrument that measures the wavelengths present in electromagnetic radiation. (A **spectrograph** is a spectrometer that produces a photograph of the spectrum.)

spectroscopic binary   An orbiting pair of stars that can be distinguished as two due to the changing Doppler shifts in their spectra.

spectroscopic parallax   The method of measuring the distance to a star by comparing its absolute magnitude to its apparent magnitude.

spectrum   The order of colors or wavelengths produced when light is dispersed.

spicule   A narrow jet of gas that is part of the chromosphere of the Sun and extends upward into the corona.

spiral galaxy   A disk-shaped galaxy with arms in a spiral pattern.

spring tide   The greatest difference between high and low tide, occurring about twice a month, when the lunar and solar tides correspond.

standard solar model   Today's generally accepted theory of solar energy production.

star   A self-luminous celestial object.

stellar parallax   The apparent annual shifting of nearby stars with respect to background stars, measured as the angle of shift.

stellar wind   The flow of particles from a star.

summer and winter solstices   The points on the celestial sphere where the Sun reaches its northernmost and southernmost positions, respectively.

sunspot   A region of the Sun's surface that is temporarily cool and dark compared to surrounding regions.

supercluster   A group of clusters of galaxies.

supergiant   The evolutionary stage of a massive star after it leaves the main sequence; at this stage, the star has a very great luminosity and size (100 to more than 1000 times the Sun's diameter).

superior planet   A planet that is more distant from the Sun than the Earth is.

superluminal motion   Motion that appears to occur at speeds faster than the speed of light.

supernova   The catastrophic explosion of a star, during which the star becomes billions of times brighter.

synodic period   The time interval between two successive similar alignments between a celestial object and the Sun, as seen from Earth.

T Tauri stars   A class of stars that show rapid and erratic changes in brightness.

tail (of comet)   The gas and/or dust swept away from a comet's head.

tangential velocity   Velocity perpendicular to the line of sight.

thermal energy   Energy that is due to the random motions of molecules and atoms of a substance.

tidal friction   Friction forces that result from tides on a rotating object.

total lunar eclipse   An eclipse of the Moon in which the Moon is completely in the umbra of the Earth's shadow.

total solar eclipse   An eclipse in which light from the normally visible portion of the Sun (the photosphere) is completely blocked by the Moon.

transit   The passage of a planet in front of its star.

triangulation   The use of parallax to determine the distance to an object.

troposphere   The lowest level of the Earth's (and some other planets') atmosphere.

Tully-Fisher relation   A relation that holds that the wider the 21-centimeter spectral line, the greater the absolute luminosity of a spiral galaxy.

tuning fork diagram   A diagram developed by Edwin Hubble to relate the various types of galaxies.

turnoff point   The point on an H-R diagram of a cluster of stars where the stars are just leaving the main sequence.

twenty-one-centimeter radiation   Radiation from atomic hydrogen, with a wavelength of 21.1 centimeters.

umbra   The portion of a shadow that receives no direct light from the light source.

universality   The property of obeying the same physical laws throughout the universe.

vernal and autumnal equinoxes   The points on the celestial sphere where the Sun crosses the celestial equator while moving north and south, respectively.

visual binary   An orbiting pair of stars that can be resolved (normally with a telescope) as two separate stars.

volatile (element)   A chemical element that can be vaporized at a relatively low temperature.

watt   A unit of power. It corresponds to a specific amount of energy each second.

wavelength   The distance from a point on a wave, such as the crest, to the next corresponding point, such as the next crest.

weight   The gravitational force between an object and the planetary/stellar body where the object is located.

white dwarf   The burnt-out relic of a low mass star. (Typical diameter is 0.01 that of the Sun—about the size of Earth.)

winter and summer solstices   The points on the celestial sphere where the Sun reaches its southernmost and northernmost positions, respectively.

zenith   The point in the sky located directly overhead.

zero-age main sequence   The main sequence of newly formed stars that just started hydrogen fusion in their cores.

zodiac   The band that lies 9° on either side of the ecliptic on the celestial sphere.

Index

Note: Page numbers in **bold** type refer to figures and tables.

Photo Credits

Starlinks icon, ©Photodisc; Advancing the Model banner art courtesy of NASA.

Prologue
Prologue Openers: NASA/STScI, and Hubble Space Telescope Comet Team and NASA. **P-1:** Fred Espenak/Science Photo Library/Photo Researchers, Inc. **P-2:** Dennis Milon/Photo by Allan E. Morton. **P-3:** T.A. Rector (NRAO/AUI/NSF and NOAO/AURA/NSF) and B.A. Wolpa (NOAO/AURA/NSF). **P-4:** Dennis Milon/Photo by Allan E. Morton. **P-6:** ©1980, Anglo-Australian Telescope Board. **P-8:** Karl Kuhn. **P-10:** NASA, ESA, A. Fruchter and the ERO Team.

Chapter 1
Chapter opener: Courtesy of the Adler Planetarium. Hand-colored engraving by Johann Bayer (1572–1625). **1-1:** AURA/NOAO/NSF. **1-3a:** Akira Fujii/NASA. **1-7:** Harvard College Observatory. **1-21:** The Granger Collection, New York. **1-22a and b:** Space.com Canada, Inc. and Starry Night Deluxe. **1-23:** Space.com Canada, Inc. and Starry Night Deluxe.

Chapter 2
Chapter opener: NASA. **2-1:** ©Enzo & Paolo Ragazzini/CORBIS. **2-2:** Crawford Collection, Royal Observatory, Edinburgh. **B2-1:** The Bettman Archives/Corbis. **B2-2:** The Granger Collection, New York. **B2-3:** By permission of the Houghton Library, Harvard University. **2-10:** The Granger Collection, New York. **2-11:** Photo Researchers, Inc.

Chapter 3
Chapter opener: NASA. **3-1:** Werner Sabo. **3-2:** William P. Sterne, Jr. **3-3c:** NASA, Voyager 2 photo. **B3-1:** The Granger Collection, New York. **B3-2:** ©Jim Sugar Photography/CORBIS. **B3-3:** The Granger Collection, New York. **B3-6:** AIP Neils Bohr Library. **3-4:** Lowell Observatory Photograph. **3-17a and b:** Andrew J. Martinez/Photo Researchers, Inc.

Chapter 4
Chapter opener: NASA. **4-3:** David Parker/Photo Researchers, Inc. **B4-6:** AIP Emilio Segré Visual Archive. **4-7c:** Karl Kuhn. **4-8c:** Karl Kuhn. **4-11:** STScI/NASA. **4-13:** Deutches Museum, Munich. **4-20a:** Karl Kuhn. **4-20b:** Education Development Center.

Chapter 5
Chapter opener a: ©NASA/CXC/SAO. **Chapter opener b:** NASA/UIT Team. **Chapter opener c:** ©European Southern Observatory. **Chapter opener d:** Image courtesy of NRAO/AUI/NSF. **5-2b:** Theo Kupelis. **5-4b:** ©Photodisc. **5-5a:** Karl Kuhn. **B5-1:** ©Lori Stiles, University of Arizona News Services. **B5-2:** Image courtesy of NRAO/AUI. **5-9:** Karl Kuhn. **5-10a:** Harvard College Observatory. **5-10b:** George C. Atamian. **5-13:** Yerkes Observatory Photograph. **5-14b:** Karl Kuhn. **5-15a:** Courtesy California Institute of Technology. **5-17 a and b:** ©European Southern Observatory. **5-18:** California Association for Research in Astronomy. **5-19:** ©1993 Roger Ressmeyer—Starlight. **5-20:** ©1993 Roger Ressmeyer—Starlight. **5-21:** University of Texas McDonald Observatory. **5-22:** LBT Project/drawing by European Industrial Engineering, Mestre, Italy. **5-23a:** AURA/NOAO/NSF. **5-23b:** NASA. **5-23c:** NASA/Galileo Project. **5-25:** NRAO/AUI/NSF. **5-26 a and b:** Courtesy of the NAIC—Arecibo Observatory, a facility of the NSF. **5-28a:** Bill Schoening, Vanessa Harvey/REU program/NOAO/AURA/NSF. **5-28b:** ©MPIfR Bonn (R. Beck, E.M. Berkhuijsen and P. Hoernes). **5-29:** NRAO/AUI. **5-33:** Image courtesy of NRAO/AUI. **5-34 a and b:** I. Gatley and R. Probst/AURA/NOAO/NSF. **5-35:** STS-82 Crew/HST/NASA. **5-36a:** Max-Planck-Institut fur Radioastronomie, Bonn (Glyn Haslam et al.). **5-36b:** NASA Goddard Space Flight Center and the COBE Science Working Group. **5-36c:** Lund Observatory, Sweden. **5-36d:** Courtesy, the ROSAT Mission and Max-Planck-Institut fur extraterretrische Physik. **5-36e:** Courtesy Tom McGlynn, NASA.

Chapter 6
Chapter opener: NASA/Mariner 10. **B6-1:** NASA. **DP6-2:** NASA. **6-2:** By permission of Johnny Hart and Creators Syndicate, Inc. **B6-2:** AURA/NOAO/NSF. **6-3a** and **b:** Karl Kuhn. **B6-3:** NASA/Galileo Imaging Team. **B6-4:** ©JPL/NASA. **DP6-5:** NASA. **6-7:** Karl Kuhn. **6-8:** Dennis Milon/Photo by Dennis Trail. **6-10a–g:** Lick Observatory. **6-16:** Photo by Jim Rouse. **6-18:** NASA Kennedy Space Center. **6-21:** Alex S. York. **6-22b:** NASA. **6-24a:** AURA/NOAO/NSF. **6-24b:** Hans Vehrenberg, Hansen Planetarium. **6-32b:** Nancy Rudger/The Exploratorium. **6-36:** Photo Researchers, Inc. **6-37b:** Photo courtesy of UCO/Lick Observatory. **6-37a:** Lick Observatory. **6-38:** NASA. **6-42:** Lick Observatory. **6-44:** William P. Sterne, Jr. **6-46:** Karl Kuhn. **6-48:** Jacques Descloitres, MODIS Rapid Response Team at NASA GSFC.

Chapter 7
B7-1: Yerkes Observatory Photograph. **7-2:** Karl Kuhn. **7-10:** ©Photodisc. **7-16a:** Royal Observatory, Edinborough/AATB/Science Photo Library. **7-16b:** ©AURA/STScI/NASA. **7-17:** D. Berry/STSCI AUL. **7-18a:** Image courtesy of C. and F. Roddier Institute for Astronomy, Honolulu, Hawaii. **7-18b:** ©T. Nakajima and S. Kulkami (CalTech), D. Durrance, and D. Golimowski (JHU), NASA. **7-19a:** NASA. **7-19b:** NASA/STScI. **Page 237:** NASA, J. Bally (University of Colorado), H. Throop (SWRI), and C.R. O'Dell (Vanderbilt University).

Chapter 8
Chapter opener: NASA, J. Bell (Cornell U.), and M. Wolff (SSI). **B8-2:** NASA/JPL. **DP8-2:** NASA. **8-3:** Courtesy of NASA/JPL/Caltech. **B8-3:** NASA/JPL. **8-5a:** Courtesy of NASA/JPL/Caltech. **8-6:** Courtesy of NASA/JPL/Caltech. **DP8-6:** NASA. **8-7:** Courtesy of NASA/JPL/Caltech. **DP8-10:** NASA. **8-11:** Courtesy of NASA/JPL/Caltech. **8-13:** TASS/Sovfoto. **8-14a and b:** NASA/JPL. **8-15:** NASA/JPL. **8-16:** NASA/JPL. **8-17:** NASA/JPL. **8-20a and b:** S. Larson, University of Arizona. **8-20c:** D. Crisp, WFPC2 team/NASA. **8-21:** Worlds in Comparison, Astronomical Society of the Pacific. **8-23:** Lowell Observatory Photograph. **8-24a:** NASA/JPL. **8-24b:** Astronomical Society of the Pacific. **8-26:** NASA. **8-27a:** Astronomical Society of the Pacific. **8-28:** NASA. **8-29:** NASA. **8-30:** Jeffrey S. Kargel, U.S. Geological Survey. **8-31a:** NASA/Johnson Space Center. **8-31b:** NASA. **8-31c:** NASA/Ames Research Center. **8-33:** D. Crisp, WFPC2 Team/NASA. **8-34:** IMP Team/NASA/JPL. **8-35a and b:** NASA/Goddard Space Flight Center Scientific Visualization Studio. **8-36:** NASA/Goddard Space Flight Center Scientific Visualization Studio. **8-37:** NASA/Goddard Space Flight Center Scientific Visualization Studio. **8-38a and b:** NASA/JPL/MSSS. **8-39a and b:** NASA/JPL/Malin Space Science Systems. **8-40a and b:** NASA/JPL/Malin Space Science Systems/Philip Christensen. **8-41a and b:** NASA/JPL/Malin Space Science Systems. **Page 281:** NASA and Lisa Frattare (STSci).

Chapter 9
Chapter opener: Erich Karkoschka (Univ. of Arizona) and NASA. **9-1:** California Institute of Technology. **B9-1:** NASA. **9-2:** NASA. **B9-2:** NASA. **DP9-2:** NASA. **9-3:** NASA. **B9-3:** NASA/JPL. **9-4:** NASA. **9-5:** NASA. **9-7b:** NASA. **9-8:** Image courtesy of NRAO/AUI. **9-9:** ©European Space Agency. All rights reserved. **9-10:** NASA. **DP9-10:** NASA/JPL. **9-12a:** Astronomical Society of the Pacific. **9-12b:** NASA. **9-13:** NASA. **DP9-14:** NASA. **9-14:** NASA. **9-15:** STScI. **9-16:** NASA. **9-17:** NASA. **9-18:** NASA. **9-19:** NASA. **9-21:** Lowell Observatory Photographs. **9-24:** J.T. Trauger (Jet Propulsion Laboratory) and NASA. **9-25:** R. Beebe (NMSU). **9-26:** NASA. **9-27:** NASA. **9-28:** Erich Karkoshcka (University of Arizona Lunar and Planetary Lab) and NASA. **9-29:** NASA. **9-32:** NASA. **9-34:** NASA/JPL. **9-35a:** NASA/JPL. **9-35b:** Erich Karkoschka (University of Arizona) and NASA. **9-39:** NASA. **9-40:** NASA. **9-41:** NASA. **9-42:** NASA. **9-44:** NASA. **9-45:** NASA. **Page 323:** NASA and E. Karkoschka (University of Arizona).

Chapter 10

Chapter opener: H. Mikuz, Crni Vrh Observatory, Slovenia. **10-1 left and right:** Lowell Observatory Photograph. **B10-1:** AP/Wide World Photos. **10-2:** NASA. **B10-2:** John Bortle, W.R. Brooks Observatory, Stormville, NY. **DP10-2:** Hubble Space Telescope/NASA. **10-4:** A. Stern (SwRI), M. Buie (Lowell Observatory), NASA, ESA. **10-5:** Yerkes Observatory Photograph. **10-6a:** NASA Galileo Imaging Team. **10-6b and c:** NASA/JPL. **10-6e:** Courtesy of the NEAR Project (JHU/APL). **PDC10-7:** NASA Galileo Imaging Team. **10-8:** Dennis Milon/Photo by George East. **PDC10-8:** ©1999 Calvin J. Hamilton/NASA. **10-9:** Photo by Nick James. **PDC10-9:** Photo by Nick James. **10-12a:** Karl Kuhn. **10-12b:** National Space Science Data Center. **10-14:** California Institute of Technology. **10-17:** David Jewitt (Institute for Astronomy, U. Hawaii) and Jane Luu (University of Leiden). **10-19:** Courtesy of Sven Kohle (www.allthesky.de). **10-20:** Kitt Peak National Observatory/AURA/NOAO/NSF. **10-24a:** Theo Kupelis. **10-24b:** NASA. **10-24c:** ©Field Museum of Natural History. **10-25:** Meteor Enterprises Inc., Flagstaff, AZ. **10-26:** Harvard-Smithsonian Institute for Astrophysics, Cambridge, MA. **Page 357:** NASA and H. Weaver (Johns Hopkins University).

Chapter 11

Chapter opener: SOHO (ESA & NASA). **11-1:** AbleStock. **B11-1:** NOAO. **DP11-1:** NASA. **11-3a:** William P. Sterne, Jr. **B11-3a:** AURA/NOAO/NSF. **11-3b:** The Observatories of the Carnegie Institution of Washington. **B11-3b:** NASA. **11-14a:** Raymond Davis, Jr., Brookhaven National Laboratory. **11-14b:** ©ICRR (Institute for Cosmic Ray Research), The University of Tokyo. **11-14c:** Courtesy of the Sudbury Neutrino Observatory. **11-15:** Courtesy Marshall Spaceflight Center. **11-16a:** NASA. **11-18:** AURA/NOAO/NSF. **11-20b and c:** NASA. **11-21a:** AURA/NOAO/NSF. **11-21b:** NASA. **11-22:** National Center for Atmospheric Research/University Corporation for Atmospheric Research/National Science Foundation. **11-23a and b:** NASA. **11-23c:** Photo by High Altitude Observatory, National Center for Atmospheric Research. The NCAR is sponsored by The National Science Foundation. **11-24a and b:** NASA. **11-25a:** NASA/NSSTC/Hathaway, 2003/05. **11-26:** NASA/MSFC/Hathaway, 2003/08. **11-28:** AURA/NOAO/NSF. **11-29:** ESA/NASA/Office of Space Science/SOHO. **11-30a:** NASA. **11-30b:** ©SOHO/CDS, SOHO/EIT (ESA & NASA); TRACE (NASA). **11-31:** NASA. **11-32:** The SOHO/LASCO data used here are produced by a consortium of the Naval Research Laboratory (USA), Max-Planck-Institut fuer Aeronomie (Germany), Laboratoire d'Astronomie (France), and the University of Birmingham (UK). SOHO is a project of international cooperation between ESA and NASA. **11-33a and b:** Courtesy of the Particle Physics and Astronomy Research Council (PPARC).

Chapter 12

Chapter opener: NOAO. **12-1:** Courtesy of Arnim D. Hummel Planetarium, Eastern Kentucky University. **B12-2:** Harvard College Observatory. **12-8a and b:** Yerkes Observatory Photograph. **12-12:** Courtesy of Chad Trujillo. **12-13:** Harvard College Observatory. **12-14:** Image supplied by Roger Bell, University of Maryland, and Michael Briley, University of Wisconsin, Oshkosh. **12-26:** Swathmore College Observatory. **12-27:** Lick Observatory. **12-39:** Harvard College Observatory. **12-41:** W. Freedman and NASA. **Page 423:** J. Newman (University of California, Berkeley) and NASA.

Chapter 13

Chapter opener: Wolfgang Brandner (JPL/IPAC), Eva K. Grebel (Univ. Washington), You-Hua Chu (Univ. of Illinois, Urbana-Champaign), and NASA. **13-1a:** Courtesy, Martin C. Germano. **13-1b:** ©European Southern Observatory. **13-1c:** ©Anglo-Australian Observatory/Royal Observatory Edinburgh. **13-1d:** ©European Southern Observatory. **13-2a:** Infrared Processing and Analysis Center, Caltech, JPL. **13-2b:** NASA. **13-5a:** AURA/NOAO/NSF. **13-5b:** C.R. O'Dell, and S.K. Wong (Rice University), NASA. **13-5c:** ©European Southern Observatory. **13-7:** ©Anglo-Australian Observatory/Royal Observatory Edinburgh. **13-8b:** Todd Bronson/NOAO/AURA/NSF. **13-8c:** ©2001 National Astronomical Observatory of Japan. **13-9:** NASA, HST. **13-10a:** Bill Shoening/AURA/NOAO/NSF. **13-10b:** J. Hester and P. Scowen (Arizona State University), NASA. **13-10c:** ESA & the ISOGAL team. **13-10d:** ©European Southern Observatory. **13-12a and b:** ©European Southern Observatory. **13-14a:** AURA/NOAO/NSF. **13-14b:** ©Anglo-Australian Observatory, photography by David Malin. **13-15:** J. Morse, STScI, NASA. **13-17:** C.R. O'Dell/Rice University and NASA. **13-18a:** NASA and C.A. Grady (Eureka Scientific, NOAO, and Goddard Space Flight Center) and the 9136 Team. **13-18b:** NASA and the Hubble Heritage Team (STScI/AURA). **13-19:** Two Armed Instability of Rotating Polytropic Star: Principle Investigators: Durisen, Richard; Yang, Shelby; Grabhorn, Robert; Department of astronomy, Indiana University. Visualization: Yost, Jeffrey; NCSA. **13-20a:** NASA. **13-20b:** Andy Steere. **13-21:** Hubble Heritage Team (AURA/STScI/NASA).

Chapter 14

Chapter opener: X-ray: NASA/UIUC/Y.Chu et al.; Optical: NASA/HST. **14-1b:** S. Kulkarni (Caltech), D. Golimowski (JHU), NASA. **DP14-1:** Robert Rubin (NASA Research Center), Reginald Dufour and Matt Browning (Rice University), Patrick Harrington (University of Maryland), and NASA. **DP14-2:** A. Hajian (USNO) et al., Hubble Heritage Team (STScI/AURA), NASA. **DP14-3:** B. Balick (U. Washington) et al., WFPC2, NASA, HST, NASA. **DP14-4:** B. Balick (U. Washington), et al, WFPC2, HST, NASA. **14-16:** ©Anglo-Australian Observatory, photography by David Malin. **14-17a:** Hubble Heritage Team (AURA/STScI/NASA). **14-17b:** NASA, NOAO, ESA, the Hubble Helix Nebula Team, M. Meixner (STScI), and T.A. Rector (NRAO). **14-19a:** ©European Southern Observatory. **14-19b:** H. Bond, STScI, NASA. **14-20a:** B. Balick (University of Washington) and NASA. **14-20b:** R. Sahai and J. Trauger (JPL), the WFPC2 Science Team, and NASA. **14-21a:** ESA & A.G.G.M Tielens (SRON/Kapteyn Astronomical Institute). **14-21b:** ESA & Valentin Bujarrabal (Observatorio Astronomico Nacional). **14-22a:** NASA/ESA. **14-22b:** NASA, ESA and the Hubble Heritage Team (STScI/AURA). **14-24:** Photo courtesy of UCO/Lick Observatory. **14-28 left and right:** Photo courtesy of UCO/Lick Observatory. **14-30:** X-ray: NASA/CXC/Rutgers/J.Hughes et al.; Optical: Rutgers Fabry-Perot. **14-31:** ©1980, Royal Observatory, Edinburgh. **Page 475:** NASA and the Hubble Heritage Team.

Chapter 15

Chapter opener: ©Anglo-Australian Observatory, photography by David Malin. **15-2a:** A. Dupree (CfA), NASA, ESA. **B15-3:** ©ESA 2002.Illustration by Medialab. **B15-4:** Illustration: CXC/M. Weiss; Spectrum: NASA/CXC/N. Butler et al. **15-5a and b:** Jack Newton. **15-6:** ©European Southern Observatory. **15-7a:** Jack Newton. **15-7b:** NASA/CXC/SAO/Rutgers/J. Hughes. **15-8:** Chris Floyd. **15-9a:** ©Anglo-Australian Observatory, photography by David Malin. **15-9b:** J. Pun (NASA/GSFC), R. Kirshner (CfA), and NASA. **15-9c and d:** NASA, P. Challis and R. Kirshner (CfA), P. Garnavich (University of Notre Dame), and the SINS Collaboration, STScI. **15-10:** University of Cambridge, Mullard Radio Astronomy Observatory. With compliments of Professor Antony Hewish. **15-15a:** NASA/CXC/SAO. **15-15b:** Jeff Hester, Paul Scowen of Arizona State University; HST, STScI. **15-15c:** N.A. Sharp/NOAO/AURA/NSF. **15-16:** European Space Agency & Francesco Ferraro (Bologna Astronomical Observatory). **15-17a:** NASA/PSU/G. Pavlov et al. **15-17b:** NASA/CXC/J. Hughes et al. **15-23:** Lois Cohen/Griffith Observatory. **15-25:** CXC/M. Weiss.

Chapter 16

Chapter opener a: NASA. **Chapter opener b:** NASA Goddard Space Flight Center. **16-1:** Photo by Akira Fujii. **B16-1:** Photo by Kim Zussman, Thousand Oaks, CA. **16-3:** Bill Keel, University of Alabama. **16-4:** AURA/NOAO/NSF. **16-6:** Yerkes Observatory Photograph. **16-7:** Yerkes Observatory Photograph. **16-8a:** AURA/NOAO/NSF. **16-8b and insert:** P. Guhathakurta (UCO/Lick Observatory, UC Santa Cruz), B. Yanny (Fermi National Accelerator Lab), D. Schneider (Pennsylvania State Univ.), J. Bahcall (Inst. for Advanced Study), and NASA. **16-15b:** NASA Goddard Space Flight Center. **16-18:** Anglo-Australian Telescope Board, Anglo-Australian Observatory. **16-22a:** A. Sharp/AURA/NOAO/NSF. **16-22b:** The Observatories of the Carnegie Institution of Washington. **16-23:** D. Figer (STScI) and NASA. **16-24a:** Produced at the Naval Research Laboratory by Dr. N.E. Kassim and collaborators from data obtained with the National Radio Astronomy's Very Large Telescope, a facility of the National Science Foundation operated under cooperative agreement with Associated Universities, Inc. Basic research in radio astronomy is supported by the Office of Naval Research. **16-24b:** National Radio Astronomy Observatory. **16-24c:** Harvard-Smithsonian Center for Astrophysics. **16-25a and b:** ©European Southern Observatory. **16-26:** EGRET All-Sky Gamma Ray

Survey Avve 100 MeV, Copyright EGRET, NASA, GSFC. **16-27a** and **b:** NASA/CXC/MIT/F.K. Baganoff et al. **16-30:** Ingrid Kallick, Lund Obsevatory, Dwingeloo Observatory, and STScI. **16-31:** Rensselaer Polytechnic Institute for the Sloan Digital Sky Survey. **Page 539:** NASA and H. Richer (University of British Columbia).

Chapter 17

Chapter opener: R. Williams, Hubble Deep Field Team (STScI), NASA. **17-1:** N. A. Sharp/NOAO/AURA/NSF. **B17-1:** California Institute of Technology. **17-2a** and **b:** NOAO/AURA/NSF. **17-2c:** ©European Southern Observatory. **17-3a:** ©European Southern Observatory. **17-3b:** NOAO/AURA/NSF. **17-3c:** Todd Boroson/NOAO/AURA/NSF. **17-4a–d:** NOAO/AURA/NSF. **17-5a:** AURA/NOAO/NSF. **17-5b:** NASA and the Hubble Heritage Team (STScI). **17-6a:** B. Whitmore (STScI) and NASA. **17-6b:** N. A. Sharp/AURA/NOAO/NSF. **17-6c:** NASA and the Hubble Heritage Team (STScI). **17-6d:** C. Grillmair (SIRTF Science Center) and NASA. **17-6e:** Hubble Heritage Team (AURA/STScI/NASA). **17-8 left:** The Observatories of the Carnegie Institution of Washington. **17-8 right:** California Institute of Technology. **17-12:** California Institute of Technology. **17-17a:** ©Anglo-Australian Observatory/Royal Observatory Edinburgh. **17-17b:** P. van Dokkum (University of Groningen), ESA, and NASA. **17-19a:** NASA/CXC/SAO. **17-19b:** NASA/CXC/SAO/A. Vikhlinin et al. **17-20:** Courtesy of Matthew Colless, Research School of Astronomy and Astrophysics, Mount Stromlo Observatory, Australian National University. **17-21a:** Image courtesy of NRAO/AUI. **17-21b:** NASA/UMD/A. Wilson et al. **17-21c:** CXC. **17-22:** X-ray (NASA/CXC/M. Karovska et al.); Radio 21-cm image (NRAO/VLA/J. Condon et al.); Optical (Digitized Sky Survey U.K. Schmidt Image/STScI). **17-23a:** NOAO/AURA/NSF. **17-23b:** STScI. **17-23c:** NASA, NRAO, and J. Biretta (STScI). **17-24a:** WFPC2 image: NASA and J. Bahcall (IAS); ACS image: NASA, A. Martel (JHU), H. Ford (JHU), M. Clampin (STScI), G. Hartig (STScI), G. Illingworth (UCO/Lick Observatory), the ACS Science Team and ESA. **17-24b:** Optical: NASA/STScI; X-ray: NASA/CXC; Radio: MERLIN. **17-26:** NASA Education and Public Outreach at Sonoma State University/Aurore Simonnet. **17-28:** NASA, Andrew S. Wilson (University of Maryland), Patrick L. Shopbell (Caltech), Chris Simpson (Subaru Telescope), Thaisa Storchi-Bergmann and F.K.B. Barbosa (UFRGS, Brazil), and Martin J. Ward (University of Leicester, UK). **17-30a:** Image courtesy NRAO/NUI. **17-30b:** K. Ratnatunga (JHU), NASA. **17-30c:** W. Couch (UNSW), NASA. **17-30d:** W. N. Colley (Princeton University), E. Turner (Princeton University), J.A. Tyson (AT&T Bell Labs), and NASA. **17-31c:** Image courtesy of R. J. Smith/Liverpool John Moores University. **17-33:** John Biretta, Space Telescope Science Institute. **17-35a:** Image courtesy of NRAO/AUI. **17-34a:** Inset photo: John Bahcall, Institute for Advanced Study; and NASA. **17-35b:** Image courtesy of NRAO/AUI and C. O'Dea & F. Owen. **17-35c:** Walter Jaffe/Leiden Observatory, Holland Ford/JHU/STScI, and NASA. **17-36a:** K. Gebhardt (U.MI), T. Lauer (NOAO), and NASA. **17-36b:** John Bahcall (Institute for Advanced Study, Princeton), Mike Disney (University of Wales), and NASA. **17-37a:** ©European Space Agency. **17-37b:** Dana Berry/NASA. **17-40:** Achut Reddy/Adam Block/NOAO/AURA/NSF. **Page 579:** ©ESO Education & Public Relations Department.

Chapter 18

Chapter opener: BOOMERANG Collaboration. **B18-2:** NASA/WMAP. **18-9:** AbleStock. **18-11:** Courtesy of AT&T archives. **18-12:** NASA. **18-20:** NASA/WMAP. **18-21:** R.J.E. Peebles. **18-22:** NASA Goddard Space Flight Center and the COBE Science Working Group. **18-23:** NASA Goddard Space Flight Center and the COBE Science Working Group. **18-24:** NASA/NSBF. **18-26:** NASA/WMAP. **18-27:** NASA/WMAP. **18-29:** NASA/WMAP. **18-30:** NASA/WMAP. **18-31:** NASA/WMAP. **18-32:** NASA/WMAP. **18-33:** BOOMERANG Collaboration. **18-34:** NASA/WMAP. **18-35:** ©European Southern Observatory. **Page 619:** NASA/WMAP Science Team.

Chapter 19

19-1: The Arecibo Observatory is part of the National Astronomy and Ionosphere Center, which is operated by Cornell University under a cooperative agreement with the National Science Foundation. **19-2:** NASA. **19-3:** NASA.

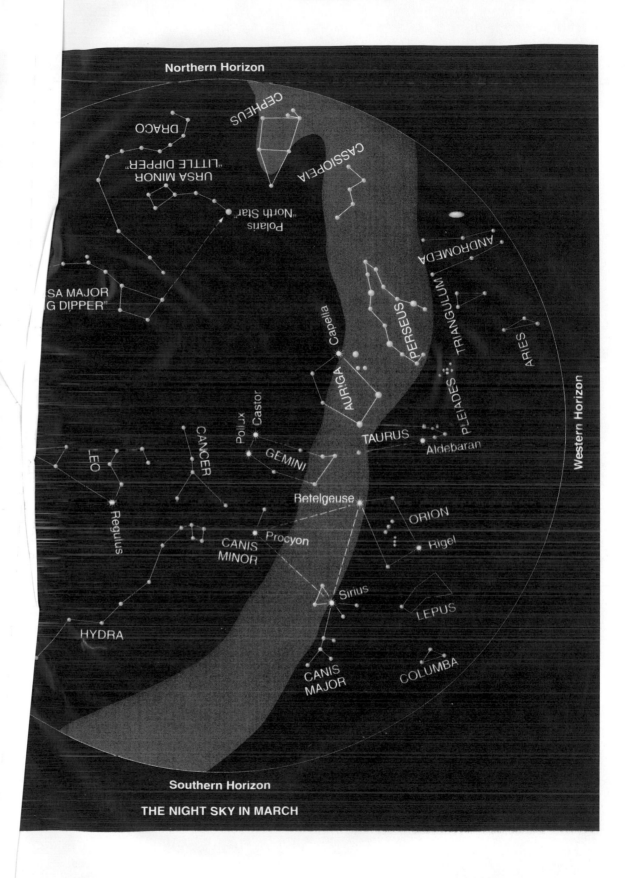

Northern Horizon

CEPHEUS

DRACO

URSA MINOR
"LITTLE DIPPER"

CASSIOPEIA

Polaris
"North Star"

ANDROMEDA

URSA MAJOR
"BIG DIPPER"

PERSEUS

TRIANGULUM

ARIES

Capella

AURIGA

PLEIADES

Castor

Pollux

CANCER

GEMINI

TAURUS

Aldebaran

LEO

Betelgeuse

ORION

Regulus

Procyon

CANIS
MINOR

Rigel

Sirius

HYDRA

LEPUS

CANIS
MAJOR

COLUMBA

Western Horizon

Western Horizon

Southern Horizon

THE NIGHT SKY IN MARCH

and turn it so the direction you are facing shows at the bottom.

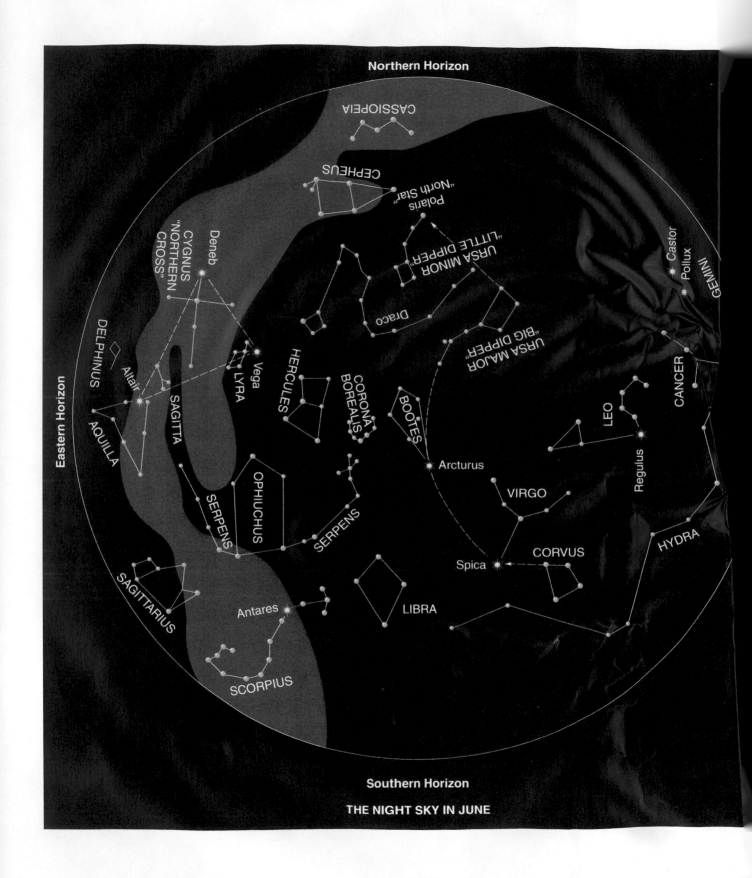

Northern Horizon

Southern Horizon

THE NIGHT SKY IN JUNE

To use: Hold chart vertically and turn it so the direction you are facing shows at the bottom.

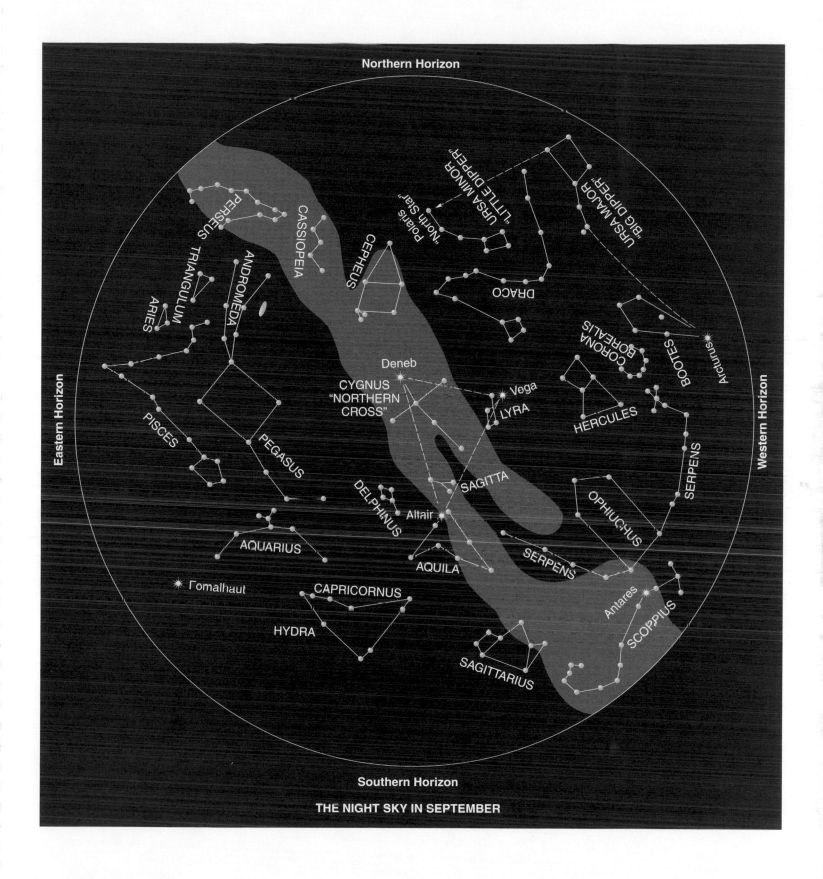

To use: Hold chart vertically and turn it so the direction you are facing shows at the bottom.

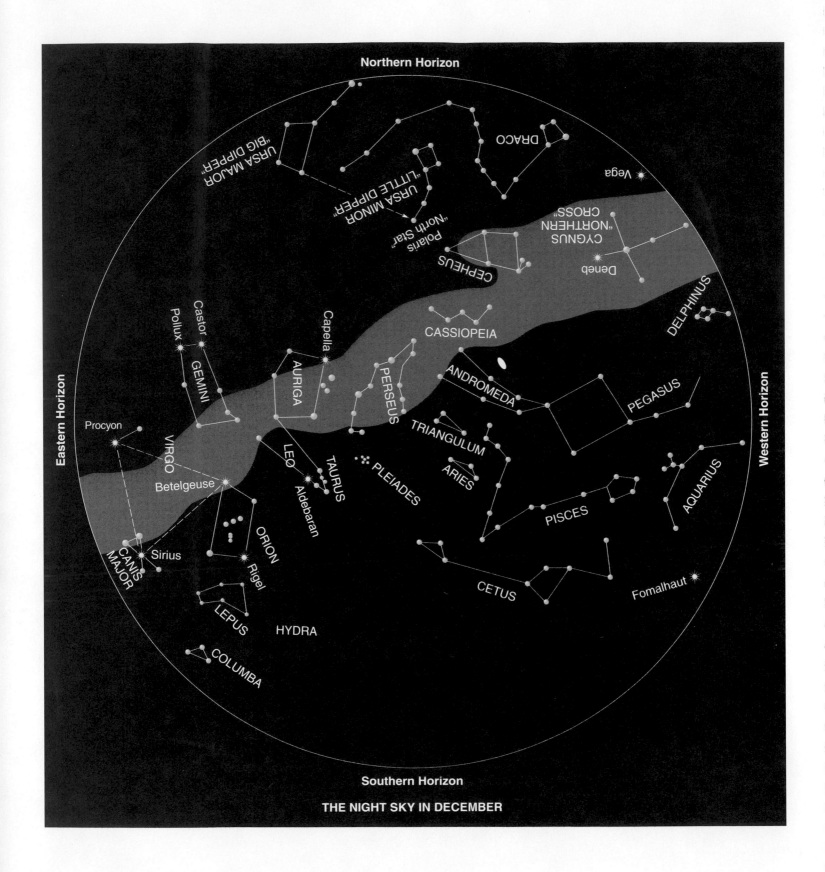

To use: Hold chart vertically and turn it so the direction you are facing shows at the bottom.